PHYSICAL GEOGRAPHY

SECOND EDITION

PHYSICAL GEOGRAPHY

RALPH C. SCOTT

Towson State University

WEST PUBLISHING COMPANY

St. Paul
New York
Los Angeles
San Francisco

WEST'S COMMITMENT TO THE ENVIRONMENT

In 1906, West Publishing Company began recycling materials left over from the production of books. This began a tradition of efficient and responsible use of resources. Today, up to 95 percent of our legal books and 70% of our college texts are printed on recycled, acid-free stock. West also recycles nearly 22 million pounds of scrap paper annually—the equivalent of 181,717 trees. Since the 1960s, West has devised ways to capture and recycle waste inks, solvents, oils, and vapors created in the printing process. We also recycle plastics of all kinds, wood, glass, corrugated cardboard, and batteries, and have eliminated the use of styrofoam book packaging. We at West are proud of the longevity and the scope of our commitment to our environment.

CREDITS

COVER PHOTO The Dauphine region of France (M. Thonig/H. Armstrong Roberts)
COVER DESIGN Paul Konsterlie
TEXT DESIGN Diane Beasley Design
ARTWORK Rolin Graphics
MAPS Alice Thiede/Carto-Graphics
COPYEDITING Mary Hoff
COMPOSITION Carlisle Communications

CHAPTER OPENING PHOTOS *Chapter 1* USGS; *Chapter 2* USGS; *Chapter 3* The Granger Collection; *Chapter 4* Rohan/Tony Stone Worldwide; *Chapter 5* Tony Stone Worldwide; *Chapter 6* Ken Fraser/FPG International; *Chapter 7* Jerry Scott; *Chapter 8* NOAA/JLM Visuals; *Chapter 9* T. Algire/H. Armstrong Roberts; *Chapter 10* © Charlie Ott/Photo Researchers, Inc.; *Chapter 11* M. Thonig/H. Armstrong Roberts; *Chapter 12* J. Irwin/H. Armstrong Roberts; *Chapter 13* USGS; *Chapter 14* Joanna McCarthy/The Image Bank; *Chapter 15* USGS; *Chapter 16* David Butler; *Chapter 17* USGS; *Chapter 18* V. Clevenger/H. Armstrong Roberts; *Chapter 19* H. Armstrong Roberts; *Chapter 20* Arthur Tress/Magnum Photos

Production, Prepress, Printing and Binding by West Publishing Company

Printed in the United States of America

99 98 97 96 95 94 93 92 8 7 6 5 4 3 2 1 0

Library of Congress Cataloging-in-Publication Data

Scott, Ralph C. (Ralph Carter), 1944–
Physical geography / Ralph C. Scott—2nd ed.
p. cm.
Includes bibliographical references and index.
ISBN–0–314–91837–X (hard)
1. Physical geography. I. Title.
GB54.5.S38 1992
551.4—dc20

91–32667
CIP

CONTENTS

C H A P T E R 3

Depicting the Earth—Maps and Remote Sensing 31

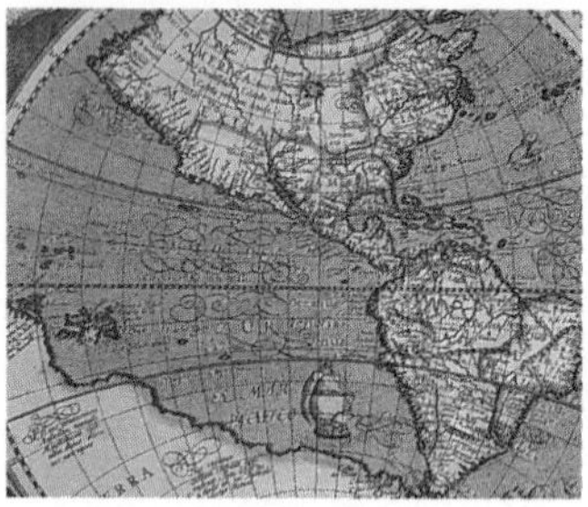

C H A P T E R 4

Earth's Atmospheric Envelope 49

C H A P T E R 5

Energy Flow and Air Temperature 65

C H A P T E R 6

Air Pressure and Wind 88

C H A P T E R 7

Atmospheric Moisture 115

C H A P T E R 8

Weather Systems 147

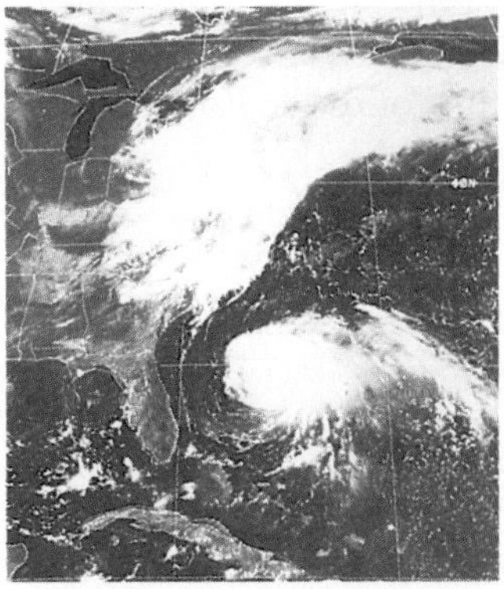

CHAPTER 9

Climates of the World 181

CHAPTER 10

Water on and Beneath the Earth's Surface 221

CHAPTER 11

Biogeography 253

C H A P T E R 15

Diastrophic and Volcanic Landforms 373

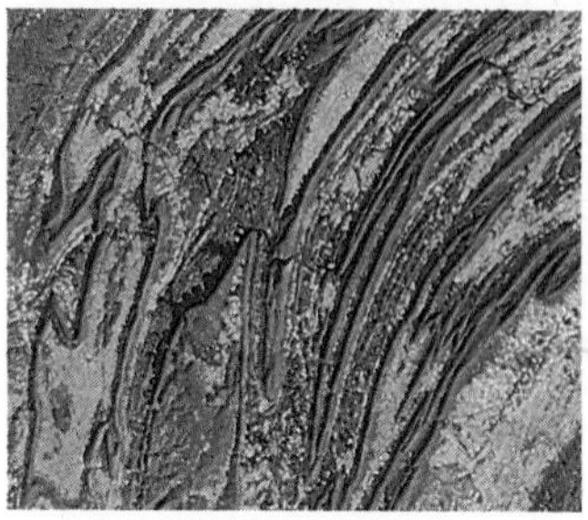

C H A P T E R 16

Weathering, Mass Wasting, and Karst Topography 405

C H A P T E R 17

Fluvial Processes and Landforms 433

CHAPTER 18

Glacial and Periglacial Landforms 463

CHAPTER 19

Eolian Processes and Desert Landscapes 492

CHAPTER 20

Coastal Landforms and Seafloor Topography 515

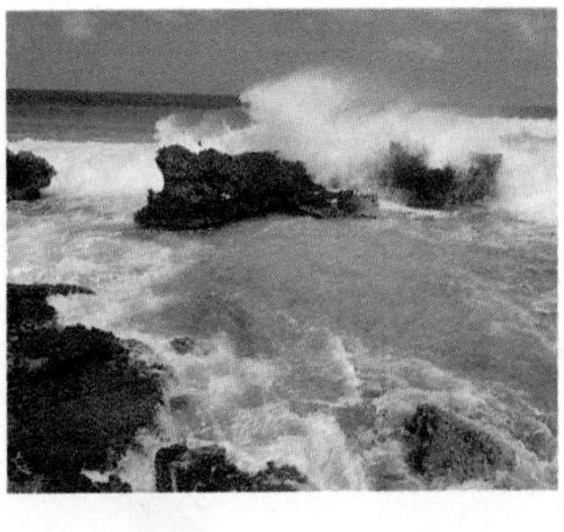

PREFACE

My primary goal in writing this book is to provide its readers with a greater understanding and appreciation of the world around them. The book is intended chiefly for individuals with limited physical science and geography backgrounds who are taking the course either to meet general university requirements or as an introductory course in a geography major program. The book provides an intermediate depth of material coverage that, I believe, is appropriate for the majority of college and university undergraduate students.

Over the past decade, geographic education in North America, and particularly within the United States, has been experiencing a long-overdue revival. Students, educators, and the general public once again are coming to realize that other places, and the events that occur in them, have an important bearing on our own lives. Especially crucial to our future well-being is the recent growth in public environmental consciousness. Our planet is a fragile one, and humanity now has a greater potential than ever before to alter permanently its physical characteristics. It is essential that the next generation of leaders display an understanding and appreciation of the environment greater than that of the leaders of the past.

APPROACH AND CONTENT OF THE BOOK

This second edition of *Physical Geography* has the same sequence of major topics as the first edition. Most material coverage is devoted to the "big four" subjects of physical geography, with weather and climate covered first, followed by treatment of natural vegetation and wildlife, soils, and landforms. Preceding these four major subjects, however, are three chapters that cover what I believe to be essential preliminary topics. Chapter 1 briefly discusses the subject of geography as an academic and intellectual discipline and presents some important basic geographical concepts. Chapter 2 discusses the Earth as a planet, treating its origin and its place in the solar system and universe, as well as covering terrestrial systems of location, direction, time, and the seasons. Chapter 3 deals with the depiction of the Earth through maps and the acquisition of geographical data by remote sensing devices.

The approach I used for major topics discussed in the book might be called the process/distribution approach. By this I mean that a certain amount of general science background appears first to explain the origin and characteristics of the topic. This is followed by a discussion of the topic's geographical distribution, and finally by the reasons for its distribution. Throughout the book, I have tried to stress the causative and geographical linkages among the various Earth phenomena.

In my discussion of world climates, I decided not to use the Köppen system so as to avoid forcing it upon instructors and students. I have always been somewhat partial to this system, however, and my climate regions correspond rather closely to Köppen's. I have also included an appendix section at the end of the book that explains this system, including its advantages and shortcomings, in some detail.

Important environmental concerns are discussed in the Case Studies found at the end of most chapters and in many of the Focus boxes appearing throughout the text. I hope that these studies will promote student interest, will illustrate the relevance of the material to contemporary issues, and will foster an increased environmental consciousness.

A number of additional learning aids have been provided in the text. At the beginning of each chapter is a subject outline as well as several Focus Questions that broadly address the core subjects of the chapter material. Within the body of the chapters, important terms appear in boldface type for easy recognition. These terms are defined in an extensive glossary at the end of the book. At the end of each chapter is a summary containing answers to the chapter-opening Focus Questions. Also appearing at the end of each chapter is a set of Review Questions and a list of Key Terms; some chapters also have a Problems section containing mathematical problems. Finally, at the end of the text is an appendix section covering the topics of scale conversions, weather map interpretation, and the Köppen climate classification system.

DIFFERENCES BETWEEN THE FIRST AND SECOND EDITIONS

The second edition of *Physical Geography* contains a number of changes from the first edition. What will be perhaps most apparent to many first edition users is that many of the illustrations, and particularly the photographs, are new. The overall quality of the illustrations has also been substantially improved. Measurements throughout the book are now given in metric, with imperial equivalents following in parentheses. The coverage of maps and remote sensing has been expanded and upgraded from an appendix section to a chapter (Chapter 3). Coverage of zoogeography has been added to the old chapter on natural vegetation and the chapter has been retitled "Biogeography." A number of new Case Studies and Focus boxes have been added; some of the new topics covered include Hurricane Hugo, the Loma Prieta earthquake, the spread of the gypsy moth, and map misinformation. A mathematical problem section has been added at the end of several of the chapters. Included are calculations involving latitude and longitude, time zone changes, temperature scale conversions, wind directions, lapse rates, and others. Answers to these problems appear in the appendix. Finally, there has been a good deal of the updating of information and refinement of writing that you would expect in a new edition.

AVAILABLE ANCILLARIES

West has provided a comprehensive package of ancillary materials that I believe will greatly aid adopters of this book. These include an instructor's manual, a laboratory manual, a student study guide, and a computerized test bank for IBM, Macintosh, and Apple computers. A set of approximately 200 color slides and 60 overhead transparencies of the more important maps and diagrams in the text are also available for teachers using the book. In addition, a free *PC Globe 3.0* program is available for adopters. Finally, a video disk containing hundreds of new illustrations as well as short videos is currently being prepared, and should be available early in 1992. Please contact the publisher or your local sales representative for further information on these materials.

ACKNOWLEDGMENTS

The preparation of a book such as this involves a great many individuals, all of whom are essential to its eventual successful completion. I would first of all like to thank the West editorial, production, and sales staffs, who enabled this book to become a reality and who have been chiefly responsible for the success of the first edition. Foremost among these individuals is my Acquiring Editor, Clark Baxter, who has overseen this project from the start and has kept me motivated through long months of work on the manuscript. Nancy Crochiere, our Developmental Editor, has provided expert analyses of the numerous manuscript reviews and has kept me on track with regard to proper procedures. Ellen Stanton, our Promotion Manager, has been responsible for the development of advertising brochures and for making sure that they reached potential users of the book. Kara Zum-Bahlen, our Photo Researcher, has managed to provide (often on very short notice) a selection of high quality photographs for the book. Lastly, the person who has been responsible for turning my manuscript and its accompanying illustrations into a book is my Production Editor, Tad Bornhoft.

Another group of individuals whose long hours of work with the manuscript have been vital to both its content and quality have been the reviewers. The work of the reviewers of my book *Essentials of Physical Geography* is also reflected in the content of this book, and I wish to thank them again for their generous assistance.

I wish particularly, however, to thank the reviewers for the first and second editions of *Physical Geography*. I found myself in agreement with most of their suggestions, and I believe they will see them incorporated in this book. These individuals are Ted J. Alsop, Utah State University; James Ashbough, Portland State University; Michael Barnhardt, Memphis State University; David Butler, University of Georgia; Gerald Brothen, El Camino College; James Bruns, Dundalk Community College; Robert F. Champlin, Fitchburg State College; Robert Churchill, Middlebury College; Percy Dougherty, Kutztown University; Richard Earl, New Mexico State University; Wayne Engstrom, California State University at Fullerton; Roberto Garza, San Antonio College; Phil Gersmehl, University of Minnesota; Richard Haiman, California State University at Chico; Martha L. Henderson, University of Minnesota–Duluth; John Hehr, University of Arkansas; David Howarth, University of Louisville; Jim Hull, Towson State University; Barbara G. Jaquay, Glendale Community College; Paul A. Knapp, University of Nevada–Reno; Clara A. Leuthart, University of Louisville; Richard Marston, University of Wyoming; Michael Mayfield, University of North Carolina at Greensboro; Bernard McGonigle, Community College of Philadelphia; Harold Meeks, University of Vermont; James R. O'Malley, West Georgia College; Don Petzold, University of Wisconsin at River Falls; Robert Quinn, Eastern Washington University; Roger Sandness, South Dakota State University; George A. Schnell, The College at New Paltz, State University of New York; Chi Sham, Boston University; Thomas Small, Frostburg State University; Stephen Thompson, University of New Mexico; Glen D. Weaver, Colorado State University; Douglas Wheeler, Utah State University and Thomas B. Williams, Western Illinois University.

Still another group of people who played a key role in this project were the artists and photo suppliers. I wish especially to thank Alice Thiede, who drew most of the maps appearing in the book, and the artists at Rolin Graphics, who produced most of the figures and graphs.

Thanks also go to my Towson State University colleagues Wayne McKim and John Morgan, who supplied photographs used in the book, and to Dr. Gregory Elmes, of West Virginia University, who supplied the text and illustrations for the Focus Box on the spread of the gypsy moth.

Finally, I owe a debt of gratitude to my wife Judi for her continuing encouragement and support for what appears to be turning into a lifelong project.

PHYSICAL GEOGRAPHY

CHAPTER 1

THE SCOPE OF GEOGRAPHY

OUTLINE

FOCUS QUESTIONS

1. What is geography? How does the geographic approach differ from other approaches to information?
2. What are the major subdivisions of the field of geography, and what topics do they cover?
3. What are regions, and why are they commonly employed in geographic studies?
4. What is a system? Why can the Earth be considered as an open system?

THE GEOGRAPHIC APPROACH

Geographers are concerned with understanding the locational aspects of Earth phenomena. In its emphasis on location, geography is unique among all the academic disciplines in its approach to its subject matter. The German philosopher Immanuel Kant (1724–1804) provided the philosophic basis for geography in the late eighteenth century when he stated that three approaches exist from which any subject can be studied. The first, the **systematic approach,** is organized by the type of thing to be studied. Most fields of knowledge are systematically defined today. They study everything about one topic. An example of a topically oriented discipline is biology, which is the study of living organisms. All life forms as well as the environmental factors that directly influence them are considered to fall within the bounds of accepted biological studies.

Kant's second approach to the study of any topic involves examining how it changes through time. This is the **historical approach.** Because virtually everything changes with time, the historical approach is not topically restricted.

The final approach in Kant's conceptual framework is the **geographical approach,** which studies topics by their differences in space or, to put it another way, by their locational characteristics. Just as nearly everything changes with time, so too does everything vary in location. This includes not only physical objects such as people, rivers, or kinds of rocks, but also less tangible things such as religious beliefs, military alliances, education levels, or tastes in music. Because geography is defined by its method of approach rather than by the topic being studied, it can be described as a field that looks at one aspect of every topic, rather than looking at every aspect of one topic, as is true of the systematic disciplines.

The distributions of earthly phenomena are not random; everything has a reason for being where it is. In nearly all cases, the distribution of one type of thing is influenced by the presence or distribution of other things. For example, human population distributions are influenced by factors such as climate, landforms, patterns of transportation routes, and the availability of housing and employment. World climates exist in patterns controlled by factors such as availability of solar energy, proximity to water bodies, and elevation. In examining the causes for the distributions of Earth phenomena, then, geographers cannot avoid noting that interrelationships exist among them. The systematic study of interrelationships among Earth phenomena therefore also has become a focus of geographic inquiry.

Suppose that you were told to go to the top of a hill, where you could get a good view of your surroundings, and to write down everything you saw. This would undoubtedly be a formidable undertaking, because a great many types of things would likely lie within your field of view. You might observe phenomena such as types of vegetation, hills and streams, animals, people, agricultural fields, roads, buildings of various types, and weather conditions. If you were to describe everything in great detail, the task could take forever. All these characteristics collectively define a **place,** or a given part of the Earth's surface. Because a large number of different types of phenomena exist in any one place, and because each phenomenon is capable of unlimited variation, every place on our planet is unique. Some geographers, rather than concentrating on the distributions of specific types of Earth features, seek to understand how the interactions of many features create differences and similarities between places.

Geographers feel that most fields of knowledge, especially those with rigidly defined topical boundaries, do not adequately emphasize the broader relationships that exist between their topics and other phenomena. The familiar expression of being "unable to see the forest for the trees" is particularly applicable in this context. Whereas many fields of study, by analogy, provide their practitioners with detailed knowledge of individual trees, geography is more concerned with the characteristics of the forest as a whole. Geography is an integrative discipline, then, that has a wide breadth of coverage (see Figure 1.1). It overlaps with most topically defined fields of knowledge because each contains spatial, or geographical, components. Like a person trying to assemble a jigsaw puzzle, the geographer's goal is to put enough pieces of knowledge about the Earth in their correct locational settings that the "big picture" begins to emerge.

As a summary of the concepts presented in the preceding paragraphs, the following definition of **geography** is offered: *Geography is the study of the distributions and interrelationships of Earth phenomena.* The geographer therefore is concerned both with the locations of things and with the causes and consequences of those locations. A basic objective of geo-

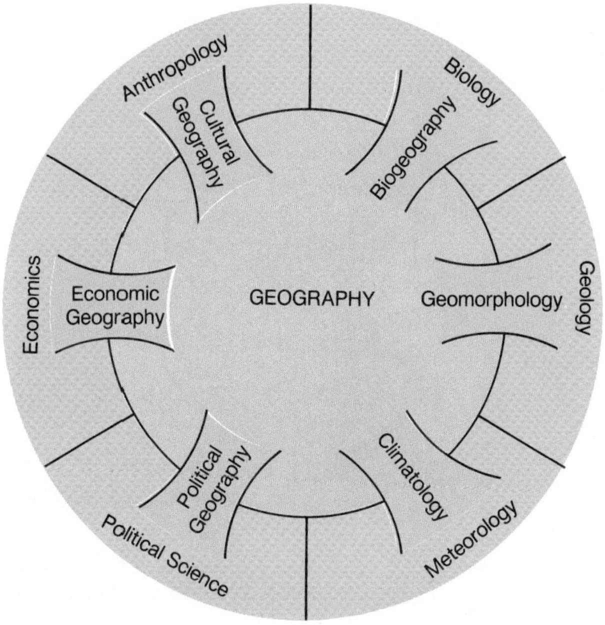

FIGURE 1.1 Geography overlaps with most topically defined fields of knowledge, because each contains spatial components. (Reprinted with permission from Julia A. Tuason, "Reconciling the Unity and Diversity of Geography," *Journal of Geography*, vol. 86, no. 5, p. 193.)

graphic research is to understand better the nature of places, and especially how distributions of terrestrial phenomena interact to cause similarities and differences between places. The ultimate goal of geography is to develop a sufficient understanding of spatial interaction to allow for a comprehension of the entire Earth as a functioning system.

Subdivisions of Geography

As the examination of a college textbook or course syllabus will show, the major academic disciplines typically are subdivided into separate units of material. This is done both for convenience and in order to permit individuals to specialize in a body of information of manageable proportions. The field of geography is no exception. Although geography *as a whole* is defined by its method of approach rather than by topic, its *subdivisions* are topically defined. While the basic spatial emphasis remains, each branch of geography deals with a different category of Earth phenomena.

There is a hierarchy of geographical subdivisions. In general, the more advanced a study, the more narrowly its particular borders are defined. At the most general level, the field of geography is partitioned into two major branches: physical geography and human geography.

Physical geography is concerned with locational aspects of natural Earth phenomena—that is, with phenomena not produced or primarily controlled by human beings. The four basic divisions of physical geography are weather and climate, surface features (including landforms and water bodies), natural vegetation, and soils. Most of this textbook is devoted to the geographical analysis of these four subjects.

Human geography, on the other hand, deals with subjects in which distributional patterns are largely or entirely controlled by people. Included are topics such as patterns of population, agricultural and industrial activities, urban areas, religions, transportation routes, political regions, and recreational facilities.

In reality, the interaction of terrestrial phenomena, both physical and human, produces a great deal of overlap between the different topical specializations, both in geography and in other disciplines. Geographers, because of their interests and training, are especially able to appreciate these interrelationships. The wide range of subjects covered in geography courses often creates difficulties in grouping this discipline with others that are topically defined. (Colleges and universities have experienced problems in this regard, because they typically contain separate academic divisions or schools of physical and social sciences.) Geography as a whole, then, is both a physical and a social science, and one of the geographer's chief goals is to emphasize the interrelationships between the two.

The Importance of Geography

From a practical standpoint, what is the advantage to a college student in gaining a knowledge of geography, and, in particular, of physical geography? The answer to this question is twofold. First, a fundamental "geographic literacy" should be a vital component of the store of knowledge of any educated individual. A knowledge of the locations and nature of places enables us to gain a fuller appreciation of the global, national, and local events that continuously shape our lives. For example, a true understanding of the causes and likely consequences of recent events in the

FIGURE 1.2 Some natural landscape changes, such as this Wisconsin stream's change in course, can occur over short periods of time. Most large-scale landscape alterations, however, occur slowly over long timespans. (JLM Visuals)

Middle East can be gained only with a basic knowledge of the nations, peoples, customs, religions, climates, land and water bodies, and natural resources of that part of the world.

A knowledge of physical geography also is needed to fully appreciate the delicate complexity of the Earth's surface environment as the only known abode of life in the universe. During the twentieth century, for the first time in history, human beings developed the technological capability of irreparably damaging the natural environment on a global scale. We will need to be far more environmentally sensitive in the future than we have been in the past. A number of case studies and focus boxes highlighting important contemporary environmental issues appear throughout this book.

FUNDAMENTAL GEOGRAPHICAL CONCEPTS

The geographic distribution of the Earth's physical features, as already noted, is not the result of chance, but exists in response to natural laws that provide reasons for the characteristics and locations of terrestrial phenomena. A basic understanding of these laws makes the study of physical geography—and of any physical science—much more logical and intellectually satisfying. Explanations for the distribution of Earth features will comprise an essential component of our geographic study throughout this book. In this section, we briefly examine a few fundamental concepts related to geographical distributions and interactions.

Regions

Although no two places on the Earth's surface are identical in every respect, different places may display considerable similarity in some of their characteristics. For example, the coastal region of southern California much more closely resembles the coast of Spain than it does the nearby Mojave Desert, even though California and Spain are some 9,600 kilometers (6,000 miles) apart. This resemblance occurs largely because of similarities in climate, landforms, and vegetation between the two sites.

An important method of geographical analysis is the formulation of **regions,** areas that are similar in selected attributes. Regionalization simplifies and organizes existing patterns of Earth phenomena. This, in turn, helps geographers to discover and understand the factors responsible for the existence of these patterns. Throughout this book, regional patterns of physical Earth phenomena will be displayed on maps and diagrams, and their causes and characteristics will be discussed.

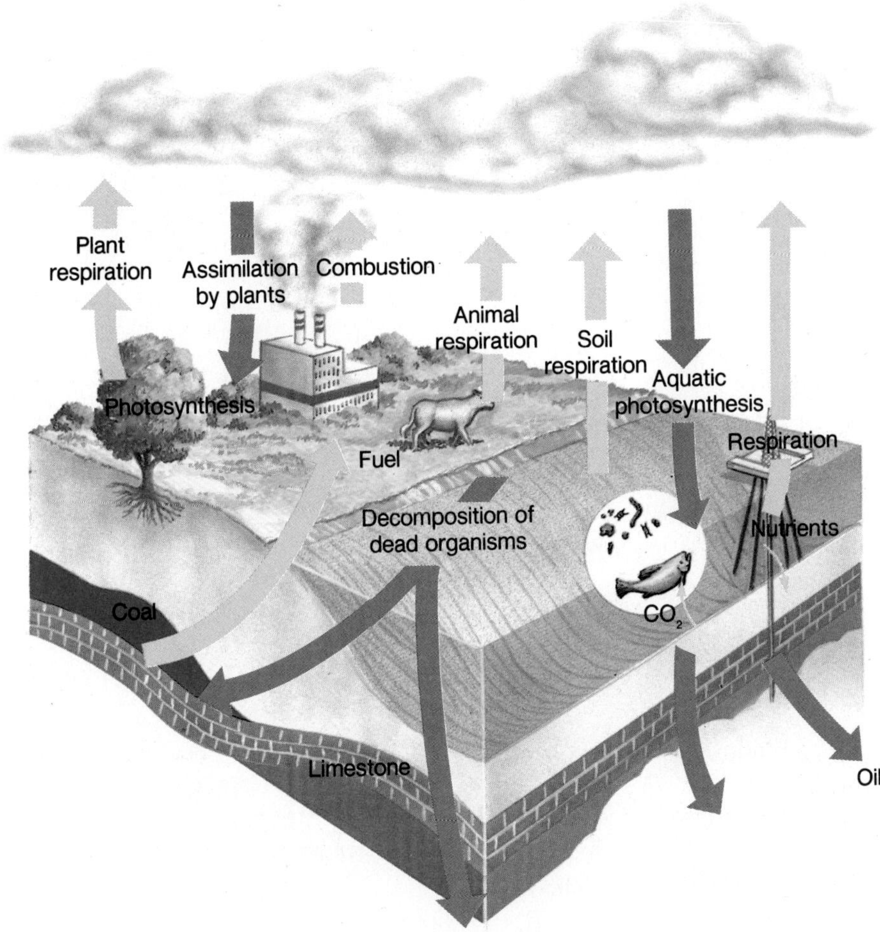

FIGURE 1.3 The Earth system contains numerous subsystems, including the carbon cycle illustrated here.

Regions—and the illustrations and text used in this book to describe them—do have limitations, however. First, descriptions of broad areas are necessarily general. Nearly every rule has its exceptions, and often these exceptions are so localized or limited in importance that they cannot be shown on a map or covered in the discussion. A second limitation of regionalization is that spatial changes in the characteristics of Earth phenomena do not usually occur abruptly, but instead take place gradually over considerable distances. As a result, regional boundary lines on maps should not be viewed as indicating sudden changes, but rather as representing zones of transition. By the same token, conditions should not be viewed as homogeneous within all portions of a given region. The climate, for example, does not normally change suddenly when one crosses a line on a climate map, nor is the climate identical in all places lying within a single climatic region.

Changes in Places through Time

Places change through time as well as over distance. Some of these changes, such as a shift in the course of a stream or the change of the seasons, are small or temporary (see Figure 1.2). Others, like the uplift of a mountain range or the onset of an ice age, are much longer in duration and of greater regional significance. Although small-scale characteristics often change relatively rapidly, broad regional patterns normally change very slowly, at least from a human perspective. Human activities, as previously noted, are proving particularly disruptive to environmental conditions in many areas and are causing rapid changes, often of an undesirable nature.

The Earth as a System

A **system** is defined as a set of interrelated components through which energy flows to produce orderly

(a)

(b)

(c)

(d)

FIGURE 1.4 The four major terrestrial subsystems, or "environmental spheres," are the (a) atmosphere (cumulonimbus cloud over South Carolina), (b) hydrosphere (Big Sur coast of California), (c) lithosphere (view from Dead Horse Point State Park, Utah), and (d) biosphere (riverbank vegetation near Ocho Rios, Jamaica). (a: Jerry Scott; b: D. Muench/H. Armstrong Roberts; c: VU/Doug Sokell; d: Ralph Scott) changes (see Figure 1.3). We commonly use many types of mechanical systems; two examples are clocks and automobiles. The fact that all terrestrial phenomena are interrelated means that any action that affects one component will ultimately influence everything else. In its association of parts, then, the Earth can be viewed as a single system of enormous complexity.

There are two basic types of systems. A **closed system** is one with a limited energy supply confined within it. Such a system can operate only as long as the energy supply lasts; then it will "run down." Familiar examples of closed systems include a wind-up clock and an automobile with one tank of gasoline.

An **open system,** on the other hand, has access to an unlimited supply of energy from an external source. The previous examples of closed systems can be altered to make them open systems by changing them to an electric clock and to an automobile that can obtain refills at a gas station whenever needed. Of course, no source of energy is truly infinite, so that all open systems, and even the universe itself, will probably eventually run down. For practical purposes, though, an open system has access to such a large energy supply that the exhaustion of this energy lies in the distant future.

The Earth system can be considered an open system powered by two different "unlimited" energy sources. These consist of solar energy and of heat from the decay of radioactive elements inside our planet. The total quantity of energy reaching the Earth system from these two sources is relatively constant, but it is capable of powering a great number of Earth processes because of the many routes it can take as it permeates the Earth system.

Like an automobile, the Earth system contains a number of interconnected subsystems, often described as "environmental spheres." The four major subsystems are the **atmosphere,** the ocean of air that overlies the entire Earth's surface; the **hydrosphere,** the water of the surface and near-surface regions of the Earth; the **lithosphere,** the massive accumulation of rock and metal that forms the solid body of the planet itself; and the **biosphere,** the layer of living organisms of which we are a part (see Figure 1.4). All four subsystems respond in various ways to the flow of energy and materials through the Earth system. The resultant distributional patterns and movements of these subsystems form the basis for the material content of this book.

SUMMARY

This chapter has provided a brief overview of the geographic approach to the study of Earth phenomena and the material content of the field of geography. It has also examined a few basic concepts that are central to geography.

As a means of summarizing the main points discussed in the chapter, we return to the Focus Questions posed on the chapter title page:

1. *What is geography? How does the geographic approach differ from other approaches to information?*

Geography is the study of the distributions and interrelationships of Earth phenomena. The distinctiveness of geography results not from the kinds of phenomena studied, but rather from the *way* they are studied, with its emphasis on spatial or locational components. A geographer generally asks three questions when studying something: "Where is it?", "Why is it there?", and "What is the significance of it being there?"

2. *What are the major subdivisions of the field of geography, and what topics do they cover?*

The field of geography generally is divided into two major branches. The emphasis of **physical geography** is on natural Earth phenomena, which have not been created or extensively modified by human actions. The chief topical subdivisions of physical geography are weather and climate, surface land and water features, natural vegetation, and soils. **Human geography** deals with Earth phenomena that are created, controlled, or modified by humans. Examples of human geography topics include economic, political, urban, and religious patterns.

3. *What are regions, and why are they commonly employed in geographic studies?*

Regions are portions of the Earth's surface that display similarity in one or more selected physical or human attributes. Regions are employed as a means of simplifying and organizing the geographic study of large areas by emphasizing selected spatial patterns.

4. *What is a system? Why can the Earth be considered an open system?*

A system is an organized set of interrelated parts through which energy flows to produce orderly changes. Closed systems contain a finite supply of internal energy, while open systems are powered by an essentially limitless supply of energy from an external source. The Earth can be considered as an enormously complex open system, powered by energy both from our planet's interior and from the Sun. One of the physical geographer's basic goals is to gain a greater understanding of the workings of the Earth system.

Review Questions

1. What is geography? In what basic way does the field of geography differ from the other physical and social sciences?
2. What are the two major subdivisions of the field of geography? Explain the basis of the difference in material content between the two.
3. What is a geographic region? What are two potential shortcomings in the use of regions to describe global patterns of physical phenomena?
4. Explain why the Earth can be described as an open system.
5. What are the four major subsystems of the Earth system?

Key Terms

geography
physical geography
region
system
closed system
open system

SELECTED REFERENCES FOR CHAPTER 1

The following books are recommended for those seeking a more comprehensive discussion of geographic methodology, philosophy, and history:

Broek, Jan O. M. *Geography: Its Scope and Spirit.* Columbus, Ohio: Merrill Publishing Co., 1965.

Gaile, Gary, and Cort Wilmott, eds. *Geography in America.* Columbus, Ohio: Merrill Publishing Co., 1989.

Holt-Jenson, Arild. *Geography: Its History and Concepts.* Totowa, N.J.: Barnes & Noble Books, 1980.

James, Preston E., and Geoffrey J. Martin. *All Possible Worlds: A History of Geographical Ideas.* 2nd ed. New York: John Wiley & Sons, Inc., 1981.

C H A P T E R 2

THE PLANETARY SETTING

OUTLINE

Planetary Motions
Large-scale Motions
Small-scale Motions
Size and Shape of the Earth
Departures from Sphericity
Direction
Magnetic North and South
Directional Systems
Latitude and Longitude
Latitude
Longitude
Determining Geographical Coordinates
Time
Local Time
Standard Time
The International Date Line
Daylight Saving Time
The Seasons
Cause of the Seasons
Lengths of Day and Night
Path of the Sun in the Sky

FOCUS QUESTIONS

1. What is the geographical setting of the Earth in space, and what movements is it making through space?
2. What are the size and shape of the Earth?
3. What basic directional and locational systems are employed on the surface of the Earth?
4. How is the global system of time zones organized?
5. What causes the seasons?

The Earth is the third of nine known planets orbiting the star we call the Sun. Sharing our **solar system** with the Sun and planets are a number of planetary satellites (moons), as well as asteroids, comets, dust, gases, and a great deal of empty space. Two opposing factors are responsible for the maintenance of stability in the solar system. One is the outward thrust of centrifugal force caused by the inertia of the planets as they orbit the Sun; the second is the inward pull of gravity. The revolution of the planets around the Sun, and of the moons around the planets, is made possible by the precise balance between these forces.

The Sun is only one relatively undistinguished member of a giant assemblage of stars called the **Milky Way galaxy.** Galaxies are large groups of stars held together by their mutual gravitational attraction. They vary tremendously in size, shape, and density, and contain anywhere from a few thousand to hundreds of billions of stars. Most galaxies also contain extensive clouds of gases and dust, out of which new stars constantly form.

The Milky Way is a large, discus-shaped galaxy consisting of more than 100 billion (or 10^{11}) stars. From end to end, the Milky Way has a diameter of 120,000 light years (a light year is the distance that light travels in a year), or about 13×10^{17} kilometers (8×10^{17} mi). Despite its enormous stellar population, the great preponderance of the galaxy consists of virtually empty space containing only the most rarefied concentrations of gas and dust.

The Milky Way galaxy revolves about its center of gravity. This motion has generated the centrifugal force responsible for its discus-like shape. The galaxy's peripheral disk is not homogeneous; rather it is organized into spiral arms of stars, giving the Milky Way something of a pinwheel appearance. Our solar system is located on a spiral arm an estimated 30,000 light years from the center of the Milky Way. This places us about two-thirds of the way from the galactic center to its outer fringe. The center of the galaxy is rendered invisible to us by vast clouds of interstellar gas and dust. The stars that so liberally sprinkle the sky on a clear night are our neighbors, occupying nearby portions of our arm of the Milky Way.

PLANETARY MOTIONS

The Earth is engaged in several motions through space, each at a differing level of magnitude. These movements can be divided into two categories—large-scale movements and small-scale movements. The large-scale movements are of limited direct significance to the Earth because millions or even billions of years are needed for them to produce major changes. The Earth's small-scale movements, however, critically affect nearly every aspect of our environment and will be discussed in greater detail. For clarity, the Earth's various motions through space will be discussed in order of decreasing magnitude.

Large-scale Motions

The largest-scale motion of all is the movement of the Earth, along with the rest of the solar system and galaxy, away from the center of the universe—that is, from the site of the Big Bang (see the Focus Box on page 12). The universe is still expanding at great speed.

The second large-scale motion is the revolution of our solar system around the center of the Milky Way galaxy. The time required for one full revolution is estimated to be 230 million years; this period is a function of our distance from the center of the galaxy.

A third motion of the Earth is associated with the drift of our solar system among the neighboring stars in our arm of the Milky Way. The positions of the various stars relative to one another are gradually changing. Over thousands of years, this movement will radically alter the present constellations of the night sky.

Small-scale Motions

The small-scale motions cause the Earth to change its position constantly with respect to the Sun. Because the Sun is our only major source of planetary heat and light, and because it powers external terrestrial processes such as the weather, plant growth, and erosion, these movements are extremely crucial to the Earth and its life forms. The small-scale motions also are responsible for the seasons and for the alternation of day and night, and form the basis for our system of keeping time.

Revolution

One of the two primary small-scale motions of the Earth is its **revolution** around the Sun. This motion occurs as a result of the combined effect of centrifugal force generated by the inertia of the Earth as it

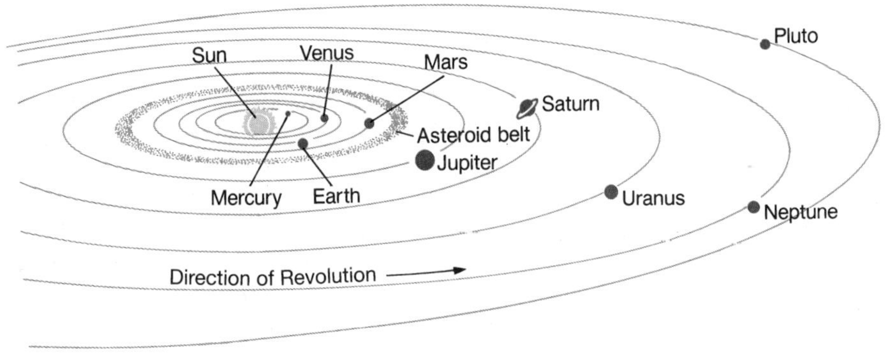

FIGURE 2.1 The planets of our solar system all orbit the Sun in the same direction and, except for Pluto, on nearly the same orbital plane.

orbits the Sun and the inward pull of the Sun's gravitational force.

The Earth's revolution is the basis of our calendar year. To be precise, the period of time needed for one complete revolution around the Sun is 365 days, 5 hours, and 49 minutes. For the sake of convenience, though, calendar years are defined in whole-day increments. Most years are therefore 365 days long. Because the extra 5 hours and 49 minutes required for a full revolution is very close to an extra quarter of a day, every fourth year has an added day (February 29) and is termed a **leap year.**

The Earth orbits the Sun on an angular plane termed the **plane of the ecliptic** in the same direction as all the other planets (see Figure 2.1). This direction appears as counterclockwise when Earth is viewed from a point in space above the northern hemisphere. (The vantage point in space is crucial, because the direction of revolution is clockwise when Earth is viewed from above the southern hemisphere.)

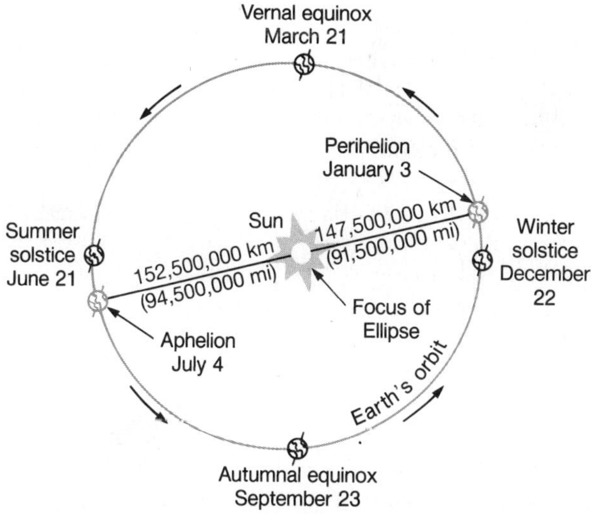

FIGURE 2.2 The Earth's orbit is slightly elliptical. As a consequence, we are approximately three million miles nearer the Sun in early January than in early July.

The Earth's orbit is roughly circular, but more precisely forms a broad ellipse. The mean radius of this ellipse, which represents the average distance between the Earth and Sun, is approximately 150 million kilometers (93 million miles). At its closest approach to the Sun, the Earth is said to be at **perihelion** (from the Greek *peri,* or "near," and *helios,* or "Sun"). At this point in its orbit, which occurs about January 3 each year, the Earth is only 147.5 million kilometers (91.5 million miles) from the Sun (see Figure 2.2). Six months later, on or about July 4, the Earth is at its farthest point from the Sun, or at **aphelion** (from the Greek *apo,* or "away from"). Our distance from the Sun is then approximately 152.5 million kilometers (94.5 million miles). Therefore, the Earth's distance from the Sun varies by about 3 percent during the course of a year.

Rotation

Rotation, the smallest-scale motion of the Earth, involves the spinning of our planet on its **axis.** The rotation of the Earth on its axis is similar to the rotation of a model globe on its axis bar except that, of course, the Earth's axis is not a physical entity and provides no support.

The rotation of the Earth allows us to define three important and vary basic locations. The **North Pole** and **South Pole,** located on opposite sides of the planet, are the two points where the axis intersects the Earth's surface. Midway between the poles, and exactly equidistant from both along its entire length, lies the **equator.** This is the line of fastest rotational speed on the Earth.

Rotation is the basis of our calendar day, and is the primary reason for the apparent westward motion of the Sun, Moon, and stars through the sky. (The situation is analogous to riding eastward in an automobile and watching the outside scenery apparently rushing by to the west.) The Earth rotates in the same direction as it revolves; this rotational direction can be described as counterclockwise when viewed from above the North Pole, clockwise when viewed from above the South Pole, or eastward if the Earth is viewed from a point in space above the equator.

Because all places on Earth complete one 360° rotation in a 24-hour period, they rotate through an angle of 1° every 4 minutes. The linear speed of rotation, however, is highly variable and depends upon distance from the equator. The speed of rotation is at its maximum value of approximately 465 meters per second (1,040 mph) at the equator, but diminishes to 0 meters per second at the poles.

FORMATION OF THE UNIVERSE, SOLAR SYSTEM, AND EARTH

Astronomical observations have revealed that the universe is composed of vast expanses of almost perfectly empty space, as well as the widely separated clouds of gases, dust, and stars we call galaxies. Within the known universe are many billions of galaxies, each containing, on the average, tens of billions of stars. The number of stars in the universe is for all practical purposes limitless. It has been said that there are more stars in the universe than grains of sand on all the beaches of the world, and this is apparently a great understatement. Moreover, a large proportion of these stars may have planetary systems similar to our own solar system.

The origin of the universe is one of the great mysteries facing humanity, and numerous theories concerning its formation have been developed over the centuries. Any modern scientific theory, however, must take into account a startling fact that has been revealed through the spectrographic analysis of the light from distant galaxies. This fact is that the galaxies are rushing away from one another. In general, the farther apart two galaxies are, the faster their rate of recession seems to be. The only way astronomers can explain this phenomenon is by theorizing that the occupied universe is expanding in all directions, much like an inflating balloon, as the result of a titanic explosion.

THE BIG BANG THEORY

Over the past few decades a theory of the origin of the universe—the so-called "Big Bang Theory"—has been developed and refined. It seems to fit most known facts, and is now generally accepted by the scientific community. According to this theory, some thirteen or so billion years ago, all matter in the known universe was concentrated into one tiny and inconceivably dense mass sometimes nicknamed the "cosmic egg." How it formed, and what events occurred prior to this time, nobody knows. The universe as we know it was initiated by the explosion of this mass. The explosion, the "Big Bang," destroyed the mass, and all the matter it contained was hurled into space in all directions, much like shrapnel from a detonating bomb. The violence of the explosion was so great that the super-dense matter of the cosmic egg was reduced to its constituent atomic particles, which soon reassembled to form only two elements—hydrogen and helium. Shortly after the explosion, then, the universe consisted of an expanding cloud of gas composed of approximately 73 percent hydrogen and 27 percent helium.

Gravity, however, began to alter the characteristics of the cloud. The expanding cloud gradually lost its homogeneity as it was drawn into many smaller gas clouds, each held together by its own internal gravitational attraction. Within these gas clouds, or proto-galaxies, gravity produced much denser and more localized concentrations of gas. As these smaller gas clouds contracted, their central cores were gradually heated by compression until thermonuclear fusion reactions were spontaneously initiated, and stars were born.

The basic thermonuclear reaction that powers the stars, including the Sun, involves the conversion of hydrogen to helium. A series of other stellar fusion reactions are be

continued on next page

The Earth's axis is not perpendicular to the plane of the ecliptic, but instead is inclined from the perpendicular at an angle of approximately 23½° (see Figures 2.4 and 2.16). Because this value does not change significantly over the short term as the Earth revolves around the Sun, the Earth's axis is described as exhibiting **parallelism,** and the poles remain directed toward approximately the same points in space for extended periods of time. This is why the North Star, or Polaris, occupies a fixed position above the North Pole.

The five motions of the Earth are essentially steady and cannot be felt. All five, however, are very slightly slowed by friction from various sources. Our planet's axial rotation will likely be the first motion to be significantly affected by friction. Sometime in the distant future the Earth will stop rotating, or, more accurately, its period of rotation will equal its period

lieved to have manufactured all the elements heavier than helium.

The earliest stars formed under highly favorable conditions and were large, extremely hot, and short-lived. These massive stars consumed their fuel at a prodigious rate, and soon became unstable as their fuel supplies were exhausted. They eventually exploded as supernovas, spewing a portion of the heavy elements they had produced throughout the galaxy. Succeeding generations of stars, including our Sun, thus formed from concentrations of gas containing not only hydrogen and helium, but the heavier elements as well. It is believed that the elements that constitute most of the Earth, and even our own bodies, originated in the cores of these supernova stars!

FORMATION OF THE SOLAR SYSTEM AND EARTH

During the contraction of a cloud of gas to form a star, any initial rotational motion is magnified because of the conservation of angular momentum. (This is the same principle that causes ice skaters to spin faster when their arms are drawn in.) The centrifugal force that is generated draws the contracting gases into a disk shape. The major concentration of gases at the center of the disk is under enough gravitational pressure to initiate thermonuclear reactions and become a star, while it is believed that smaller nodes of gases form on the rotating rim of the disk to produce planets. This would explain why the planets of our solar system all orbit the Sun on roughly the same angular plane and why they revolve around the Sun in the same direction. If this system operates everywhere, most stars would have planetary systems and, of course, the chances of earthlike planets and the existence of life elsewhere in the universe would be greatly enhanced.

In the case of our solar system, it is believed that the weaker gravity of the Earth and the other inner planets, coupled with intense radiation emanating from the nearby Sun (the so-called "solar wind"), drove off the hydrogen and helium that formed the initial bulk of the masses of these planets. The inner planets therefore consist primarily of the heavier residual elements that originated in supernova stars. The massive outer planets, though, were able to retain most of their light gases. They still consist largely of hydrogen and helium, which probably surround small solid cores of heavier elements.

FIGURE 2.3 The Andromeda galaxy, pictured here, is a spiral galaxy similar to the Milky Way. (Palomar Observatory, California Institute of Technology)

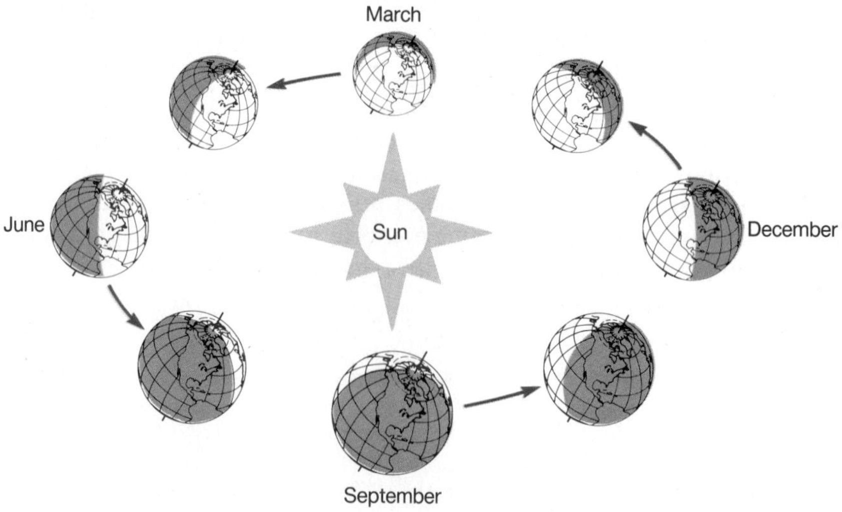

FIGURE 2.4 The Earth's axis is tilted at a constant angle with respect to the plane of the ecliptic as it revolves around the Sun during the course of the year.

of revolution, so that one side of the planet always faces the Sun while the other side faces the cold and darkness of outer space. (The Moon currently is in this position with respect to the Earth and always keeps its same face toward us.) If and when this occurs, perhaps billions of years from now, only the "twilight zone" along the boundary between day and night will likely be habitable by life forms as we know them.

SIZE AND SHAPE OF THE EARTH

It is common knowledge today that the Earth is roughly spherical. The shape of the Earth is not readily apparent from observation, however, and was the subject of speculation and debate for centuries. Most early scientists and philosophers believed that the Earth was flat. This view was based on the physical appearance of the Earth's surface, just as the early belief that the rest of the universe revolved around the Earth was based on the apparent motion of the Sun, Moon, and stars.

Modern science has produced many proofs of the sphericity of the Earth. These include the virtual equality of gravitational attraction at all points on the Earth's surface, numerous precise surface and astronomical measurements of distances and directions, and the visual evidence provided by high altitude and space photography.

Earth's spherical shape is produced by gravity. This incompletely understood force causes all matter to have an attraction for, and to be attracted by, all other matter. The mutual gravitational attraction of the Earth's matter has caused our planet to assume a spherical form—the most compact form that exists. Gravity also has produced a layering of Earth materials by density, with lighter materials tending to overlie heavier ones. (Fortunately for us, this layering is far from perfect.) Almost all other known astronomical objects of large size are also essentially spherical, indicating that gravity is apparently a universal force that operates wherever a sufficient mass of matter has assembled.

The three basic measurements of the size of a sphere are its **radius, diameter,** and **circumference.** For a given sphere, if any one of these measurements is known, the other two can be readily calculated. The radius of the Earth, or the distance from its surface to its center, is approximately 6,350 kilometers (3,950 mi). Earth's diameter, the linear distance from any point on the surface through the center of the Earth to its opposite (**antipodal**) point, is equal to two radii, or 12,700 kilometers (7,900 mi). Finally, Earth's circumference, the length of a surface line that bisects the planet, is 40,000 kilometers (24,900 mi).

Departures from Sphericity

Although gravity causes the Earth to be nearly spherical, it is not precisely so because other forces also influence the shape of our planet. The effects of two of these forces are sufficiently important to merit discussion. They are the outward thrust of centrifugal force generated by the Earth's rotation on its axis, and the tectonic forces produced in the planet's deep interior.

The centrifugal force of rotation is, fortunately for us, much too weak to overcome the binding force of gravity and cause the Earth to fly apart. It is, however, sufficient to slightly distort our planet's shape. In the vicinity of the equator, where the speed of rotation is greatest, the Earth bulges out from its spherical form. In contrast, the slowly rotating polar areas are somewhat flattened as the Earth's rotational inertia displaces their mass toward the equator (see Figure 2.5). The Earth thus assumes the shape of an **oblate ellipsoid,** a three-dimensional ellipse that is flattened at the poles. The amount of bulging and flattening is too small to be visible, and from space the Earth looks perfectly spherical. The departure from exact sphericity, however, affects the precise calculations required for surveying and mapping and also influences the force of gravity on different parts of the Earth.

The second, more important departure from sphericity (discussed at length later in this book), results from the Earth's variable surface features. If our planet were a true sphere, or even a true oblate ellipsoid, its surface would be smooth and would appear flat to an observer at ground level. Instead, the surface is highly wrinkled and contorted in places, exhibiting a great variety of landforms of many sizes and shapes. Total surface irregularity increases if we consider the sea floor, rather than the ocean surface, as comprising more than 71 percent of our planet's surface. Thus the Earth, when examined in sufficient detail, has a unique shape, undoubtedly duplicated by no other planet in the universe.

How great a departure from true sphericity do Earth's surface features represent? One way of judging this is by determining the total vertical distance between the world's highest mountain and its greatest ocean depth. Mt. Everest in the Himalayas is the highest peak, reaching an elevation of 8,847 meters (29,028 ft) above sea level. Challenger Deep in the western Pacific is the deepest known point of any ocean, with a maximum depth of 11,033 meters (36,198 ft) below sea level. The total vertical difference between these two extreme points is 19,880 meters (65,226 ft), or about 20 kilometers (12.3 mi). This value is less than 0.3 percent of the Earth's radius.

Despite the relatively diminutive size of the Earth's surface features when compared to the size of the entire planet, they are extremely important to humankind because we happen to live just where these features are located—at the surface. As a result, the Earth's surface features have strongly influenced most major aspects of human society, including population distributions, transportation routes, and agricultural patterns. In addition, Earth's landforms profoundly influence global patterns of climate, vegetation, and soils.

DIRECTION

The concept of direction is vital to the field of geography because of the discipline's emphasis on location. Unfortunately, no natural "universal" directions exist in space due to a lack of reference lines or surfaces. In outer space, in fact, even the words "up" and "down" are meaningless.

The basis for the horizontal directional systems used on Earth is our planet's rotational orientation. Rotation has established the positions of the North and South Poles and the equator. These locations enable us to construct reference lines for stating horizontal directions. Vertical directions are based on

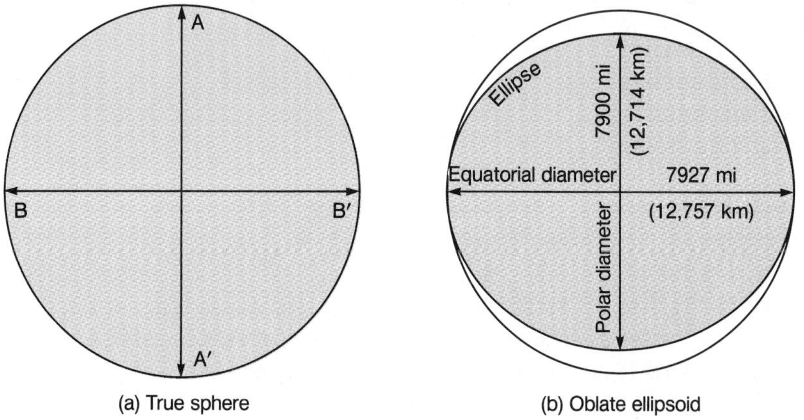

FIGURE 2.5 The Earth's shape is somewhat distorted from that of a true sphere (Figure a) because of the centrifugal force of the Earth's rotation on its axis. This motion causes our planet to assume the shape of an oblate ellipsoid (Figure b), with its equatorial diameter slightly larger than its polar diameter.

the orientation of the Earth's gravitational force, which is considered to pull directly downward toward the center of the Earth.

North, often considered the most basic of all directions, is by definition the straight-line direction from any point toward the North Pole. Likewise, **south** is the direction from any point toward the South Pole. These two directions, of course, are always 180° apart with respect to a given starting point because the poles are on opposite sides of the Earth. **East** and **west** may be defined either as those directions lying perpendicular to north and south, or those lying parallel to the equator. The Earth rotates in an eastward direction, and when an observer stands facing north, east is on his right and west on his left. Any other direction can be determined by reference to these four cardinal directions.

It is important to note that the directions north, south, east, and west are all equally horizontal. Expressions such as "up North" and "down South" are strictly figures of speech and are probably based on the fact that many maps are displayed vertically, as on classroom walls.

Magnetic North and South

The exact determination of direction is difficult or impossible without an accurate map or surveying equipment, although estimates often can be made from the position of the Sun, the North Star, or other celestial objects (see the Focus Box on page 20.) The most commonly employed instrument for determining direction in the field is the magnetic compass. A compass does not point toward the geographic (rotational) North Pole, however, but instead aligns itself with the magnetic field extending outward from the **magnetic North Pole.**

The magnetic North and South Poles, unlike the geographic poles, are not fixed in position, but slowly drift about. The magnetic North Pole is presently located in the Canadian Arctic, north of Hudson Bay, at a point some 1,600 kilometers (1,000 mi) south of the geographic North Pole. The magnetic South Pole is located near the coast of Antarctica south of Sydney, Australia, at a point about 2,900 kilometers (1,800 mi) from the geographic South Pole.

The origin of the Earth's magnetic field is incompletely understood. It seems to be related to the flow of electrically charged particles, both within the Earth and approaching Earth from the Sun and other points in space, that have been set into circular motion by the Earth's rotation. Because the magnetic poles are at considerable distances from the geographic poles, compass directions can differ greatly from geographical directions, and corrections are generally needed.

Directional Systems

A number of systems have been devised for stating direction on the Earth's surface. The ideal directional system is one that is uncomplicated, allows for easy directional computations, and is capable of accurately describing any direction. At present, two systems are widely employed in the United States. One, the **azimuth system,** has the advantages just noted; the other, sometimes called the **Mariner's Compass,** has serious shortcomings but has long been in use and is familiar to the general public. These two systems are illustrated in Figures 2.6 and 2.7.

LATITUDE AND LONGITUDE

The analysis of spatial data necessitates the development of an orderly system of reference points, distances, and directions. Geographers are concerned

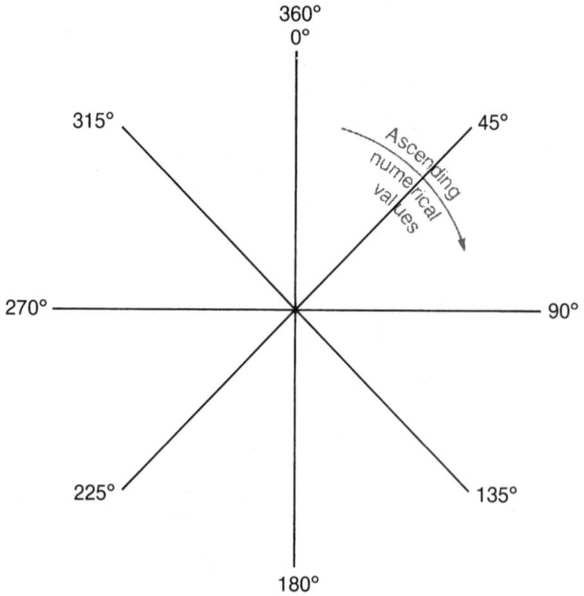

FIGURE 2.6 The azimuth system states directions in degrees, progressing in a clockwise direction from due north (0°).

with understanding spatial phenomena and interactions over the entire Earth. They must therefore employ locational systems that can accurately describe the geographic position of any point on the Earth's surface.

A number of global locational systems have been devised, but the most widely used, by far, is the system of **latitude** and **longitude.** This system is based on the locations of the North and South Poles and the equator. Related to these locations are the directions established by the Earth's rotational orientation. All lines in the system of latitude and longitude are oriented either due north-south or due east-west. We will examine the system by discussing the characteristics first of lines of latitude, then of lines of longitude. Finally, we will put the two together to show how these lines form a system for describing locations.

Latitude

Lines of latitude have the following characteristics, as illustrated in Figure 2.8:

1. They extend in an east-west direction but are used to state locations in a north-south reference frame.
2. Lines of latitude are really circles of latitude if looked at in their entirety, because they form full circles on the globe.

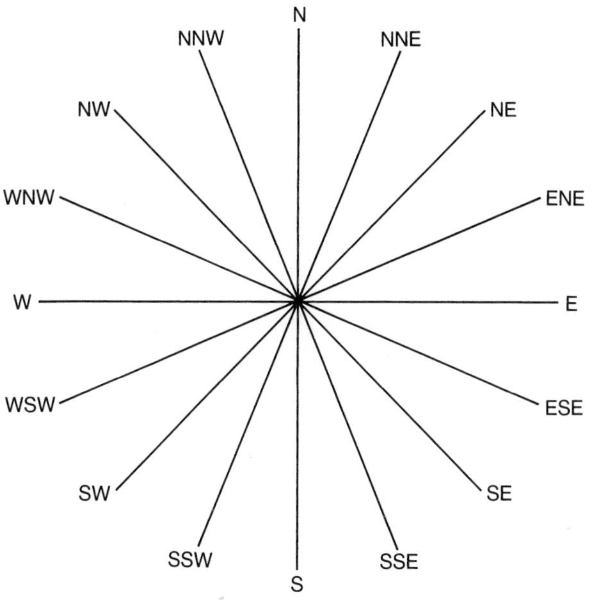

FIGURE 2.7 The 16 basic points of the Mariner's Compass.

3. Lines of latitude are parallel, or equidistant to one another, along their entire lengths. For this reason, they are commonly termed **parallels.**
4. Lines (circles) of latitude vary greatly in length (circumference), but do so in an orderly fashion. The longest circle of latitude is the equator. Progressing poleward, the circumferences of the circles of latitude become progressively shorter, until at the poles they have shrunk to a point.
5. An unlimited number of parallels exist, but on a map they are usually spaced at set intervals, decided upon by the **cartographer** (mapmaker) in view of the map's intended purpose.

By international agreement the equator is used as the starting line for the determination of latitude values. The equator is unique in that it is the only individual line in the system that is a full circumference of the Earth. A circle that is a circumference on a sphere bisects the sphere and is termed a **great circle.** Any circle drawn on a sphere that is not a circumference is termed a **small circle.** All circles of latitude other than the equator are small circles.

Parallels of latitude are identified by their angular distance, in degrees, north or south of the equator. As the starting line, the equator is assigned a value of 0° latitude. The half of the Earth's surface located north of the equator is termed the **northern hemisphere,** and all lines of latitude here contain an *N* after their numerical value. Likewise, the half of the Earth's surface located south of the equator is termed the **southern hemisphere,** and all latitude values are followed by an *S.* If we took a large globe and drew a parallel for each degree of curvature north of the equator, we would find that the fortieth line we drew would pass through Philadelphia, Pennsylvania. Philadelphia therefore has a latitude of 40°N. So do Columbus, Ohio; Boulder, Colorado; Ankara, Turkey; Beijing (Peking), China; and many other cities and towns. It is evident, then, that a statement of latitude is insufficient by itself to pinpoint a location.

The higher the numerical value of a line of latitude, the farther it is from the equator. The most northerly point on Earth is the North Pole, which is 90° north of any point on the equator. Its latitude is therefore 90°N. By the same token, the latitude of the South Pole is 90°S. Because all low-numbered parallels are near the equator, areas within about 30° of the equator are frequently referred to as the **low latitudes.** Areas between about 30° and 60°N and S

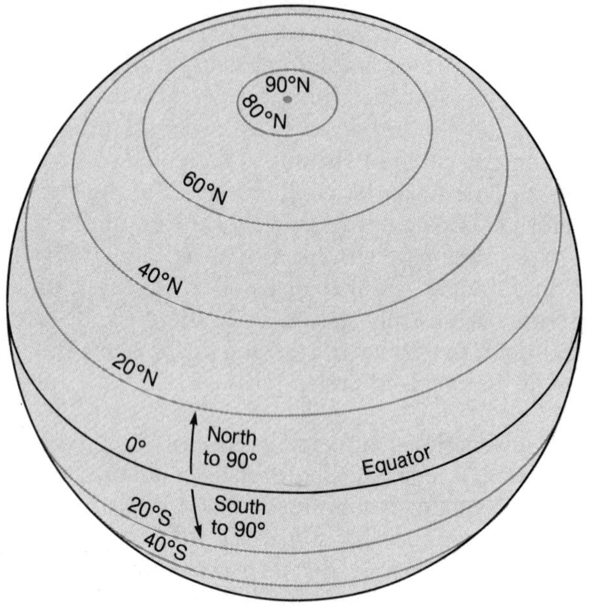

FIGURE 2.8 The globe with selected lines of latitude. These lines, termed parallels, actually take the form of concentric circles, numbered from the equator.

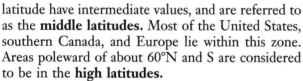

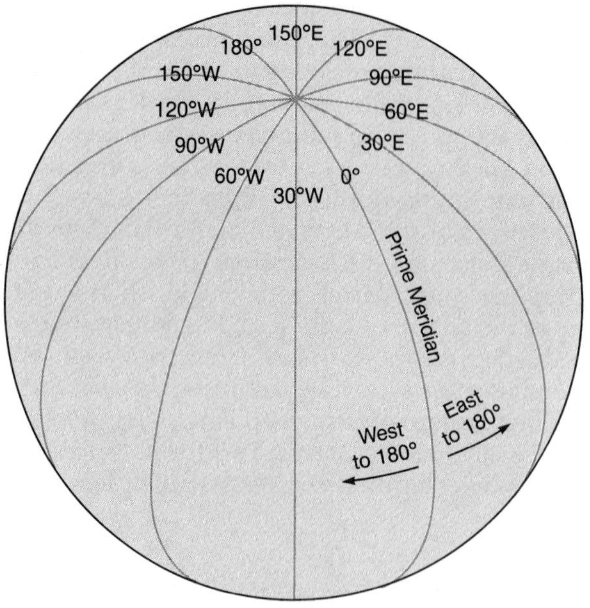

FIGURE 2.9 Lines of longitude, or meridians, are half circles with their end points at the poles. They are spaced farthest apart at the equator and are numbered in degrees east and west of the prime meridian.

latitude have intermediate values, and are referred to as the **middle latitudes.** Most of the United States, southern Canada, and Europe lie within this zone. Areas poleward of about 60°N and S are considered to be in the **high latitudes.**

Because lines of latitude are parallel to one another, if the Earth is considered to be a sphere with a circumference of 40,000 kilometers (24,900 mi), the length of one degree of latitude can be calculated readily. A circle contains 360° of arc. The length of one degree of latitude must therefore be 40,000/360 or 111 kilometers (69 mi). Actually, since the Earth is not precisely spherical, a degree of latitude is slightly longer at the poles (111.7 kilometers or 69.4 mi) than it is at the equator (110.6 kilometers or 68.7 mi).

For greater precision, a degree of either latitude or longitude can be subdivided into sixtieths, termed **minutes** (′), and each minute, in turn, can be divided into sixtieths, termed **seconds** (″). Minutes and seconds of curvature therefore subdivide degrees as minutes and seconds of time subdivide hours. One minute of latitude is approximately 2.85 kilometers (1.15 mi) long, and one second of latitude is about 31 meters (100 ft) long. A latitudinal reading is given from left to right in progressively smaller units, for example, 39°07′15″S.

Longitude

Lines of longitude, or **meridians,** are similar in some respects to lines of latitude, but they also have a number of unique characteristics (see Figure 2.9). The term *meridian* means "noon" or "midday" and, as we shall see, longitude is a basis of our system of time.

Lines of longitude have the following characteristics:

1. They extend in a north-south direction but are used to state locations in an east-west reference frame.
2. Each meridian is exactly half a great circle with its end points at the North and South Poles and its midpoint at the equator.
3. Meridians are *not* parallel to one another; they are spaced farthest apart at the equator and converge northward and southward to meet at the poles. Because meridians converge, the surface length of a degree, minute, or second of longitude depends on its latitude. The length of one degree

of longitude at the equator is approximately the same as that of a degree of latitude— 111.2 kilometers (69.2 mi). This length decreases to zero at the poles. The length of a degree of longitude at any given latitude is equal to 69 multiplied by the cosine of the latitude.

4. Meridians always intersect parallels at 90° angles.

5. No limitation exists as to the number of meridians that may be placed on a map, but, like parallels, they usually are spaced at set intervals.

The numbering of meridians is accomplished in the same way as is the numbering of parallels—by choosing a starting 0° line and then by measuring subsequent lines in degrees of surface curvature, this time to the east and west. A problem long existed, though, in choosing an appropriate starting meridian because no one meridian differs from the rest in any geometrical manner. As a result, many countries that made maps established 0° meridians through their own capital cities or other points of national importance. This practice caused a great deal of confusion because longitude values for the entire world depended upon the country of origin of the map or book being consulted. It was finally decided at an international meeting in 1884 to adopt the English system, in which the starting or **prime meridian** passed through Greenwich Observatory on the outskirts of London.

From the prime meridian (also called the **Greenwich meridian**) westward, west longitude values increase from 0° to a maximum of 180° halfway around the world. Similarly, going eastward, values of east longitude increase from 0° to 180°. The numerical values of meridians therefore go twice as high as they do for parallels.

Longitude divides the world into the **eastern hemisphere** and the **western hemisphere.** These terms are not as frequently employed as are the terms northern hemisphere and southern hemisphere, but such longitudinally derived expressions as the Western World and the Far East are in common usage.

By convention, the 0° and 180° meridians are not followed by the letters *E* or *W*, because these two lines separate the eastern and western longitudinal hemispheres. The numerical values of all other meridians, however, are followed by an *E* or *W*, just as all lines of latitude except the equator are followed by an *N* or *S*.

Determining Geographical Coordinates

Taken together, the latitude and longitude of any location describe its geographical position. A given parallel intersects a given meridian at only one point on the Earth, so if the correct value of each of these lines is provided, the position of any place can be pinpointed. The accuracy of placement of any point depends on the accuracy with which its position is stated. Locations stated to the nearest whole degree contain a potential margin of error of many miles, while those stated to the nearest second of latitude and longitude are accurate to within 15 meters (50 ft).

When the geographical coordinates of a site are given, latitude is always stated first. An example of a location stated to the nearest second is as follows:

39°22′18″N 76°38′05″W.

It must be remembered that, with the exception of the equator, the prime meridian, and the 180th meridian, the numerical value of each parallel or meridian is followed by the letter *N*, *S*, *E*, or *W*. It should also be kept in mind that both minutes and seconds are sixtieths, not hundredths. Most maps have selected lines of latitude and longitude either drawn on them or indicated by tick marks on their margins. The numerical spacing of the lines of latitude drawn on a map is not necessarily the same as that for its lines of longitude.

TIME

Our system of keeping time is based on the east-west (or longitudinal) position of the Sun. The Sun's position and apparent motion during the course of a day are associated primarily with the Earth's axial rotation by also are slightly affected by Earth's revolution around the Sun. The basic unit of time is the **solar day,** which is the period needed for the Sun to make one apparent 360° circuit of the sky. The solar day is, of course, subdivided into the familiar units of hours, minutes, and seconds.

Local Time

The event historically used for establishing the time of day is the occurrence of **solar noon.** Solar noon occurs the instant the Sun reaches its daily high point in the sky. For centuries, until the late 1800s, communities made the time of the mean occurrence of

GEOGRAPHIC POSITIONING AT SEA

Determining the location of a ship at sea—and out of sight of any identifiable terrestrial landmarks—has long been a crucial problem in long-distance oceanic travel. For this reason, the navigator is one of the most essential members of a ship's crew. Over the centuries, navigation techniques have evolved from rough "guesstimates," based on a combination of luck, intuition, and visual observations of such phenomena as the Sun and stars, waves, and clouds, into a highly sophisticated and accurate science.

The basic problem in oceanic navigation is to determine both the ship's north-south (latitudinal) and east-west (longitudinal) position in the absence of visual landmarks. In ancient times, the latitudinal position was the easier of the two to determine with reasonable accuracy. This was done by observing the altitude of various celestial objects. Polaris, the North Star, was best in this regard, because its position in the sky does not change appreciably with time at any one location. In addition, the altitude of Polaris above the horizon corresponds to the ship's angular distance, or latitude, north of the equator. For example, at a latitude of 30°N, Polaris will appear at an altitude of 30° above the northern horizon. If the latitude of one's destination was known, it was a fairly easy matter to use Polaris as a guide to reach this latitude and then to sail due east or west along the parallel to the destination.

FIGURE 2.10 The astrolabe, an ancient instrument for determining geographic location. The altitude of the Sun or stars above the horizon was determined by sighting through pinholes on the alidade, or pointer bar (crossing the center of the astrolabe). From this, one's latitude could be determined. (The Bettmann Archive)

When sailing in the southern hemisphere where Polaris is not visible, navigators used the position of southern hemisphere stars. Other stars and planets also were used to determine positions at sea, and numerous instruments were developed to measure the precise locations of these objects in the sky (see Figure 2.10). Most measurements were taken during the late twilight period before sunrise or after sunset, when the horizontal reference plane of the horizon was still visible.

The accurate determination of longitudinal positions remained a problem into the late eighteenth century, and it was the development of the chronometer—a highly precise clock set to Greenwich mean time—that provided the eventual solution. The Earth's rotation, as we have seen, causes the Sun's position in the sky to shift westward at a rate of exactly 15° each hour. By comparing the east-west position of the Sun at one's own location to the position of the Sun at Greenwich as calculated from the chronometer, the longitude could be readily determined.

In recent decades, navigational instruments have been developed that provide highly accurate locational data. Among these is **RDF** (radio direction finder), which employs radio beams from land-based transmitters to allow the determination of lines of position at sea. **Loran** (long-range navigation) uses synchronized, intersecting radio signals that provide precise positions. Even more recent is the widespread use of shipboard computers that receive and process navigational satellite data in order to provide highly accurate positional information in less than a second.

solar noon the official instant of noon and then simply subdivided the periods between solar noons into the appropriate hours, minutes, and seconds. This system of **local time** was highly accurate with respect to the position of the Sun, because the time was individually adjusted to each locality. For example, when it was noon in Washington, D.C., it was 12:24 P.M. in Boston and 11:17 A.M. in Chicago.

During the nineteenth century, however, a major shortcoming in the local time system became increasingly evident, and eventually led to its discontinuance over most of the world. The problem was that different places, even those located near one another, inevitably had somewhat different times. Because the local time system used the east-west position of the Sun in the sky, only places on precisely the same meridian had exactly the same time at any one instant. For example, at 40°N, the Earth's rotation causes the position of the noon Sun to shift westward over the Earth's surface at a speed of approximately 300 meters (1,000 ft) each second. This means that a location only 300 meters (1,000 ft) to the east of a reference point would, when strictly observing local time, be a second later, and a place 300 meters (1,000 ft) to its west would be a second earlier than at the reference point. Two towns located 66 kilometers (35 mi) from one another along an east-west highway would have times differing by more than three minutes.

Prior to the nineteenth century, time differences between nearby localities using the local time system were only a minor inconvenience. Timepieces were rare and rather inaccurate, appointments were not generally made for exact times, and means of transport were so slow that it took a long period to reach a destination far enough away to have a significantly different time. In the nineteenth century, however, the advent of the railroads provided a rapid and long distance means of transportation, while the telegraph offered a means of instantaneous long distance communication. The railroad and telegraph companies, as might be imagined, both experienced major difficulties because of the multiplicity of times existing throughout the world. As a result, they brought increasing pressure to bear on government agencies to standardize time over large areas.

Standard Time

Our present **standard time system** was officially established for the United Sates in November 1883 and soon was adopted by most other countries. The standard time system, like the local time system, is based on the position of the noon Sun, but only at selected meridians of longitude rather than at each specific site. The areas surrounding these standard meridians all have the same official time. Because the Earth rotates 15° in one hour, the standard time zones differ by exactly one-hour intervals. Thus, 24 time zones cover the Earth (see Figure 2.11). The basic decision made in changing from local time to standard time, therefore, was to sacrifice accuracy of time determination for much greater convenience.

The meridians used as the centers of their respective time zones are multiples of 15°. The meridian that provides the basis for all the time zones is the Greenwich meridian, and the time zone established by this meridian is called Greenwich mean time (GMT). Many people employed in fields such as transportation and communications base their activities directly on Greenwich mean time.

If the standard time system were rigorously followed, each time zone would be centered exactly on its own standard meridian and would extend 7°30′ east and west from the meridian to the borders of the adjacent time zones. For example, the 75°W meridian is used as the basis for the eastern time zone of the United States. In theory, eastern time should extend westward to 82°30′W, changing to central time from that line westward to 97°30′W. In reality, though, considerable liberties have been taken with the placement of the time zone boundaries; they are generally drawn along state or county lines and are sometimes quite distant from the theoretical 7°30′ boundaries (see Figure 2.12). The rationale for this, of course, is to avoid the problems that arise when time lines cut through the middle of a county, town, or other political unit.

Time zone boundaries sometimes are changed for convenience. For example, several years ago, most of Alaska changed to Yukon Time, one hour later than the time zone it had been in previously.

Internationally, time zone boundaries also have been adjusted considerably, as is evident from Figure 2.11. In many cases entire countries use the same time, and the time changes at the international boundary. Most large countries, especially those of great longitudinal extent, have multiple time zones. The United States, for example, has seven different time zones (four in the coterminous United States), and the Soviet Union has eleven! Because time zones

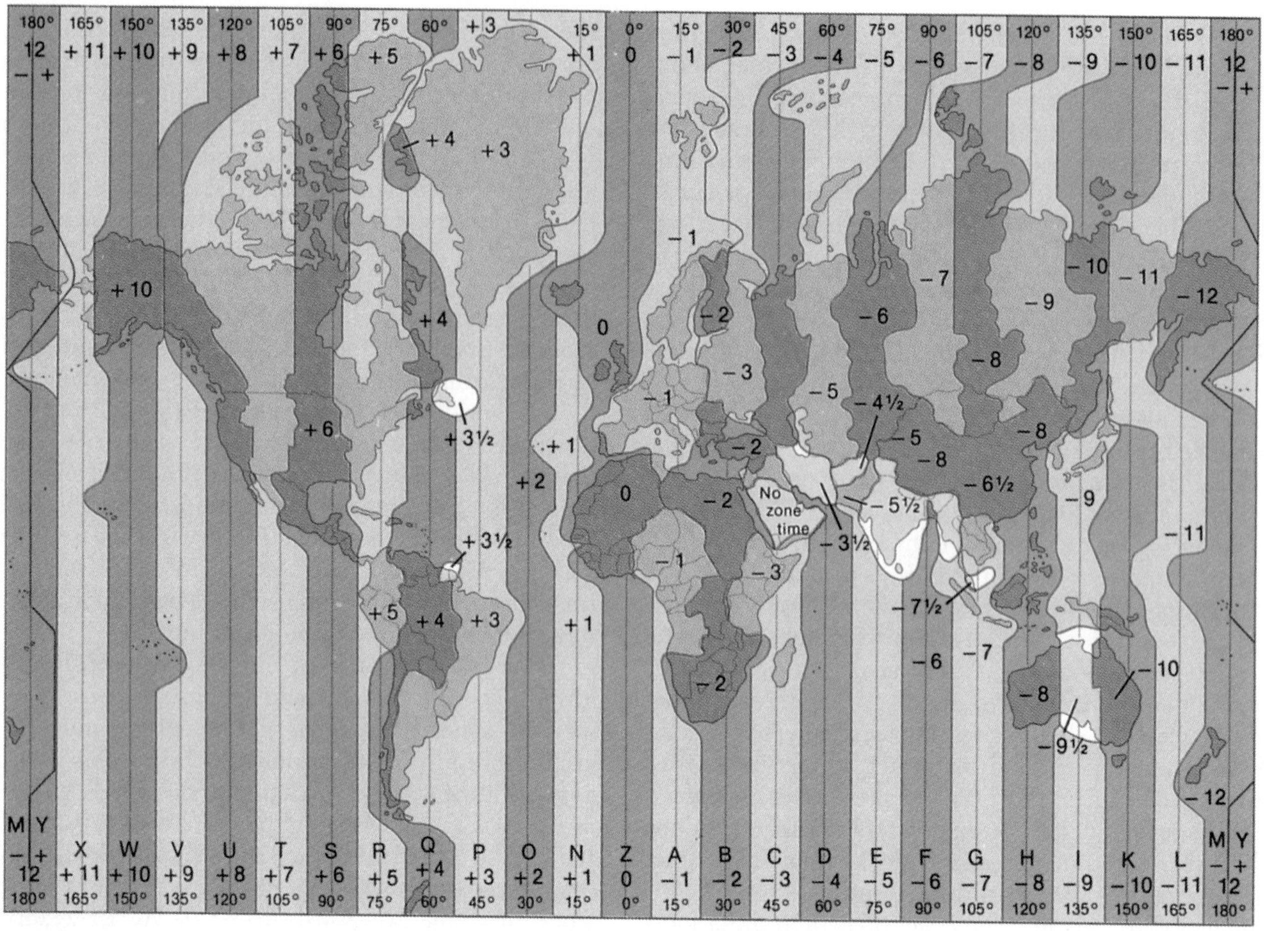

FIGURE 2.11 The global system of time zones. The International Date Line appears near the left and right margins of the map.

FIGURE 2.12 The standard time zones of the coterminous United States. The central meridian of each time zone is indicated.

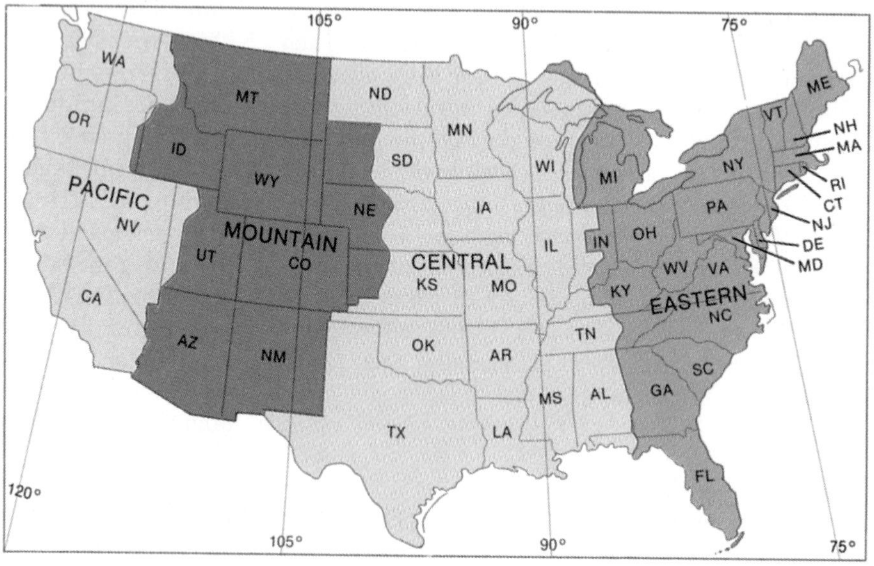

are based on 15° increments of longitude, they are wedge-shaped, and taper poleward. High-latitude countries such as Canada and the Soviet Union therefore tend to have narrower and frequently more numerous time zones than do countries of comparable size in the lower latitudes.

Some countries, such as India, Iran, and Burma, as well as the Canadian island of Newfoundland, have adopted times that are a half hour off the world standard times. The reason for doing so is that these countries are centered near boundaries between two time zones, and this variation makes their times more accurate with respect to the Sun. A few countries, such as Saudi Arabia, have not adopted standard time and still officially use local time. Over the open ocean, the official time boundaries generally are used because no practical advantage is gained from shifting them.

The International Date Line

The **International Date Line** is a unique line in the world standard time system. When it is crossed, time changes by a full 24 hours. Therefore, although the hour of the day remains the same, the calendar date and day of the week both change. In addition, the direction of the time change is the opposite of all those established by the 24 one-hour time zone boundaries. Crossing the date line from east to west causes the time to become one day later, and crossing it from west to east causes the time to become one day earlier.

The necessity for a line with these characteristics results from the existence of the 24 time zones. In effect, the International Date Line is the line that counterbalances the total changes in time produced by the time zone boundaries. The necessity arises from the fact that it is possible to travel either eastward or westward indefinitely around the Earth. Let us imagine that we have the means to travel instantaneously, and that we choose to circle the planet in a westward direction. For every 15° of longitude we travel, we cross a time zone boundary and the time becomes one hour earlier. Upon returning, if no date line existed, we would find that we had crossed 24 time lines and that time had been set back by 24 hours. We would thus be returning from our trip the day before we started it! Likewise, if we traveled eastward around the globe, the time would become later by 24 hours. Somewhere, then, a line must exist that compensates for all the gains or losses experienced when crossing the hourly time zone boundaries. The International Date Line is this line.

The International Date Line is closely associated with the 180th meridian, and the two lines run together through the Pacific along much of their lengths (see Figure 2.11). For the sake of convenience, though, the Date Line contains three major longitudinal digressions that allow continents or island groups crossed by the 180th meridian to remain entirely on one side or the other. The 1883 decision to extend the 0° longitude line through England was at least partially made so as to place the International Date Line, with its potentially confusing date changes, in as remote a portion of the Earth as possible.

Daylight Saving Time

For most people, the daylight hours are the hours around which most activities, especially work-related ones, are centered. Major advantages exist in concentrating our waking periods and activities during the day, including the presence of light to see by and the economic benefits derived from decreased fuel needs for lighting and heating. Yet, in many societies, especially in more technologically developed countries, the waking hours of most individuals are considerably offset from the daylight period. To illustrate this point, consider that on a typical day the Sun is above the horizon from 6 A.M. to 6 P.M. The average adult in Europe or North America, however, is likely to awaken at perhaps 7 A.M. or later, and to retire at 11 P.M. or later. Thus, an hour or more of daylight is wasted in the morning, while nearly half the night is spent in wakefulness. **Daylight saving time** represents an attempt to achieve a greater correlation between the daylight hours and the human activity period.

In order to convert from standard time to daylight saving time, we simply move the official time forward by one hour. This obliges individuals with set time schedules to do everything an hour earlier with respect to solar time. The implementation of daylight saving time changes no time zone boundaries; only the time within each time zone is changed.

Although it had long been in use in some parts of the United States, daylight saving time was adopted nationwide in 1966 for six months of the year, and this period was increased slightly in 1986. (A few states and localities have elected not to use the system.) Daylight saving time begins at 2 A.M. on the first Sunday in April (when the time is moved ahead

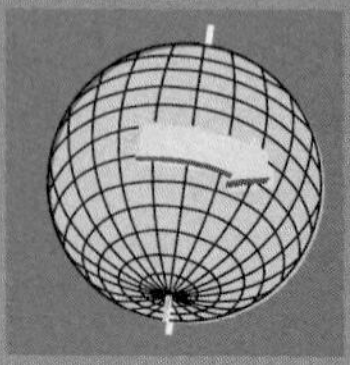

DETERMINING TIMES

Because time is based on the apparent westward movement of the Sun in the sky, time zones to one's east are progressively later and those to one's west are progressively earlier, regardless of the longitude. For example, the time in California is three hours earlier than on the East Coast. (Announcements concerning the times of television programs have familiarized most Americans with this fact.) Conversely, standard time is five hours later in Great Britain than in the eastern United States.

The simplest method of calculating the time difference in hours between any two points is to determine the difference in longitude and then divide this difference by 15. The time in the location with the most easterly longitude is always the later.

When determining the difference in longitude between two places in opposite longitudinal hemispheres for the purpose of comparing their times, it is recommended that you measure by way of the prime meridian in order to avoid the potential confusion encountered when crossing the International Date Line. The following two time problems should serve to clarify the methodology involved:

Problem 1: If it is 10 P.M., Sunday, at 135°W, what are the time and day at that same instant at 15°W?

Analysis: In going from 135°W to 15°W, we are traveling eastward; therefore the time gets later. Since both points are in the same longitudinal hemisphere, we subtract their numerical values to determine their longitudinal separation. Since 135° − 15° = 120°, the time must be 120/15 or 8 hours later at 15°W. Eight hours later than 10 P.M. Sunday, is 6 A.M. Monday.

Problem 2: If it is 9 A.M., Tuesday, at 105°E, what are the time and day at that same instant at 150°W?

Analysis: In going from 105°E to 150°W, we are traveling far west in longitude, so the time must become much earlier. Since the two given longitudes are in opposite longitudinal hemispheres, their total longitudinal separation is calculated by adding their numbers. Thus, we must go 105° + 150°, or 255°, westward from 105°E to reach 150°W. The time at 150°W must therefore be 255/15 or 17 hours earlier than 9 A.M. Tuesday. Subtracting, our new time and day becomes 4 P.M. Monday.

In this second problem, a much shorter distance would be involved if we traveled *eastward* past 180° longitude to reach 150°W. Traveling in this direction, however, includes the complication of crossing the International Date Line. Figure 2.13 shows diagrammatically the time in each time zone between 105°E and 150°W as one crosses the International Date Line.

105°E	120°E	135°E	150°E	165°E	180°		165°W	150°W
9 AM Tues	10 AM Tues	11 AM Tues	Noon Tues	1 PM Tues	2 PM Tues	2 PM Mon	3 PM Mon	4 PM Mon

FIGURE 2.13 Diagram for Problem 2.

to 3 A.M.) and ends at 2 A.M. on the last Sunday in October (when the time is moved back to 1 A.M. and standard time resumes). A phrase sometimes employed to help keep track of the resulting clock adjustments is "spring forward and fall back."

We use daylight saving time only during what is essentially the summer half of the year because the Sun rises earliest during this period and, correspondingly, the greatest loss of daylight hours is likely to occur. A number of countries in western Europe, as well as the Soviet Union and most of Alaska, now use daylight saving time throughout the year.

THE SEASONS

Over most of the Earth's land surface, and including nearly all areas lying outside the tropics, the seasonal climatic changes are among the most important characteristics of the natural environment. All life forms, both plant and animal, have had to adjust to the seasonal rhythm of the climate in these areas, and most species have evolved biological cycles that cause them to depend on the orderly seasonal progression of climatic changes.

Cause of the Seasons

The seasonal climatic changes are caused by the Earth's axial inclination of 23½° from perpendicular to the plane of the ecliptic as our planet revolves around the Sun (see Figure 2.16). This results in a yearly cycle of change in both the angle and duration of sunlight for all places on the Earth. The resulting changes in solar energy produce annual cycles of higher and lower temperatures over most of our planet.

Before we examine Earth-Sun relationships associated with each of the four seasons, we must discuss the key factor responsible for the relationship between the Sun's angle and its heating ability. The angle of the Sun in degrees above the horizon is termed its **altitude.** The altitude of the Sun is the most important factor controlling the amount of solar heating received at the Earth's surface in all but the high latitudes. When the Sun is directly overhead, or at an altitude of 90°, its heating power is greatest; as its altitude decreases, its rays gradually lose intensity. This loss of intensity occurs because a decrease in the Sun's altitude causes an equivalent amount of solar energy to strike a larger Earth surface area, spreading and reducing the energy received per unit area (see Figure 2.14). The solar energy reaching the Earth's surface is further reduced when the Sun is at a low altitude because of increased scattering and absorption of sunlight resulting from its oblique passage through the atmosphere.

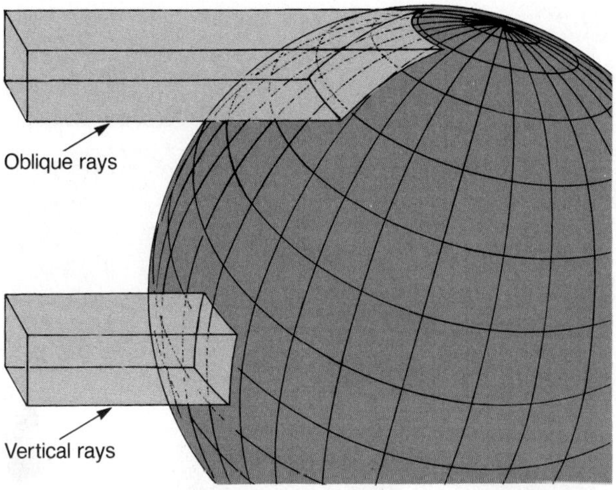

FIGURE 2.14 The intensity and heating power of the Sun's rays are directly related to the Sun's altitude. The same amount of energy is contained in both "boxes" of incoming radiation, but the oblique rays striking the higher latitudes are spread over a larger area, resulting in lower temperatures.

The tilt of the Earth's axis causes the noon Sun to vary in altitude at any place during the course of the year. This results in a considerable variation in the concentration and heating ability of the solar energy received. As the Earth revolves around the Sun, the North and South Poles constantly point toward the same two locations in space. The inclination of the poles, or of any other location on Earth, is *not* constant, though, with respect to the Sun. In fact, the

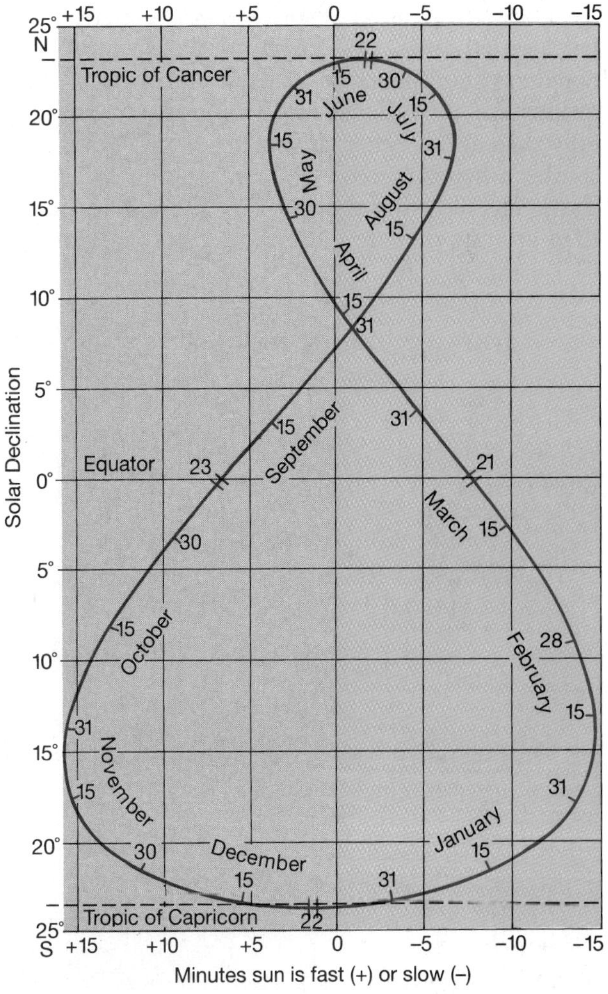

FIGURE 2.15 The analemma indicates both latitudinal and longitudinal changes in the Sun's position with respect to the Earth at different times of the year.

combination of rotation and revolution causes a constant shift in the position of the **subsolar point,** the point on the Earth's surface that has the Sun directly overhead (see Figure 2.15). Through the course of a year this point shifts alternately northward and southward because of the parallelism of the Earth's axial inclination and also moves constantly westward because of the Earth's rotation.

The four seasons are based on the changing orientation of the Earth's axis with respect to the Sun during the course of a year. The official beginning of each season marks the instant that a key Earth-Sun positional relationship occurs.

Summer officially begins in the northern hemisphere when our hemisphere reaches its maximum inclination of 23½° toward the Sun (see Figure 2.16). When this occurs, the Sun is at its **summer solstice** position, and at solar noon is directly overhead at a point 23½° north of the equator. The northern hemisphere at this time receives a larger proportion of the Earth's total allotment of solar energy than at any other time of the year. Because the southern hemisphere is on the opposite side of the Earth, it is receiving a smaller amount of solar energy than at any other time during the year. Therefore, the northern hemisphere's summer solstice also marks the beginning of winter in the southern hemisphere, where the Sun is at its **winter solstice** position. All the seasons are exactly reversed between the two hemispheres, so that when any seasonal event occurs in one, the opposite event occurs in the other.

In the northern hemisphere, the summer solstice occurs on, or within two days of, June 21. In fact, each of the four seasonal positions always occurs between the nineteenth and twenty-third day of its respective month. The exact time and date vary in different years because of leap years as well as slight variations in the Earth's orbit from year to year, caused by external gravitational forces.

The decision to begin each of the four seasons at the four "quarter positions" of the Earth's orbit around the Sun was an arbitrary one; no natural law mandates this. As a consequence, the change of seasons does not lead immediately to the changes in the weather that some people seem to expect. In fact, in large parts of the tropics, day-to-day weather conditions remain highly similar throughout the year.

During the three months of summer that follow the northern hemisphere summer solstice, the revo-

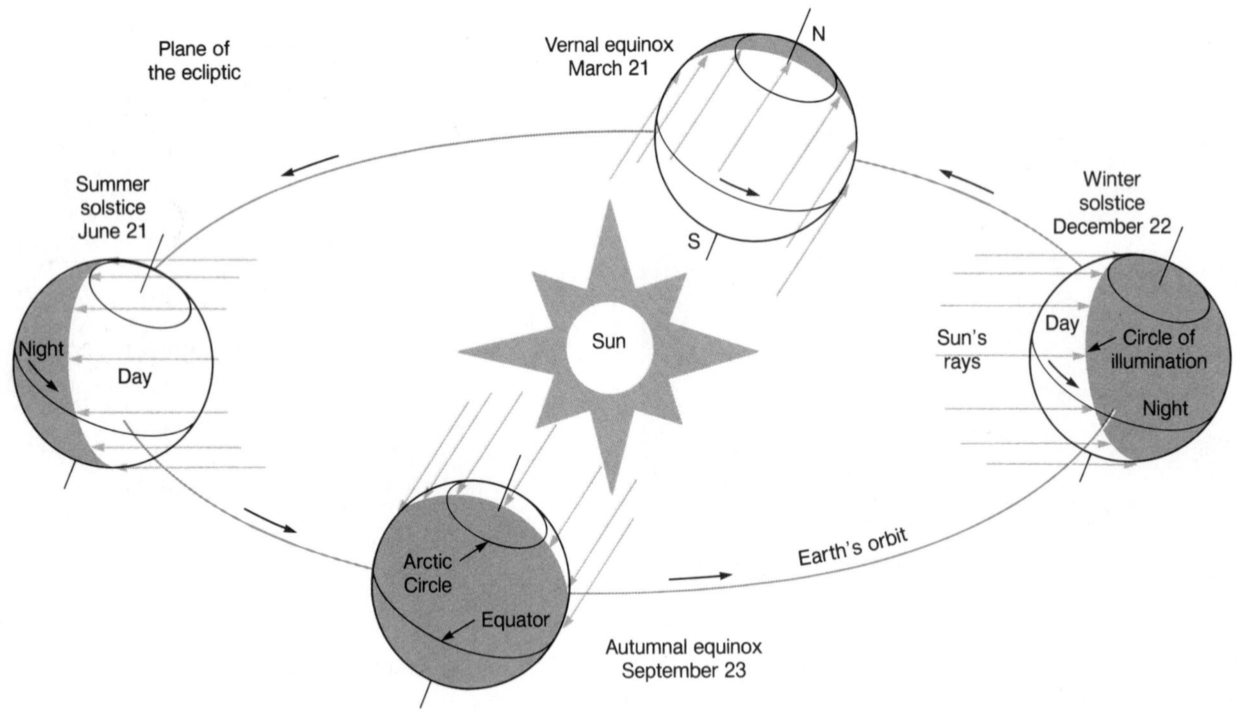

FIGURE 2.16 The four seasonal positions of the Earth. Note the parallelism of the Earth's axis in all four seasons.

lution of the Earth causes the inclination of the North Pole toward the Sun to be progressively reduced. Consequently, the Sun's vertical rays gradually shift southward from 23½°N until they reach the equator. When this occurs, between September 19 and 23, the Sun is at its **autumnal equinox** position for the northern hemisphere, and the fall season begins. In the southern hemisphere, where spring is beginning, the Sun is in its **spring** (or **vernal**) **equinox** position. At this time the axial tilt of the Earth is precisely perpendicular to the Sun's vertical rays, and the amount of solar energy reaching both hemispheres is equal.

As the fall season progresses, the continued revolution of the Earth causes the northern hemisphere to incline more and more away from the Sun. Finally, between December 19 and 23, the entire 23 ½° tilt of the axis is directed away from the Sun in the northern hemisphere (and toward the Sun in the southern hemisphere), and the northern hemisphere winter begins (see Figure 2.17b). The situation is now exactly the same for the northern hemisphere as it was six months earlier for the southern hemisphere. At this time, the Sun follows its lowest course through the southern sky and provides less energy to the northern hemisphere than at any other time of the year.

During the three months of the winter season, the Sun's vertical rays gradually progress northward from their initial position at 23½°S to reach the equator as the Earth's inclination again becomes perpendicular to the Sun's rays. Between March 19 and 23, the northern hemisphere spring equinox occurs as the Sun crosses the equator, and both hemispheres again receive equal energy. The following three months of spring witness the increasing tilt of the northern hemisphere toward the Sun until once again the summer solstice position is reached.

Two lines of latitude achieve importance as a result of the Earth-Sun relationships involved in the changing of the seasons. The **Tropic of Cancer** (23½°N) and the **Tropic of Capricorn** (23½°S) mark, respectively, the highest latitudes north and south that the Sun's vertical rays ever reach. Stated differently, the Sun is always directly overhead somewhere between 23½°N and 23½°S. Places between these two parallels have the Sun directly overhead on two occasions during the year—once while the Sun is shifting northward toward the Tropic of Cancer, and once while it is shifting southward toward the Tropic

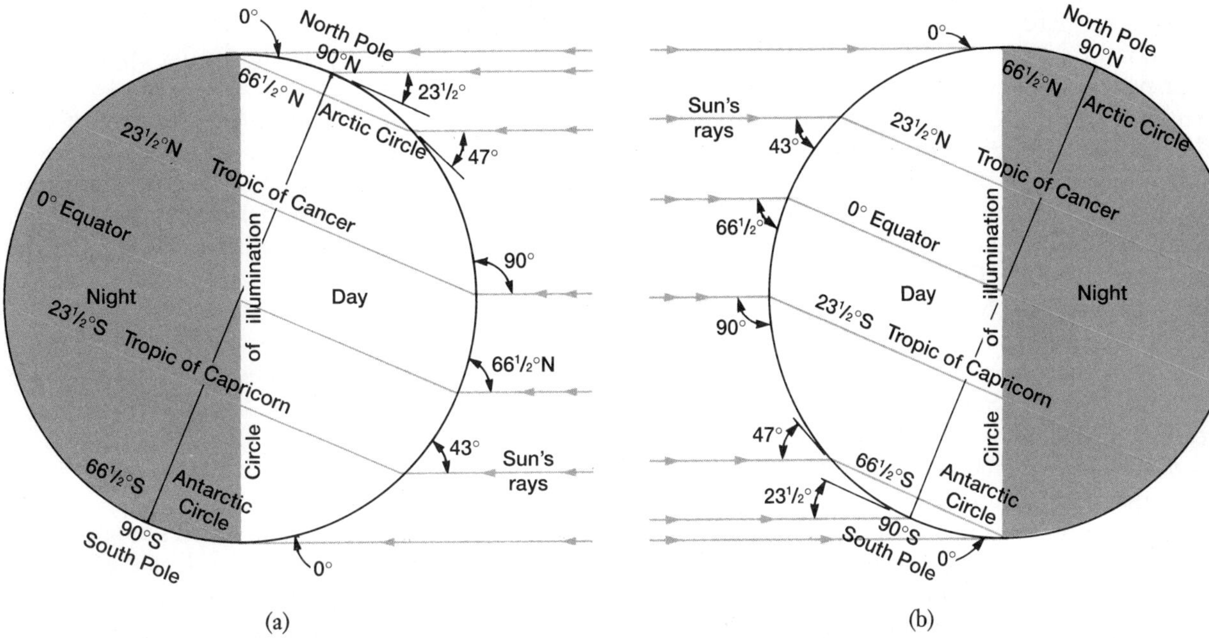

FIGURE 2.17 The angle of incidence of the Sun's rays at the times of the solstices. At the time of the northern hemisphere summer solstice (a), the Sun is directly overhead at the Tropic of Cancer. Days are longer than nights in the northern hemisphere, while nights are longer than days in the southern hemisphere. The situation is reversed during the winter solstice (b), at which time the Sun is directly overhead at the Tropic of Capricorn.

of Capricorn. More than half the world is poleward of these lines (or outside the **tropics**) and never has the Sun directly overhead. For nontropical locations, the Sun reaches its highest altitude at solar noon on the date of the summer solstice.

Lengths of Day and Night

The changing of the seasons involves not only constant changes in the Sun's noon altitude, but also variations in the lengths of day and night for all places on Earth except the equator. All locations in the northern hemisphere experience longer days than nights during the spring and summer, when the northern hemisphere is inclined toward the Sun. Days are shorter than nights during the fall and winter, when the northern hemisphere is tilted away from the Sun. The lengths of day and night at equivalent latitudes in the southern hemisphere are exactly reversed from those in the northern hemisphere because the seasons are reversed. Only the equator, the line separating the two hemispheres, experiences equal-length (12-hour) days and nights throughout the year.

The hemisphere inclined toward the Sun always occupies the major part of the daylit half of the Earth. The greater the inclination, the greater the proportion of the hemisphere that is in daylight. On the two solstice dates, when the Earth's inclination is at a maximum with respect to the Sun's rays, the hemisphere experiencing the summer solstice is 63 percent in daylight and only 37 percent in darkness.

The length of daylight is not evenly distributed by latitude, but becomes progressively less equal poleward. Figure 2.17a shows that more of each line of latitude north of the equator is on the day side than on the night side of the Earth. At the time of summer solstice, all parallels north of the **Arctic Circle** (66 ½° N) are entirely in daylight. The Sun therefore does not set in this area (sometimes called "The Land of the Midnight Sun") at this time. Conversely, increasing proportions of parallels progressively south of the equator (and at their winter solstice) are in darkness until, south of the **Antarctic Circle** (66 ½°S), continuous night prevails. Of course, this situation is exactly reversed six months later (see Figure 2.17b) so that both hemispheres (and indeed all places on Earth) receive six months of day and six months of night over the course of a year.

The Arctic and Antarctic Circles therefore mark the equatorward limits of the two portions of our planet in which day and night exceed 24 hours at the times of the summer and winter solstices, respectively. All other areas experience both day and night each calendar day. The situation is most extreme at the North and South Poles, where day and night each last for six months. The Sun at these two locations remains constantly below the horizon during the fall and winter but is constantly above the horizon during the spring and summer. On the dates of the equinoxes all places on Earth experience 12 hours of day and 12 hours of night.

Path of the Sun in the Sky

Contrary to popular opinion, the Sun does not rise and set vertically except at the equator. It actually describes a curved path through the sky, with the horizontal component of apparent motion increasing directly with latitude. At the poles, the Sun's motion is essentially horizontal as it circles the sky in a clockwise direction at the North Pole and a counterclockwise direction at the South Pole.

The horizontal component of the Sun's motion in the higher latitudes causes a substantial proportion of the night hours to not be completely dark, but instead to experience **twilight.** Twilight is the familiar condition of semidarkness that occurs before sunrise and after sunset as sunlight is scattered by air molecules, clouds, and dust particles in the upper atmosphere. It occurs when the Sun is less than 18° below the horizon. Just as the Sun never rises more than 23½° above the horizon at the poles, it is never more than 23½° below the horizon. The polar regions therefore experience substantial periods of twilight, especially in early autumn and late winter. On the other hand, the vertical motion of the Sun at the equator is the basis of the often expressed observation that "night falls swiftly in the tropics," because the Sun moves directly and rapidly past the 18° mark below the horizon.

SUMMARY

This chapter has examined the Earth's origin, position, and movements in space, as well as the size and shape of our planet. These factors have enabled us to devise systems for describing locations and determining times on the Earth's surface; they also are responsible for the changing of the seasons. We return to the Focus Questions posed at the beginning of the chapter to summarize the most important concepts presented:

1. *What is the geographical setting of the Earth in space, and what movements is it making through space?*

The Earth is the third planet from the Sun in a solar system that contains nine known planets. The solar system is located in the outer portion of a spiral arm of the Milky Way galaxy—a rotating collection of many billions of stars. The Earth is involved in several movements through space, occurring at greatly differing orders of magnitude. At the largest scale, we, along with the entire galaxy, are moving away from the center of the universe as an apparent consequence of the Big Bang. We are also revolving around the center of the Milky Way galaxy. On a smaller scale, the Earth revolves around the Sun once each year and rotates on its axis once each day.

2. *What are the size and shape of the Earth?*

The Earth is roughly spherical because of the inward pull of gravity, and has a diameter of approximately 12,700 kilometers (7,900 mi). The centrifugal force of the Earth's axial rotation, however, has caused the equatorial regions to bulge outward and the polar regions to become flattened, so that the Earth is more accurately described as an oblate ellipsoid. Forces above and below the Earth's surface also have produced landforms of many sizes and shapes, so that in detail the shape of the Earth is complex and undoubtedly unique.

3. *What basic directional and locational systems are employed on the surface of the Earth?*

Most directional and locational systems in common use are based on the positions of the North and South Poles and of the equator. Two directional systems widely used in the United States are the Mariner's Compass and the azimuth system. Mariner's Compass directions are stated as various combinations of the words *north*, *east*, *south*, and *west*, or letters *N*, *E*, *S*, and *W*. Azimuths are stated in degrees in a clockwise direction from north (0°).

The system of latitude and longitude is the most important locational system currently in use. Degrees of latitude are measured north and south of the equator and degrees of longitude are measured east and west of the prime meridian.

4. *How is the global system of time zones organized?*

The standard time zones ideally form 15°-wide swaths of longitude, each centered on a meridian divisible by fifteen. Each of the twenty-four time zones differs in time by one hour from adjacent zones. Over land areas, the zones usually are adjusted to coincide with political boundaries. The international date line, located in the central Pacific, is a line over which a 24-hour time difference exists.

5. *What causes the seasons?*

The seasons occur because the Earth's axis is inclined at 23 ½° from a perpendicular angle to the plane of the ecliptic. As the Earth revolves around the Sun during the course of the year, the northern and southern hemispheres alternately are inclined at varying angles toward and away from the Sun. Spring and summer occur when a hemisphere is inclined toward the Sun; at this time the Sun is higher in the sky, periods of daylight are longer, and total energy receipts are greater. Conversely, fall and winter occur when the hemisphere is inclined away from the Sun.

Review Questions

1. Describe the appearance and characteristics of the Milky Way galaxy. Explain in general terms the location of our solar system within the Milky Way.
2. Briefly describe the Big Bang theory of the origin of the universe. Include an explanation of why the universe is expanding and how the galaxies and stars were formed.
3. Draw diagrams illustrating the Earth's revolution around the Sun and the rotation of the Earth on its axis. Be sure to indicate the correct directions of motion.
4. Why is the Earth spherical? What two major departures from a precise spherical shape exist? Why do they occur?
5. Explain the differences in the nature and locations of Earth's geographic and magnetic poles.
6. What is the purpose of the system of latitude and longitude? What similarities and differences exist between parallels and meridians? How are degrees of latitude and longitude subdivided for greater precision?
7. Why was the local time system discontinued? What advantages and disadvantages does the standard time system offer? Explain how this system is organized.
8. In what time zone are you located? Based on your longitude, determine by how many minutes your local time differs from standard time.

9. Where is the International Date Line located? Explain what change in time occurs as one crosses the International Date Line, and why this line is necessary.
10. What is the justification for using daylight saving time? Why is it used in the United States during only part of the year?
11. Draw a single diagram illustrating the parallelism of the Earth's axis and the position of the Earth with respect to the Sun at the times of the solstices and the equinoxes. Label each position.
12. Describe the locations and explain the significance of the Arctic and Antarctic Circles and of the Tropics of Cancer and Capricorn.

Problems

1. If you travel 60° south of 45°N, at what latitude will you be located?
2. If you travel 50° west of 140°W, on what meridian will you be located?
3. If it is 11 P.M. Monday at 45°W what are the time and day at the same instant at 135°W?
4. If it is 9 A.M. Thursday at 120°W, what are the time and day at the same instant at 165°E?
5. Is the Sun's noon altitude higher in Maine or in Florida during the summer? Does Maine or Florida have the longest daylight period during the summer? Which U.S. state experiences the greatest number of hours of twilight during the year?

Key Terms

solar system
Milky Way galaxy
revolution
rotation
geographic North and South Poles
equator
parallelism
oblate ellipsoid
latitude
longitude
parallel
meridian
great circle
small circle
prime meridian
local time
standard time
International Date Line
daylight saving time
solstice
equinox
Tropics of Cancer and Capricorn
Arctic and Antarctic Circles

C H A P T E R 3

DEPICTING THE EARTH—MAPS AND REMOTE SENSING

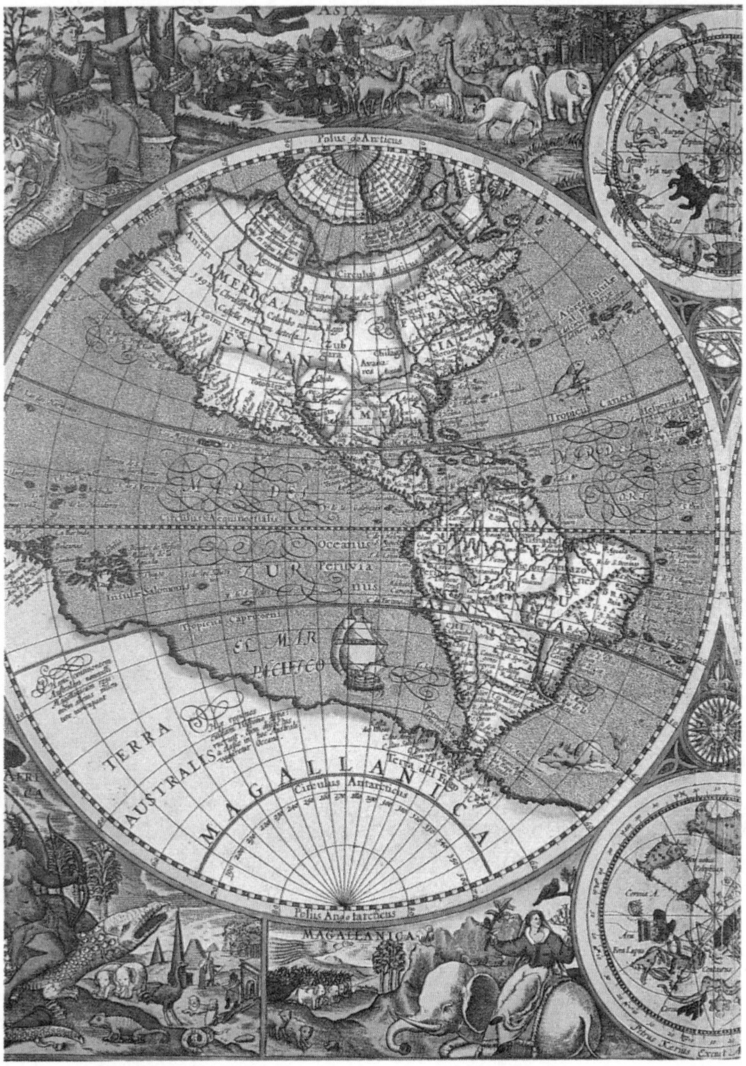

OUTLINE

FOCUS QUESTIONS

1. What is a map? With what basic map information must the user become familiar before the map can be employed to full advantage?
2. How is locational information about the Earth transferred onto a flat surface to make a map? Why are all flat maps of the Earth distorted?
3. How is information concerning differing surface elevations displayed on a flat map?
4. How is map information collected and compiled?

In order to conduct geographical research, it is generally necessary to collect, display, and analyze large quantities of spatial data. In this chapter, we examine the remote sensing techniques by which most spatial data are currently gathered, the properties of the maps that permit us to effectively display this information, and the geographic information systems that are increasingly being used in its analysis. Strictly speaking, maps, remote sensing techniques, and geographic information systems are not in themselves topics in physical geography, as they are human constructs rather than natural phenomena. Because they are essential tools for the effective geographic analysis of both physical and human phenomena, however, a basic understanding of these tools and techniques is of great importance to students of geography.

MAPS

A **map** is a pictorial representation of the geographic locations of selected surface features at a reduced scale. Maps perform two basic functions: (1) they reduce the size of an area of interest so that it can be shown in its entirety in the space available on the map; and (2) they simplify spatial patterns by displaying only those phenomena of importance to the map user.

The origins of maps are lost in the mists of antiquity. There can be little doubt that, even before the emergence of civilized societies, crude maps were being scratched with sticks on the ground or drawn with pieces of chalk or charcoal on cliff walls. Moreover, the need to convey, by means of a diagram, the locations of important phenomena such as a trail or a source of water was so basic that the use of maps apparently began independently in several early societies.

Maps currently are employed for a great variety of purposes by individuals in many different occupations. The subject of maps and mapmaking is one of the few associated more closely with geography than with any other academic field. This close association is a natural one; geographers are interested in the geographic distributions of phenomena, and this is precisely what maps are designed to depict. In fact, mapping is the only practical method by which large amounts of data can be presented for many purposes.

Modern maps, such as that shown in Figure 3.10, contain a great deal of information. The verbal description of the locations and relationships of all the features shown in this figure would take many pages of text and could never be communicated as effectively as it is on this one map sheet. Nearly all geographical studies employ maps, and maps have rightfully been called the chief research tool of the geographer. By displaying the locational patterns of the phenomena under study, maps often aid in geographic understanding by shedding light on the reasons for these patterns.

Cartography, the art and science of mapmaking, has developed continually over the past several centuries, and currently is highly automated and precise. Much of the information contained on modern maps is obtained through the remote sensing techniques discussed later in this chapter.

Basic Map Information

Maps have been devised to show many different types of spatial information, and this information can be displayed in a great variety of forms. Therefore, before employing a map, map users should take the time to acquaint themselves with the types of information contained on the map and the ways it is shown. Among the most basic types of map information are the title, date, location, directional orientation, and legend.

Title

The map title, which should be prominently displayed, indicates the area depicted as well as the subject matter of the map. It should be clear and concise, because many people obtain maps, sight unseen, based on their titles.

Date

The date of the map is important because most types of map information change over time, and the information depicted will likely become outdated. In many cases the year is sufficient, but in some instances, as with weather maps, information changes so rapidly that the map should be dated to the hour or even to the minute. If the information on the map is substantially older than the date of publication of the map itself, this too should be indicated.

Location

Many maps indicate locations by the use of numbered parallels and meridians (see Chapter 2), although other locational systems also are employed. A map that depicts a small area whose location may not be easily recognizable by its shape or surroundings may contain a small inset map showing the location of the mapped area within its larger geographical setting.

Directional Orientation

The compass orientation of the area shown on a map is commonly indicated by the use of parallels and meridians. Not all maps maintain constant directions everywhere, so these lines, although actually straight in a two-dimensional sense, may curve on the map. A north directional arrow also is frequently used. Most maps are oriented so that north is toward the top, south toward the bottom, east toward the right, and west toward the left.

Legend

The legend is the information section of a map. The primary purpose of the legend is to explain the major map properties and symbols. **Symbols** are the devices employed to represent the phenomena displayed on the map. They may be point symbols, such as dots to represent towns; line symbols, such as those used to represent the roads on a highway map; or area symbols, such as colors that indicate elevations or shading patterns that represent types of land use.

Most geography textbooks employ numerous maps to show the spatial distributions of the phenomena being studied. Two frequently used methods for presenting this information are the familiar **dot maps,** where each dot represents the presence of a certain number of objects, and **isarithmic maps,** which use lines connecting places of equal numerical value. A number of isarithmic maps appear in this book. For either type of map, the values represented by the dots or lines are indicated in the title, in the legend, or in both.

Scale

The **map scale** is the ratio between distances on a map and the corresponding distances on the Earth's surface. All maps are reduced from reality, and the scale indicates the extent of this reduction. Scale information is especially vital when one of the chief purposes of the map is to indicate distances between places, as on a road map. Unfortunately, for reasons to be explained shortly, the scale of a map showing a large area cannot be made consistent over its entire surface. The scales of maps depicting restricted areas such as states or cities may, for practical purposes, be considered constant, however.

Three basic methods of stating map scale are commonly employed, and one or more of these should appear in the map legend. **Verbal scales** often are provided on maps designed for use by people with limited map experience. They frequently appear, for example, on road maps. In this case, the scale is stated in a sentence format, as in the following example:

1 inch equals 8 miles.

By convention, the first figure given (1 inch) is the map distance, and the second (8 miles) is the corresponding Earth distance.

A **graphic scale** generally takes the form of a bar or line graph on which the map distances corresponding to real Earth units are shown. On some maps, multiple bars may be used to show either different units of measure (see Figure 3.1) or the scale on different parts of the map. Graphic scales have the advantage of providing a visual indication of the scale and of allowing for direct measurements to be taken. Unlike the other two methods of stating scale, they remain correct if the map is enlarged or reduced.

A **representative fraction (RF) scale** is stated as a direct ratio between map distances and corresponding Earth distances. It may be written in either of the following fashions:

1:63,360 or 1/63,360
(values, of course, vary)

This fraction is a dimensionless quantity that can be thought of as the "degree of reduction" used to dis-

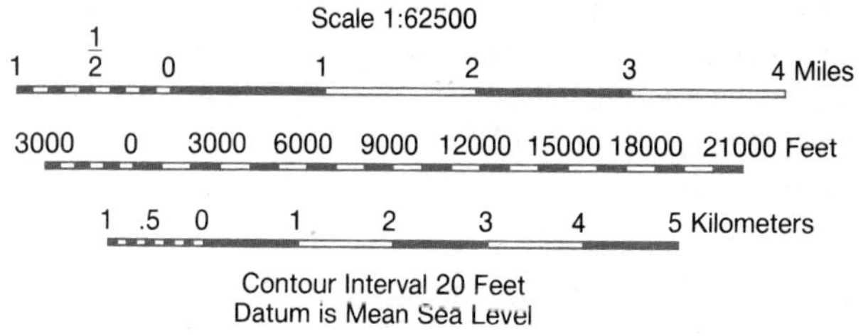

FIGURE 3.1 The map scale information section in a United States Geological Survey (USGS) topographic map legend. The scale appears both in representative fraction and in multiple graphic scale formats.

play the portion of the Earth's surface depicted on the map surface. In both cases the first number (the fraction numerator) is the map distance, which is always set at a value of one; the second figure (the fraction denominator) is the corresponding Earth distance. A major advantage of the RF scale is that any unit of measure can be employed, as long as the same unit is used for both map distances and Earth distances. In the example above, one inch on the map equals 63,360 inches on the Earth, one centimeter on the map equal 63,360 centimeters on the Earth, and so forth. Because there are 63,360 inches in one statute mile, the RF scale of 1:63,360 is equivalent to the verbal scale "one inch equals one mile."

The cartographer (mapmaker) chooses the scale of a map based largely on the intended use of the map, the amount of information to be shown, and the amount of space available on the map sheet. An unlimited number of scales are possible, but, for general descriptive purposes, maps are often described as being small-scale or large-scale.

A **small-scale map** is greatly reduced from reality. This large reduction enables the map to show a large surface area but has the disadvantage of showing little surface detail. Most classroom wall maps are small-scale maps, even though the maps themselves are made large in size for easy viewing from a distance. **Large-scale maps** are less reduced than are small-scale maps. They therefore cannot show as much surface area, but are capable of showing more detail. An example would be a road map of a city or county. No precise definitional boundary between small-scale and large-scale maps has been agreed upon. Generally speaking, though, maps displaying areas the size of states of countries are considered small-scale maps, while those of cities or counties are considered large-scale maps.

It is important to note that the amount of surface area depicted on maps differs by the inverse square of their difference in linear scale. For example, if one of two equal-sized maps is constructed at a scale three times larger than the other, it will show only one-ninth the total area of the smaller-scale map.

Map Properties and Distortion

Maps, as we have seen, are representations of the spherical surface of the Earth. The great majority of maps, however, are drawn on plane (flat) surfaces, usually sheets of paper. The planar nature of these maps produces the major problem in cartographic depiction—that of map distortion. Distortion in this context refers to inconsistencies in the spatial properties of different portions of the map. The crucial map properties of size, shape, distance, and direction may differ from one place to another, and it is not possible to eliminate this problem completely.

Map distortion exists because it is geometrically impossible to depict a spherical surface, or any substantial part of it, accurately on a plane surface. This dilemma can be illustrated by painting a map of the Earth on a large rubber ball such as a basketball, and then attempting to flatten it out. This obviously cannot be done without considerable stretching, cutting, tearing, or compression. Even pieces of the ball (map) will be curved and cannot be flattened without stretching or tearing. Yet, this transformation from a spherical surface to a planar one is exactly what the cartographer must accomplish in producing a flat map. In the process, some of the basic map properties of size, shape, distance, and direction must be distorted, and on many maps all of these properties are distorted.

Because map distortion results from the flattening of the Earth's curved surface, maps depicting larger surface areas are more distorted than maps depicting limited areas. Maps of the entire Earth, therefore, are subject to the greatest distortion. On the other hand, large-scale maps generally show such a small geographical area that distortion can be considered negligible.

The one way to map the Earth without distortion is to use a globe, which duplicates the three-dimensional shape of our planet. The geometrical accuracy of globes is their greatest advantage over flat maps. Because globes are not distorted, relative sizes, shapes, distances, and directions can all be shown as they actually exist on the Earth.

If flat maps are distorted, and globes are not, why are flat maps so much more commonly used? The reason is rather obvious: the handling and use of globes presents numerous practical difficulties. Undoubtedly the biggest problem is their very small scale and resulting lack of detail. There is no way, for example, that a globe can be used as a practical road map. Other major shortcomings are the great bulkiness and high cost of globes, difficulties in storing them (see how many you can fit in your car's glove compartment!), the difficulty of making measurements on a curved surface, and the impossibility of seeing more than half the Earth at any one time. Because of these problems, globes are used primarily for display and general reference purposes. They provide an accurate depiction of the Earth's surface, but once the correct geographical relationships are mastered, practical work then normally proceeds by use of the necessarily distorted flat maps. On the other hand, it is undoubtedly true that globes are not used enough for educational purposes, and that many students have developed erroneous impressions about the geography of our planet by taking literally the views of the Earth appearing on flat maps.

Although flat map distortion cannot be eliminated, it can be manipulated. Maps can be made to have their greatest distortion in what are generally considered the less important parts of the Earth's surface, such as the polar regions or oceans. It is also possible to construct maps that have their distortion confined to certain spatial characteristics, while other characteristics remain undistorted. The types and amounts of distortion on an unfamiliar map can be estimated by comparing it to a globe or by examining the behavior of the map's parallels and meridians.

Correct map location is a property of vital importance. All maps, regardless of distortion, should show the absolute locations of mapped features correctly with respect to latitude and longitude (or other reference system). Two additional basic map properties are their method of depicting an area's size and shape. These two properties are strongly influenced by the problem of distortion; in fact, it is impossible for a single flat map to display correctly both an area's size and its shape. It is possible, however, to keep either one or the other of these map properties from being distorted.

A map that has the property of consistency in relative size is called an **equal-area map.** All places on such a map are shown in their proper proportional sizes—that is, subject to the same total areal reduction. Proportional sizes can be maintained only by distorting lengths, and therefore shapes; lengthening in one direction is exactly compensated for by shortening in another. Shape is therefore badly represented on an equal-area map. Equal-area maps are commonly used to show areal distributions of specific phenomena. They are generally used, for example, for dot maps, because the spacing of the dots indicates the density of the subject being represented.

A map that displays shapes correctly is termed a **conformal map.** Actually, only shapes of small areas can be shown without distortion on conformal maps; over large distances, shapes and directions remain distorted. Conformality can be achieved only by greatly distorting sizes, and high-latitude regions usually appear at a much larger scale than low-latitude regions. An important use for conformal maps is in determining directions, and they are valuable as navigational charts. The parallels and meridians on a conformal map cross at right angles, just as they do on a globe.

On conformal maps, directions from any point start out correctly, but gradually become distorted with distance. It is impossible to devise a flat map on which all directions remain correct over the entire map surface. However, a third type of map, the **azimuthal map,** retains true directions outward from one central point, regardless of distance.

Many maps are neither equal-area, conformal, nor azimuthal, Instead, they represent some degree of compromise among these basic types. Their distortion is distributed among the various map properties, usually producing a better appearing map. Such maps are commonly used for general reference or classroom purposes.

Map Projections

A geometric figure such as a sphere that cannot be flattened without distortion is said to be **undevelopable.** The basic goal in mapmaking is to transfer the Earth's undevelopable surface onto a **developable** surface—one that can be flattened. A **map projection** of the Earth is produced by systematically transferring or projecting the image of all or part of the Earth's surface onto a developable surface, which is then developed, or flattened. Three particular developable surfaces—the plane, the cone, and the cylinder—are commonly used for map projections (see Figure 3.2).

In order to visualize the map projection principle, imagine that a light bulb has been placed in the center of a transparent globe that has drawn on it parallels, meridians, and the outlines of the continents. Place in turn a plane, a cone, and a cylinder over the globe as shown in Figure 3.2. The shadows of the features on the globe will then be projected onto these surfaces to produce maps that, in the cases of the cone and cylinder, can then be cut and flattened, as shown in Figure 3.3.

When map projections are made in this fashion, no distortion of any type occurs at the points or lines

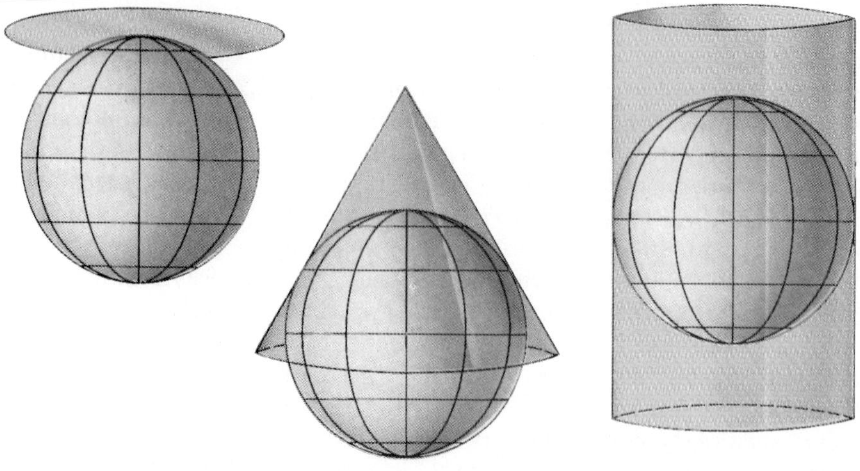

FIGURE 3.2 The projection principles for plane, conic, and cylindrical map projections. To envision how the projections are accomplished, imagine a light placed at the centers of each of the globes projecting the outlines of the Earth features onto the three geometrical surfaces.

where the developable surface rested in contact with the globe. These places are termed **standard points** or **standard lines.** Map projections often are constructed so that standard lines are particular parallels or meridians. Distortion occurs everywhere else on the map, increasing as the developable surface lies progressively farther from the globe's surface. Because the projected images expand outward in all directions after they leave the globe, locations on portions of the developable surfaces that are situated well away from the globe will appear at a larger scale, and angular relationships in these locations will also be distorted.

Another option is for the developable surface to cut through the globe, rather than lie tangent to it, as shown in Figure 3.4. For a plane projection, this will produce a standard line rather than a standard point. For both conic and cylindrical projections, the result will be two standard lines rather than one. Because the amount of contact between the globe and the developable surface has been increased, the total map area that has little or no distortion is also increased. The portions of the map that pass through the globe will now appear at a smaller scale than those at the standard points or lines.

While the preceding paragraphs describe the principles of construction of plane, conic, and cylindrical projections, many modern projections are constructed mathematically and often are altered so as to achieve some desired property. For any projection,

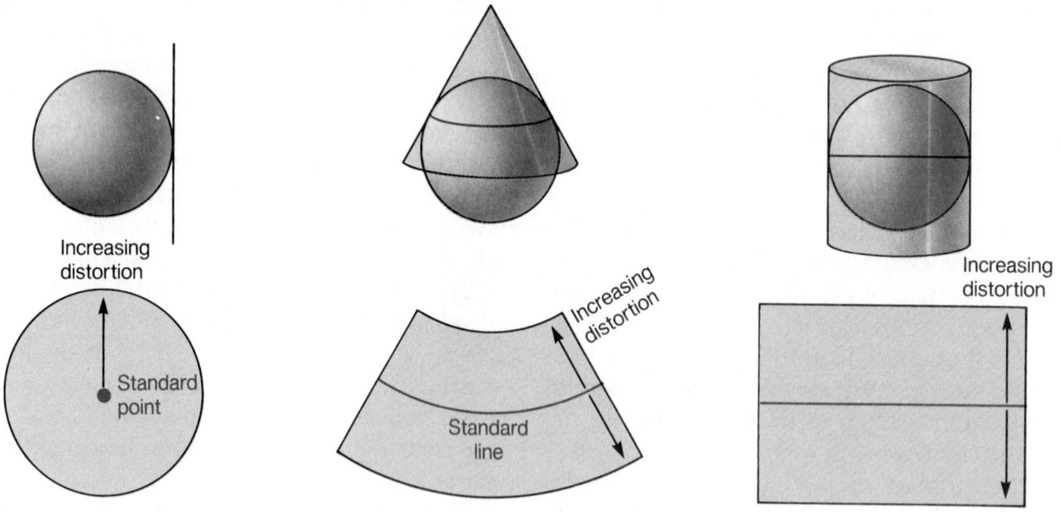

FIGURE 3.3 A map projection formed by a plane, cone, or cylinder lying tangent to the Earth's surface has only a single standard point or standard line, with distortion increasing outward from it.

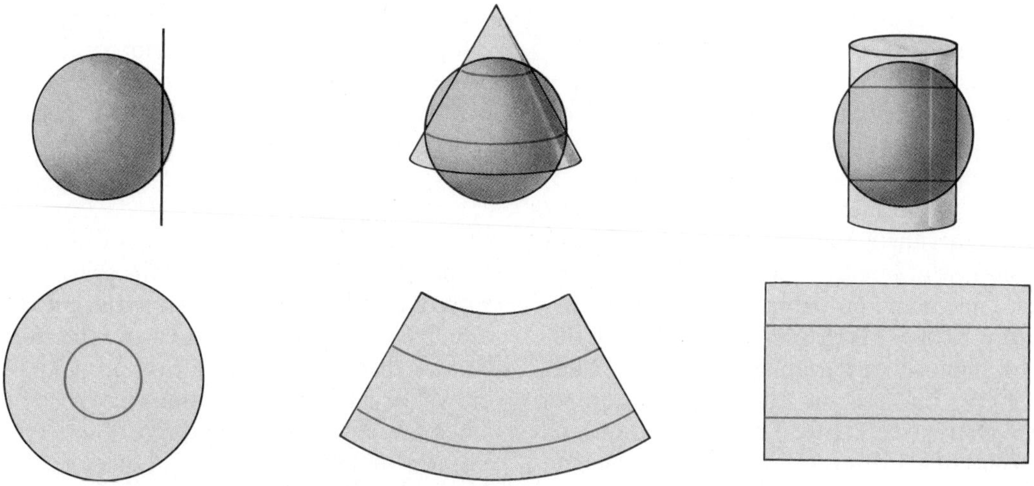

FIGURE 3.4 A map projection formed by a plane passing through the Earth's surface has a circular standard line of no distortion. A map projection formed by a cone or cylinder passing through the Earth's surface has two standard lines, allowing a larger map area to display little or no distortion.

however, the transfer of map information must be accomplished in a systematic fashion. Because many types of developable surfaces exist and because they can be mathematically altered in varying ways, an unlimited number of types of map projections are possible.

Characteristics of Major Map Projections

Most commonly used map projections are members of the plane, conic, and cylindrical projection families. Each family has characteristic strengths and weaknesses. Because the uses of maps are so varied, no single projection group or type can be considered as best; the type of map projection chosen should depend on the map properties that will allow the map to display the desired information most clearly and accurately.

Plane Projections

A plane projection is a direct projection of a portion of the Earth's surface onto a plane surface. The resulting maps are circular and are capable of showing half of the Earth at most. Most plane projections are constructed so that they have a single point of contact with the globe. If so, they are azimuthal (correct in direction) from that point, with distortion increasing symmetrically outward.

Most plane projections are centered on the North or South Pole, and are termed **polar projections.** They may, however, be centered on any point, such as a city, airport, or park, to show directions and distances to other locations. Any straight line drawn from the standard point to any other point will represent the line of shortest Earth distance between the two. Such lines are termed **great circle routes** because they form arcs of great circles on the globe. Because of the ease of plotting great circle routes on them, plane projections are commonly used for the preparation of navigational charts for ships and aircraft.

Conic Projections

Conic projections are generally constructed so that the apex of the cone is located directly above the North and South Pole. This produces a standard parallel of no distortion in the middle latitudes. Conic projections therefore are well suited for depicting middle latitude areas with long east-west dimensions, such as the United States, Canada, Europe, and Australia. If the cone cuts through the globe so as to produce two standard parallels, the width of the zone of little distortion can be expanded. On conic projections, parallels appear as arcs of circles, while meridians appear as straight lines that gradually converge poleward.

Cylindrical Projections

Cylindrical projections are almost always oriented so that the equator forms the standard parallel. As a result, they display the low latitudes most accurately, with scale—and distortion—increasing poleward. Because only the poles cannot be shown, such projections are frequently used for world maps. On a cylindrical projection, both parallels and meridians appear as straight lines, giving the map a rectangular shape (see Figures 6.10 and 6.11, for example). Unless they have been mathematically adjusted, the spacing of the meridians remains constant in a poleward direction, while that of the parallels increases.

The best known and probably the oldest cylindrical projection is the **Mercator projection,** developed by the Flemish cartographer Gerhardus Mercator in 1569 for use in the construction of navigational charts (see Figure 3.6a). The unique attribute that has made this projection so valuable and durable is the fact that any straight line drawn on a Mercator map is a **rhumb line,** or line of constant compass bearing. In plotting a course, on a Mercator map, a navigator simply draws a straight line on the map between his current position and his destination and measures the compass orientation of the line. Even though such a course does not normally follow the shortest (or great circle) route, the ability to easily determine an accurate compass heading to one's destination was a major breakthrough at a time when navigational techniques were very imprecise. Unfortunately, the convenient rectangular shape of a Mercator map has led to its frequent use as a classroom wall map. As a consequence, many students have developed an inaccurate impression of the relative sizes of land and water bodies in different parts of the world. This problem is explored further in the Focus Box on pages 40–41.

Elliptical Projections

None of the three basic projection types just discussed is capable of showing the entire Earth. In order to meet this need and avoid the excessive high latitude distortion of cylindrical projections, a variety of oval or elliptical projections have been developed. They are currently in widespread use for wall maps and textbooks, including this book (see, for example, Figure 11.11). Elliptical projections are not based on the projection of light, but instead are mathematically derived. Most types contain straight parallels and have meridians that bow outward in opposite directions from a straight central meridian. Distortion is generally greatest near the margins, especially in the high latitudes.

Interrupted and Condensed Projections

A problem shared by most types of projections is that they have only one, or at most two, standard lines where no distortion exists. Interrupted projections

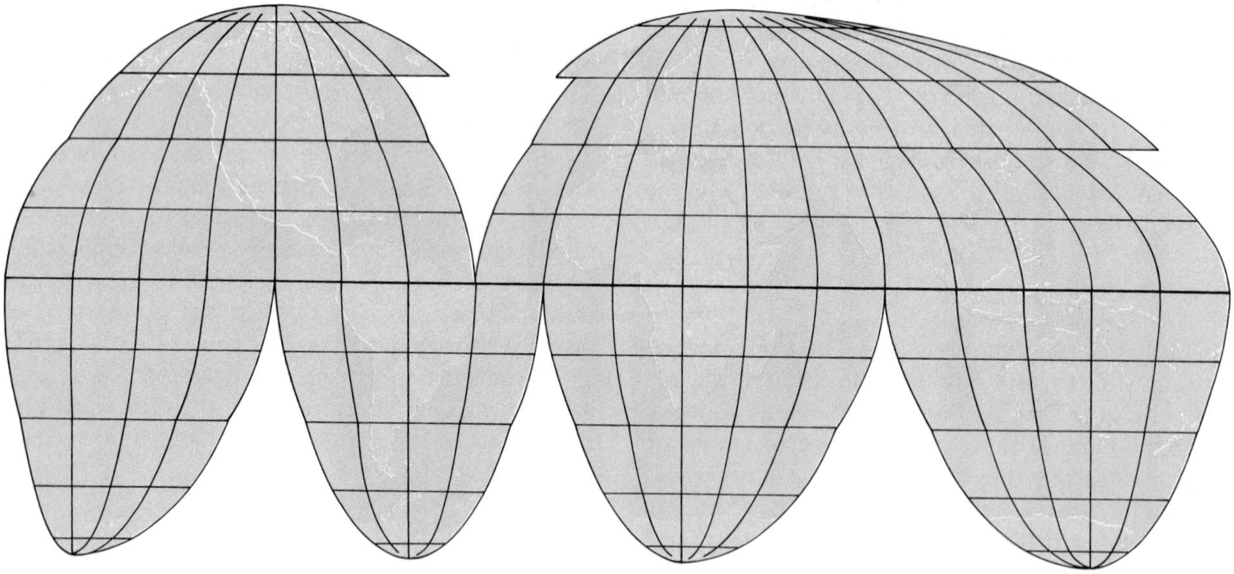

FIGURE 3.5 Goode's interrupted projection of the Earth. The separation of the ocean segments allows the more important land segments to be depicted with less distortion.

use several standard lines. By separating the projection surface in various places (as illustrated in Figure 3.5), they increase the amount of surface area that can be depicted accurately. In most cases, map users are more interested in the continents than the oceans, so the interruptions are normally made in ocean areas.

On **condensed projections,** portions of the Earth's surface are omitted from the map in order to save space and show the remaining areas at a larger scale. Again, the missing portions are usually ocean areas, especially in the Atlantic and Pacific. Figure 10.6, for example, deletes most of the Atlantic Ocean in order to save space.

Topographic Maps

Despite the problem of distortion, maps are intrinsically well suited for displaying horizontal locational information because map sheets are flat surfaces. The Earth's surface, however, is far from flat. Cartographically depicting the vertical dimensions of surface features therefore presents a major challenge.

Some maps contain no information about the vertical characteristics of the surface, or do not treat the subject systematically. Such a map is called a **planimetric map,** because it presents a plane or two-dimensional view of the surface. A planimetric map may contain the written names of features such as mountain ranges, may use symbols such as miniature peaks for mountains, or may even provide **spot elevations** of notable points for reference, but it does not allow for the systematic determination of elevations over the mapped area as a whole. A good example of a much-used type of planimetric map is a road map.

A map that, in addition to its other data, displays information on elevations in a consistent fashion for the entire area covered, is termed a **topographic map.** It contains three-dimensional information on a two-dimensional surface. Topographic maps can employ any of several methods to depict the vertical characteristics of the landscape, and two or more methods sometimes may be combined for greater effectiveness. The five most common methods involve the use of hachures, color, shading, raised relief, and contour lines.

Hachures

Hachures are short line segments that are drawn directly downslope (see Figure 3.7). They appear somewhat like gullies and indicate the direction water would take as it flows downhill. Hachure maps were produced in quantity by several European countries during the first half of the twentieth century, but are no longer in widespread use.

Color

The use of colors to indicate elevations is a technique familiar to those who use classroom wall maps. Each color represents an elevation range that is described in the legend, and a logical gradational sequence of colors is normally employed. The particular color sequence of dark green, light green, yellow, orange, red, and brown for increasing elevations is especially common. Many maps use a sequence of progressively darker shades of blue for increasing ocean depths as well.

Shading

On maps employing shading, the topography is shown by differences in surface brightness simulating those resulting from oblique sunlight (see Figure 3.8). A realistic and sometimes almost three-dimensional impression can be produced by the skillful use of this technique. Shading is frequently combined with other methods of elevation depiction to enhance the visual impression of topographic relief.

Raised Relief

A raised relief map differs from all other types because the map itself is three-dimensional, producing a miniaturized model of the landscape it portrays. In most instances, the vertical scale is exaggerated to stress the topographic characteristics of the surface. Raised relief maps are attractive and are popular for displaying surface features in a highly realistic fashion. They have the disadvantage of being expensive, because they must be molded from stiff material (usually plastic). In addition, like the other methods so far discussed, they do not permit precise measurement of elevations.

Contour Lines

Contour lines are lines that connect points of equal elevation above mean sea level (see Figure 3.9). In following a contour line, one remains at the same elevation, while in crossing contour lines, one travels uphill or downhill. Contour lines give a less realistic visual impression of the topography than do some of the other methods discussed. However, they provide by far the most useful method of elevation depiction for physical geographers and others engaged in fieldwork because they are well suited to large-scale maps and they usually permit a reasonably accurate determination of elevations at any point on the map.

Some of the more important characteristics of contour lines are as follows:

1. Each contour line has a specific elevation. The elevation value of a given contour line may be written on it, or the value may have to be determined from the elevations of nearby contour lines.

2. A contour line separates areas with elevations higher than its value from areas with elevations lower than its value.

3. The map distance between contour lines indicates slope steepness. Widely spaced contour lines

MAP MISINFORMATION

We use maps in order to gain an accurate understanding of the sizes, shapes, and locations of places around the world. Because flat maps of the Earth are necessarily distorted, however, the geographical concepts we derive from them may be distorted as well. Of particular concern to geographic educators are the erroneous impressions that elementary school students get of our planet's geography through the use of rectangular wall maps. When students who are learning world geography for the first time study these highly distorted maps, they can gain lasting impressions that are difficult to later correct.

The worst offender of all is the popular Mercator projection, which excessively enlarges the sizes of high latitude regions at the expense of the tropics (see Figure 3.6a). The lines of latitude on a Mercator map are all the same length, although these lines actually become shorter poleward, and they are spaced progressively farther apart toward the poles, even though they are virtually equidistant on the globe. In addition, lines of longitude are parallel, while in reality they converge poleward. Finally, many Mercator maps display more of the northern hemisphere than the southern hemisphere, thus placing the equator below the center of the map.

In 1989, six major professional geographical and cartographic organizations urged book and map publishers, the media, and government agencies to stop using rectangular world maps. Recommended instead were some of the more than 100 nonrectangular world maps—many of them oval—created to overcome the problem of rectangular map distortion. Of course, these map types are also distorted, but their distortion in the relative sizes of mapped features is generally not nearly as severe as that of most rectangular maps. In December 1988 the prestigious National Geographic Society adopted the Robinson projection for its world maps, and many publishing companies followed suit (see Figure 3.6b). The Robinson projection is an oval projection, but is flattened in the polar regions. Its developer is Arthur Robinson, Professor Emeritus at the University of Wisconsin-Madison, who is often considered the dean of American cartographers.

Another recent but highly controversial contender as a replacement for the Mercator projection is the Peters projection, developed in the late 1970s by German historian Arno Peters (see Figure 3.6c). Billed when it first appeared as "the most honest map of the world yet devised," it is an equal area rectangular map. In order to avoid the high latitude distortion typical of rectangular projections, longitudes on the Peters projection are compressed in the tropics and latitudes are greatly compressed in the polar regions. The result is the preservation of correct relative sizes at the expense of a great distortion in shapes.

Compared to conventional rectangular projections, the Peters projection makes Africa and southern Asia look very large and greatly reduces the size of high-latitude countries such as Canada and the Soviet Union. Its use was promoted enthusiastically by many tropical developing nations, especially in Africa, and it has been used on United Nations publications dealing with international development.

continued on next page

indicate gentle slopes, while closely spaced contour lines indicate steep slopes.

4. Contour lines do not meet or cross one another, except in the rare cases of vertical or overhanging topographic features.

5. The bends in contour lines point uphill when crossing streams or other linear lowlands and point downhill when crossing ridges.

6. A closed contour line normally indicates the presence of a high point of land, such as a hilltop. Closed depressions, which are less common, are indicated by special **depression contours,** which have inward-facing (i.e., downhill-extending) hachure marks.

The number and elevation values of the contour lines used on a given topographic map are decided upon by the cartographer. The vertical interval used, called the **contour interval,** is normally constant over the entire map. Some topographic maps, such as

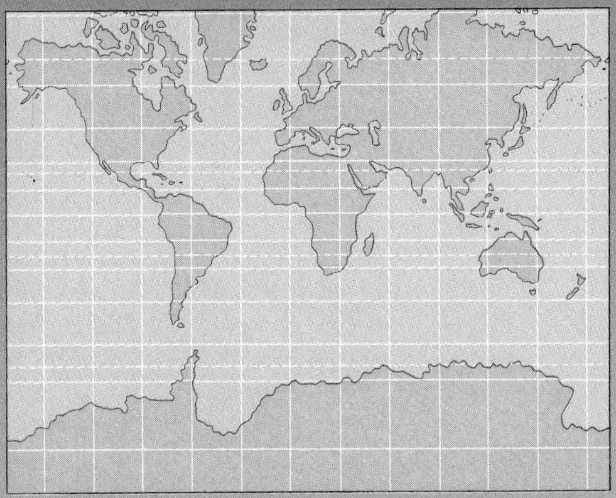

(a)

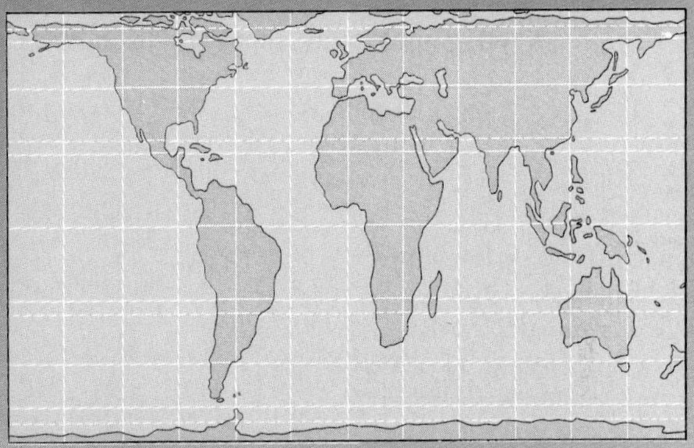

(c)

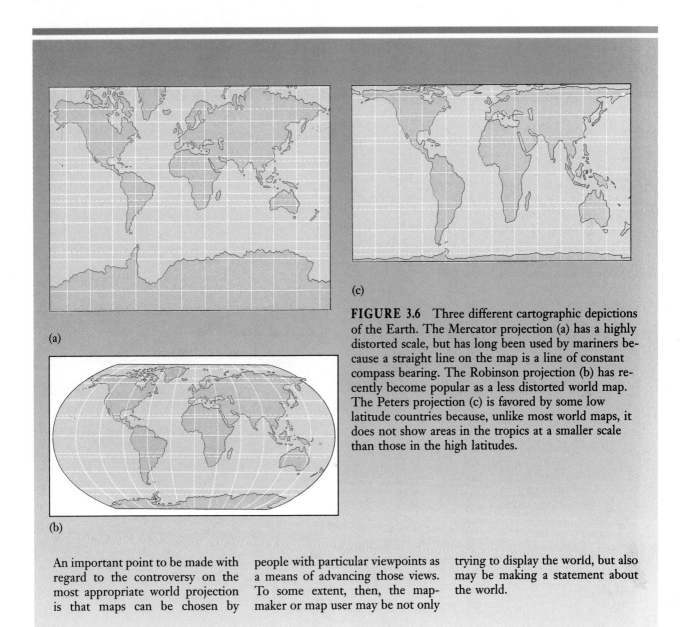

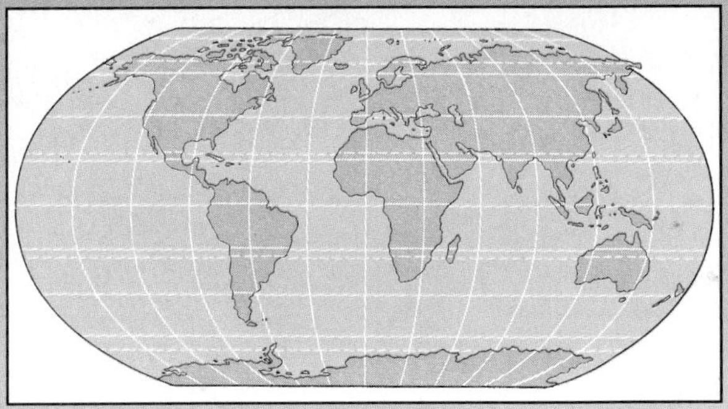

(b)

FIGURE 3.6 Three different cartographic depictions of the Earth. The Mercator projection (a) has a highly distorted scale, but has long been used by mariners because a straight line on the map is a line of constant compass bearing. The Robinson projection (b) has recently become popular as a less distorted world map. The Peters projection (c) is favored by some low latitude countries because, unlike most world maps, it does not show areas in the tropics at a smaller scale than those in the high latitudes.

An important point to be made with regard to the controversy on the most appropriate world projection is that maps can be chosen by people with particular viewpoints as a means of advancing those views. To some extent, then, the mapmaker or map user may be not only trying to display the world, but also may be making a statement about the world.

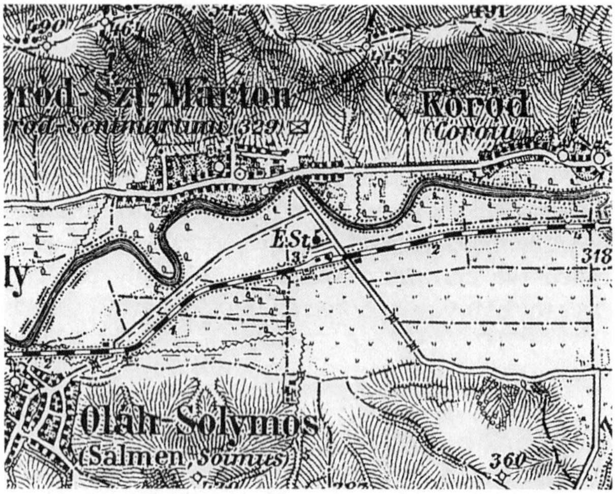

FIGURE 3.7 A topographic map employing hachures.

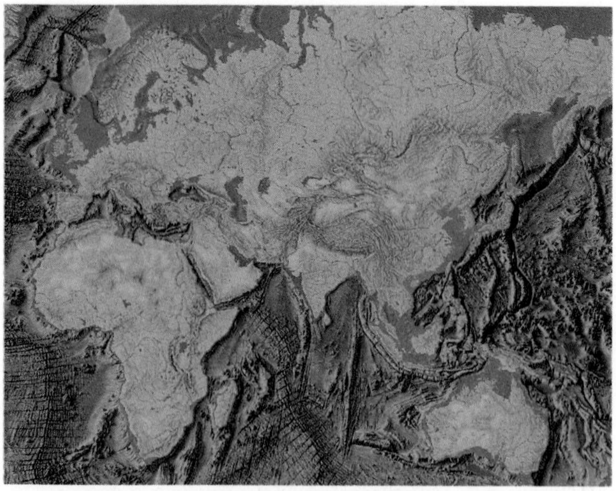

FIGURE 3.8 The skillful use of shading gives the surface of a map a three-dimensional appearance.

FIGURE 3.9 Contour lines can be visualized as the locations in a mapped area that would be intersected by plane surfaces drawn at set elevation levels. The top portion of the illustration shows a mountainous island with contour lines drawn at intervals of 20 units (usually the units are meters or feet). The bottom view depicts the resulting contour map.

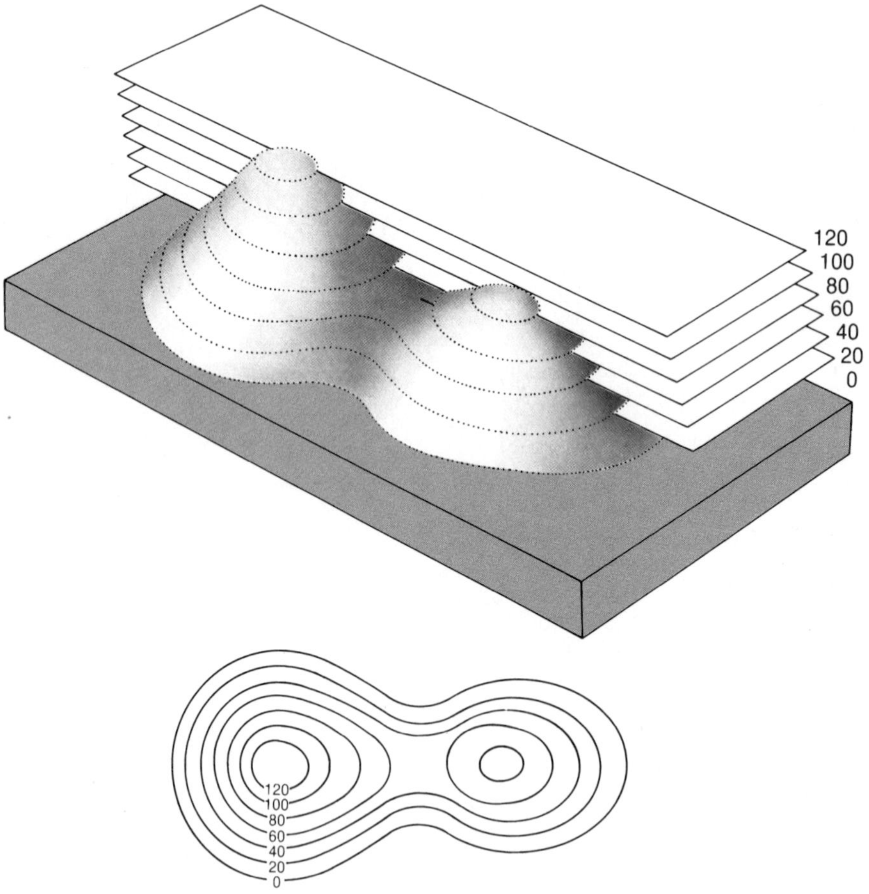

those prepared by the U.S. Geological Survey (USGS), also use thicker **index contours** with a larger contour interval for faster elevation determination. For example, on a map with a 20-foot contour interval, every fifth line might be thickened to produce a 100-foot index contour interval.

The use of contour lines allows for the exact determination of elevations only on the lines themselves. Elevations of areas between the lines are estimated by the process of interpolation, or proportional spacing. For example, if you wish to estimate the elevation of a house located on a contour map three-fourths of the way between the 320-foot and 340-foot contour lines, you would calculate the number that is three-fourths of the way between 320 and 340. This number is 335, the best estimate of the elevation of the house. Slopes may not be smooth, so there is no guarantee that you will be exactly correct; given no evidence to the contrary, however, a relatively smooth slope is the most reasonable assumption.

Many countries have government agencies responsible for the preparation of topographic maps. In the United States, the chief mapping agency is the USGS. The USGS has been systematically mapping the country since 1879, and the maps it produces are revised periodically. Most USGS map information is now derived from aerial photographs, but the maps are also locationally controlled by carefully surveyed ground sites called **bench marks.**

USGS topographic maps are available for purchase by the public and frequently are used for fieldwork and lab exercises in Earth science and engineering courses. In addition to elevations, which are indicated by the use of brown contour lines, these maps contain a great deal of other information (see Figure 3.10). The larger scale maps in particular contain a wealth of surface detail pertaining to both natural and cultural features, and more than 100 standardized symbols have been developed to identify these features (see Figure 3.11).

FIGURE 3.10 A portion of a modern USGS topographic map. Contour lines are colored brown.

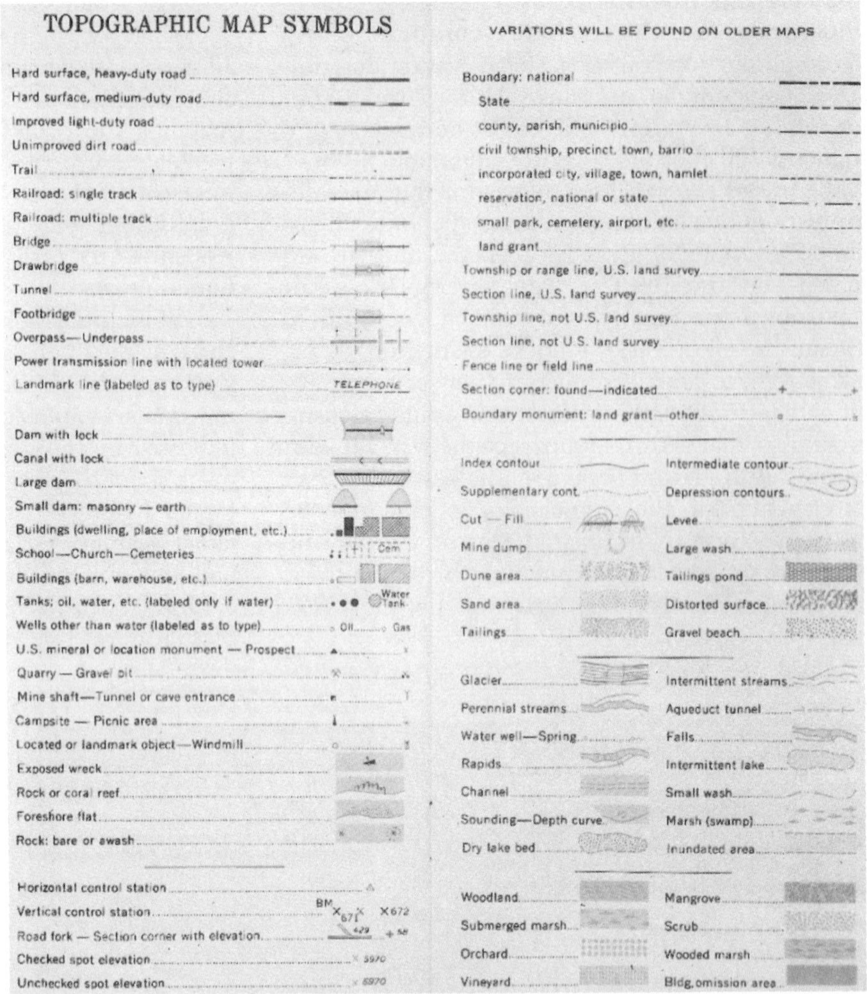

FIGURE 3.11 USGS standardized topographic map symbols.

REMOTE SENSING AND COMPUTER MAPPING

As we have seen, sophisticated topographic maps such as those prepared by the USGS contain a tremendous quantity of spatial data. How is this information collected, and how is it precisely located on maps? In contrast to the laborious manual ground surveying and cartographic techniques of the past, most map information currently is gathered by remote sensing instruments, and the maps themselves are largely computer generated.

Every object or substance on the Earth's surface absorbs, reflects, and emits energy in the form of electromagnetic radiation. The quantity and wavelengths of this radiation depend on the nature and temperature of the surface materials. In the rapidly evolving field of **remote sensing,** information is acquired from a distance by mechanical devices that receive portions of this electromagnetic energy. This information enables us to gain more knowledge of the nature or characteristics of the radiation sources.

A large number of remote sensing devices are used at the Earth's surface for a great variety of purposes. Some familiar examples are radios, cameras, binoculars, telescopes, and radar. Of greatest importance to physical geographers, however, are remote sensing devices that are sent aloft to gather data on the nature of the Earth's surface and lower atmosphere. These devices, which operate on a variety of electromagnetic wavelengths, have made possible the accumulation of vast quantities of spatial data.

The oldest remote sensing tool is black-and-white photography. Surface photography was developed in the mid-nineteenth century, and the earliest aerial photographs were taken in the late 1800s by cameras carried aloft by balloons. With the development of the airplane, it became possible to control the area being photographed, and systematic air photo coverage began in the United States and portions of Europe after World War I. This led to the growth of the field of **photogrammetry**—the making of accurate measurements from photographs.

Following World War II, the use of color film, which more realistically portrays the surface and provides a greater variety of surface information, gradually increased. At about the same time, remote sensing devices that used radiation wavelengths other than visible light were developed. Of particular importance were infrared sensors, which recorded patterns of heat rather than light. These sensors can readily penetrate clouds, fog, and darkness.

In the 1960s, the deployment of orbiting satellites started a revolution in remote sensing technology equivalent to that resulting from the development of the camera and the airplane. From their high altitudes, satellites have greatly expanded the field of view of the Earth below. In addition, unlike aircraft, satellites can remain aloft and continuously collect and transmit surface data for months and even years. Satellites have been able to acquire data from all parts of the world because they are affected neither by adverse weather conditions nor by international boundaries. They therefore are critically important in gathering military intelligence, and many of their data-collecting capabilities remain classified.

Weather satellites, among the first types developed, have been operating ever since the United States launched the first TIROS satellite in 1960. The current generation of GOES (Geostationary Operational Environmental Satellite) satellites remain at fixed positions 36,000 kilometers (22,300 mi) above the Earth's surface, permitting continuous monitoring of weather systems affecting North America. Infrared sensors in these satellites allow cloud patterns to be photographed both during the day and at night.

Other satellites, including the LANDSAT series, provide high resolution color imagery of surface features. These satellites have permitted the accurate mapping of remote areas, such as mountainous and desert regions, and have also been extensively employed in monitoring air and water pollution patterns, in identifying areas of diseased crops and forests, and in exploring for minerals (see Table 3.1).

At present, literally hundreds of satellites serving a great variety of purposes are orbiting the Earth, and additional satellites of progressively increasing sophistication are constantly being deployed. The U.S. government makes aircraft and satellite imagery of all parts of the Earth's surface available to the public, and a number of these photos have been employed as illustrations in this book.

Another recent and ongoing revolution with major applications to both cartography and remote sensing lies in the areas of **computer mapping.** Computers can be used to store, organize, and display huge quantities of data in a tiny fraction of the time that such tasks could be accomplished manually. Modern cartography and remote sensing technology now depend upon the use of computers for preparing maps. For example, plotting machines have been developed that can make accurate topographic maps directly from overlapping series of aerial photographs (see Figure 3.12).

As an example of computer technology, the Soil Conservation Service of the U.S. Department of Ag-

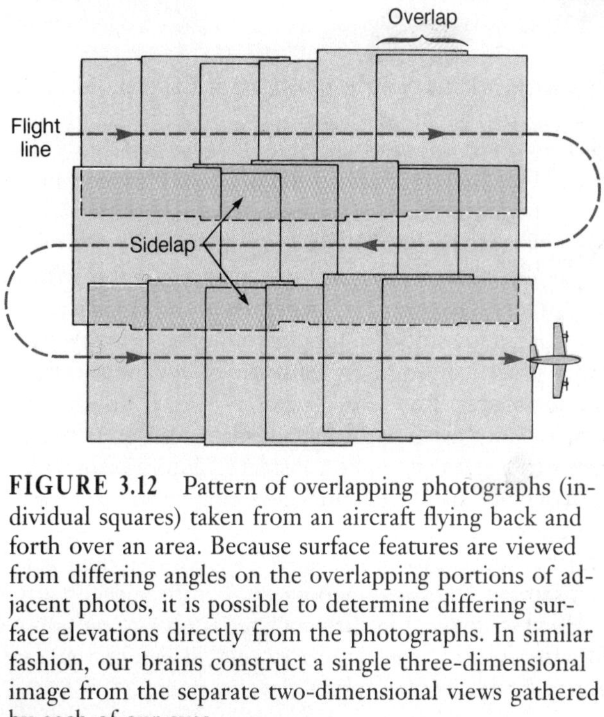

FIGURE 3.12 Pattern of overlapping photographs (individual squares) taken from an aircraft flying back and forth over an area. Because surface features are viewed from differing angles on the overlapping portions of adjacent photos, it is possible to determine differing surface elevations directly from the photographs. In similar fashion, our brains construct a single three-dimensional image from the separate two-dimensional views gathered by each of our eyes.

TABLE 3.1 Applications of Landsat and Weather Satellites

ATMO-SPHERE	AGRICULTURE, FORESTRY, AND RANGE RESOURCES	CARTOGRAPHY AND LAND USE/COVER	GEOLOGY	WATER RESOURCES	OCEANOGRAPHY AND MARINE RESOURCES	ENVIRONMENT
Weather data Pollution mapping Cloud types Water vapor content Fog Haze conditions	Discrimination of vegetative types: Crop types Timber types Range vegetation Measurement of crop acreage by species Measurement of timber acreage and volume by species Determination of range readiness and biomass Determination of vegetation stress Determination of soil associations Assessment of grass and forest fire damage	Classification of land uses/cover Mapping and map updating Categorization of land capability Separation of urban and rural categories Preparation of regional plans Mapping of transportation networks Mapping of land-water boundaries Mapping of wetlands	Identification of rock types Mapping of major geologic units Revising geologic maps Delineation of unconsolidated rock and soils Mapping igneous intrusions Mapping recent volcanic surface deposits Mapping landforms Search for surface guides to mineralization Determination of regional structures Mapping lineaments (fractures)	Determination of water boundaries and surface water area and volume Mapping of floods and floodplains Determination of areal extent of snow and snow boundaries Measurement of glacial features Measurement of sediment and turbidity patterns Determination of water depth Delineation of irrigated fields Inventory of lakes and water impoundments	Determination of turbidity patterns and circulation Mapping shoreline changes Mapping of shoals and shallow areas Mapping of ice for shipping Study of eddies and waves	Monitoring surface mining and reclamation Mapping and monitoring of water pollution Determining effects of natural disasters Monitoring environmental effects of human activities (lake eutrophication, defoliation, etc.)

SOURCE: The Landsat Story (Washington, D.C.: Earth Resources Satellite Data Application Series, Module U-2. 1980), p. 14.

riculture, which compiles soil maps and related data for the nation, is in the process of digitizing (converting to computer-readable form) maps of soils data. This information can be processed by computers to produce maps of important soil characteristics, such as soil fertility or erosion potential.

In addition to assisting in map preparation, some computers store, retrieve, manipulate, and display a variety of geographic data for portions of the Earth's surface. These computer systems, or **geographic information systems,** are revolutionizing the way geographers work with mapped data. Geographic information systems can be considered automated atlases of mapped data. They allow geographers to synthesize geographical information by superimposing several maps in the computer in order to identify spatial patterns of Earth surface phenomena. For instance, county maps of soil characteristics, rock types, and topographic features might be combined to aid in the search for areas suitable for residential development.

SUMMARY

Maps are essential geographical tools, and are extensively employed to display the distributions of phenomena of all types. In order to make full use of maps, it is necessary to understand something of their characteristics and principles of production, as well as their advantages and shortcomings. We return to the Focus Questions:

1. *What is a map? With what basic map information must the user become familiar before a map can be employed to full advantage?*

A map is a pictorial representation of the geographic locations of selected surface features at a reduced scale. The following categories of basic map information should be examined before a map is selected or used: (1) the map title, which should clearly indicate the purpose of the map; (2) the date of the map or of the information contained on the map; (3) the locations covered by the map; (4)

the directional orientation of the map with respect to the map sheet; (5) the symbols used on the map, as explained in the map legend; and (6) the map scale, indicating the degree of reduction of surface features appearing on the map. The scale may be stated verbally, graphically, or as a representative fraction.

2. *How is locational information about the Earth transferred onto a flat surface to make a map? Why are all flat maps of the Earth distorted?*

In order to make a map, locational information about the Earth is commonly projected onto a developable surface, such as a plane, a cone, or a cylinder, that has been placed on or around the globe. The Earth's image is projected onto the developable surface in much the same way that the images on a movie film are projected onto a screen. Once the projection has been made, the surface is "developed," or flattened, to produce a flat map. Unfortunately, it is not possible to project the Earth's spherical surface onto a flat surface without creating distortion—inconsistencies in the relative sizes or shapes of the areas depicted.

3. *How is information concerning differing surface elevations displayed on a flat map?*

A map that depicts surface elevations in a systematic fashion for the entire mapped surface is termed a topographic map. The five methods discussed in this chapter for displaying elevational information are hachures, color, shading, raised relief, and contour lines. Contour lines, which are lines connecting places of equal elevation, are the most frequently used method of elevation depiction on large-scale maps.

4. *How is map information collected and compiled?*

In the past, map information generally was collected by laborious ground surveys and was drafted onto the map by hand. A variety of remote sensing techniques now enable map data to be collected and assembled much more quickly and easily than before. Most surface information is currently obtained by aircraft and satellites, which often use electromagnetic wavelengths other than visible light to obtain their imagery.

Computers can generate maps directly from aerial photographs. Sophisticated computer programs termed *geographic information systems* are revolutionizing the geographical analysis of mapped data. These programs permit researchers to superimpose several different categories of map data so as to identify complex spatial patterns resulting from the interaction of multiple factors.

Review Questions

1. What is a map? Why are maps particularly associated with the field of geography?

2. What is map scale? What are the three basic methods of stating map scale? What major advantage does a large-scale map have over a small-scale map? What major advantage does a small-scale map have over a large-scale map?

3. Why are all flat maps distorted? Is the problem of map distortion greater with small-scale maps or with large-scale maps? Why? Why aren't globes more widely employed as a means of avoiding the problem of map distortion?

4. Explain the basic principle of construction of a map projection. What are the three basic types of map projections? Why are conic projections especially well suited for depicting the United States?

5. Briefly describe five different methods for indicating the vertical dimensions of the Earth's surface on a topographic map. Which of these methods allows for the most accurate determination of elevations?

6. What is remote sensing? What remote sensing devices have been most important in the acquisition of information about the Earth's surface during the twentieth century?

7. What role do computers play in the mapping of spatial data? How do geographic information systems allow researchers to identify interrelationships among different Earth phenomena?

Problems

1. Change the verbal map scale of "1 inch equals 8 miles" to the equivalent RF scale. How many kilometers would one centimeter represent at this scale?

2. Change the RF map scale of 1:1,000,000 to the equivalent miles-per-inch scale. Change it to the equivalent kilometers-per-centimeter scale.

3. If a map with an RF scale of 1:126,720 measures 50 centimeters (20 in) from side to side and 20 centimeters (10 in) from top to bottom, how many square miles of Earth surface area does it depict? How many square kilometers?

4. If the mapped area described in the previous problem were remapped at an RF scale of 1:63,360, how many times as much paper would be needed? What would be the dimensions of the new map?
5. If a topographic map had a contour interval of 25 feet, how many contour lines would encircle a mountainous oceanic island with a summit elevation of 673 feet? What would be the elevation of the uppermost contour line?

Key Terms

map
map scale
map projection
topographic map
contour lines
remote sensing
geographic information systems

CHAPTER 4

EARTH'S ATMOSPHERIC ENVELOPE

OUTLINE

FOCUS QUESTIONS

1. What basic influences does the atmosphere have on the Earth's surface environment?
2. What is the composition of the atmosphere?
3. How do the characteristics of the atmosphere change with altitude?
4. How are human activities influencing the composition of the atmosphere?

The **atmosphere** is the gaseous envelope that surrounds the Earth. Although relatively transparent and seemingly insubstantial, it nonetheless profoundly affects nearly every aspect of our physical environment. Life on Earth could not exist without it.

The atmosphere is the outermost of the four environmental spheres that interface at the surface of the Earth. In essence, it forms a canopy over the others—the hydrosphere, the lithosphere, and the biosphere. The cycles of development and change that characterize the four environmental spheres require tremendous amounts of energy. Nearly all this energy is derived from the Sun, although a small quantity comes from the interior of the Earth. Because the atmosphere is the outermost of the four spheres, it is the first to receive the solar energy on which it and the others depend. As solar energy passes downward through the atmosphere, it is partially absorbed and redirected before it reaches the Earth's surface and is converted to the thermal and chemical energy that largely powers the other spheres.

ATMOSPHERIC ORIGIN AND FUNCTIONS

Earth's atmosphere is believed to have been derived primarily from volcanic gases released from our planet's interior. Volcanic eruptions not only produce large quantities of lava and ash, they also liberate great volumes of gases. Scientists believe that more than four billion years of volcanic eruptions have released not only the gases that have formed our atmosphere, but also the water vapor that has formed all the Earth's water bodies and glaciers. Present-day eruptions continue to liberate gases, originally dissolved under great pressure in rock deep beneath the surface. Evidence for the volcanic origin of the atmosphere is provided by measuring the composition of gases currently being released from active volcanoes. These gases consist of approximately 85 percent water vapor, 10 percent carbon dioxide, and 1 or 2 percent nitrogen. These three gases, and their derivatives (including oxygen), comprise about 99 percent of our atmosphere.

The atmosphere performs a number of vital functions that profoundly affect the nature of the Earth's surface and its ability to support life. The most obvious and vital function is the supplying of oxygen to animals and carbon dioxide to plants. These gases are essential to their biological processes.

Earth's atmosphere also protects our planet from the impact of small meteors. If the atmosphere did not exist, the Earth's surface would probably closely resemble that of the Moon (see Figure 4.1). The Moon is the same distance from the Sun as the Earth, and is composed of similar geologic materials. What the Moon lacks, however, is sufficient mass to generate enough gravity to have maintained an atmosphere. One of the striking differences between the Earth and Moon is the nature of their major surface features. The Moon's surface is pockmarked by meteorite craters, some of them billions of years old. Earth's atmosphere, in contrast, protects our planet from the impacts of small meteors by causing them to be incinerated by friction before reaching the surface. The large meteorites that do reach the Earth's surface produce craters like those on the Moon. These craters have relatively short spans of existence, however, because the forces of erosion associated with atmospheric phenomena, especially wind and water, cause them gradually to disappear. In fact, all terrestrial landforms exist for limited periods of time because of erosional and depositional processes. The Moon's surface, in contrast, is almost frozen in time,

FIGURE 4.1 The gray, pockmarked surface of the Moon stands in stark contrast to the brilliance of the distant Earth in this photo taken by the Apollo astronauts. If the Earth had no atmosphere, its surface would closely resemble the Moon's surface. (Department of the Interior, U. S. Geological Survey)

and most of its large-scale surface features are well over a billion years old.

The blanketing influence of the atmosphere also protects us from most cosmic radiation. **Cosmic radiation** consists of charged particles, mostly hydrogen and helium nuclei, that enter Earth's atmosphere at nearly the speed of light. Because of their great kinetic energy, these particles are capable of destroying or mutating plant and animal cells and would be lethal to Earth's life forms in the concentrations that bathe the Moon's surface. Fortunately, the atmosphere acts to reduce greatly the influx of cosmic radiation.

The resistance of the atmosphere to the passage of energy greatly reduces the range of Earth surface temperatures. On the Moon's surface, daytime temperatures can exceed 94°C (200°F), only to fall to −128°C (−200°F) at night. Earth's atmosphere reflects some solar energy and distributes the remainder through a vertical distance of many kilometers, so that temperature ranges are much less extreme. (Temperature ranges on the Moon are further increased by that satellite's slow rotation, which causes a full day/night cycle to take 29 Earth days, and by the lack of surface water to evaporate—a process that absorbs large amounts of terrestrial solar energy.) Earth's atmosphere also has a substantial capacity to absorb solar radiation (see Chapter 5), permitting it to escape only slowly into space. This causes mean temperatures at the Earth's surface to be much higher than they would be in the absence of an atmosphere.

Finally, the atmosphere makes possible the cycle of evaporation and precipitation that provides water to the land areas of the Earth. Without this cycle (discussed in Chapter 7), there would be no clouds, no rain or snow, no rivers or lakes, little or no land vegetation or soil, and little animal life either on the land or in the sea. In addition, many important minerals could not form, and most landforms with which we are familiar would not develop.

COMPOSITION OF THE ATMOSPHERE

The atmosphere is a physical mixture of gaseous elements and compounds. Most of these gases exist in constant proportions in all but the atmosphere's outermost layers (above 80 kilometers—50 mi—altitude). The reason for the homogeneity of the atmosphere is the mixing action of the wind. If it were not for the turbulent transfer of air molecules from one level to another by wind currents, the various atmospheric gases would become stratified into horizontal layers on the basis of density. A few gases, however, have characteristics that prevent the wind from mixing them evenly, and consequently they vary in their atmospheric concentrations.

In addition, small quantities of liquids and solids are suspended in the atmosphere. Atmospheric liquids and solids tend to vary greatly in quantity at different times and in different places because they are much denser then the atmospheric gases and have a tendency to settle to the surface of the Earth. The distribution of these substances depends on the size and location of their sources, the rate at which they enter the atmosphere, and the speed and direction of the wind currents.

In the following discussion, the components of the atmosphere are divided into three categories: nonvariable gases, variable gases, and particulates (liquids and solids).

Nonvariable Gases

The nonvariable gases comprise over 98 percent of the total atmosphere by volume (see Figure 4.2). These gases are sufficiently stable, both physically and chemically, to remain in the atmosphere long enough to be dispersed evenly by wind currents.

As Table 4.1 indicates, the most abundant gas by far is **nitrogen,** which by volume comprises 78 per-

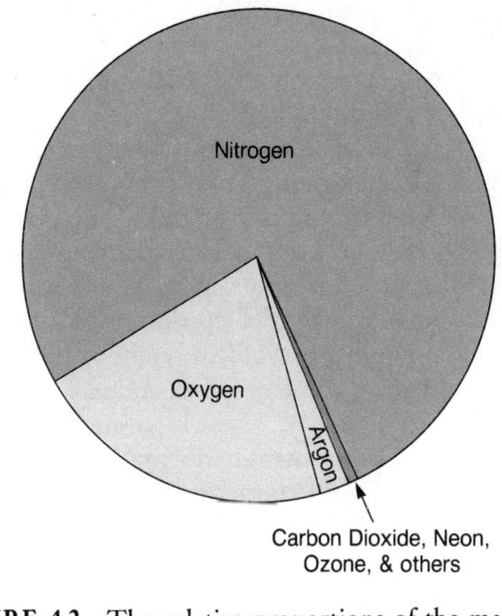

FIGURE 4.2 The relative proportions of the major atmospheric gases.

TABLE 4.1 Composition of the Atmosphere

NAME	SYMBOL	PERCENT BY VOLUME*
Nonvariable Gases		
Nitrogen	N_2	78.08
Oxygen	O_2	20.95
Argon	Ar	0.93
Neon	Ne	0.002
Others		0.001
Variable Gases and Particulates		
Water vapor	H_2O	0.1 to 4.0
Carbon dioxide	CO_2	0.035
Ozone	O_3	0.00006
Other gases		trace
Particulates		normally trace

*Percentage figures, except for water vapor, are for dry air.

cent of the atmospheric total of nonvariable gases. Although it is only the third most abundant gas released by the volcanic eruptions that formed the atmosphere, nitrogen has achieved its position of dominance because of its physical and chemical stability. Gaseous nitrogen is diatomic; that is, two atoms of nitrogen combine to form one nitrogen molecule (N_2). In this form, nitrogen is relatively inactive and does not participate directly in the respiratory processes of either plants or animals. Small amounts of atomic nitrogen released by lightning or bacterial action, however, are incorporated into compounds that serve as essential plant nutrients.

Most of the remainder of the atmosphere consists of **oxygen** (O_2), another diatomic molecule, which is essential to the respiratory processes of plants and animals. Oxygen also enters into many chemical reactions with rock materials at or near the Earth's surface to form a variety of products, including soil. Free oxygen is not liberated in quantity by volcanic eruptions. Instead, these eruptions release carbon dioxide (CO_2), which plants later separate into its constituent carbon and oxygen during the process of photosynthesis. Plants therefore have been responsible for placing nearly all the free oxygen in the atmosphere. Nitrogen and oxygen together comprise slightly over 99 percent, by volume, of dry air.

One other nonvariable atmospheric gas exists in more than a trace amount. This gas is **argon** (from the Greek **argos**, or "inactive"), which comprises 0.93 percent of dry air. It is an inert gas and is virtually incapable of entering into any chemical reactions. Argon enters the atmosphere both during volcanic eruptions and by slow diffusion from near-surface rocks.

Variable Gases

Several gases exist in variable quantities in the atmosphere. Although they collectively comprise only a very small proportion of the atmosphere's total mass and volume, three of these **variable gases**—carbon dioxide, water vapor, and ozone—are crucial to the existence of life.

Carbon Dioxide

Carbon dioxide is essential to the photosynthetic processes of plants and is the source of the carbon that is the basic element of Earth's plant and animal life. The amount of carbon dioxide in the atmosphere varies slightly during the year because of variations in its rate of absorption by plants (see Figure 5.0). It declines slightly in quantity each northern hemisphere summer, when the vegetation of the vast northern hemisphere land masses is growing most vigorously, and increases again in quantity during the winter as it is released by organic decay. The atmosphere's carbon dioxide content also has been slowly increasing, especially during the twentieth century, as a result of human activities. This phenomenon and its potential environmental consequences are discussed in the Case Study at the end of Chapter 5.

Carbon dioxide and oxygen are involved in a constant cycle of interchange. Because the dissociation of a carbon dioxide molecule into its component carbon and oxygen is an endothermic (energy-absorbing) reaction, plants require sunlight to supply the energy for this process. Conversely, when animals recombine carbon from the food they have ingested with atmospheric oxygen, an exothermic (energy-producing) reaction occurs. This reaction enables animals to perform all their activities. In a sense, then, we are powered by stored sunlight, which has been "fixed" by plants in the carbohydrate molecules we digest.

It was noted earlier that about 10 percent of the gas emitted by volcanic eruptions is carbon dioxide, yet only a very small percentage (0.035 percent) of the atmosphere currently is made up of this gas. This

ATMOSPHERES OF THE OTHER PLANETS

The Earth is unique among the planets of our solar system in having an atmosphere that contains enough free oxygen to support life as we know it. Our planet, however, is not the only one in the solar system with an atmosphere; six of the other eight planets also have one. Let's take a brief tour of the solar system to examine the atmospheres of our neighbors in space.

- *Mercury.* The innermost planet contains no appreciable atmosphere, both because of its small size and resulting weak gravity (it is only slightly larger than Earth's Moon), and because of its extremely hot surface temperatures.
- *Venus.* Venus, in contrast, has a very dense atmosphere, with surface air pressures about 90 times those at the Earth's surface. It consists of about 97 percent carbon dioxide and 2 percent nitrogen, and has only trace amounts of water and oxygen. Because of its great density and high carbon dioxide content, Venus's atmosphere is very effective in absorbing solar energy; consequently, surface temperatures on Venus average about 477°C (890°F).
- *Mars.* The fourth planet from the Sun has a highly rarefied atmosphere, with a surface pressure about 0.006 times that at the Earth's surface. Like the atmosphere of Venus, it is dominated by carbon dioxide (95 percent), with about 3 percent nitrogen, 1.5 percent argon, and 0.25 percent oxygen. Although there are only trace amounts of water vapor in the Martian atmosphere, the planet has substantial polar icecaps of frozen water and possibly carbon dioxide that cannot melt because of perennially subfreezing temperatures.

FIGURE 4.3 Jupiter, as photographed by Voyager 1 from a distance of 28.4 million kilometers (17.5 million mi). The gas giant's colorfully banded atmosphere displays complex patterns highlighted by the Great Red Spot, a large, circulating atmospheric disturbance. Jupiter's innermost large satellite, Io, can be seen against the planet's disk. (NASA)

- *Jupiter.* The giant planet Jupiter has an atmosphere about 1,000 kilometers (620 mi) thick, below which the pressure is so great that materials must exist in a liquid or solid state. The combination of Jupiter's large mass and consequent high gravity (2.5 times that of the Earth's), coupled with its cold temperatures, have allowed the light gases to be retained. Jupiter's outer atmosphere therefore closely reflects the compositional mix of gases in the universe as a whole. It consists of 80 percent hydrogen and 19 percent helium, with the remaining 1 percent composed largely of methane, ammonia, and water vapor. Jupiter's outer atmosphere exhibits a complex turbulent flow pattern, giving the planet a beautiful orange and white banded appearance when viewed from space (see Figure 4.3).
- *Saturn.* Like Jupiter, Saturn's atmosphere consists mostly of hydrogen (89 percent) and helium (11 percent), with traces of methane, ammonia, and other gases. Wind velocities in the outer atmosphere are the highest of any planet—about 500 meters per second (1,000 mph) near the equator. Saturn's largest moon, Titan, also has a substantial atmosphere. Its density is 1.6 times that of Earth's atmosphere and it consists of 98 percent nitrogen and 2 percent methane.
- *Uranus* and *Neptune.* So far little is known about the atmospheres of these two large planets, except that they are extensive, extremely cold, and contain hydrogen and methane, and probably also helium and ammonia.
- *Pluto.* Remote Pluto, the outermost known planet of our solar system, is probably too small and cold to have a gaseous atmosphere, although frozen gases may exist on its surface.

invites the question of where the remainder has gone. A small portion of the carbon dioxide forms the bodies of plants, especially trees, that are alive or have recently died. A considerably larger quantity is stored beneath the Earth's surface in the form of peat, coal, oil, and gas deposits. These deposits represent the remains of plants and animals, most of which lived more than one hundred million years ago. The bulk of the carbon dioxide, though, has dissolved into the ocean waters. Carbon dioxide is highly soluble in water (a fact that the carbonated beverage industry uses to its advantage), and, over the course of geologic time, most atmospheric carbon dioxide has dissolved into the oceans, later to be precipitated in the form of carbonate rocks such as limestone and dolomite.

Water Vapor

Water vapor (H_2O), or water in the invisible gaseous state, is one of the four most abundant gases in the atmosphere, but in highly variable in amount. Cold, dry air may contain less than 0.1 percent water vapor by volume, while very warm, moist air in the tropics may contain as much as 4 percent. Worldwide, air just above the Earth's surface contains an estimated average of 1.4 percent water vapor. This proportion declines rapidly with altitude, so that approximately half the atmosphere's water vapor content exists within 1.6 kilometers (1 mi) of sea level.

The major reason that the atmosphere's water vapor content varies so greatly is that the water vapor holding capacity of the air is highly dependent upon its temperature. Warm air can hold much more water vapor than can cold air. In addition, the availability of water vapor differs greatly in different areas because of the nature of the Earth's surface. For example, air over land areas is likely to be drier than air over large bodies of water. As a result of the continuous temperature and locational changes to which the air is subjected, its water vapor content alternately increases through the process of evaporation and decreases through the processes of condensation (conversion to liquid water) and sublimation (conversion to ice).

Atmospheric water vapor is critically important to life on Earth because it is the source of all of the Earth's precipitation. If the atmosphere could not alternately gain and lose huge quantities of this gas, the land areas of our planet would be sterile deserts. Atmospheric moisture is of sufficient importance to be treated in its own chapter (Chapter 7), and its

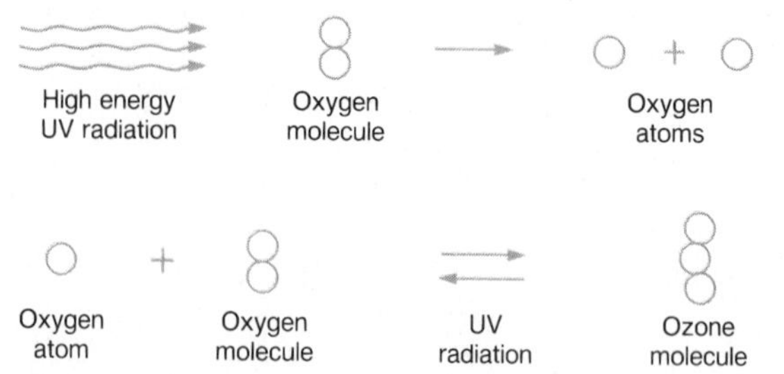

FIGURE 4.4 Ozone forms when a free oxygen atom, released by the splitting of an oxygen molecule by cosmic radiation, combines chemically with another oxygen molecule.

effects also are dealt with in subsequent chapters covering the subjects of climate, the Earth's surface waters, fluvial landforms, and glacial landforms.

Ozone

The third important variable gas is ozone. Ozone (O_3), a triatomic form of oxygen, has three chemically bonded oxygen atoms instead of the normal two. Most ozone exists between an altitude of about 15 kilometers (10 mi) and 65 kilometers (40 mi), a portion of the atmosphere sometimes referred to as the **ozone layer.** Concentrations are quite small; even at an altitude of 24 kilometers (15 mi), where ozone is most abundant, it typically comprises only about 0.0012 percent of the total gases present.

Ozone is produced by the bombardment of normal oxygen (O_2) molecules by cosmic radiation, which causes some to split and to recombine as ozone (see Figure 4.4). The ozone molecule is not particularly stable chemically and has a tendency to revert to the diatomic oxygen (O_2) form.

The minute amounts of ozone produced in the upper atmosphere are important because of their ability to absorb the potentially lethal quantities of cosmic and ultraviolet radiation that enter the atmosphere. The absorption of this short-wave radiation releases heat, producing a zone of relatively warm temperatures centered at an altitude of about 48 kilometers (30 mi). Considerable concern has developed in recent years that human activities may be depleting the ozone layer. This subject is explored further in the Focus Box on pages 56–57.

The atmosphere also contains trace amounts of several other variable gases. Among the more significant in quantity are carbon monoxide (CO), methane (CH_4), sulfur dioxide (SO_2), and nitrogen dioxide (NO_2). A substantial proportion of these gases is produced by human activities, although they are formed by natural processes as well.

Particulates

Atmospheric particulates consist of liquids and solids (with the exception of water droplets or ice crystals) that are suspended in the atmosphere. They vary considerably in size, and the smallest are capable of remaining aloft almost indefinitely. Airborne particulates are often considered to be atmospheric "impurities," because the atmosphere is essentially a mixture of gases, and particulates represent temporary additions derived largely from the Earth's surface. Nonetheless, these particulates play a vital role in atmospheric processes. Most particulates are solids, rather than liquids, and are collectively referred to as **dust.**

A considerable variety of sources of dust exist, both as a result of natural processes and, increasingly, human activities (see Figure 4.6). Natural sources may be divided into six categories. First, there is a continuous sifting of meteoric dust through the atmosphere. A second source, volcanic dust and ash, comes from the opposite direction—the interior of the Earth. Although major volcanic eruptions are highly sporadic occurrences, they have the short-term capacity to release more dust into the atmosphere than any other source. A third source consists of wind-blown surface materials, usually soil, derived from areas not protected by vegetation. A fourth source is smoke from fires, which may attain impressive dimensions, particularly in the lower latitudes. Fifth, and affecting primarily oceanic and coastal areas, are minute salt crystals released from the evaporation of sea spray. Finally, many "dust" particles of biological origin exist, including pollen, spores, seeds, and bacteria.

Particulates released by human activities are increasing rapidly in many areas, especially in the so-called developing nations. They originate primarily from urban sources such as factory and automobile exhausts and smoke from heating and cooking, as well as from agricultural areas unprotected by vegetation (see the Case Study at the end of this chapter). The well-known hazy conditions that prevail in summer over much of North America and Europe result largely from the combination of dust-generating human activities and the occurrence of relatively light winds at this time of year.

Excessive quantities of airborne particulates can produce unpleasant and occasionally unhealthful atmospheric pollution. On the other hand, dust particles play a vital role in the formation of clouds and precipitation by serving as **condensation nuclei.** Water vapor normally requires solid or liquid surfaces on which to condense or sublime in order to produce water droplets or ice crystals, and airborne particulates provide these surfaces. Every raindrop is composed of many thousands of cloud droplets, each of which has formed around a mote of dust. If no atmospheric dust particles were present, the air near the surface would become saturated with water vapor, but there would be no way for this moisture to be precipitated.

VERTICAL CHARACTERISTICS OF THE ATMOSPHERE

In the next few chapters, we examine the horizontal distribution of such atmospheric phenomena as temperature, air pressure, winds, and moisture. The horizontal (or at least relatively horizontal) level in which we are most interested is the base of the atmosphere; that is, the interface between the atmosphere and the Earth's surface. It is in the vertical direction, however, that the greatest variations in the atmosphere occur. These variations are produced largely by the force of gravity, which causes the air to differ both in pressure and in composition at different altitudes. An additional major influence on the characteristics of the lowest portion of the atmosphere is the surface of the Earth itself.

Weight, Density, and Vertical Extent

The air may appear weightless, but in reality it has mass and weight, just as does any other substance. The total weight of Earth's atmosphere is enormous; it is estimated to be 5.44 quintillion kilograms, normally written as 5.44×10^{18} kg (or as 12×10^{18} lb). The weight of the overlying air subjects each square inch of the Earth's surface at sea level to an average pressure of 6.7 kilograms (14.7 pounds). Because the mean elevation of the Earth's land and water surface, taken as a whole, is about 273 meters (900 ft) above sea level, the surface air pressure for the entire Earth averages approximately 1 kilogram per square centimeter (14.2 lbs/in^2).

This weight is very unevenly distributed throughout the vertical extent of the atmosphere because the

ARE HUMAN ACTIVITIES DEPLETING THE OZONE LAYER?

Considerable evidence has accumulated in recent years that human activities are seriously depleting Earth's ozone layer. It has been found that the average global concentration of stratospheric ozone declined by about 3 percent between 1969 and 1989. In a study released by the U.S. Environmental Protection Agency in April 1991, ozone depletion of 4 to 5 percent was found to exist over the United States. These statistics are of concern because ozone is the only gas in the atmosphere capable of limiting the amount of ultraviolet radiation reaching the surface. Because this radiation, in sufficient quantity, can destroy or mutate the cells of exposed plants and animals, continued ozone layer depletion could eventually have a devastating impact on Earth's life forms. For example, the yields of many important agricultural crops would be greatly diminished, and there would likely be a major increase in the incidence of skin cancer among humans.

The chief culprit responsible for the loss of ozone is the release of chlorofluorocarbon gases (CFCs), which currently are widely employed as refrigerants and aerosol propellants. When CFCs break down, they release chlorine atoms that act as catalysts to trigger the conversion of ozone molecules into normal oxygen molecules. The chlorine atoms, unaffected by this transformation, act over and over again, eventually destroying great quantities of ozone.

CFCs are relatively stable gases, with chemical half-lives of 75 to 100 years. They also take several years to diffuse upward into the ozone layer following their release. Because these factors mean that there is a considerable period between the time CFCs are produced and the time they begin to affect the ozone layer, even a worldwide ban on their use would be slow to show results. Currently, about 5 percent more ultraviolet radiation reaches the Earth's surface than in 1969, and recent computer models indicate that the continued erosion of the ozone layer will likely cause a further increase in ultraviolet radiation of from 5 to 20 percent within the next 40 years.

The scientific concern regarding possible human effects on the ozone shield became a headline issue in 1985, when the British Antarctic Survey discovered a hole in the ozone layer over Antarctica (see Figure 4.5). This hole apparently first appeared in 1975, and has been returning each Antarctic spring. The ozone-depleted zone extends from 12 to 20 kilometers (7 to 12 mi) in altitude and exhibits a maximum overall ozone loss of about 45 percent. In the center of this zone, between 13 and 18 kilometers (8 and 11 mi) in altitude, ozone losses of 70 to 90 percent have been registered.

The hole in the ozone layer over Antarctica is apparently caused by the extremely low air temperatures that develop over the continent during the sunless winter season. This leads to the formation of ice crystal clouds in the normally clear lower stratosphere. The ice crystals react with CFC gases to release chlorine molecules. When sunlight returns to Antarctica during August and September, these molecules are photochemically separated into individual chlorine atoms which then react with the ozone molecules and

continued on next page

air is highly compacted by gravity. The great vertical extent of the atmosphere and the imperceptibility with which it merges with the scattered molecules of interplanetary space make it impossible to locate precisely the top of the atmosphere, but it extends several thousand miles above the Earth's surface. Half the atmosphere's total mass, however, lies below an altitude of only 5.6 kilometers (3.5 mi, or 18,500 ft). At this altitude, the air pressure averages just half its value at sea level. Many of the world's higher mountain peaks extend well beyond this altitude; Mt. Everest, for example, has a summit elevation of 8.8 kilometers (5.5 mi). Because the air pressure is so low, many mountain climbers employ oxygen masks when climbing above 6,200 meters (20,000 ft).

Compositional Layers

The lowest 80 kilometers (50 mi) of the atmosphere, which contains more than 99.9 percent of the total

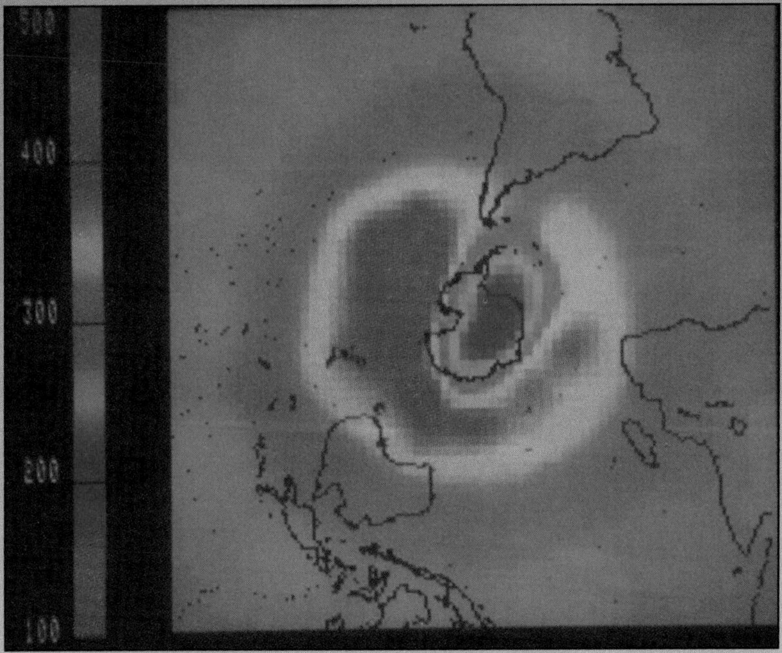

FIGURE 4.5 A color-coded satellite view of the ozone hole over Antarctica. The purple area has the greatest ozone depletion. (National Oceanic and Atmospheric Administration/National Environmental Satellite Data and Information Service)

convert them to normal oxygen. The strength of the ozone hole in any given year appears to depend on temperatures in the Antarctic stratosphere. For example, the hole was strongly developed in 1987, was much weaker during the relatively mild winter of 1988, and appeared strongly again in 1989. The hole disappears each summer as temperatures in the stratosphere rise.

Within the past few years, a second, weaker ozone hole has begun to form over the North Pole. Early in 1989, a scientific team measured 25 percent reductions in ozone concentrations at altitudes between 22 and 26 kilometers (13 and 16 mi) over the Arctic Ocean.

Aerosol cans currently are the world's largest single source of ozone-destroying CFCs. Although both the United States and Canada banned the nonessential use of CFCs as aerosol propellants in 1978, the resulting decline in emissions has been largely offset by an increase in their use in foreign countries. A significant international agreement to limit future releases of CFCs was reached in September 1987, when 47 nations signed the Montreal Protocol, pledging to reduce CFC production by 50 percent over a ten-year period. In June 1990, more than 70 countries agreed to totally discontinue CFC use by the year 2000. This agreement was especially significant because it included several developing countries such as India that had, until recently, been rapidly expanding their production of CFCs.

air, is kept mixed by wind currents so that the proportions of the nonvariable gases remain constant. Because of its relatively homogeneous composition, this portion of the atmosphere is sometimes termed the **homosphere.** The ability of the homosphere's gases to remain mixed actually results from both the mixing action of vertical wind currents and the numerous collisions between air molecules in this dense portion of the atmosphere. When air molecules collide, they tend to bounce in random directions. This action thoroughly mixes the various gases, regardless of their masses.

Above an altitude of 80 kilometers (50 mi), the atmosphere's density is so low that the number of air molecule collisions is greatly reduced. The wind flow at these higher altitudes also is predominantly horizontal, so that the air currents are subjected to very little vertical mixing. As a result, even though strong winds are common in this low-friction environment, no effective mechanisms exist to mix gases of differ-

(a)

(b)

(c)

FIGURE 4.6 Among the chief sources of atmospheric dust particles are (a) erupting volcanoes (in this case, Washington's Mount St. Helens), (b) urban and industrial activities (here, a copper refinery in Nevada), and (c) forest fires. (a: U.S. Geological Survey/JLM Visuals; b: © Paolo Koch/Photo Researchers, Inc.; c: Charles Phelps Cushing/H. Armstrong Roberts) ing densities. This leads to an increasing tendency, with altitude, for the various gases to exhibit a density layering. Scientists have named this zone the **heterosphere,** because it is vertically heterogeneous in its composition.

Thermal Layers

The atmosphere not only contains layers of varying composition, it also is stratified with respect to temperature (see Figure 4.7). The lowest and most important of these thermal layers is the **troposphere,** which contains nearly 80 percent of the atmosphere's total mass. This is the zone in which most observable weather phenomena occur and in which we live. The troposphere contains most of the atmosphere's water vapor supply. Consequently, it contains nearly all clouds and is the origin of all precipitation reaching the surface. Winds in the troposphere are highly variable in both speed and direction, and these variations are directly responsible for many of the changing weather conditions at the surface. In fact, the term *troposphere* was derived from the Greek word *tropos,* which means "turning" or "changing."

The dominant vertical characteristic of the troposphere is a rather rapid decrease in temperature with increasing altitude. The actual rate of temperature change at any given time and place is quite variable, but over long periods the mean rate of decline is approximately 6.4 C° per 1,000 meters (3.5 F° per 1,000 ft). This value holds true as a global statistical mean at all levels in the troposphere.

The sharply defined upper boundary of the troposphere, the **tropopause,** is the level at which the temperature stops falling with increasing altitude. The tropopause bulges outward above the tropics because of their high surface temperatures. Over the equator it is located at an altitude of about 18 kilo-

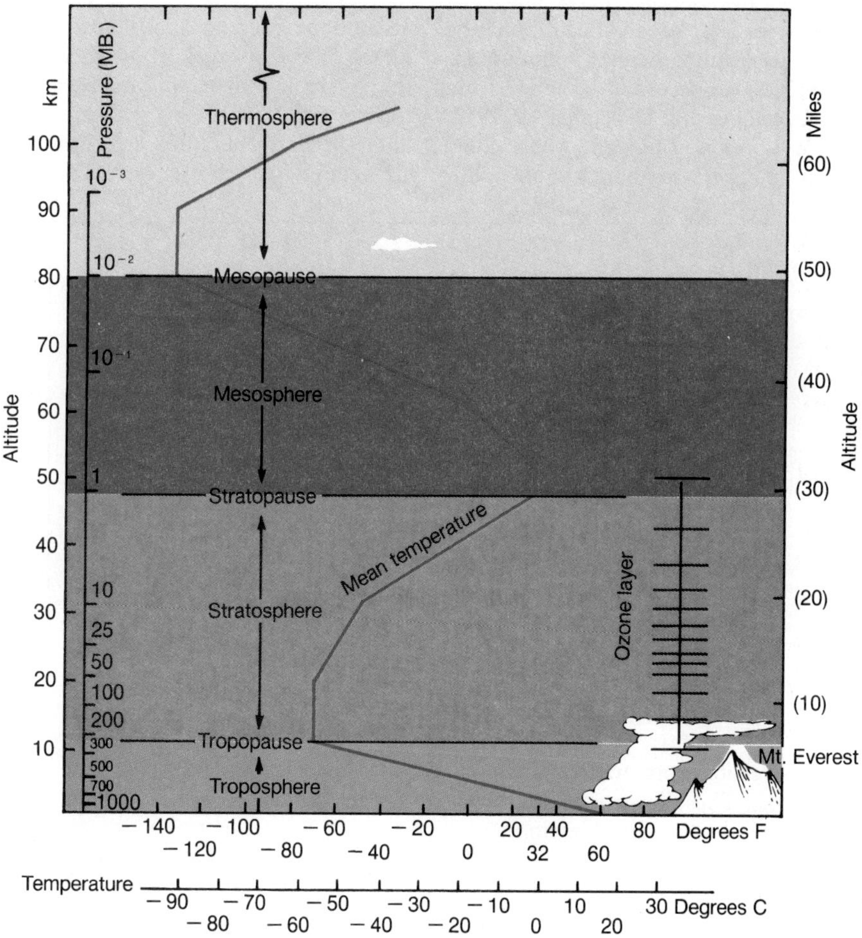

FIGURE 4.7 The average vertical temperature profile of the atmosphere, and its relationship to the four thermal atmospheric layers.

CASE STUDY

AIR POLLUTION: A GLOBAL ENVIRONMENTAL PROBLEM

The problem of air pollution has faced humanity for thousands of years. Within the past century, though, air pollution has increased in magnitude from a local source of annoyance into an important global environmental concern. While air pollution is produced by both natural and human sources, the focus here is on the human sources.

Two general categories of human activities are major producers of polluted air: urban-related activities and agricultural activities. Urban air pollution is associated primarily with motor vehicles, factory emissions, and home heating. The leading agricultural air pollutant is wind-blown soil resulting from the removal of protective vegetation, plowing, and other soil-loosening activities. Airborne pesticides and the burning of grassland areas to clear land for crops also create problems in some areas.

URBAN AIR POLLUTION

A combination of gaseous, liquid, and solid pollutants lower the quality of life in many large urban areas today. These substances produce unsightly deposits, reduce visibilities, irritate eyes and respiratory systems, and under some conditions can cause serious illness and even loss of life. As an illustration, the most deadly urban air pollution episode ever recorded occurred from December 5 to December 10, 1952, in London, England, when 4,000 deaths were attributed to air pollution resulting primarily from the household burning of low-quality soft coal. London has since enacted legislation that has curtailed this practice and brought about a noteworthy improvement in air quality.

Atmospheric conditions most commonly associated with severe urban air pollution are a lack of precipitation, light winds, and a near-surface temperature inversion. A **temperature inversion** occurs when a shallow surface layer of dense, cool air is trapped beneath warmer air. It is termed an inversion because it represents a localized reversal of the normal tropospheric characteristic of declining temperatures with increasing altitude. Short-lived temperature inversions are common occurrences. They

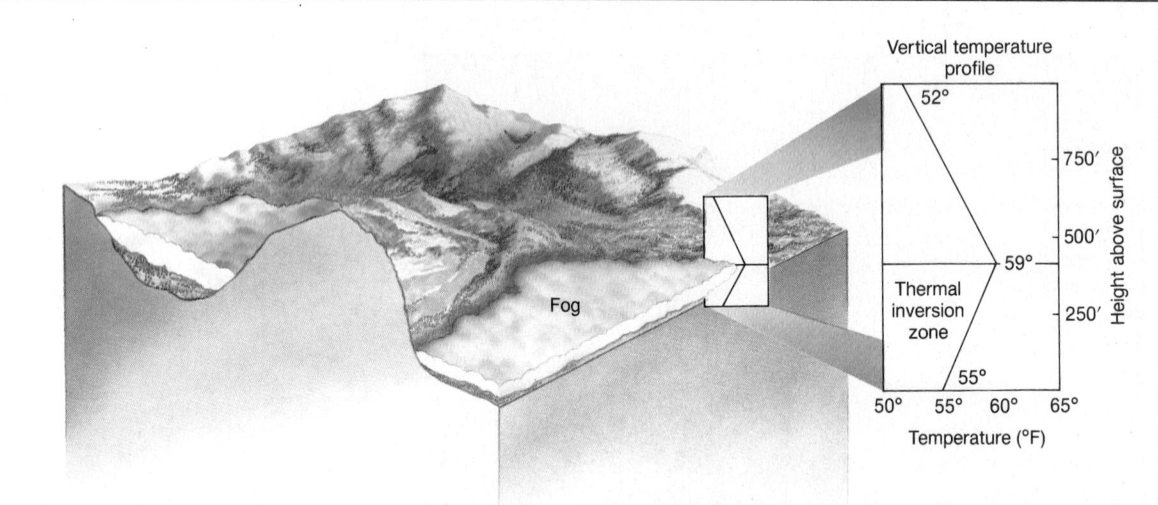

FIGURE 4.8 A low-level temperature inversion exists when a shallow layer of cool, relatively dense air lies just above the ground, with warmer air aloft. These inversions are frequently associated with fog that hinders the ability of sunlight to heat the surface and destroy the inversion.

continued on next page

form frequently in the late night and early morning as a result of the nightly radiation of surface heat into space, only to dissipate by midday because of solar heating. Occasionally, though, foggy weather within a stagnant air mass helps maintain a temperature inversion for several days. This enables pollutants to accumulate to potentially dangerous levels in the trapped layer of cool air at the surface.

Conditions favoring air pollution occur most frequently in late summer and autumn in much of North America and western Europe. A prime site for severe air pollution in the United States is the Los Angeles Basin. Its eastern rim of mountains helps trap the pollutants produced by a major urban area, while the cold waters of the offshore California Current generate low-level temperature inversions and frequent fog. As a consequence, Los Angeles experiences more days with moderate to severe air pollution than any other large city in the nation.

Fortunately, a combination of increased concern and technological advances has brought about an improvement in the urban air quality of most developed countries during the past decade. In the United States and Canada, marked improvements have been recorded recently as a result of a number of factors. These include the desulfurization of coal and oil, the increased use of solar and nuclear power, the trend toward smaller cars with better-designed engines, the mandatory use of vehicle emissions control systems, and a substantial increase in the quantity and quality of industrial gaseous effluent treatment equipment. These changes have resulted largely from a growing public concern over the problem of urban air pollution and the consequent enactment of a great deal of anti-pollution legislation. As a result, within the United States, the emission of particulates declined by 62 percent between 1970 and 1987, while sulfur dioxide emissions declined by 28 percent.

Improved air pollution control technology, however, has been largely offset by the rapid population growth and industrial development of many developing countries of the world, particularly those in Africa, Latin America, and Southeast Asia, which are for the first time producing significant quantities of air pollution. Large-scale industrialization of the countries of Eastern Europe since World War II also has been carried out with little regard to air pollution controls. In 1991, oil field fires in Kuwait were injecting tremendous quantities of gases and particulates into the air over the Middle East. The geographic consequence of these events is that urban air pollution is becoming less restricted to established industrial areas such as North America, Europe, and Japan, and is increasingly a global phenomenon.

AGRICULTURAL AIR POLLUTION

While the severity of urban air pollution for the world as a whole probably is remaining nearly steady or even declining slightly, agricultural air pollution continues to increase rapidly. The chief air pollutant of agricultural origin is windblown dust from unprotected soil. Dust storms occur in many parts of the world, especially where semiarid conditions prevail. These areas, despite their dryness, contain some of the world's most fertile soils, and extensive areas have been cleared for agriculture.

As in the case of urban air pollution, the erosion of soil by the wind is being reduced gradually in more

FIGURE 4.9 Smog in the Los Angeles Basin sometimes gives the sky and setting Sun an orange cast. (R. Krubner/H. Armstrong Roberts)

continued on next page

technologically advanced countries such as Canada and the United States. During the Dust Bowl era of the 1930s, tremendous dust storms carried off millions of tons of fertile soil from the American Great Plains (see Figure 19.1), reducing visibilities in the cities of the East Coast and creating spectacular sunsets in Europe. Improved agricultural techniques and changes in land use patterns have reduced the frequency and severity of dust storms in this area, although continuing soil erosion remains a serious problem (see the Case Study at the end of Chapter 11).

In contrast, high population growth rates in many of the developing countries of the subtropics, especially in Africa and Asia, have necessitated a rapid expansion in agricultural land use. Much of this land is marginal in quality, and the removal of protective vegetation, plowing operations, and frequent droughts combine to augment the potential for major dust storms. Dust storms not only temporarily add great quantities of dust to the air, they also remove vast amounts of irreplaceable soil. Much of this soil is deposited in the oceans, where it is irretrievably lost. The quantities of dust involved are so large that some scientists theorize they may have reduced the influx of solar energy sufficiently to account for the estimated worldwide drop in temperature of about 0.3C° (0.5F°) over the northern hemisphere between 1940 and the mid-1980s. Vegetation removal and soil erosion also are leading to the formation of desert-like conditions over large areas. This process, called **desertification**, is discussed further in the Case Study at the end of Chapter 19. It represents one of the more serious long-range environmental problems faced by humanity.

meters (11 mi); it gradually lowers poleward, until at the North and South Poles it is only about 8 kilometers (5 mi) above sea level. The troposphere also varies somewhat in altitude with the season of the year, being about 1.5 kilometers (1 mi) higher in summer than in winter over the middle and high latitudes. Temperatures at the tropopause are usually between −50°C and −80°C (−60°F and −110°F).

The tropopause can be visualized as a boundary surface separating the troposphere from the **stratosphere**. This second thermal layer extends from the tropopause to an altitude of 50 kilometers (30 mi). The lower stratosphere, up to about 32 kilometers (20 mi) in altitude, remains fairly constant in temperature, but above this level a gradual temperature rise occurs. This is caused by the absorption of solar ultraviolet energy by ozone, which converts this energy into heat. Temperature readings near the top of the stratosphere are quite variable, depending on the quantity of ultraviolet emissions from the Sun. Normally, they are near or slightly below freezing, but on occasion they can reach 7°C (45°F). Because temperatures increase with altitude in the stratosphere, there is much less tendency for the vertical air movements that form clouds to develop. The stratosphere is therefore characterized by cloudless, thin, dry air, and often is subject to strong horizontal winds.

The upper edge of the stratosphere is marked by the **stratopause**, another well-defined thermal boundary. Crossing the stratopause, the temperature stops rising and begins falling as the atmosphere's third thermal layer, the **mesosphere**, is entered. The mesosphere is located between 50 and 80 kilometers (30 and 50 mi) above the Earth's surface. Because little solar energy is absorbed at this level, temperatures within the upper mesosphere fall to the lowest values found anywhere in the atmosphere. At the upper boundary of the mesosphere, termed the **mesopause**, readings are in the vicinity of −90°C (−130°F).

Above the mesosphere, a final reversal in temperature occurs as the **thermosphere** is entered. In this zone, located at altitudes above 80 kilometers (50 mi), the temperature gradually rises to eventually reach the highest readings occurring anywhere in the atmosphere. In the upper thermosphere, temperatures may exceed 1,100°C (2,000°F) because of the absorption of ultraviolet radiation from the Sun. Although this is a very high temperature, the air is so rarefied at these great heights that very little thermal energy is actually represented.

THE ATMOSPHERE, WEATHER, AND CLIMATE

The variability of the atmosphere is made evident by the constantly changing array of weather conditions with which we are all familiar. This chapter, in fact, serves as a preface to Chapters 5 through 9, which deal with global weather and climate characteristics.

Weather is the short-term condition of the atmosphere. Variations in the weather occur both over time at any given location, and between different locations at any one time. As a result, the weather becomes more complex when examined over progressively larger segments of time and space.

When discussing long-term weather conditions, it becomes necessary to generalize. These long-term conditions constitute an area's climate. The **climate** of a region is sometimes defined as its average weather because mean (or statistically average) conditions are usually stressed. Also of great importance, though, are extreme conditions, because they can produce disastrous consequences. People therefore should expect conditions frequently near the climatic means, but also should be prepared to cope occasionally with the extremes.

Weather and climate therefore are closely related but not identical concepts. The branch of science having the weather as its subject area is termed **meteorology** (from the Greek *meteora*, or "atmospheric phenomena," and *logos* or "description"). Meteorology today is a highly sophisticated field that is considered a branch of applied physics. Forecasting meteorology involves the acquisition, rapid transmittal, and computer analysis of large volumes of surface and upper air data. **Climatology**, the science that deals with the subject of climate, is concerned with the causes and global distributions of climatic characteristics.

The atmospheric phenomena studied in weather and climate can be divided into six discrete segments or elements: temperature, air pressure, wind, humidity, clouds, and precipitation. Each element varies almost continuously over time and space, but, when analyzed in terms of its permanent or average condition, each displays an organized and logical geographical pattern. The next several chapters examine each weather element in turn in order to provide an understanding of what it is, how and why it operates, and how it is distributed around the world. Temperature is discussed first because it is not only one of the most important weather elements, but because, in a sense, it is also a measure of the energy that powers all the rest.

SUMMARY

This chapter has examined the physical and chemical properties of the Earth's atmosphere and has briefly discussed its influence on the surface environment. The atmosphere originated from gases released from our planet's interior by volcanic eruptions over the course of several billion years. The mixture of atmospheric gases has been altered gradually by physical, chemical, and biological activities. The Earth's atmosphere is unique in its properties among the atmospheres of the planets of our solar system, and is the only one capable of supporting life as we know it.

In order to organize our summary of the significance and characteristics of the atmosphere, we return to the chapter-opening Focus Questions:

1. *What basic influences does the atmosphere have on the Earth's surface environment?*

The atmosphere performs several vital functions that profoundly affect the nature of the Earth's surface and enable plant and animal life to exist. First, it provides the oxygen and carbon dioxide needed for animal and plant respiratory processes. Second, it produces frictional resistance to incoming meteors, incinerating all but the largest before they reach the surface. Third, it protects the Earth's surface from temperature extremes and from potentially lethal cosmic and ultraviolet radiation. Finally, it makes possible the weather phenomena that water the Earth's surface and produce a great variety of distinctive landforms and surface materials.

2. *What is the composition of the atmosphere?*

The atmosphere is composed of a physical mixture of gases. The most abundant are nitrogen (N_2), oxygen (O_2), argon (Ar), water vapor (H_2O), and carbon dioxide (CO_2). The proportions of the first three of these gases remain relatively constant, while those of the last two vary. In addition, small quantities of a number of liquids and solids exist within the atmosphere.

3. *How do the characteristics of the atmosphere change with altitude?*

The most important vertical changes in the atmosphere are in density and temperature. The atmosphere rapidly becomes less dense with altitude, and half of the atmospheric gases are concentrated within 5.6 kilometers (3.5 mi) of sea level. Four distinct thermal layers exist, each with vertical temperature characteristics reversed from adjacent layers. From the Earth's surface to the top of the atmosphere, these layers are the troposphere, the stratosphere, the mesosphere, and the thermosphere. The composition of the atmosphere also changes with altitude. Water vapor is concentrated near the Earth's surface, ozone (O_3) is relatively more abundant at middle altitudes, and hydrogen is more abundant in the upper atmosphere.

4. *How are human activities influencing the composition of the atmosphere?*

Human activities add a variety of substances to the air, especially in urban areas. The three substances of greatest concern are carbon dioxide, chlorofluorocarbons, and dust. Carbon dioxide, which is released in large quantities by the burning of fossil fuels, may be altering the Earth's climate. (This situation is explored in more detail in Chapter 5.) Chlorofluorocarbons, which are used as propellants and refrigerants, apparently are depleting the ozone layer, especially in the polar regions. Dust, especially windblown dust from agricultural land, may be significantly reducing the quantity of incoming solar energy.

Review Questions

1. How does the atmosphere affect the Earth's surface environment? List several factors.

2. Explain how the atmosphere is believed to have originated.

3. What five gases comprise most of the atmosphere? What is the direct source of atmospheric oxygen?

4. Why are most gases distributed evenly throughout the lower atmosphere? Explain why the atmospheric distribution of water vapor varies greatly from time to time and from place to place.

5. What is ozone, and how is it produced? What important function does it serve with respect to life on Earth?

6. What are the major natural and human sources of atmospheric dust particles? What vital function do atmospheric dust particles perform?

7. Describe how and why the atmosphere varies in density at increasing altitudes. Why is it difficult to determine the location of the "top" of the atmosphere?

8. Why is the upper atmosphere less homogeneous in its constituency than the lower atmosphere?

9. What are the four thermal layers of the atmosphere, and what are the altitudes and temperature characteristics of each?

10. Why do temperatures vary with altitude in the atmosphere? What are the two warmest portions of the atmosphere, and what are their respective sources of heat?

Key Terms

nonvariable gases
variable gases
ozone
homosphere
heterosphere
troposphere
tropopause
stratosphere
stratopause
mesosphere
mesopause
thermosphere

CHAPTER 5

ENERGY FLOW AND AIR TEMPERATURE

OUTLINE

FOCUS QUESTIONS

1. How does solar energy get to the Earth, and what happens to it once it reaches the Earth?
2. What are the major factors controlling air temperatures at the Earth's surface?
3. What is the geographic pattern of temperatures that results from the action of these controls?
4. Do human activities have the potential to significantly alter global temperatures?

Most people spend a substantial amount of time outdoors. Many, in fact, are employed in occupations, such as construction, agriculture, recreation, and transportation, that require several hours of outside work each day. A major consideration for individuals about to spend an extended period outside and, indeed, one that may partially dictate the nature of their outside activities, is the temperature.

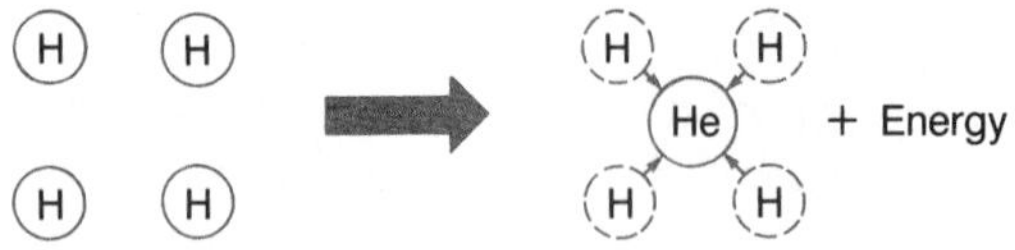

FIGURE 5.1 Hydrogen fusion involves the fusing of four hydrogen atoms into one helium atom. The attendant release of energy powers the Sun and most other stars.

ENERGY AND TEMPERATURE

The constant cycles of change that occur in the Earth system,[1] including its various subsystems—the atmosphere, hydrosphere, lithosphere, and biosphere—are driven by energy derived from the Sun and from the Earth's interior. By definition, **energy** is the capacity to do work on some form of matter. **Work,** in turn, is defined as the transfer of force from one body to another. The amount of work accomplished is the product of the amount of force applied and the amount of displacement produced by that force. For example, the amount of work you accomplish in lifting a stone is determined by both the mass of the stone and the height to which you lift it.

Energy can exist in a number of forms, and it is capable of changing from one form to another. Any moving substance possesses energy of motion, or **kinetic energy.** The faster an object or substance moves, the more kinetic energy it contains. A strong wind, for instance, has greater kinetic energy than does a gentle breeze. The kinetic energy of an object also is determined by its mass. As an illustration, equal volumes of water and air may be traveling at the same speed, but because the water has a much greater mass, it contains much more kinetic energy.

Energy also can be stored in any object or substance. This energy is termed **potential energy,** because it represents the potential of the substance to perform work in the future. For instance, the stone you lifted contains potential energy because it will perform work if you drop it on your foot. A log contains chemical potential energy that will be released as heat if it is burned. A pot of water will gain potential energy if it is heated on a stove, and the water molecules will gain further potential energy (or **latent heat**) if they evaporate to form water vapor as the water boils away. In fact, large amounts of latent heat are absorbed and later released as actual heat when surface or atmospheric water undergoes transformations between the solid, liquid, and gaseous states. We will examine this topic and its implications, for the Earth system in greater detail in Chapter 7.

When we measure the **temperature** of any object or substance, we are actually making a quantitative measure of its thermal energy. **Thermal energy** is a form of kinetic energy in which the temperature of the substance involved is related to the rate of motion of its molecules. As energy is absorbed (gained from another energy source and converted to molecular kinetic energy), the temperature rises. If enough heating takes place, the substance involved may undergo phase changes from a solid, to a liquid, and finally to a gaseous (vapor) form. Because no limiting maximum temperature exists, the temperature attained by any substance is restricted only by the limits of its ability to continue to absorb more energy than it loses. The Earth's air temperature therefore is a measure of the rate of motion of the air molecules that results from the absorption of energy from all sources.

Of the several energy sources that heat the atmosphere, the Sun is overwhelmingly dominant. It is estimated to supply at least 99.97 percent of the atmosphere's total heat. A second source, heat from the interior of the Earth, provides nearly all the remainder.[2] The Sun's energy is produced by atomic fusion: within the Sun's core, atoms of hydrogen are fused into helium under conditions of tremendous heat and pressure (see Figure 5.1). The solar rate of helium production is approximately 3.6×10^{11} kilo-

1. The Earth and its atmosphere, taken together, will be referred to as the **Earth system**.

2. Heat from the Earth's interior, termed **geothermal heat**, sometimes manifests itself spectacularly in the form of earthquakes, geysers, and volcanic eruptions. This energy source, although unimportant in providing atmospheric heat, plays a crucial role in the development of the Earth's surface features and will be examined in detail later in this book.

grams (400 million tons) per second. The mass of the helium produced is slightly less than that of the original hydrogen, and this lost mass is converted to the energy that is radiated in all directions from the Sun. The Earth is such a small target and is so far away from the Sun that it intercepts only about .0000000005 (or one/two billionth) of the Sun's total energy output. This small proportion of energy, however, is the dominant power source for the Earth system.

ENERGY TRANSFER PROCESSES

How does solar energy cross the great void separating the Earth from the Sun? Once it has reached the Earth system, how is this energy distributed to all parts of our planet? Three energy transport processes—radiation, conduction, and convection—collectively perform these tasks.

Radiation is the process of energy transmission by electromagnetic waves. These waves, so-named because they display both electric and magnetic properties, have the ability to pass through a vacuum such as the 1.5×10^8 kilometers (93 million miles) of space separating the Earth from the Sun. When it meets no resistance, radiation travels at the speed of light (light itself being a form of electromagnetic radiation), 300,000 kilometers per second (186,000 mi/sec). Radiation is divided into several major categories on the basis of wavelength to form an **electromagnetic spectrum** (see Figure 5.2). As will be explained shortly, when solar radiation is absorbed, it is converted to the heat energy that warms the Earth's surface and atmosphere.

Once solar radiation reaches the Earth system, the processes of conduction and convection play major roles in its subsequent distribution. **Conduction** involves heat transfer by physical contact between ob-

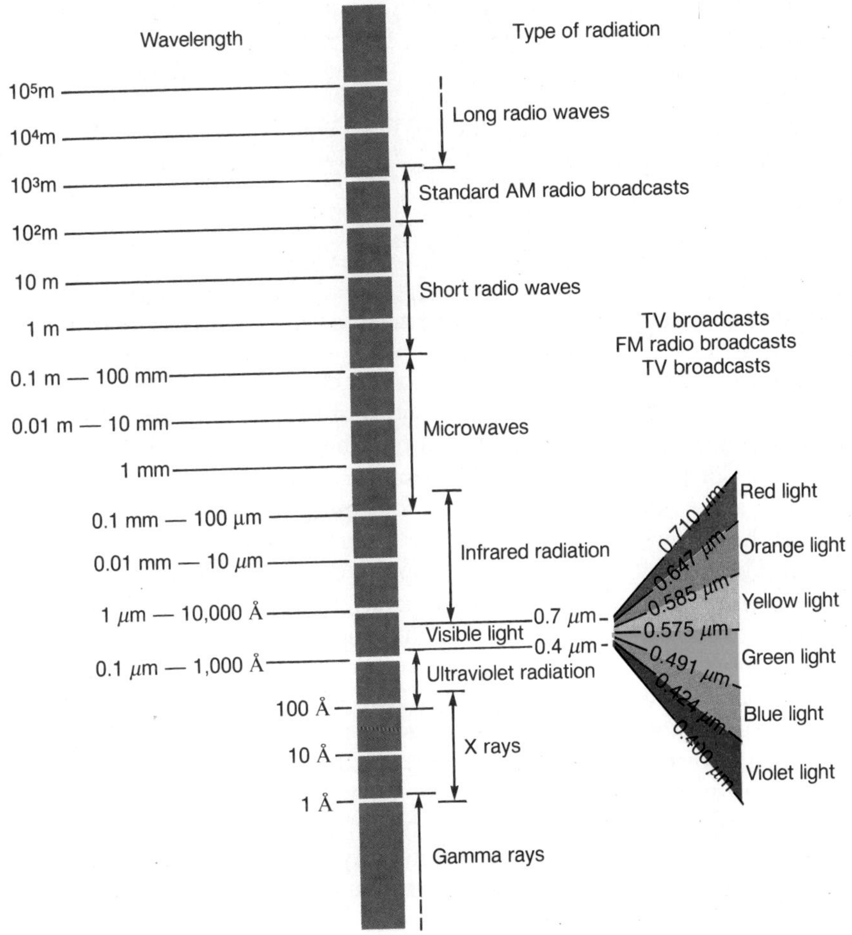

FIGURE 5.2 The electromagnetic spectrum.

CHARACTERISTICS OF ELECTROMAGNETIC RADIATION

Although electromagnetic radiation constantly surrounds us and all objects or substances absorb and emit radiation, we still have a very imperfect understanding of this phenomenon. The radiation spectrum commonly is divided into categories based on wavelength, but all forms of electromagnetic radiation are essentially similar in most respects.

The quantity and wavelengths of radiation emitted by any object depend on its temperature and **emissivity**, which is a measure of its efficiency as an emitter of radiation. A perfect emitter is termed a "black body," and has an emissivity of 100 percent. Good emitters of radiation are also good absorbers of the same radiation wavelength, and vice versa. The Sun, the Earth (both land and water), and even clouds approach black body characteristics with respect to emissivity, but the variation in the emissivity of the atmosphere is quite complex, depending on the wavelength of the radiation reaching it (see Figure 5.8). At some wavelengths, especially between 0.3 and 1.0 micrometers, the air's emissivity is near zero. Consequently, virtually no radiation of these wavelengths passing through the atmosphere is absorbed. Much of the Sun's radiation falls within this wavelength range; it therefore readily penetrates the Earth's atmosphere to warm the surface. At other wavelengths, such as those shorter than 0.3 micrometers, the atmosphere's absorption rate is nearly 100 percent, so that this energy is effectively blocked from reaching the surface.

The hotter an object is, the greater the quantity of radiation it emits per unit area, and the shorter the wavelength of the emitted energy. The Sun, with a surface temperature of approximately 6,000°C (11,000°F), radiates maximum energy in the vicinity of 0.5 micrometers, which is in the visible portion of the electromagnetic spectrum (see Figure 5.3). Because it is so much hotter than the Earth, a given area on the surface of the Sun emits about 160,000 times the radiation of an area of similar size on the Earth's surface. The maximum radiation for the Earth occurs at a wavelength of about 10 micrometers, well within the portion of the infrared range that we sense as heat (see Figure 5.4). Because wavelengths in the zone of maximum terrestrial radiation are roughly 20 times longer than wavelengths in the zone of maximum solar radiation, terrestrial radiation is generally referred to as **longwave radiation** and solar radiation as **shortwave radiation**.

jects or substances of different temperatures. Heat always flows from the warmer body to the colder one, and, if contact is maintained, will continue to do so until their temperatures have equalized. Conduction is the basis of our ability to sense heat and cold. For example, when you hold an ice cube, heat flows from your hand to the much colder ice. The loss of heat makes your hand cold, while the ice may gain enough energy from your hand to melt.

With respect to weather and climate, most conduction occurs between the Earth's surface (both land and water) and the atmosphere. Because air, water, and land are all rather poor conductors, the direct effects of this energy transfer process are restricted largely to the lowest few meters of the atmosphere.

Convection is the process of energy transfer through the physical movement of a fluid (a liquid or gas) from one place to another. At the Earth's surface, winds and ocean currents are the major convectors. (If the motions of these two fluids are predominantly horizontal, the energy transfer process is termed **advection.**) A given quantity of air has a lower capacity for transporting heat than does the same quantity of water, but wind typically travels at a much greater speed than do ocean currents. As a result, approximately twice as much energy is convected by the atmosphere as by the ocean waters.

On a global basis, convection produces a poleward net transport of heat (see Figure 5.5). If it were not for this exchange of energy between the lower and higher latitudes, the tropics would be significantly hotter, and the polar regions substantially colder, than they already are. Such a situation would greatly reduce the habitability of our planet. As we shall see, the latitudinal imbalance in solar energy reaching the Earth system is also largely responsible for driving

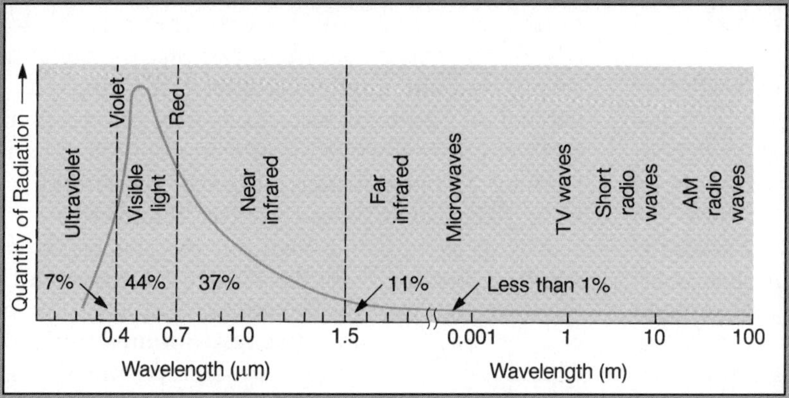

FIGURE 5.3 The solar radiation spectrum. Most energy radiated by the Sun is visible light and near-infrared radiation.

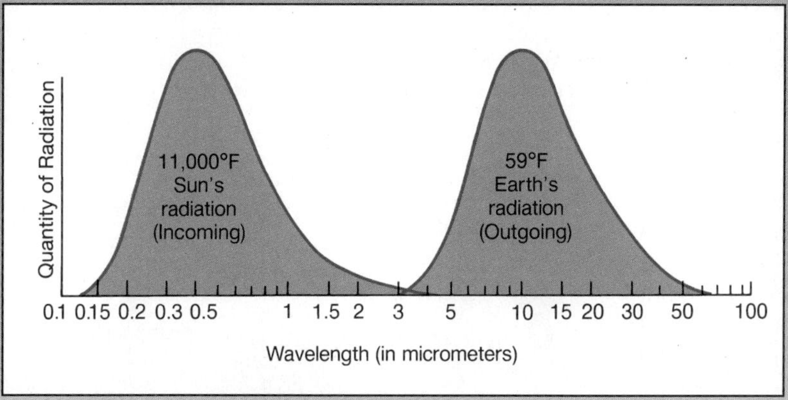

FIGURE 5.4 Comparison of solar and terrestrial radiation spectra. Although the total quantities of incoming and outgoing energy are equivalent, the peak radiation wavelength for outgoing terrestrial radiation is about twenty times longer than that of incoming solar radiation (insolation).

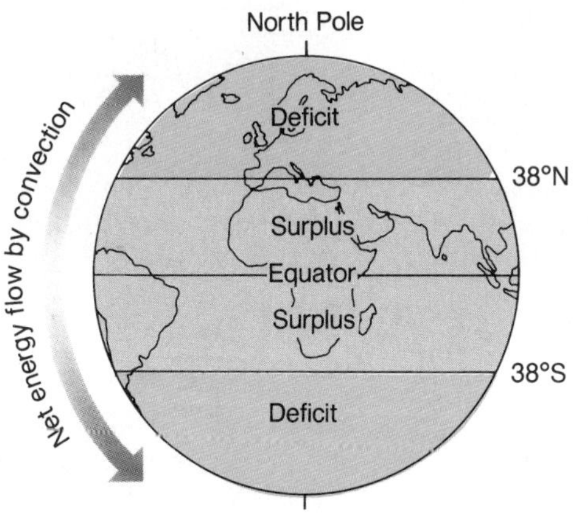

FIGURE 5.5 Latitudinal variation in the terrestrial energy budget. The Earth receives more energy from the Sun than it reradiates to space equatorward of 38°; this situation is reversed poleward of 38°.

our planet's atmospheric and hydrologic (water) subsystems.

The Global Energy Balance

Solar shortwave radiation passes through the 1.5×10^8 kilometers (93 million mi) of space between the Sun and the Earth system without being appreciably altered in any way. During its comparatively short passage through the atmosphere to the Earth's surface, however, solar radiation is subjected to a number of redirections and transformations of great importance to life on Earth.

Solar energy can be visualized as reaching the Earth system in a steady stream of shortwave radiation. As this **insolation** (a shortening of the term ***incoming solar radiation***) enters the atmosphere, it is effectively divided into three segments, as illustrated in Figure 5.6. Approximately 30 percent of the energy is redirected back to space. While most of this redirected energy is reflected from the upper surfaces of clouds, some is scattered by air molecules and dust particles, and some passes through the atmosphere to be reflected from the Earth's surface.

Energy that is scattered is broken up and reflected in all directions by air molecules and dust particles. Without the diffuse light produced by scattering, objects not exposed to a direct light source would be totally black, and therefore invisible. Because the wavelengths that produce the color blue are scattered more effectively than are those of the longer visual wavelengths, a clear sky in the presence of sunlight takes on a bright blue color. The sky as viewed from the Moon, which has no atmosphere, is black.

The ability of different materials to reflect or scatter solar energy varies greatly; this ability is indicated by their visual brightness in sunlight. For example, the almost blinding brightness of ice and snow or of a cloud deck when viewed from above is produced by reflected sunlight and shows that these objects are excellent reflectors. In contrast, the dark green color of a dense forest or the black color of an asphalt road indicates that they are poor reflectors, and therefore good absorbers, of insolation. The reflection rate, or **albedo**, of the Earth system as a whole is 30 percent, but local albedos display considerable variation, both temporally and spatially (see Table 5.1). In general, albedos are higher in the high latitudes than in the lower latitudes because of the presence of ice and snow and the oblique passage of sunlight through the atmosphere in the high latitudes. For the same reasons, albedos are higher, on average, in winter than in summer outside the tropics.

TABLE 5.1 Typical Albedos of Terrestrial Materials

MATERIAL	ALBEDO (%)
Earth system	30
Sand	25
Dense forest	8
Field crops	25
Water (Sun near horizon)	40–75
Water (overhead Sun)	6
Fresh snow	85–95
Concrete	20–30

The 30 percent of insolation that is reflected or scattered upward is not used in terrestrial energy processes because it is redirected back into space. The remaining 70 percent is absorbed to heat the Earth's surface and atmosphere. The actual absorption of this energy occurs in two places—the atmosphere and the Earth's surface. Approximately 19 percent of the total insolation does not reach the surface, but is absorbed in the atmosphere. The remaining 51 percent passes through the atmosphere to be absorbed at the surface (see Figure 5.6).

Energy is neither gained nor lost during the absorption process. What does occur is a major shift in the wavelengths of the radiation. Most insolation arrives as visible light and as near-infrared radiation, with an average wavelength of about 0.5 micrometers (μm) (see the Focus Box on pages 68 and 69).[3] The

3. Microscopic dimensions are generally measured in micrometers, commonly written μm. One micrometer (1μm) equals 10^{-6} meters.

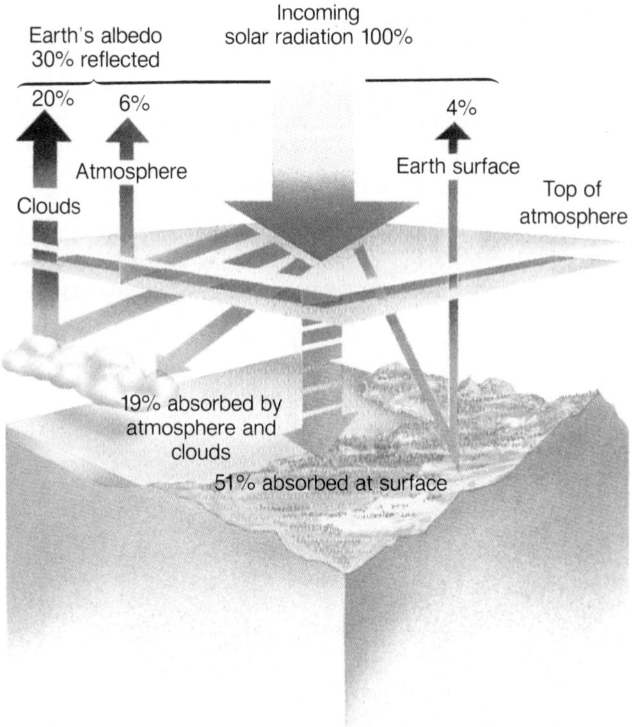

FIGURE 5.6 Approximately 30 percent of the solar radiation reaching the Earth system is reflected or scattered back to space, mostly from the atmosphere. The remaining 70 percent is absorbed within the atmosphere or at the surface.

much cooler Earth system, though, reradiates the absorbed energy in the form of long-wave **terrestrial radiation**, which has an average wavelength of about 10μm. We sense this long-wave radiation as heat.

The radiation absorption process not only generates the sensible heat that warms the Earth's surface and atmosphere, it also powers a number of vital terrestrial subsystems. Some of the radiation is converted to kinetic energy that powers the winds and ocean currents; some becomes biochemical energy used for the manufacture of plant tissue (therefore enabling terrestrial life forms to exist); and some is used in the evaporation of water, thereby becoming latent heat, and making precipitation possible.

All insolation absorbed by the Earth system is eventually reradiated to space. This process results in the maintenance of a global energy balance (see Figure 5.7). A similar energy balance, however, does not exist at the local level over short periods of time. Whenever the local influx of energy exceeds its rate of loss, as during the day, the temperature rises. When a net loss of energy occurs, as at night, the temperature falls.

Most insolation that passes through the atmosphere is absorbed by the Earth's surface and converted to either latent or sensible heat. This explains why, for example, an asphalt driveway or a closed automobile that has been in sunlight is so much hotter than the surrounding air. The heated surface sub-

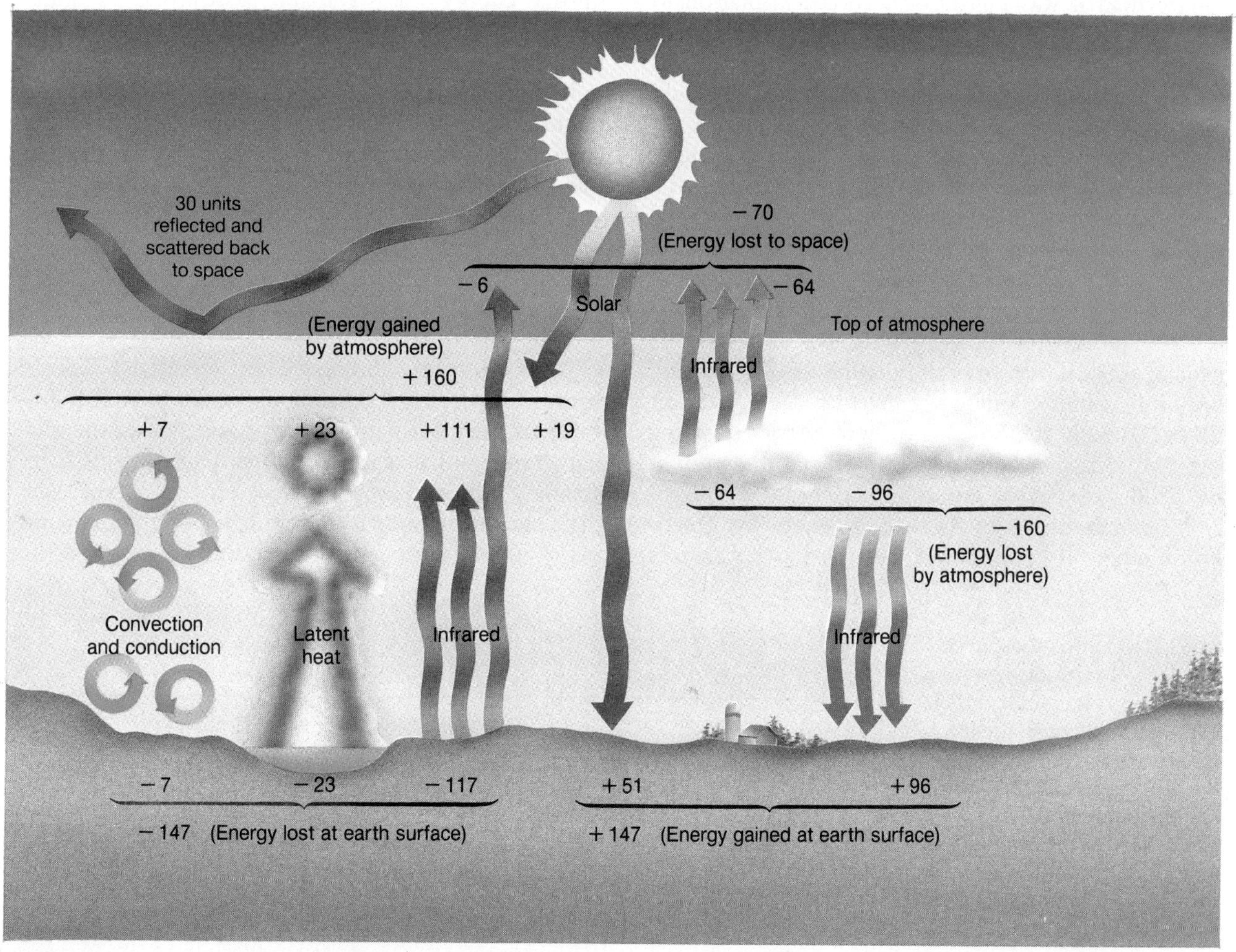

FIGURE 5.7 Terrestrial energy balance diagram. Numbers are energy units based on the receipt by the Earth system of 100 units of insolation in any given period of time. Incoming and outgoing quantities of energy to both the Earth's surface and atmosphere are equal. The total energy flux in either direction exceeds 100 units because most of the terrestrial long-wave radiation is absorbed by the atmosphere and reradiated back to the surface.

sequently transfers the energy by reradiation, conduction, and evaporation to the surrounding atmosphere, which can then transport it over great distances. The atmosphere therefore is heated primarily from below; it absorbs much more heat from the Earth's surface than from direct insolation. This is why the warmest portion of the atmosphere is the lower troposphere and why tropospheric temperatures decline with increasing altitude.

The wavelengths of most of the terrestrial radiation that warms the atmosphere are between 3 and 40 μm. Figure 5.8 indicates that most wavelengths in this range are absorbed by atmospheric water vapor and carbon dioxide. As a result, it is more difficult for long-wave terrestrial radiation to leave the atmosphere than it was for the short-wave insolation to enter it. This long-wave heat energy therefore is temporarily trapped in the atmosphere, and atmospheric temperatures become much warmer than they would otherwise be at our distance from the Sun.

The heat-trapping ability of the atmosphere, which is analogous to the heating of the inside of a greenhouse or a closed car on a sunny day, is called the **greenhouse effect**. Glass, like our atmosphere, is relatively transparent to short-wave solar radiation, but not to long-wave heat. As a result, solar energy easily enters a greenhouse, where it is converted to heat energy. This causes the greenhouse to heat up until it becomes so much warmer than the outside air that eventually the amount of heat energy escaping through the glass by conduction equals the amount of short-wave energy entering, and a raised equilibrium temperature is established. Without the greenhouse effect, it is estimated that the Earth's surface temperature would average about −20°C (−4°F), rather than its present value of 15°C (59°F). Because long-wave terrestrial radiation escapes so slowly, the Earth's surface and lower atmosphere cool only gradually during the nighttime hours, when no new insolation is available to offset radiation losses. Cooling is especially slow when the air is moist or a cloud cover is present to radiate energy back to the surface.

The global radiation balance, illustrated diagrammatically in Figure 5.7, can be written in the form of the following equation:

$$Q = K\downarrow - K\uparrow + L\downarrow - L\uparrow$$

In this equation, Q represents the net quantity of radiative energy at the Earth's surface. It is the amount of solar energy that is available to do work, including heating the air, melting snow, evaporating water, and powering plant photosynthesis. $K\downarrow$ is incoming solar radiation (insolation); $K\uparrow$ is outgoing (reflected and scattered) short-wave solar radiation; $L\downarrow$ is incoming (reradiated) long-wave terrestrial radiation; and $L\uparrow$ is outgoing long-wave terrestrial radiation.

WORLD TEMPERATURE PATTERN AND CONTROLS

Geographers are interested in terrestrial energy sources and energy transfer processes primarily because of the information they provide about the global temperature pattern. This pattern, in turn, strongly influences vegetative patterns, soil types, the distribution of human and animal populations, and many other factors. In this section we examine the

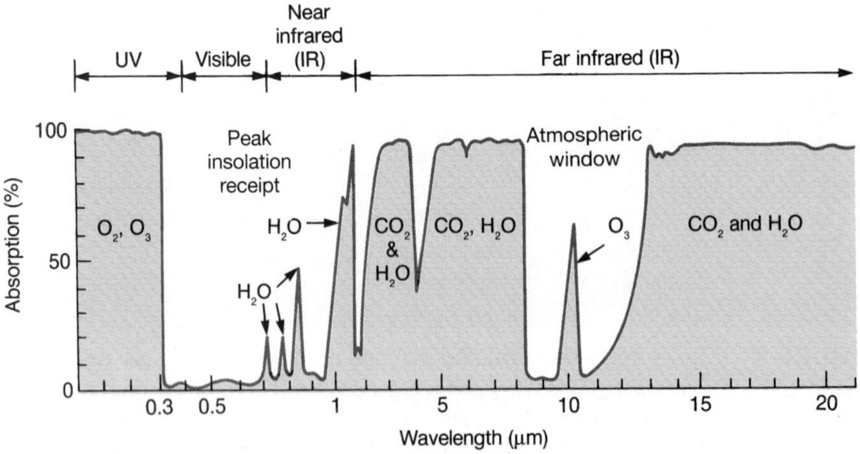

FIGURE 5.8 Atmospheric emission/absorption spectrum. Insolation readily enters the atmosphere because the atmospheric gases are relatively transparent to radiation wavelengths between 0.3 and 1.0 μm. Much of the outgoing terrestrial radiation escapes through the "atmospheric window" between 8 and 13 μm. The atmospheric gases that are the chief absorbers of radiation at various wavelengths are indicated.

influence of the physical factors primarily responsible for the Earth's surface air temperature pattern. Before proceeding, a few basic terms used in geographic temperature analysis must be introduced.

Temperature patterns usually are mapped by the use of **isotherms**, lines connecting places with the same temperature. Figures 5.9 and 5.10 depict the mean temperatures for the world in January and July, the months that represent the extremes of summer heat and winter cold for most land areas.

Weather reporting stations record a variety of temperature data, but the daily high and low temperatures provide the basis for calculating several basic temperature statistics. The **daily mean temperature** is calculated by adding the high and low temperatures and dividing the total by two. The **monthly mean temperature** is the average of the daily means for the entire month. The **daily temperature range** is the number of degrees between the day's high and low temperatures.

Last, and probably most important in terms of world temperature analysis, is the **annual temperature range,** defined as the number of degrees between the warmest and coldest monthly mean temperatures. For example, the warmest month in Baltimore, Maryland, is July, with a long-term mean of 25°C (77°F). Baltimore's coldest month is January, with a monthly mean of 1°C (33°F). Its annual temperature range is thus 24 Celsius degrees (44 Fahrenheit degrees). This figure gives a good indication of the "seasonality" of an area's climate, with higher values indicating a climate with a more pronounced summer/winter temperature variation. The global distribution of annual temperature ranges is displayed in Figure 5.12.

Primary Temperature Controls

Although a number of factors influence the Earth's temperature pattern, five stand out as the primary

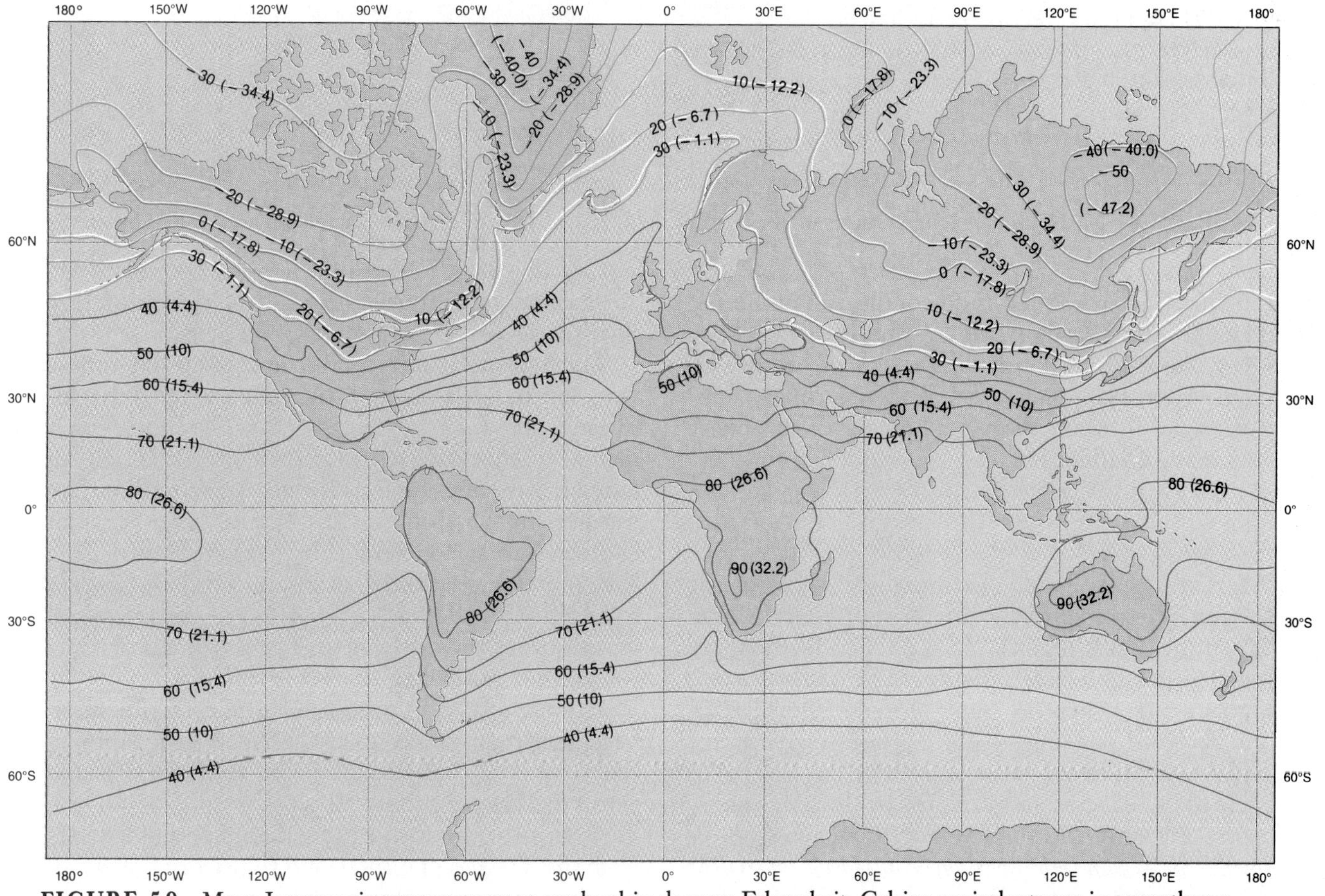

FIGURE 5.9 Mean January air temperatures at sea level in degrees Fahrenheit. Celsius equivalents are in parentheses.

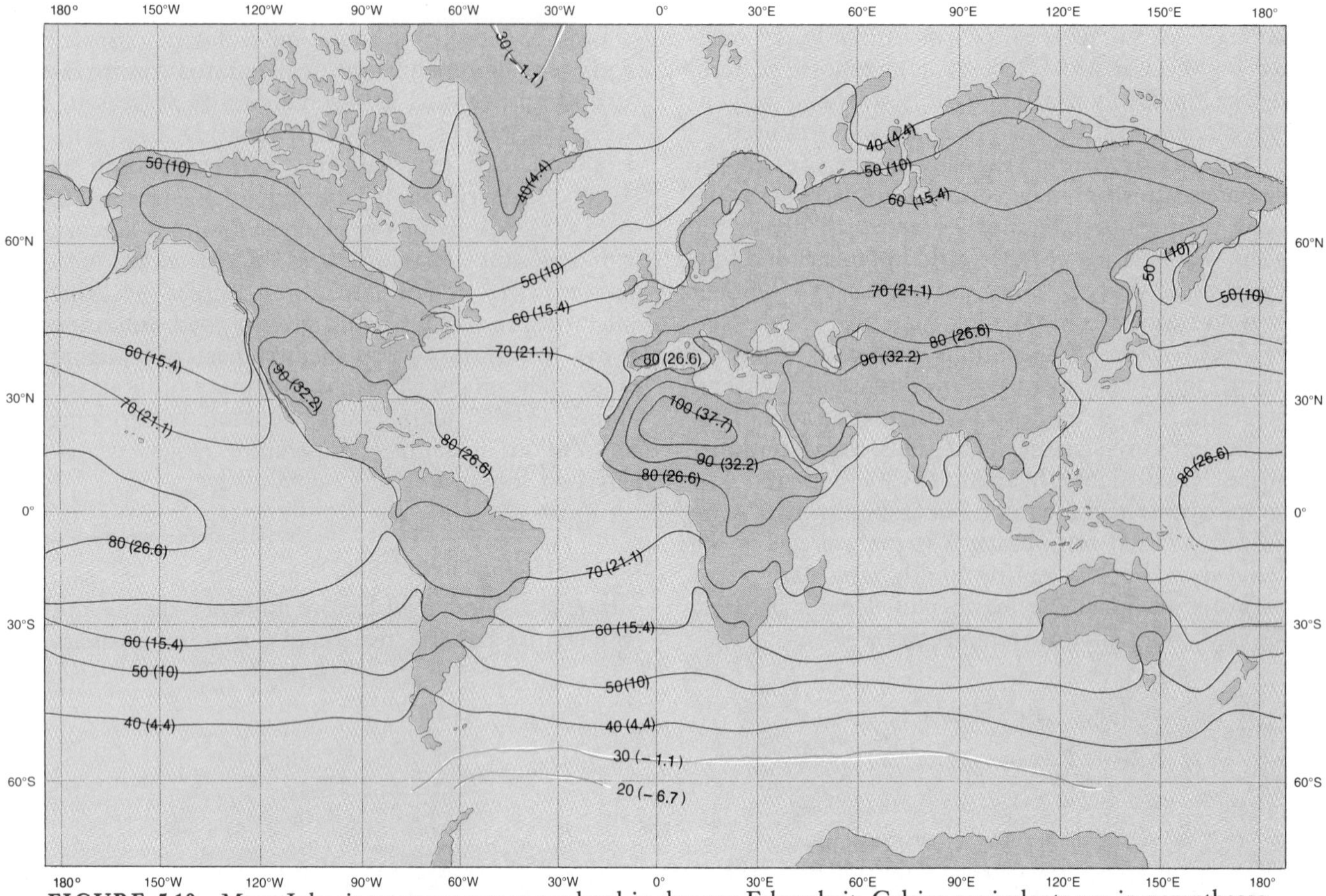

FIGURE 5.10 Mean July air temperatures at sea level in degrees Fahrenheit. Celsius equivalents are in parentheses.

controls. These are the altitude of the Sun in the sky, the duration of daylight, the distribution of land and water, the elevation, and the pattern of ocean currents. After examining each of these primary controls, we will briefly discuss a number of secondary temperature influences.

Altitude of the Sun and Duration of Daylight

We have already noted that nearly all atmospheric heating is derived directly or indirectly from the Sun. When the Sun is high in the sky, the radiant energy received per unit area of surface is maximized (see Figure 2.14). When the Sun's altitude is low, radiation received per unit area of surface is greatly reduced, and more of the energy that does arrive is likely to be reflected and scattered by the atmosphere before reaching the surface. For these reasons, *the altitude of the Sun is the most important factor controlling air temperatures at the Earth's surface.* This control affects air temperature patterns over three differing timespans—long-term, annual, and daily.

The long-term pattern, the most basic and important, can be seen on both the January and July isotherm maps (Figures 5.9 and 5.10). The most fundamental long-term characteristic of the global temperature distribution is that temperatures are warmest in the tropics and decrease poleward. Put another way, as a general rule, temperatures decline with increasing latitude because of decreasing mean solar altitudes. Because isotherms are lines of equal temperature, they display an overall zonal, or east-west trend, paralleling lines of latitude.

Although the Sun's average altitude at solar noon becomes progressively lower with increasing latitude, it is subject to seasonal variations in all parts of the world because of the changing inclination of the Earth's axis with respect to the Sun. For all locations from the Tropics of Cancer and Capricorn poleward, the Sun gets progressively higher in the noon sky

TEMPERATURE SCALES

Two temperature scales are in common use in the United States. Both have values based on the boiling and freezing points of water under standard atmospheric pressure conditions. The **Fahrenheit** scale, developed in 1714 by the German physicist Gabriel Daniel Fahrenheit, has been losing international popularity but is still the scale most frequently used in the United States. This scale has values of 32°F and 212°F, respectively, for the freezing and boiling points of water. The metric system temperature scale is the **Celsius** (or centigrade) scale, named in honor of Anders Celsius, a Swedish astronomer who devised it in 1742. The freezing and boiling points for water on the Celsius scale are 0°C and 100°C, respectively.

Where both temperature scales are commonly employed, as in the United States, it is often necessary to convert from one to the other. The formulas used are as follows:

$$°C = \frac{°F - 32}{1.8}$$

$$°F = 1.8°C + 32$$

A third temperature scale, used primarily in the laboratory sciences, is the **Kelvin** (or absolute) scale. Like the Celsius scale, it is calibrated to have 100 equal temperature increments between the freezing and boiling points of water. Numerical values, however, begin at absolute zero, the temperature at which all molecular vibration ceases. The freezing point of water is 273°K, and the boiling point is 373°K. Any Kelvin temperature has a numerical value 273° higher than its Celsius equivalent. Stated as a formula, this relationship is as follows:

$$°K = °C + 273$$

The Fahrenheit, Celsius, and Kelvin scales are illustrated in Figure 5.11.

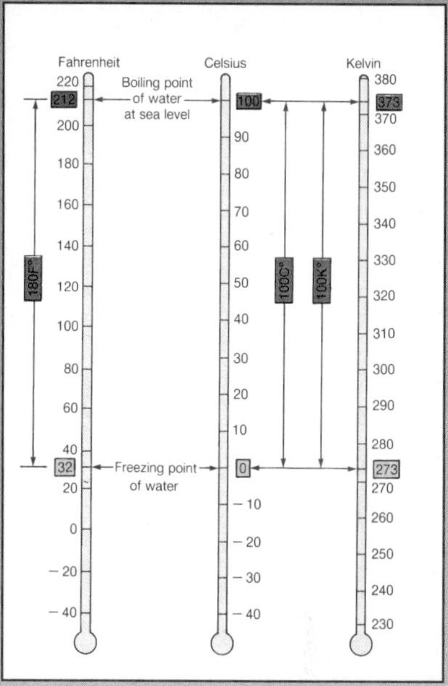

FIGURE 5.11 Comparison of the Fahrenheit, Celsius, and Kelvin temperature scales.

during the six-month period from the winter solstice to the summer solstice. It then decreases in altitude during the following six months. Locations within the tropics have two periods of increasing and decreasing solar noon altitudes during the course of a year. At all locations except the equator itself, however, the Sun's average noon altitude is higher in the spring and summer than in the fall and winter. In addition, all places on Earth (again excepting the equator) have longer days than nights in spring and summer and longer nights than days in fall and winter. These two related factors are responsible for the seasonal variations in temperature distributions that can be observed between the January and July isotherm maps. For example, the combination of vertical Sun angles, limited cloud cover, dry surface conditions, and 13- to 14-hour periods of sunlight contribute to the fact that land areas in the subtropics, rather than those in the equatorial regions, experience the hottest summer temperatures in the world.

The greater intensity and duration of insolation during the spring and summer typically cause these two seasons to be warmest. This tendency becomes progressively stronger poleward, because seasonal differences in the receipt of insolation increase with latitude. In the high latitudes, poleward of the Arctic and Antarctic Circles, seasonal differences in the daily duration of sunlight reach extremes, with

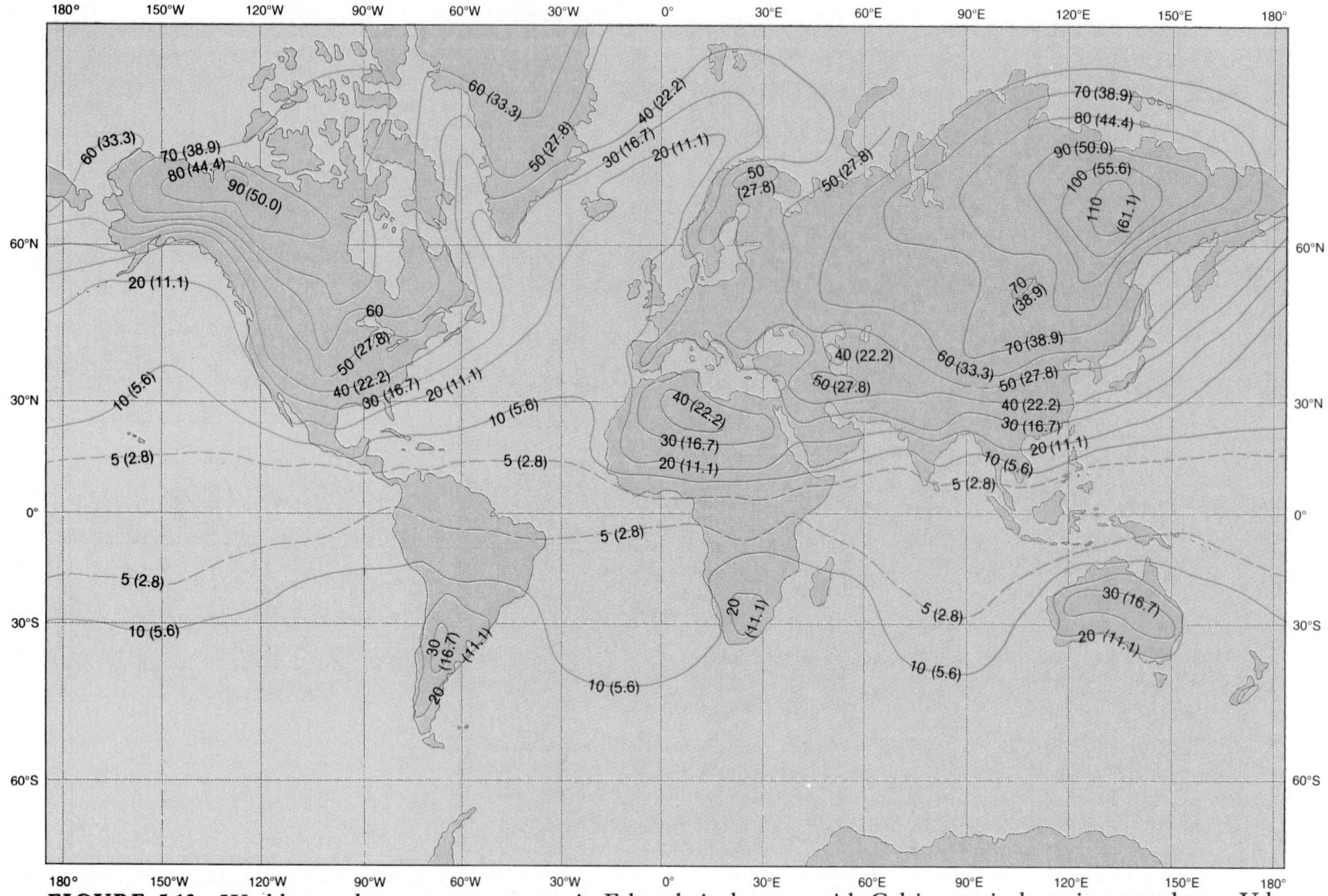

FIGURE 5.12 World annual temperature ranges in Fahrenheit degrees, with Celsius equivalents in parentheses. Values indicate the differences between the warmest and coldest monthly mean temperatures. The largest ranges occur over continental interiors in the high latitudes.

twenty-four hours of sunlight at the time of the summer solstice and zero hours at the time of the winter solstice. Annual temperature ranges therefore are very large.

Because the Earth's surface heats and cools rather slowly, seasonal temperature lags exist. The hottest and coldest times of the year actually occur a month or more after the dates of the summer and winter solstices. For example, July and August are the two warmest months in most of North America, even though the greatest insolation receipts occur during the summer solstice month of June. Likewise, January and February typically are colder than the winter solstice month of December. As a result of the seasonal temperature lag, summer is warmer than spring, and winter is colder than fall in most places.

The familiar day/night cycle of air temperature is also caused by differing Sun altitudes due to the rotation of the Earth on its axis. The resulting continuous change in the Sun's altitude rarely permits temperatures to remain steady for very long, especially when skies are clear. Although the atmosphere restricts the magnitude of daily temperature ranges, variations of from 3C° to 22C° (5F° to 40F°) are common.

Daily heating and cooling lags also occur. Although the altitude of the Sun begins to decrease after solar noon, the Sun normally remains high enough in the sky for total insolation amounts to exceed outgoing terrestrial radiation totals until midafternoon. This causes maximum temperatures to typically occur between 2 P.M. and 4 P.M. By the same token, minimum temperatures more commonly occur near sunrise than near midnight.

Land and Water Distribution

The complex distribution of land and water over the Earth's surface acts as a highly important world temperature control. Whereas the altitude of the Sun

APPLYING THE LAPSE RATE

It is easy to estimate expected temperatures at different elevations in mountainous regions if you keep in mind the fact that the air temperature can be expected to fall approximately 6.4C° for each 1,000 meters of elevation. (The equivalent value is 3.5F° for each 1,000 feet.) Let's try a couple of problems applying this temperature lapse rate:

1. If the temperature near the equator averages 27°C at sea level, what should be the approximate mean freezing level in the Andes Mountains of Ecuador? (These mountains reach elevations exceeding 6,100 meters, or 20,000 feet.)

Solution: We know that the temperature of 27°C is expected to fall by 6.4C° for each 1,000 meters of elevation. Dividing, we find that 6.4 goes into 27 approximately 4.2 times. Therefore, the freezing level should be at an elevation of about 4,200 meters.

2. If the temperature in Denver, Colorado, at 5,200 feet elevation is 60°F, what should be the temperature atop nearby Mt. Evans, at an elevation of 14,200 feet?

Solution: We know that the average environmental lapse rate on the Fahrenheit scale is 3.5F° for each 1,000 feet of elevation. In traveling from Denver to the top of Mt. Evans, we need to ascend a total of 14,200 − 5,200 or 9,000 feet. The temperature should therefore decline by 9 × 3.5, or by 31.5F°. Subtracting 31.5F° from 60°F gives an estimated temperature of 28.5°F for the summit of Mt. Evans.

largely affects the *mean* temperatures of different parts of the world, the distribution of land and water most greatly influences temperature *ranges*, both on a daily and an annual basis.

This factor's importance stems from the ability of land to change temperature far more readily than water. The following factors are responsible for the differences in the thermal behavior of land and water:

1. Water has a specific heat five times greater than that of dry land. The **specific heat** of a substance is the quantity of heat energy that must be absorbed or released in order to produce a given temperature change. It is therefore necessary for a given mass of water to absorb or release five times as much heat in order to undergo the same temperature change as a similar mass of dry land.
2. Water is transparent while land is opaque. Insolation striking a water surface therefore can penetrate to a considerable depth before being completely absorbed; consequently, solar energy is distributed through a large volume of water. This is particularly true when the Sun is high in the sky. On land, insolation is converted to heat at the immediate surface; as a result, the surface is heated much more intensely.
3. Unlike land, water is horizontally and vertically mobile. Heat absorbed near the water surface is distributed in all directions by waves and currents.
4. Water loses large amounts of energy as latent heat through the process of evaporation, because considerable energy is required to convert water from the liquid to the vapor state. Water, of course, will also evaporate from a damp land surface, but a dry surface cannot lose heat to evaporation, so that more energy is instead converted to sensible heat. This factor is largely responsible for the fact that the world's highest temperatures occur in arid desert regions.

The net effect of these factors is that land areas tend to have larger daily and annual temperature ranges and shorter seasonal and daily heating and cooling lags than do water bodies at similar latitudes. In fact, the influence of the surrounding land or water surface is so marked that climates often are described as being either continental or maritime (see Figures 5.13 and 5.14).

Continental climate characteristics include large daily and annual temperature ranges, one-month seasonal heating and cooling lags, and a tendency toward relatively low amounts of precipitation, fog,

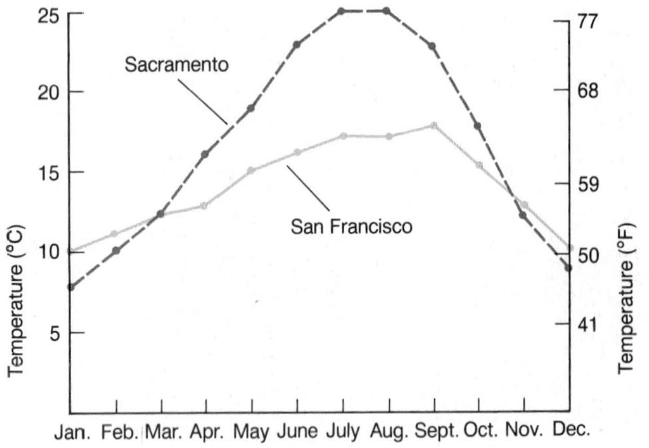

FIGURE 5.13 The differences between the annual temperature graphs of the coastal city of San Francisco and of Sacramento, located about 145 kilometers (90 mi) inland, illustrate the influence of continentality.

cloudiness, and humidity. Continental climates are best developed in the interiors of large continents such as Asia and North America.

Maritime climate characteristics, conversely, include relatively small daily and annual temperature ranges, two-month seasonal heating and cooling lags, and a greater quantity of atmospheric moisture. Maritime climates are best developed over the oceans, but also occur in many coastal areas that have a predominantly onshore windflow. They even develop to some extent along the shores of large lakes such as the Great Lakes. In reality, it is oversimplifying matters to describe the climate of any location as being just continental or just maritime, because all degrees of continentality exist between, for example, the supercontinental conditions of the central Sahara and the ultramaritime conditions of Easter Island in the South Pacific.

The influence of the Earth's land and water distribution on temperature ranges is clearly apparent from the world map of annual temperature ranges (Figure 5.12). A striking variation can be seen between the temperature ranges of large land and water masses at similar latitudes. Land areas have much larger ranges than do adjacent water bodies, especially in the higher latitudes. The world's largest annual temperature ranges occur on large continents in the high latitudes, where the influences of large seasonal variations of insolation and continentality are combined. In northeastern Asia, the monthly means of January and July differ by as much as 64 C° (115F°). Ranges of more than 45C° (80F°) occur in northwestern Canada and northeastern Alaska, on the somewhat less massive continent of North America.

Elevation

Another important influence on the global temperature pattern is elevation. This factor primarily affects temperature means (see Figure 5.15). For locations

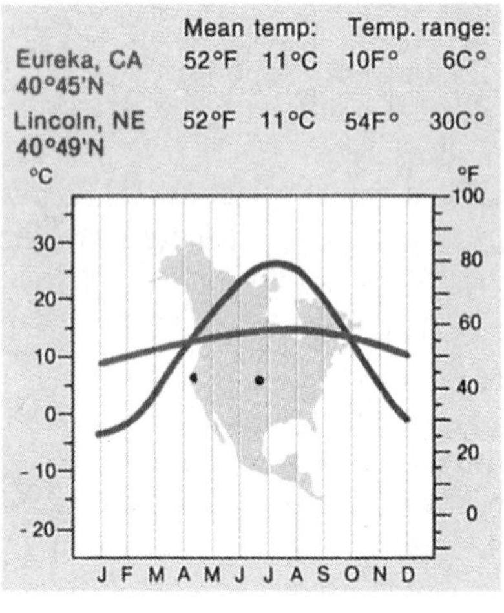

FIGURE 5.14 Eureka, California, and Lincoln, Nebraska, are at the same latitude and have identical annual mean temperatures. The difference in their continentality, however, causes their annual temperature ranges to differ greatly.

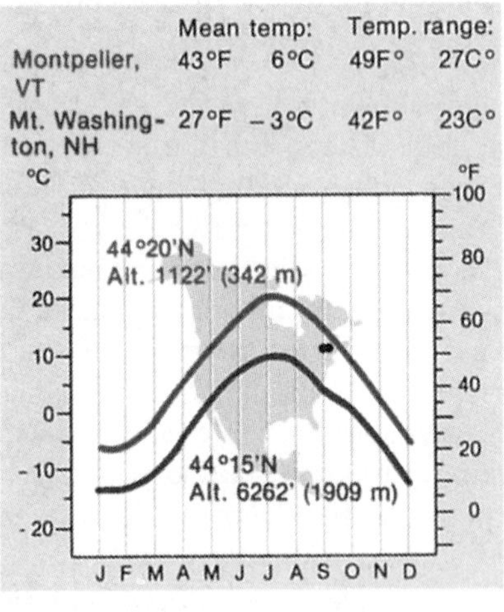

FIGURE 5.15 Temperature graphs for Montpelier, Vermont, and Mt. Washington, New Hampshire, illustrate the influence of elevation on air temperature.

above approximately 2,100 meters (7,000 ft), elevation becomes a dominant temperature influence. Peaks above 5,500 meters (18,000 ft) are perpetually covered with snow and ice even at the equator. Only a small proportion of the world's surface area, however, is situated at such high elevations. The importance of this control is therefore reduced by the limited areal extent of the Earth's highland surfaces.

Elevation is capable of producing substantial differences in both temperature and precipitation means over smaller distances than is any other major climatic control. Climatic characteristics in mountainous areas therefore vary greatly; a hike or a short drive often brings travelers to places with climatic conditions very different from those they left. In lowland areas, in contrast, it often is necessary to travel hundreds of kilometers to reach locations with substantially different climates. Travel to both higher elevations and higher latitudes results in colder temperatures. Depending on the location and season of the year, however, temperatures decrease some 500 to 1,000 times faster with elevation than with latitude, so that an increase in elevation of 300 meters (1,000 ft) may be equivalent, in thermal terms, to traveling 160 to 320 kilometers (100 to 200 mi) poleward.

The decline in temperature with altitude in the troposphere is a function of distance from a direct source of heat. High elevation locations are simply farther away from sea level, the level at which most insolation is converted to heat. In addition, the air is less dense at high elevations and has a reduced ability to absorb or conduct heat energy from either above or below. As mentioned in Chapter 4, the worldwide average rate of vertical temperature decrease is 6.4C° per 1,000 meters (3.5F°/1,000 ft) in the troposphere. This value is termed the **average environmental lapse rate.**[4] It should be stressed that this is a mean value, and that the existing lapse rate for a location at any given time probably will differ from this figure.

Ocean Currents

The final major world temperature control is ocean currents (see Figure 5.16). We have already seen that water can absorb and transport tremendous quantities of heat, and that ocean currents are responsible for about a third of the net poleward transport of thermal energy over the Earth's surface.

Ocean currents usually are classified as warm or cold. Warm currents form when water that has absorbed heat in the low latitudes is transported poleward. Cold currents, conversely, have spent an extended period in the higher latitudes before flowing equatorward.

Ocean currents can exert a strong effect on air temperatures, but are reduced in their overall geographical significance by the limited areal extent of well-developed warm and cold currents. This is especially true with respect to land areas. Many areas are far inland, out of reach of the effects of ocean currents. In addition, **leeward coasts**, which have a predominantly land-to-water wind flow, often do not experience the inland penetration of maritime air masses. Those land areas with temperatures most strongly influenced by ocean currents are associated with **windward coasts**, which have a predominantly water-to-land windflow.

The ocean currents that most strongly influence the world temperature pattern flow northward and southward along the east and west margins of the ocean basins in the middle latitudes and subtropics. In general, the east coasts of the continents are paralleled by warm ocean currents, while the west coasts are paralleled by cold currents (see Figure 10.19). The thermal effects of these currents, as might be expected, are most pronounced along the immediate

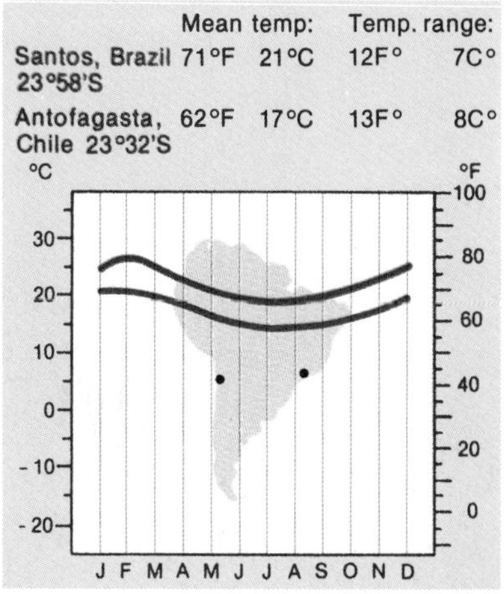

FIGURE 5.16 Ocean current effects on temperature. Santos, Brazil, is on the east coast of South America and is warmed by the Brazil Current. Antofagasta, Chile, on the west coast, is chilled by the Humboldt (Peru) Current.

4. This value also is often referred to as the **normal lapse rate**.

WORLD RECORDS OF HEAT AND COLD

World weather records of many kinds have been established, and this type of information is so popular that several books on the subject have been published. Among the most basic of weather records are extremes of heat and cold.

The world's highest generally accepted air temperature was 58°C (136.4°F), re-corded in El Azizia, Libya, in the northern Sahara, on September 13, 1922. The hottest annual mean temperature reported from a regular weather station is 34.4°C (94°F), recorded in Dallol, Ethiopia.

The world's coldest temperatures occur in Antarctica. At Vostok, a Soviet research station located on the Antarctic Plateau at 78°S 96°E and at an elevation of 3,420 meters (11,220 ft), annual temperatures average -57.8°C (-72°F), and a world record low temperature of −89.2°C (−128.6°F) was set on July 21, 1983. The coldest temperature ever officially recorded in any permanently inhabited town was −71°C (−96°F) in Oymyakon, Siberian USSR, in 1964.

coasts and diminish inland. The distance to which oceanic influences penetrate inland depends primarily on the direction and strength of the prevailing winds and the presence or absence of mountain barriers.

The January and July isotherm maps (Figures 5.9 and 5.10) illustrate, in places, the thermal effects of warm and cold currents. As a general rule, the influence of warm currents is most marked in winter, when the water is warmest relative to the land. For example, the pronounced poleward bulge of isotherms over the North Atlantic on the January map is produced by the warm North Atlantic Drift, a continuation of the Gulf Stream, which flows along the southeastern coast of the United States. Near Iceland, this current keeps January air temperatures more than 22C° (40F°) warmer than the world average for that latitude at that time of year. Mild air from the North Atlantic makes the climate of Europe much more comfortable than would be expected on the basis of latitude alone. The climate of London, for example, is often considered to be rather similar to that of Boston, Massachusetts. London, however, has a latitude of 52°N, 10 degrees farther north than Boston. In fact, London's latitude is the same as that of southern Labrador, where snow lies on the ground from September through May.

Cold ocean currents are most clearly indicated on summer isotherm maps. The pronounced equatorward bulge of isotherms off the California coast on the July map, and off the west coasts of South America, South Africa, and Australia on the January map, are quite evident.

Secondary Temperature Controls

In addition to the five primary controls, a number of secondary influences affect the global distribution of temperature. The influence of these factors often is not obvious on world temperature maps, but collectively they modify the global temperature pattern. The most important of these influences are the seasonally changing distance between the Earth and the Sun, geographical variations in cloud cover, variations in wind patterns, topographic influences, and, increasingly, the effects of human activities, which are discussed in the Case Study at the end of the chapter.

Changing Distance between Earth and Sun

The Earth's orbit around the Sun is somewhat elliptical (see Chapter 2), causing the distance between the Earth and the Sun to vary by about 4.8 million kilometers (3 million mi) during the course of a year. When the Earth is at perihelion in January, it intercepts approximately 7 percent more insolation than it does at aphelion in July.

Because of its varying distance from the Sun, the Earth might be expected to be somewhat warmer as a whole near the time of perihelion than it is near the time of aphelion. With some time lag factor, this may be true; the difficulty in assessing the overall significance of this influence is that it is masked by two much more powerful temperature controls—seasonal differences in Sun angle and the effect of land and water distribution.

Geographical Variations in Cloud Cover

The amount of cloudiness and dust cause atmospheric transparency to vary both spatially and temporally. As a consequence, the proportion of total insolation that is absorbed often varies greatly on a short-term basis. Even when the sky is clear and the air appears transparent, insolation must penetrate many miles of atmosphere to reach the Earth's surface. As a result, about half the total insolation entering the outer fringes of the atmosphere does not reach the surface because of scattering by air molecules and dust particles. When the Sun is low in the sky, its oblique rays must penetrate a much greater thickness of atmosphere before reaching the ground. The receipt of energy is therefore reduced substantially in the early morning, late afternoon, and winter. The greatest overall reduction occurs in the polar regions, where Sun angles are always low.

Clouds are even more effective than dust at reducing insolation by scattering and reflection. Satellite photographs have shown that, on the average, slightly more than half the Earth's surface is covered by clouds at any given time, and that some areas are much cloudier than others. Because clouds are excellent reflectors of insolation, it might be expected that the temperature means of these cloudy zones would be substantially lowered. A cloud cover, however, is nearly as effective at keeping terrestrial radiation from escaping the atmosphere as it is at keeping solar radiation from entering. As a consequence, the daily and annual temperature ranges, much more than the means, are lowered by cloudy conditions. For this reason, not only are daytime temperatures held down by a cloud cover, but nighttime temperatures are also raised. The abundant cloudiness associated with many maritime locations is responsible for strengthening the effect of the nearby water in maintaining relatively constant temperatures, while the clear skies of desert regions help produce large daily and annual temperature ranges.

Variations in Wind Patterns

Wind currents are another secondary temperature influence. The origin of the wind largely determines the day-to-day temperatures experienced throughout most of the middle and higher latitudes. Daily temperature variations can be quite striking, especially in the winter half of the year. The United States, for example, is frequently invaded by cold air masses from Canada and by warm air masses from Mexico and the Gulf of Mexico. In general, though, the effects of warm and cold air masses largely offset one another when it comes to computing long-term temperature means. This is especially true since the prevailing global wind belts are zonal (west to east, or east to west) in direction.

Many people believe that the speed of the wind, as well as its direction, strongly influences temperature. It is well known that the air normally feels much colder on a windy day than on a calm day. Wind-chill values, which indicate just how cold it feels due to the combined effect of air temperature and wind, have become popular statistics with radio and television weather broadcasters (see Table 5.2). In reality, the wind does not lower the air temperature, but merely blows away the insulating layer of warm air next to our bodies and replaces it with outside air. With sufficiently thick or impermeable clothing, this insulating layer stays intact and we are not chilled by a strong wind. However, an important influence of wind on air temperatures is to reduce daily temperature ranges by mixing air at different levels.

Topographic Influences

The *topography*, or the spatial distribution of the land surface features, also has an effect on temperatures. As is true of elevation, it most strongly influences temperatures in mountainous regions, because mountain barriers act to block or redirect air masses. Just as high mountain ranges restrict the movement of people across them, so too do they impede the passage of air masses. Especially affected are cold, dense air masses, which resist the lifting necessary to cross mountain barriers, and air masses that are moving slowly because of the presence of weak upper-level winds. The most effective mountain barriers are ones that are exceptionally high and continuous.

Probably the best example of the topographic influence is provided by the Himalayas along the India-Tibet border in south-central Asia. The January isotherm map (Figure 5.9) shows that the isotherms are compressed in this area, indicating a very large or "steep" temperature gradient. Part of this gradient results from the ability of the Himalayas to block the

TABLE 5.2 Windchill Equivalent Temperatures

WIND SPEED (M/SEC)	AIR TEMPERATURE (°C) 6	3	0	−3	−6	−9	−12	−15	−18	−21	−24	−27	−30	−33	−36	−39
3	3	−1	−4	−7	−11	−14	−18	−21	−24	−28	−31	−34	−38	−41	−45	−48
6	−2	−6	−10	−14	−18	−22	−26	−30	−34	−38	−42	−46	−50	−54	−58	−62
9	−6	−10	−14	−18	−23	−27	−31	−35	−40	−44	−48	−53	−57	−61	−65	−70
12	−8	−12	−17	−21	−26	−30	−35	−39	−44	−48	−53	−57	−62	−66	−71	−75
15	−9	−14	−18	−23	−27	−32	−37	−41	−46	−51	−55	−60	−65	−69	−74	−79
18	−10	−14	−19	−24	−29	−33	−38	−43	−48	−52	−57	−62	−67	−71	−76	−81
21	−10	−15	−20	−25	−29	−34	−39	−44	−49	−53	−58	−63	−68	−73	−77	−82
24	−10	−15	−20	−25	−30	−35	−39	−44	−49	−54	−59	−63	−68	−73	−78	−83

WIND SPEED (MPH)	AIR TEMPERATURE (°F) 45	40	35	30	25	20	15	10	5	0	−5	−10	−15	−20	−25	−30	−35	−40	−45
5	43	37	32	27	22	16	11	6	1	−5	−10	−15	−20	−26	−31	−36	−41	−47	−52
10	34	28	22	16	10	4	−3	−9	−15	−21	−27	−33	−40	−46	−52	−58	−64	−70	−76
15	29	22	16	9	2	−5	−11	−18	−25	−32	−38	−45	−52	−58	−65	−72	−79	−85	−92
20	25	18	11	4	−3	−10	−17	−25	−32	−39	−46	−53	−60	−67	−74	−82	−89	−96	−103
25	23	15	8	0	−7	−15	−22	−29	−37	−44	−52	−59	−66	−74	−81	−89	−96	−104	−111
30	21	13	5	−2	−10	−18	−25	−33	−41	−48	−56	−63	−71	−79	−86	−94	−102	−109	−117
35	19	11	3	−4	−12	−20	−28	−35	−43	−51	−59	−67	−74	−82	−90	−98	−106	−113	−121
40	18	10	2	−6	−14	−22	−29	−37	−45	−53	−61	−69	−77	−85	−93	−101	−108	−116	−124
45	17	9	1	−7	−15	−23	−31	−39	−47	−55	−62	−70	−78	−86	−94	−102	−110	−118	−126

southward penetration of cold central Asian air masses into India. Similarly, the west coast of the United States, especially California, normally is spared incursions of cold Canadian air masses in winter by the blocking effects of the Rockies and the Sierra Nevada. The mountains of California also are quite effective in restricting the inland penetration of cool, damp air masses from the Pacific, especially during the summer months (see Figure 5.17). This helps make Death Valley the hottest place on the North American continent.

Additional topographic influences on temperature result from the effects of slope angle and slope **aspect**, or compass orientation, on the receipt of insolation. In the northern hemisphere, particularly in the higher latitudes, steep south-facing slopes are inclined toward the Sun and therefore have substantially warmer daytime temperatures than do slopes facing in other directions, especially north. Slope aspect can produce marked differences in vegetation types and soil moisture characteristics over small distances and is an important microclimatic influence in mountainous regions. In the southern hemisphere, of course, the situation is reversed; north-facing slopes receive the most insolation and have the warmest temperatures.

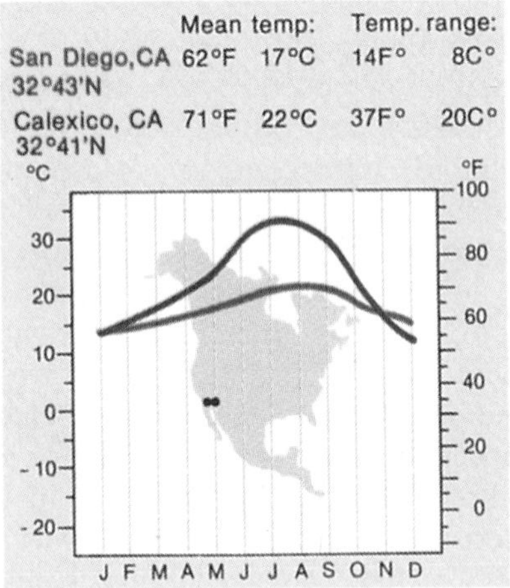

FIGURE 5.17 Mountain barrier effect on temperature. San Diego, California, is situated on the coast and frequently experiences onshore winds from the cool Pacific. Calexico is located about 145 kilometers (90 mi) inland over several low mountain ranges.

SUMMARY

This chapter has discussed both the causes and the geographic distribution of the world temperature pattern. Temperature is the quantitative measurement of heat energy; the heat energy that drives physical processes at the Earth's surface comes primarily from the Sun, which is a giant hydrogen fusion reactor. Most terrestrial processes are ultimately powered by this energy source.

We now return to the Focus Questions at the beginning of the chapter:

1. *How does solar energy get to the Earth, and what happens to it once it reaches the Earth?*

Solar energy is radiated to the Earth in the form of electromagnetic waves that fall mostly within the visible light and near-infrared portions of the electromagnetic spectrum. Once this energy reaches the Earth system, much is converted to long-wave terrestrial radiation, which we sense as heat. The heat energy is distributed around the Earth's surface and atmosphere by the combined processes of radiation, conduction, and convection, before finally being reradiated to space. The resulting distribution of this energy is responsible for the global temperature pattern.

2. *What are the major factors controlling air temperatures at the Earth's surface?*

Numerous factors are responsible for the global pattern of air temperatures. The most important are the altitude of the noon Sun, the duration of daylight, the distribution of land and water, surface elevation, and ocean currents. Additional factors of more limited importance are the varying distance between the Earth and Sun, the amount of atmospheric cloudiness and dust, the wind direction, topographic influences, and the influence of human activities.

3. *What is the geographic pattern of temperatures that results from the action of these controls?*

Because each of the world temperature controls operates with differing intensities in different regions, the global pattern of temperatures is complex. The Sun's altitude, the most dominant of all the controls, causes the overall temperature pattern to be latitudinal in nature, with mean temperatures generally decreasing as latitude increases. The duration of daylight is a seasonally varying factor which tends to cause progressively greater seasonal temperature contrasts with increasing latitude. Much of the thermal influence of the Earth's land and water distribution also is seasonal. Areas with continental climates, which are dominated by winds from the land, tend to have large temperature ranges on both a daily and an annual basis. Maritime climate areas, which receive the majority of their air from large water bodies, tend to have small daily and annual temperature ranges. Elevation reduces mean temperature values at an average worldwide rate of 6.4C° per 1,000 meters (3.5F° per 1,000 ft), so that high mountain and plateau regions are noted for their cold temperatures. Ocean currents may be either warm or cold, depending on their origins, and correspondingly act to warm or cool regions in their vicinities. Especially affected by ocean currents are windward coastal regions in the middle latitudes and subtropics.

4. *Do human activities have the potential to significantly alter global temperatures?*

There is increasing scientific evidence that the answer to this question is "yes." It already is well documented that temperatures within urban areas have been raised substantially by human activities. Of greater present concern, though, are human activities that may have a global impact. Most frequently discussed is the large-scale burning of fossil fuels, which, it is feared, may cause a global warming trend because of the vast quantities of heat-retaining carbon dioxide they release to the atmosphere.

Review Questions

1. What is energy? What forms can energy take, and how do these forms of energy differ from one another?

2. Describe the three energy transfer processes of radiation, conduction, and convection. In which portions of the Earth system does each primarily operate?

3. Explain what happens to the insolation reaching the Earth system. What proportion is absorbed, and where? What happens to the energy when it is absorbed? What proportion is reflected and scattered, and from where?

4. What is the greenhouse effect, and how does it influence world temperatures? Explain the origin of the term.

5. What are the five primary world temperature controls? Briefly explain how each is able to influence temperatures.

6. Explain the relationship between mean Sun altitudes and the long-term global temperature pattern. What more short-term temperature characteristics also are controlled by the angle of the Sun?

7. Does land and water distribution more greatly affect temperature means or temperature ranges? Why? What are the basic characteristics of continental climates and of maritime climates?

8. How rapidly do temperature means fall with increasing altitude? Why does this occur?

9. Which side of a continent is most commonly paralleled by warm ocean currents? By cold currents? Explain why windward coasts are more strongly influenced by ocean currents than are leeward coasts.

10. Briefly discuss the influence of each of the world temperature controls in your hometown or campus area. How does your geographic location influence the effects or importance of each control?

11. Cloud cover varies greatly over different parts of the Earth. Are predominantly cloudy areas substantially cooler than predominantly clear areas? Explain.

12. In addition to the effects of elevation, how does the topography influence temperatures within an area?

13. Explain how human activities that involve the burning of fossil fuels might affect global temperatures. What other types of human activities may be countering this tendency?

Problems

1. Convert 50°F to the equivalent Celsius reading. Convert it to the equivalent reading in degrees Kelvin.

2. What is the Fahrenheit equivalent of absolute zero (0°K)?

3. Convert 32°C to the equivalent temperature in degrees Fahrenheit.

4. If the average environmental lapse rate happens to exist and the temperature at sea level is 67°F, what is the temperature at an elevation of 8,000 feet?

5. If the temperature at an elevation of 6,500 feet is 37°F, and the existing lapse rate is 2.5F° per thousand feet, what is the temperature at an elevation of 2,500 feet?

6. Compute the average environmental lapse rate in degrees Celsius per 1,000 meters. If the temperature at an elevation of 1,500 meters is 12°C and the average environmental lapse rate happens to exist, what is the temperature at an elevation of 4,500 meters?

Key Terms

energy
radiation
conduction
convection
insolation
greenhouse effect
continental climate
maritime climate
environmental lapse rate

CASE STUDY

ARE HUMAN ACTIVITIES ALTERING THE GLOBAL TEMPERATURE PATTERN?

Until recently, the direct human impact on the global temperature pattern was minor. Human-derived sources of energy, primarily the burning of fossil fuels, still produce only about 0.007 percent of the Earth's total heat supply. This addition of heat is negligible on a global basis. A widespread concern, however, has developed in recent years that human activities may be bringing about a global-scale warming of the climate.

POTENTIAL CAUSES OF GLOBAL WARMING

We have seen that carbon dioxide (CO_2) gas, through its influence on the greenhouse effect, plays a vital role in enabling the Earth's atmosphere to absorb enough long-wave radiation to make our planet habitable. Unfortunately, a combination of human activities is producing a rapid increase in CO_2 levels. Between 1850 and the present, atmospheric concentrations of CO_2 have risen from approximately 280 parts per million to 355 parts per million. The rate of increase is accelerating, and it has been estimated that the amount of atmospheric carbon dioxide could double from its present concentration within the next 60 to 75 years (see Figure 5.18).

The increase in CO_2 levels is attributed primarily to the large-scale burning of fossil fuels, notably coal, oil, and natural gas, which releases carbon dioxide as the chief by-product of combustion. Recent studies have indicated that another major cause of increasing atmospheric CO_2 levels is the rapid worldwide removal of vegetation, especially in the tropics (see the Case Study at the end of Chapter 11). The decrease in the global vegetation cover is reducing the amount of CO_2 withdrawn from the air by plants. In addition, although carbon dioxide is highly soluble in water and is absorbed in large quantities by the oceans, the rate of oceanic assimilation is not keeping pace with the rate of production. As a re-

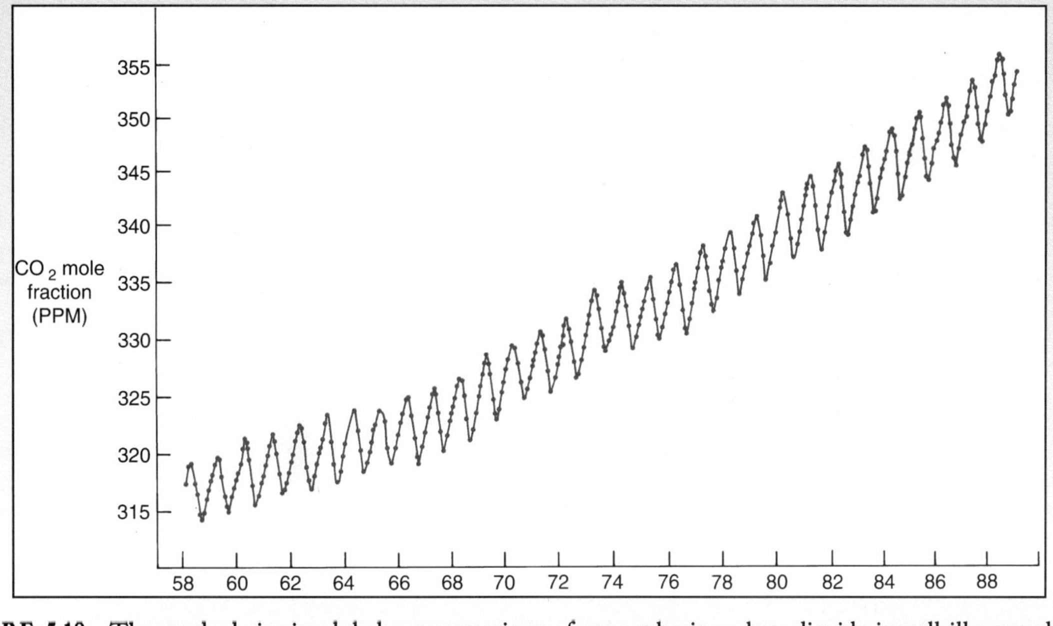

FIGURE 5.18 The gradual rise in global concentrations of atmospheric carbon dioxide is well illustrated by this graph of CO_2 levels monitored at the isolated measuring station atop Mauna Loa, Hawaii. CO_2 levels fall each northern hemisphere summer due to vegetative absorption, but then rise each winter to new record highs.

continued on next page

sult, the amount of atmospheric CO_2 is increasing at nearly half the rate at which it is currently being added to the air through the burning of fossil fuels.

Carbon dioxide is not the only gas contributing to the intensifying greenhouse effect. In fact, it is believed to be responsible for only about 50 percent of the human share of the global warming potential. Methane (CH_4) is a gas that, on a per molecule basis, is 20 to 30 times as effective as CO_2 in absorbing heat. Culturally produced methane is derived from the decomposition of crops (particularly rice) under oxygen-deficient conditions and is also produced by the digestive processes of livestock. The atmospheric concentrations of methane currently are double those of the pre-industrial period, and it is believed responsible for 15 to 20 percent of the human greenhouse potential. Chlorofluorocarbon (CFC) gases, whose effects on the ozone layer we have already examined, are at least 10,000 times as effective as carbon dioxide, on a per-molecule basis, in promoting global warming. These gases are believed responsible for another 20 percent of the global warming potential.

Some countries are much greater contributors to the global warming potential than others. The United States, with its large industrial base and great numbers of automotive vehicles, has a world-leading 17.6 percent share of greenhouse gas production. It is followed by the combined members of the European Community, the Soviet Union, Brazil, and China. Collectively, these countries account for half of total net greenhouse gas emissions. On a per-capita basis, Canada was the leading greenhouse gas emitter in 1987, followed closely by Brazil, the Ivory Coast, the United States, and Australia.

THE GEOGRAPHY OF A WARMER EARTH

It has been known for some time that the human-produced greenhouse gases have the potential to produce a global warming trend, but only within the past decade has it become apparent that this warming trend may already have begun. Global mean temperatures since the late 1800s have risen approximately 0.6C° (1F°), and the six warmest years of the twentieth century all were recorded during the 1980s. It is predicted that carbon dioxide levels could reach 550 to 600 parts per million by the middle of the next century. World population is expected to double during this same period, leading to a doubling of greenhouse gases even if the per-capita production of these gases remains constant. Currently, per- capita production is rising steadily. The likely improvement in the standard of living in the developing countries of Africa, Asia, and Latin America could raise greenhouse gas emissions still more, as these countries industrialize and their rapidly growing populations increasingly obtain automobiles, air conditioners, and other major energy-using devices.

Most computer models predict that the expected doubling of greenhouse gases over the next 75 or so years will warm the Earth as a whole by 2C° to 5C° (4F° to 8F°), but that the pattern of temperature change will be highly uneven. Interactions and feedback mechanisms among different components of the environment are so complex, however, that the predictive limits of even the most advanced computers are strained by the attempt to forecast detailed temperature patterns. For example, increasing global temperatures probably would also increase the Earth's cloud cover, which would act to reduce the warming trend in some areas by reflecting more sunlight. Rising temperatures also would alter patterns of air pressure, wind, and ocean currents, perhaps drastically, warming some areas and cooling others. Melting glaciers and the thermally induced expansion of seawater also would cause ocean levels to rise, adversely affecting coastal population centers as well as affecting coastal temperature patterns.

Most computer models indicate that the greenhouse effect warming is likely to be most intense in the higher latitudes (see Figure 5.19). High-latitude areas such as Scandinavia, Canada, and Siberia would likely benefit from such a climate change, because their growing seasons would undoubtedly become longer and their precipitation heavier. In the major grain-producing areas of the world such as the American Midwest, however, the advent of hotter summers would greatly in-tensify the frequency and severity of the droughts that already periodically plague these regions. Indeed, they might even become semi-deserts, unsuitable for agriculture without irrigation. A major change in climate also would place a severe strain on many plants and animals, especially those occupying limited geographical ranges, and would undoubtedly lead to the extinction of numerous species.

Humanity has a tremendous investment in the existing world pattern of climates. Virtually all economic development has taken place under current conditions, and a significant alteration in the global temperature pattern would doubtless produce a major worldwide calamity. More precise analyses of the potential effects of human activities on global climatic patterns are needed before these effects become only too painfully self-evident.

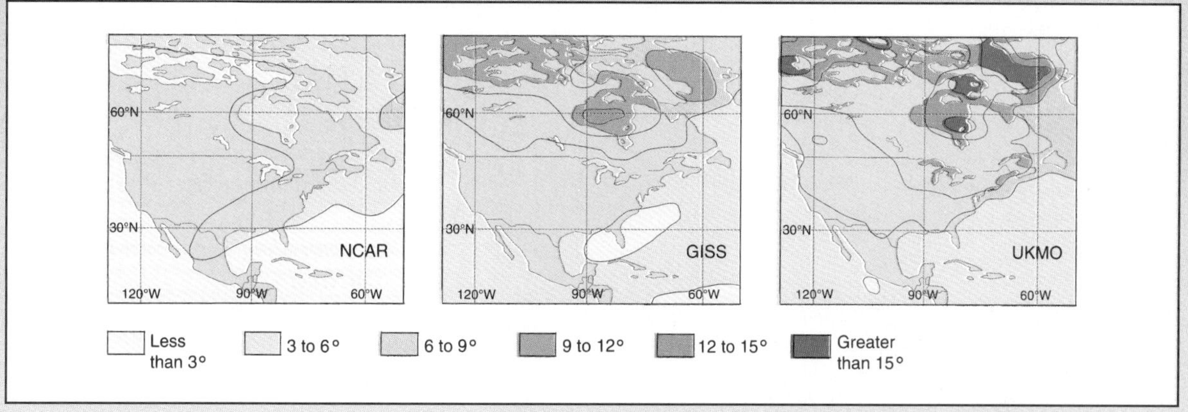

FIGURE 5.19

CHAPTER 6

AIR PRESSURE AND WIND

OUTLINE

FOCUS QUESTIONS

1. What causes variations in air pressure, and how do these variations affect the weather?
2. What is the basic pattern of global air pressure belts?
3. What factors control wind speed and direction?
4. What is the basic pattern of global wind belts?
5. What are the jet streams, and how do they influence surface weather conditions?

This chapter examines the causes and global distributions of two important weather elements: air pressure and wind. Day-to-day variations in air pressure and wind, like those of temperature, are complex and subject to constant change, but both elements display relatively simple patterns when examined at the global scale. Air pressure and wind are covered in a single chapter because they are more highly interrelated than any other pair of weather elements. They exist in a cause-and-effect relationship, with variations in air pressure causing changes in wind direction and speed.

AIR PRESSURE

Air pressure is the force that air exerts on its surroundings because of its weight. The amount of pressure exerted on any point at the surface or in the atmosphere is proportional to the weight of the overlying air from that point directly upward to the outer edge of the atmosphere. Because the atmosphere is gaseous, this pressure is applied equally in all directions. The air pressure at sea level averages 1,013.2 millibars (14.7 pounds/in^2). The pressure at any given site, however, fluctuates constantly and rarely equals this exact value.

Analyzed at the molecular level, air pressure is produced by air molecule collisions. Air molecules at room temperature travel in random directions at speeds of about 450 meters per second (1,000 mph) and, near sea level, collide elastically with one another or with surrounding objects billions of times each second. The air pressure is actually determined by the number and force of these molecular collisions.

Air pressure varies both horizontally and vertically. Because of the dominating influence of gravity on the vertical distribution of the atmosphere, however, the vertical **pressure gradient,** or the vertical variation in air pressure per unit of distance, is typically several thousand times greater than is the horizontal pressure gradient.

Measured through time at any one place, air pressure variations generally are very small, and can be neither seen nor felt. The practical importance of this weather element results largely from the fact that air pressure variations greatly influence, directly or indirectly, the global distributions of each of the other five weather elements. The air pressure pattern controls both the speed and direction of the wind; the wind, in turn, strongly influences the global patterns of temperature and moisture.

Causes of Air Pressure Variations

Why do changes in air pressure occur? The constant variation in the atmospheric pressure pattern is largely related to two factors: temperature changes and the rotation of the Earth. As discussed in Chapter 5, temperatures at the Earth's surface and in the lower atmosphere vary continuously because of the interplay of a variety of factors. When the air is heated, the air molecules increase their rate of vibration, resulting in more energetic collisions that push the molecules farther apart. This causes the heated air to expand in all directions, reducing the total mass of the air molecules within any given area and causing a reduction of air pressure (see Figure 6.1). Similarly, when air cools, its molecular activity decreases. This allows the air molecules to crowd more closely together, increasing the surface air pressure. In summary, heating causes the air pressure to decrease, while cooling causes it to increase.

The temperature/air pressure relationship produces a minor daily cycle of air pressure change in many areas, especially over land, with somewhat

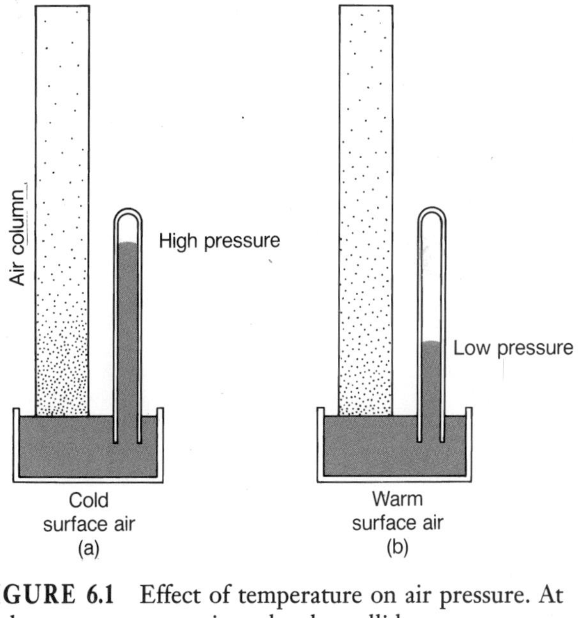

FIGURE 6.1 Effect of temperature on air pressure. At higher temperatures, air molecules collide more energetically, causing the air to expand outward and upward. This reduces air pressure values near the surface, but results in higher pressures in the upper atmosphere.

lower pressures during the warmer daytime hours and higher pressures at night. Over a longer timespan cold surfaces, such as high-latitude continental interiors during the winter, tend to develop shallow, thermally induced areas of high air pressure known as **thermal highs.** Similarly, the summertime heating of some land surfaces tends to foster the development of **thermal lows.** These thermal pressure centers greatly influence seasonal weather conditions in large parts of the world.

The second major cause of air pressure change is variations in air density resulting from the deflection of air into or out of certain latitudinal zones because of the Earth's rotation. This factor is rather complex and its effects are not as yet entirely understood, but it is related to the Coriolis effect, discussed later in this chapter. At this point in our discussion, it is sufficient to note that the Earth's rotation causes air to accumulate in certain areas, forming high pressure systems, and deflects air away from other areas, producing low pressure systems.

Global Distribution of Air Pressure

The Earth's rotation, coupled with the influence of warmer and colder surface temperatures, produces constantly changing patterns of air pressure in the lower troposphere. The resulting high and low pressure systems, often referred to simply as "highs" and "lows," are each associated with characteristic weather conditions. The high pressure in a "high" is produced by descending air that is compressed as it approaches the Earth's surface (see Figure 6.2). The compression of the air makes it warmer and relatively drier. Highs therefore are normally associated with fair weather, and a rising barometer is viewed as a sign of good weather to come. Lows operate in precisely the opposite manner (see Figure 6.3). They are areas of rising air, which expands and cools, thus reducing its ability to hold moisture (see Chapter 7). Lows therefore are normally associated with clouds and precipitation. A falling barometer, by signaling the approach of a low, often indicates that inclement weather is on its way. The alternating passage of highs and lows at intervals of a few days is largely responsible for the changing weather of the middle and high latitudes.

Global Surface Pressure Belts

Although the precise distribution of pressure systems around the world varies daily, the global pattern of air pressure is not entirely random. On a long-term basis, some parts of the world are dominated by high pressure; others are dominated by low pressure. In addition, some regions tend to experience marked seasonal differences in air pressure patterns, while others do not. The whole situation is somewhat analogous to the pattern of traffic on a busy highway. Although a limitless potential for variability exists in terms of the number, types, and placement of the

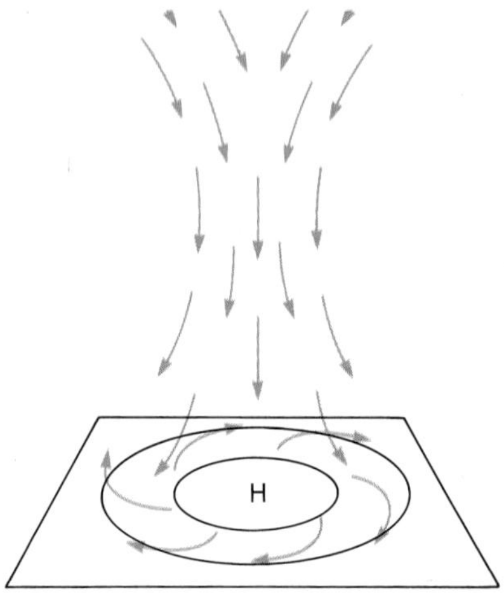

FIGURE 6.2 Three-dimensional wind-flow pattern in the northern hemisphere associated with a high pressure center.

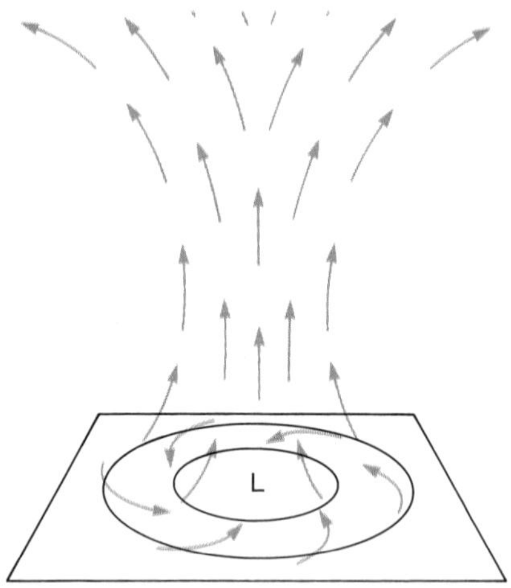

FIGURE 6.3 Three dimensional wind-flow pattern in the northern hemisphere associated with a low pressure center.

AIR PRESSURE MEASUREMENT

The most frequently measured weather element other than temperature is air pressure, and the instrument used to measure it is the **barometer.** Barometers may be grouped into two basic categories: mercurial barometers and aneroid barometers.

The mercurial barometer was invented by the Italian physicist Evangelista Torricelli in 1643. Its operating principle is quite simple; changes in air pressure control the height of the mercury column in a graduated glass tube from which the air has been removed (Figure 6.4). Despite their basic simplicity and high potential accuracy, mercurial barometers are not widely used because they are rather awkward, fragile, and expensive.

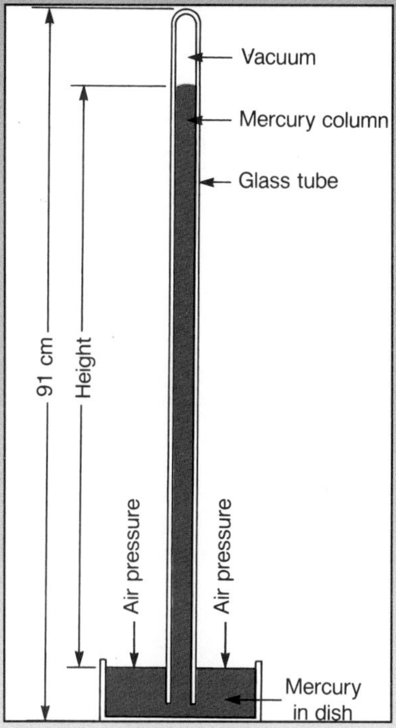

FIGURE 6.4 Operating principle of the mercurial barometer. Air pressure exerted on the mercury in the bowl is transmitted to the mercury inside the glass tube. The mercury in the tube rises to the level at which its weight exactly counterbalances the air pressure.

The second type, the aneroid barometer, is much more popular and is found in many homes. It generally takes the form of a dial-type instrument that is compact, sturdy, adjustable, and accurate (Figure 6.5). The heart of an aneroid barometer is the sylphon cell—a collapsible, accordion-shaped box from which the air has been partially evacuated. An increase in air pressure causes the sylphon cell to be compressed, while a decrease in air pressure allows it to expand. It is connected through a series of chains, levers, and springs to a pointer on a dial.

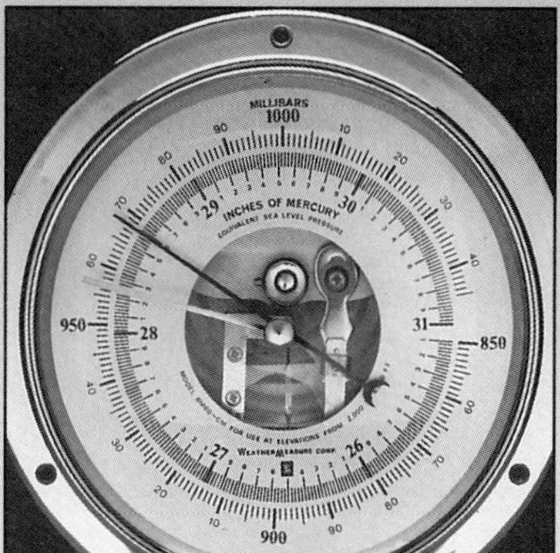

FIGURE 6.5 The aneroid barometer (pictured here) contains a sealed box that expands or contracts in response to changes in air pressure. (Courtesy of Qualimetrics, Inc.; Sacramento, CA)

The two standard scales used for barometric pressure readings in the United States are the inches of mercury and the millibar scales (see Figure 6.6). The older **inches of mercury** scale simply measures the height to which the mercury column in a mercurial barometer rises. Standard sea level air pressure of 1 kilogram per square centimeter (14.7 pounds/in^2) is equivalent to 76 centimeters (29.92 in) of mercury. Both aneroid barometers and mercurial barometers often use this scale. The second scale, which is currently employed by meteorologists throughout the world, uses units of pressure termed **millibars.**[1] Standard sea level air pressure is 1,013.2 millibars.

All surface air pressure measurements must be adjusted to eliminate the influence of elevation. Although air pressure varies vertically and horizontally, the horizontal pressure gradient is of greater significance. Vertical pressure gradients are maintained by the force of gravity and therefore do not greatly affect the weather. Horizontal pressure gradients, however, have no force maintaining them and therefore produce wind currents that, in turn, affect other weather phenomena.

[1] By definition, a millibar is 0.001 bar. A bar is a force of one million dynes per square centimeter. A dyne is the force required to accelerate a one-gram object at a rate of 1 centimeter per second each second.

continued on next page

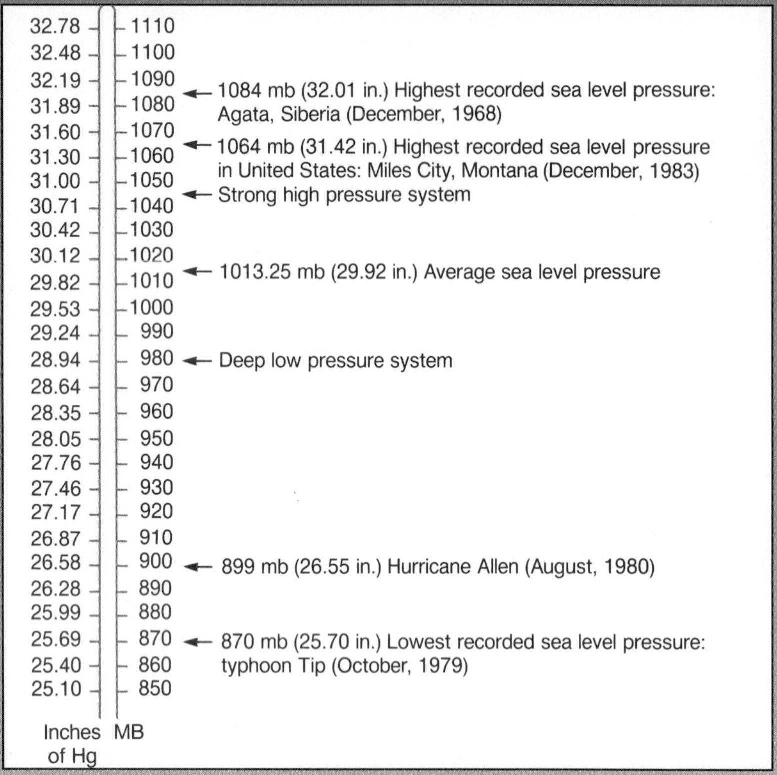

FIGURE 6.6 A comparison of the inches of mercury and the millibar pressure scales.

Because of the steepness of the vertical pressure gradient, air pressure values are greatly influenced by the elevation of the measuring station. In order to eliminate the influence of elevation, barometric readings are normally "corrected" to sea level. In other words, it is not the actual air pressure at a station that is given, but rather the air pressure that the station would have if it were situated at sea level.

On weather maps prepared by the U.S. National Weather Service (see Figure 6.7), air pressure patterns are depicted by the use of **isobars.** These are lines connecting points with the same air pressure; they are typically drawn at four-millibar intervals.

vehicles, an underlying logic and order governs the pattern. Likewise, pressure systems—and weather systems in general—tend to follow certain principles that control their development and movement.

Computing long-term averages of air pressure reveals a belted pattern of high and low pressure systems. These pressure belts encircle the Earth in an east-west direction. In a general sense, the arrangement of the world air pressure belts can be considered symmetrical with respect to the equator, with the northern hemisphere pattern being a mirror image of the southern hemisphere pattern. Each hemisphere contains three complete pressure belts, while a fourth straddles the equator and is shared by both hemispheres (see Figure 6.8). There are thus seven belts in all. The belts are in reality far from having symmetrical beltlike shapes, especially in the northern hemisphere, and vary considerably from summer to winter in many areas. The reasons for this will be explained shortly.

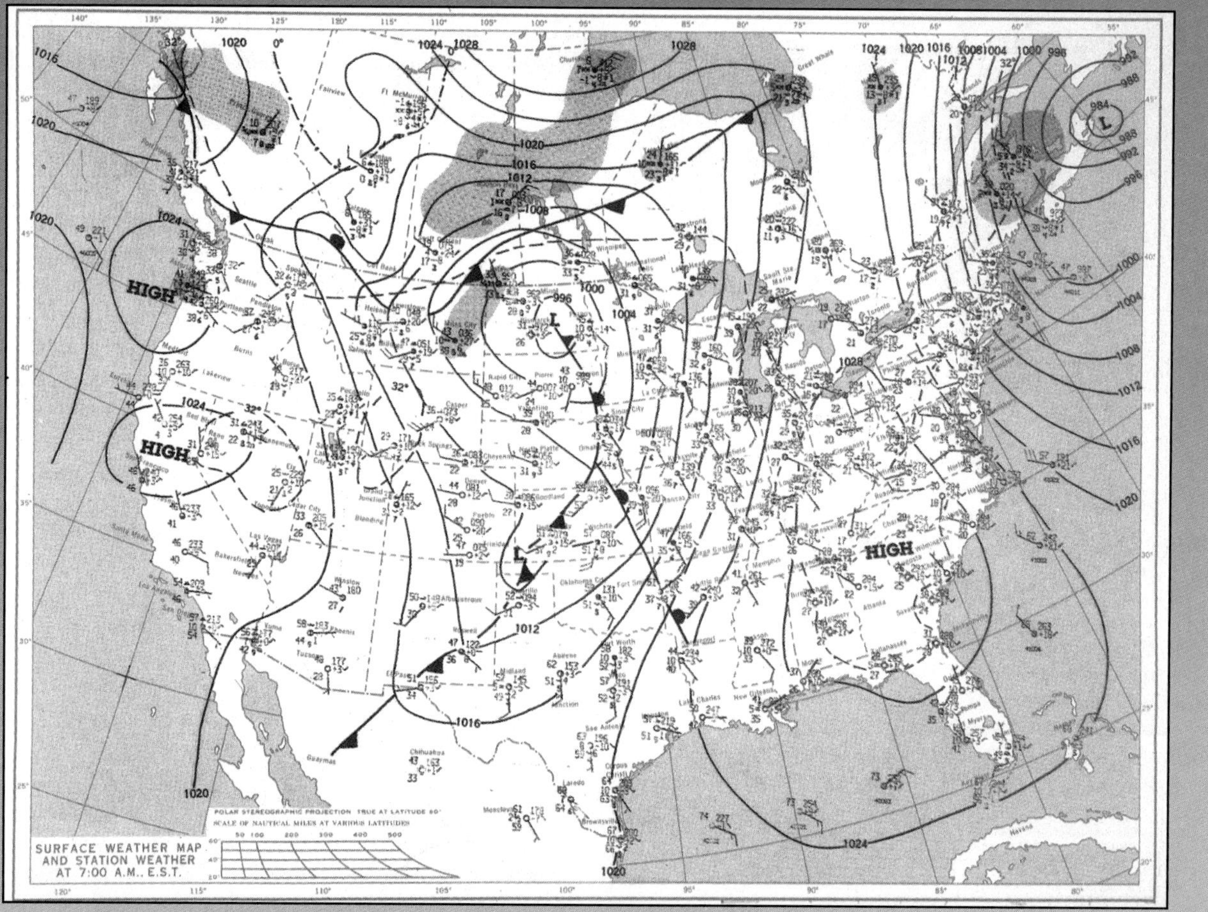

FIGURE 6.7 Surface weather maps prepared by the U.S. National Weather Service depict air pressure values by the use of isobars drawn at 4 millibar intervals. The air pressure at each reporting station is also given to the nearest tenth of a millibar. Additional information on the data contained on this map appears in Appendix B.

The most equatorward of the global pressure belts dominates the region between about 5°N and 5°S. It is a broad but rather weak trough of low pressure termed the **Intertropical Convergence Zone (ITCZ).** This low pressure zone is thermally induced by the high Sun angles and resulting year-round warmth of the equatorial region. It is associated with some of the world's rainiest climates.

Proceeding poleward in either direction, air pressures gradually rise until the centers of the **subtropical highs** are reached at latitudes between 25° and 40°. Unlike the ITCZ, the two subtropical highs are produced largely by air at high altitudes that is deflected into the subtropics by the Earth's rotation and subsequently sinks to the surface. These highs therefore are dynamically, or rotationally, induced. The subtropical highs vary regionally in the effectiveness with which they suppress precipitation, but in many places they cause arid or semiarid conditions. As a result, most of the world's major deserts are located in the subtropics.

Air pressures typically decline poleward through the middle latitudes, until the **subpolar lows** are reached at latitudes of approximately 55° to 70°. These lower pressures are generated both by the Earth's rotation, which deflects air away from this latitudinal zone, and by the lifting of warm air from the lower latitudes as it encounters denser cold air from the polar regions along weather boundaries, or **fronts.** Although often described as pressure belts, the subpolar lows are less beltlike than either the ITCZ or the subtropical highs. Instead, especially in the northern hemisphere, the subpolar lows tend to form a zone in which storms occur frequently. These storms can become very powerful, especially over ocean areas in the winter. A semipermanent storm center that develops over the subpolar North Atlantic is termed the Icelandic Low, while its counterpart in the North Pacific is the Aleutian Low (see Figure 6.9).

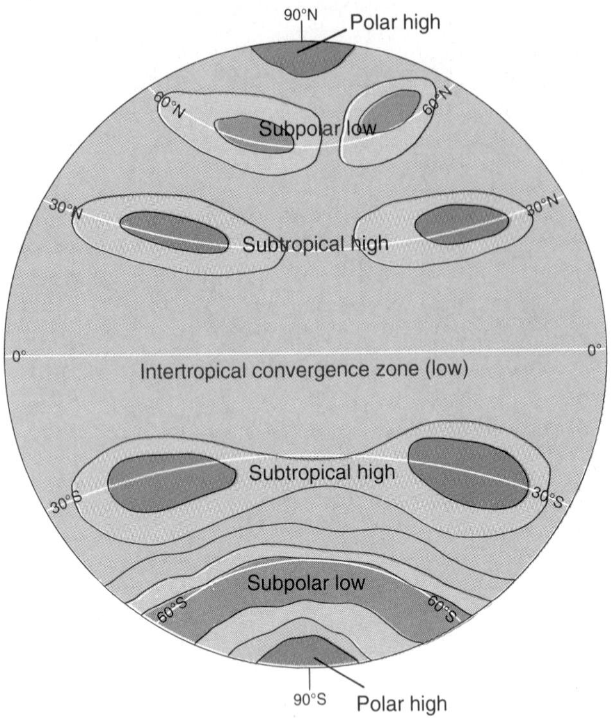

FIGURE 6.8 The idealized latitudinal distribution of the global air pressure belts.

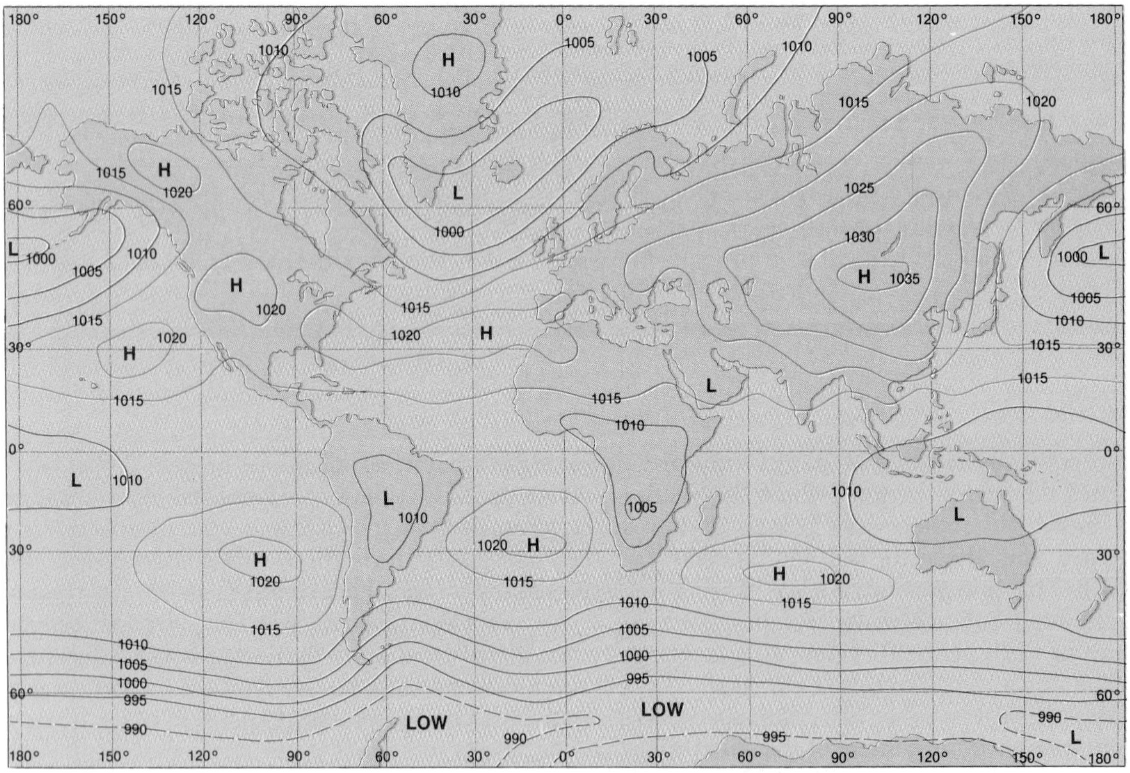

FIGURE 6.9 January global mean sea level air pressure values in millibars.

The most poleward (and typically the most weakly developed) of the four global pressure belts are the *polar highs*. In actuality, they are not belts at all, but ideally form roughly circular caps over the polar regions. They are induced by the cold temperatures that prevail in this energy-deficient region, and are quite shallow. The South Polar high is much the better developed of the two because Antarctica is considerably colder than is the Arctic Ocean location of the North Polar high.

It should be stressed that the world pressure belts merely represent average pressure conditions. They display considerable daily and seasonal variation in location and strength. In addition, numerous traveling high and low pressure centers exist independent of the pressure belts. These pressure systems are especially common in the middle latitudes, where they strongly control daily weather conditions. Their characteristics and influences on the weather will be explored more fully in Chapter 8.

Seasonal Pressure Variations

As we have seen, the combination of latitudinal temperature differences and the Earth's rotation produces a beltlike world pressure pattern. If these two factors were the only controlling influences on air pressure, the seven pressure belts would be strongly developed and would probably appear much as they do in Figure 6.8. Two additional factors, however, distort the global pressure pattern and produce marked seasonal variations in air pressures over much of the Earth's surface. These two factors are the seasonal shift in latitude of the Sun's rays and the uneven distribution of land and water surfaces.

The entire pattern of world pressure belts experiences a seasonal shift in latitude that mirrors, on a reduced scale, the annual latitudinal shift of the Sun's vertical rays between the Tropics of Cancer and Capricorn (see Chapter 2). The belts shift north or south from their mean positions only a few degrees over mid-ocean areas, but up to 10 degrees or more over large continents.

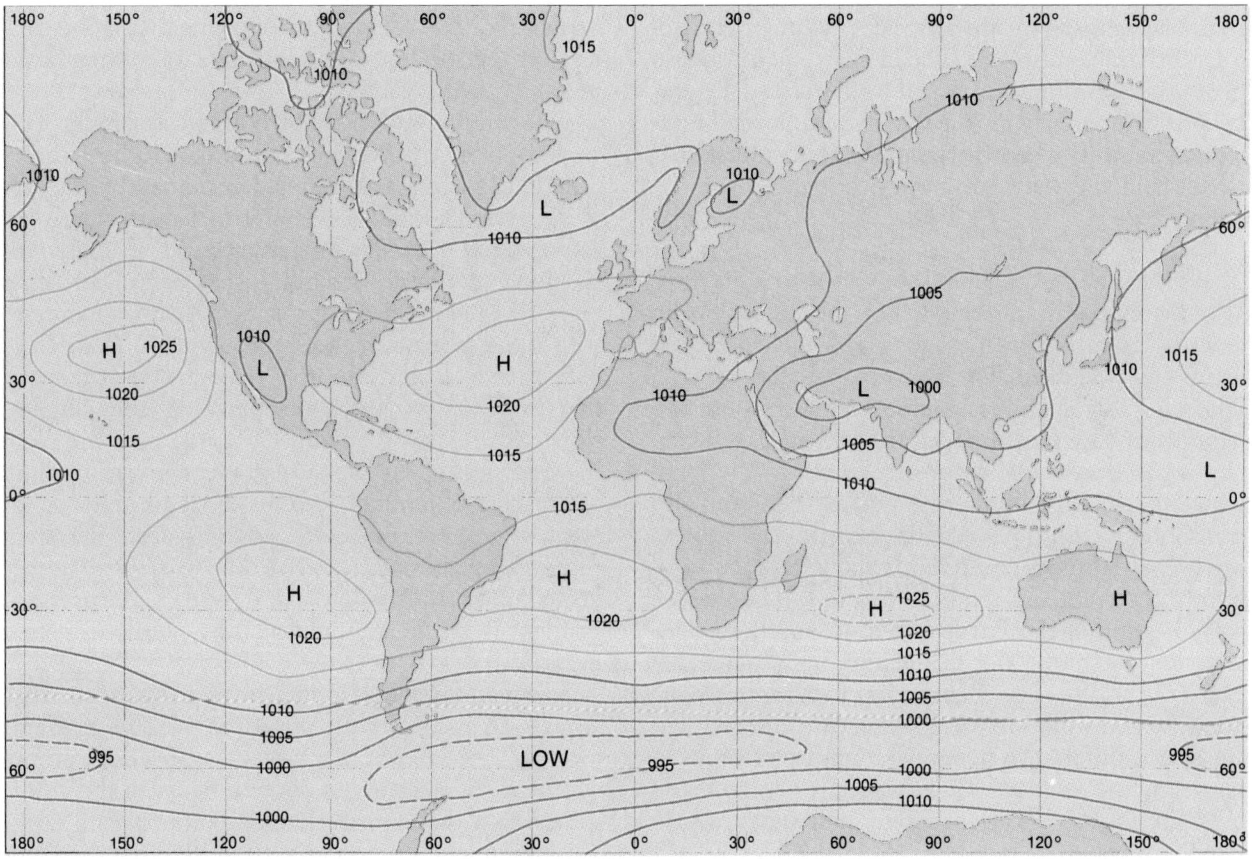

FIGURE 6.10 July global mean sea level air pressure values in millibars.

The shift of the pressure belts is thermally generated by the seasonal movement of Earth's surface temperature pattern with the Sun. As already noted, the uneven distribution of land and water bodies produces marked seasonal differences in temperature over areas at similar latitudes. The greater pressure belt displacement over land is caused by the ability of land to heat and cool more readily than water. These temperature differences strongly influence air pressure patterns, resulting in the formation of thermal highs and lows. Because land areas heat and cool more readily than do water bodies, land areas typically experience greater seasonal pressure fluctuations. Specifically, land areas usually are hotter than water bodies during the summer months and therefore tend to have lower surface pressures; during the winter months, land areas are colder and tend to have higher pressures. Seasonal heating and cooling variations are the dominant factor responsible for disrupting the orderly pattern of global pressure belts. Because the southern hemisphere has relatively little land, it heats and cools much more evenly than does the northern hemisphere. As a result, the air pressure belts of the southern hemisphere are better defined than those of the northern hemisphere. The January and July world pressure maps (Figures 6.9 and 6.10) portray the global pressure patterns during the midsummer and midwinter months, when the temperature contrasts between land and water are most pronounced.

WIND

Wind is simply air in motion. This motion may be in any direction, but in most instances the horizontal component of wind flow far surpasses the vertical. The wind is tremendously important as a global distributor of both moisture and thermal energy. It carries water evaporated from the oceans to the land, where much of it falls as precipitation. The wind also transports tremendous quantities of tropical heat to the higher latitudes, moderating the temperatures of both the tropics and the polar regions. The uneven solar heating of the Earth system is largely responsible for generating the air pressure imbalances that cause the winds. The relationship between air temperature and wind therefore is complex and reciprocal: spatial differences in temperatures cause winds to blow, and the resulting winds transport heat and cold over the Earth's surface, thereby influencing world temperatures.

Factors Controlling Wind Speed and Direction

The wind has two distinct components: speed and direction. These components not only are measured and stated separately, but also are produced by differing combinations of controlling forces.

Wind Speed Controls

The wind blows because of differences in air pressure. Wind results from the tendency of the atmosphere, in the absence of offsetting forces, to achieve a pressure equilibrium between different points at the same altitude. As we have already seen, a large or "steep" vertical atmospheric pressure gradient exists permanently because of the surfaceward pull of gravity. At most times, the downward force of gravity is counterbalanced by the upward reduction in air pressure. When the vertical pressure gradient is not balanced, the wind has an upward or downward directional component. The steepest unbalanced pressure gradients, however, normally exist in a horizontal direction. This causes most winds, both at the surface and aloft, to blow essentially parallel to the Earth's surface. For this reason, the following discussion centers on the influence of horizontal variations in air pressure on the wind.

The wind always blows from areas of high pressure to areas of low pressure and, if other factors remain constant, its speed is determined by the rate of air pressure change per unit of distance between the pressure systems. The situation is analogous to the flow of water down a hill. Just as water flows from higher to lower elevations at a speed that increases with the steepness of the slope, air flows from areas of higher to lower atmospheric pressure at a speed that increases with the steepness of the pressure slope, or pressure gradient. On an air pressure map, the steepness of the pressure gradient is indicated by the spacing of **isobars,** lines of equal air pressure. In areas where the isobars are closely spaced, the pressure gradient is steep and strong winds can be expected. In areas of widely spaced isobars, the pressure gradient is gentle and light winds typically exist (see Figure 6.11).

A second force that influences wind speed and that always serves to partially offset the pressure gradient force is friction. Some molecular friction occurs within an air stream, but friction between the atmosphere and the Earth's surface is more significant. The amount of friction generated by the surface is a

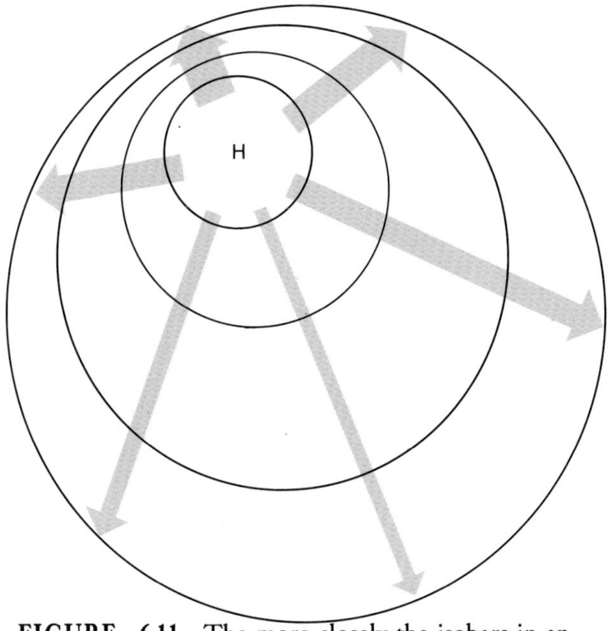

FIGURE 6.11 The more closely the isobars in an area are spaced, the steeper the pressure gradient and the greater the wind speed. In this diagram, wind speed corresponds to arrow thickness. Wind directions are shown as they would be with no Coriolis deflection.

function of both wind speed and surface roughness. With increasing wind speeds, friction between the air and the nonmoving surface increases rapidly. For this reason, a steepening of the pressure gradient will produce a less-than-proportional increase in the speed of the wind.

The frictional resistance provided by the surface is determined largely by its roughness. Land areas vary greatly in this respect. At one extreme, a smooth, ice-covered surface such as the interior of Antarctica offers very little frictional resistance to the passage of air, and winds there frequently attain very high velocities. Conversely, a forested, mountainous surface generates a high degree of friction, and wind speeds are correspondingly much reduced.

Surface friction not only reduces wind speeds, it also produces wind **turbulence.** This refers to the eddying or swirling motions generated in near-surface air currents by the uneven blocking and redirecting influences of frictional barriers such as trees, hills, and buildings. Turbulence is responsible for the gustiness of a windy day. The generally smoother surfaces of water bodies cause average wind speeds to be greater over water than over land and also reduce turbulence. This often is quite evident on the seashore or the shore of a large lake. The highest average wind speeds are encountered in the upper atmosphere, above the frictional influence of the surface. Here, friction is produced only by collisions between air molecules, and the wind normally flows in a smooth, steady stream that is almost free of turbulence. A pressure gradient sufficient to produce a gusty 18 m/sec (40 mph) wind over a land surface typically will produce a smoother wind of approximately 30 m/sec (65 mph) over water, and a steady 45 m/sec (100 mph) wind above a height of 900 meters (3,000 ft).

Wind Direction Controls

The wind direction is always stated as the direction from which the wind is coming. The direction of the wind is determined by three factors: the orientation of the pressure gradient, the Coriolis effect produced by the Earth's rotation, and the frictional resistance to the flow of the wind.

The primary factor controlling wind direction is the orientation of the pressure gradient. Air flowing from high to low pressure tends to do so along the most direct route down the steepest pressure gradient. Again, an analogy can be drawn with the flow of a stream. Just as a stream tends to flow down the steepest slope from high to low elevation, wind tends to flow down the steepest pressure gradient from high to low pressure (see Figure 6.15).

If the pressure gradient were the only factor controlling the wind direction, the wind at any given location would blow perpendicularly across the isobars, just as a stream tends to flow perpendicularly across contour lines. In the time it takes the wind to flow from a high pressure system to a low, however, both pressure systems are moving in circular paths because of the rotation of the Earth. The Earth's rotation constantly reorients the direction of the steepest pressure gradient. If the wind is to continue to flow down the steepest pressure gradient, it must gradually turn in a similar arc. Its angular momentum, though, keeps it from fully doing so. As a result, to an observer on the Earth's surface, it appears that the wind starts by blowing directly down the pressure gradient, but, with time, gradually changes its direction to a path that increasingly parallels the pressure gradient (see Figure 6.16). In reality, however, it is the underlying Earth, not the wind, that is turning.

WIND MEASUREMENT

The wind varies in both direction and speed. Therefore, two different systems of measurement and two types of meteorological instruments must be employed to fully describe this important weather element. A crucial and sometimes misunderstood aspect of wind description is that its direction is always stated as the *direction from which the wind comes.* Hence, a west wind is traveling eastward, and a southeasterly wind is blowing toward the northwest. This practice is sometimes confusing to people not familiar with wind measurement because we tend, when traveling, to give directions in terms of where we are going, rather than in terms of where we have been. A very good reason, however, exists for emphasizing the wind's direction of origin. The air's temperature and moisture are determined by its previous location, not its future destination. Thus, a north wind tends to bring colder than normal temperatures to the middle latitudes of the northern hemisphere, while a southerly breeze is typically associated with warm weather.

The Mariner's Compass is usually employed for stating wind directions in weather reports intended for the general public because of its widespread familiarity. Pilots and navigators, however, normally use the more precise azimuth system. The instrument commonly used to measure wind directions is the well-known **wind vane,** sometimes imprecisely termed a weather vane. This simple instrument consists of a flattened metal design mounted on a shaft about which it can rotate freely to point in the direction from which the wind is blowing (see Figure 6.12).

Wind speeds in the United States normally are stated in miles per hour in public broadcasts, but are measured in nautical miles per hour, or **knots,** by the National Weather Service and other official agencies. A nautical mile is the mean length of one minute of latitude, or 1,853 meters (6,080 ft). It is equal to 1.15 statute miles, so that a knot equals 1.15 statute miles per hour.

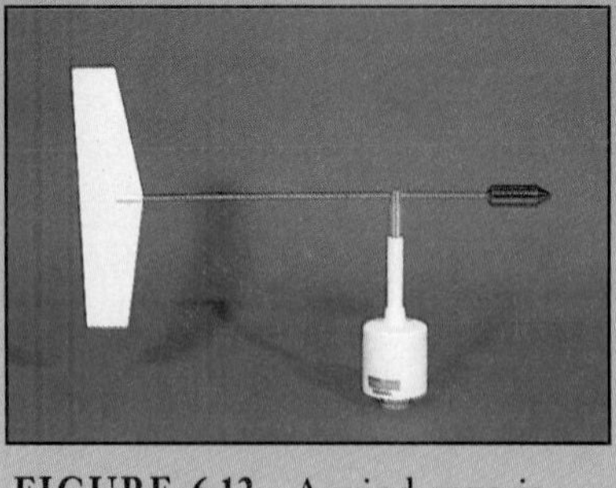

FIGURE 6.12 A wind vane is designed to point toward the direction from which the wind is blowing. (Courtesy of Qualimetrics, Inc.; Sacramento, CA)

The instrument most commonly used to measure wind speeds is the **anemometer.** It normally consists of three or four hemispherical cups mounted on a metal shaft so that they are free to rotate with the wind (see Figure 6.13). While an ob-

This apparent deflection of moving fluids on the Earth's rotating surface, which affects ocean currents as well as winds, is termed the **Coriolis effect.**

The counterclockwise rotation of the northern hemisphere, as observed from above the North Pole, deflects winds in the northern hemisphere to the right of their original paths, regardless of their directions of flow. Conversely, the clockwise rotation of the southern hemisphere deflects all southern hemisphere winds to the left. The Coriolis effect at the equator is zero, so no wind deflection exists there. The Coriolis effect increases poleward because the linear component of planetary rotation progressively decreases with increasing latitude—in other words, the rotational paths of high latitude locations form "tighter" circles. Even in the high latitudes where the Coriolis effect is strongest, though, the deflection is gradual, and local winds are not significantly affected.

The Coriolis effect tends to continue turning the wind from its current direction, regardless of how much it already has been deflected. This tendency is resisted, however, by the inability of the wind to blow up the pressure gradient from lower to higher pressure. The ultimate tendency of the Coriolis effect therefore is to deflect the wind by 90°, so that it flows parallel to the isobars (see Figure 6.17). When this server can easily determine wind direction from a wind vane, the speed of the wind cannot be readily determined merely by looking at a whirling anemometer. For this reason, the typical anemometer powers a small electrical generator, which is wired to an instrument panel inside the building on which the anemometer is mounted. The generated electricity moves a pointer on a calibrated dial to indicate the correct wind speed.

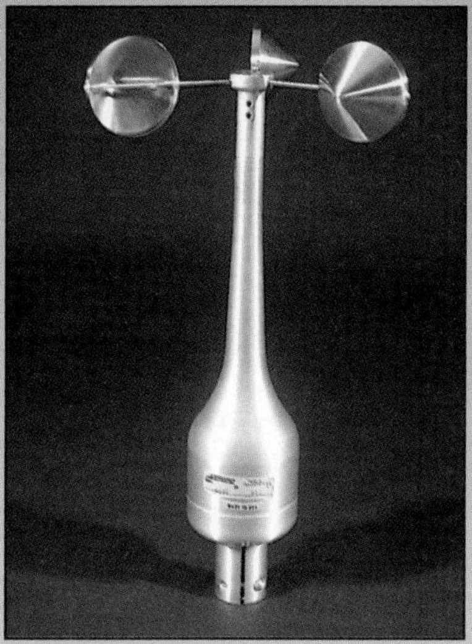

FIGURE 6.13 An anemometer. As the wind blows, air is trapped within the cups, causing them to spin at a speed proportional to the wind speed. (Courtesy of Qualimetrics, Inc.; Sacramento, CA)

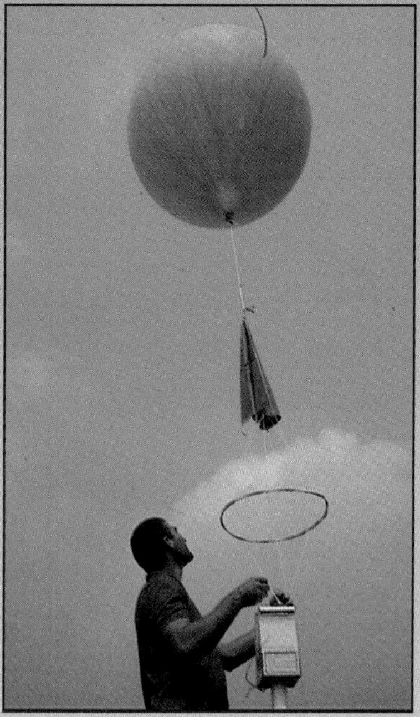

FIGURE 6.14 The launching of a weather balloon carrying a radiosonde. As it ascends, the radiosonde automatically measures and transmits back information on temperature, air pressure, and humidity. (NOAA)

Upper-level measurements also are of vital importance, not only to aircraft, but also to weather forecasters. Systematic observations are provided by the release of large hydrogen- or helium-filled balloons with a known rate of ascent. As they rise, these balloons also are carried downwind, so that the horizontal component of their motion indicates the wind directions and speeds for the various levels through which they ascend (see Figure 6.14). Instrument packages called **radiosondes** may be carried aloft by these balloons. They transmit temperature, humidity, and air pressure readings as they ascend, and their radio signals enable them to be tracked in order to determine wind direction and speed.

occurs, the pressure gradient force exactly balances the Coriolis effect, and the net force on the wind's direction becomes zero.

The third influence on wind direction is friction. Friction interferes with the Coriolis effect, causing it to deflect the wind less than a full 90°. With increasing friction, the angle of deflection is progressively reduced. The resulting wind therefore blows at some intermediate angle between that of the pressure gradient force and the angle that would be produced by a full Coriolis effect. Friction reduces the deflection of the wind because, when wind speed is reduced, its momentum in that direction also is diminished, allowing it to turn more easily toward the new direction of the pressure gradient orientation. A similar reduction in momentum, for example, makes it much easier for an automobile to navigate a sharp turn when it is traveling slowly than when it is speeding.

Because most land surfaces provide more frictional resistance to the wind than do water surfaces, the wind is generally deflected less over land than over water. Deflection angles average about 60° from the orientation of the steepest pressure gradient over land areas in general, but can be reduced to as little as 45° over extremely rough terrain. Over water, deflection angles range between 70° and 75°. In the

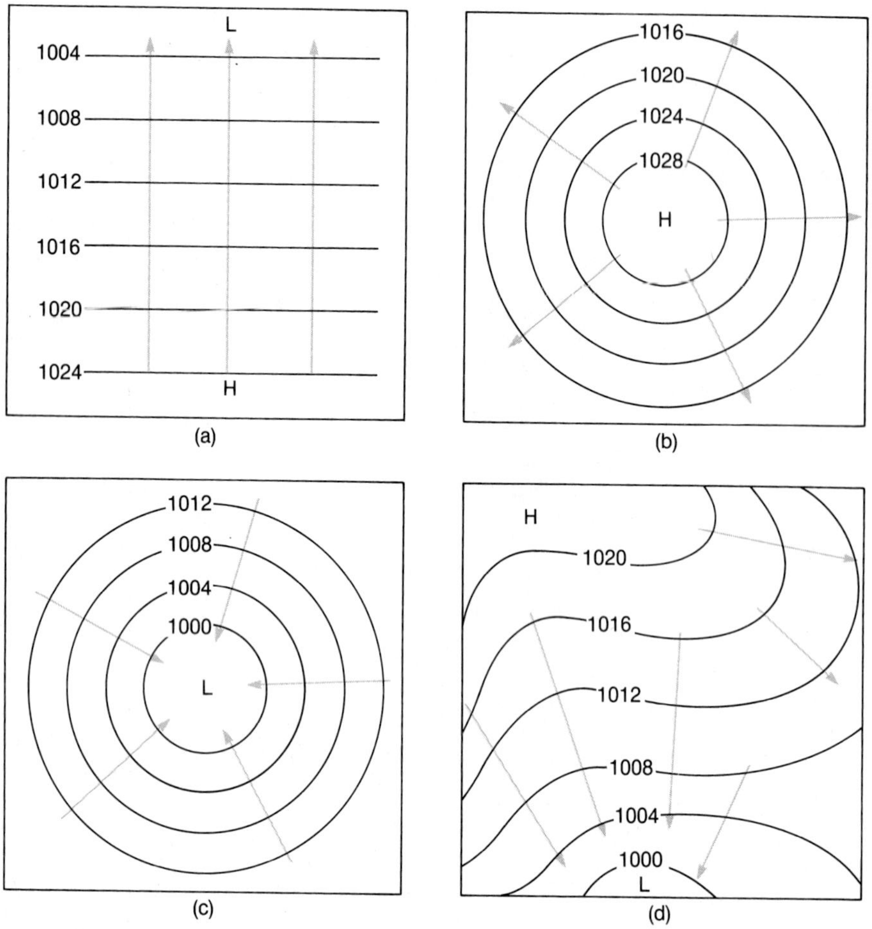

FIGURE 6.15 Wind directions as they would be if controlled solely by the orientation of the pressure gradient. In each case, the wind blows perpendicularly across the isobars from high to low pressure. The Earth's rotation prevents this from actually occurring.

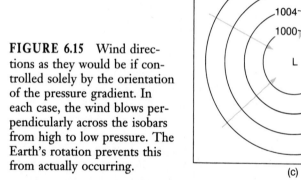

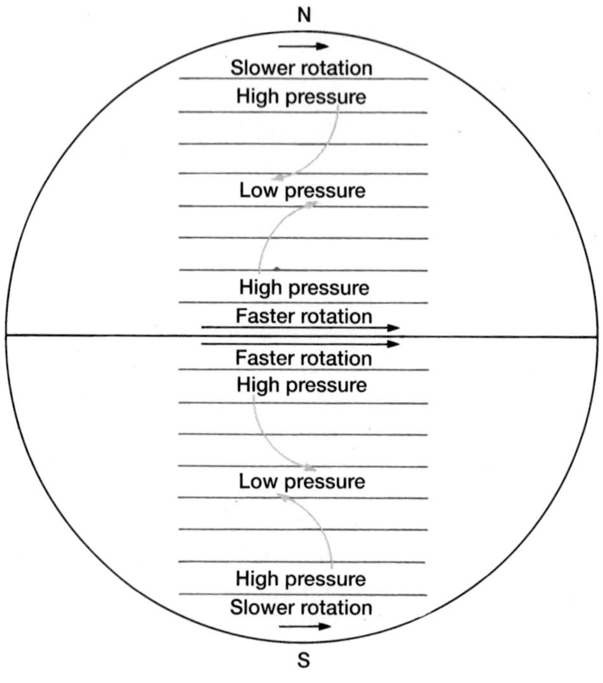

FIGURE 6.16 Coriolis effect deflection of the wind caused by the Earth's rotation. Deflection is to the right in the northern hemisphere, and to the left in the southern hemisphere. Deflection occurs largely because the Earth's rotational speed differs at different latitudes. In the diagram, the wind currents maintain some of their initial rotational momentum as they flow poleward to areas of slower rotation or equatorward to areas of faster rotation. They therefore end up behind or ahead of their original targets.

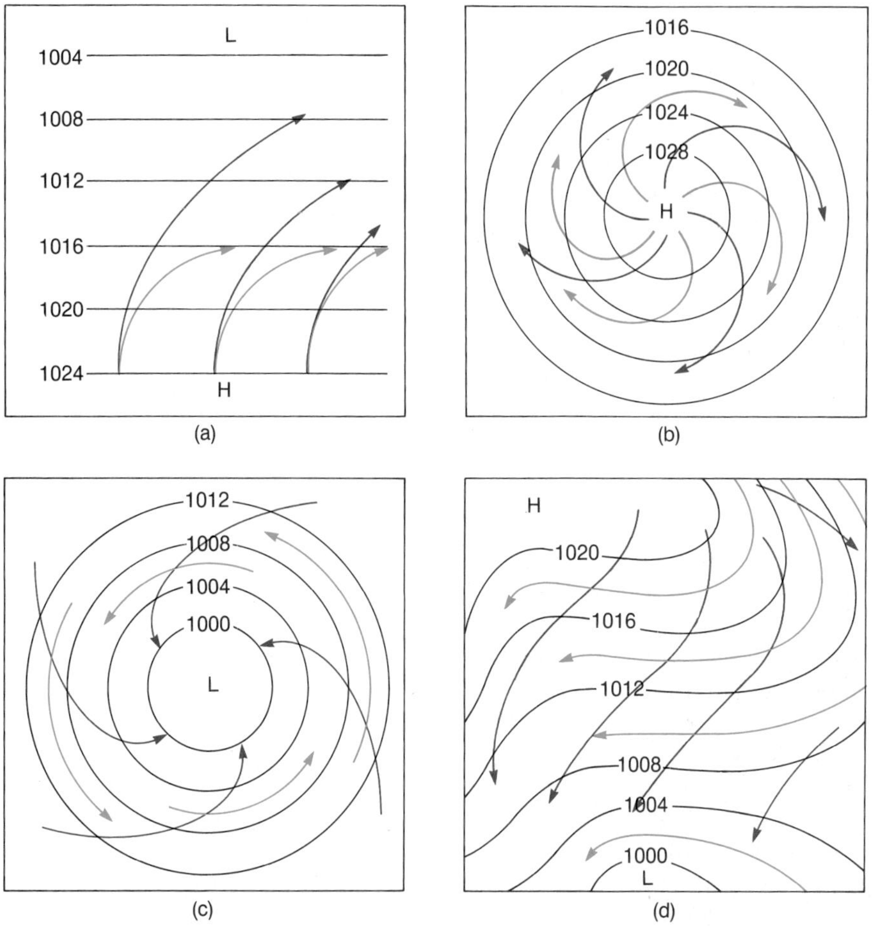

FIGURE 6.17 Northern hemisphere wind-flow patterns after a full 90° Coriolis effect deflection are shown by the blue arrows. Wind directions here parallel the isobars. Frictional resistance prevents a full 90° deflection from occurring near the Earth's surface, so the actual wind directions are shown by the red arrows. Air pressure values are in millibars.

upper atmosphere, where friction is minimal, the wind is deflected by almost 90° so that it flows nearly parallel to the isobars. The procedure for estimating wind directions for both land and water surfaces on a pressure map is illustrated in Figure 6.18.

General Circulation of the Atmosphere

Because differences in air pressure cause the wind to blow, the global pattern of wind currents is closely related to the world air pressure pattern. Therefore, keeping the air pressure pattern (Figure 6.8) in mind will greatly facilitate the understanding of the major wind currents.

Two factors are primarily responsible for maintaining the pressure imbalances that produce the global wind systems: the uneven latitudinal heating of the surface and the Coriolis effect resulting from the Earth's rotation. Before discussing the actual global wind pattern, it is desirable to examine the influence of each of these factors, which combine to produce what might be considered as a simplified theoretical global pattern of wind currents.

We might begin by imagining that the Earth is not rotating, but that instead the Sun is circling the Earth. With no rotation there would be no Coriolis effect, and winds would blow directly across the isobars from high to low pressure. In addition, only the thermally induced pressure belts would be present, and global air pressure and wind patterns would be considerably simplified. As they do now, high Sun angles would produce a zone of rising air and low surface pressures near the equator. This air, after rising, would flow poleward through the upper troposphere. Meanwhile, limited insolation in each polar region would produce a mound of dense, cold air at the surface, as occurs to some extent on the rotating Earth. This air would flow equatorward because of its weight and would continually be replaced by air descending from the upper troposphere. The linking of these two systems would produce in each hemisphere a sausage-shaped pattern of air flowing equa-

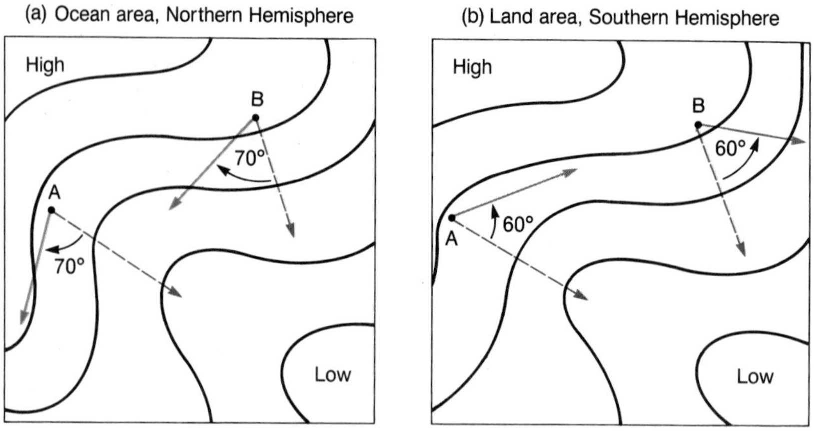

FIGURE 6.18 Illustration of the method for estimating surface wind directions from the isobar pattern on a weather map. The left map represents an ocean area in the northern hemisphere. In order to determine the approximate wind directions at Points A and B, first draw arrows (shown here as dashed red lines) perpendicularly across the isobars from high to low pressure. Then rotate the arrows 70° in a *clockwise* direction to approximate the actual wind directions. On the right-hand map, representing a land area in the southern hemisphere, draw arrows directly across the isobars as before, then rotate them 60° *counterclockwise* to approximate the wind directions.

torward at the surface, counterbalanced by a poleward flow of air in the upper troposphere.

Such a pattern does not actually form because of the deflection of wind currents toward the east and west by the Earth's rotation. As the air heated over the tropics rises and begins moving poleward in each hemisphere, it is turned gradually toward the east by the Coriolis effect, thus becoming a westerly wind. As a result, it can no longer move farther poleward to make way for new air that continues to flow from the equatorial regions. The air therefore accumulates over the subtropics, gradually subsiding to the surface to form the subtropical highs of each hemisphere. This air then flows outward at the surface, both poleward and equatorward, in the process again being considerably deflected by the Coriolis effect.

The air flowing equatorward eventually returns to the equatorial region, by which time it has once more been heated sufficiently to rise and flow poleward; thus, the cycle is perpetuated. Meanwhile, the air flowing poleward at the surface from the subtropical highs travels across the middle latitudes at an angle, because the major component of motion has been redirected to a west-to-east flow. This air eventually encounters the equatorward outflow of dense, cold air from the polar highs somewhere in the upper middle latitudes or subpolar regions. Because the air originating in the subtropical highs is much warmer, and therefore lighter, than the polar air, it is forced to rise. The boundary along which these two masses of air collide is termed the **polar front.** The lifting of the warmer air along this front helps generate the storms of the subpolar low pressure belts. This lifted air eventually flows outward in the middle troposphere in both a poleward direction, providing new air for the polar highs, and equatorward, to feed into the subtropical highs.

The net result of all this is, in effect, a triple cell pattern of winds when observed three-dimensionally (see the left side of Figure 6.19). The three circulation cells are anchored to the four world pressure belts. The two high pressure belts have an airflow pattern of upper-level convergence, followed by the descent and divergence of the air near the surface. The two low pressure belts have a reverse pattern of surface air convergence, ascent, and upper level divergence. The entire pattern, powered by the Sun and redirected by the Earth's rotation, helps to distribute the global supply of insolation so as to reduce latitudinal extremes of heating and cooling.

The foregoing discussion of the idealized pattern of global wind and pressure belts satisfactorily explains the general characteristics of the major wind systems of the lower troposphere. In detail, however, the global wind pattern is made much more complex because of the influence of additional factors such as land/water heating differences, the development and movement of weather fronts and storms, daily and seasonal variations in heating and cooling, and topographic influences. These factors, when coupled with the Sun angle and rotational influences already discussed, create a complex and ever-changing pattern of winds over the face of the Earth.

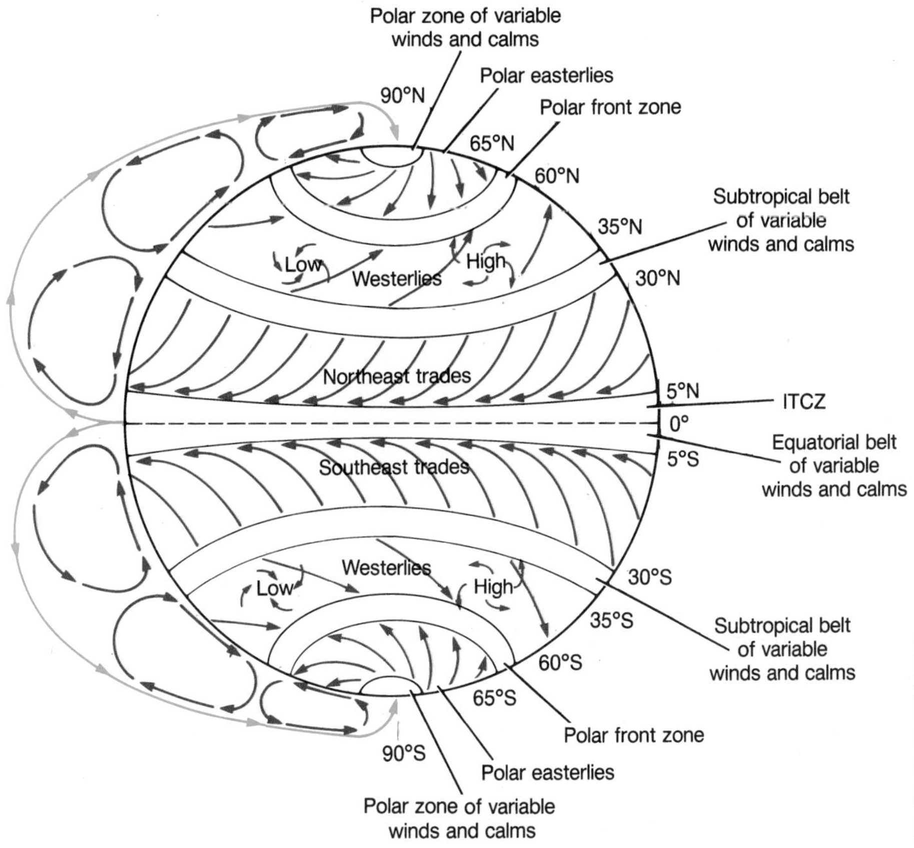

FIGURE 6.19 The generalized pattern of global wind belts, both at the surface and aloft. Air is exchanged between the surface and the upper atmosphere in the vicinity of the pressure belts.

The remaining portion of this chapter explores the basic characteristics of the wind-flow pattern of the troposphere. The discussion is centered on four subjects: the global surface wind belts, winds associated with traveling high and low pressure systems, local and regional winds, and upper-level winds.

Global Surface Wind Belts

The global surface wind belts are the largest units of the Earth's surface wind system. They are closely associated with the world pressure belts, and have many features in common with them. Altogether, seven east/west-oriented wind belts are located in each hemisphere. They are zones of prevailing wind conditions; the wind pattern on any given day may differ considerably from the norm. The "pumping mechanisms" that power the wind circulation pattern both within and between the wind belts often are considered to be the descending air currents within the subtropical highs and, to a lesser extent, the polar highs.

The two hemispheres, ideally speaking, have a mirror-image reversal of their wind belt patterns, just as they do with their pressure belt patterns. As is true of the pressure belts, the wind belts of the southern hemisphere are better defined than those of the northern hemisphere. A seasonal shift in latitude of the wind belts also occurs along with, and because of, the thermally induced seasonal shift of the pressure belts.

Our examination of the wind belts begins at the equator and progresses poleward. Figure 6.19 illustrates the positional relationships of the various belts.

Centered near the equator, and extending to about 5° N and S, is the **equatorial belt of variable winds and calms.** This is a weak wind belt associated with the low pressure center of the ITCZ. Winds here normally are light and variable in direction, with wind variations depending on small changes in air pressure patterns from day to day. The calm central portion of this wind belt is sometimes referred to as the **doldrums.** Locally strong, gusty winds, however, often are associated with rain squalls and thunderstorms in this unsettled area.

Just poleward of the equatorial belt of variable winds and calms, and surrounding it to both the north and south, are the broad expanses of the **trade winds,** which dominate the zones between 5° and 25° N and S. The trade winds received their name because they aided the European sailing ships in their trading voyages from Europe to the West Indies during the colonial period. They are generated by the pressure gradient between the subtropical highs of each hemisphere and the low pressure of the ITCZ. The trade winds are noted for both their persistence and their steadiness of direction. In the northern hemisphere, the winds blow from a northeast or east-northeast direction, producing the **northeast trades.** In the southern hemisphere, the winds blow from the southeast or east-southeast to produce the **southeast trades.**

Within the centers of the subtropical highs, the winds again become light and variable due to the absence of a significant pressure gradient. The **subtropical belts of variable winds and calms,** as they are called, are, from a meteorological standpoint, the most tranquil latitudinal zones on Earth. They are characterized by clear skies, calm or light winds, dry air, and warm temperatures. They cover at most five degrees of latitude and are usually located between approximately 30° and 35° N and S. Their seasonal shift in latitude, however, typically carries them several degrees poleward in summer and equatorward in winter. Legend has it that, on occasion, the crews of ships bound for the New World with cargoes of horses exhausted their supplies of drinking water and were forced to throw the horses overboard when the ships were becalmed for extended periods in the subtropical highs. As a result, mariners have named this wind belt the "horse latitudes."

Occupying the zone between the subtropical highs and the subpolar lows of each hemisphere are the **westerlies,** the major wind belts of the middle latitudes. They dominate the zones from 35° to 60° N and S, making them second only to the trade winds in total area. Although the westerlies are well developed wind systems, they are neither as steady nor as persistent as the trade winds because the middle latitudes are much more prone to the development of traveling high and low pressure systems than are the tropics. Each such pressure system, while being carried eastward by the flow of the westerlies, maintains its own wind circulation in much the same fashion that an eddy of water in a river maintains a circular flow pattern while being borne downstream.

Within the latitudinal zones dominated by the subpolar lows, centered at about 60° to 65° N and S, wind directions again become variable. In this region, sometimes termed the **polar front zone,** prolonged periods of light winds are uncommon. In fact, stormy conditions often prevail because the subpolar low pressure zone tends to contain well-developed traveling storm centers. The daily positions and intensities of these storms largely determine both the speed and direction of the winds. The storm centers, and their associated winds, are especially intense over ocean areas in the winter because the strongest latitudinal temperature contrasts exist at that time of year.

Proceeding farther poleward, a poorly developed wind belt termed the **polar easterlies** is encountered. Dominating the latitudinal zone between 65° and 80° N and S, it exists because of the pressure gradient between the polar highs and the subpolar lows. Winds of the polar easterlies theoretically come from a direction slightly poleward of due east, but the constantly changing positions and intensities of high and low pressure centers in these zones actually produce considerable short-term variation in wind direction and speed.

A final pair of wind systems exist within the centers of the two polar highs. These systems, which can be termed the **polar zones of variable winds and calms,** are not beltlike in shape but ideally form roughly circular caps over both polar regions in the vicinity of 80° to 90° N and S. In reality, they are poorly developed because the position and strength of the polar highs vary and also because the ice-covered, low-friction polar surfaces promote strong winds.

Traveling High and Low Pressure Systems

The wind-flow patterns of the middle and high latitudes of both hemispheres are modified by the development and general eastward movement of traveling high and low pressure systems that exist independent of the global pressure belts. Low pressure systems also are termed **cyclones** because they are characterized by a cyclonic pattern of converging and rising air (see Figure 6.3). High pressure systems, conversely, are termed **anticyclones** because of their reverse pattern of diverging and descending air

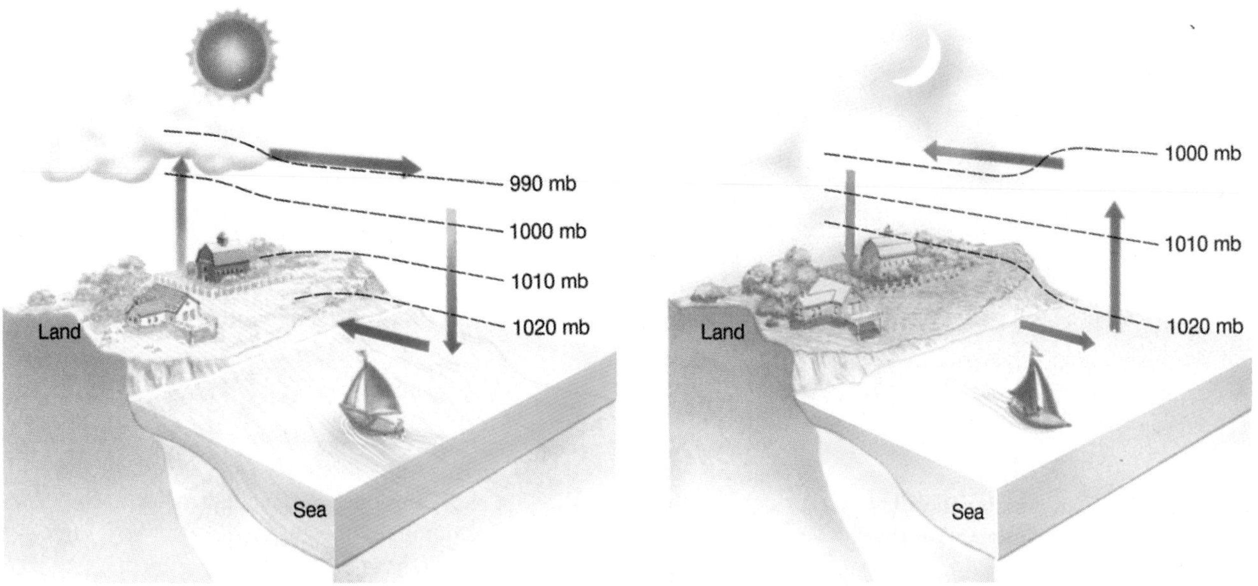

FIGURE 6.20 Land and sea breezes are produced by air pressure differences resulting from land/water temperature variations. During the day, air heated over the land expands and rises, producing an onshore sea breeze. At night, the cooler and denser air from the land spreads over the adjacent water, producing a land breeze.

(see Figure 6.2). The Coriolis effect deflects the wind fields associated with these pressure systems, giving clockwise or counterclockwise twists to their circulations. The development and characteristics of these weather systems, which play a vital role in our day-to-day weather, are discussed further in Chapter 8.

Local and Regional Winds

Further contributing to the overall complexity of the Earth's total wind pattern are various local and regional winds. Most of these winds are produced directly or indirectly by air pressure differences resulting from the differential heating or cooling of the Earth's surface. Surface temperature differences, in turn, usually result from variations in surface materials, elevation, or slope orientation. In this section we will briefly examine some of the most important of these winds.

Land and Sea Breezes

Land and sea breezes are familiar to residents of coastal areas in many parts of the world. They are produced by daily differences in heating and cooling over adjacent land and water areas (see Figure 6.20). During the day, the land warms rapidly, heating the overlying air and causing it to expand and rise. The resulting reduction in surface air pressure causes somewhat cooler and denser air, called a **sea breeze,** to flow in from the adjacent water. (The same process also occurs along the shores of large lakes such as the Great Lakes, where the resulting breeze is known as a **lake breeze.**) During a sunny afternoon, a sea breeze may penetrate inland as far as 15 to 25 kilometers (10 to 15 miles) before losing its identity. At night, a reverse process occurs, with the more quickly cooling land fostering the formation of an overlying mass of relatively cool, dense air. By the late night hours, a **land breeze** may develop as the cooler and denser air from the land spreads over the coastal waters and displaces the lighter, milder oceanic air. Land and sea breezes develop best when skies are clear, midday Sun angles are high, and the regional pressure gradient is weak. Consequently, they occur most frequently in summer, especially in the lower latitudes, where they are a daily phenomenon in many coastal areas.

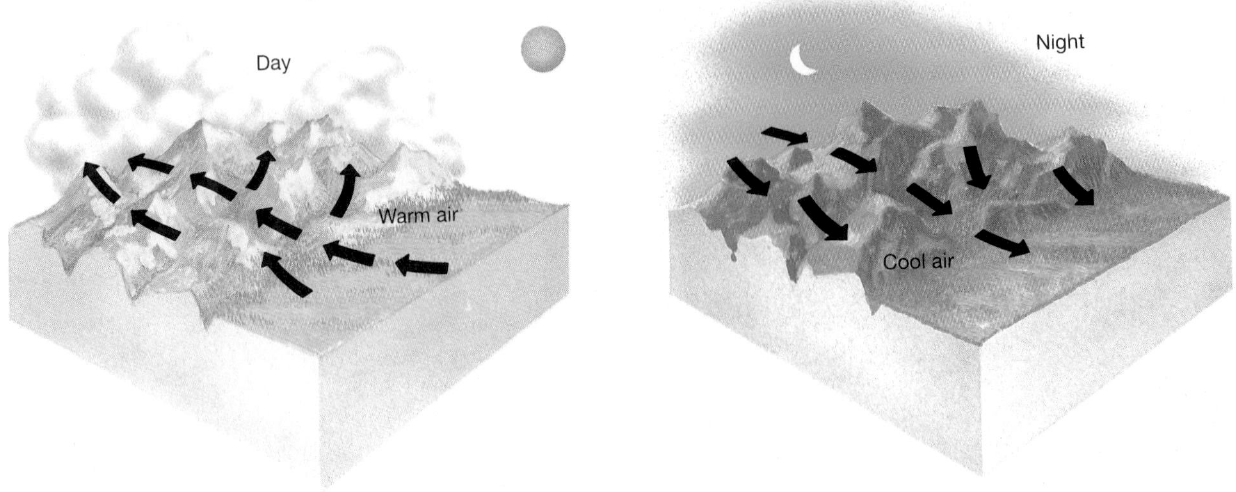

FIGURE 6.21 Mountain and valley breezes. During the day, heated air slides upslope, producing a valley breeze. At night, air chilled by contact with the surface drifts downslope, producing a mountain breeze.

Mountain and Valley Breezes

Mountain and valley breezes occur in areas of rugged terrain under conditions of fair weather and weak pressure gradients. Like land and sea breezes, they are thermally generated local winds that undergo a daily reversal in direction. In this instance, the temperature contrasts producing them occur between air near the surface and air farther aloft (see Figure 6.21). During the day, the slopes, especially those inclined toward the Sun's rays, are heated strongly. This warms the overlying air, which becomes buoyant and begins to drift upslope, producing a **valley breeze.** At night the slopes cool quickly, chilling the adjacent overlying air. Late at night and into the early morning this air becomes sufficiently cool and dense to flow downslope under the influence of gravity, producing a **mountain breeze.**

Chinook and Foehn Winds

Another group of mountain-related winds is known by such names as **chinook** in the Rocky Mountains and **foehn** in the Alps. These winds are not produced in the mountain areas they influence; rather, they result from the regional pressure gradient. However, they are modified by their passage over the mountains and subsequent descent into the adjacent lowlands. Typically, mild, moist air on the windward side of the mountains is carried over the mountain crests, where uplift causes precipitation and warming by the released heat of condensation (see Chapter 7). Upon descending the leeward flanks of the mountains, the air is further heated and dried by compression so that it enters the lowlands as a relatively warm, dry wind. The **Santa Ana** of southern California is an example of a related wind type that also has its characteristics altered by descent. It is discussed in the Focus Box on pages 108-109.

Monsoons

The most important of the world regional winds not associated with the semipermanent global wind belts are the monsoons. A **monsoon** is a seasonally reversing wind system produced by the differing thermal characteristics of continents and oceans. In terms of formation, it is the large-scale equivalent of a land and sea breeze system, where summer and winter are substituted for day and night (see Figure 6.22). During the summer, land areas that experience a monsoon are strongly heated by the Sun. This causes the overlying air to expand, producing a thermal low pressure area. The existence of higher pressure over the adjacent ocean induces an onshore flow of cooler, denser air to produce the **summer monsoon.** In winter, the reverse process occurs as the land and the air overlying it become cooler than the adjacent ocean. This results in the formation of a thermal

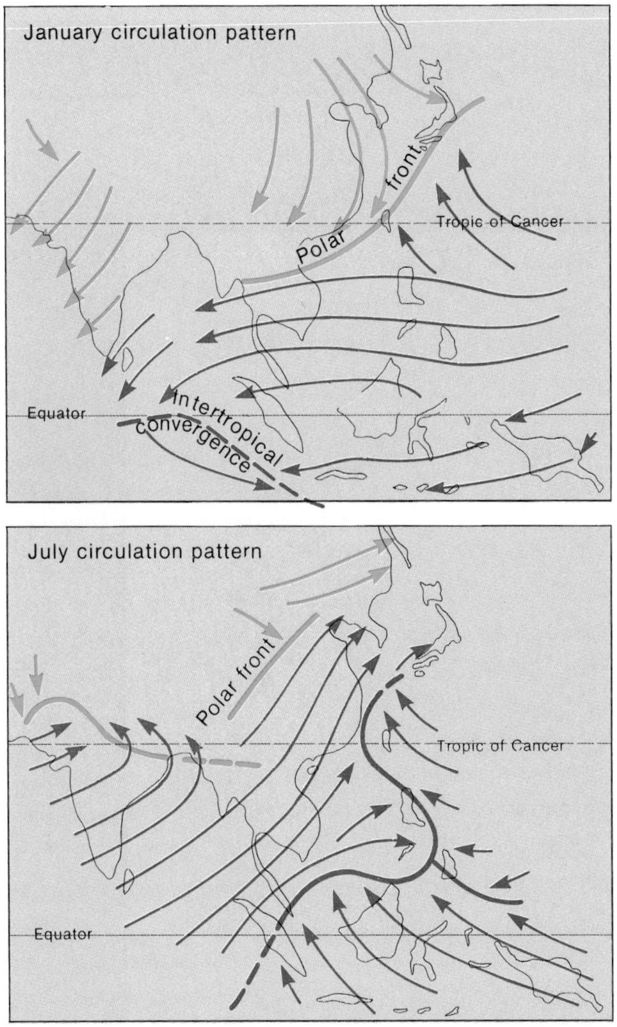

FIGURE 6.22 The summer and winter wind-flow pattern of Southeast Asia. The blue arrows represent continental air from the interior of Asia, and the pink arrows indicate maritime air from the Indian and Pacific Oceans.

high over the land, producing the seaward flow of dry air that characterizes the **winter monsoon.**

The monsoon circulation tends to be most strongly developed over large land masses in the lower latitudes, which seasonally heat and cool to greater extremes than do small land masses. In addition, large land masses develop larger and stronger thermal lows and highs as a result of their seasonal heating and cooling. It is the size and strength of these thermal pressure centers that produce the pressure gradients that set the monsoonal winds in motion. Asia, by far the largest continent, has the best developed monsoon system. Monsoonal circulations also are well developed in northern Australia and in sub-Saharan North Africa. Most other large land masses, including North America, have weak monsoonal tendencies that alter somewhat their predominant wind directions between summer and winter.

The importance of the monsoon results primarily from its influence on the moisture content of the air. In areas where it is well developed, the summer monsoon causes the inland transport of vast quantities of moisture-laden oceanic air. The lifting of this air due to heating over land and the possible influence of mountain barriers generates clouds and often copious precipitation. (The resulting association of the summer monsoon with the rainy season in Southeast Asia and other areas has led many people to the erroneous belief that the monsoon is a rain system rather than a wind system.) In winter, the reversal of the monsoonal flow produces conditions opposite to those that prevail in summer. Skies tend to be clear, humidities low, and daily temperature ranges large. This generally produces the dry season for areas dominated by the monsoon. The Case Study at the end of this chapter further explores the characteristics and human impact of the Asiatic monsoon.

Upper-Level Winds

Before the development of high-altitude aircraft during the World War II period, relatively little was known about the circulation patterns of the upper troposphere. In the succeeding decades, though, we have gained a great deal of knowledge about the upper-level winds, and it has become apparent that they exert a controlling influence on the development and movement of surface weather systems.

General Circulation

The air pressure and wind patterns of the upper troposphere are considerably simpler than those of the lower troposphere because of the absence of an adjacent surface. Because no land and water heating differences are present, and because local variations in wind direction and speed caused by frictional and blocking effects are absent, the winds in the upper troposphere normally flow as smooth, steady currents. In the absence of significant friction, moreover, the Coriolis effect is able to exert its full influ-

CALIFORNIA'S SANTA ANA WINDS

The Santa Ana is a desert-derived wind that periodically visits the lowlands of coastal southern California, where it causes considerable discomfort and sometimes major destruction. The Santa Ana typically develops in the winter half of the year when a strong high pressure system forms over the Great Basin region northeast of California (see Figure 6.23). The clockwise flow of air around this high produces a flow of mild, dry air from the east that can attain great velocities as it is funneled through passes in the San Gabriel and San Bernardino Mountains east of Los Angeles. The air is heated and further dried by compression during its descent of 1,000 meters (3,000 ft) or more to the Los Angeles Basin, pro-

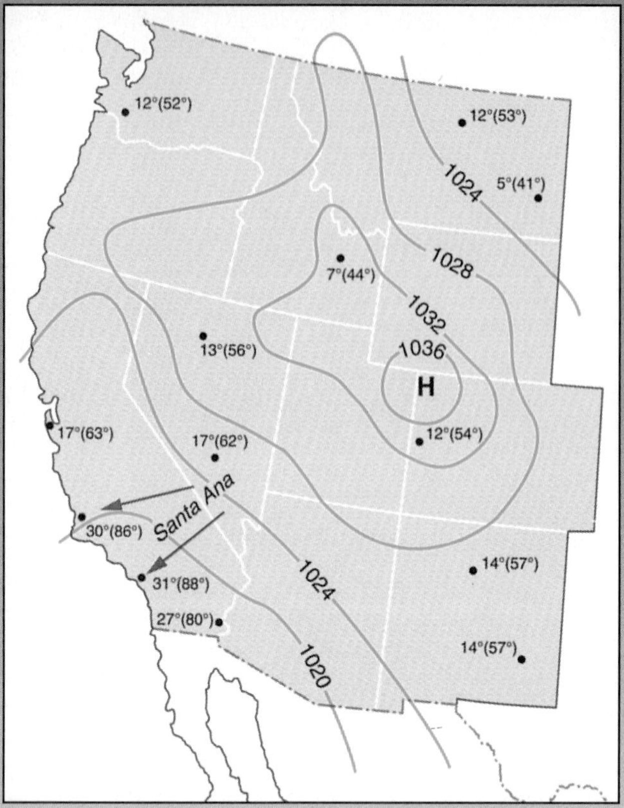

FIGURE 6.23 The development of a high pressure system over the Great Basin produces a pressure gradient favoring the development of Santa Ana winds.

ence in deflecting the winds so that they blow at right angles to the pressure gradient. This causes the upper-level winds to be much more zonal than are the near-surface winds. It should be stressed, however, that the upper-level winds are closely associated with the low-level wind systems and that they also are powered largely by the absorption of surface insolation. Air is exchanged constantly between the upper and lower levels by the rising air columns within low pressure systems and the descending air columns within highs.

Two wind systems dominate the upper tropospheric flow patterns of each hemisphere. The **tropical easterlies** form a broad belt of relatively light easterly winds that occupy the zone from about 15°N to 15°S. They actually are the upper-level continuation of the surface trade winds. The latitudinal extent of the tropical easterlies, however, is not nearly as great as that of the trade winds, which extend to approximately 25°N and S.

Poleward of approximately 20° latitude in each hemisphere, and extending all the way to the polar regions, is a vast belt of westerly winds termed the **upper-level westerlies.** This wind belt not only covers the majority of each hemisphere, it also contains by far the strongest winds in the upper tropo-

ducing a hot, dust-laden wind that can blow for days. Wind gusts sometimes reach speeds of 130 to 140 kilometers per hour (80 to 90 mph), causing property damage and desiccating crops. The unsettling impact of this wind on the populace of southern California is suggested by police statistics that typically show a marked increase in crime rates and suicides during Santa Ana events. Satellite photos taken during a Santa Ana sometimes show plumes of smoke and dust extending well out over the adjacent Pacific Ocean (see Figure 6.24).

Perhaps the most dangerous aspect of a major Santa Ana event is the accompanying high fire danger resulting from the combination of hot temperatures, strong winds, and extremely low relative humidities. The danger is particularly great when the Santa Ana develops during the fall, before the onset of the annual winter rains. One of the worst such fires was the infamous Bel Air fire of November 1961. It burned for three days, destroying 484 homes and causing more than $25 million in damage. In 1977, another fire associated with a Santa Ana burned more than 25, 000 acres in the Santa Monica Mountains, destroying 91 homes. Hillsides denuded of their protective vegetation by such fires are particularly vulnerable to serious erosion and potentially deadly mudflows during ensuing rainstorms. Thus, although the Santa Ana is a short-term phenomenon, its effects on the landscape may remain for years.

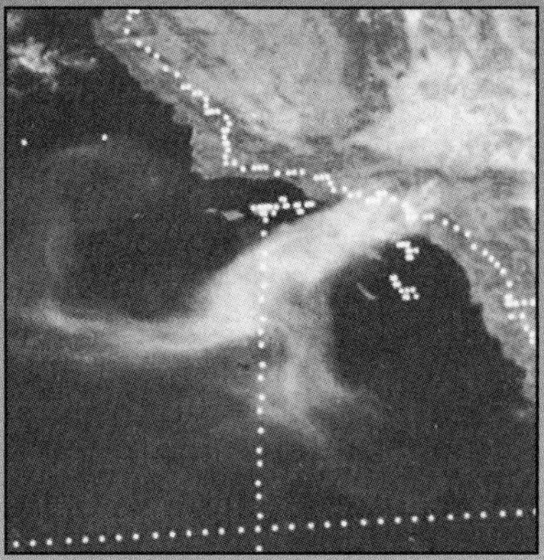

FIGURE 6.24 Satellite photo showing the smoke from Santa Ana-induced forest fires in the hills east of Los Angeles being blown far out over the Pacific. (NOAA)

sphere. Although the upper-level westerlies have a mean direction of flow that is almost due west to east, they typically display a wavy pattern because they tend to develop a series of eastward migrating ridges and troughs. These irregularities frequently cause winds to blow from the northwest or from the southwest. In the vicinity of the North and South Poles, but commonly offset from them by several hundred kilometers, are giant low pressure centers.

The Jet Streams

The fastest-flowing upper level winds, and those with the greatest impact on the surface weather, are the **jet streams.** They consist of relatively narrow bands of very strong winds centered in the upper troposphere at altitudes of approximately 9 to 13 kilometers (6 to 9 mi). When viewed three-dimensionally, a jet stream is tubular in shape, with wind speeds increasing toward the center from all directions (see Figure 6.25). Maximum wind speeds vary considerably along the path of a jet stream but usually range between 80 and 240 kilometers per hour (50 to 150 mph). The winds of the jet streams generally flow from west to east in both hemispheres but, like the upper-level westerlies as a whole, tend to develop large-scale north/south-oriented waves termed **Rossby waves** (see Figure 6.26). The entire undulating pattern of a jet stream

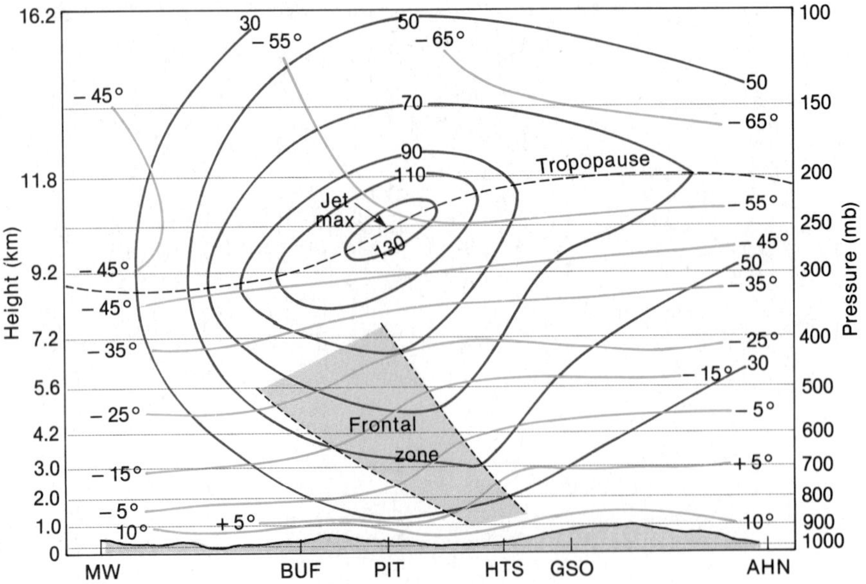

FIGURE 6.25 Vertical cross-section through the atmosphere along a transect from Ottawa, Ontario (MW) to Athens, Georgia (AHN) at 7 a.m. on October 16, 1973. The red lines are lines of equal wind speed, and the blue lines are lines of equal temperature (isotherms). Wind speeds are given in knots, and temperatures in degrees Celsius. The jet stream is centered over Pittsburgh at a height of about 10.5 kilometers.

typically shifts slowly eastward at about 25 to 40 kilometers per hour (15 to 25 mph). As this shift occurs, existing waves change their amplitude, new waves may form, and old ones disappear. A jet stream thus exhibits a complex, dynamic flow pattern.

At most times there are two separate jet streams in each hemisphere. These are termed the subpolar jet and the subtropical jet. The **subpolar jet** is produced by the temperature contrasts between the high and low latitudes, and its position tends to coincide closely with the surface location of the polar front. It is strongest during the winter because the temperature contrast between high and low latitude air masses is greatest then. The jet stream also shifts equatorward during the winter because the frontal boundary between warm and cold air masses is located at a lower latitude. In North America, for example, the subpolar jet in midwinter typically extends over the southern United States; by midsummer it usually is positioned over southern or central Canada.

The **subtropical jets** are typically centered above the subtropical highs at a somewhat higher altitude than the subpolar jets. Their origin is less certain than that of the subpolar jets. Although they may be thermally generated to some extent, they are also apparently produced in part by the Earth's rotation.

The practical significance of the jet streams is associated primarily with their influence on weather conditions at the Earth's surface. They act as upper-level steering currents or "highways," largely controlling both the speed and the direction of movement of the migratory highs, lows, and fronts that influence the surface weather. These surface weather systems extend high enough into the atmosphere to be carried along by the jet streams. As a result, weather systems located beneath a jet stream tend to travel along its path, and their forward speed is determined by the jet stream's strength. The speed of movement of surface weather systems is normally between one-half and one-third the maximum speed of the jet stream because of the restraining influence of surface friction.

The subpolar jet is particularly important to the weather of the middle latitudes because it is located above the surface boundary between warm and cold air masses. An equatorward loop in this jet, for example, will draw a mass of air from the higher latitudes unusually far equatorward, producing an outbreak of cold weather (see Figure 6.26). A poleward loop, conversely, will result in the poleward flow of a mass of warm air. In addition, the loops of the subpolar jet influence the locations of storm development. Because the subpolar jet is critically important to surface weather conditions, information on the jet streams is regularly gathered and mapped, and a great deal of research has been conducted to better understand the jet streams and predict their behavior.

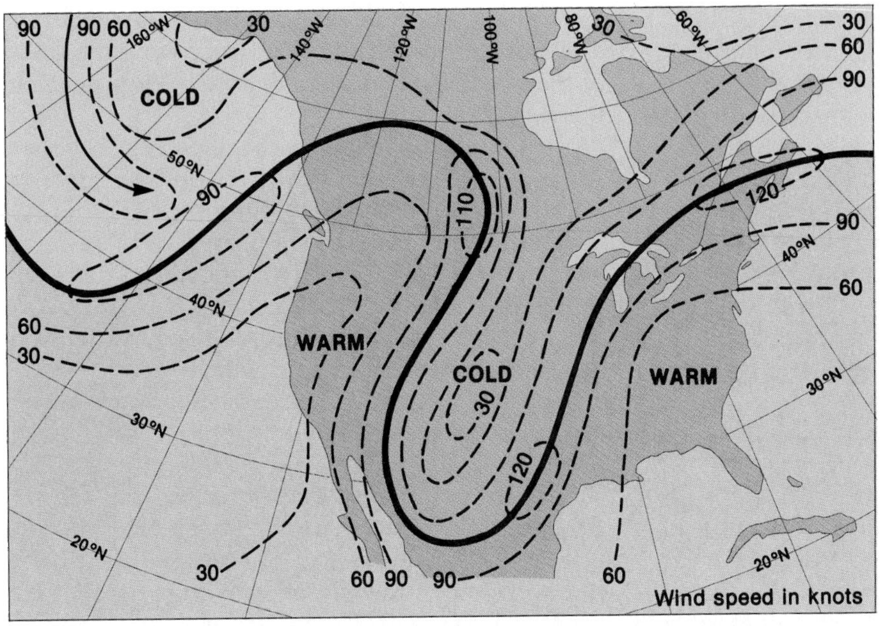

FIGURE 6.26 A deep trough in the subpolar jet stream associated with a cold air outbreak over the central United States appears on this map of February 22, 1975, showing windflow patterns at the 300 millibar pressure level. Dashed lines indicate wind speeds in knots. The jet stream maximum (thick line) separates warm surface areas (shaded pink) from cold surface areas (shaded blue).

SUMMARY

This chapter has examined the causes and global distributions of two weather elements—air pressure and wind—that are perhaps most important because of their influence on other aspects of the weather. Air pressure is the only weather element to undergo changes that can be neither seen nor felt, yet is of great significance because of its effect on wind direction and speed. The wind, in turn, is important both because of its direct effects on the surface and because it transports energy and moisture. Daily changes in temperature and the potential for precipitation are strongly influenced by short-term changes in wind patterns. On a global scale, regions of high and low precipitation are associated, respectively, with zones of rising air within lows and descending air within highs.

For a more complete summary, we return to the Focus Questions appearing at the beginning of the chapter:

1. *What causes variations in air pressure, and how do these variations affect the weather?*

Air pressure varies largely because of global heating differences and the Earth's rotation. Heating causes air to expand and the air pressure to fall, while cooling causes air to contract and the pressure to rise. The Earth's rotation produces the Coriolis effect. This causes the air to be deflected into some latitudinal zones, producing high pressure, and to be deflected away from other latitudinal zones, producing low pressure.

2. *What is the basic pattern of global air pressure belts?*

In general terms, a symmetrical pattern of global pressure belts exists. The Intertropical Convergence Zone (ITCZ), a zone of low pressure that is produced primarily by high temperatures, straddles the equator. Surface air pressures rise poleward until the centers of the subtropical highs are reached at about 30° to 35°N and S. Mean air pressure values then decline poleward to about 60° to 65°N and S, where the centers of the subpolar lows are found. Finally, polar highs are located in the vicinities of the North and South Poles. This beltlike pattern is somewhat idealized and is generally best developed in the southern hemisphere, where land/water temperature differences are less pronounced than in the northern hemisphere.

3. *What factors control wind speed and direction?*

The wind is produced by differences in air pressure. Air tends to flow from high to low pressure at speeds proportional to the steepness of the pressure gradient existing between the two pressure systems. The wind speed is greatly reduced, however, by the frictional resistance it encounters at the Earth's surface.

Wind direction is controlled by three factors: the orientation of the pressure gradient, the rotation of the Earth, and surface friction. The net result of these factors is that surface winds in the northern hemisphere are deflected clockwise as they flow from high to low pressure. The angles of deflection vary from about 45° to 80° from the orientation of the steepest pressure gradient. In the southern hemisphere, a similar counterclockwise deflection occurs.

4. *What is the basic pattern of global wind belts?*

The pattern of global wind belts closely resembles that of the global pressure belts. Each pressure belt is associated with a central zone of variable winds. This wind variability results from a changeable, and often weak, pressure gradient. The broad zones of relatively steep pressure gradients between the pressure belts are occupied by belts of stronger and steadier winds. The trade winds occupy the latitudinal zone between the ITCZ and subtropical highs; the westerlies occupy the zone between the subtropical highs and the subpolar lows; and the polar easterlies occupy the zone between the subpolar lows and the polar highs.

5. *What are the jet streams, and how do they influence surface weather conditions?*

The jet streams are narrow, meandering belts of very strong westerly winds centered in the upper troposphere. They result largely from the surface temperature contrasts between the low and high latitudes. The jet streams act as steering currents that influence the direction and speed of movement of surface weather systems.

Review Questions

1. Explain why air pressure exists. How and why do temperature changes affect air pressure?

2. What characteristic weather conditions are associated with high and low pressure systems, respectively?

3. Describe the locations of the global pressure belts. In which hemisphere is the beltlike pattern best developed, and why?

4. Explain why a seasonal shift occurs in the positions of the global pressure belts. How large a shift takes place, and in what directions?

5. What causes the wind to blow? What two factors must be considered when forecasting the speed of the wind? Explain the influence of each on wind speed.

6. What factors influence the direction of the wind? What is the Coriolis effect, and how does its influence on wind direction differ between the northern and southern hemispheres? Why is the wind deflected more greatly in the upper atmosphere than near the Earth's surface?

7. Draw a profile diagram of the basic pattern of wind-flow cells connecting the upper and lower troposphere along a 90° arc extending from the equator to the North Pole.

8. Draw a diagram of the Earth showing the idealized pattern of global surface air pressure belts. Label the wind belts that exist within each pressure belt and between each pair of pressure belts. Indicate with arrows the wind-flow pattern for those wind belts that have a prevailing direction.

9. Explain the cause, seasonal flow pattern, and attendant climatic characteristics of the monsoon winds. In what parts of the Earth are they best developed, and why?

10. Why do coastal land and sea breezes form, and in what geographical areas are they best developed? How are they similar in origin to the monsoon winds?

11. What are the jet streams? Where and why do they form? Explain why a knowledge of the pattern of the jet streams is essential for accurate surface weather forecasting.

Problems

1. If you were aboard a ship in the North Atlantic, with high pressure centered due north of you and low pressure centered due south, from what direction would you expect the wind to come? (Give your answer in both the Mariner's Compass and the azimuth system.)

2. If you were in a New Zealand forest, with high pressure centered due west of you and low pressure centered due east, what would be your most probable wind direction?

3. If the wind is coming from the south, in what compass direction from your location is a high pressure center most likely to be found?

Key Terms

air pressure
pressure gradient
Intertropical Convergence Zone (ITCZ)
subtropical highs
subpolar lows
polar highs
wind
Coriolis effect
trade winds
westerlies
polar easterlies
land and sea breezes
mountain and valley breezes
summer and winter monsoons
jet streams

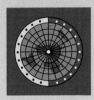

CASE STUDY

THE IMPACT OF THE ASIATIC MONSOON

The Asiatic monsoon is a thermally induced wind system. Climatically, however, it is most important as the mechanism controlling the annual precipitation pattern of southern and eastern Asia. The "monsoon lands" of Asia, where the effects of the monsoon are especially dominant, occupy a broad arc from Pakistan to extreme eastern Siberia. Approximately half the world's population lives within this area.

The Asiatic monsoon is caused by the seasonal temperature differences that develop between the air over the continent of Asia and that over the adjacent Pacific and Indian Oceans. During the late fall and early winter, the Asian land mass cools rapidly, causing surface air pressures to increase. The coldest temperatures occur in eastern Siberia, producing the giant Siberian High. The clockwise outflow of air

FIGURE 6.27 Southeast Asian rice paddies are terraced both to control soil erosion on sloping land and to make the most of the summer monsoon rains. (A. Tovy/H. Armstrong Roberts)

from this pressure center produces cool, dry, north-to-northeasterly winds over most of monsoon Asia. The relative dryness of this continental air is increased by its descent from the land toward the sea, so that in many areas no rain may fall for several months.

As the Sun climbs higher in the sky and the length of daylight increases during the late winter and early spring, a rapid warming occurs. This causes a reduction in air pressures over eastern Asia and weakens the northeast monsoon winds. By late spring, temperatures have become hot over southern Asia, and inland air pressures have begun to fall to values lower than those over the cooler adjacent oceans. This pressure gradient initiates an onshore flow of tropical maritime air over extreme southern and eastern Asia, which gradually strengthens and expands northward and westward as the summer progresses (see Figure 6.22).

In India, rains begin on the Bay of Bengal coast in early June and spread northwestward to the Pakistani border by mid-July. The onset of the rains associated with the summer monsoon is often very sudden and is sometimes referred to as the "burst" of the monsoon. In Bombay, for example, mean monthly rainfall totals increase from 18 millimeters (0.7 in) in May to 485 millimeters (19.1 in) in June (see Table 6.1). It is estimated that 85 percent of India's annual rainfall occurs during the three-month period from mid-June until mid-September. Rainfall in most areas is not steady, but instead takes the form of frequent, often torrential, showers and thunderstorms. By September the rains begin to retreat southeastward, ending where they began on the Bengali coast in November as high pressure again becomes dominant over Asia.

Several factors account for the intensity of the summer monsoon rains of Southeast Asia. One is the large water vapor content of the tropical oceanic air masses that are drawn into the continent by the summer monsoon airflow. A second factor is that the ITCZ, with its associated wind convergence and uplift, is drawn northward deep into Asia during the summer. Third, and probably most important, is the forced lifting of the air as it passes from the ocean at sea level to the Asian land mass, much of which is at a high elevation. Rainfalls can be extremely heavy where mountain ranges lie directly across the path of the monsoon winds. For example, the town of Cherrapunji, located on the southern slopes of the Khasi Hills in eastern India, has a mean annual rainfall of 1,087 centimeters (428 in). Cherrapunji holds the world's record for the most rainfall ever recorded in a single twelve-month span. From August 1860 through July 1861, it received an amazing 2,645 centimeters (1,042 in) of rain! This is about thirty times the world average annual total.

The monsoon precipitation cycle is especially critical to the inhabitants of Southeast Asia because of the dominance of subsistence agriculture in that part of the world. In India, for example, two-thirds of the population is engaged in agriculture; in Pakistan, three-fourths of the population is agricultural. Land holdings are small, and high population growth rates continue to increase the pressure on the land to provide ever-larger crop yields.

The agricultural calendar of events is dictated by the rainy season. Little can be done in the hot, dry spring months, when the land lies parched, cracked, and bare. The onset of the monsoon rains brings about a frenzy of planting and plowing, as the crops must have maximum exposure to the available moisture. Most harvesting takes place in the fall at the beginning of the winter monsoon dry season. As winter arrives, the land again lies fallow until the spring and the annual return of the rains.

TABLE 6.1 Monthly Mean Precipitation Totals for Bombay, India

	JAN.	FEB.	MAR.	APR.	MAY	JUNE	JULY	AUG.	SEPT.	OCT.	NOV.	DEC.	YEAR
Inches	0.1	0.1	0.1	0.0	0.7	19.1	24.3	13.4	10.4	2.5	0.5	0.1	71
Centimeters	0.3	0.3	0.3	0.0	1.8	48.5	61.7	34.0	26.4	6.3	1.3	0.3	180

CHAPTER 7

ATMOSPHERIC MOISTURE

OUTLINE

FOCUS QUESTIONS

1. Under what conditions do water phase changes occur, and what role do they play in the global distribution of energy?
2. What is the hydrologic cycle?
3. How do clouds and precipitation form?
4. What is the geographic pattern of precipitation, and why does it exist?

Water covers most of the Earth's surface and also is a vital component of the atmosphere. Atmospheric moisture is, however, highly variable in amount, comprising less than 0.1 percent by volume of very cold, dry air and as much as 4 percent of very warm, moist air. This variability exists because of the unique ability of water to enter or leave the atmosphere in large quantities over comparatively short time spans. The continuous cycle of water between the Earth's surface and atmosphere not only supplies lifegiving moisture to the land, but also plays a crucial role in the processes of energy transfer throughout the Earth system.

PHASE CHANGES OF WATER

The phase changes that water readily undergoes are the key to its ability to enter and leave the atmosphere. Matter can exist in three physical states, or phases: solid, liquid, or gaseous (vapor). Although any substance, with sufficient gain or loss of energy, can theoretically change from any one of these states to any other, water is the only atmospheric component that undergoes phase changes within the existing range of atmospheric temperatures and pressures.

Energy is absorbed or released when water changes temperature or phase. This energy is commonly measured in terms of calories. A **calorie** is the amount of heat needed to raise the temperature of one gram of liquid water by one Celsius degree (1.8 F°). Heat is absorbed by water when it is warmed and released when it cools. Large quantities of energy are involved in the six phase changes in which water can participate. Energy is absorbed as latent heat by the water when it changes from a lower to a higher phase, and is released as sensible heat to the environment when it changes from a higher to a lower phase. Figure 7.1 illustrates these changes. They consist of three pairs of opposing processes, which we shall now examine.

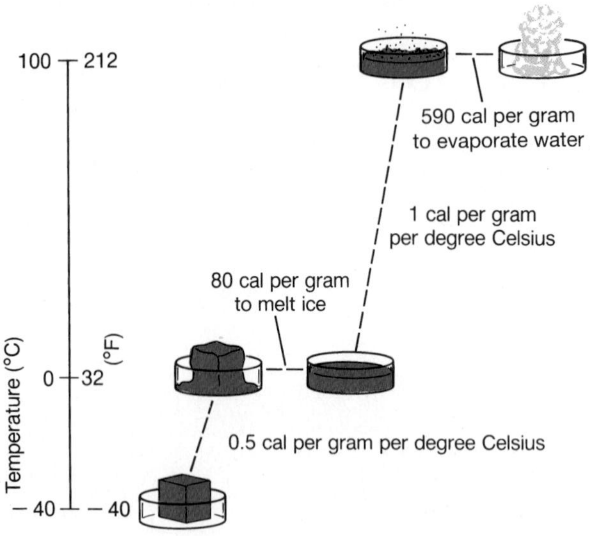

FIGURE 7.1 Temperature and phase change energy requirements for water.

■ *Evaporation and condensation.* **Evaporation,** the change of liquid water to water vapor, occurs as individual molecules in a water body acquire sufficient kinetic energy through random molecular collisions to escape into the atmosphere. For every gram of water that evaporates at normal Earth temperatures, approximately 590 calories of energy are transferred from the liquid water to the atmosphere.[1] Because most evaporating water is located on the Earth's surface, this process produces a major transfer of energy from the surface to the atmosphere. The addition of this water vapor does not directly heat the atmosphere, however, because the absorbed energy is used to change the physical state of the water molecules rather than to raise their temperature.

Condensation, the change of water vapor back to liquid water, is the exact opposite of evaporation. When condensation occurs, the energy absorbed by the water molecules during evaporation is released as the *heat of condensation.* Most condensation takes place well above the Earth's surface during the process of cloud formation; it therefore is a major heating process for the troposphere.

■ *Melting and freezing.* A much smaller but still important quantity of energy is associated with the melting of ice and the freezing of water. Termed the **heat of fusion,** it involves the transfer of 80 calories per gram of water. (It is the transfer of this energy that makes your hand cold when you hold an ice cube as it melts.) When water freezes this energy is released as sensible heat to the atmo-

[1]The lower the temperature of the evaporating water, the greater the quantity of energy required for evaporation. At the boiling point, 540 calories are required to evaporate a gram of water; at the freezing point, 600 calories are required.

sphere. At the surface this often is a slow process because, even at temperatures well below the freezing point, considerable time is required for a large water body to radiate enough energy into the atmosphere to freeze. However, freezing can occur rapidly in the atmosphere as water droplet clouds are lifted to higher and colder levels, where they are converted to ice crystal clouds in a process termed **glaciation.**

- *Sublimation.* Under certain conditions ice can change directly to water vapor, or vice versa, without passing through the liquid state. Both changes are known as **sublimation.** Sublimation involves the transfer of a greater amount of energy per unit of mass than does any other change of state: 670 calories per gram. (This is the sum of the heats of condensation and fusion.) In the atmosphere, sublimation occurs primarily when ice crystal clouds form directly from water vapor at temperatures well below freezing. Although it liberates large amounts of energy per gram of water involved, sublimation is much less important as a source of atmospheric heating than is condensation because it occurs only at cold temperatures, which greatly reduce the capacity of the air to hold water vapor.

Taken together, the phase change processes perform three vital atmospheric functions. First, they transfer from the Earth's surface to the atmosphere much of the energy that powers the atmospheric processes, especially storms. Second, they add water vapor to the atmosphere. Third, they make possible the conversion of water vapor to the liquid and solid states, thereby enabling precipitation to occur.

THE HYDROLOGIC CYCLE

Earth's surface water supply is involved in an endless cycle of movement, called the **hydrologic cycle,** which is powered by solar energy and made possible by water's ability to change phase. As a result of this cycle, water is transported from the oceans to the continents, the land surface is sculptured by erosion, and the Earth's supply of solar energy is spread more evenly over the face of our planet. Most of the topics discussed in the remainder of this book are influenced at least indirectly by this water circulation system.

The hydrologic cycle is diagramed in Figure 7.2. The cycle actually is more complex than it appears in the figure because a number of alternative routes exist for the water depending on local conditions. Basically, though, the cycle consists of five key steps that, when completed, return the water to its point of origin and set the stage for a repetition of the cycle.

The hydrologic cycle is essentially a closed material system, because it involves the continuous recycling of the same water. It therefore is possible to begin the discussion with any of the cycle's five steps. Perhaps the best starting point, however, is with the ocean waters. The oceans contain an estimated 96.7 percent of the Earth's near-surface water supply, and they represent by far the longest stop that water normally takes in moving through the hydrologic cycle (see Table 7.1).

The evaporation of water from the oceans and its subsequent upward diffusion into the atmosphere represent, in our illustration, the first step of the hydrologic cycle.[2] Rates of evaporation from oceans in different parts of the world vary greatly for reasons that will be discussed shortly, but a mean annual rate of 100 cm (40 in) can be used as an approximation. At any one time, the atmosphere contains about 14,200 cubic kilometers (3,400 mi^3) of water, mostly in the vapor state.

The second step in the hydrologic cycle is the condensation (and/or sublimation) of atmospheric water vapor to form clouds. When condensation occurs, the water returns to its original liquid physical state, but is far from having its original appearance. Where before it was part of an extensive, interconnected body of water covering approximately 71 percent of the Earth's surface, it is now dispersed high in the atmosphere in myriad tiny water droplets centered around dust particles that serve as condensation nuclei.

Most clouds do not gain enough moisture through condensation or sublimation to produce precipitation, but instead simply evaporate back into the air. Clouds that do generate precipitation, though, return water to the surface. In contrast to its long stay of perhaps thousands of years in the oceans, water remains in the atmosphere only a short time before precipitating out. Its estimated average residency time in the atmosphere is 10 days. If the clouds are still above the ocean, a three-step hydrologic cycle of

[2]Large amounts of water also evaporate from the land, but the total quantity is much less than that from the oceans. We shall temporarily ignore this source in order to concentrate on the basic steps of the hydrologic cycle.

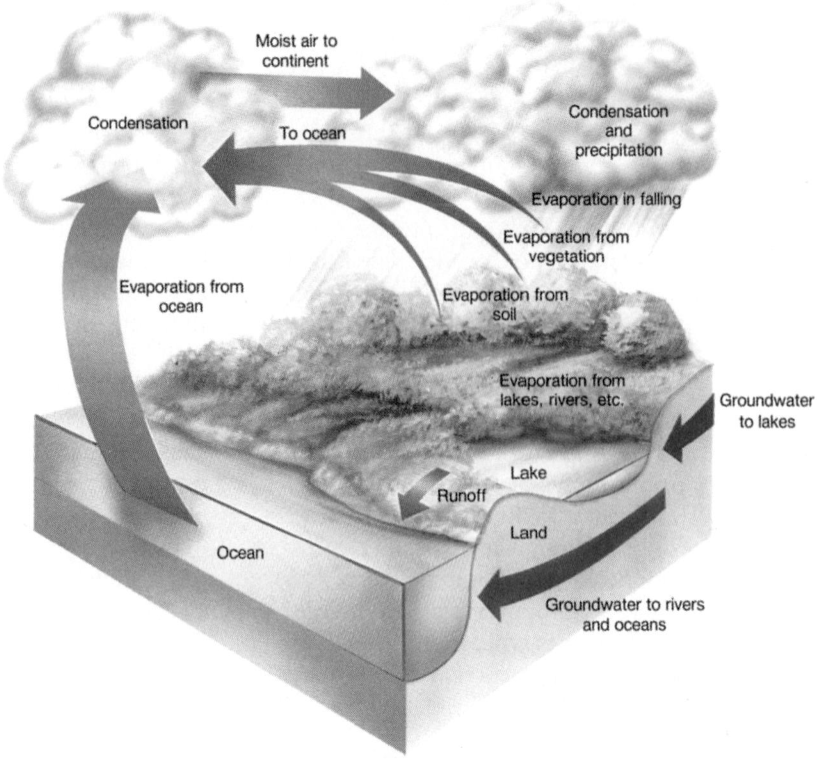

FIGURE 7.2 The hydrologic cycle involves the continuous cycling of water between the oceans, the land, and the atmosphere.

TABLE 7.1 Distribution of Global Near-Surface Water Supplies

LOCATION	VOLUME		PERCENT
	Cubic Miles	Cubic Kilometers	
Oceans	286,230,000	1,192,625,000	96.7%
Fresh water	9,769,600	40,707,000	3.3%
	DISTRIBUTION OF FRESH WATER		
	Cubic Miles	Cubic Kilometers	
Ice caps and glaciers	7,327,500	30,531,000	75.00%
Ground water below 800 meters (2,500 ft)	1,327,000	5,529,000	13.60%
Ground water above 800 meters (2,500 ft)	1,074,700	4,478,000	11.00%
Rivers and lakes	32,000	133,000	0.32%
Atmospheric moisture	3,400	14,000	0.03%
Soil moisture	5,000	20,800	0.05%
Total	9,769,600	40,707,000	100.00%

evaporation, condensation, and precipitation results. Indeed, since 80 percent of the world's precipitation falls into the oceans, this is the most common pathway within the hydrologic cycle. However, we are usually much more concerned with the 20 percent of the precipitation that falls on land. The third step of the hydrologic cycle therefore involves the transport of the moisture from sea to land by wind currents either before or after condensation has occurred.

The fourth step is the precipitation of the water onto the land surface. Land areas in different parts of the world receive greatly differing amounts of precipitation, but the average is approximately 76 centimeters (30 in) of liquid water per year. Much of this moisture has been transported hundreds or even thousands of kilometers from the places where it was evaporated.

The fifth and final step of the hydrologic cycle is runoff, which eventually returns the water to the oceans. Once again, complications can arise, with the likelihood that much of the water will evaporate before it reaches the ocean. Flow to the sea may be on or beneath the surface. The water also can be stored for varying periods in rivers, lakes, and glaciers, or as ground water. Eventually, though, the water completes its journey and is available to begin the cycle anew.

The hydrologic cycle, as described here, provides an outline for the material covered in the remainder of this chapter, as well as that treated in several subsequent chapters of this book. Each of the final three weather elements to be discussed—humidity, clouds, and precipitation—first appears at a particular point in the hydrologic cycle. Their analyses in a sense constitute studies of specific phases of the cycle. Later material on surface and subsurface waters (Chapter 10), weathering and mass wasting (Chapter 16), fluvial processes and landforms (Chapter 17), glacial processes and landforms (Chapter 18), and coastal processes and landforms (Chapter 20) covers the surface stages of the hydrologic cycle as well as its direct effects on the Earth's surface.

Evaporation

Evaporation obviously is a critical step in the hydrologic cycle, since it is the means by which the atmosphere obtains nearly all of its water vapor. For the Earth as a whole, the amount of moisture entering the atmosphere by evaporation must, in the long run, approximate that leaving it as precipitation. This is to be expected because precipitation is a product of the processes of condensation and sublimation, which are opposite phase changes from evaporation. However, although evaporation and precipitation may be considered equal in quantity on a global basis, their geographic distributions are very different.

A number of factors control the rate of evaporation from any point on the Earth's surface. One is the availability of water. In many land locations, the lack of surface moisture results in low evaporation rates. Water that evaporates from land areas comes from the surface (including the soil), from water bodies such as rivers and lakes, and from water evaporated from plants—a process known as **transpiration.** In well-vegetated regions, water losses from transpiration can considerably exceed those from direct surface evaporation. The combined water loss from both evaporation and transpiration is termed **evapotranspiration.** Globally, it is estimated that 15 percent of the total atmospheric water vapor supply evaporates from the continents, including inland water bodies such as lakes and rivers, while 85 percent evaporates from the oceans. This means that, as a world average, the evaporation rates over land areas are about half those over the oceans per unit of area (see Figure 7.3).

A second factor affecting evaporation rates is the temperature of the evaporating water. Warm water evaporates more quickly than cold water because its molecules are in more rapid motion and thus more are able to overcome the surface tension of the water body and enter the vapor state. The amount and intensity of solar energy receipts are important in determining the supply of available heat.

Additional factors include the temperature and water vapor content of the air overlying the surface. Warm air has a much larger water vapor retention ability than does cold air. Water therefore evaporates much more rapidly into warm air than into cold air, if other factors are equal. Evaporation rates at any given temperature also tend to be greater when relative humidity values are low.

A final influence on evaporation rates is wind speed. Under calm conditions, evaporation produces a shallow layer of moistened air overlying the surface. The wind either removes this layer or mixes it with drier air from farther aloft, allowing evaporation to proceed at an increased rate. Well-known mechanical devices that use a combination of heat and

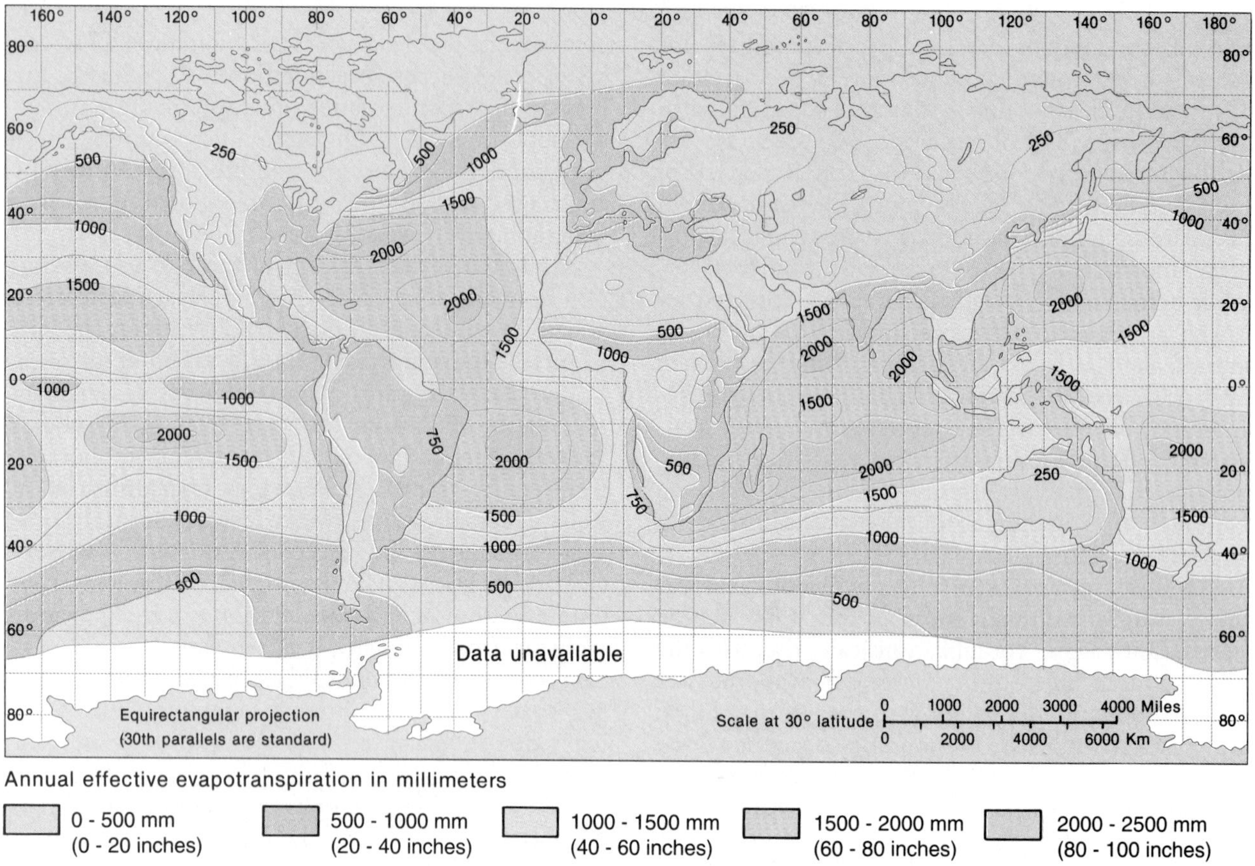

FIGURE 7.3 World distribution of mean actual annual evapotranspiration rates. Globally, it is estimated that 15 percent of the atmosphere's water vapor is derived from the continents (including inland water bodies), while 85 percent evaporates from the ocean.

airflow to promote rapid evaporation include electric hair driers and clothes driers.

Taking all these factors and relating them to what we know about the world distribution of land and water, temperature, pressure, and winds, we can make some generalizations concerning the global pattern of evaporation rates. One is that most evaporation occurs from the warm oceans of the lower latitudes (see Figure 7.4). In fact, the leading source of the moisture received as precipitation for much of the world is the ocean surface in areas dominated by the subtropical highs. With their combination of clear skies, warm temperatures, and relatively dry air, the subtropics lose vast quantities of water to evaporation, and many subtropical areas receive little back as precipitation. On a global basis, evaporation greatly exceeds precipitation at latitudes between 10° and 40° in both hemispheres, while precipitation exceeds evaporation in all other latitudinal zones largely as a result of the transport of subtropical moisture into those zones, where it is released by storms. Because of temperature and vegetation factors, evaporation rates outside the tropics also are generally higher in summer than in winter and during the day than at night.

Humidity

When water evaporates, it contributes to the air's **humidity,** the amount of water vapor contained by the air. The atmosphere's water vapor content can be calculated in a variety of ways, so several different types of humidity measurements exist. The two most commonly used in the atmospheric sciences are specific humidity and relative humidity.

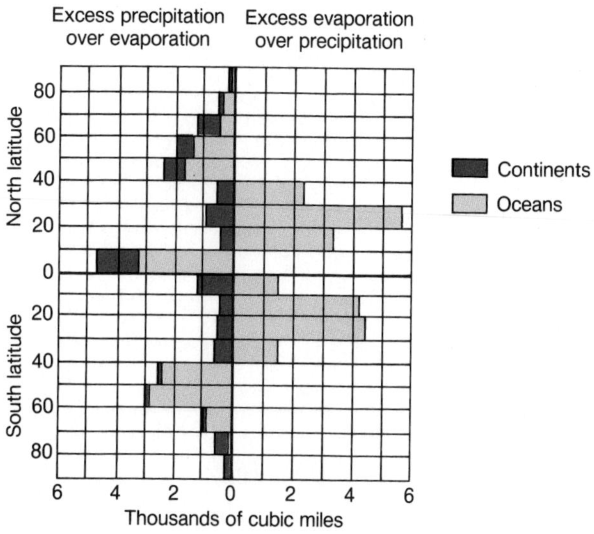

FIGURE 7.4 Continental and oceanic pattern of annual evaporation and precipitation totals by ten degree increments of latitude. For the world as a whole, evaporation and precipitation totals are equal. Note the influence of the global air pressure belts; evaporation exceeds precipitation only in the subtropics.

Specific Humidity

Specific humidity is the ratio of the mass of water vapor in a parcel of air to the total mass of the air, including the water vapor. It usually is given in grams per kilogram, so that the numerical value is in easily calculable parts per thousand. Specific humidity values give the actual water vapor content of the air. They are unaffected by changes in temperature or air pressure—factors that influence the values of most other humidity measurements, including relative humidity. The geographic distribution of specific humidity by latitude is shown in Figure 7.5.

Relative Humidity

Relative humidity is the most commonly used measure of atmospheric water vapor; it also is the one most familiar to the general public. Its popularity probably results from the fact that relative humidity readings correspond rather well to the sensible moisture content of the air; that is, to the apparent moisture content felt by people. It is, however, a poor indicator of the air's total water vapor content.

By definition, the **relative humidity** is the percent ratio of the amount of water vapor actually in the air to the maximum amount of water vapor the air could hold at that temperature and pressure. Put another way, the relative humidity indicates what percent of its capacity of water vapor the air is currently holding. The mean relative humidity of the air near the Earth's surface is about 78 percent. The atmosphere therefore is holding an average of 78 percent of its capacity of water vapor at its existing temperature and pressure.

Like most weather conditions, the relative humidity is subject to continual changes in value at any given location. Two factors are chiefly responsible for these changes. The first is variations in the water vapor content of the air. As in the case of the specific humidity, if surface water evaporates into the air, the relative humidity will increase, assuming other conditions remain constant. Similarly, if **hygroscopic** (water-attracting) materials are present to remove water vapor from the air, the relative humidity will decrease. Unlike the specific humidity, the relative humidity is directly affected by changes in air temperature (see Figure 7.6). Because the air's capacity for holding water vapor increases with a rise in temperature, heating the air causes the relative humidity to fall. This occurs because the amount of water va-

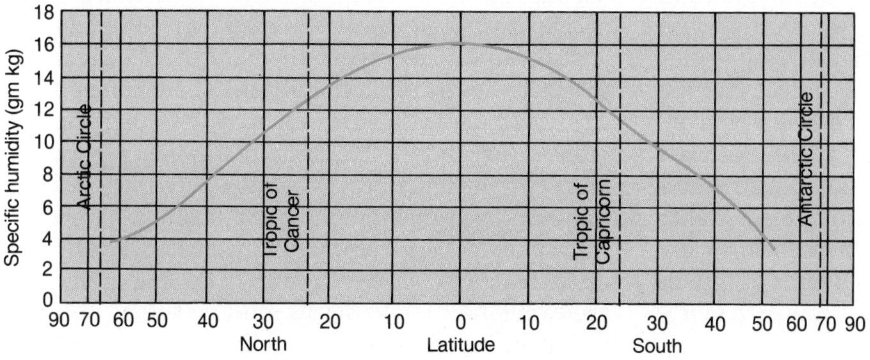

FIGURE 7.5 Mean specific humidity values by latitude. The dominant influence on this pattern is the effect of temperature on the water vapor holding capacity of the atmosphere.

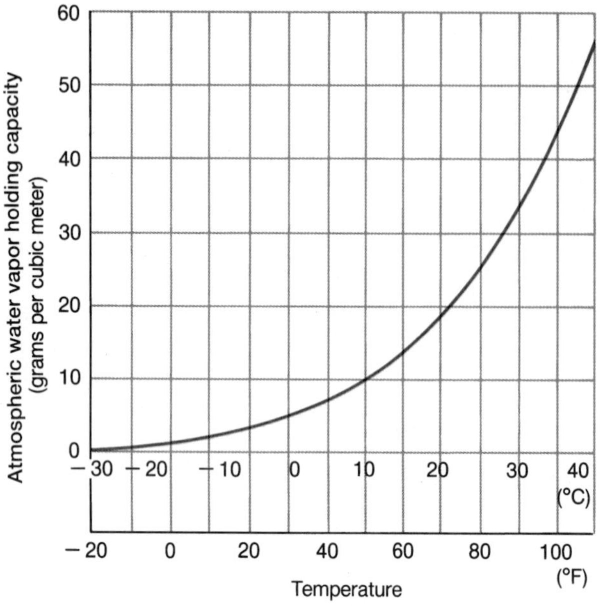

FIGURE 7.6 The relationship between temperature and the maximum water vapor holding capacity of air at sea level.

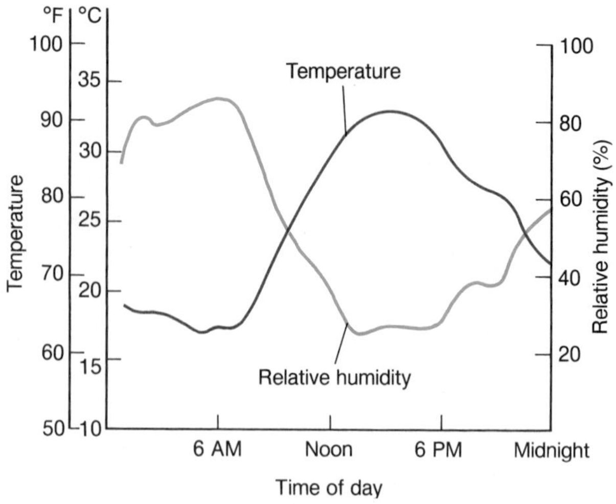

FIGURE 7.7 Air temperature and relative humidity readings at Tucson, Arizona, during September 18, 1982. Note the inverse relationship between the two graph patterns.

por contained by the air, though unchanged, now constitutes a smaller percentage of the air's increased capacity to hold this moisture. Similarly, if the air becomes cooler, its water vapor holding capacity is reduced, and the relative humidity rises, even though no new water vapor is added to the air. As a consequence, relative humidity readings usually fall during the day as the air is heated by insolation, and rise during the night as the atmosphere cools (see Figure 7.7).

Figure 7.8 depicts the mean distribution of relative humidity values on a latitudinal basis. While the global distribution of specific humidity values is closely correlated with temperature, the distribution of relative humidity values is strongly correlated with the global air pressure belts. The highest relative humidity values are associated with the ITCZ and the subpolar lows, where cloudiness and precipitation are prevalent. The lowest values are associated with the subsiding air of the subtropical highs. Relative humidity values in the southern hemisphere are higher on average than those in the northern hemisphere because of the presence of greater amounts of surface water available for evaporation in the southern hemisphere.

Condensation and Sublimation: Processes and Products

Water that evaporates from the Earth's surface eventually returns as precipitation. In order for this to occur, the water vapor, one of the atmosphere's lightest gases, must first be converted to the much denser liquid or solid states. The phase change processes involved in these conversions are condensation and sublimation.

Under what conditions do condensation and sublimation occur? Obviously, they do not continually remove water vapor from the atmosphere, because the weather often is fair and dry. At other times, though, the occurrence of precipitation or the formation of dew or frost makes it apparent that water vapor is indeed being removed from the air. The key to these processes is cooling. It was noted earlier that cooling the air reduces its water vapor holding capacity, causing an increase in the relative humidity. If it is cooled sufficiently, a temperature will be reached at which the relative humidity is 100 percent and the air is saturated with water vapor. Any further cooling reduces the ability of the air to hold water vapor even more, producing an excess of moisture that must be removed by condensation or sublimation.

The temperature at which a given parcel of air will be saturated is its **dew point.** The dew point, by indicating how much the air must be cooled in order to initiate condensation or sublimation, is important

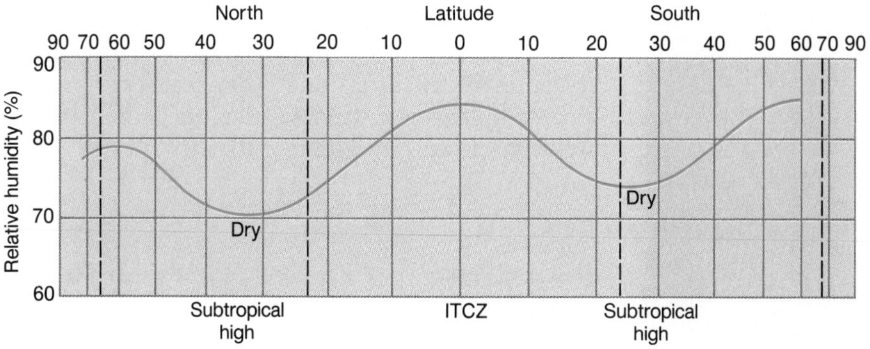

FIGURE 7.8 Mean relative humidity values by latitude. The dominant influence on this pattern is the distribution of the global air pressure belts.

in predicting the likelihood of precipitation, dew, frost, or fog, and the amount and height of certain types of clouds. When the relative humidity is low, the dew point is much lower than the air temperature, and a large amount of cooling must occur before the air becomes saturated. When the air is saturated, the dew point and air temperature values are identical.

The forms taken by the condensed or sublimed moisture vary with temperature and location. Condensation occurs when the air is cooled below a dew point that is near or above freezing. The products of condensation are dew, fog, water droplet clouds, drizzle, and rain. Sublimation normally occurs when the dew point is well below freezing. Its three major products are frost, ice crystal clouds, and snow. The remainder of this chapter is devoted to an examination of the causes and global distributions of these forms of atmospheric moisture.

Adiabatic Processes and Atmospheric Stability

Atmospheric cooling that leads to the formation of clouds and precipitation nearly always results from the rising of air. As a parcel of air rises, it enters areas of reduced atmospheric pressure and consequently expands. The expansion of the rising air requires energy, which is taken from the internal kinetic (or thermal) energy of the air molecules, making the air parcel cooler. When this cooling has reduced the air temperature to the dew point, condensation and cloud formation begin. Clouds thus are the visible tops of rising air currents. The cooling of air resulting from its expansion is termed **adiabatic cooling.** This cooling process involves no actual loss of energy, because the energy has not left the air but has merely changed form from sensible heat to mechanical energy. As long as the air keeps rising, it continues to cool, causing the condensation or sublimation of additional water vapor. This allows the cloud to grow larger and its constituent water droplets and ice crystals to increase in size, so that eventually they may become large enough to fall as precipitation. Conversely, when the air descends, it undergoes **adiabatic heating** because the surrounding air forces the descending parcel to contract, returning the energy it used in expansion.[3] As the descending air warms, its water vapor holding capacity increases, so that a cloud within a descending airflow will soon evaporate. This is why the descending air associated with high pressure systems produces fair, dry weather.

The concept of atmospheric stability also is important in understanding cloud development. Several different factors can cause air to rise sufficiently to produce clouds and precipitation. If the air is **stable,** however, it tends to resist any lifting influence; if forced to rise, it will return to its original altitude when this becomes possible. **Unstable** air, however, tends to rise; once given an initial lift, it continues to rise on its own. The implications of atmospheric stability on cloud types and precipitation characteristics will be explored shortly.

Dew and Frost

Dew and frost both form on the Earth's surface. Because they do not fall from the atmosphere, they are not technically considered forms of precipitation. **Dew** consists of water droplets that have condensed

[3]An analogy might be made to a compressed spring, which uses energy when it is released and expands, but regains this energy when it is again compressed.

FIGURE 7.9 The crystalline patterns formed on a window pane by a heavy frost indicate that it is a product of sublimation. (Edward F. D'Arms, Jr./Photo Researchers, Inc.)

on surface objects because the overlying air has cooled below the dew point. It nearly always occurs at night and is most likely when the sky is clear, the wind light or calm, and the air moist. Under these conditions, terrestrial radiation causes the land surface to cool rapidly. Conduction with the cold surface then chills the overlying layer of air, forcing the excess atmospheric moisture to condense on available surfaces such as leaves and blades of grass. Dew normally evaporates in the morning as the air is warmed and its water vapor holding capacity increases. It is a relatively minor source of surface moisture, but typically provides land areas in the middle latitudes with 12 to 50 millimeters (0.5 to 2.0") of water per year.

Frost forms under nearly identical conditions, but when the dew point is below freezing. Frost is not frozen dew, but instead forms by sublimation directly from water vapor. All products of sublimation have a crystalline structure, and the feathery patterns formed by frost on windowpanes and other surfaces are well known (see Figure 7.9).

Fog

Dew and frost, as we have seen, form when a layer of air just above the surface is cooled below its dew point, and the resulting moisture is deposited on objects at the surface. When a deeper layer of air is cooled, however, condensation or sublimation occurs on atmospheric dust particles. The result is fog. **Fog** is a visible accumulation of minute water droplets or ice crystals suspended in air immediately overlying the surface. It is actually a cloud at ground level.

The most common fog type over land areas is **radiation fog** (see Figure 7.10). It forms under the same clear, calm, moist nighttime conditions that result in the formation of dew and frost. Its presence indicates that conditions for terrestrial radiation favor the cooling of an unusually deep layer of near-surface air, and it is associated with heavy dew or frost. Local topographic features often control the distribution of radiation fog. Over flat surfaces, it can form an extensive blanket covering many square kilometers. In more rugged areas, though, the downhill drainage of air chilled by contact with the cold surface often causes radiation fog to accumulate in valley bottoms, while the surrounding uplands remain clear.

Another important fog type, **advection fog,** forms when moist, mild air flows over a colder surface. Al-

FIGURE 7.10 Radiation fog in the Appalachians is typically concentrated in valley locations. (J. Irwin/H. Armstrong Roberts)

though it occurs less frequently worldwide, it typically develops much more extensively than radiation fog, and sometimes covers hundreds or even thousands of square kilometers. It also is generally deeper than a radiation fog and is likely to persist for an extended period.

The most favorable locations for advection fogs are ocean areas where warm and cold waters lie adjacent to one another. One notorious site is the Grand Banks area southeast of Newfoundland, where air that has traveled over the warm Gulf Stream passes over the adjacent, much colder water of the Labrador Current. Another well-known site is the California coast, where fog forms as mild Pacific air is chilled by its passage over cold water that has upwelled near the coast (see Figure 7.11). Extensive advection fogs also can be produced when mild maritime air is transported over a cold, perhaps snow-covered, land surface in winter or a glacier surface in summer. A southerly flow of air from the Gulf of Mexico often produces dense winter fogs over the central United States in this manner.

The geographical pattern of fog frequency in the United States is depicted in Figure 7.12. Fog is particularly prevalent in two types of settings: areas near large bodies of cold water, and mountainous regions. The cold-water coasts notably include the entire West Coast from Washington to southern California, the northern Great Lakes states, and coastal New England. In each of these areas, fog typically is produced by damp air that is chilled below the dew point as it passes over the cold water. The chief mountain region frequented by fog is the Appalachians, where the air is cooled by being lifted by the elevated surface. It is interesting to note that even though the mountains of the western United States are much higher than the Appalachians, the western mountains typically experience much less fog because of their continental climates and dry air.

FIGURE 7.11 Advection fog bank approaching the California coast near La Jolla. (John S. Shelton)

Clouds

Clouds, with their endless variety and constantly changing patterns of form and color, add substantially to the beauty of the natural environment. They also provide valuable information about the operation of often complex atmospheric processes, and most important of all, they produce virtually all of the Earth's precipitation.

Like fog, **clouds** are accumulations of minute water droplets or ice crystals, suspended in the air, that have condensed or sublimed around dust particles because of atmospheric cooling. Clouds may be differentiated from fog by the fact that they are not in contact with the surface. The diameter of a typical cloud droplet is on the order of 0.01 millimeters (0.0004 inches), and its mass is about one-millionth of a gram. Particles of this size and mass can remain suspended in the air almost indefinitely.

Most clouds consist largely or entirely of water droplets. This is true both because warmer air typically contains more water vapor for cloud formation and because tiny cloud droplets, with their high surface tension values, can remain in the liquid state at temperatures well below freezing. Clouds are composed mostly of water droplets down to a temperature of about −7°C (20°F), a mixture of supercooled water droplets and ice crystals between −7°C and −39°C (20°F and −38°F), and entirely of ice crystals at temperatures below −39°C (−38°F). Because very cold temperatures are needed for their formation, ice crystal clouds generally are high clouds, located in the upper troposphere.

CLOUD CLASSIFICATION Although an unlimited variety of clouds exists, ten basic types are commonly recognized on the basis of their altitude and their appearance. The names of the types are derived from five Latin root words: *cumulus, stratus, cirrus, nimbus, and altus.*

Cumulus means "pile" or "heap." All cumuliform clouds have an uneven, puffy appearance produced by individual parcels of unstable air that rise to form

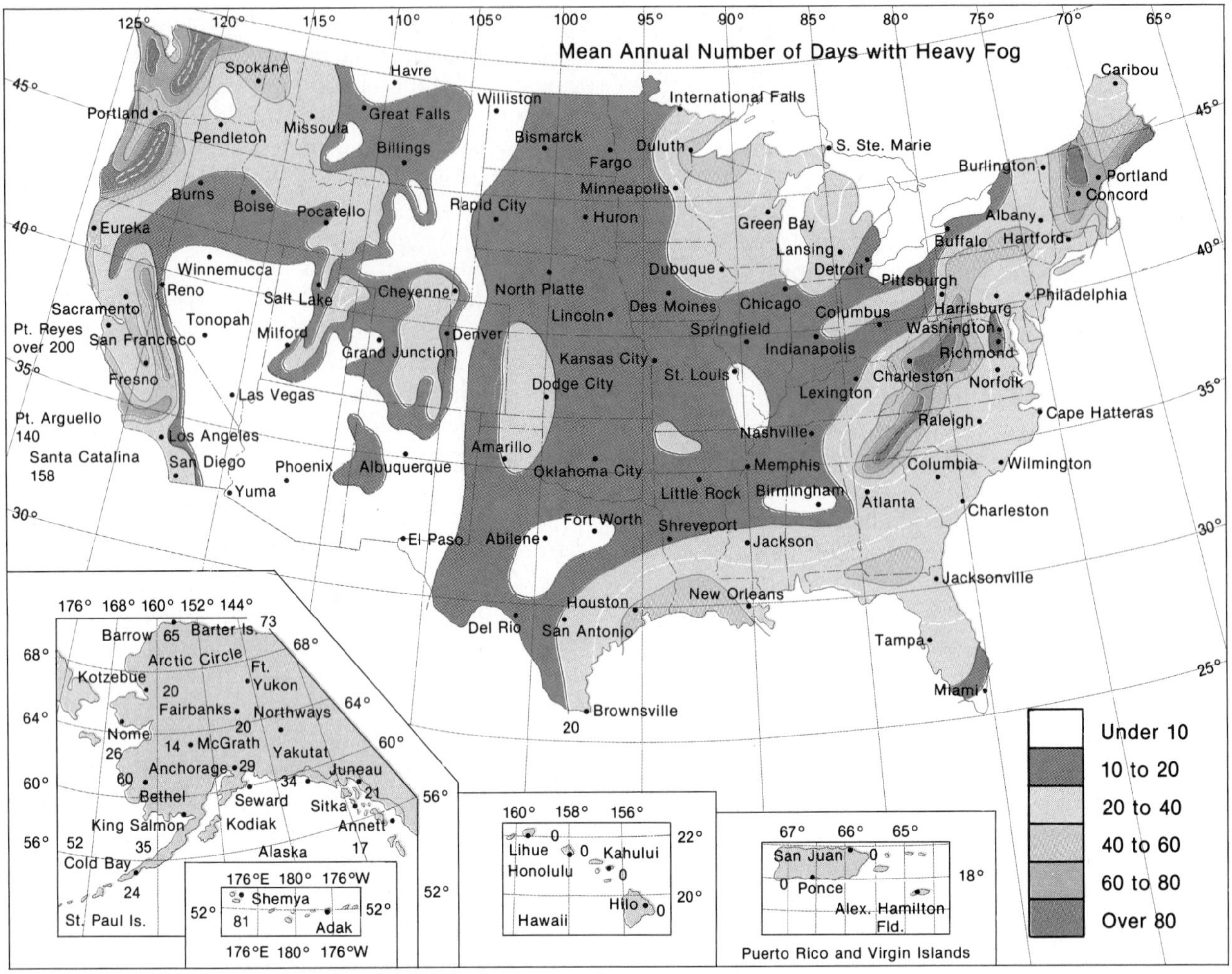

FIGURE 7.12 Average number of days per year with dense fog in the United States.

discrete cloud masses. *Stratus* means "layer." Stratiform clouds have a smooth, sheetlike appearance produced by the forced gentle uplift of stable air. They often are relatively thin, but may cover vast areas. *Cirrus*, which literally means "curl of hair," refers to high, ice-crystal clouds, often with a wispy appearance. *Nimbus*, which means "rain cloud," is used in conjunction with the two cloud types that produce nearly all of the Earth's precipitation. Finally, *altus*, which means "high," is used as the prefix "alto" in the names of two middle-level relatives of common low cloud types.

The ten cloud types often are separated into four categories on the basis of height, as indicated in Table 7.2. It is difficult to provide adequate verbal descriptions of clouds, so you should carefully examine the accompanying photographs of each cloud type as you read about it in the following paragraphs (see Figure 7.13).

Three high cloud types are recognized. All are relatively thin, whitish in color in sunlight, and composed entirely of ice crystals. **Cirrus** clouds are the highest and most common of the three. They have a distinctive wispy, hairlike, or "broom-swept" appearance. **Cirrostratus** clouds cover most or all of the sky in a smooth, thin white veil. The Sun and Moon shine through these clouds fairly brightly, but often are surrounded by a halo. **Cirrocumulus** clouds are probably the least common of the ten cloud types. They consist of a large number of very small,

TABLE 7.2 Major Cloud Groups, Types, and Heights*

HIGH CLOUDS 5,000 - 18,300 meters (16,000 - 60,000 ft)	
Cirrus	
Cirrostratus	
Cirrocumulus	
MIDDLE CLOUDS 2,000 - 7,600 meters (6,500 - 25,000 ft)	
Altostratus	
Altocumulus	
LOW CLOUDS 0 - 2,000 meters (0 - 6,500 ft)	
Stratus	
Stratocumulus	
Nimbostratus	
CLOUDS OF VERTICAL EXTENT	
Cumulus	600 - 6,000 meters (2,000 - 20,000 ft)
Cumulonimbus	600 - 18,300 meters (2,000 - 60,000 ft)

*Height values are approximate and relate to the low and middle latitudes only. High-latitude clouds are considerably lower.

rounded, and sometimes iridescent cloud masses arranged in groups or lines. Cirriform clouds are often forerunners of storms in the middle latitudes, especially during the colder half of the year.

Two types of middle clouds exist, both with the prefix "alto." **Altocumulus** clouds are rather similar to cirrocumulus in that they consist of small, circular cloud masses arranged in a linear pattern, but the individual masses appear considerably larger, darker in their centers, and have more distinct margins. **Altostratus** typically forms a smooth, textureless layer of gray cloud over the entire sky. The Sun may be visible as a diffuse bright spot, and a light, steady rain or snow may occur.

There are three types of low clouds, all stratiform in nature. **Stratus** is a dense, uniformly gray cloud, usually covering the entire sky, and capable of producing a light drizzle. Common on dull, damp mornings, it is the lowest cloud type and often is formed by the lifting of a fog layer. **Stratocumulus** clouds form a whitish or grayish layer with distinct cloud masses, usually in the form of rolls, between which patches of clear sky may show. They are intermediate between stratus and cumulus and are often produced by the lifting and breakup of a stratus overcast. Stratocumulus clouds sometimes produce light rain showers or snow flurries. **Nimbostratus** clouds are the important winter precipitation producers of the middle and high latitudes; they are associated with the large-scale forced uplift of stable air. They form a dark, generally textureless cloud layer often partially obscured by precipitation. From the surface, nimbostratus clouds often appear identical to stratus, but may be distinguished from them by their continuously falling precipitation.

The final two cloud types differ from the others in that they display a considerable degree of vertical development. Both are cumuliform and are associated with unstable air. **Cumulus** clouds are the world's most familiar cloud type. A common sight on summer afternoons, they consist of distinct individual cloud masses with flat bases and uneven sides and tops. The flat bases of cumulus clouds indicate the level at which rising and expanding masses of warm, near-surface air have cooled below the dew point, thereby initiating condensation. Cumulus clouds typically have sharply defined outlines and can assume a great variety of shapes. If especially large, they can generate showers. The **cumulonimbus** is the most visually impressive cloud type. It is the chief precipitation producer of the tropics and, probably, of the world as a whole. It is associated with heavy rain, lightning and thunder, and occasionally hail, destructive winds, and even tornadoes. Cumulonimbus clouds form in moist air that is unstable to great altitudes. They look like overgrown cumulus clouds, which is exactly what they are. A large cumulus cloud (cumulus congestus) becomes a cumulonimbus when its top grows high enough for the supercooled water droplets to freeze, or glaciate. Cumulonimbus clouds are by far the deepest cloud types. They are usually at least 6,000 meters (20,000 ft) thick and can occasionally attain depths of 15,000 or even 18,000 meters (50,000 to 60,000 ft), as they extend from perhaps 1,000 meters (3,000 ft) above the surface up to the tropopause. Because their great thickness prevents most light from penetrating to the surface, their bases can appear ominously black even during the midday.

FIGURE 7.13a Cirrus (Camerique/H. Armstrong Roberts)

FIGURE 7.13b Cirrostratus (Thomas M. Conrow/Educational Images)

FIGURE 7.13c Cirrocumulus (Richard Jacobs/© JLM Visuals)

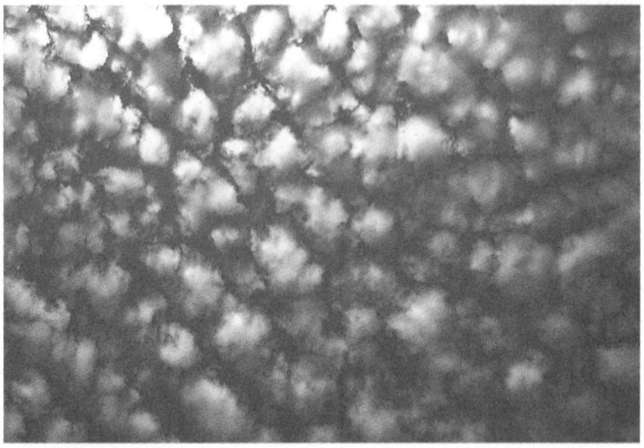

FIGURE 7.13d Altocumulus (Graham Macfarlane/Tony Stone Worldwide)

FIGURE 7.13e Altostratus (Thomas M. Conrow/Educational Images)

FIGURE 7.13f Stratus (Carl Wilmington/San Francisco Convention and Visitors Bureau)

FIGURE 7.13g Stratocumulus (B. F. Molnia)

FIGURE 7.13h Nimbostratus (Thomas M. Conrow/ Educational Images)

FIGURE 7.13i Cumulus (JLM Visuals)

FIGURE 7.13j Cumulonimbus (Jerry Scott)

CLOUD DISTRIBUTION Clouds have a complex and constantly changing geographic pattern that is influenced by a number of factors. These include surface land and water features, pressure and wind systems, fronts, and daily and seasonal variations in temperature. Satellite observations have indicated that an average of about 52 percent of the world is covered by clouds at any one time. Distinctive cloud bands typically demarcate the locations of the ITCZ and the subpolar lows. Areas of cold ocean waters in the subtropics and the tops of high mountain ranges also are frequently cloud covered. In addition, the middle and higher latitudes experience a constant progression of distinctive cloud patterns attending traveling cyclonic storms and their associated frontal systems. These are examined in more detail in the next chapter. As a general rule, atmospheric heating causes cumuliform clouds to be more common in the lower latitudes, during the day, and in the summer. The existence of cooler, more stable conditions causes stratiform clouds to predominate in the higher latitudes, at night, and during the winter.

The tropopause forms a "lid" on vertical cloud development. Tropospheric temperature conditions favor cloud formation because temperatures decline with altitude, thereby encouraging the uplift of relatively warm and buoyant near-surface air. Above the tropopause, however, the temperature trend is reversed (see Chapter 4). The characteristic flattened "anvil" shape associated with thunderheads often develops as the rapidly rising tops of cumulonimbus clouds encounter the tropopause.

PRECIPITATION MEASUREMENT

The accurate measurement of precipitation is vital because of the influence of precipitation on soil moisture levels and crop growth, water supplies, and flood potentials. As a result, precipitation is measured at thousands of meteorological field stations around the world.

The instrument most commonly used for collecting and measuring precipitation in North America is the standard 20-centimeter (8-in) *rain gage* (see Figure 7.14). It is a simple apparatus consisting of a funnel-shaped collector with a diameter of 20 centimeters (8 in), that feeds into a 50-centimeter (20-in) long measuring tube.

A much more sophisticated instrument that is increasingly used to estimate the intensity of falling precipitation is **radar.** Radars transmit pulses of microwave energy that are reflected from falling precipitation. The returning radar "echo" is plotted on a screen to depict the size, shape, and intensity of the precipitation area (see Figure 7.15). Observed over a period of time, changes in the radar echoes indicate the direction and speed of movement of storms.

FIGURE 7.14 A standard 20 centimeter (8 inch) rain gage consists of a narrow collecting tank with a mouth 20 centimeters in diameter. The barrel-shaped outer container reduces evaporation and provides basal support. (NOAA)

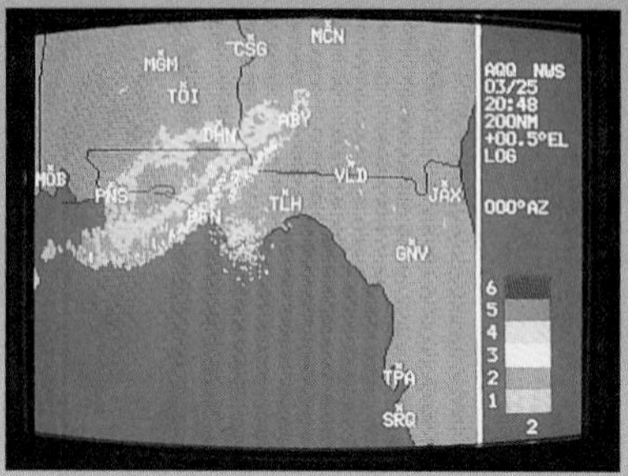

FIGURE 7.15 Color radar depiction of a rain area fronted by a line of thunderstorms extending from Southwestern Georgia into the Gulf of Mexico. (Courtesy of Alden Electronics, Inc.; Westboro, MA)

Precipitation

Precipitation is water, in either liquid or solid form, falling through the atmosphere toward the surface of the Earth. If it did not occur, life on land likely would never have evolved. Without the physical and chemical influences of water, Earth's landforms would be strikingly different in appearance, and a true soil cover would not have formed. Even the oceans, without the receipt of nutrients eroded from the land, would be far less capable of supporting life than they are today. The condensation and sublimation from which precipitation is derived are also associated with the release of huge amounts of heat energy to the atmosphere. For example, a rainfall of one centimeter (0.4 in) on one square kilometer (0.4 sq mi) of surface is associated with the release of a quantity of heat equivalent to that produced by burning approximately 1,400,000 kilograms (3,000 tons) of coal. Much of this energy escapes into space, but the rest is widely distributed to the Earth's atmosphere and surface.

Precipitation Formation and Types

Precipitation usually is produced in deep clouds that have been uplifted far enough for the air to have cooled adiabatically to well below the original dew point, forcing the condensation or sublimation of large quantities of water vapor. Because the upper portions of clouds have risen and cooled the most, their water droplets and ice crystals are much larger than those of the lower portions. Most precipitation therefore is initiated near the tops of clouds.

Although condensation and sublimation supply the moisture for precipitation, these two processes by themselves cannot produce raindrops or snowflakes large enough to precipitate. Two other processes are believed to play a critical role in enabling the myriad tiny cloud droplets or ice crystals to grow large enough to fall to the surface. The first of these is **coalescence.** This is the merging of cloud droplets either by incidental contact or because the updrafts in a cloud blow smaller droplets upward to collide with more slowly rising larger droplets. When a droplet grows large enough to begin falling, it collides and coalesces with still more droplets as it descends through the cloud. Coalescence is believed to be the chief mechanism by which "warm" clouds of the tropics (that is, clouds formed entirely of liquid water droplets) are able to produce precipitation.

In the middle and high latitudes, even in the summer, clouds deep enough to produce precipitation usually are "cold" clouds, with temperatures well below freezing in their upper portions. In this environment, a cloud typically contains a mixture of ice crystals and supercooled water droplets. The ice crystals in such a mixed cloud grow at the expense of the water droplets by a mechanism known as the **Bergeron process.** The vapor pressure, or pressure to evaporate, is slightly higher for water than for ice at the same temperature. As a result, the water droplets in a mixed cloud tend to evaporate, and this water vapor is then sublimed onto the growing ice crystals, which eventually fall as snow. Because of the Bergeron process, a cloud composed almost entirely of water droplets, and with only a small amount of ice crystals, will produce snow rather than rain. In fact, nearly all precipitation outside the tropics begins falling as snow and simply melts to become rain as it passes through warmer air on its way to the surface.

Precipitation can assume a variety of forms, depending on atmospheric conditions both within the clouds and between the clouds and the Earth's surface. The two types of liquid precipitation are rain and drizzle. **Rain** is by far the most common and important form of precipitation; it comprises about 90 percent of the Earth's precipitation. By definition, it consists of liquid drops with diameters of at least 0.5 millimeters (0.02 inches). **Drizzle** consists of droplets with diameters less than 0.5 millimeters (0.02 inches). It falls only from stratus clouds developed from weak updrafts.

FIGURE 7.16 Heavily weighted tree branches following an ice storm. (John S. Shelton)

There are four major forms of frozen precipitation. Rain falling onto a subfreezing surface will freeze quickly on contact with whatever object it strikes to form a solid coating of ice. This makes **freezing rain** one of the most potentially dangerous forms of precipitation. Severe ice storms, although they may transform the landscape into a crystalline wonderland, are capable of causing widespread damage to power lines and trees because of the sheer weight of the accumulated ice (see Figure 7.16).

When a considerable depth of the lower atmosphere has subfreezing temperatures, raindrops will freeze while falling to form sleet. **Sleet** consists of small, hard pellets of ice that collect to form a crunchy, granular surface.[4] Both freezing rain and sleet indicate the presence of a temperature inversion and form when above-freezing air overlies subfreezing air (see Figure 7.17).

The most common form of solid precipitation, comprising nearly a tenth of the Earth's precipitation total, is snow. **Snow** consists of crystalline ice formed directly from water vapor by sublimation. Its occurrence indicates that subfreezing (or near-freezing) conditions exist from the upper cloud levels all the way to the surface.

The most spectacular of the major precipitation forms is hail. **Hail** consists of relatively large, rounded particles of ice, which typically have a lay-

[4]In English-speaking countries outside of North America, sleet is commonly considered to be a mixture of rain and snow.

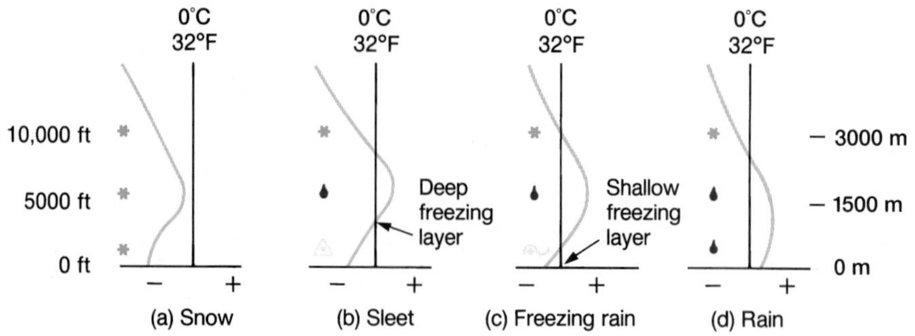

FIGURE 7.17 Vertical temperature profiles associated with various forms of precipitation.

ered structure and are associated with severe thunderstorms. Hailstones form as the strong updrafts and downdrafts of a thunderstorm carry ice pellets alternatively above and below the freezing level in a cumulonimbus cloud. Water and ice crystals adhere to the outside of the hailstone and freeze onto it when it is carried above the freezing level, causing it to grow. Hailstorms are most common in continental climate areas of the middle latitudes, especially the central United States (see Figure 7.18). They occur most frequently during the spring and early summer, when the atmosphere may be highly unstable and the freezing level is lower than it is later in the summer. Most hailstones have diameters of a centimeter (0.4 inches) or less, but some have been recorded that were larger than baseballs (see Figure 7.19). Severe hailstorms can do great damage to crops, livestock, and buildings. Hail depths can occasionally be impressive; individual storms have deposited more than 30 centimeters (1 ft) of hail in several localities in the United States.

Causes of Precipitation: Atmospheric Lifting Processes

We have noted that clouds and precipitation are produced by the rising and adiabatic cooling of air, but we have not yet examined the factors causing air to rise. Four basic atmospheric mechanisms can cause sufficient uplift to produce precipitation. They are convectional lifting, orographic lifting, frontal lifting, and cyclonic lifting.

Convectional lifting refers to the rising of unstable air because it is warmer, and therefore lighter, than the surrounding air (see Figure 7.20). The chief way by which the air is warmed is by absorbing energy from a warm surface. Over land, this frequently occurs during the midday or afternoon hours because of surface heating by the Sun. Over water, convectional lifting may occur at almost any hour of the day or night because of the constancy of surface temperatures, but it is restricted largely to areas of relatively warm water. Convectional lifting is associated with cumuliform clouds and localized, showery precipita-

FIGURE 7.18 The average number of days each year on which hail occurs in the coterminous United States.

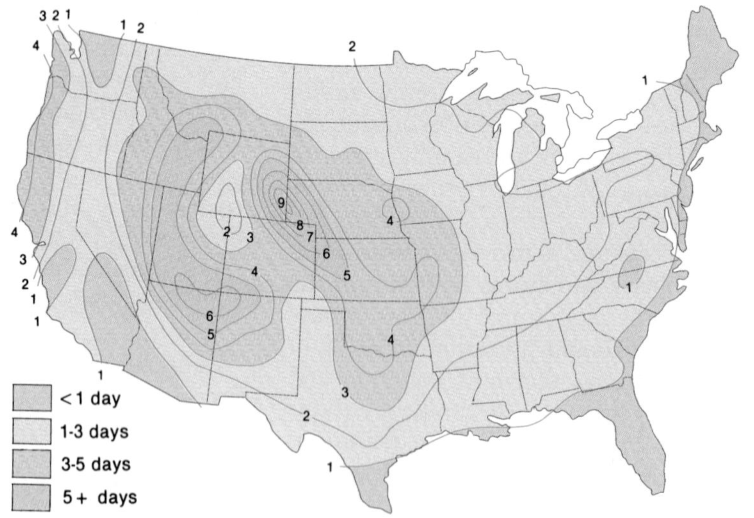

FIGURE 7.19 This hailstone—the largest ever reported—fell on the community of Coffeyville, Kansas, on September 3, 1970.

tion. The summertime afternoon showers and thunderstorms commonly experienced in many areas in the middle latitudes are mostly products of this atmospheric lifting process. The brevity of a convectional shower results largely from the fact that the shower cools and stabilizes the air, thereby reducing its ability to continue to rise and prolong the precipitation (see the Focus Box on the next page). The other three lifting processes are produced by the forced uplift of air; they are typically characterized by lighter and more prolonged precipitation.

Because it is generally associated with warm, unstable air, convectional lifting occurs most frequently in the tropics, where it is the leading cause of precipitation in most areas. In the middle and especially the higher latitudes, it is less common and is restricted largely to the warm season. Convectional precipitation, however, is not always associated with warm weather. This is illustrated by the heavy convectional snowfalls that may be triggered by the passage of extremely cold arctic air over an unfrozen water body. (As an example, see the Case Study at the end of the chapter.)

Orographic lifting is the lifting of air over a mountain range or other elevated surface. In contrast to convectional lifting, it involves the forced ascent

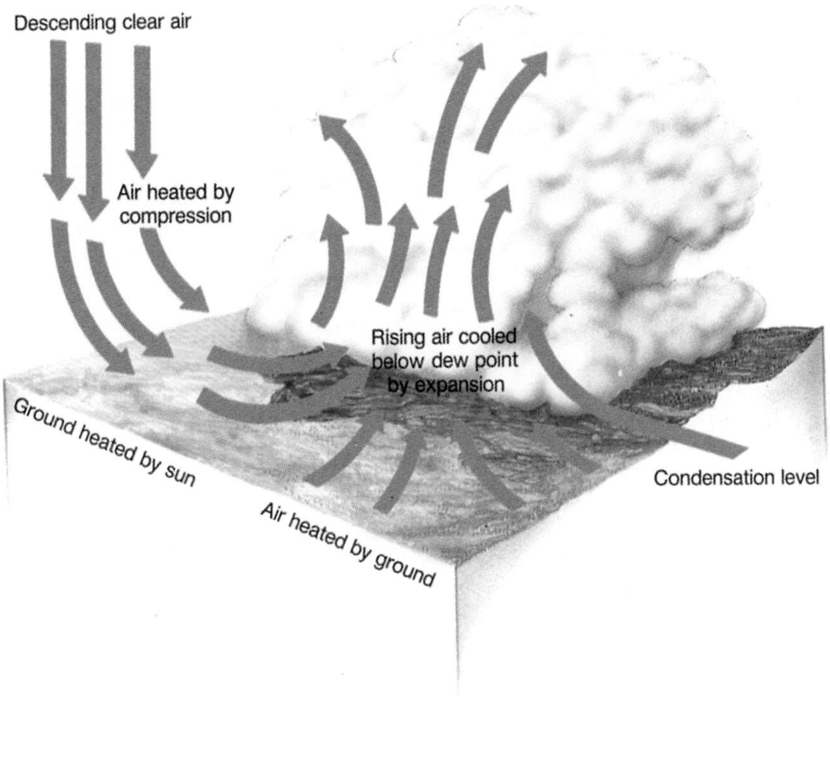

FIGURE 7.20 Convectional lifting is initiated by surface heating, which causes the air to expand, become buoyant, and rise. Adiabatic cooling then causes the rising air to cool below the dew point, resulting in cloud formation.

ATMOSPHERIC STABILITY

Convectional lifting is closely associated with the vertical distribution of temperatures within the troposphere. This factor controls the stability of the atmosphere. We saw earlier that the average environmental lapse rate within the troposphere is 6.4°C per 1,000 meters (3.5°F per 1,000 ft). We also noted that rising air cools adiabatically as it enters areas of decreased air pressure aloft. If the rising air is unsaturated, it cools at a constant rate of approximately 10°C per 1,000 meters (5.5°F per 1,000 ft). This value is termed the **dry adiabatic rate.**

Let us imagine that a parcel of unsaturated air at sea level with a temperature of 70°F is forced to rise to an altitude of 4,000 feet on a day when the average environmental lapse rate exists, and is then released to test its stability (see Figure 7.21). What will happen? Recall that if it is stable, the air will descend to its original level, but if it is unstable, it will continue to rise and perhaps produce clouds and precipitation. While being lifted, the air expands and cools adiabatically by a total of 5.5F° × 4, or 22F°, so that its temperature at 4,000 feet is 48°F. The surrounding air, however, is only becoming cooler at a rate of 3.5°/1,000 feet, so that at a height of 4,000 feet, the outside air temperature is 56°F. The parcel of air that was lifted is thus substantially cooler and denser than the air into which it was lifted. It therefore is stable. When released, it will return to the surface, warming at the dry adiabatic rate, so that when it reaches its original level, its temperature is once again 70°F.

The experiment can be repeated with a similar air parcel on a day when the environmental lapse rate

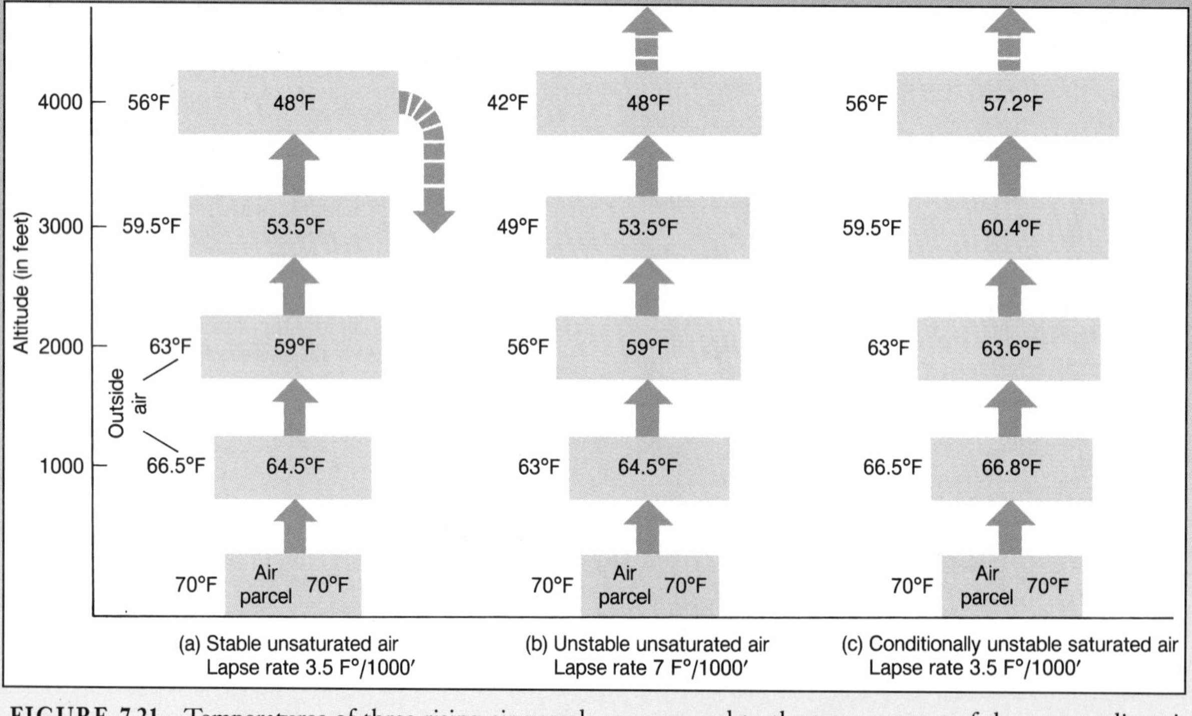

FIGURE 7.21 Temperatures of three rising air parcels as compared to the temperatures of the surrounding air. The air parcels are unstable if they are warmer than the surrounding air, and will continue to rise on their own. If they are cooler than the surrounding air, they are stable and will have a tendency to descend.

of 7F°/1,000 feet. Once again the air parcel is lifted a total of 4,000 feet and cools adiabatically to 48°F. This time, however, the surrounding air at the 4,000-foot level is only 42°F. As a result, the released air parcel is warmer and lighter than the surrounding air. It therefore is unstable, and will continue to rise on its own.

A complication occurs if the air parcel is saturated or becomes saturated while rising. In this case, adiabatic cooling causes the condensation or sublimation of some of the water vapor, releasing heat to the air. The released heat slows the cooling process to an average rate for air in the lower troposphere of 6C°/1,000 meters (3.2F°/1,000 ft). This reduced rate of cooling for rising, saturated air is termed the **wet adiabatic rate.** Unlike the dry adiabatic rate, the value of the wet adiabatic rate varies with the rate of condensation or sublimation. More condensation occurs per unit of cooling for warm air than for cold air (see Figure 7.6); therefore, the wet adiabatic rate becomes progressively higher as the air becomes colder. In order to avoid undue complication, however, we will use a constant wet adiabatic rate of 6C°/1,000 meters (3.2F°/1,000 ft).

If we now repeat our first rising air experiment with saturated air, we find that the outcome has changed. At a level of 4,000 feet, the surrounding air temperature is still 56°F, but because of the released heat of condensation, our rising air parcel has cooled only to 57.2°F. It therefore is unstable and will continue to rise. This illustrates the fact that saturated air is considerably less stable than unsaturated air.

The conclusion that can be drawn from these examples is that *air is unstable if the environmental lapse rate exceeds the operative adiabatic rate.* More specifically, unsaturated air is unstable if the lapse rate exceeds 10C°/1,000 meters (5.5F°/1,000 ft) and saturated air is unstable if the lapse rate exceeds the operative wet adiabatic rate. If the air in either case has a lapse rate lower than these values, it is stable (see Figure 7.22).

A point to note is that the lapse rate is likely to vary considerably at different altitudes, and that a parcel of air will continue to remain unstable and to rise only as long as it remains warmer than the surrounding air. For this reason, cumulonimbus clouds and heavy convectional precipitation, caused by air that has risen for many kilometers, can occur only if the air is unstable to great heights.

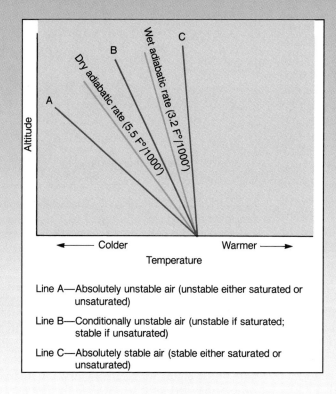

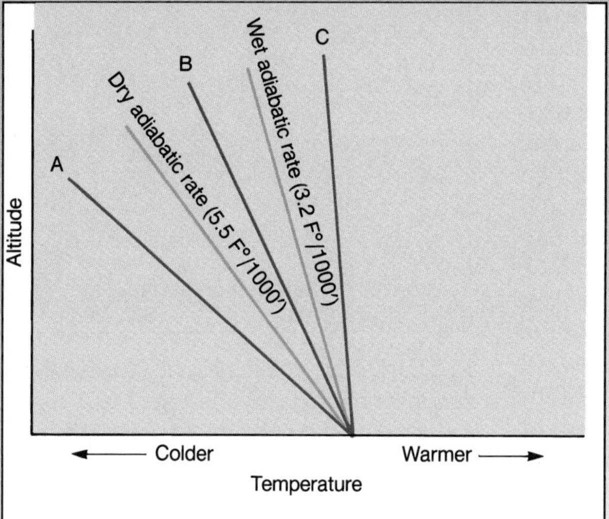

Line A—Absolutely unstable air (unstable either saturated or unsaturated)

Line B—Conditionally unstable air (unstable if saturated; stable if unsaturated)

Line C—Absolutely stable air (stable either saturated or unsaturated)

FIGURE 7.22 Graphic depiction of the three categories of relative atmospheric stability.

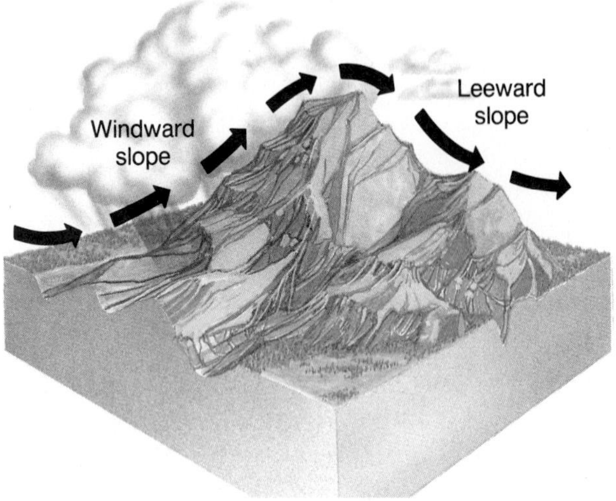

FIGURE 7.23 The orographic effect. The windward sides of mountains are associated with rising air, clouds, and precipitation; the leeward sides with descending air and dry conditions.

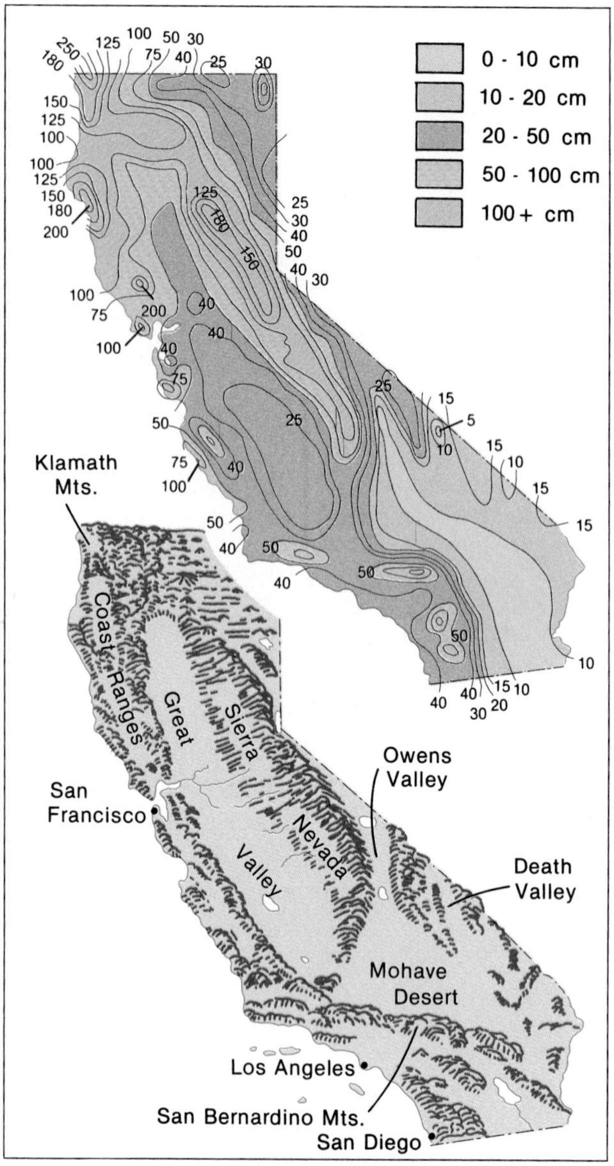

FIGURE 7.24 The effect of surface topography on precipitation is clearly illustrated by a comparison of topographic and mean annual precipitation maps of California. The highest precipitation totals are associated with the Coast Ranges and Sierras.

of air because the topography imposes a physical barrier to its movement. As a result, the stability of the air is not a crucial factor, although unstable air is more effectively lifted by any lifting mechanism than is stable air.

As an air mass approaches a mountain barrier (Figure 7.23), it begins to rise and to cool adiabatically. The amount of cloudiness and precipitation produced by this uplift depends on several factors. The most favorable conditions for heavy orographic precipitation include a high, continuous mountain range, the presence of warm, moist, unstable air, and strong winds blowing directly across the mountain range.

Most clouds and precipitation produced by the orographic effect are concentrated on the side of the mountain range facing the wind (the **windward** side), because the forced uplift occurs here. As the air passes over the crest and descends the downwind, or **leeward,** side of the mountains, it is adiabatically heated, and progressively drier conditions prevail. An area that has a dry climate because it is on the leeward side of an orographic barrier is said to experience a **rainshadow effect.** The orographic effect can therefore either increase or decrease precipitation amounts.

Figure 7.24 illustrates the importance of the orographic effect in the mountainous state of California. A comparison of the two maps indicates that California's precipitation pattern is strongly controlled by the state's topography. It can be seen that each of California's major mountain ranges has higher precipitation means than the surrounding regions, and that lowland areas, especially those in the eastern part of the state, are relatively dry. On a larger scale, orographic lifting is responsible for the narrow band

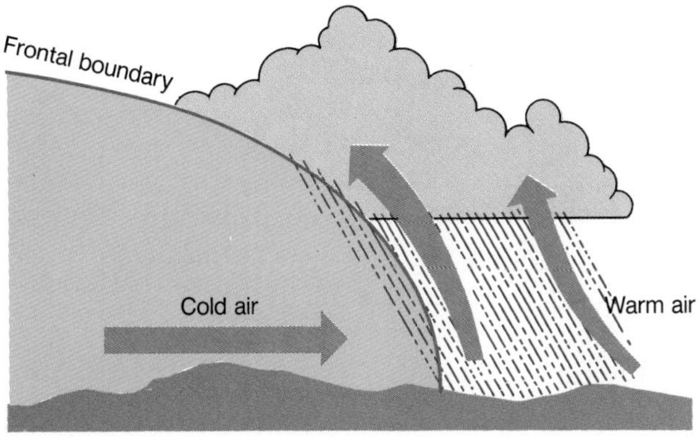

FIGURE 7.25 Frontal lifting occurs as a warm air mass is forced to rise over a denser cold air mass.

of heavy precipitation that occurs along the mountainous west coast of North America from southern Alaska to northern California, while the rainshadow effect to the east of these same mountains produces the aridity of the Great Basin region of the interior western United States and the interior valleys of western Canada (see Figure 7.26).

Although orographic lifting can be an extremely effective precipitation generator, it is the most geographically restricted of the four atmospheric lifting mechanisms. It is normally strongly developed only in the vicinity of relatively high mountain ranges or plateaus, which comprise at most 10 percent of the Earth's surface. Because it is tied to the Earth's surface features, it is incapable of movement. The other three lifting mechanisms are readily transported by the wind currents, greatly increasing the total surface area they influence.

Frontal lifting is associated with the lifting of relatively light warm air by denser cold air along a weather front (see Figure 7.25). It is the primary cause of precipitation in most of the middle and higher latitudes. In the middle latitudes frontal lifting is most prevalent in winter and early spring, when fronts are much stronger and more numerous than they are during the summer. In the high latitudes fronts are fairly numerous throughout the year, but the lowered water vapor holding capacity of the cold air limits the amount of precipitation they produce. Fronts and frontal precipitation are generally unimportant in the tropics because cool air masses from the higher latitudes usually weaken and lose their identity before penetrating far into this area. The different types of fronts and their characteristics are examined in Chapter 8.

The final atmospheric lifting mechanism, **cyclonic lifting,** is produced by the rising spiral of converging air associated with a low pressure system. In the middle and high latitudes, low pressure systems usually form on weather fronts, so that frontal and cyclonic lifting act together. Because cyclonic storms of the middle and high latitudes almost always are associated with weather fronts, the precipitation accompanying these storms is produced by the combined influences of frontal and cyclonic lifting—a combination sometimes described as **convergent lifting.** Low pressure systems of the tropics such as hurricanes do not have fronts. Their cloudiness and precipitation are induced primarily by cyclonic uplift acting alone.

It should be noted that various combinations of lifting mechanisms may occur at the same time. Heavy precipitation in the Pacific Northwest, for example, may be generated when a cyclonic storm with its associated weather fronts crosses mountain ranges that orographically intensify the rainfall or snowfall. Similarly, precipitation along a weather front is typically heavier during the day than at night, because the daytime air is warmer and less stable.

World Precipitation Distribution

The total amount of precipitation received at any given location depends upon the moisture content of the atmosphere and the frequency and intensity of atmospheric lifting.The estimated average annual precipitation for the Earth as a whole is 90 centimeters (35 inches). This value provides a useful point of comparison for assessing the annual total at any given site.

In the next several paragraphs, the basic characteristics and causes of the global pattern of precipitation are examined. As you read this material, it is a good idea to make frequent reference to Figure 7.26. More detailed information on regional precipitation patterns is provided in Chapter 9.

The equatorial zone, located between approximately 15°N and 15°S, is the world's rainiest zone of latitude. Most locations receive 125 to 300 centimeters (50 to 120 inches) of rainfall annually. The air in the tropics is very warm and unstable, and has an exceptionally large water vapor content, giving it a high rainfall potential. Rainfalls are further aug-

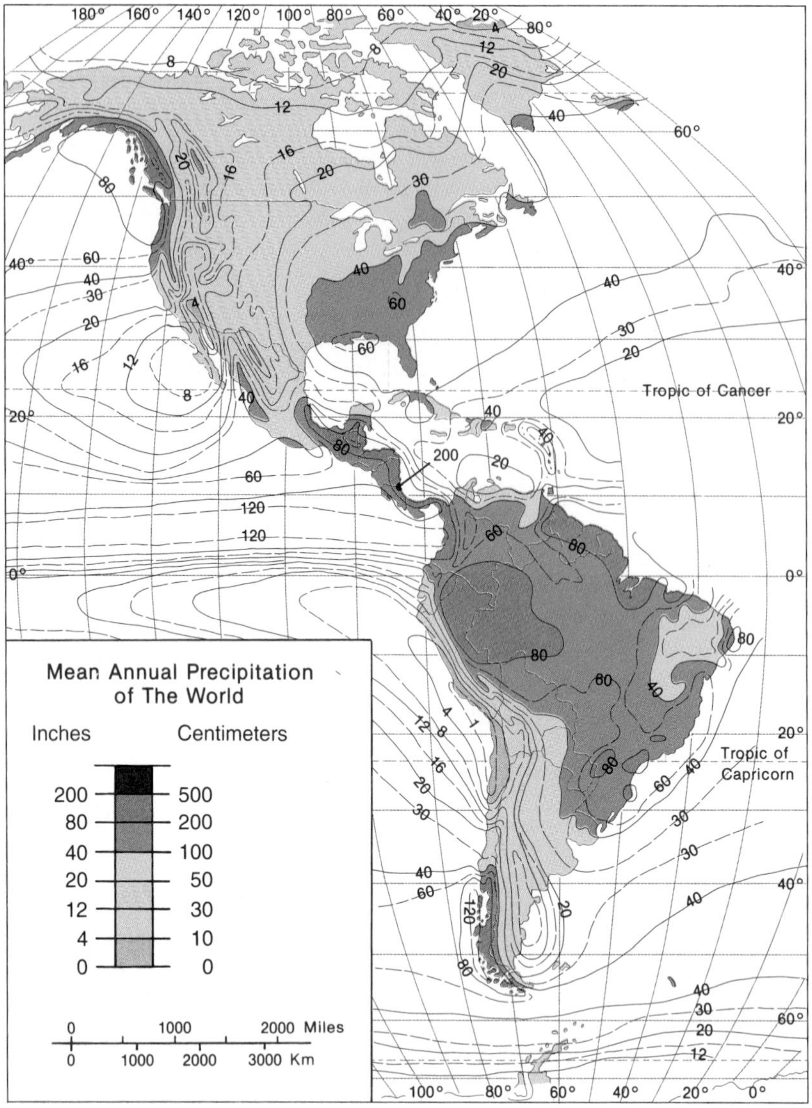

FIGURE 7.26 Mean annual world precipitation totals. Values are in centimeters (or inches) of liquid water.

mented by the convergence of air from the trade wind zones of both hemispheres into the ITCZ. The heaviest rainfall averages occur in areas where high mountains produce substantial orographic uplift.

The subtropics, comprising those areas between 15° and 35°N and S, are substantially drier as a whole than the equatorial zone, although great regional variations in precipitation means exist. The predominant precipitation influences here are the subtropical belts of high pressure with their subsiding air. In places where the highs are especially well developed, precipitation totals are so greatly reduced that arid or semiarid conditions prevail. Along continental west coasts paralleled by cold ocean currents, atmospheric lifting is further discouraged by the chilling of the air at the surface, and the world's driest deserts occur here. In contrast, the warm ocean currents that flow along the east coasts of continents in the subtropics serve to heat and destabilize the air as well as to add large quantities of water vapor. Precipitation amounts in these areas generally are substantial, especially if orographic or monsoonal influences are present.

The middle latitudes, which include areas between 35° and 65° N and S, are characterized by moderate precipitation amounts. Many areas receive annual totals near the global mean of 90 centimeters (35 inches) per year. This is due largely to the influence of two factors. First, temperatures in the middle lat-

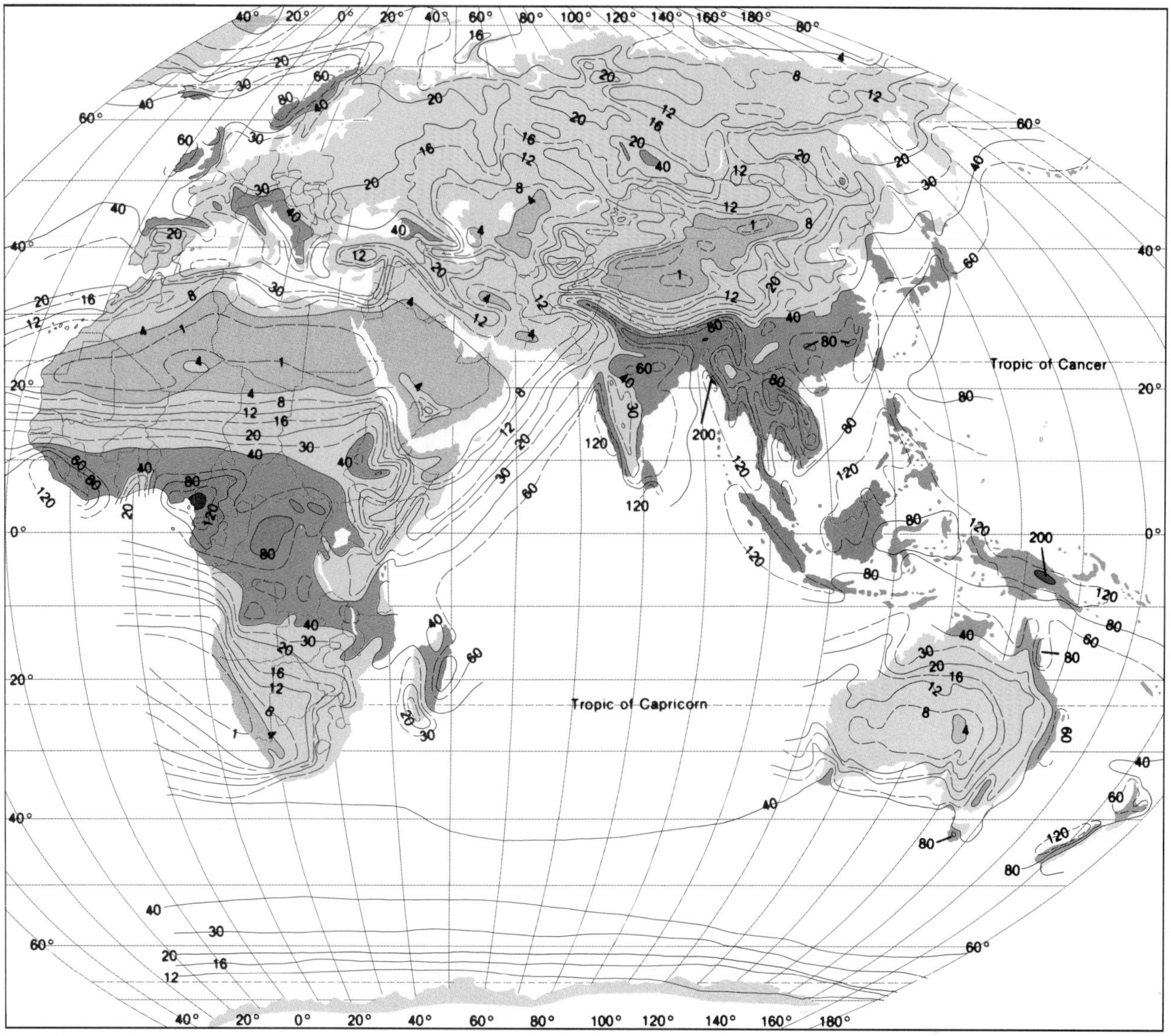

itudes are mild because of the intermediate quantities of insolation received. This causes the air's water vapor holding ability, and therefore its potential for precipitation, to be moderate. In addition, the middle latitudes are located between high and low pressure belts—specifically, the subtropical highs and subpolar lows. Because the effect of highs is to decrease precipitation amounts and that of lows is to increase them, the tendency exists for moderate totals. As is true of all latitudinal zones, however, orographic factors and proximity to water can have a profound influence on precipitation receipts. The heaviest precipitation totals are typically experienced along windward (west-facing) coasts backed by high mountains.

In the high latitudes, extending from 65°N and S to the poles, precipitation means are lower than in any other latitudinal zone. This is primarily a function of temperature. The cold air overlying these areas simply does not contain enough water vapor to generate large precipitation totals.

Other Important Precipitation Variables

Up to this point, the discussion of precipitation patterns has centered on the precipitation amounts re-

WORLD PRECIPITATION RECORDS

Precipitation records of various types have been set at a large number of meteorological stations around the world. Despite their widespread distribution, a certain geographic similarity is evident in the physical settings of these stations. For example, most rainfall and snowfall records were established at sites with a strong orographic effect. Most rainfall records were set in the tropics and subtropics. Most records for dryness have been established in subtropical deserts under the dominance of the subtropical highs.

A sampling of especially significant precipitation records includes the following:

- Greatest amount of rain received in one minute was 3.8 centimeters (1.50 in) at Barst, Guadeloupe, in the Lesser Antilles (November 26, 1970).
- Greatest 24-hour rainfall total was 187.0 centimeters (73.62 in) at Cilaos, Réunion, in the Indian Ocean (March 15–16, 1952).
- Greatest rainfall total received in a one-month period was 930.0 centimeters (366.14 in) at Cherrapunji, India (July 1861).
- Greatest rainfall total received in a period of one year was 2,646.1 centimeters (1,041.78 in) at Cherrapunji, India (August 1860–July 1861).
- Heaviest mean annual precipitation is 1,146 centimeters (451 in) at Mt. Waialeale, Kauai, Hawaii (1920–1972).
- Greatest 24-hour snowfall was 193 centimeters (76 in), at Silver Lake, Colorado (April 14–15, 1921).
- Greatest one-year snowfall was 3,110 centimeters (1,224.5 in), at Mt. Rainier, Washington (February 19, 1971–February 18, 1972).

ceived in different parts of the world. This is undoubtedly the most important of the geographic precipitation variables, but several others also are noteworthy.

The *form* of the precipitation is of great practical significance. Most of the world's precipitation is rain, but large quantities fall in solid forms, especially as snow. Because of the greater availability of atmospheric moisture, annual snowfall totals typically are heavier in parts of the upper middle latitudes and subpolar regions than near the poles.

The *intensity* and *duration* of precipitation play major roles in determining how much moisture can be absorbed by the soil and how much will become runoff. High-intensity rainfalls may occur at rates exceeding the absorption capacity of the soil, especially in arid areas or in areas disturbed by human activities. This will lead to erosion and possible flooding. On the other hand, steady, gentle rainfalls allow the soil to absorb more moisture, causing runoff to be reduced and spread over a greater time span. Intense rainfalls occur only when substantial amounts of water vapor are available, and are usually produced by the convergence of unstable air. They are most common in the low latitudes and during the summer months in the middle latitudes.

Another factor that has a great effect on water supplies and agricultural activities is the **seasonal distribution** of precipitation. In many areas the precipitation is relatively evenly distributed throughout the year, but other areas have distinctive wetter and drier seasons. Where precipitation is highly seasonal, the maintenance of a continuous water supply is a problem, and cycles of human activities, especially those relating to agriculture, are strongly affected.

The **annual variability** of precipitation refers to the average percentage deviation of an area's precipitation from the mean. For example, if a location has a mean annual precipitation variability of 30 percent, a 50 percent chance exists that the total precipitation it receives in any given year will be at least 30 percent higher or lower than its long-term annual mean. A low annual variability of precipitation means that the area can more confidently expect to receive a quantity of precipitation close to its mean value. A high variability, on the other hand, means that much less

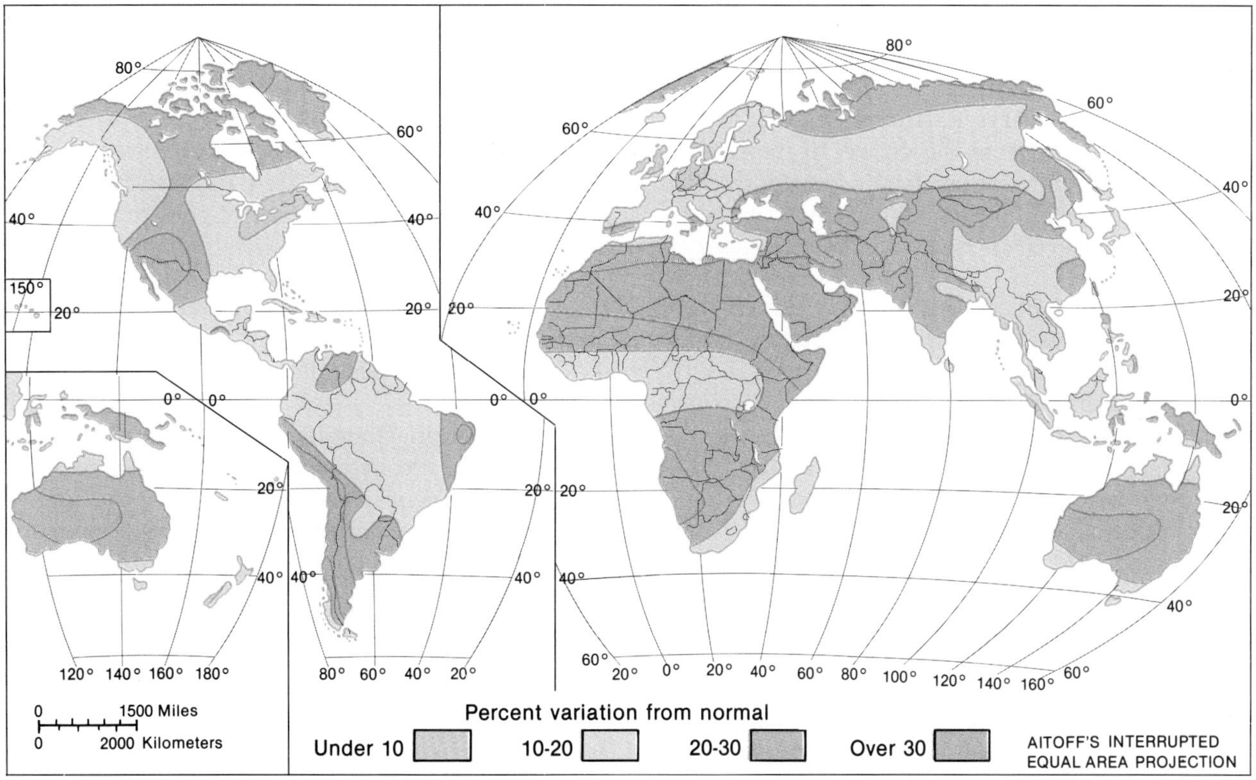

FIGURE 7.27 Mean annual precipitation variability. Values represent the average departure of a given year's precipitation from the long-term mean.

certainty exists as to the amount of precipitation to be received, and floods or droughts are more likely to occur. Figure 7.27 depicts the pattern of the mean annual variability of precipitation for the world.

A final factor that is not actually a precipitation factor, but nonetheless has a tremendous effect on surface water supplies, is the **potential evapotranspiration rate.** It may be defined as the rate at which water would be lost through the combined processes of evaporation and transpiration if the surface were constantly supplied with an unrestricted quantity of water. The balance between precipitation and evapotranspiration largely determines the "wetness" of an area. Of the total precipitation falling on land areas of the world, approximately 75 percent is evaporated or transpired; only 25 percent reaches the oceans as runoff. Therefore, while precipitation is the chief means by which the land gains its moisture supply, evapotranspiration is the most important mechanism by which this moisture is lost. Potential evapotranspiration rates are influenced primarily by the air temperature, and to a lesser extent by the relative humidity. In general, they are highest in the tropics and subtropics, especially in dry areas, and become progressively lower through the middle and high latitudes.

The necessity of examining both precipitation amounts and evapotranspiration rates in order to determine the moisture balance of an area is well illustrated in the case of the interior of Antarctica. This area annually receives the liquid equivalent of only a few centimeters of precipitation, but is currently buried under approximately 3 kilometers (2 miles) of ice. The Antarctic ice has formed from the accumulation of precipitation over many thousands of years, coupled with the virtual absence of losses by evaporation or sublimation. Conversely, areas with high annual or seasonal temperatures require substantial amounts of precipitation to support a continuous vegetation cover.

SUMMARY

This chapter has examined the role of water in the atmosphere. Water is a unique atmospheric component because of its ability to undergo phase changes between solid, liquid, and gaseous states. In so doing, it produces a variety of atmospheric phenomena and, more importantly, makes possible the transfer of great quantities of water from the oceans to the land. In order to review more systematically some of the most important information presented in the chapter, we now return to the chapter-opening Focus Questions:

1. *Under what conditions do water phase changes occur, and what role do they play in the global distribution of energy?*

Water undergoes phase changes when environmental conditions favor a gain or loss of energy. Energy is gained when ice melts or sublimes to water vapor, or when liquid water evaporates. Energy is released as heat to the environment when water vapor condenses or sublimes to ice, or when liquid water freezes. Because most surface water exists as liquid water and most atmospheric water exists as water vapor, the net result of the phase changes is the transfer of great amounts of energy from the surface to the atmosphere. This energy provides heat to the atmosphere, powers atmospheric processes, and is widely distributed around the Earth by wind currents.

2. *What is the hydrologic cycle?*

The hydrologic cycle is the continuous closed cycle of water transport between the surface and atmosphere. Although a number of alternate routes can be employed, the following five steps comprise the basic elements of the hydrologic cycle as it involves land areas: (1) evaporation of ocean water; (2) condensation of this water vapor to form clouds; (3) transport of the clouds to the land by wind currents; (4) precipitation of the moisture onto the land; and (5) surface and subsurface runoff of the moisture back to the ocean.

3. *How do clouds and precipitation form?*

Clouds form when air is cooled below the temperature at which it can hold all its moisture as water vapor. When this happens, the excess moisture condenses or sublimes onto dust particles. Most cooling that results in cloud formation occurs adiabatically as the air rises and expands. Expansion is an energy-using operation that removes heat from the air. The air rises because of one or more of the following four factors: atmospheric heating and resulting instability (convectional lifting); lifting along a weather front (frontal lifting); lifting caused by passage over a topographic barrier (orographic lifting); or lifting within the rising air spiral of a low pressure system (cyclonic lifting).

Precipitation forms as cloud water droplets or ice crystals become large enough for gravity to cause them to fall toward the Earth's surface. Cloud droplets grow by means of collision and coalescence. Ice crystals grow by sublimation, which is encouraged by the relatively low vapor pressure over ice as compared to that over water droplets within a mixed cloud.

4. *What is the geographic pattern of precipitation, and why does it exist?*

The global precipitation pattern is complex because it is affected by many factors. In general, the largest totals occur in the tropics, where the air is very moist and low pressure prevails. Most portions of the subtropics experience varying degrees of dryness because of the influence of the subtropical highs. The middle latitudes typically receive moderate precipitation amounts caused largely by traveling frontal cyclones and summertime convectional showers. The high latitudes are the driest latitudinal zone because of the limited water vapor holding capacity of the cold air. In general, precipitation is highest near warm ocean currents and on the windward sides of mountain ranges.

Review Questions

1. List and define each of the six water phase change processes. How many calories of energy are transferred in each process per gram of water involved? What is the meteorological significance of the energy released when water undergoes a phase change to a lower energy state?

2. Draw a diagram of the hydrologic cycle and label the various steps. In which portion of the cycle does water remain longest? In which portion does it remain for the briefest period, on the average?

3. Describe the basic global pattern of evaporation rates on a latitudinal basis, and briefly explain why this pattern exists.

4. What is evapotranspiration? Why might you expect evapotranspiration losses to be considerably higher in a well-vegetated region than losses by pure evaporation in a vegetation-free region under similarly moist surface conditions?

5. Define the terms *specific humidity* and *relative humidity.* Explain why specific humidity varies more by latitude than relative humidity, and why relative humidity varies more during the day than specific humidity.

6. What must normally happen to the air temperature in order for condensation or sublimation to occur? (Use the term *dew point* in your explanation.) List and briefly describe the major products formed by the condensation and sublimation of atmospheric water vapor.

7. What happens to the temperature of rising air? Why? What happens to the temperature of the air as it descends? Why?

8. What environmental conditions favor the formation of radiation fog? What conditions favor the formation of advection fog? List three geographical areas where advection fogs frequently occur, and explain why.

9. What are clouds, and why do they form? How are clouds classified? What are the two primary precipitation-producing cloud types, and how do they differ in their precipitation characteristics?

10. Define the term *precipitation.* Why is precipitation most likely to be produced by large, deep clouds?

11. Describe the six basic forms of precipitation and briefly explain the atmospheric and/or surface conditions under which each occurs.

12. List and briefly describe the four atmospheric lifting processes. Which one is predominant in the tropics, and why? Which ones are predominant in the middle latitudes, and why? Which one is the most geographically restricted, and why?

13. What is the difference in meaning between the environmental lapse rate and the adiabatic rate? How are these two values compared in order to determine the stability of the atmosphere?

14. What factors account for the large precipitation amounts received in most equatorial regions? Why are the high latitudes and much of the subtropics dry? Why do large areas of the middle latitudes receive intermediate precipitation amounts?

15. What precipitation factors other than mean amounts received have important imp tions to the inhabitants of a region? Why is precipitation generally needed for agricultura poses in the low latitudes than in the high latitudes?

16. Explain why the Great Lakes snow belts develop. Would there be a tendency for heavier snow belt snowfalls in early winter or in late winter under equally favorable atmospheric conditions? Why?

Problems

1. How many calories of heat are required to evaporate 25 grams of water from the Earth's surface?

2. How many calories of energy are released to the atmosphere by the formation of 40 grams of frost?

3. If a kilogram of air near the surface has the capacity to hold 2.3 grams of water vapor but is currently holding only 1.4 grams of water vapor, what is its relative humidity?

4. Is unsaturated air with a lapse rate of 7.5C°/1000 meters (4F°/1,000 ft) stable or unstable? Is saturated air with this same lapse rate stable or unstable?

Key Terms

evaporation
condensation
sublimation
hydrologic cycle
evapotranspiration
specific humidity
relative humidity
dew point
adiabatic temperature changes
stable and unstable air
dew
frost
radiation fog
advection fog
cloud types (ten major types)
convectional lifting
orographic lifting
frontal lifting
cyclonic lifting

CASE STUDY

SNOW BELTS OF THE GREAT LAKES

The Great Lakes moderate the climate of the areas that surround them, and because the prevailing winds blow from the west, the influence of the lakes is especially strong along their eastern shores. During the spring and summer the lake waters are colder than the land, and the shorelines receive cooling lake breezes. During the fall and winter, the lakes are warmer than the surrounding land and warm the areas to their east. This has the beneficial effect of delaying autumn frosts and has permitted the planting of extensive orchards of cold-sensitive fruit trees. The same warming influence, however, is responsible for the formation of the heavy lake-effect snowstorms that plague the lee shores of the Great Lakes each winter. Each of the five lakes has a distinctive snow belt along its southern and eastern shores.

The lake-effect snowstorms are highly localized storms of a predominantly convectional nature. They form following the passage of strong cold fronts, when frigid arctic air is carried on gusty west or northwest winds over the unfrozen waters of the Great Lakes. The air picks up large quantities of water vapor in its lower levels during its passage over the lakes, providing the moisture for heavy snow squalls.

The key factor responsible for lifting this moistened air is the air's great temperature contrast with the water of the Great Lakes. Arctic air, as it reaches the lakes, may have temperatures well below −18°C (0°F). As it passes over the water, which usually is a few degrees above freezing, it suddenly encounters a surface 15 to 30C° (30 to 60F°) or more warmer than it is. Large quantities of this heat are conducted to the overlying air, making it much lighter and more buoyant than the unmodified air farther aloft. This produces conditions of extreme instability. Convectional uplift of the air results, producing cumuliform clouds and heavy snow squalls. Latent heat energy, released as the water vapor condenses and sublimes, enhances conditions for uplift.

When the air reaches the hilly terrain along the eastern and southern shores of the lakes, orographic and frictional influences come into play to produce further lifting. The low hills, trees, and buildings act as barriers to reduce the speed of the strong winds that have developed over the open waters of the lakes, causing the air to pile up and rise. Higher hills are encountered farther inland, especially leeward of Lakes Erie and Ontario, and produce a further significant orographic uplift.

The configuration of the topography and the wind direction seem to be the two factors that exert the greatest control on the distributional pattern of snow squalls at any given time. If the wind blows across the narrow axis of a lake, the amounts of heat and water vapor it picks up are limited, and the resulting snow showers tend to be widely

FIGURE 7.28 A view of snow depths in Adams, New York, on February 2, 1977. (NOAA)

scattered and of light to moderate intensity. When the wind blows down the long axis of a lake, however, large quantities of heat and moisture are absorbed, and narrow bands of heavy snowfall typically result. Along the eastern shores of Lakes Erie and Ontario in upstate New York, these bands usually are 15 to 50 kilometers (10 to 30 mi) wide and extend inland from 80 to 160 kilometers (50 to 100 mi). Their edges are often very sharply defined, so that one site may experience a heavy snowfall, while another site a short distance away receives little or nothing. These localized snowstorms are sometimes accompanied by lightning and thunder, and on occasion funnel clouds may form.

One of the most remarkable aspects of the lake-effect snowstorms is their persistence in any given area, despite their localized nature and the rapid movement of the clouds that produce them. In most parts of the world a convectional storm will influence a given location for only a short period, perhaps an hour or less. This occurs largely because these storms are produced by the surface heating of the air, and the heat is soon dissipated by the cooling effect of the storm. The waters of the Great Lakes, however, generate a continuous supply of heat, and as long as the wind direction is constant, new snow clouds tend to form along the same band to replace those that are streaming inland and dissipating. Under extreme conditions, this can produce continuous heavy snowfall for a period of several days, resulting in amazingly large snow depths.

The winter of 1976–77 was a record-setting year for lake-effect snowstorms in many portions of the eastern Great Lakes region. It was caused by a jet stream pattern that produced a nearly continuous flow of arctic air over the Great Lakes for most of the winter. The season was capped by the great snowstorm of January 28 through February 1, which produced snowfall totals exceeding 250 centimeters (100 in) in some locations southeast of Watertown, New York (see Figure 7.28). This storm, which paralyzed the city of Buffalo, produced snowdrifts more than 7.5 meters (25 ft) high and resulted in 29 deaths. Winds gusted to 137 kilometers per hour (85 mph), creating zero visibilities in blowing snow for much of the five-day period, and producing wind-chill readings as low as −51°C (−60°F). President Carter declared several counties in western New York a disaster area after the storm. For the entire 1976–77 winter season, Buffalo received a record 506.5 centimeters (199.4 in) of snowfall, but Hooker, New York—a small town east of Lake Ontario—was buried under 1,185.9 centimeters (466.9 in) of snow! Figure 7.29 depicts the 1976–77 snowfall totals for the snowbelt areas of Lakes Erie and Ontario. The much higher totals to the east of Lake Ontario resulted from the fact that relatively shallow Lake Erie froze over in midwinter, cutting off its moisture supply, while Lake Ontario, which is much deeper, remained unfrozen.

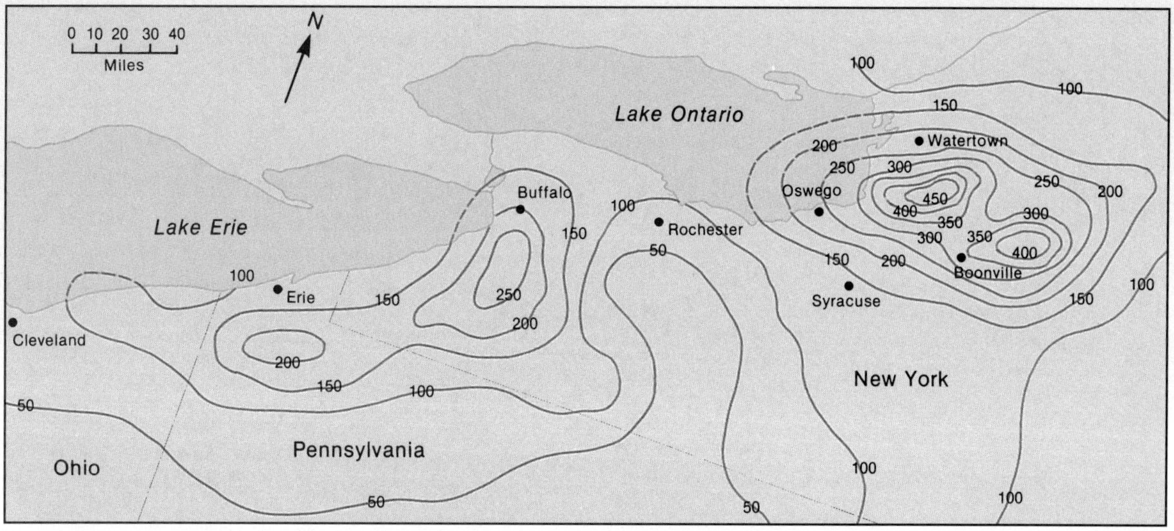

FIGURE 7.29 Snowfall totals in inches for the 1976–77 winter season in the snow belt region of Lakes Erie and Ontario. (*Weatherwise*, "Lake-effect Snowstorms. . ." by Kenneth F. Dewey, Dec. 1977. Reprinted with permission of the Helen Dwight Reid Educational Foundation. Published by Heldref Publications, 4000 Albemarle St., N.W., Washington, D.C. © 1977.)

CHAPTER 8

WEATHER SYSTEMS

OUTLINE

FOCUS QUESTIONS

1. What are air masses? How and why do they form, and what influence do they have on our weather?

2. What are the most important weather systems of the middle and higher latitudes, and what types of weather are associated with them?

3. How do thunderstorms and tornadoes form, and what areas are most affected by them?

4. How and why do weather conditions of the tropics differ from those of the higher latitudes?

5. What are hurricanes? How and where do they form, and what land areas do they most affect?

The preceding three chapters have discussed the characteristics, causes, and world distributions of the six basic elements of weather and climate. The weather elements, however, do not occur separately. At any location and at any given moment, the air has a temperature, pressure, and relative humidity. Additionally, in most cases, wind is blowing from a given direction at a certain speed, clouds of one or more types are present, and precipitation may be falling. Over large areas, these weather elements are organized into patterns that form **weather systems** such as air masses, fronts, and storms. Now that the components of the weather have been discussed, it is time to put them together to discover how and why these weather systems develop and what areas of the world they influence.

Some weather systems, notably air masses, occur in all parts of the world. Because they influence weather conditions everywhere, they will be examined first. Most other weather systems are restricted primarily to certain zones of latitude. For example, weather fronts, frontal cyclonic storms, and tornadoes are predominantly nontropical phenomena; easterly waves and hurricanes originate within the tropics. Following the discussion of air masses, the weather systems of the middle and high latitudes will be treated; this will be followed by a discussion of the systems of the tropics.

AIR MASSES

Television weather forecasters make it obvious that air masses are important controllers of present or future weather conditions. Statements such as "A frigid Canadian air mass should reach our area by tonight . . ." or "A continued onshore flow of cool, damp Pacific air can be expected . . ." are the forecasters' stock in trade. But what exactly are air masses, and why do they form?

An **air mass** is simply a large body of air with relatively uniform temperature and moisture characteristics. A typical air mass has a diameter of 800 to 2,400 kilometers (500 to 1,500 miles). Its properties develop gradually over a period of several days as the air lies relatively stationary over a large land or water body. During this period, the temperature and moisture characteristics of this surface area, termed the **source region** of the air mass, are gradually imparted to the overlying air.

Air Mass Formation and Types

Air masses form with much greater frequency over some areas than others. An important source region has several attributes. Like the air masses that form over it, it must be large in size and homogeneous in its temperature and moisture characteristics. For this reason, it must be either a land or a water body, not a combination of both. The source region also should frequently experience light upper-level winds, because strong winds will transport air near the surface out of the region too rapidly for the air mass to attain its characteristics. Finally, a source region should be an area that favors air subsidence and the development of high pressure. The centers of most air masses, in fact, appear on weather maps as high pressure systems. The diverging airflow associated with a high pressure system helps an air mass maintain its homogeneity because this flow pattern constantly pumps air from the center of the air mass over an expanding area. Because the air originates from one central location, it has similar temperature and moisture values throughout. Conversely, the converging winds associated with a low pressure system draw air of dissimilar characteristics inward toward a common central area. For this reason, lows are frequently associated with two or more air masses.

The principal source regions for the world's air masses are in the low and high latitudes, not the middle latitudes (see Figure 8.1). The middle latitudes lack some of the basic criteria for air mass formation. For example, the large differences in insolation experienced over relatively short north-south distances cause the middle latitudes to lack the thermal homogeneity needed for air mass formation. In addition, the frequent presence of strong upper-level winds, especially the subpolar jet stream, does not often permit air to remain over a single location long enough to acquire the characteristics of the surface.

The virtual elimination of middle latitude source regions means that air masses must form over large land or water bodies in the high and low latitudes. This has led to the development of a simple four-type classification system for air masses based on the nature of their source regions; the classifications also implicitly describe the temperature and moisture characteristics of the air masses themselves. Each classification consists of two words. The first indicates the surface nature of the source region. An air mass that forms over land is called *continental* and can be expected to contain dry air. Conversely, a *maritime*

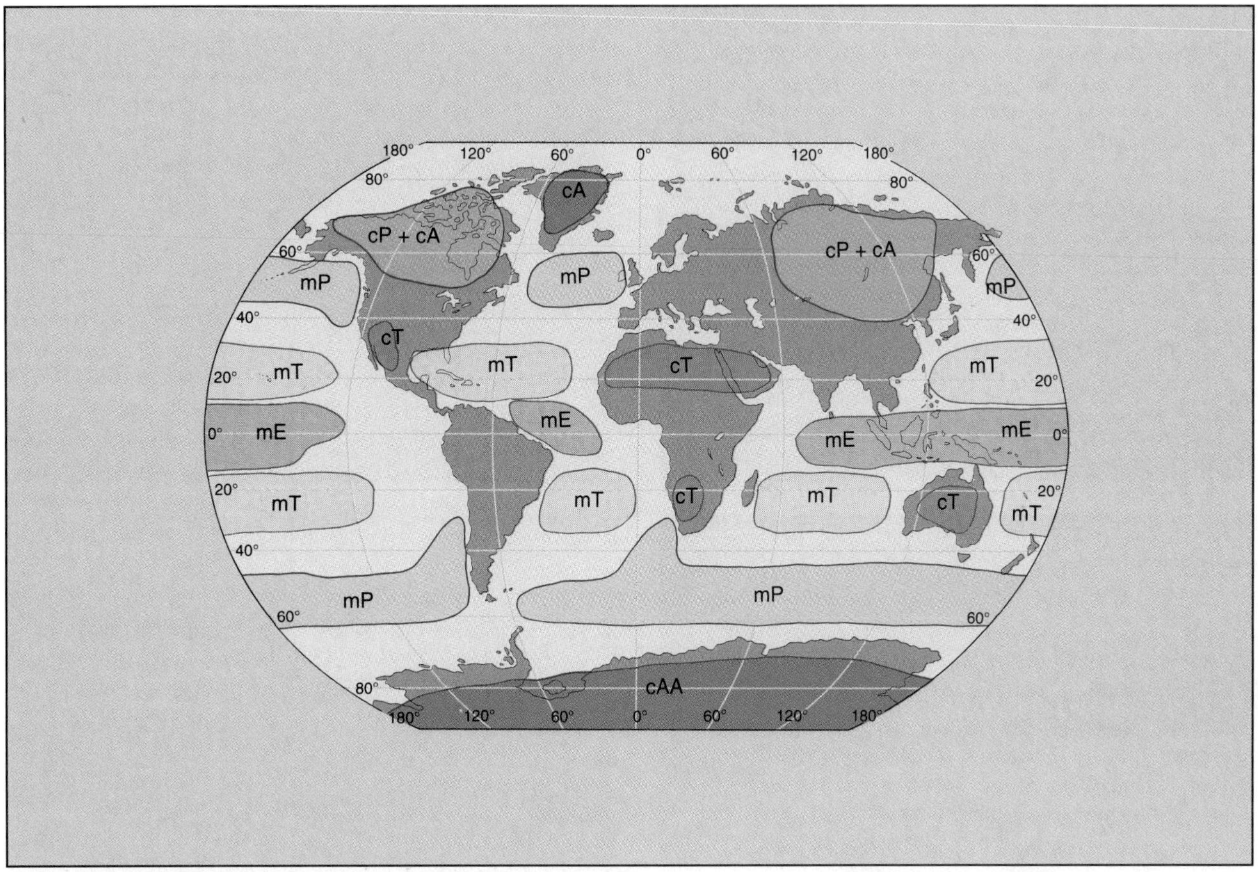

FIGURE 8.1 Major air mass source regions of the world.

air mass forms over water and can be expected to have a high water vapor content, at least in the lower levels. The second word indicates the general latitudinal zone in which the air mass forms. A *polar* air mass has formed over a polar or subpolar source region and is relatively cold. A *tropical* air mass has formed over a tropical or subtropical source region and is characterized by warm temperatures.

Combining these terms, we have an air mass classification system that consists of four types (see Figure 8.2). On weather maps, these air masses often are abbreviated by using lowercase *c* and *m* for continental and maritime, and uppercase *P* and *T* for polar and tropical as follows: continental polar (cP), maritime polar (mP), maritime tropical (mT), and continental tropical (cT). The source regions and the summer and winter characteristics of each of these air masses are summarized in Table 8.1.

Meteorologists employ two other temperature terms for additional accuracy in classifying air masses. They are *arctic* (abbreviated *A*) and *equatorial* (abbreviated *E*). Use of these terms gives rise to two additional air masses. One is continental arctic (cA) air, which is extremely cold and dry. In the northern hemisphere, it forms over northern Canada, Alaska, and Siberia only in the winter season. It is responsible for the occasional winter cold waves that affect most of the United States and Europe. In the southern hemisphere, it forms over the continent of Antarctica, where it is labeled *continental Antarctic* (cAA) air. The second air mass type is maritime equatorial (mE) air, which forms near the equator in the vicinity of the Intertropical Convergence Zone (ITCZ) and does not affect the middle latitudes. It is similar to maritime tropical air, but typically is somewhat warmer, moister, and less stable.

Air Mass Movement and Modification

Air masses are frequently transported out of their source regions by the upper-level winds. As the air

FIGURE 8.2 Typical views of the source regions of the four major air mass types, clockwise from upper left: cP—black spruce forest in interior Alaska; cT—sandy desert near Parachilna, South Australia; mP—the coast of Scotland near Glenborrodale; mT—beach scene at Grande Case, French Saint Martin (Caribbean). (Visuals Unlimited/© Steve McCutcheon; © Bill Bachman/Photo Researchers; R. Stottlemeyer, Michigan Technological Univ.; © Dick Davis/ Photo Researchers)

masses move, they bring their weather characteristics to new areas. Successive passages of air masses from diverse source regions are largely responsible for day-to-day changes in our weather conditions, and a basic task of weather forecasters is to predict accurately the direction and speed of these air mass movements. Although air masses rarely form in the middle latitudes, those latitudes are constantly invaded by air masses from both high and low latitudes. Indeed, the conflict between dissimilar air masses makes the weather of the middle latitudes more variable on a day-to-day basis than that of either the high or low latitudes. The boundaries between air masses are indicated on maps by weather fronts, and the differences in temperature and humidity conditions attending a frontal passage result largely from the change in the controlling air mass.

An air mass moving out of its source region gradually is modified as it passes over surfaces of differing temperature and moisture characteristics. Topographic barriers also may remove moisture through the orographic lifting process and may generate vertical wind currents that mix air in the lower levels. Just as the air mass took several days to gain its char-

TABLE 8.1 Air Mass Characteristics

	CONTINENTAL POLAR (cP)	MARITIME POLAR (mP)	MARITIME TROPICAL (mT)	CONTINENTAL TROPICAL (cT)
Source Regions	Continental interiors in high latitudes	Oceans in subpolar regions	Oceans in tropics and subtropics	Continental interiors in dry portions of tropics and subtropics
Basic Attributes	Cold and dry	Cool and moist	Warm and moist	Warm to hot and very dry
Associated Summer Weather in Middle Latitudes	Pleasantly cool, low humidities, excellent visibilities, scattered cumulus	Cool to mild, moderately high humidities, mostly sunny in lowlands, showers likely in mountains	Uncomfortably warm and humid, hazy conditions inland, considerable cumulus cloudiness, scattered showers and thunderstorms	Very hot and dry, good visibilities, scattered daytime cumulus, very rare showers or thunderstorms
Associated Winter Weather in Middle Latitudes	Cold, dry, and often windy, excellent visibilities; clear to scattered or broken cumuliform cloudiness, snow showers likely in mountains or leeward shores of warm water bodies	Chilly and damp, frequently windy, considerable cloudiness, rain or snow showers especially in mountains	Mild and moist, fog or low stratiform cloudiness common especially night and morning hours; substantial precipitation along frontal boundaries	Does not reach middle latitudes

acteristics, however, it normally takes several days to lose them. As a result, when it arrives over a new location, an air mass has temperature and moisture values that reflect the conditions both of the source region and of the areas over which it has passed during its journey. The degree of modification that takes place depends on the speed at which the air mass moves, the distance it travels, and the extent of the dissimilarity between the areas over which it passes and its source region. As an illustration of the modification of a southward moving cP air mass, note the progressively warmer temperatures that exist from Canada southward to Texas in Figure 8.10 on page 159. (The temperature in degrees Fahrenheit for each reporting station is to the upper left of the circle that indicates the station.)

Eventually an air mass is so extensively modified that it becomes in effect a different air mass. The formation, movement, modification, and dissolution of air masses is a continuous process that produces a constantly changing and never exactly repeated pattern of weather conditions over the Earth's surface.

SECONDARY CIRCULATION SYSTEMS OF THE MIDDLE AND HIGH LATITUDES

The global air pressure and wind belts, which play a crucial role in the distribution of the Earth's long-term climatic characteristics, are sometimes referred to as primary circulation systems of the atmosphere. While the primary systems exert their control over the climates of the world, a traveling and constantly changing array of smaller secondary circulation systems is responsible for the variations in daily weather conditions at a given site.

The atmosphere at any one time contains a large number of high and low pressure systems that vary in size, shape, and intensity. On occasion, pressure systems remain nearly stationary over one area for an extended period. More commonly, they actively move, or "build," across the Earth's surface at speeds ranging up to approximately 55 kilometers per hour (35 mph). In general, the pressure systems of the

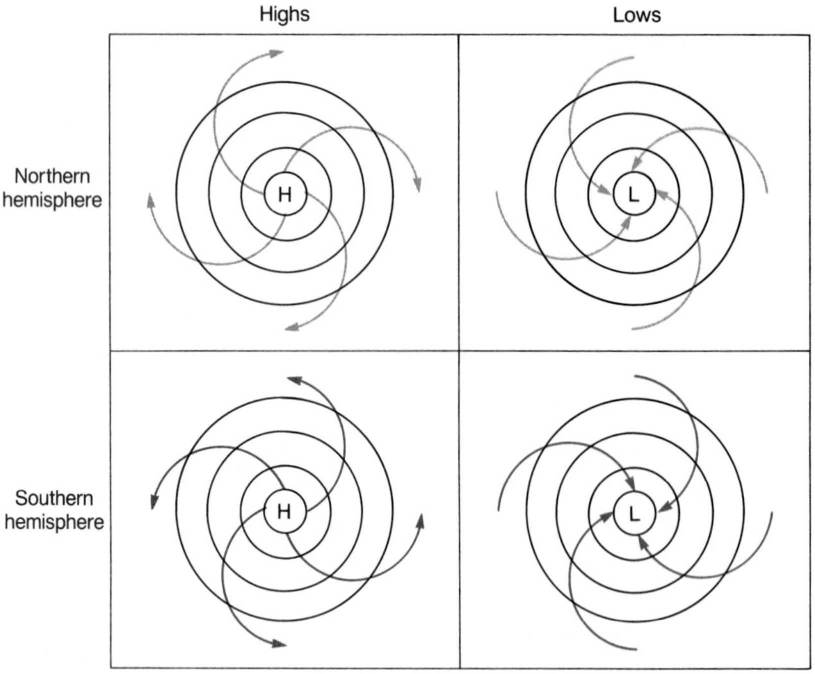

FIGURE 8.3 Wind-flow patterns for high and low pressure systems in the northern and southern hemispheres.

middle and high latitudes are better defined and move more rapidly than those of the low latitudes.

The direction and speed of motion of the pressure systems are largely controlled by wind currents in the upper atmosphere. With the passage of time, pressure systems form and disappear, strengthen and weaken, change size and shape, and vary in their speed and direction of motion. The Earth is thus covered by an ever-changing kaleidoscope of pressure patterns—one that contains an unlimited potential for variation and that never exactly repeats itself. The great potential for change as well as our lack of full understanding of the complex forces controlling not only pressure, but all weather systems, makes accurate weather forecasting a highly challenging task.

The changing patterns of secondary circulation systems are especially important in controlling the weather of the middle and high latitudes, which tends to be much more variable than that of the tropics. There are three major types of systems: traveling anticyclones (high pressure systems), weather fronts, and frontal cyclones (low pressure systems). Outside the tropics, all three are transported in a generally eastward direction by the upper-level steering currents, and each is associated with its own sequence of weather conditions. The causes, characteristics, and distributions of these weather systems will now be examined.

Traveling Anticyclones

A high pressure system, or **anticyclone**, can be described as a mound of air under higher pressure than the surrounding air. Highs develop where atmospheric conditions favor the upper-level convergence and subsidence of air. Because highs are generally located in the centers of air masses, they normally contain air that is relatively homogeneous in temperature and water vapor content.

High pressure systems have distinctive wind-flow patterns that differ somewhat between the northern and southern hemispheres because of the influence of the Coriolis effect. Highs in both hemispheres contain air that converges in the upper levels, subsides, and then diverges in the lower levels. In the northern hemisphere, the Coriolis effect imparts a clockwise spiral to these diverging winds, while in the southern hemisphere, it gives them a counterclockwise spiral (see Figure 8.3). The descent of the air results in the compression and adiabatic heating of the atmosphere. This, in turn, is responsible for both the high barometric pressure and for the fair weather normally associated with highs.

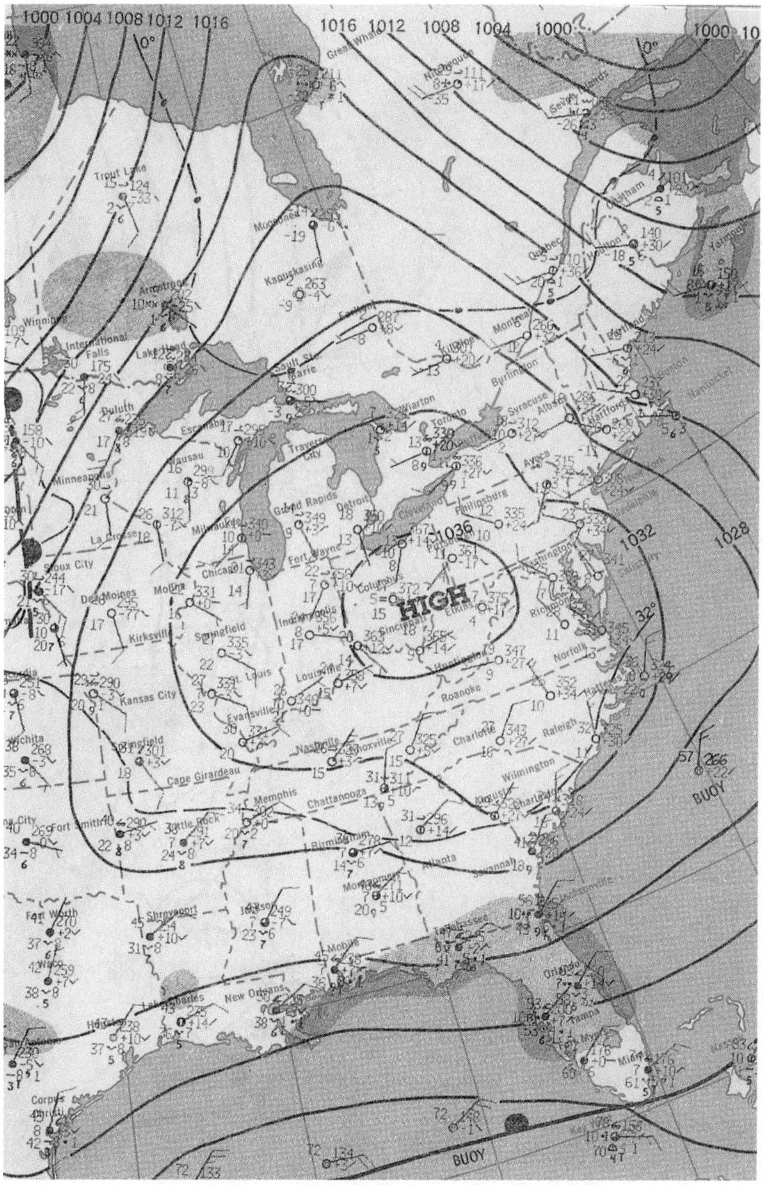

FIGURE 8.4 A portion of the surface weather map for March 28, 1982, showing a large Canadian anticyclone consisting of cP air centered over the central Appalachians. (NOAA)

Anticyclones affecting the United States generally come from one of three source regions. During the colder portion of the year, when weather systems typically are in active movement across the country, high pressure systems containing Pacific mP air masses frequently enter the western United States. As they cross the western mountains they dry out and are reclassified as cP air masses, but their air remains relatively mild for the time of year. Southward loops in the jet stream frequently cause cold cP or cA anticyclones from central or northern Canada to invade the United States during the winter season (see Figure 8.4). They may bring subfreezing conditions deep into the South and East. During the summer, the southeastern and south-central portions of the country normally come under the domination of a quasi-stationary poleward extension of the northern subtropical high. This Bermuda High, as it is called, contains warm, humid mT air derived largely from the Gulf of Mexico (see Figure 8.5).

An orderly sequence of weather conditions typically is associated with the approach and passage of a

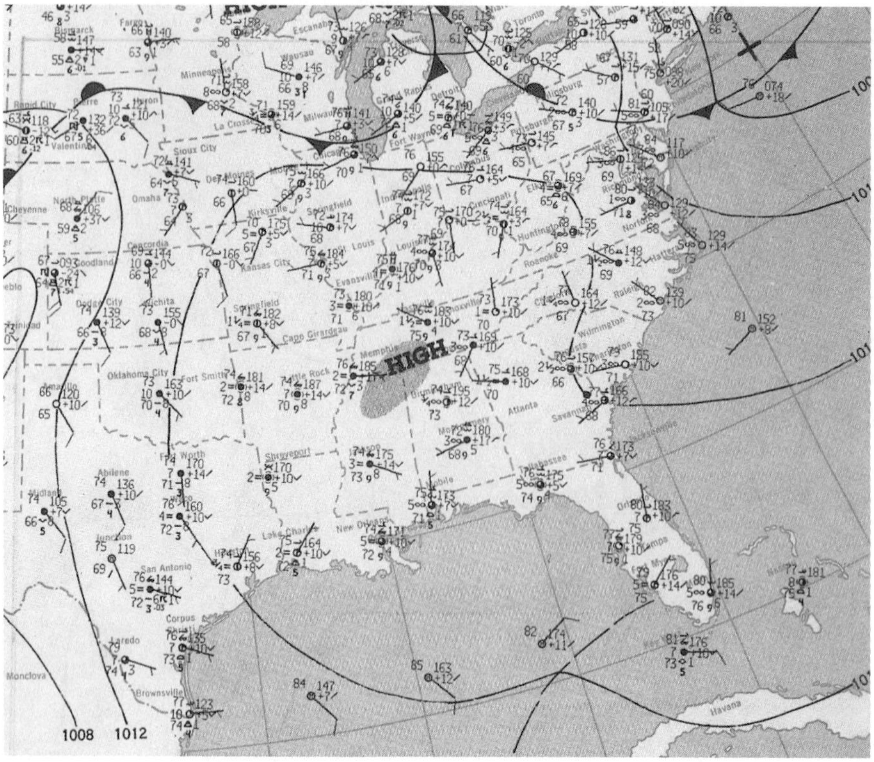

FIGURE 8.5 The surface weather map for July 17, 1983, showing a typical summertime "Bermuda high," composed of mT air, blanketing the Southeast. (NOAA)

cold Canadian anticyclone. As the cold front that precedes the high pressure system passes through the area, the weather turns cold, with blustery northwest winds and considerable cumuliform cloudiness. As the high pressure center approaches and the pressure gradient relaxes, the winds decrease in speed, the skies clear, temperatures remain cold, and visibilities are excellent. If the center of the high passes nearby, the winds become light and variable for a time, and then gradually turn to the south or southwest as the center moves to the east of the area. The change in wind direction, coupled with the gradual modification of the cP or cA air within the system, causes a warming trend. As the high continues moving eastward, an increase in cloudiness is likely as the next low pressure system, with its associated fronts, approaches.

Weather Fronts

A **weather front** is a boundary between adjacent air masses. It is a relatively narrow zone where air masses of different temperature and moisture characteristics come into contact. Because it is situated between the high pressure centers of two air masses, a front is a trough of low pressure. This low pressure causes air to flow from both air masses into the frontal zone. As the air converges, the warmer and lighter air mass is forced to rise over the cooler and denser air mass; this frequently leads to the formation of clouds and frontal precipitation.

The surface position of a front appears as a line on a weather map, but a front is actually a three-dimensional surface with height and width as well as linear extent. From its surface location, the frontal boundary extends aloft at a very gentle angle from the horizontal, so that it resembles an inclined ramp (see Figure 8.8). As a result, a few kilometers above the surface the front may be located several hundred horizontal kilometers from its surface position. This frontal inclination is caused by the differing densities of the two air masses, with the warmer air, which is lighter, always lying above the colder air. Fronts also have width because some mixing of the adjacent air masses takes place, and several hours may be required for the full transition of air masses to occur at a specific site following a frontal passage.

Fronts are especially numerous and important in the middle latitudes because this portion of the world

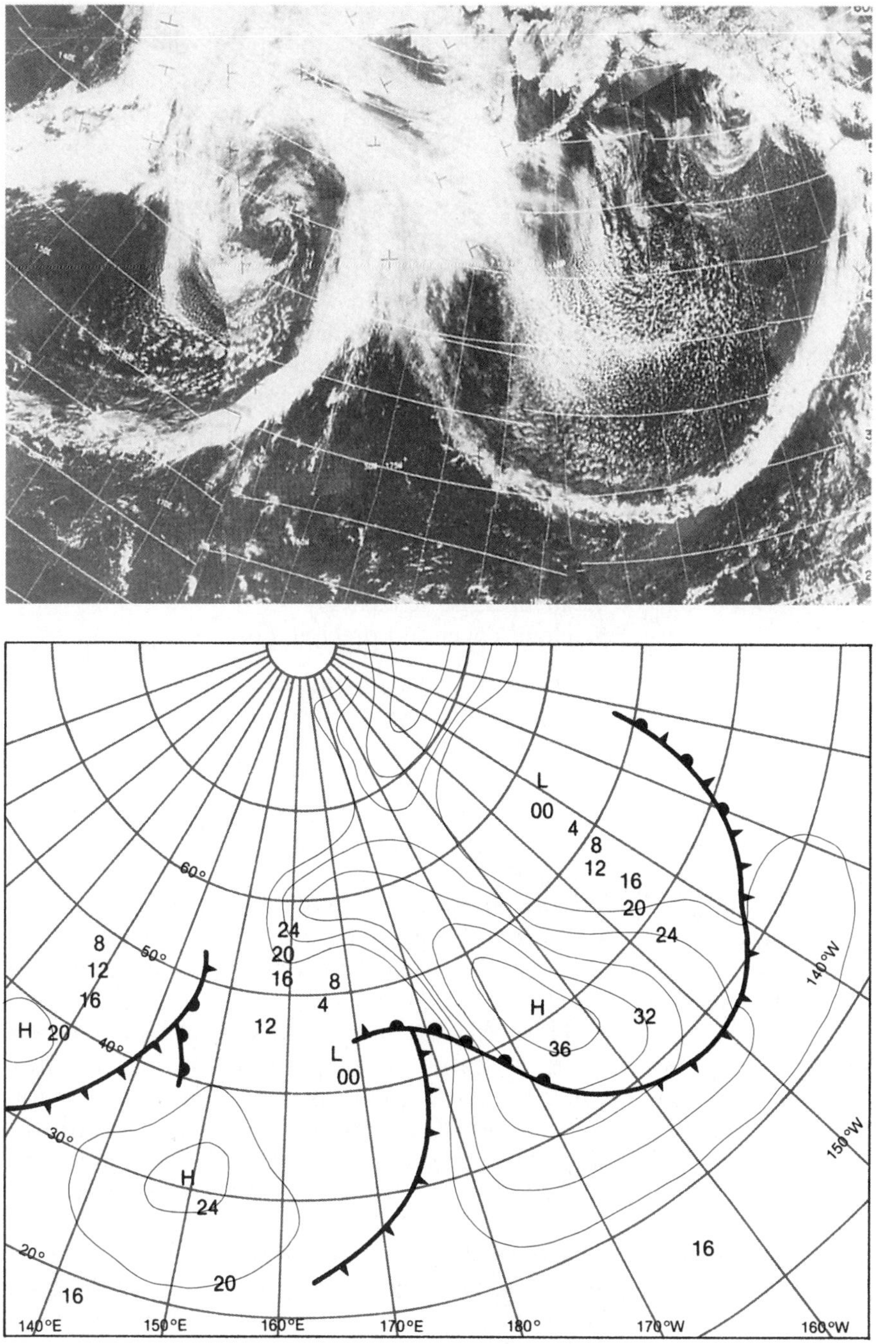

FIGURE 8.6 A pair of low pressure systems with their associated fronts over the North Pacific as depicted both by satellite and by surface weather map. The clouds in the upper photo are white, while the ocean surface appears black. (Photo courtesy of NOAA)

is within reach of both the cold air masses of the polar and subpolar regions and the warm air masses of the subtropics. The rapid weather changes for which the middle latitudes are noted usually are associated with frontal passages. These weather changes are most dramatic from late autumn through early spring, when the latitudinal variation in insolation, and therefore in temperature, is especially large. The stronger upper-level wind currents of the winter also transport fronts faster than they travel in the summer, allowing less time for modification and mixing of the opposing air masses.

TABLE 8.2 Idealized Frontal Weather Conditions*

COLD FRONT
Temperature: Warm before passage; progressively colder after passage.
Moisture: High humidity and dew point before passage; progressive lowering after passage.
Wind direction: Southwest winds before passage; northwest winds after passage.
Air pressure: Slow fall before passage; rapid rise after passage.
Clouds and precipitation: Brief period of heavy showers or a thunderstorm; rapid clearing after passage.
WARM FRONT
Temperature: Cold before passage; warmer after passage.
Moisture: High relative humidity but low dew point before passage; high relative humidity and high dew point after passage.
Wind direction: Easterly winds before passage; southwest winds after passage.
Air pressure: Rapid fall before passage; steady readings after passage.
Clouds and precipitation: Gradual increase and lowering of cloudiness followed by extended period of steady precipitation; slow clearing after passage.
STATIONARY FRONT
Weather: Extended period of cloudiness and light precipitation; temperature, moisture, wind, and air pressure remaining relatively steady.
OCCLUDED FRONT
Temperature: Cold before and after passage.
Moisture: High relative humidity but moderate dew point before passage; lowering relative humidity and low dew point after passage.
Wind direction: Southerly before passage; west to northwest after passage.
Air pressure: Falling before passage; rising after passage.
Clouds and precipitation: Gradual increase and lowering of clouds, followed by extended period of precipitation; short period of heavy precipitation at end, followed by clearing.

*Conditions described are typical, but do not occur in all cases. For example, weak fronts or fronts separating dry air masses may pass with no precipitation and little cloudiness. Wind directions are those typical of the northern hemisphere; the north-south directional component should be reversed for southern hemisphere fronts. Terms such as warm and cold are *relative* to those that normally occur.

The primary frontal boundary separating air masses of subpolar and subtropical origin (usually cP and mT air) is called the *polar front.* Although subject to large daily and seasonal fluctuations in position, it is generally located within the middle latitudes. This front is not straight, but develops large-scale waves that travel eastward along it, causing it to shift alternately northward and southward.

The passage of a front is usually attended by a number of changes in weather conditions. The extent and speed of the changes are subject to considerable variation, depending on factors such as the temperature contrast on opposite sides of the front and its speed of motion. These weather changes can be grouped into the following five categories:

1. temperature
2. atmospheric water vapor content
3. wind direction and speed
4. air pressure tendency
5. cloud cover and precipitation

By observing changes in these weather elements, a knowledgeable weather observer, even without sophisticated meteorological instruments, usually can tell when a front has passed (see Table 8.2).

Types of Fronts

Weather fronts, which separate warmer and colder air masses, are named according to which air mass, if either, is advancing. Four types of fronts are recognized.

The **cold front** is the most common and typically the most sharply defined of the four. It is a surface along which advancing cold air is displacing warm air. The cold air, being denser, wedges beneath the warm air, lifting it well above the surface (see Figure 8.7). If the warm air has sufficient water vapor, this uplift will produce frontal clouds and precipitation. Surface friction tends to slow the advance of the cold air near the surface, steepening the slope of the front to an angle that is usually between 1:40 (that is, 1 vertical unit to 40 horizontal units) and 1:80. This relatively steep frontal slope causes the warm air to rise rapidly, typically producing a brief period of heavy precipitation followed by rapid clearing. Fair, cooler, and drier weather may be expected over the next few days.

Along a **warm front,** advancing warm air is displacing cold air. Warm fronts tend to move consid-

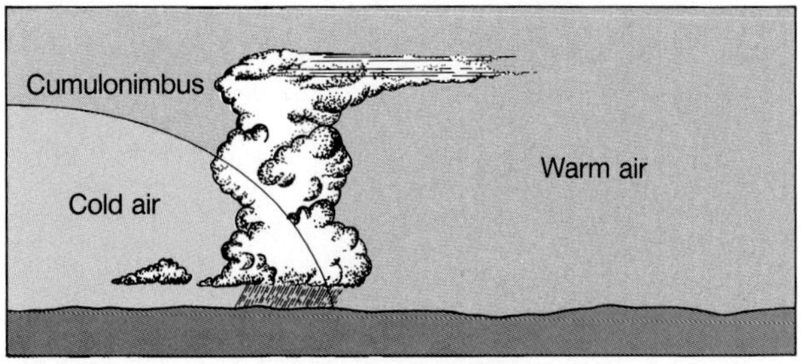

FIGURE 8.7 Profile (side) view of a cold front, showing the relative positions of the warm air, cold air, clouds, and precipitation. The direction of frontal motion is from left to right.

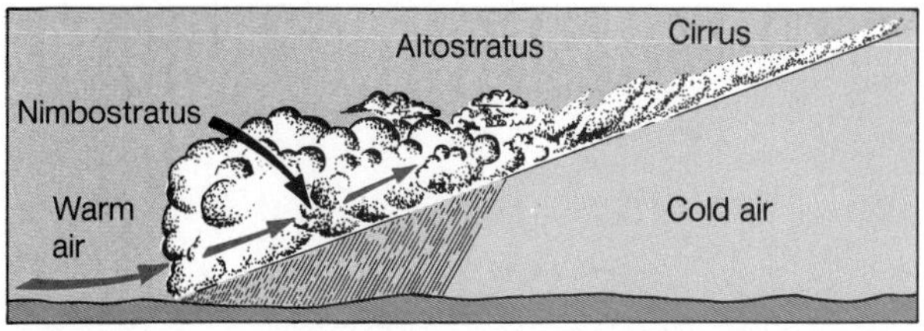

FIGURE 8.8 Profile view of a warm front. Note that a warm front, unlike a cold front, is tilted forward from its surface position. Most precipitation occurs on the poleward (cold) side of the front.

erably more slowly than cold fronts, because in this case the advancing air is lighter than the air it is displacing. The warm air rides up over the retreating wedge of cold air, causing the front aloft to be tilted in the direction of motion. (See Figure 8.8. Note that a cold front aloft is tilted in the opposite direction.) As the warm air rises, it is cooled adiabatically and, with sufficient moisture, clouds and precipitation again develop. By retarding the movement of the lower air, surface friction stretches and flattens the front, rather than steepening it. This produces a frontal slope of only 1:100 to 1:200. Because the warm air gradually slides up this gentle incline, warm frontal precipitation usually falls from nimbostratus clouds and tends to be lighter in intensity and longer in duration than cold frontal precipitation. If, however, the rising warm air is unstable, heavier showers and possibly thunderstorms may be interspersed within the area of precipitation.

The mean forward speed of cold fronts is about 40 kilometers per hour (25 mph). That of warm fronts is about 25 kilometers per hour (15 mph). Speeds average somewhat faster in winter than in summer because of the greater strength of the upper-level steering currents.

A **stationary front,** as the name implies, shows little or no present movement. Although the front itself may be stationary, warm air continues to drift up the frontal surface, producing the associated cloudiness and precipitation. Fronts that remain stationary for several days tend to weaken and eventually lose their identity as the opposing air masses gradually mix and moderate. On the other hand, a front that has been stationary frequently will begin to move as an imbalance between the opposing forces develops; as it moves, it becomes either a cold front or a warm front.

It should be noted that any of these three types of fronts can, by a change in its motion, become any of the other types. In fact, different portions of the same front can, because of their motions, be classified as different types of fronts. This situation is analogous to that of a battlefront between two opposing armies. On one day, the battlefront may move in one direction, and on the next day, a counterattack may cause it to move in the opposite direction. In addition, an army may be making gains by advancing the front in one location, while at the same time it is losing ground in another. Weather "fronts" received their name because these air mass boundaries, like battlefronts of World War I, are linear zones of conflict between opposing forces.

The fourth type of front differs from the others in that it is formed by the meeting of three, rather than two, air masses. It was noted earlier that cold fronts tend to move somewhat faster than warm fronts. On occasion, a cold front will overtake a warm front, lifting the intervening mass of warm air entirely above the surface (see Figure 8.9). The merging of the warm and cold fronts forms an **occluded front.** Although a newly formed occluded front may pro-

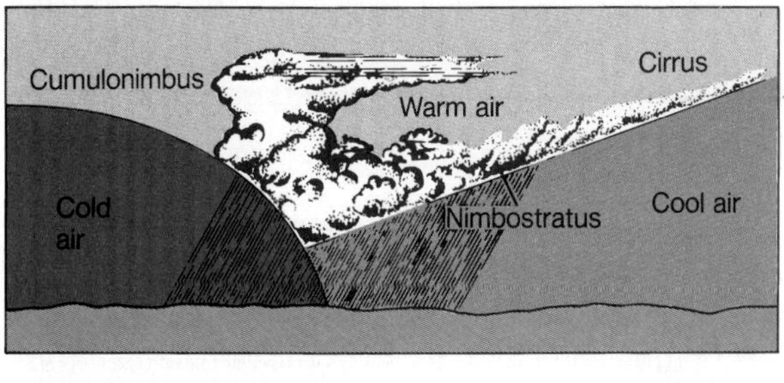

FIGURE 8.9 Profile view of an occluded front. These fronts are formed by the meeting of three air masses.

THE WEATHER MAP AS A FORECASTING TOOL

The modern science of forecasting meteorology is highly complex, and meteorologists employ a variety of sophisticated instruments to gather and analyze the tremendous quantity of weather data needed for accurate weather forecasting. Most weather analysis and prediction currently is performed by high-speed computers that can simultaneously digest and synthesize data from a large number of sources.

Despite this fact, a relatively old forecasting tool—the surface "synoptic" weather map—is still widely used by forecasters. Although it appears complex, a map such as that shown in Figure 8.10 chiefly indicates the current weather conditions at a large number of reporting stations in the United States and Canada. From this combined station data, large-scale weather patterns can be discerned and drawn on the map. These include the surface air pressure pattern (indicated by isobars with pressure units in millibars), the positions of highs, lows, and weather fronts, areas of current precipitation, and the locations of the freezing and 0° F lines. A person who can translate the weather information shown on such a map and who has a basic knowledge of weather systems often can make reasonably accurate short-term forecasts using only this source of data. Appendix B provides an explanation of the various map symbols used.

Before we try to forecast weather using Figure 8.10, note the following points:

1. Each reporting station is indicated by a circle. The amount of cloudiness is indicated by the proportion of the circle that is blackened in. The temperature in degrees Fahrenheit is given to the upper left of the station circle. The direction the wind is *coming from* is indicated by the "shaft" protruding from each station circle and the wind speed in knots is indicated by the number of "feathers" at the end of the shaft.
2. Areas of current precipitation on the map are shaded gray, and precipitation at any station is indicated to the left of each station circle. Dots indicate rain and stars indicate snow.
3. Weather fronts are shown as thick lines. Triangles on a front indicate a cold front, half-circles indicate a warm front, and an alternation of these symbols indicates a stationary front.
4. Weather systems in the middle latitudes typically travel eastward at an average speed of 30 to 45 kilometers per hour (20 to 30 miles per hour), or roughly 800 to 950 kilometers (500 to 600 miles) per day.

Let's now try to forecast the weather for the next 24 hours at a couple of representative cities:

- *Oklahoma City.* We see that Oklahoma City lies to the west of the cold front cutting through the center of the country. The sky is cloudy, the temperature is 46° F, and the wind is from the northwest at about 15 knots. We would expect the cold front to continue to move eastward, away from the city, and the large high pressure area containing cold cP air to approach from the northwest. Accordingly, we would forecast that the sky will gradually clear and the temperature turn colder.
- *Pittsburgh.* Pittsburgh lies in the warm mT air mass to the east of the cold front. The sky is mostly cloudy, the temperature is 58° F, and the wind is light and southerly. We would expect the cold front to pass through the Pittsburgh area within the next 24 hours, placing the city in the cP air mass. The forecast, then, should go something like this:

duce substantial precipitation, the warm air that supplies most or all of this moisture has been cut off from its surface source of supply. As a consequence, precipitation soon diminishes as the remaining moisture is exhausted. Occluded fronts are weak fronts, since at the surface they separate two cold air masses. Weak fronts tend to become weaker as the opposing air masses mix and modify; they usually lose their identity within a few days.

Frontal Cyclones

The importance of weather fronts is not solely the result of their associated precipitation and the

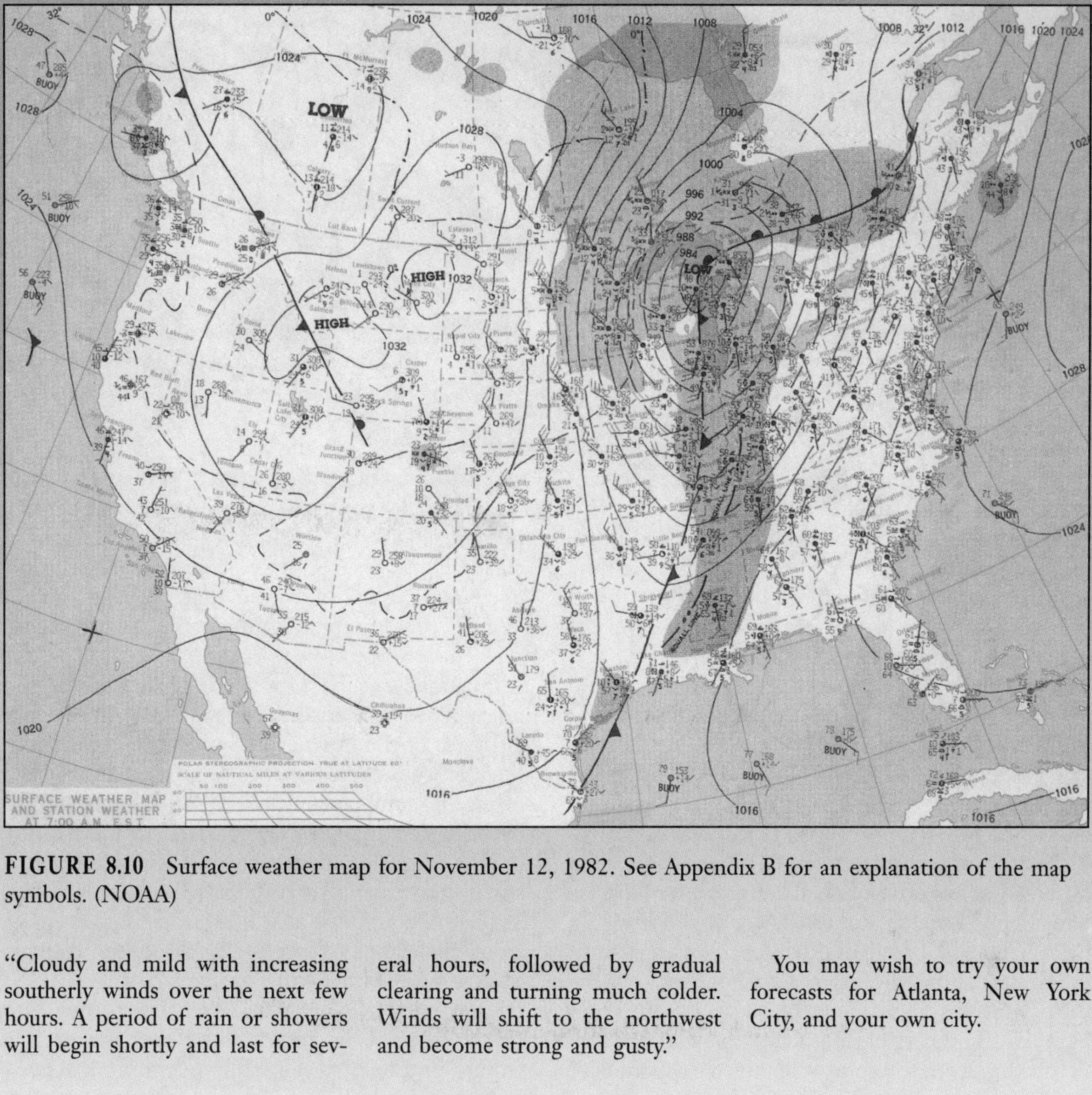

FIGURE 8.10 Surface weather map for November 12, 1982. See Appendix B for an explanation of the map symbols. (NOAA)

"Cloudy and mild with increasing southerly winds over the next few hours. A period of rain or showers will begin shortly and last for several hours, followed by gradual clearing and turning much colder. Winds will shift to the northwest and become strong and gusty."

You may wish to try your own forecasts for Atlanta, New York City, and your own city.

changes in temperature and humidity that accompany their passage. They also produce conditions favoring the development of **frontal cyclones.** These storms are among the most important secondary circulation features of the atmosphere and are the dominant type of weather disturbance in both size and significance for areas outside the tropics. Frontal cyclones produce most of the large storms of rain and snow in the middle and high latitudes. Their low air pressures cause diverse air masses to converge, often setting the stage for severe weather. In addition, they transport thermal energy from the surface to the upper atmosphere and help to carry oceanic moisture to the continents.

Formation and Developmental Cycle

A weather front provides a favorable setting for the development of cyclonic disturbances because it is a trough of low pressure and therefore a zone of air convergence. Low pressure centers can form only in areas where convergence occurs. In addition, the presence of adjacent air masses of differing densities on opposite sides of a front provides a mechanism for releasing the heat of condensation of atmospheric water vapor through the process of frontal lifting.

The critical factor in determining whether a frontal cyclone will develop is the pressure and wind-flow pattern of the upper troposphere. Specifically, an area of wind divergence must exist above the area where the cyclone is forming. This is necessary because near the surface, any low consists of converging and rising air. As the air rises, it must have a place to go. If the air in the upper troposphere is not diverging, the rising air cannot spread out to make room for the continued upward flow of air. The updraft therefore will be blocked, and the surface low will fill and disappear. If, on the other hand, a diverging flow exists in the upper levels, air rising into the low will be able to flow outward to make room for more air from the lower levels. In order for a low to deepen, upper-level divergence must exceed surface convergence.

A zone of air convergence that forms a favorable site for the formation and development of frontal cyclones is the eastern flank of a trough in the subpolar jet stream (see Figure 8.11). The flow of the jet stream above the middle latitudes tends to develop equatorward-projecting troughs that move eastward. The pressure and wind patterns are generally most compressed at the axis of such a trough, creating a zone of upper-level air convergence on the trough's western flank and a zone of divergence on its eastern flank. Surface lows therefore tend to form on the eastern flanks of these troughs, ahead of the main trough axis. The converging airflow on the western flank of the trough causes the air to accumulate there. It can then subside to the surface and diverge to produce a high pressure system.

The cycle of formation and dissolution of a frontal cyclone can be compared to that of an ocean wave, because both tend to build, break, and dissipate. In Figure 8.12, various stages of the cycle are diagrammed as they would appear on a weather map. The entire process normally takes about four or five days to complete, so that the time interval between each diagram is approximately 20 to 24 hours. It should be stressed that both the diagrams and the description of the cycle in the next several paragraphs pertain to "model" cyclones, and that no two frontal cyclones are exactly alike. Nevertheless, the differences are more in degree than in kind, and most frontal cyclones go through the basic stages described here.

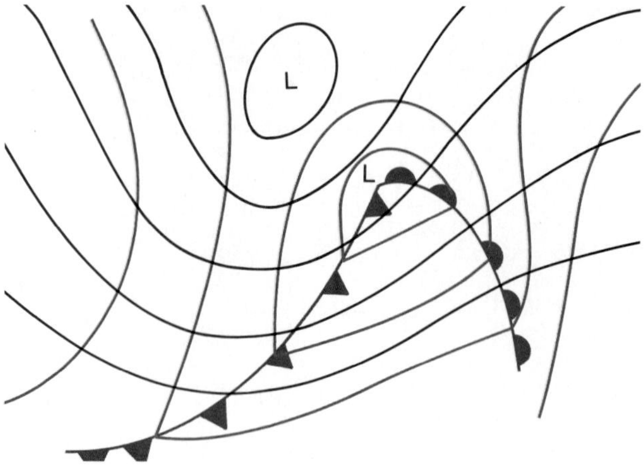

FIGURE 8.11 The typical position of a surface frontal cyclone in relation to an upper-level low pressure trough. Red lines are surface isobars and black lines are upper-level isobars.

In panel (a), a portion of the polar front extends in an east-west direction through an area in the middle latitudes of the northern hemisphere. This front delineates the boundary between cold cP air to the north and warm mT air to the south. The wind in the opposing air masses flows parallel to the front, with only a slight tendency toward convergence. This slight convergence is responsible for the presence of a zone of cloudiness and light precipitation along the front. A balance of forces exists between the two air masses, so that the front is stationary.

In panel (b), a bend or **wave** has formed on the front, perhaps in response to the approach of a trough in the jet stream. The position of the wave is indicated by both the actual bend or wave on the front and by the formation of a center of slightly lower barometric pressure. This low pressure begins to cause the winds on opposite sides of the front to blow inward, and the influence of the Coriolis effect on this converging air initiates a counterclockwise spiral. Because warm air to the east of the center of the wave flows northward into the front, it begins to move northward as a warm front. At the same time cold air to the west of the center is being directed

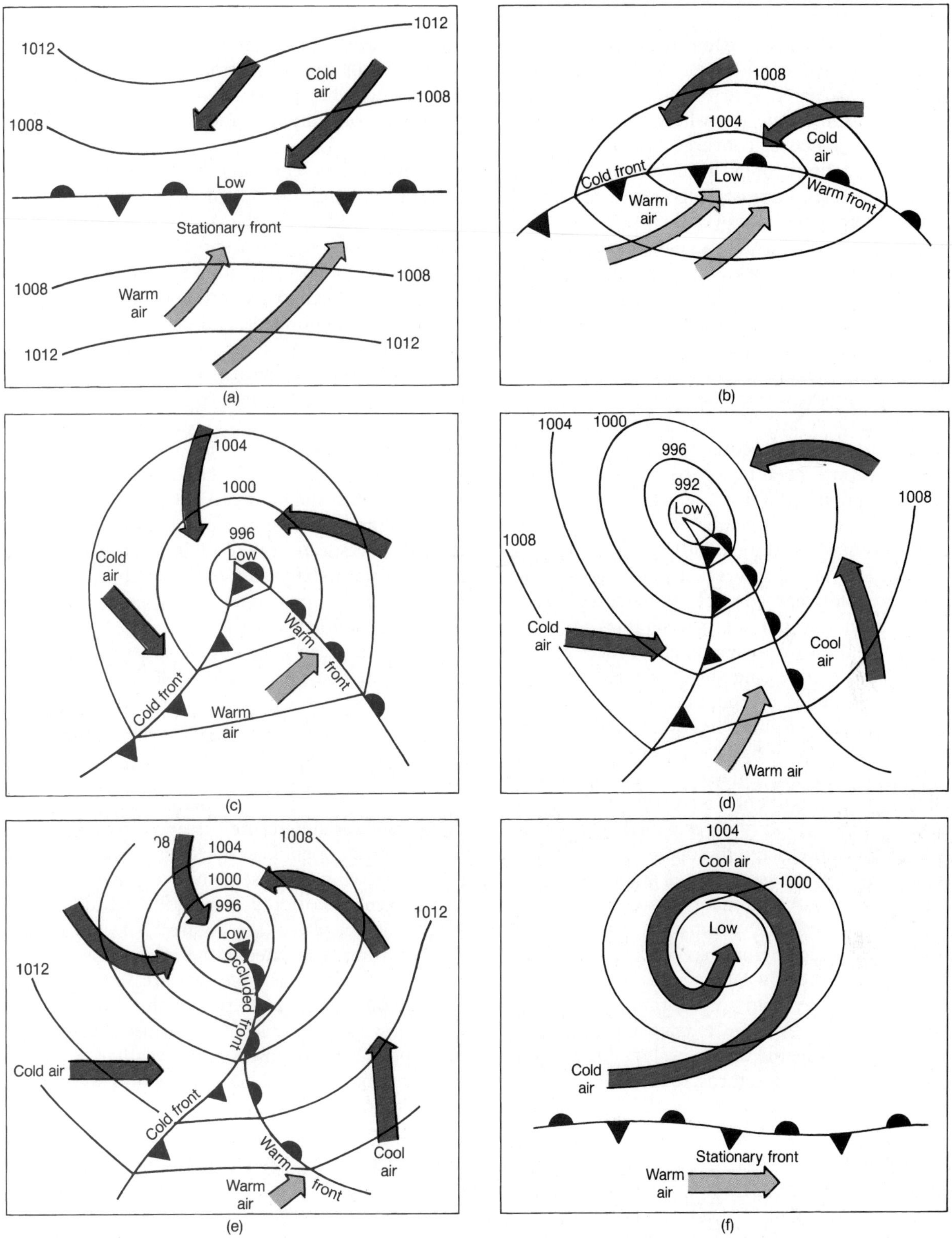

FIGURE 8.12 Six stages in the "life cycle" of a frontal cyclone. (See text for a discussion of each panel.) Air pressures are indicated in millibars.

southward, causing that portion of the front to advance as a cold front. The area of cloudiness and precipitation associated with the front undergoes a marked expansion as the wind-flow patterns of the two opposing air masses converge more directly on the front, increasing frontal lifting of the warm air mass. This, in turn, increases the release of heat of condensation available to power the low.

Panels (c) and (d) illustrate stages in the development of a mature open-wave cyclone (one that has not yet occluded). The developing wave is carried northeastward by the upper-level winds and continues to intensify as more and more heat energy is released by condensation and sublimation. This causes air pressures within the cyclone to fall and wind speeds to increase. The warm front, extending southeastward from the center of the low, advances northeastward, preceded by a broad band of precipitation. The trailing cold front, which extends southwestward from the low's center, advances southeastward at a somewhat more rapid rate, producing a narrow band of heavy showers and thunderstorms. Because the cold front advances more rapidly than the warm front, the mT air that supplies moisture and energy to the storm is gradually reduced in size to a narrowing wedge, and the storm center is increasingly surrounded by cold air.

In panel (e), the cold front has overtaken the warm front in the vicinity of the low pressure center to form an occluded front, and the process of occlusion is continuing at the point of intersection between the warm and cold fronts. As this occurs, the warm air is lifted from the surface, and its supply of water vapor is cut off. The low pressure center gradually begins to weaken as its source of energy is slowly depleted.

Panel (f) depicts the dissipating stage of the cyclone. The occluded front has weakened and lost its identity, and the low pressure center, now entirely surrounded by cold air, continues to fill and weaken. The polar front once again has stabilized, and as the last remnants of the cyclonic circulation die away, the situation returns to that depicted in panel (a). The stage is now set for the formation of another frontal cyclone that will repeat the pattern.

Frontal Cyclone Distribution and Movement

Frontal cyclones develop primarily within the middle latitudes. Their specific areas of formation vary during the course of the year because they are closely associated with the position of the polar front. The most vigorous storms develop during the winter and early spring, when the temperature contrast between air masses of subpolar and subtropical origins is greatest.

Frontal cyclones affecting the United States in winter most commonly develop in the North Pacific, in the Gulf of Mexico, along the Atlantic coast, and in the eastern Rockies and adjacent sections of the Great Plains (see Figure 8.13). During the summer, the major storm track lies well to the north, near the border of the United States and Canada, and the storms typically are much weaker due to the reduced temperature contrast between the opposing air masses. The frontal cyclones that develop in winter over the central and southern United States or along the Atlantic coast typically deepen and strengthen as they move northeastward. They eventually reach their peak development over the subpolar North Atlantic south of Greenland. In similar fashion, storms developing along the east coast of Asia move northeastward, often attaining massive proportions over the waters south of Alaska. In the southern hemisphere, the great temperature contrast between the frigid continent of Antarctica and the relatively mild adjacent ocean waters often causes that land mass to be surrounded by a ring of well-developed frontal cyclones.

Sequence of Frontal Cyclone Weather Conditions

The following paragraphs describe the weather conditions associated with the passage through our area of an idealized open-wave cyclone. It is assumed that we are viewing this sequence of meteorological events from a location in the middle latitudes of the northern hemisphere (see Figure 8.14). Because the weather fronts associated with such storms generally extend southward from the low pressure center, we will have the storm pass to our north.[1] This will allow both the warm and cold fronts to pass by, giving us the opportunity to observe the weather conditions associated both with their passage and with the air masses they separate.

We begin in Position 1 with respect to the storm. At this time we are still under the domination of a cP

[1] When a low passes by to the south (the equatorward side) of a site, the site remains in the cool air and typically experiences a prolonged period of precipitation and cool easterly winds.

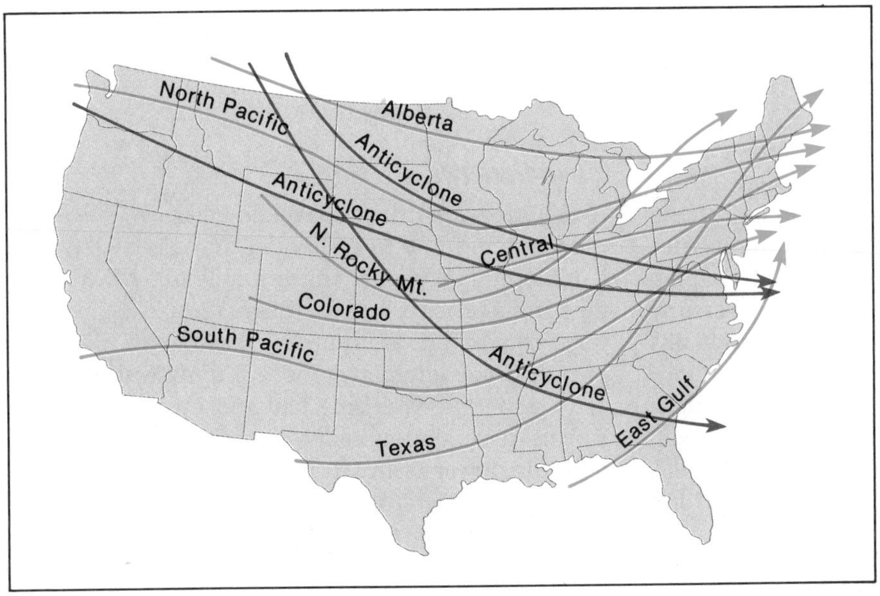

FIGURE 8.13 Typical winter tracks of cyclones (in blue) and of anticyclones (in red) across the United States. Note that while all systems track generally eastward, most cyclones have a poleward directional component, while anticyclones have an equatorward component of motion.

air mass, with high pressure centered to our northeast. The weather is clear, cool, and dry, with light easterly winds. As the low approaches us at a speed of about 40 kilometers per hour (25 mph), thin cirrus clouds gradually overspread the sky from the southwest, signaling the approach of the warm front, and the barometer begins to fall. During the following 12 hours, the clouds gradually thicken and become lower as they change first to cirrostratus and then to altostratus. At the end of this period a light rain begins to fall. (If the temperatures are cold enough, the precipitation may begin as snow and later change to sleet and finally to rain.)

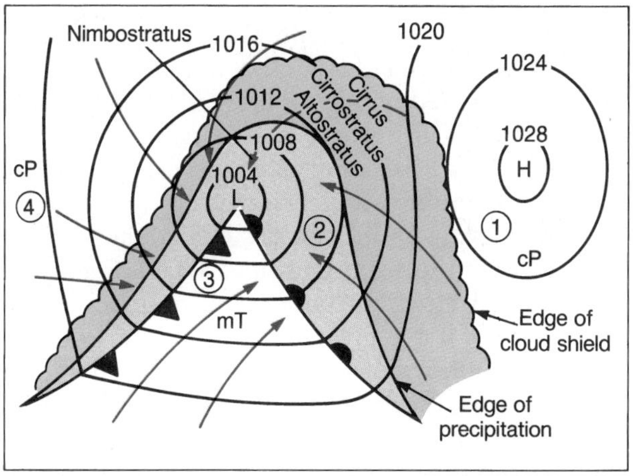

FIGURE 8.14 Idealized frontal cyclone, showing areas of cloud cover, and of precipitation (shaded area). The four station positions described in the text are indicated.

By the time the low has moved so that we are in Position 2, a cold, steady rain has been occurring for several hours and has become moderate in intensity. The wind has shifted gradually to the southeast and has become stronger. Visibilities have been reduced to a kilometer or less in fog. It is interesting to note that, although we are still in the cold air, the rain is actually coming from the overrunning warm air and is falling through the front into the colder air before reaching the surface.

A few hours later the rain tapers off and ends, followed within an hour or two by the gradual lifting of the fog and partial clearing of the skies. The wind slowly becomes southwesterly, and the temperatures become decidedly warmer. The barometric pressure stops falling and remains low and steady. By the time we are in Position 3, we are well within the warm mT air mass. Skies, although rather hazy, are clear except for scattered cumulus clouds, and winds are gusty from the southwest.

Within several hours the appearance of a line of towering cumulus and cumulonimbus clouds on the northwest horizon signals the approach of the cold front. These clouds soon cover the northern and western sky, which turns very dark and ominous. Distant lightning flickers, and the rumble of thunder can be heard. Then, with great rapidity, a line of very low, black, turbulent clouds approaches; the wind becomes northwesterly and very gusty; and the thunderstorm breaks upon us with heavy rain, vivid lightning, and possibly hail. The rain and winds soon slacken, however, and after an hour or so the rain ends completely. The sky clears rapidly, and in a few hours we are in Position 4, well within the new cP air mass. Only a few cumulus clouds are now present, visibilities have become excellent, and strong northwesterly winds prevail. The temperatures are falling steadily, humidities are quite low, and the air pressure is rising rapidly. Fair, cold weather with gradually diminishing winds may be expected over the next few days as the high pressure center of the cold air mass approaches and passes through the area. This will likely be followed by another frontal cyclone, bringing a repetition or variation of the preceding weather sequence.

Localized Storm Types of the Middle and High Latitudes

Frontal cyclones have a profound impact on the middle and high latitudes because of their large size, widespread geographical distribution, and frequency of occurrence. For most areas outside the tropics, they and their attendant fronts produce more precipitation than do orographic and convectional sources combined. In terms of their effects on people and property, though, two much smaller and generally more intense storm types, the thunderstorm and the tornado, are especially noteworthy. In actuality, all three storms are interrelated, because thunderstorms outside the tropics often are associated with frontal cyclones, and tornadoes are produced only by severe thunderstorms.

FIGURE 8.15 Time-lapse view of multiple lightning strokes illuminating the interior of an intense thunderstorm at night. (Dave Baumhefner/NCAR)

Thunderstorms

A **thunderstorm** is a convectional storm accompanied by lightning and thunder and associated with cumulonimbus clouds and unstable air. Most thunderstorms produce heavy precipitation, usually in the form of rain, but are quite localized, with diameters of only a few kilometers. Severe thunderstorms also are capable of producing hail and tornadoes. These storms have undoubtedly impressed more people with the power of the forces of nature and have created more general interest in the weather than has any other single meteorological phenomenon (see Figure 8.15).

The thunderstorm is by far the world's most common storm type. The key factor required for thunderstorm development is the vigorous convectional lifting of a deep layer of moist, unstable air. Conditions favoring thunderstorm formation occur frequently, and it is estimated that, at any given moment, an average of 1,800 thunderstorms are in progress over the Earth's surface.

Two general categories of thunderstorms are commonly recognized by meteorologists. **Air mass thunderstorms** are strictly convectional in origin and may form anywhere when the atmosphere is sufficiently warm, moist, and unstable. Over land areas, they are most numerous in the afternoon and early evening, when insolation has increased the temperature and lapse rate of the lower troposphere to its daily maximum values. Air mass thunderstorms are usually distributed within an area in a poorly organized fashion; currently it is impossible to predict accurately the precise localities they will affect on any given day. **Frontal thunderstorms** are triggered by the frontal lifting of unstable air. Unlike the air mass type, they tend to form in a line. This line may be located either along the front itself or at some distance ahead of it as a prefrontal **squall line.** Some of the world's most violent and destructive thunderstorms develop along squall lines that form about 240 kilometers (150 miles) in advance of strong cold fronts in the middle latitudes. The causes of squall line formation are complex, but are basically associated with the compression or "crumpling" of the warm air by the approaching cold air.

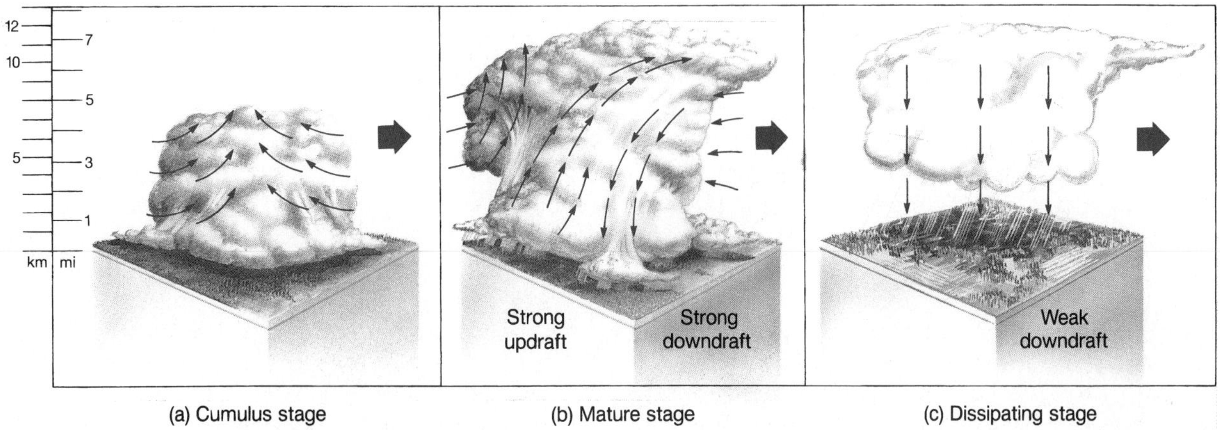

FIGURE 8.16 The three stages of a single-cell thunderstorm "life cycle." During the *cumulus stage*, the developing cell consists of a single strong updraft of air that has been heated by both the ground and by the released heat of condensation. No rain has yet reached the surface. During the *mature stage*, lightning and thunder begin, and heavy rain begins to reach the ground. The frictional drag of the heavy rainfall columns produces strong downdrafts that gradually choke off the initial updraft feeding the cloud. In the *dissipating stage*, the downdraft has spread throughout the base of the cloud. This causes the cloud's lower portion to evaporate as its upper ice-crystal portion gradually spreads out. Rainfall tapers off and eventually ends.

Both air mass and frontal thunderstorms are similar in structure. Each is composed of one or more **cells** that originally form from a roughly circular updraft of air several kilometers in diameter. Larger thunderstorms or those that have formed into a line are composed of multiple cells located adjacent to one another. The life cycle of a thunderstorm cell, which typically lasts less than an hour, is illustrated in Figure 8.16.

Thunderstorms occur in nearly all parts of the world, but vary greatly in frequency in different areas. Figure 8.17 illustrates two major aspects of their global distribution. The first is that the frequency of thunderstorms is correlated with latitude. They are most numerous near the equator and become progressively less common poleward. This distribution is related to that of both global mean temperatures and specific humidity values—an indication of the great importance of heat and moisture to thunderstorm development. The second distributional aspect indicated by this figure is that thunderstorms are far more common over land than over water. This is primarily a thermal factor. Land areas attain maximum temperatures considerably higher than those experienced over water, making the air correspondingly lighter and more unstable. Orographic lifting also can trigger thunderstorms when the air is unstable. This lifting mechanism is, of course, restricted to land areas.

The global pattern of thunderstorms is shown in more detail in Figure 8.18. As can be seen, the continent of Africa leads the world in the frequency of thunderstorms, followed closely by the world's wettest continent—South America. Thunderstorms oc-

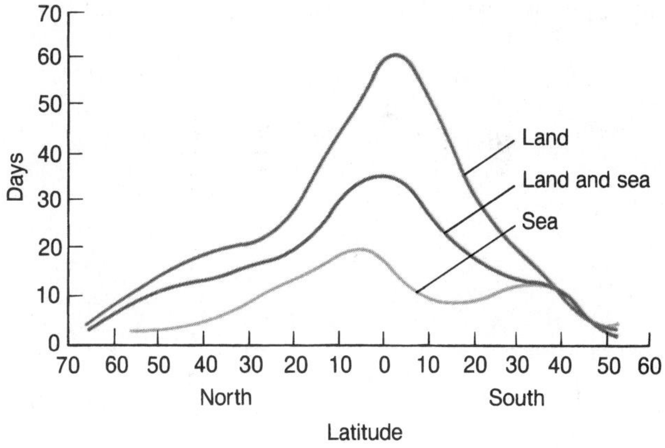

FIGURE 8.17 Latitudinal distribution of mean annual thunderstorm days for land areas, water areas, and the Earth's surface as a whole. Note that thunderstorms are most common near the equator and are also far more frequent over land than water.

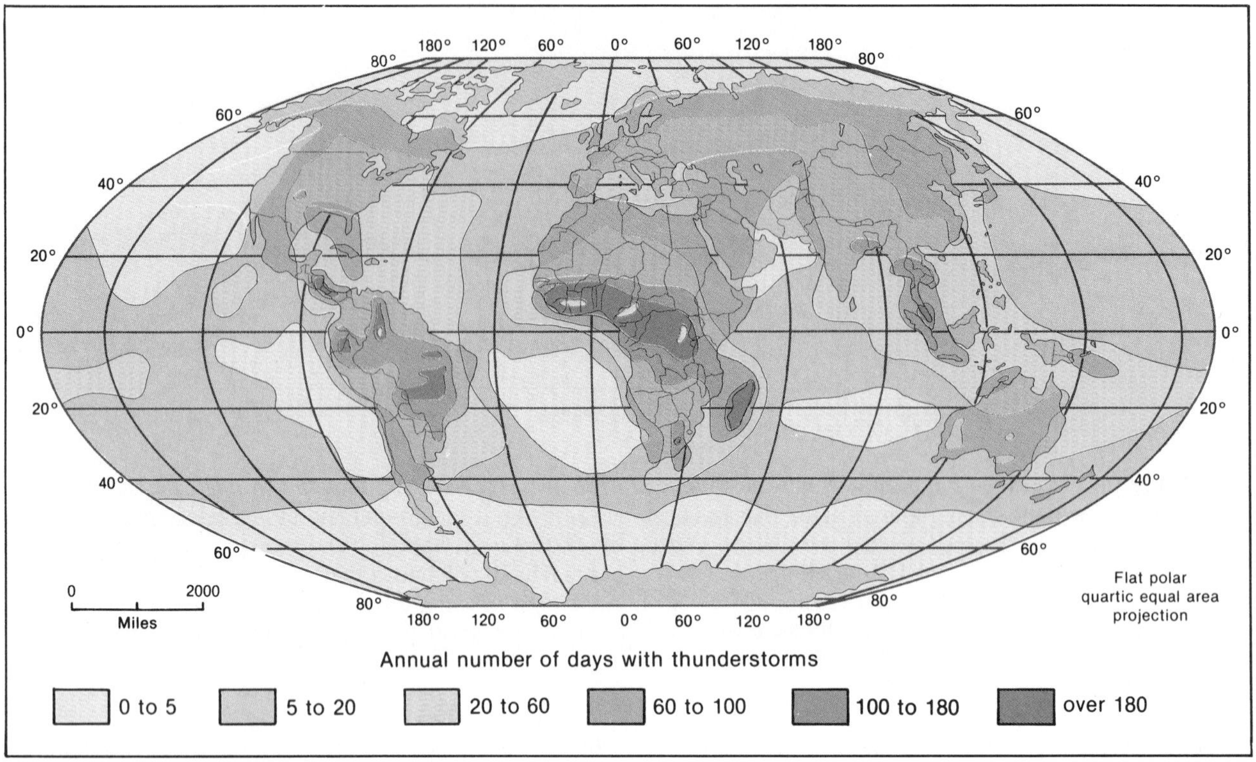

FIGURE 8.18 Global distribution of thunderstorm days.

FIGURE 8.19 Mean annual number of thunderstorm days in the United States.

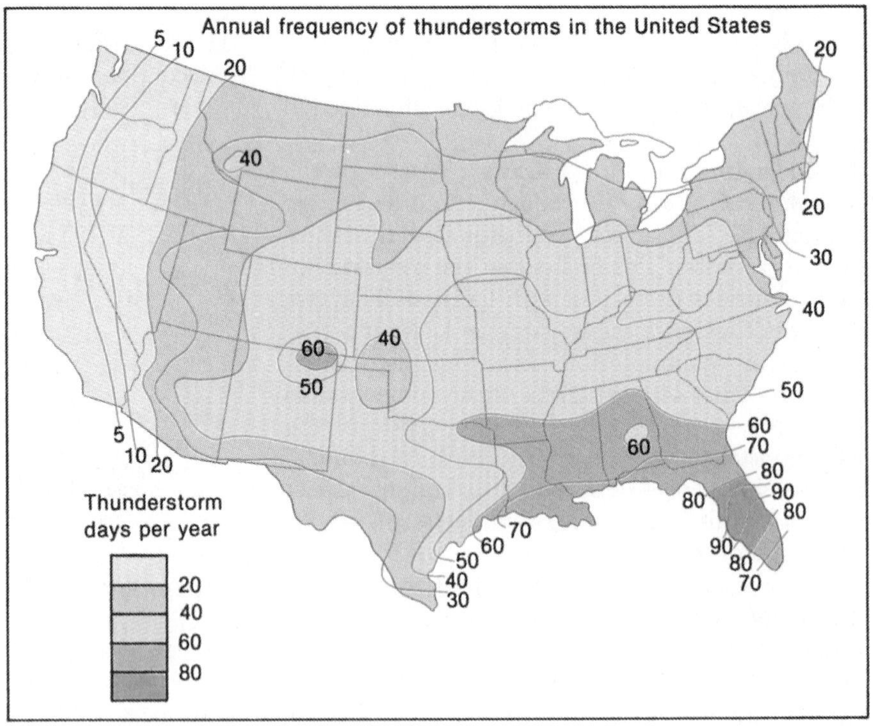

cur almost daily in many parts of the tropics and provide much of the rainfall received in that part of the world. In the middle and higher latitudes, thunderstorms occur primarily during the late spring and summer, when insolation is sufficiently strong to make the air unstable. Most of the middle latitudes have an intermediate frequency of thunderstorms, while the high latitudes, some desert regions, and areas dominated by cold ocean currents have the lowest numbers.

In the United States, thunderstorms are most common in the Gulf Coast area, especially in central Florida, where strong solar heating in summer induces the convergence of unstable mT air from both the Atlantic and the Gulf of Mexico (see Figure 8.19). They are least frequent along the Pacific coast, where the cold California current keeps the air cool and stable throughout most of the year.

Tornadoes

The **tornado** is by far the smallest of the world's major storm types, but it also is the most violent. It takes the form of a tubular vortex of whirling air surrounding a central core of extremely low pressure that extends downward from the base of a cumulonimbus cloud (see Figure 8.20). In a sense, the tornado should not be considered a separate storm type. Along with lightning, thunder, and hail, it is a phenomenon uniquely associated with thunderstorms. Because the specific causes, characteristics, and world distributions of tornadoes differ from those of thunderstorms, however, they are treated here as a distinctive storm type.

Most tornadoes form in thunderstorms associated with squall lines that have developed in advance of strong cold fronts in the middle latitudes (see Figure 8.21). The atmosphere must be highly unstable, giving the air a tendency to rise rapidly. For this reason, tornadoes are most frequent during the spring, when a large temperature contrast exists between opposing mT and cP air masses. In addition, air in spring heats much more rapidly near the surface than it does in the upper troposphere, producing steeper lapse rates than are typical during the summer. The tornado season begins in March in the Gulf Coast area and advances slowly northward with the retreat of the polar front, reaching the vicinity of the Canadian border in June or July.

A tornado apparently is initiated in the middle levels of a mature cumulonimbus cloud when a vigorous updraft begins to rotate, producing a **mesocyclonic** circulation. This occurs as updrafts feeding into the cloud are given a cyclonic spiral by wind-shear conditions (opposing directions of wind flow) at different altitudes. The mesocyclone gradually extends downward through the cloud and narrows, producing an increase in its rotational speed. Eventually, the cyclonic vortex protrudes from the base of the cloud as a **funnel cloud,** which becomes a tornado if it reaches the ground.

FIGURE 8.20 A tornado swirls up a cloud of soil at its base as it moves across North Dakota farmland in July, 1978. (Edi Ann Otto)

A tornado funnel generally tapers from cloud to ground somewhat like an elephant's trunk. It moves along the surface at speeds usually of 30 to 70 kilometers per hour (20 to 45 miles per hour). Typical diameters of tornado funnels at the surface are 100 to 600 meters (300 to 2,000 ft), although diameters of more than 1,500 meters (1 mile) have been observed

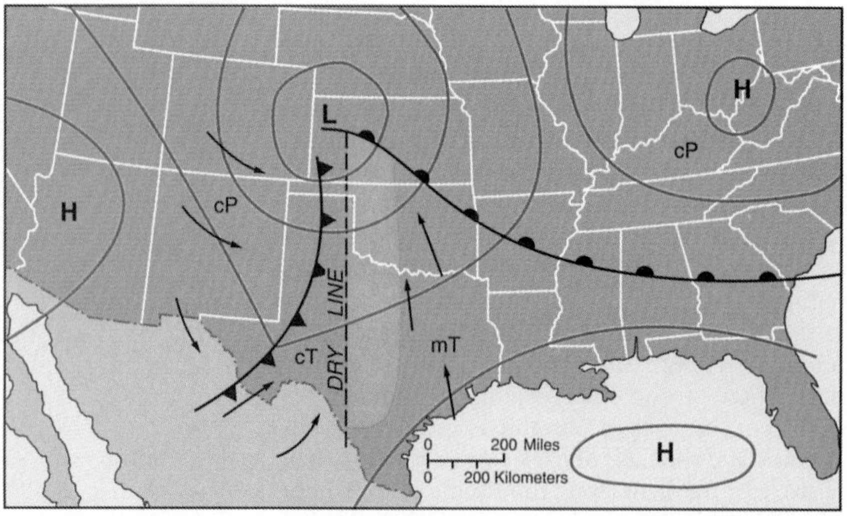

FIGURE 8.21 An outbreak of severe weather, including possible tornadic activity, is most likely to occur in the area shown here in orange. It precedes the "dry line" (the boundary between mT and cT air masses) ahead of a cold front in the southern Great Plains.

(see Figure 8.22). Wind speeds associated with tornadoes are difficult to determine with any accuracy, but currently are believed to range from about 22 to 135 meters per second (50 to 300 miles per hour).

The paths of tornadoes frequently are highly erratic and vary greatly in length. On the average, tornadoes are on the ground for a distance of only about 7 kilometers (4 miles) over a time span of perhaps 10 minutes. The record in this respect is held by a tornado that, on May 26, 1917, carved a path through Illinois and Indiana that was 470 kilometers (292 miles) long. It was on the ground for more than seven hours.

The great destructiveness of tornadoes is attributable to several factors. One is the winds of the tornado funnel, which are the most powerful of any storm type on Earth. These winds are capable of blowing down buildings, leveling swaths of trees, and even lifting railroad locomotives from their tracks. Another tornado attribute that can have a deadly destructive effect on large structures is the wind shear within the funnel, which can literally rip buildings apart. A third factor, which is largely responsible for the substantial death toll annually associated with these storms, is their unpredictability. The extremely localized nature of tornadoes, coupled with their erratic movement, rapid formation, and generally short spans of existence, make accurate tornado forecasts extremely difficult. Although the capability of reliably pinpointing specific areas that will experience tornadoes lies well in the future, the recent development of Doppler radar holds considerable promise. Doppler radar, which can remotely measure and dis-

FIGURE 8.22 The narrow but destructive path of a tornado that sliced through Edmond, Oklahoma, on May 8, 1986. (Chris Johns © 1987 National Geographic Society)

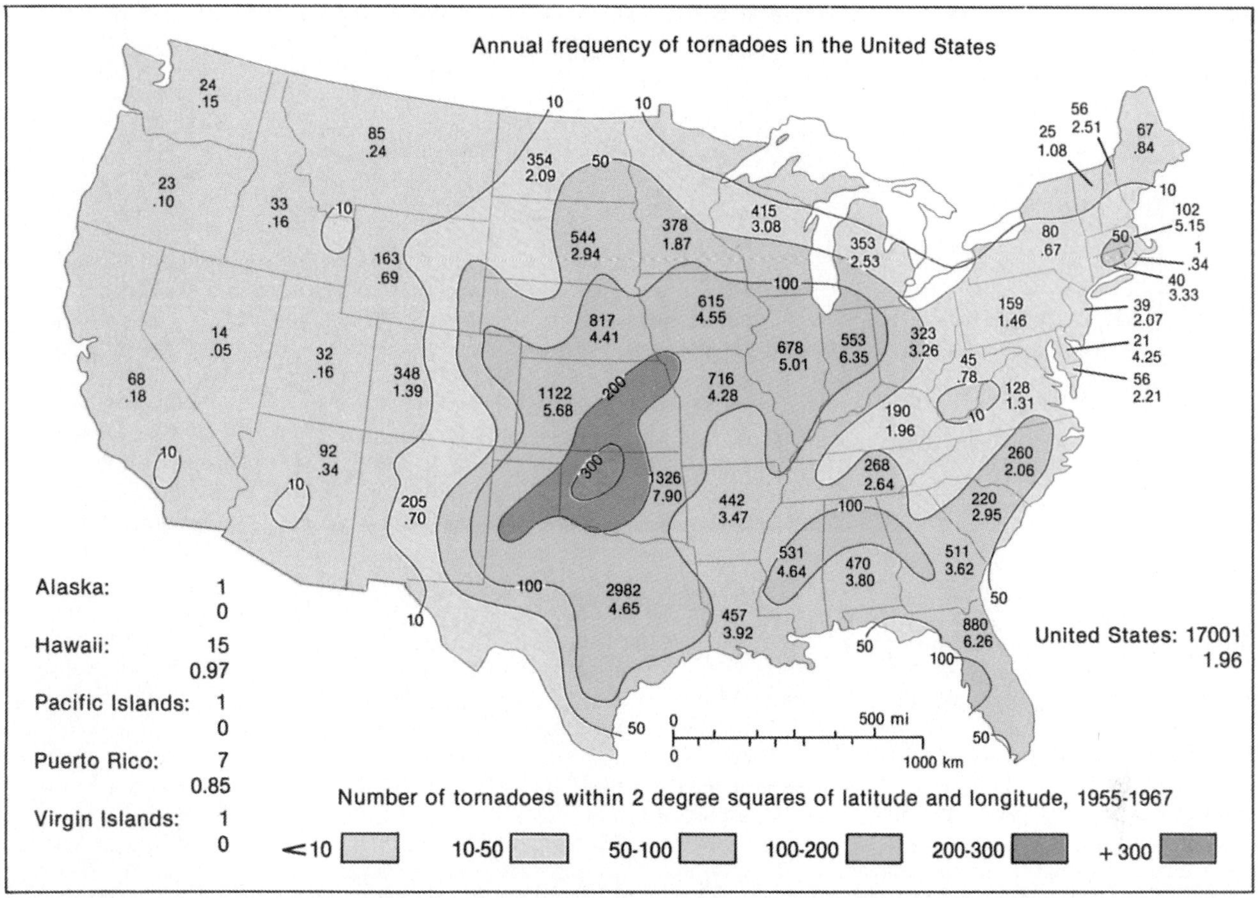

FIGURE 8.23 Distribution of tornadoes within the United States. The upper number in each pair indicates the total number of tornadoes within 2-degree squares of latitude and longitude between 1955 and 1967. The lower number represents the average annual number of tornadoes per 10,000 square miles.

play the speed of winds moving toward or away from the radar site, can detect larger-scale (mesocyclonic) rotary circulations developing within severe thunderstorms up to 20 minutes before they form tornadoes.

Tornadoes are largely restricted to land areas and are experienced primarily within the middle latitudes, where air masses of subtropical and subpolar origin are most likely to collide. Most tornadoes occur in the United States, which, with an annual average of over 700 tornadoes, has the dubious distinction of experiencing more of these storms than the rest of the world combined. The concentration of tornadoes within the United States results largely from its geographic location between major source regions of both cP and mT air, coupled with an absence of east/west-oriented mountain barriers. These two contrasting air masses are often drawn into contact by the circulation systems of traveling frontal cyclones, sometimes producing major outbreaks of dozens of tornadoes within a one- or two-day period.

The distribution of tornadoes within the United States is shown in Figure 8.23. The area of greatest frequency is the southern Great Plains, especially the states of Texas, Oklahoma, and Kansas. This area is sometimes referred to as "Tornado Alley." Tornadoes also are quite common throughout the southeastern states. Within any given region, tornadoes tend to be more common in flat, treeless areas than in hilly or forested topography that offers higher surface friction.

TROPICAL WEATHER

The subtropical belts of high pressure serve as a sort of meteorological barrier, dividing the Earth into two zones that are similar in size, but that display marked

THE TORNADO SUPEROUTBREAK OF APRIL 3–4, 1974

When large-scale clashes between maritime tropical and continental polar air masses take place over the central United States, the stage may be set for major tornado outbreaks. One such outbreak of at least 60 tornadoes that occurred on February 19, 1884, from Mississippi through Kentucky to Virginia, took as many as 800 lives. Another, the so-called Palm Sunday outbreak of 31 tornadoes on April 11, 1965, produced 256 fatalities. In terms of sheer magnitude, however, the tornado "superoutbreak" of April 3–4, 1974, is the largest on record.

During the 16-hour period from 2 P.M., April 3, to 6 A.M., April 4, a total of 148 tornadoes formed in advance of a strong cold front moving through the Ohio and Tennessee River valleys. Most of the storms developed along three squall lines well in advance of the front and, under the influence of strong upper-level winds, raced toward the northeast at 65 to 100 kilometers per hour (40 to 60 miles per hour). Not only did this outbreak contain more tornadoes than any other on record, it affected a larger geographical area than had any before. The total path length of all tornadoes combined was estimated at 3,200 kilometers (2,000 miles), about half the average annual total path length for the entire country in a year.

Hardest hit was the town of Xenia, Ohio (population 27,000). At approximately 4:40 P.M. a tornado struck the town and, in a five-minute period, carved a path a kilometer (0.6 mile) wide and 5 kilometers (3 miles) long. The storm damaged or destroyed 2,400 homes as well as most of the business district, killed 32 people, and injured nearly 600 others.

The paths of the 148 tornadoes that are known to have occurred during the superoutbreak were mapped by University of Chicago meteorologist T. Theodore Fujita (see Figure 8.24). The Xenia tornado has path number 37. The very small numbers appearing along the paths of each of the tornadoes are their estimated **Fujita Scale** intensities at those locations. The Fujita scale is a numerical scale that rates the severity of a tornado from a minimum of 0 (weak) to a maximum of 5 (devastating). Tornado 37 had a Fujita rating of 4 as it passed through Xenia.

differences in weather and climate. On the equatorward side of the subtropical highs, the weather is much less susceptible to dramatic change, especially with respect to temperatures. The tropics also experience their own storm types, which are triggered much more by the instability of the intensely heated atmosphere than by the conflict between opposing air masses.

General Weather Characteristics of the Tropics

When examined over an extended period, several important differences in weather and climate between tropical and nontropical regions become apparent. Nearly all the differences in the following list result from the year-round receipt by the tropics of large quantities of insolation.

1. Temperatures are warmer on average for the year as a whole in the tropics than in the higher latitudes. No cold season occurs, so annual temperature ranges tend to be small and often are exceeded by daily temperature ranges.

2. Day-to-day temperature fluctuations are small and usually are related to variations in cloud cover. The constancy of temperatures exists largely because cold fronts do not normally penetrate into the low latitudes. In fact, genuine weather fronts of any type are rare within the tropics, and mT, mE, or cT air masses permanently control most areas.

3. The chief difference between summer and winter, if indeed there is any significant difference at all, is in cloudiness and precipitation, rather than in temperature. Stated another way, while temperature is the leading seasonal climatic variable in the higher latitudes, precipitation is the leading sea-

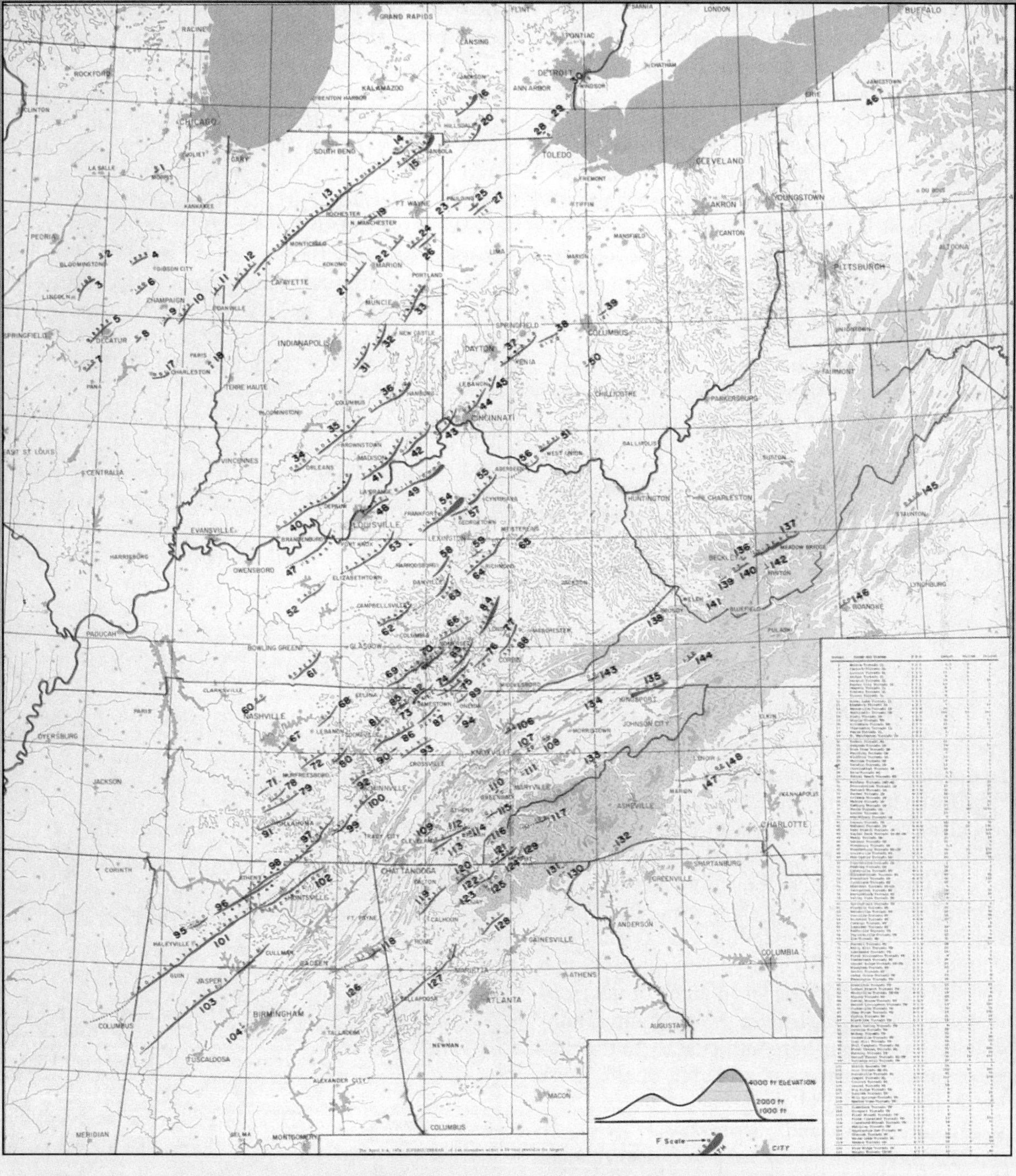

FIGURE 8.24 Paths of the 148 tornadoes that formed during the "superoutbreak" of April 3–4, 1974. (Reprinted by permission of T. Theodore Fujita)

sonal climatic variable in the tropics. In most locations, summer is the wettest season and winter is the driest.

4. While most precipitation in the higher latitudes is generated by frontal lifting, most tropical precipitation is convectional. As a result, it tends to be showery, localized and often intense.

5. Frozen moisture forms, such as frost, snow, and sleet, do not occur in most tropical areas because of a lack of subfreezing temperatures. The only natural form of ice ever seen by many residents of the tropics is hail.

6. Although the prevailing direction of movement of wind and weather systems is from west to east in most nontropical areas, it is from east to west within the tropics. The weather systems of the tropics tend to move relatively slowly as well, because the upper tropospheric steering currents are weaker, on the average, than those of the higher latitudes.

All in all, the weather of the tropics is much more constant than that of the higher latitudes. Most areas tend to experience a well-defined, repetitious daily cycle of morning sunshine and heating, afternoon clouds and scattered showers, and nighttime clearing and cooling.

Weak Tropical Weather Disturbances

The predominant tropical weather disturbances are localized convectional showers and thunderstorms. Convectional activity in general and thunderstorms in particular are more common in the tropics than anywhere else on Earth. Large-scale weather disturbances also develop within the lower latitudes, but the relative lack of conflicting air masses causes them to be less numerous and generally weaker than those of the higher latitudes. Because of the instability and high moisture content of the air in much of the tropics, however, these disturbances are substantial rainfall producers and contribute significantly to the precipitation totals of many areas. Three of the more important larger tropical disturbances are easterly waves, equatorial lows, and hurricanes.

Easterly waves are weak weather disturbances that are most likely to develop during the summer and early autumn over ocean areas. They consist of north/south-oriented troughs in the trade wind belts of each hemisphere that drift slowly westward (see Figure 8.25). Their western sides contain diverging air and generally fair weather. To the east of the trough line, the inflowing air converges as it approaches the axis of the wave, reducing its stability and producing numerous showers and thunderstorms. Easterly waves are important producers of precipitation for the islands and east coasts of continents that lie between approximately 10° and 30° N and S. They are watched closely during the summer and early autumn because of their occasional tendency to develop into hurricanes.

Weak centers of low pressure, termed **equatorial lows,** frequently develop along the ITCZ and drift westward. Although they normally are situated too near the equator to develop into hurricanes, they usually are associated with extensive areas of cloudiness and heavy convectional precipitation.

Hurricanes

The hurricane, the world's most destructive storm, is the only large, strong storm type that develops in the low latitudes. By definition, a **hurricane** is a tropical cyclonic storm with sustained winds of at least 33.5 meters per second (75 miles per hour or 65 (knots).

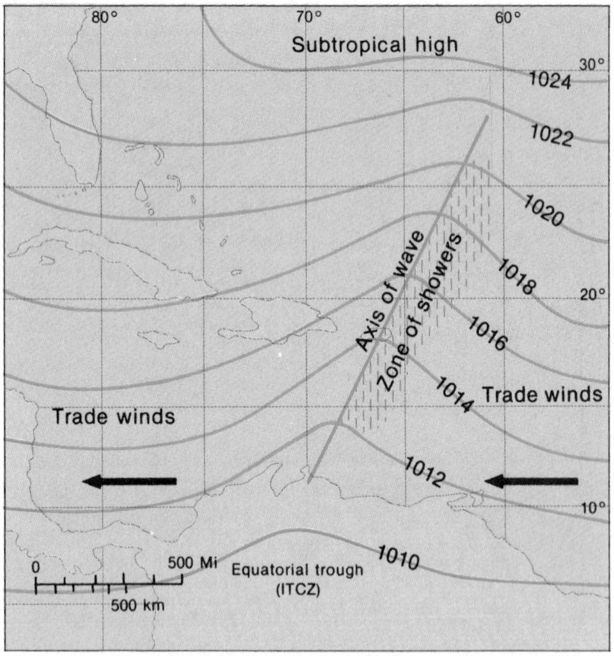

FIGURE 8.25 Weather map depiction of an easterly wave in the western Atlantic. Isobars are drawn at 2 millibar intervals.

In the western North Pacific, where they are most numerous and intense, they are termed **typhoons.** Other names for this storm type include "cyclone" in the Indian Ocean and "chubasco" off the west coast of Mexico.

Hurricanes gain energy and moisture through a combination of convectional and cyclonic lifting. Tremendous quantities of warm, moisture-laden air that spiral into these storms because of their very low central barometric pressures are lifted into the upper troposphere and are cooled adiabatically to release torrential rainfalls.[2] The heat of condensation released during this process powers the storms and helps to maintain the global heat balance by transferring excess thermal energy from the surface to the upper atmosphere and eventually, as the storms progress poleward, to the higher latitudes.

Development and Characteristics

Hurricanes form only over tropical or subtropical oceans with surface water temperatures of 26° C (79° F) or more. Unlike the frontal cyclones of the middle and higher latitudes, a hurricane does not develop on the frontal boundary between opposing air masses, but instead forms within a single mass of maritime tropical air. In its earliest stages of development it is little more than a large, unorganized mass of convectional showers and thunderstorms. The rising of the air within the showers reduces surface pressures and initiates a weak inflow of air, which is given a cyclonic rotation by the Coriolis effect. The development of a cyclonic wind flow raises the disturbance to the category of a **tropical depression.** When sustained winds reach 18 meters per second (40 mph), the developing storm is given a name and is classified as a **tropical storm.**[3] Upon reaching sustained winds of 34 meters per second (75 mph), the storm officially becomes a hurricane (see Figures 8.26 and 8.27). If conditions remain favorable for intensification, the hurricane eventually may attain wind speeds exceeding 67 meters per second (150 mph); the most powerful storms on record have had sustained winds in the vicinity of 90 meters per second (200 mph).[4]

Hurricanes develop over warm tropical oceans at latitudes between 5° and 30° N and S (see Figure 8.28). They occur with greatest frequency on the west sides of oceans, both because of the prevalence of warmer waters and because of the general westward movement of weather systems in the tropics; they are most frequent of all in the vast reaches of the western tropical and subtropical North Pacific. Cool water inhibits hurricane formation poleward of 30° latitude, although already developed storms often are able to move well poleward of this limit before losing intensity. Oddly, the very center of the tropics—the area within five degrees of latitude of the equator—is immune from hurricanes. This is because the Coriolis effect is not strong enough here to impart the necessary rotary motion to disturbances.

The paths followed by most hurricanes are rather similar and are controlled by the upper-level windflow patterns. Most hurricanes form within the trade wind belt. Here the upper-level winds that steer the surface weather systems blow in a direction slightly poleward of due west (see Figure 8.28). As a result, most hurricanes take an initial course toward the west-northwest in the northern hemisphere and toward the west-southwest in the southern hemisphere at speeds of perhaps 15 to 30 kilometers per hour (10–20 mph). As they move poleward, they approach the subtropical highs. The steering currents in the subtropics are weak and often erratic, but most hurricanes gradually turn more poleward and reduce their forward speed to 10 or 15 kilometers per hour (5–10 mph). Eventually, the storms enter the westerly wind belt and recurve toward the northeast in the northern hemisphere or toward the southeast in the southern hemisphere. Their forward speed grad-

[2] World records for both low barometric pressures and high rainfall amounts have been set in hurricanes. The world's lowest sea level air pressure of 870 millibars (25.69 inches of mercury) was recorded during a 1979 typhoon in the Philippine Sea. A record 24-hour rainfall total of 187 centimeters (73.62″) was established during a 1952 storm on the island of Réunion in the Indian Ocean.

[3] The practice of naming tropical storms and hurricanes according to alphabetical lists was begun in 1953 by the U.S. National Weather Service. Separate lists are maintained for the western Atlantic area, the eastern Pacific, and the western Pacific. Originally, only female names were used. This, however, was denounced as sexist, and since the late 1970s, male and female names have been alternately employed.

[4] Only one storm this powerful is known to have occurred during the twentieth century in the Atlantic Ocean. This was the Labor Day Storm of 1935 that hit the Florida Keys. Two relatively recent Gulf of Mexico hurricanes, however—Camille in 1969 and Allen in 1980—had peak winds of at least 83 meters per second (185 miles per hour).

FIGURE 8.26 Satellite photo taken on July 11, 1978, showing four tropical systems in differing stages of development in the eastern Pacific. (NOAA)

ually accelerates to perhaps 50 or 60 kilometers per hour (30–40 mph), carrying them rapidly poleward. Eventually they pass over colder waters, which causes them to lose strength and eventually to dissipate because of the loss of the water vapor that supplies their energy.

Although most hurricanes reach colder waters and dissipate without ever encountering land, others strike coastal areas, often with disastrous consequences. The great destruction and loss of life associated with the landfall of these storms results from three factors. The best known, but often least destructive of these, are the hurricane-force winds which are capable of uprooting trees, collapsing weaker buildings, and turning loose objects such as boards and strips of sheet metal into lethal missiles. A second factor is the extremely heavy rainfall associated with hurricanes. Rainfall totals, which may exceed 25 centimeters (10in) with slow-moving storms, frequently produce extensive flooding, landslides, and loss of life. By far the most deadly feature of hurricanes, however, is the **storm surge** that accompanies their landfall. The very low air pressure within a hurricane, coupled with strong air convergence, causes the ocean surface to bulge upward as much as 10 meters (30 feet) above normal. This water may then be blown onshore by winds well in excess of 45 meters per second (100 miles per hour), inundating the coast. Flooding is most extensive on low-lying coastlines and on those occasions when a hurricane crosses the coast during the time of normal high tide. The greatest loss of life caused by a hurricane in the United States occurred in Galveston, Texas, in 1900, when 6,000 people were drowned by a storm surge. The largest numbers of human casualties on record are associated with hurricanes (locally called "cyclones") that have struck the Bay of Bengal coast in what is now Bangladesh in 1737 and in November 1970. The loss of life in each of these storms was

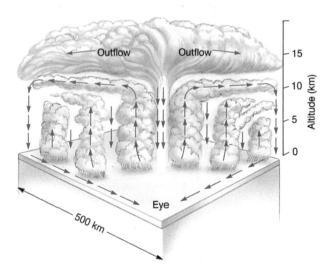

FIGURE 8.27 Cross-sectional view through the center of a hurricane. Note the bands of cumulonimbus clouds alternating with clear air.

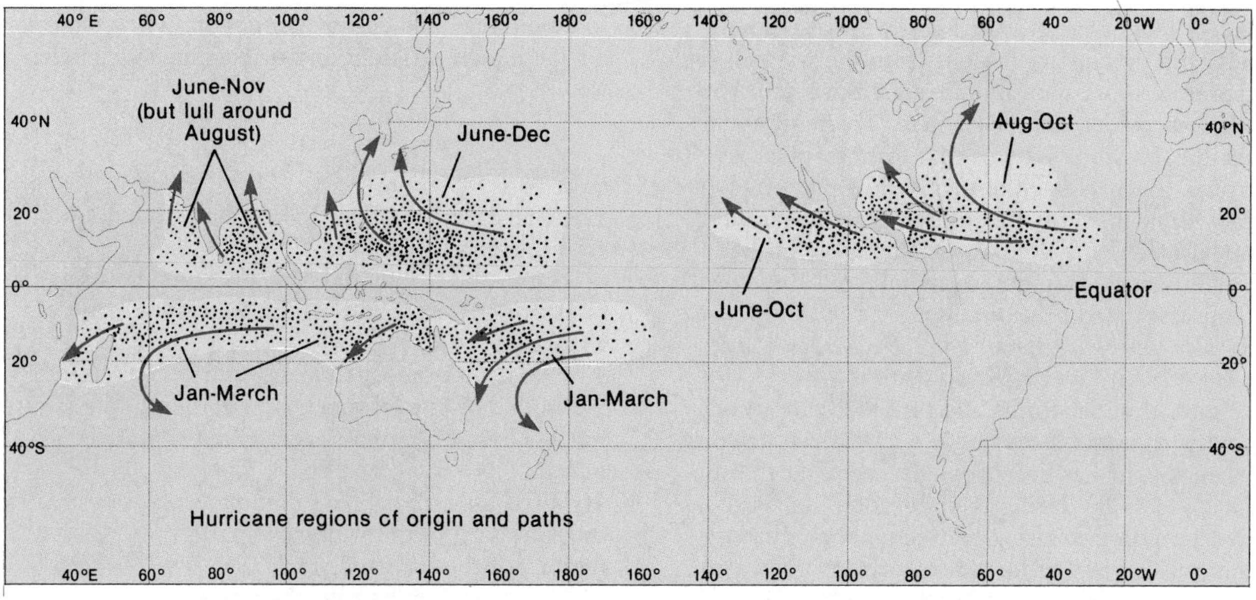

FIGURE 8.28 Areas of hurricane formation and typical storm tracks. Each dot indicates the point where a tropical storm initially reached hurricane intensity.

estimated at a staggering 300,000! Recently, in spring 1991, another hurricane in the same area was responsible for 180,000 deaths. Most of the lives lost in all three storms resulted from the storm surge, which in the 1737 storm was said to have raised water levels by 13 meters (37 ft), causing the ocean to advance inland for many kilometers.

SUMMARY

This chapter has examined the Earth's major weather systems. These systems are in a continuous state of flux—forming; changing size, shape, and intensity; and dissipating—as they drift across the face of our planet largely in response to the upper-level wind currents. As they pass over any given location, they bring their attendant weather conditions to that location for a period of a few hours or days.

Each weather system is an organized assemblage of the weather elements discussed in the previous chapters. Each exists because of the differential receipt of insolation in various portions of the Earth system. The seven weather systems discussed in greatest detail in this chapter are air masses, anticyclones, weather fronts, frontal cyclones, thunderstorms, tornadoes, and hurricanes. In order to review the formation, characteristics, and distribution of these systems, we return to the Focus Questions posed at the beginning of the chapter:

1. *What are air masses? How and why do they form, and what influence do they have on our weather?*

Air masses are large bodies of air characterized by internal homogeneity in temperature and water vapor content. They form over large land or water bodies that gradually impart their surface temperature and moisture conditions to the overlying air. After they have formed, air masses may be transported to other parts of the world by the upper-level winds. Most air masses form in either the high or low latitudes. The middle latitudes experience frequent incursions of both polar and tropical air masses.

2. *What are the most important weather systems of the middle and higher latitudes, and what types of weather are associated with them?*

The major large-scale weather systems of the middle and higher latitudes are anticyclones (high pressure systems), weather fronts, and frontal cyclones. Anticyclones generally are associated with the centers of air masses. They are areas of descending and diverging air and generally produce fair weather. Weather fronts are linear boundaries where air masses from different source regions meet. They typically are associated with cloudiness and precipitation as air from the warmer, and therefore more buoyant, air mass flows over the adjacent colder, denser air mass. Frontal cyclones are low pressure systems that form on weather fronts. Their convergent wind-flow patterns typically draw in warm air on their equatorward and eastern sides and cold air on their poleward and western sides, forming warm and cold fronts, with their associated precipitation.

3. *How do thunderstorms and tornadoes form, and what areas are most affected by them?*

Both thunderstorms and tornadoes are associated with unstable air and are most likely to form when the air has been strongly heated at the surface, especially during the afternoon. This produces vigorously rising convectional air currents that, in the case of tornadoes, are apparently given a cyclonic spiral by wind-shear conditions (opposing directions of wind flow) at differing altitudes. Thunderstorms are most abundant over land areas in humid tropical regions and generally decrease in frequency poleward. Tornadoes, however, are most frequent in the middle latitudes, especially in the central United States, where they typically form in advance of strong cold fronts.

4. *How and why do weather conditions of the tropics differ from those of the higher latitudes?*

Tropical weather differs from that of the higher latitudes largely because the tropics experience large quantities of insolation throughout the year. This makes the air warmer and more buoyant and increases its water vapor holding capacity. As a consequence, temperatures are warm throughout the year, fronts generally are absent, and precipitation is often heavy, coming largely in the form of afternoon convectional showers.

5. *What are hurricanes? How and where do they form, and what land areas do they most affect?*

Hurricanes are powerful cyclonic storms that form over ocean areas at latitudes between approximately 5° and 30°. They form when a region of moist, unstable air containing numerous showers and thunderstorms develops a cyclonic wind-flow pattern that gradually intensifies. Like most tropical weather systems, hurricanes tend to move in a westerly direction, so that the east coasts of land masses are most likely to be affected. If they enter the middle latitudes, however, they usually recurve to a poleward and easterly direction. They typically dissipate after striking land or passing over colder water.

Review Questions

1. What is an air mass? Where and how does it attain its characteristics?
2. Name and briefly describe the temperature and moisture characteristics of the four basic types of air masses.
3. Explain why most air mass source regions are in the high and low latitudes, not in the middle latitudes. How do air masses reach the middle latitudes?
4. What type of air mass dominates in your area in the summer? In the winter? What are the source regions of these air masses?
5. Describe the wind-flow patterns associated with anticyclones and cyclones in each hemisphere. Why are the patterns different in the northern and southern hemispheres?
6. What is a weather front? Explain how weather fronts are classified, and briefly describe each of the four types of fronts.
7. Describe the sequence of weather events associated with the approach and passage of an idealized cold front and an idealized warm front. Why does the average duration of precipitation associated with these two types of fronts differ?
8. Briefly describe the "life cycle" of a typical frontal cyclone. Include an explanation of why the cyclone normally begins to lose intensity after frontal occlusion takes place.
9. During what time of year are thunderstorms most frequent in your area? Explain why. What essential characteristics must the atmosphere exhibit in order for thunderstorms to develop?
10. What are tornadoes? Under what meteorological conditions are they likely to occur? Why do they occur more frequently in the United States than in other parts of the world?
11. Discuss the basic ways in which you would expect weather conditions to differ from those typical of your own area if you lived near the equator.
12. What is a hurricane? Where do these storms form, and why? Why do you think they are experienced frequently on the East Coast of the United States, but not on the West Coast?

13. By what methods can hurricanes cause destruction in coastal regions? Which method has caused the greatest loss of life, and why?

Key Terms

air mass
anticyclone
cold front
warm front
stationary front
occluded front
frontal cyclone
thunderstorm
tornado
hurricane

CASE STUDY

HURRICANE HUGO OF 1989

The entire Atlantic and Gulf Coast of the United States, from New England to the Mexican border, is vulnerable to the devastation produced by hurricanes. Despite continuing advances in hurricane-forecasting technology, the unprecedented pace of human development of this coastal region continues to increase the risk to lives and property. For the past few decades, hurricanes have been especially numerous and intense within the Gulf of Mexico, while the Atlantic coast has received relatively few major storms. This situation, however, changed dramatically in September 1989, when powerful Hurricane Hugo slammed ashore near Charleston, South Carolina.

Hurricane Hugo began as a cluster of thunderstorms off the coast of Senegal in West Africa on September 9. The rising air currents within these thunderstorms lowered sea level air pressures, causing an inflow of air at the surface. This inflow, given a counterclockwise twist by the Earth's rotation, organized the disturbance into a tropical depression on September 10. Upper-level winds carried the strengthening depression toward the west-northwest into the open Atlantic, where it attained tropical storm status the next day. The new storm was named Hugo by forecasters at the National Hurricane Center in Miami. By September 12, the rapidly strengthening storm reached hurricane intensity as its peak winds topped 34 meters per second (75 miles per hour).

The first land in Hugo's path was the Leeward Islands chain separating the open Atlantic from the Caribbean Sea (see Figure 8.29). In the predawn hours of Sunday, September 17, the now-powerful storm raked the French island of Guadeloupe with winds of 65 meters per second (145 mph). Shortly afterward, the island of Montserrat felt the storm's full fury; the famed "Emerald Isle" turned brown as most trees were defoliated by the fierce winds, and 10 people lost their lives. Antigua and Barbuda next received glancing blows, while St. Kitts and Nevis sustained more serious damage as the storm passed near them.

The next islands directly in the hurricane's path were the U.S. Virgin Islands of St. Croix and St. Thomas. St. Croix was devastated by the storm, with nearly all houses damaged or destroyed and most of its 53,000 inhabitants left without shelter. Damage estimates on the island exceeded $2 billion; further destruction resulted from the widespread looting that followed, necessitating the deployment of 1,200 U.S. troops to the island.

Hugo's next target was Puerto Rico, the first sizable land mass to be encountered by the storm. Hugo passed over the northeast portion of the island, which received extensive damage from the storm's 140 mile-per-hour winds. More than 30,000 people were left homeless and the

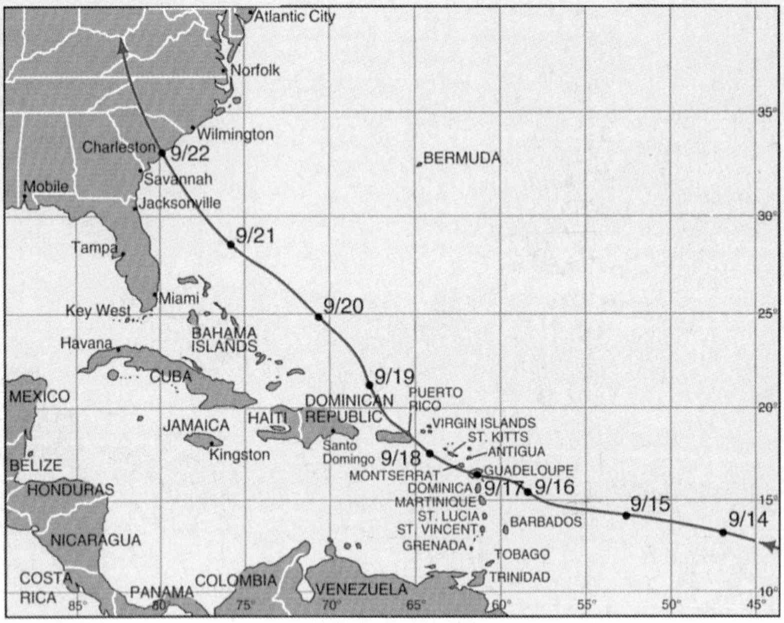

FIGURE 8.29 Path of Hurricane Hugo.

continued on next page

majority of trees in the Caribbean National Forest—the only tropical rainforest in the United States—were blown down. Hugo itself did not escape unscathed from its Puerto Rican encounter; friction caused by its passage over the mountainous island reduced it to the status of a tropical storm. However, Hugo gradually regained strength as it passed over the warm waters of the Gulf Stream on a northwesterly course toward the United States mainland.

By Wednesday, September 20, Hugo again attained hurricane status, with winds reaching 45 meters per second (100 mph) as it accelerated toward the west-northwest. At 6:00 A.M. on September 21, hurricane warnings were posted along the southeast coast from Fernandina Beach, Florida, to Cape Lookout, North Carolina. Shortly afterward, the governor of South Carolina issued an evacuation order for all peninsulas, barrier islands, and beachfront property in the state, with the exception of the city of Charleston. In the ensuing 10 hours, over a quarter of a million people successfully evacuated this area as the storm continued to bear down on the coast.

Hugo made landfall in the vicinity of Charleston just after midnight on Friday, September 22 (see Figure 8.30). It was accompanied by winds of 58 meters per second (130 mph) and a storm surge of 5 to 6.5 meters (15 to 20 feet). Despite the friction encountered over land, the storm

FIGURE 8.30 Radar depiction of Hurricane Hugo as its eye passed over Charleston, South Carolina. Rainfall intensities are color-coded, with the most intense rain appearing in red. (NOAA/Satellite Data Services Division)

retained its hurricane intensity for several hours after moving inland, with winds of 45 meters per second (100 mph) recorded in Columbia, in central South Carolina, and 38 meters per second (85 mph) in Charlotte, North Carolina. Moving rapidly on a northwesterly and then more northerly path, the storm crossed the Appalachians, passing through West Virginia, Pennsylvania, and New York before finally carrying dwindling winds and rains into eastern Canada.

Hugo's winds and the storm surge and high waves that accompanied its landfall had a devastating impact on the central South Carolina coast. Destruction was especially severe on the state's low-lying coastal barrier islands, which were lined with beach homes and resorts (see Figure 8.31). Dozens of these structures were destroyed by the fury of the storm. Charleston itself suffered its worst natural disaster since an 1886 earthquake leveled much of the city. Winds also destroyed half the mature trees in the extensive Francis Marion National Forest in the eastern part of the state. In addition, nearly a million homes and businesses in the southeastern United States were left without electricity, some for several weeks.

FIGURE 8.31 Destruction to ocean-front homes in Myrtle Beach, South Carolina, caused by Hurricane Hugo. (Reuters/Bettmann Newsphotos)

Hurricane Hugo will be remembered as one of the great hurricanes of the twentieth century. Total property damage caused by the storm was estimated at nearly $10 billion, of which $6 billion occurred in South Carolina alone. However, only 54 lives were lost (24 in the Carolinas). In retrospect, this figure is surprisingly small and is a tribute to both the accuracy of the warnings and the efficiency of the evacuation that preceded the landfall of this great storm.

C H A P T E R 9

CLIMATES OF THE WORLD

OUTLINE

FOCUS QUESTIONS

1. What are the chief advantages and shortcomings of climate classification systems?
2. What are the major climatic characteristics of each of the global climate regions, and why do these characteristics occur?

In this chapter, the global pattern of climates is examined on a regional basis. A region's climate consists of its characteristic long-term weather conditions. Separating weather and climate often is difficult, however, because only the time factor differentiates the two subjects. Much of the material presented in the previous four chapters, especially the discussions of the global distributions of the various weather elements, was actually climatic information. In contrast to the approach of the previous four chapters, however, where weather and climate phenomena were discussed in a topical fashion, we now examine the long-term condition of the atmosphere in a spatial context by dividing the Earth into a number of climatic regions.

A basic goal in physical geography, as noted in Chapter 1, is to develop an understanding of the world as a single functioning system. Such an understanding is developed largely through a knowledge of the relationships that exist among the Earth's natural phenomena. The role of climate in these relationships is vital because climate is the leading independent environmental variable and exerts a powerful and sometimes controlling influence on the other components of the environment. For example, climate is the primary factor controlling the world distribution of natural vegetation. Climate, vegetation, and rock type together largely determine the type of soil that will develop within a given area. Climate also exerts a major influence on landform characteristics.

The distribution of Earth's human population also is strongly influenced by climate. Especially important in this respect are climatic characteristics, such as extremes of temperature and precipitation, that restrict the growth of the cultivated plants forming the basis of our food supply. Continuing rapid advances in science and technology are reducing the degree to which climate influences the distributional pattern of the human population, but its continuing importance cannot be denied.

CLIMATE CLASSIFICATION

Examined in detail, the climate of any locality is unique because its climatic characteristics will never be precisely duplicated anywhere else. The key to understanding regional climate patterns and to developing climatic classification systems lies in generalization. Climate types are of necessity based not on specific values, but on ranges of values. All areas having conditions falling within the stated ranges are, perhaps somewhat arbitrarily, considered to have the same type of climate. Climatic systems are devised to simplify the study of climate by reducing the number of climate types to a manageable total.

Mean values normally are stressed in climate descriptions because they can be readily calculated and easily described, and generally are indicative of the most common atmospheric conditions. Regional climatic analyses are thus based largely on the compilation of past weather data by weather reporting stations within the region. It is assumed that no significant changes in climate are occurring and therefore that the weather conditions of the past will occur with statistically similar frequencies in the future. As a result, the longer that climatic data have been gathered, and the more comprehensive these data are, the more accurately a region's climatic characteristics may be determined.

The division of the Earth into climatic zones allows regional associations to emerge. It becomes apparent that the global distribution of climates is not random; rather, climates are spatially organized, with similar climatic conditions occurring in widely separated parts of the Earth. This, in turn, points to the existence of unifying controls that operate in similar fashions in areas that may be distant from one another.

The use of classification systems to regionalize climates has limitations that must be understood by those employing them. One of the most important shortcomings is loss of detail. The more that climatic characteristics are generalized, the larger the individual climate regions tend to become, and the more detail is sacrificed. A second shortcoming is the danger of taking climatic regions too literally. It should be understood that the regions and boundaries shown are correct only for the system being used, and that other, equally valid systems may divide the area under consideration into quite different regions. It must also be realized that climates do not change suddenly as boundary lines are crossed, but tend to vary gradually over large distances.

The data employed to develop climate classification systems also have limitations. Climatic data are inadequate for large portions of the Earth. This includes the 71 percent of our planet covered by oceans as well as extensive areas in the polar regions, subtropical deserts, and tropical forests. As data from these areas become increasingly available, climatic boundaries must on occasion be revised.

Because of the complexity of world climate patterns and the diversity of human interests, no single classification system can be satisfactory to everyone. A variety of systems has been developed through the years, and the choice of the system to be utilized in any specific instance depends largely on its intended purpose.

The most widely known climatic classification system was first published in 1918 by the German botanist and climatologist Wladimir Köppen. Köppen, who was interested in gaining a better understanding of the causes of the world pattern of natural vegetation, realized that vegetative distribution was more closely tied to climate than to any other environmental control. His climate system therefore stressed annual means and seasonal variations of temperature and precipitation, the climatic factors that have the greatest influence on global vegetative characteristics. The system has been revised and refined on several occasions. Details of the Köppen climatic classification system, as well as a brief discussion of the system's chief advantages and shortcomings, are provided in Appendix C.

Our need in this chapter is for a global climate classification system that is relatively simple, stresses geographic patterns of climates, and corresponds to the world patterns of natural vegetation and soils examined in subsequent chapters. The system used is primarily descriptive and is based closely on the Köppen classification system. It divides the world into 11 climatic regions defined largely on the basis of their temperature and moisture characteristics. For each region, stress is placed on the following factors:

1. geographic distribution
2. basic atmospheric temperature and moisture characteristics
3. the reasons for these characteristics

Following the precedent established in the discussion of air pressure and wind belts, our examination of world climatic regions begins at the equator and progresses poleward. The locations of the various climate types are mapped in Figure 9.1.

THE LOW LATITUDE CLIMATES

Approximately half of the Earth's surface is located within the low latitudes. The basic weather characteristics of this part of the world were discussed at the end of the preceding chapter; in essence they consist of continuously warm temperatures, along with showery precipitation that is predominantly convectional in nature. The placement of the subtropical highs and the ITCZ are largely responsible for the differences among the three major climates of the low latitudes. Near the equator, the ITCZ produces year-round cloudy and humid conditions, with generally abundant rainfalls. Within the subtropical highs, clear skies, low relative humidities, and scanty rainfalls generally prevail. In much of the low latitudes, however, climatic conditions are controlled by the seasonal migration of these two pressure belts. This produces a pattern of alternating rainy and dry seasons.

Tropical Wet Climate (Köppen Af and Am)

The Tropical Wet Climate is the most equatorward of the Earth's climatic types. It occurs in extensive areas between 10° N and S, and in some coastal areas it extends to the vicinity of the Tropics of Cancer and Capricorn (see Figure 9.1). Three large land areas are contained within the Tropical Wet Climate: the Amazon Basin in South America, the northern Congo Basin in western Africa, and the majority of the islands of the western Pacific "East Indies," including most of Malaysia and Indonesia and the Philippines. Additional smaller areas include coastal strips in eastern Central America and Brazil, the Guinea coast of West Africa, southwestern India, the eastern coastal margin of the Bay of Bengal, central Vietnam, and northeastern Australia.

The Tropical Wet Climate may be characterized as warm and rainy. Most locations near sea level have monthly mean temperatures near 27°C (80°F) throughout the year. Constantly high midday Sun angles and daylight periods of nearly equal length throughout the year are responsible for the warm temperatures as well as for the very small annual temperature ranges. Ranges are 3C° (5F°) or less for most locations, and even those areas situated near the poleward margins of the climate almost never experience annual ranges of more than 6C° (10F°).

Not only are seasonal differences in temperature very small, but day-to-day variations are minor and result primarily from differences in cloud cover. Air masses of nontropical origin rarely if ever penetrate the inner tropics to produce frontal temperature contrasts, and maritime equatorial (mE) air dominates throughout the year. The result is that one day

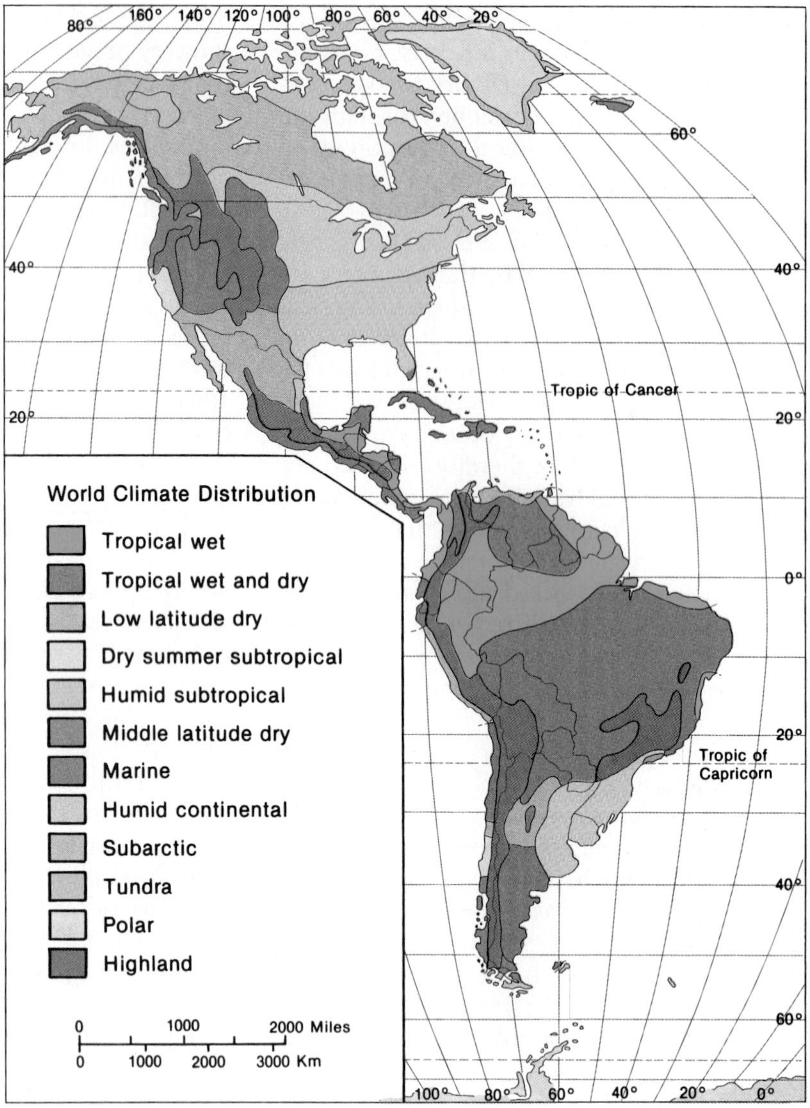

FIGURE 9.1 Distribution of world climate types.

is very much like another with respect to temperature. Only the constant rhythm of daily heating and nightly cooling provides thermal variety, leading some writers to describe day and night as the "seasons" of the tropics. In all portions of the Tropical Wet Climate, daily temperature ranges substantially exceed annual ranges.

Yearly rainfall totals usually average 125 to 500 centimeters (50 to 200 inches) a year, making the Tropical Wet Climate the world's rainiest climate type. Large variations in amount frequently occur over short distances, with the heaviest totals in mountainous regions and along windward coasts. Annual means for the region as a whole range from about 200 to 230 centimeters (80 to 90 inches).

The heavy precipitation totals result from the high moisture content of the air, intense surface heating, and the proximity of the ITCZ. Most rainfalls occur as brief heavy showers and thunderstorms interspersed with periods of sunshine. Along coastal areas, showers are likely to occur at any hour of the day or night. Inland areas, on the other hand, generally experience distinct afternoon precipitation maxima caused by surface heating, which results in the convective uplift of the unstable air. Especially heavy rainfalls occur where mountains provide an addi-

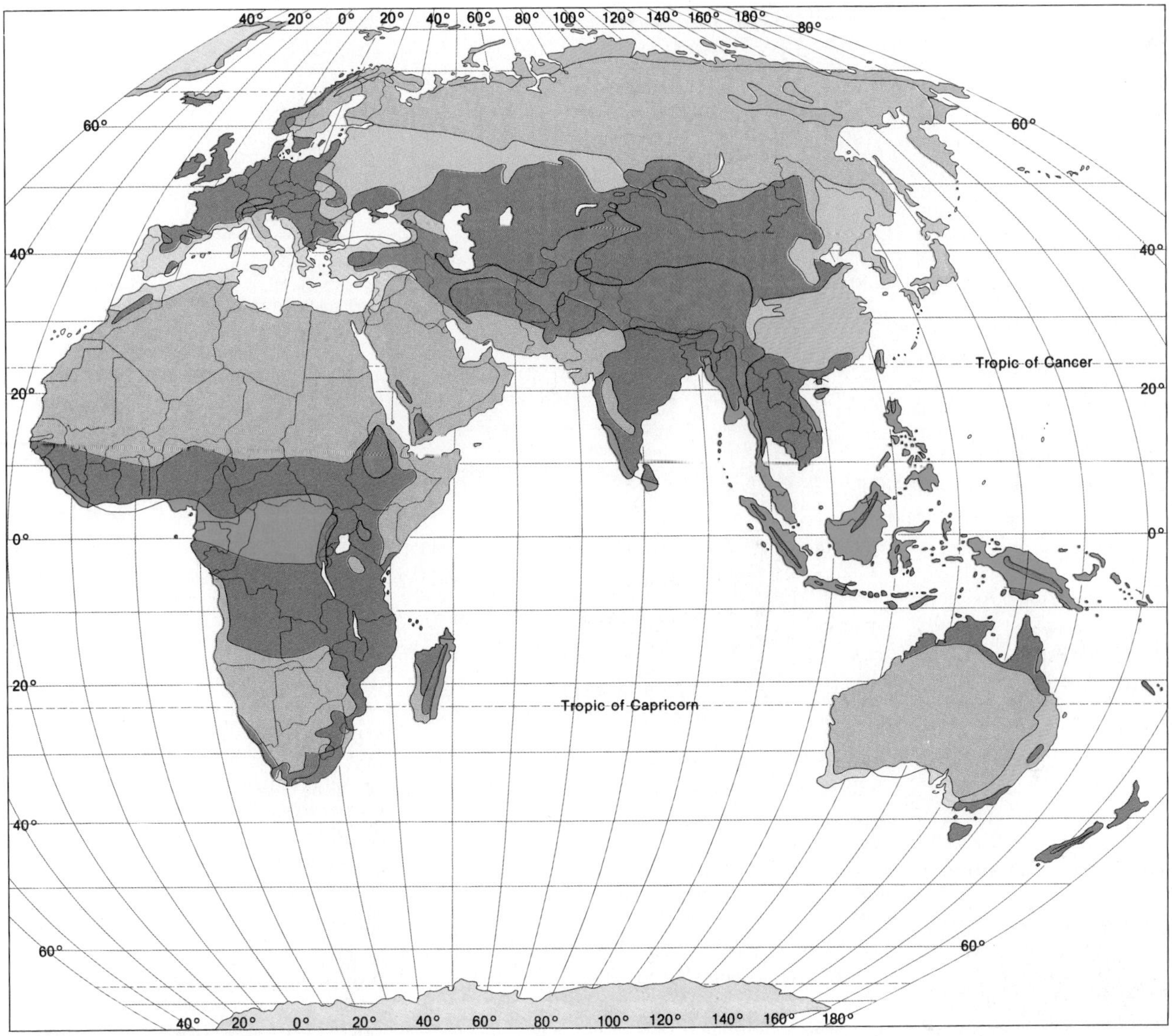

FIGURE 9.1 *continued*

tional lifting mechanism. In fact, it is the orographic lifting of the trade winds along mountainous windward coasts such as eastern Central America and Brazil that results in the poleward extensions of the Tropical Wet Climate in those areas.

Most locations within the Tropical Wet Climate experience distinct periods of increased and decreased rainfall during the year. Indeed, many of the more poleward and coastal locations have a short dry season, with little or no rainfall over a two- or three-month period. The annual precipitation distribution pattern of Jakarta, Indonesia, depicted in Figure 9.5, serves as a representative example.

The chief cause of the seasonal variability in precipitation is the proximity of the ITCZ. This zone of converging air is associated with unstable atmospheric conditions and heavy rainfalls, and its seasonal shift in latitude causes the area of heavy rains to oscillate alternately northward and southward, affecting different portions of the Tropical Wet Climate. Locations some distance from the equator typically experience a summer precipitation maximum because the ITCZ tends to shift into the hemisphere

THE COLLECTION AND DISPLAY OF WEATHER DATA

Weather data such as temperature, precipitation, air pressure, and humidity are gathered daily at hundreds of locations throughout North America. The more sophisticated weather collection centers have automated sensing and recording instruments and often are located at airports. Most weather data, however, are collected by cooperative weather observers located on farms, in suburban neighborhoods, on college campuses, and in a variety of other settings.

Outdoor weather instruments typically are housed in shelters like that shown in Figure 9.2. The standard weather instrument shelter is constructed of white painted wooden slats that are louvered to allow for free air flow. It has a double-layered roof and is set on a long-legged stand about 1.5 meters (5 feet) above the ground. The shelter generally is located in an open grassy area well away from trees, buildings, or other obstacles to the wind. These characteristics are all designed to allow air temperatures inside the shelter to remain the same as those on the outside while at the same time protecting the instruments. A weather instrument shelter normally contains a variety of thermometers, including maximum and minimum thermometers, and a sling psychrometer. A rain gage is usually located nearby. Readings are taken once or twice a day at established times.

FIGURE 9.2 A standard weather instrument shelter. (Ross DePaola)

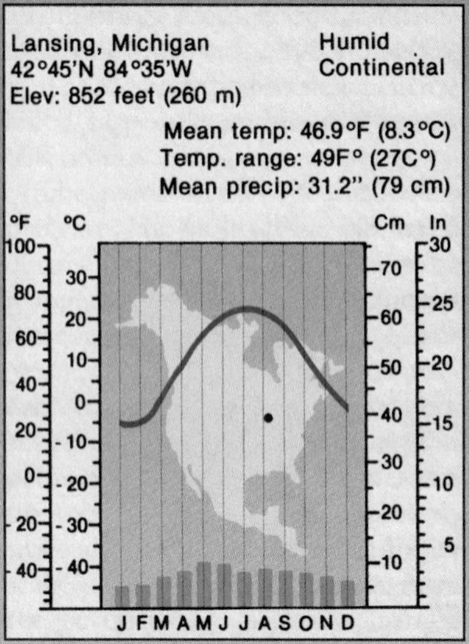

FIGURE 9.3 Climagraph for Lansing, Michigan. Mean temperatures throughout the year are depicted by a line graph, and mean precipitation amounts for each month are indicated by bar graphs.

In order to graphically depict a station's climatic characteristics, a **climagraph** such as the one in Figure 9.3 is often employed. Most climagraphs depict monthly temperature and precipitation means. For clarity, the temperature data usually are indicated by points that are connected to form a line graph, while the precipitation values are indicated by the heights of twelve bars. Additional information that may be provided on climagraphs includes station latitude and longitude, elevation, a location map, annual temperature means and ranges, and annual precipitation totals. Climagraphs providing this information are used throughout this chapter to illustrate representative conditions in each of the climatic regions discussed.

receiving the direct rays of the Sun. Such is the case for Jakarta. Some locations near the equator, particularly in Africa, experience two periods of increased rainfalls, one in the spring and one in the fall. These rainy periods are caused by passages of the ITCZ as it shifts alternately between its extreme positions in the northern and southern hemispheres. Such a double maximum pattern occurs in Kisangani, Zaire (see Figure 9.6).

Except for local gusts in heavier showers, winds in many areas within the Tropical Wet Climate are

FIGURE 9.4 Virgin tropical rainforest in the highlands of interior Costa Rica. (J. Carmichael, Jr./The Image Bank)

light and variable. Daily sea breezes develop in most coastal locations, providing welcomed relief from the heat. Locations in Southeast Asia experience the strongest winds because of their monsoonal wind-flow pattern.

Tropical Wet and Dry Climate (Köppen Aw)

Proceeding poleward in either direction from the Tropical Wet Climate, one enters regions dominated by alternating wet and dry seasons. This climate type, commonly known as the Tropical Wet and Dry Climate, exists in most areas between 5° and 20° N and S. It is the chief climate of the trade wind belt and is located between the ITCZ and the subtropical highs of each hemisphere.

From both a geographic and a human standpoint, the Tropical Wet and Dry Climate is one of the world's most important climate types because it oc-

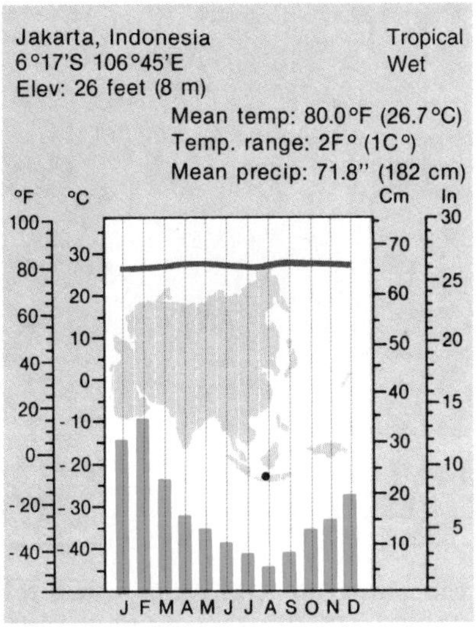

FIGURE 9.5 Climagraph for Jakarta, Indonesia. Jakarta is situated well south of the equator, and experiences a single well-defined summer rainfall peak.

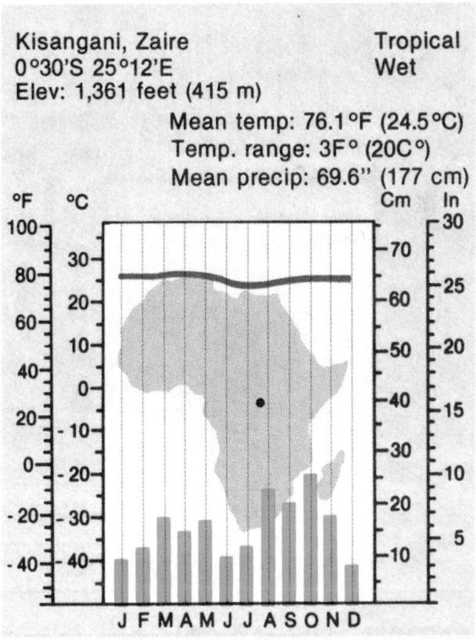

FIGURE 9.6 Climagraph for Kisangani, Zaire. This station's double precipitation maximum is caused by the semiannual passage of the ITCZ through the area as it follows the Sun.

cupies an extensive and heavily populated segment of the Earth. The largest areas occur in Latin America and Africa. In the Americas, it dominates a vast area of central and southern Brazil as well as most of Bolivia. It also occupies much of northern South America and coastal Central America and most of the islands of the Caribbean area, as well as the southern tip of Florida. It covers nearly 40 percent of the African continent, occurring in a broad arc that nearly surrounds the relatively small central core of Tropical Wet Climate. Also within the Tropical Wet and Dry Climate are densely populated areas of South and East Asia, including most of India and extending eastward to Vietnam and extreme southern China. Finally, northern Australia and the southernmost islands of the East Indies are located within this climate type.

The annual precipitation pattern of the Tropical Wet and Dry Climate is its dominant characteristic and also is largely responsible for the annual temperature distribution. The well-defined alternation of wet and dry seasons is caused by the thermally induced seasonal migration of the ITCZ and the subtropical belts of high pressure; in many areas this alternation is further enhanced by a monsoonal tendency. The wet season nearly always occurs during the summer half of the year and usually centers on the midsummer months, when the ITCZ makes its deepest penetration into the hemisphere. Conversely, the driest season in most locations occurs during midwinter, when the subtropical high has shifted to its most equatorward position. The resulting pattern can be readily seen on the climagraphs for Bombay, India, and Antananarivo, Madagascar (see Figures 9.7 and 9.8).

The Tropical Wet and Dry Climate is transitional between the constant wetness of the equatorial regions and the aridity of the subtropics. Annual precipitation means are mostly 90 to 180 centimeters (35 to 70 inches). Precipitation amounts decline poleward as the subtropical high is approached, and the climate ends at the ill-defined line where semiarid conditions and an annual water deficit begin. During the rainy season, conditions in the Tropical Wet and Dry Climate are much like those occurring throughout the year in the Tropical Wet Climate. Skies are mostly cloudy, the air is very warm and humid, and convectional showers and thunderstorms are numerous. Especially numerous showers, and sometimes even steady rains, are associated with the occasional passage of weak equatorial lows or easterly waves, and in some areas tropical storms or hurricanes may bring torrential rainfalls in the late summer and autumn. During the dry season, conversely,

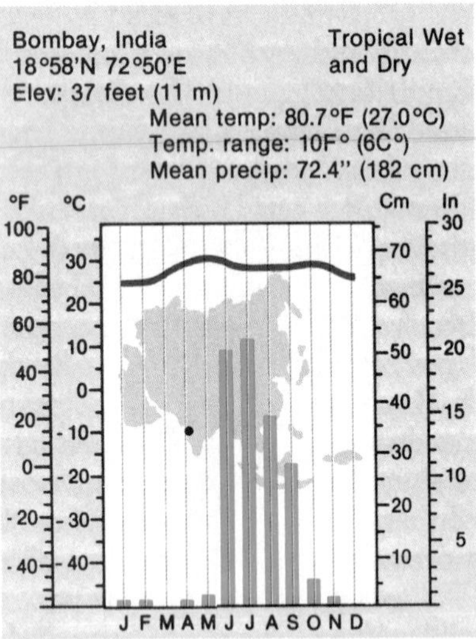

FIGURE 9.7 Climagraph for Bombay, India. The city's strong summertime precipitation maximum is caused by a combination of the shift of the ITCZ and the summer monsoon.

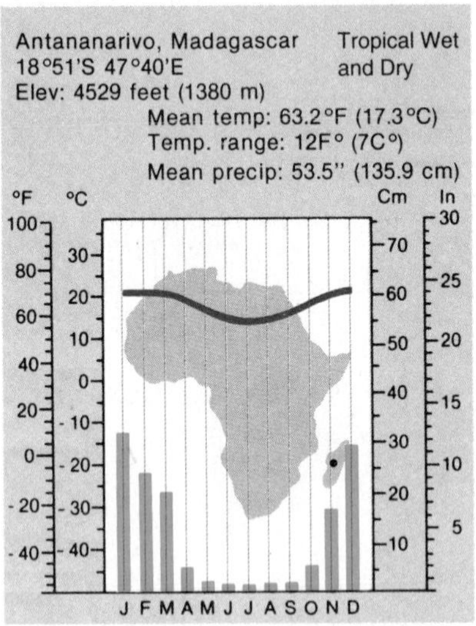

FIGURE 9.8 Antananarivo, Madagascar, has a pattern of rainy summers dominated by the ITCZ and dry winters under the influence of the southern subtropical high.

FIGURE 9.9 Wet and dry season contrasts within Mikumi National Park, Tanzania, located within the Tropical Wet and Dry Climate. (Wayne McKim)

sunshine prevails, humidities are lower, and no rain may fall for periods of weeks or even months (see Figure 9.9).

The annual temperature pattern is controlled by a combination of seasonal Sun angle variations and the alternation of wet and dry seasons. Winter usually is the coolest season, especially in the more poleward locations where Sun angles are lowest. Temperatures rise rapidly during spring as the Sun moves higher in the sky, the length of daylight increases, and the weather stays dry and often virtually cloudless. The hottest temperatures of the year commonly occur during middle or late spring, just before the onset of the summer rains. Note, for example, that Bombay's warmest month is May. Perhaps surprisingly, the hottest spring temperatures tend to occur in more poleward locations because the rainy season begins later in these areas.

Temperatures usually fall somewhat during the summer rainy season because of the reduction of sunshine and the increase in evaporational cooling. Despite the somewhat lower temperatures, the combination of continued warmth and very high humidity produces sticky, uncomfortable conditions. If the rainy season ends soon enough, a secondary period of warmer temperatures may occur during the autumn, when clearing skies once again allow for an increase in insolation and Sun angles are still fairly high. (This is the cause of Bombay's slight rise in temperatures in October.) The disruption of the normal summer temperature maximum by the rainy season is the most notable thermal characteristic of the Tropical Wet and Dry Climate.

Low Latitude Dry Climate (Köppen BWh and BSh)

In the dry climates, potential evapotranspiration rates exceed precipitation amounts, producing long-term conditions of water shortage. In assessing the dryness of an area, net radiation and resulting temperatures are crucial because they are the primary factors affecting water loss from vegetation and soil. The warmer a region, the more precipitation it can receive and still be "dry." For this reason, the world's dry climates are found in the low and middle latitudes—not in the high latitudes.

The dryness of an area exerts a major influence on such factors as the extent of soil development, vegetation type, and the potential for many types of human use. It is common practice to divide the dry lands into two categories (see Figure 9.10). Areas that receive significant amounts of precipitation and are only moderately dry are termed **semiarid**. They commonly have well-developed and often fertile soils, support a natural grass cover, and offer considerable potential for certain types of agricultural development. The very dry areas are termed **arid** and generally can be described as deserts. Arid regions typically have poorly developed soils, contain at most a scattering of drought-resistant natural vegetation and are normally capable of supporting only limited human populations. (The dry climate areas of the

FIGURE 9.10 Arid and semiarid portions of the Low Latitude Dry Climate differ greatly in appearance, largely because of differences in vegetation. The left photo shows a semidesert region dominated by scattered eucalyptus trees in Flinders Ranges National Park, South Australia. The right photo is of semiarid grassland in Kamberg Nature Reserve, Natal, South Africa. (*left*: © Bill Bachman/ Photo Researchers; *right*: Visuals Unlimited/© David L. Pearson)

low and middle latitudes are treated in separate sections of this chapter. Although their common characteristic of dryness causes these areas to share some important attributes, differences in the causes of their aridity, temperature characteristics, and geographic distribution give them sufficient regional distinctiveness to warrant separate discussions.)

The Low Latitude Dry Climate is centered between 20° and 30° N and S, but in places extends equatorward to 15° and poleward to 35°. It is the chief climate of the subtropics, although it should be noted that large portions of the subtropics have humid climates. The climate is especially extensive on the west sides of the continents and adjacent east sides of the oceans where the subtropical highs are most strongly developed and cold ocean currents help stabilize the atmosphere.

In North America, the areas with a Low Latitude Dry Climate include most of the southwestern United States and northern Mexico. In South America, this climate is restricted primarily to the narrow Atacama Desert of coastal Peru and northern Chile. By far the most extensive land areas with this climate are to be found in Africa, Asia, and Australia. In Africa, it includes the vast expanse of the Sahara and its borderlands, the Kalahari region of Southwest Africa, and the easternmost African "horn," located mostly within Somalia. The low latitude dry lands of Asia are in the southwestern part of the continent and include the Arabian Peninsula, most of Iraq, Iran, Pakistan, and western India. In Australia, sometimes called the "desert continent" because a greater proportion of this continent is desert than any other, all of the interior and much of the western and southern margins are dry. Finally, it should be noted that vast ocean areas within the tropics and subtropics experience a dry climate.

The relative lack of precipitation in Low Latitude Dry Climate areas results primarily from the year-round dominance of the subtropical highs. The dryness is further accentuated in most areas by high temperatures, high potential evapotranspiration rates, continentality, and, in some cases, mountain barriers. Arid areas, which typically have annual precipitation means averaging less than 25 centimeters (10 inches), form the central or core areas within the regions having this climate. They are surrounded by broad peripheral rims of semiarid conditions that generally have annual precipitation means of 25 to 75 centimeters (10 to 30 inches).

Three distinct precipitation patterns exist within the Low Latitude Dry Climate. These correspond with the three bands of differing dryness shown in Figure 9.11. On the climate's equatorward margins, semiarid conditions with a pronounced midsummer rainfall maximum and almost totally dry winter predominate. Areas with this rainfall pattern are adjacent to the Tropical Wet and Dry Climate, and their precipitation regime represents a poleward extension of the rainfall pattern of this climate. During most of the year in this zone, the subtropical high dominates and suppresses precipitation, but for two or three

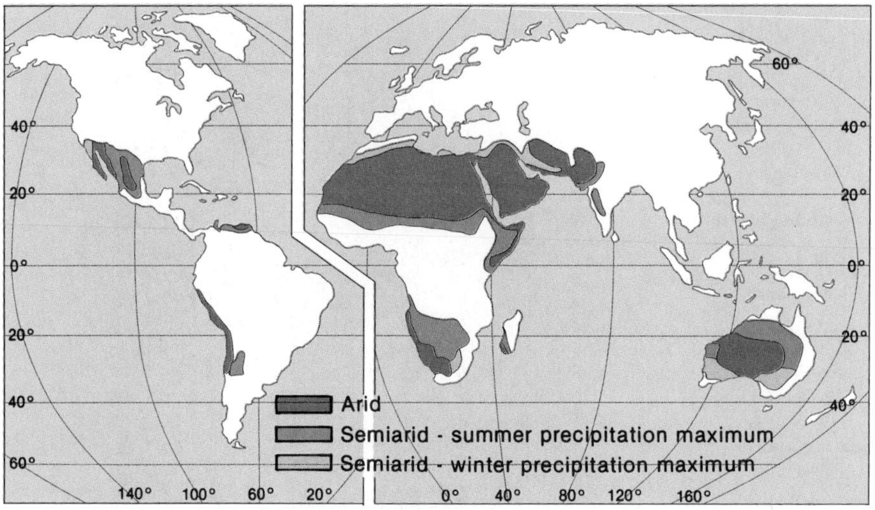

FIGURE 9.11 Distribution of arid and semiarid subtypes of the Low Latitude Dry Climate.

months during the summer, the poleward shift of the ITCZ brings substantial convectional rainfall. Rainfall is insufficient to offset the dryness of the other seasons, however, so that a semiarid climate prevails. This pattern is illustrated by the climagraph of N'Djamena, Chad (see Figure 9.12).

The arid core of this climate occupies the largest total area. Rainfalls here are very scanty and no distinctive pattern may be evident. The passage of time commonly produces an endless succession of days with deep blue skies, perhaps with scattered cirrus or afternoon cumulus clouds, and excellent visibilities. On occasion, however, visibilities may be reduced by airborne dust particles or even full-fledged dust storms. During the summer, intense heating may be sufficient to produce scattered convectional showers and thunderstorms, especially over upland regions. Often the air between the clouds and ground is so dry that the rainfall evaporates as it descends. By contrast, if rain falls in winter, it generally is produced by the unusually deep equatorward penetration of frontal cyclonic storms originating in the middle latitudes. The winter rainfalls typically are lighter and more widespread than those produced by summer season convectional storms. The climagraph for Yuma, Arizona (see Figure 9.13), is typical of a low latitude desert station.

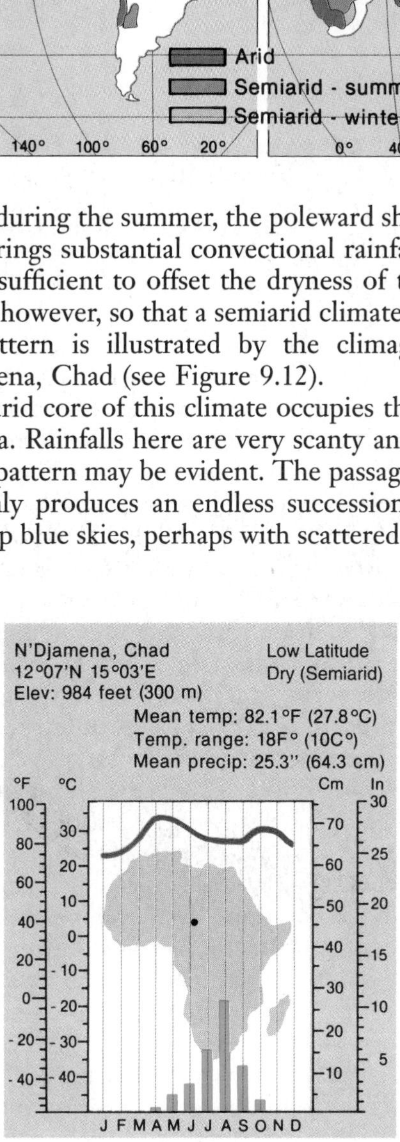

FIGURE 9.12 Climagraph for N'Djamena, Chad, with its double maximum temperature pattern caused by the advent of the rainy season.

Semiarid regions on the poleward margins of the Low Latitude Dry Climate have a reverse precipitation pattern from that of the climate's equatorward margins. In these areas, which are transitional between the low and middle latitudes, summers generally are dry because the subtropical high shifts poleward at this time of year to blanket the area. During the winter, however, the high shifts equatorward, allowing the westerlies with their eastward-traveling cyclonic storms to penetrate. Occasional one- or two-day periods of light to moderate precipitation can result, and in some of the cooler locations significant snowfalls are sometimes experienced. The climagraph for Tripoli, Libya (see Figure 9.14), illustrates the pattern.

The Low Latitude Dry Climate is as well known for its heat as it is for its dryness. The subtropical

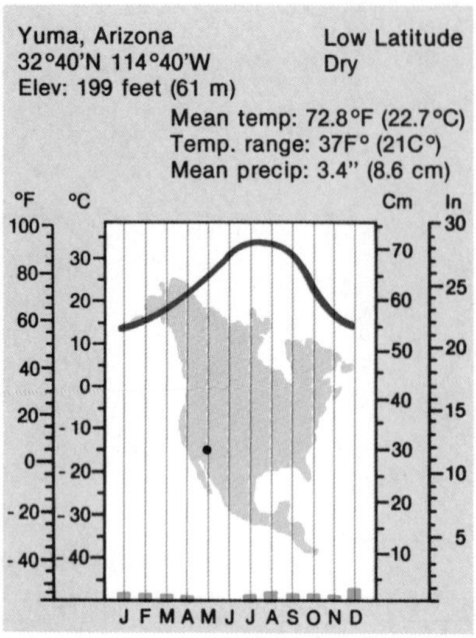

FIGURE 9.13 Climagraph for Yuma, Arizona, a station dominated by the northern hemisphere subtropical high.

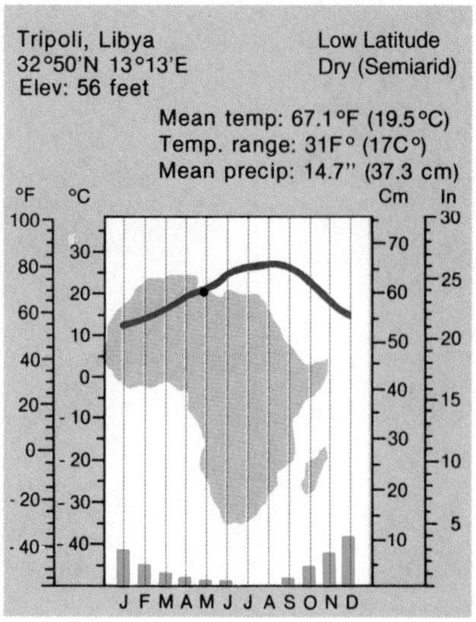

FIGURE 9.14 Tripoli, Libya, is a semiarid station located on the poleward margin of the northern hemisphere subtropical high.

deserts are the source regions for continental tropical air masses, and their summer temperatures are the hottest in the world. Many Low Latitude Dry Climate stations have means in the mid 30s Celsius (the 90s Fahrenheit) for several consecutive summer months, and a few reach a mean of 38°C (100°F) in their hottest month. Daytime highs of 46° to 50°C (115° to 120°F) are not uncommon in some inland desert areas at low elevations, and these daytime maximum readings often are remarkably consistent for weeks at a time.

Associated with the extreme summer heat are very low relative humidity values. Low humidities are caused not so much by a lack of atmospheric water vapor as by the very high water vapor holding capacity of the air at such hot temperatures. Daytime relative humidity readings in the desert often fall below 10 percent in the afternoon, and a low of 2 percent has been recorded. The relatively dry air may make the heat more bearable, but it results in extremely high evaporation rates from moist surfaces such as reservoirs and irrigated fields and causes inhabitants to need large quantities of liquids.

Semiarid areas, with their somewhat larger amounts of evaporable surface moisture, experience slightly less extreme heat. This is especially true for areas on the equatorward side of the subtropical highs that receive substantial convectional summer rainfalls (such as N'Djamena, Chad) and for coastal regions with a maritime influence (such as Tripoli, Libya).

In contrast to the excessive heat of the summer, winter temperatures within the Low Latitude Dry Climate tend to be surprisingly cool. The combination of clear skies and dry air makes the atmosphere relatively transparent to outgoing terrestrial radiation, while the lower Sun angles and shorter periods of daylight considerably reduce insolation receipts. As a result, midwinter temperatures average 10° to 18°C (the 50's to low 60's Fahrenheit) in all but the equatorward portions of this climate. (Compare, for example, the winter temperatures of Yuma and Tripoli with those of N'Djamena, which lies 20° closer to the equator.) Winter days generally are pleasantly warm, especially in the sunlight, but nighttime lows can be quite cold, and frost and freezing temperatures are occasionally experienced in many of the more poleward locations. Daily temperature ranges within subtropical deserts during the winter half of the year are among the highest in the world. These ranges usually are between 18C° and 35C° (30 and 60F°), and temperatures have been known to rise from below freezing to highs exceeding 38°C (100°F) in a single day. Little wonder, then, that stories of the desert sometimes describe it as "broiling during the day and freezing at night"!

Strong surface heating during the summer half of the year produces very steep near-surface lapse rates, resulting in strong, gusty daytime winds and the occasional formation of dust devils. These winds are capable of picking up clouds of dust and stinging fine sand, adding one more discomfort for the desert traveler. Hot, dust-laden winds from the Sahara, for example, are periodically experienced all along that desert's margins. The problems of atmospheric dust have been compounded in recent decades by the large-scale human removal of protective vegetation from low latitude semiarid areas for agriculture and for fuel (see the Chapter 19 Case Study).

THE MIDDLE LATITUDE CLIMATES

Between the zones dominated by the subtropical highs and the subpolar lows lie the broad expanses of the middle latitudes. This area often is subject to fast-changing weather conditions as both high pressure systems and cyclonic storms with their associated fronts pass by in endless succession in the westerly wind flow. Climatic conditions are determined largely by continentality and latitude and in some places are strongly modified by the topography. In general, more poleward locations are farther from the subtropical highs and experience increased cyclonic and frontal activity. The decline of temperatures in a poleward direction, however, eventually significantly limits the water vapor content and therefore the precipitation-generating potential of the atmosphere.

Isolation receipts in the middle latitudes are sufficiently low in winter to produce a cold season that substantially lowers annual temperature means. The net result of the lowered temperatures and increased storminess is that annual precipitation totals exceed potential evapotranspiration rates within most of the middle latitudes, bringing a return to humid or subhumid climatic conditions.

Dry Summer Subtropical Climate (Köppen Csa and Csb)

The Dry Summer Subtropical Climate (also called the Mediterranean Climate) is found in the western portions of continents in the lower middle latitudes, generally between 30° and 45° N and S. This climate is transitional in location between the subtropical deserts and the humid middle latitudes; in effect, the climatic characteristics of each region dominate for a portion of the year. The mountainous nature of most land areas in which this climate occurs typically restricts it to rather narrow coastal zones. As a result, the Dry Summer Subtropical Climate occupies the smallest land area of any climatic type. Its human significance, however, is great since most areas are densely populated and economically prosperous (see Figure 9.15).

Portions of five continents have areas with a Dry Summer Subtropical Climate. Although separated by vast distances, these areas have similar relative geographical positions and remarkably similar climatic conditions. In North America, the climate extends from just south of Los Angeles north to near the Oregon border. In general, it is restricted to the coastal zone west of the Sierra Nevada and the Cascade Mountains, but it does extend inland up the Columbia River and its tributaries into southern Idaho. In South America, it is located in central Chile west of the high Andes. It occurs throughout the Mediterranean Basin region of southern Europe and northwestern Africa, east into parts of the Middle East. This is the only area where there are both enough land and a sufficient lack of restricting mountain barriers to allow the climate to cover a large area. Finally, the Dry Summer Subtropical Climate occupies the coastal portions of southern and southwestern Australia and brushes the southwestern tip of Africa in the vicinity of Cape Town.

Temperatures within the Dry Summer Subtropics generally are held within moderate limits by the presence of nearby water bodies. The maritime influence lowers both daily and annual ranges. During the summer, large differences in temperature frequently exist between locations near the coast and those farther inland. Coastal areas often are kept quite cool by the adjacent water, but readings rise rapidly as one proceeds inland, and interior locations usually have hot temperatures with warmest monthly means in the upper 20s Celsius (upper 70s or lower 80s Fahrenheit). Temperature differences are especially striking along coasts paralleled by cold ocean currents and backed by hills or low mountains that block the inland penetration of cool ocean breezes. Such is the case in California and to a lesser extent in Chile and Portugal. (Compare, for example, the climagraphs for San Francisco and Red Bluff in Figures 9.16 and 9.17.)

The Mediterranean coast has much warmer summer means than do locations such as San Francisco

FIGURE 9.15 The town of Capileira, in Andalusia, southern Spain, is located in the Dry Summer Subtropical Climate. (Stephen Johnson/Tony Stone Worldwide)

that are situated on an oceanic coast. This is because the Mediterranean Sea is a restricted interior body of warm water that experiences intense summer insolation and receives only a very limited inflow of cooler Atlantic water. As a result, most locations experience summer means in the mid 20s Celsius (the high 70s Fahrenheit). Split, a city on the Adriatic coast of Yugoslavia, is a typical example (see Figure 9.18).

Although daytime highs in the summer commonly rise into the mid 30s Celsius (the 90s Fahrenheit) and sometimes exceed 38°C (100°F) in interior locations, nighttime low temperatures in most areas are pleasantly cool, with readings in the low to mid teens Celsius (the 50s to lower 60s Fahrenheit). This results from the generally clear skies of summer and the fact that the combination of subsiding air in the subtropical highs and relatively cool water produces

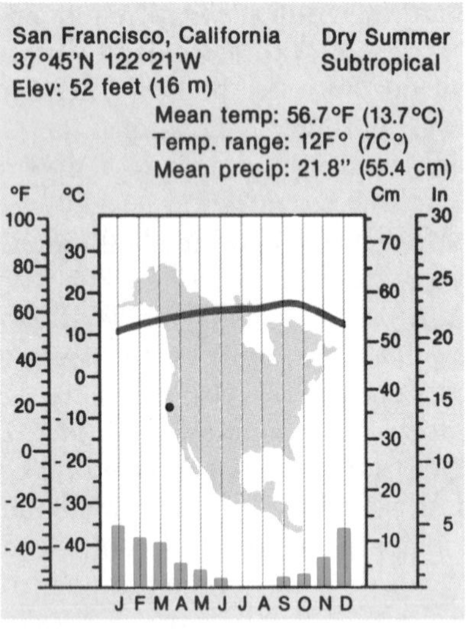

FIGURE 9.16 Climagraph for San Francisco, California. San Francisco's exposed coastal location causes it to have unusually cool summers and mild winters for its latitude.

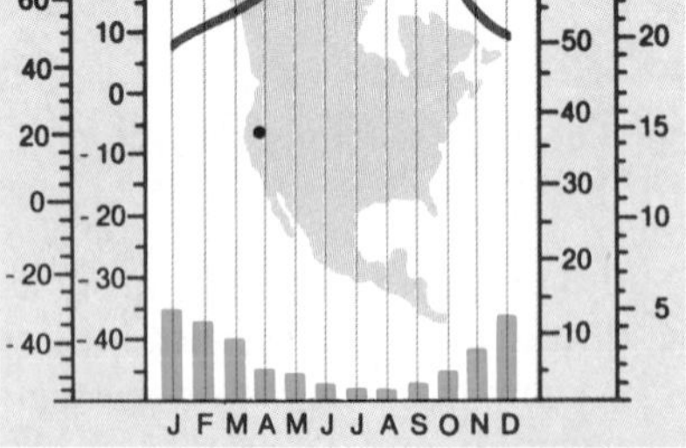

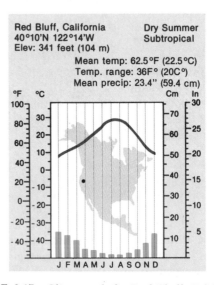

FIGURE 9.17 Climagraph for Red Bluff, California. Red Bluff's interior location causes it to have three times the annual temperature range of San Francisco.

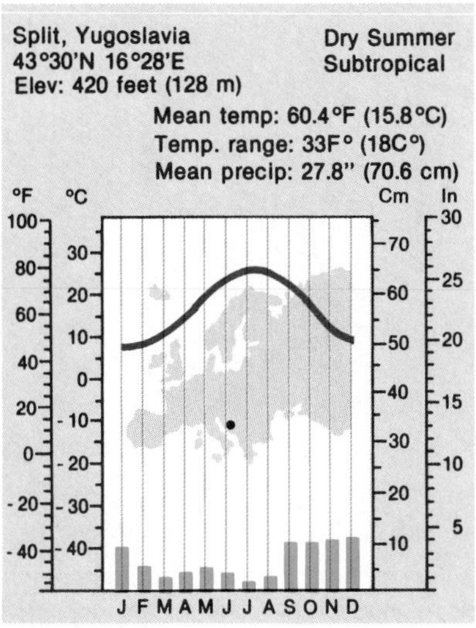

FIGURE 9.18 Climagraph for Split, Yugoslavia.

a reasonably low atmospheric water vapor content even along the coasts. These factors promote radiational cooling.

Winter temperatures within the Dry Summer Subtropics average mostly between 5° and 12°C (the 40s and lower 50s Fahrenheit). Temperature variations at this time of year between coastal and inland locations are much smaller than they are in summer, and generally the interiors are now cooler. (Compare winter values for San Francisco and Red Bluff.) Winter daily temperature ranges are also considerably higher inland than near the water. As a result, frosts and freezes, though rare in coastal areas, are common inland. This is especially true for locations situated well above sea level. Maritime polar air masses dominate all areas during the winter, but on the west coast of North America and in the northern Mediterranean Basin area, occasional invasions of continental polar air from the northeast can bring killing frosts even to the coasts.

The most distinguishing characteristic of the Dry Summer Subtropical Climate is its annual precipitation pattern of dry summers and humid winters. During the summer season, the subtropical high in each hemisphere shifts poleward with the Sun to blanket the Dry Summer Subtropics. The summer months are almost rain-free along cold water coastal regions, such as California and Chile, where cool temperatures further reduce the already low lapse rates. The climagraph for San Francisco illustrates this situation. Summer advection fogs prevail along these coasts because of the cold water. The dry season also is exceptionally long and dry along the equatorward margins of the climate, because these areas lie closest to the centers of the subtropical highs. In more poleward locations, the influence of the subtropical highs is not as strong, and occasional rainfalls from convectional showers or weak frontal passages may occur. Monthly precipitation means, however, are generally less than 2.5 centimeters (1 inch) during midsummer. The climagraph for Split illustrates this situation.

During the winter, the subtropical highs retreat equatorward, and a moderate onshore flow of less stable maritime polar air develops. Fronts and frontal cyclones also extend their influence into the subtropics at this time of year, providing the lifting needed to trigger periods of precipitation. Storms typically occur every three or four days, producing moderate to occasionally heavy rainfall. The periods between storms usually consist of sunny, mild weather, and most areas receive 50 percent or more of their maximum possible winter sunshine. Monthly precipitation means during the winter are commonly close to 7 to 10 centimeters (3 to 4 inches), with amounts increasing poleward (away from the subtropical highs) and with increasing elevation. Most areas receive at least 75 percent of their total precipitation during the winter half of the year.

Annual precipitation totals in most cases average 40 to 90 centimeters (15 to 35 inches), with the lower totals most common in the more equatorward locations. These amounts are quite low, and some of the drier areas would be classified as semiarid were it not for the fact that most of the precipitation occurs in the winter, thus maximizing its effectiveness in supplying moisture to the surface by reducing evapotranspirational losses. Yearly variations in the position of the subtropical high and the storm tracks result in a large annual variability in amounts received in all areas. As a result, droughts and floods are not uncommon (see the Focus Box on page 196).

Humid Subtropical Climate (Köppen Cfa)

The Humid Subtropical Climate is the east coast counterpart of the Dry Summer Subtropics. It occupies the eastern sides of continents in the lower middle latitudes between approximately 25° and 43° N and S. The leeward position of these areas results in

THE CALIFORNIA DROUGHT

In 1991 California entered the fifth year of a drought that has produced serious water shortages throughout the southern half of the state. California's precipitation pattern is notoriously fickle because the state is situated in the area of transition between the westerlies, with their traveling cyclonic storms, and the arid climates associated with the eastern Pacific center of the subtropical high. In order for California's winter rains to arrive on schedule, the subtropical high must shift southward over Mexico, and the jet stream also must shift southward so that it enters North America in California, rather than in Canada or the Pacific Northwest.

Major precipitation events are related to the pattern of Rossby waves that develop within the flow of the jet stream. In some years, deep wintertime troughs propagate inland at regular intervals over California from the Pacific, bringing copious rains and mountain snows. In other years the jet stream remains largely north of the state; this has been the case for most of the late 1980s and early 1990s.

FIGURE 9.19 The California drought resulted in drastically reduced water levels in many reservoirs, such as the one pictured here. (Visuals Unlimited/© Walt Anderson)

a moderate continental influence, but this is tempered by large adjacent bodies of warm water (usually oceans with warm currents) that supply sufficient water vapor to produce humid conditions.

The most extensive land area within the Humid Subtropics is the southeastern one-third of the United States, extending from extreme southeastern New England south through most of Florida and west to Texas and eastern Kansas. Another large area occupies southeastern South America from southern Brazil through Uruguay and northeastern Argentina. In Asia, much of eastern China, southern Japan, and extreme southern Korea experience a Humid Subtropical Climate. A fourth relatively extensive area is found in eastern Australia. Finally, small patches of Humid Subtropics occur in southern and southeastern Europe, especially around the northern portions of the Adriatic Sea and the Black Sea, and in Africa, along a portion of the southeast coast of the Republic of South Africa.

Temperatures within this climate are relatively similar to those experienced in inland portions of the Dry Summer Subtropics, with warm to hot summers and cool to mild winters, depending upon latitude and elevation. Higher humidities, however, make the Humid Subtropics less comfortable, especially in summer. Poleward areas also generally have colder winters and larger annual temperature ranges than do comparable west coastal locations.

The long period of summer heat and humidity is one of the most notable characteristics of the Humid Subtropics. Oceanic centers of the subtropical high

The human impact of the California drought has been greatly augmented by the state's continued rapid economic growth. California currently has a population of 30.3 million—by far the largest of any state—and is expected to reach 39 million by the year 2010. This population is distributed very differently than is water availability; three-fourths of the people live in southern California, while three-fourths of the precipitation falls in the northern half of the state. California has developed a sophisticated water collection and distribution system to make use of its limited water resources; it has 1,200 reservoirs, a system of deep wells, and long-distance aqueducts to carry water from sources such as the Colorado River and the Sierra Nevada to the parched south. Unfortunately, however, the state is running out of expansion potential for its water supply system because there are few rivers left to dam and ground water levels in some areas are declining as rapidly as 0.6 meters (2 feet) a year.

Another basic problem affecting future water availability in California is the fact that 80 to 85 percent of the state's water is used for irrigation agriculture. Although California is the leading agricultural state in the nation and its farmers grow half the fruits and vegetables consumed in the United States, agriculture generates only 3 percent of the state's revenues. Water allocation laws passed decades ago allow farmers to purchase vast quantities of water at bargain rates. They also make it difficult or impossible in many cases to transfer agricultural water to urban supply systems.

Water supply problems are certain to continue to face Californians for many years because no end is in sight for the state's rapid economic and population growth. Any significant global warming would likely compound the situation by producing a northward shift of both the subtropical high and the westerlies. This would reduce winter precipitation while at the same time increasing potential evapotranspiration rates.

A variety of solutions to the water supply problem are being explored. One is the large-scale construction of plants to desalinate sea water; unfortunately, this option is very expensive, costing an estimated $1,000 to supply an average household with water for a year. An even more expensive option that has been discussed is importing water from British Columbia in supertankers. More realistic in the short term is a policy of water conservation and reallocation. It appears that the large-scale use of vast quantities of water for low-value agricultural purposes such as keeping livestock pastures green and growing alfalfa will have to be phased back. California, like several other western states, must come to grips with the fact that its water supplies are finite and are now approaching the limits of their potential for regional development.

far to the east pump in a nearly continuous flow of maritime tropical air. (In the case of southeastern Europe, moisture is derived from the Atlantic Ocean and the Mediterranean and Black Seas.) Adding to the discomfort produced by the humidity is heat resulting from high Sun angles and long daylight periods. Although weak incursions of continental polar air bring occasional interludes of pleasant weather to the more poleward areas, the heat and humidity are almost unbroken equatorward of approximately 35° N and S for a two- or three-month period. Coastal locations are slightly cooler, and daily sea breezes frequently develop, but humidities are typically higher near the coast than inland.

Summer monthly means largely depend upon latitude and range through the 20s Celsius (the lower 70s to mid 80s Fahrenheit). Daily highs usually average from the upper 20s to middle 30s Celsius (middle 80s to lower 90s Fahrenheit), but in some areas the hottest summer days have readings in excess of 38°C (100°F). Nights are warm and sultry, with overnight lows frequently remaining in the mid or high 20s Celsius (middle 70s to about 80°F). The climagraphs for Baltimore, Maryland, and Jacksonville, Florida (see Figures 9.21 and 9.22), illustrate the influence of latitude on the length and intensity of the warm season.

During the last few weeks of summer and extending into early autumn, invasions of cool continental polar air masses become increasingly frequent, and by late autumn and winter, cP air dominates in most areas. Autumn is a delightful time of year in the Humid Subtropics, with extended periods of fair, pleasantly mild weather. Temperatures fall rapidly in the more poleward areas and more slowly farther equatorward to midwinter means ranging from about −3° to 13° C (the upper 20s to the mid 50s Fahrenheit). Weather fronts are active during the winter half of the year, resulting in substantial daily temperature variations. Northern hemisphere areas are especially prone to sharp temperature fluctuations because they are within reach of arctic (cA) air masses, and most locations experience winter days in which temperature readings remain near or below freezing. Temperatures normally begin to become milder a few weeks before the vernal equinox, and the spring brings another period of pleasant temperatures before the return of summer heat and humidity.

The most pronounced climatic difference between the Dry Summer Subtropics and the Humid Subtropics is in their annual precipitation patterns. While the former climate experiences a well-defined dry season in summer, the latter receives substantial amounts of precipitation throughout the year, with summer most commonly the season with the heaviest totals. Annual means within the Humid Subtropics are about twice those of the Dry Summer Subtropics, with most areas receiving 75 to 165 centimeters (30 to 65 inches). Also in contrast to the Dry Summer Subtropics, precipitation amounts are greatest near

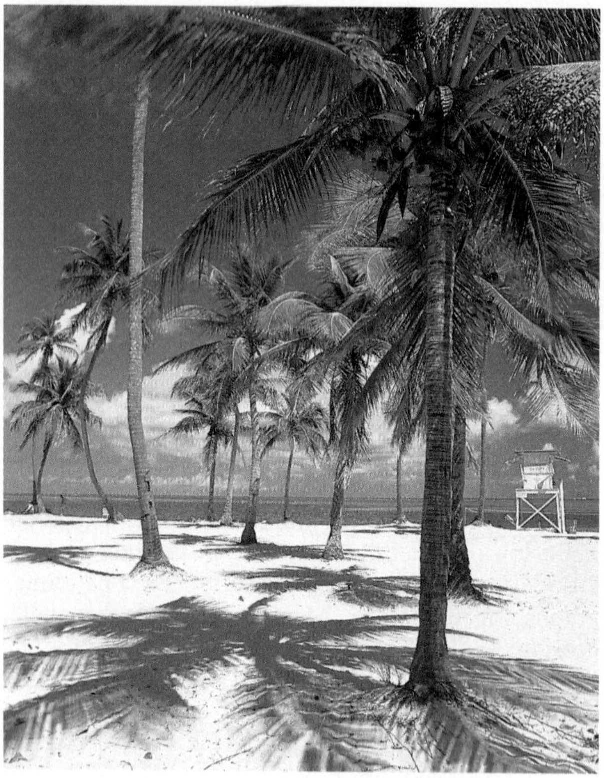

FIGURE 9.20 Key Biscayne, Florida, is a popular beach resort in the Humid Subtropics. (F. Sieb/H. Armstrong Roberts)

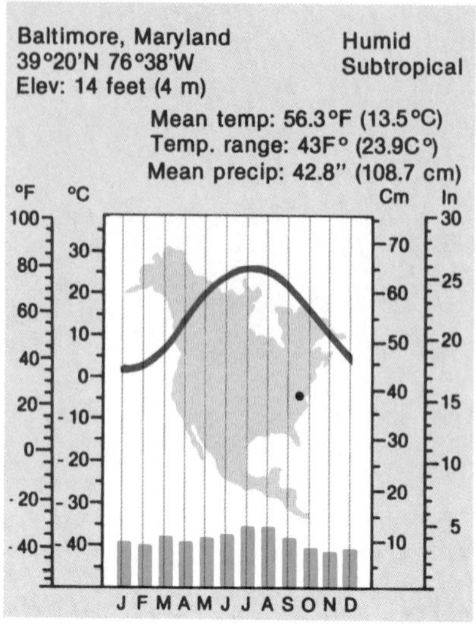

FIGURE 9.21 Climagraph for Baltimore, Maryland, near the northern border of the Humid Subtropical Climate in the United States.

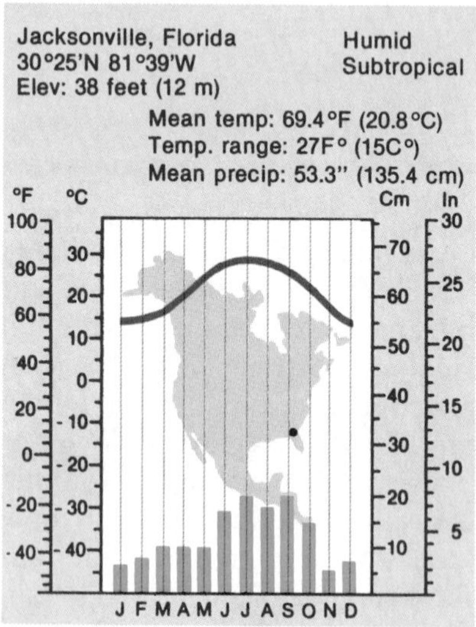

FIGURE 9.22 Climagraph for Jacksonville, Florida. The differences between Balitmore's and Jacksonville's temperature and precipitation means result largely from their differing latitudes.

the climate's equatorward margins. They are lightest poleward and inland, and interior portions of the Humid Subtropics frequently adjoin semiarid areas.

During the summer, the subtropical highs dominate, but their centers are situated well to the east, producing a weak onshore flow of unstable mT air. The monsoonal tendency serves to reinforce this wind-flow pattern, especially in eastern Asia. The resulting weather is warm, humid, and mostly sunny, but daytime heating frequently produces scattered afternoon and evening convectional showers and thunderstorms. Cumulonimbus clouds are observable almost every summer afternoon in more equatorward portions of the Humid Subtropics, such as the U.S. Gulf coastal states, and are responsible for heavy downpours. Rainfalls are highly variable in amount because these storms are localized and are slow moving due to the very weak upper-level winds that prevail at this time of year. During the winter, most precipitation is frontal and cyclonic in nature, causing it to be lighter, steadier, and much more widespread than the convectional rains of summer. Most winter precipitation falls as rain, but occasional snowstorms are experienced, especially in the more poleward areas.

Along the coasts of the United States, Asia, and Australia, summer and early autumn bring the threat of hurricanes. These tropical storms occasionally produce great destruction and loss of life within the Humid Subtropics, as illustrated by the Case Study at the end of Chapter 8.

Mid Latitude Dry Climate (Köppen BWk and BSk)

Nearly three-fourths of the world's dry climates—those areas where potential evapotranspiration rates exceed precipitation amounts on a yearly basis—are in the low latitudes. The remaining one-fourth of the dry climates occur in the middle latitudes, mostly within continental interiors in the northern hemisphere. Their dryness is not associated with the influence of the subtropical highs, but instead results from the presence of orographic barriers on their windward sides. A secondary factor in many cases is their remoteness from oceanic sources of maritime air.

Three widely separated areas are included within the Mid Latitude Dry Climate. The largest is the central portion of Eurasia extending from just north of the Black Sea eastward some 8,000 kilometers (5,000 mi) to north-central China. This vast region is cut off from moisture-bearing winds from the Indian Ocean and Mediterranean Sea by an imposing rim of mountains extending along its southern and southwestern margins. In addition, it is separated from Atlantic moisture by vast distances. Central Asia, in the heart of this area, is more remote from oceans than any other place in the world. In North America, the Great Plains and Great Basin are largely blocked from Pacific moisture by the mountains to their west, and thus experience a dry climate. These areas adjoin the drylands dominated by the subtropical high in the southwestern United States. The third, and smallest, area having this climate type is the Patagonian region of southern and western Argentina. It is geographically distinct from the other two in that it is located on a relatively narrow body of land. Its dryness is associated primarily with the rainshadow effect of the Andes, but the potential for precipitation is further reduced by the stabilizing influence of the cold Falkland Island Current off the east coast.

Most portions of the Mid Latitude Dry Climate experience highly continental conditions because maritime air masses do not normally enter these areas. Annual mean temperatures vary considerably from place to place, as illustrated by the climagraphs for Saratov, USSR; Reno, Nevada; and Mendoza, Argentina (see Figures 9.24–9.26). Readings depend most upon latitude, degree of continentality, and elevation.

Summer temperatures in the Mid Latitude Dry Climate generally are warm, with occasional hot spells. Warmest monthly means are as low as the low teens Celsius (the mid 50s Fahrenheit) in southern

FIGURE 9.23 The short grasslands of southeastern Montana are typical of those of many Mid Latitude Dry Climate regions. (Richard Jacobs/© JLM Visuals)

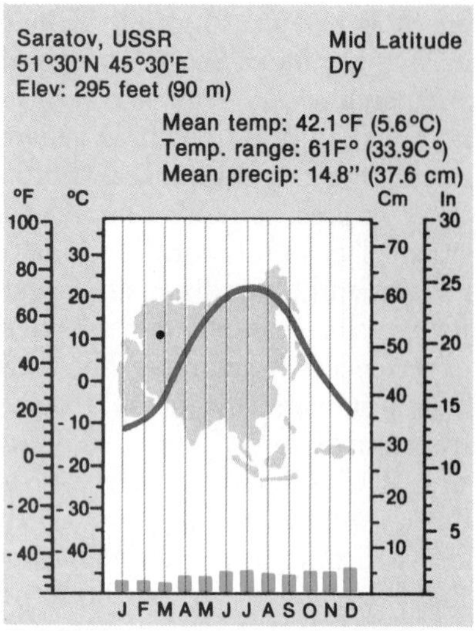

FIGURE 9.24 Climagraph for Saratov, USSR.

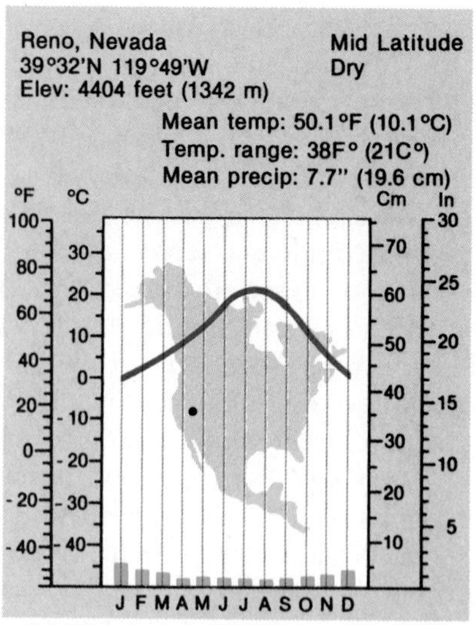

FIGURE 9.25 Climagraph for Reno, Nevada. Reno is located close enough to the Dry Summer Subtropical Climate of California to experience a modest winter precipitation maximum.

Argentina, and as high as 30°C (86°F) in some of the desert basins of eastern Asia. Clear skies and dry air allow daytime readings to soar well above 38°C (100°F) on occasion in the hottest areas. Nighttime lows, however, generally fall to comfortable readings in the upper teens to low 20s Celsius (the 60s or lower 70s Fahrenheit) in the warmer areas and to between about 7° and 15°C (the upper 40s and 50s Fahrenheit) in cooler locations at higher latitudes and elevations. Even in the warmest locations, the summer heat is not constant since these areas, unlike the dry climates of the low latitudes, are within reach of occasional air masses of subpolar origin.

Winter temperatures are cold in most places, but the degree of coldness is highly dependent upon location. The mildest of the three regions is Argentina, where virtually all areas have means that remain above freezing. In contrast, most stations in North America and virtually all those in Eurasia have subfreezing temperatures during the winter, with January means in portions of eastern Asia below −18° C (0° F). Both North America and Eurasia are within reach of wintertime arctic air masses and experience cold waves when this frigid air moves south from its source regions. Temperatures during these cold air outbreaks can get as low as −40° C (−40° F) in the northern Great Plains and −45° C (−50° F) in eastern Asia. Coupled with the bone-chilling cold are winds that frequently reach very high velocities over these flat, treeless surfaces, producing potentially deadly wind-chill values for humans and unprotected livestock. Interspersed with the cold periods, however, are intervals of milder weather as air masses from the south or west penetrate into the regions. Sudden onsets of Chinook winds in the northern Great Plains, for example, are capable of producing dramatic rises of temperature as they displace arctic air masses. An extreme example of a Chinook oc-

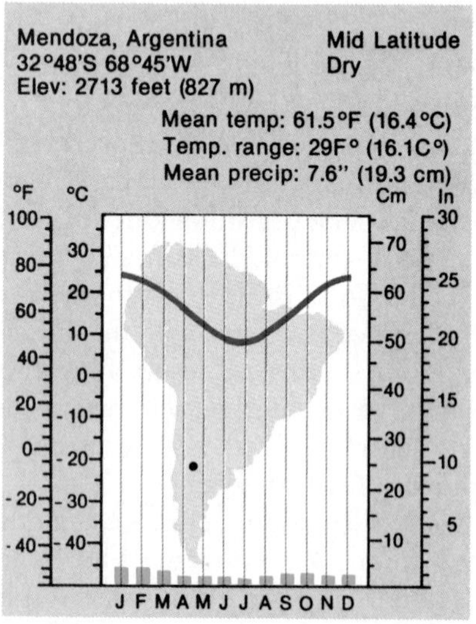

FIGURE 9.26 Climagraph for Mendoza, Argentina.

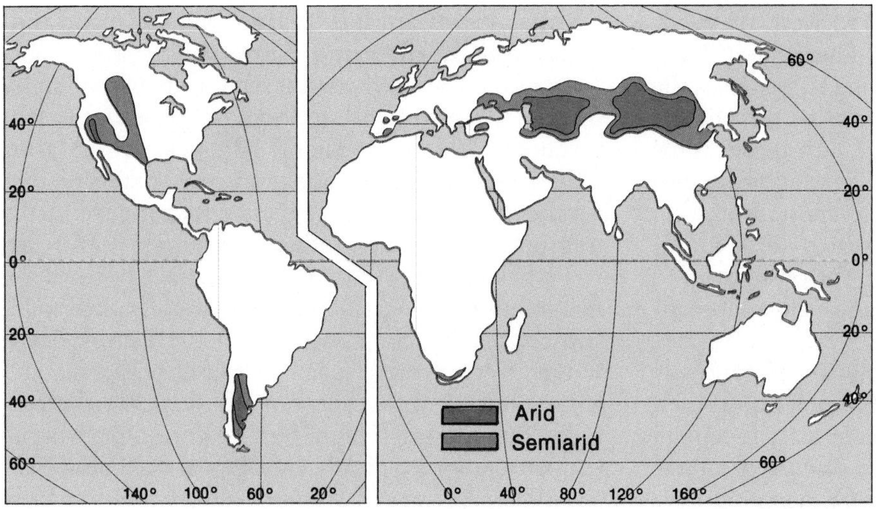

FIGURE 9.27 Distribution of arid and semiarid regions within the Mid Latitude Dry Climate.

curred in Spearfish, South Dakota, on January 22, 1943. At 7:30 A.M., the temperature was −20° C (−4° F). Two minutes later, at 7:32 A.M., it had risen to 7.2° C (45°F)!

Precipitation amounts within different portions of the Mid Latitude Dry Climate depend largely on the strength of the rainshadow effect. Both arid and semiarid conditions occur. As an approximation, arid areas receive less than 20 centimeters (8 inches) of precipitation annually, and semiarid areas, 20 to 50 centimeters (8 to 20 inches). These amounts are somewhat lower than the equivalent values for the arid and semiarid climates of the low latitudes because the cooler temperatures of the middle latitudes reduce evapotranspiration rates and thus increase precipitation effectiveness. (Cold year-round temperatures and the resulting very low evapotranspiration rates are, incidentally, the reason why no dry climates exist in the high latitudes. Deserts need not always be hot—most, in fact, are not—but they do need to be sufficiently warm during at least part of the year so that potential evapotranspiration rates exceed precipitation totals.) As is true in the subtropics, the Mid Latitude Dry Climate contains arid core areas surrounded by extensive semiarid rims. Each of the three continents has one or more arid cores (see Figure 9.27).

The seasonal distribution of precipitation is highly variable and generally is similar to that of the nearest humid climate area. For example, Reno, located near the margin of the Dry Summer Subtropics, receives a winter precipitation maximum from Pacific storms that retain some moisture as they cross the Sierra Nevada. Saratov, near the border of the Humid Subtropical Climate, has a weak summer maximum. Most locations receive a warm season precipitation maximum largely because the air contains more water vapor at this time of year.

As is typical of the middle and higher latitudes, most precipitation is associated with traveling frontal cyclones. These storms usually are capable of generating only light amounts of rain or snow because they do not have access to major sources of moisture. Winter snowstorms, however, occasionally produce substantial accumulations, with accompanying strong winds creating blizzard conditions.[1] Spring sometimes brings turbulent weather resulting from the strong temperature contrasts that develop at this time of year between opposing warm and cold air masses. In the U.S. Great Plains, especially, this conflict can produce severe hailstorms and tornadoes. During the summer, weather fronts are weaker and less numerous, but the air may be sufficiently moist and unstable to trigger convectional showers and thunderstorms.

Marine Climate (Köppen Cfb and Cfc)

Moving poleward into the upper middle latitudes, we enter the heart of the westerly wind belt. Here, frontal cyclonic storms are both numerous and well developed, and weather systems tend to move rapidly eastward under the influence of strong upper-level

[1] A **blizzard** is a heavy snowstorm accompanied by winds of more than 16 meters per second (35 miles per hour) and visibilities that are reduced to below 150 meters (500 feet) by falling and blowing snow.

winds. Situated in this zone on the west, or windward, sides of continents generally between 40° and 60° N and S, the aptly named Marine Climate has the strongest maritime influence on any of the climate types. An almost constant onshore flow of maritime polar air provides adequate moisture to allow precipitation totals to exceed potential evapotranspiration rates through most or all of the year so that no dry season is experienced. In addition, the onshore wind flow produces cool summers but unusually mild winters.

Geographically, the Marine Climate is generally located just poleward of the Dry Summer Subtropics. Like that climate, it is restricted in size, except in Europe, by the blocking influence of coastal mountain ranges and, in the southern hemisphere, by a lack of land masses at the proper latitude. In North America, the Marine Climate occupies a narrow coastal strip beginning in extreme northern California and extending northward and westward along the west coast of Canada and the south coast of Alaska to the Aleutian Islands. In South America, it occurs in a similarly restricted zone in the portion of Chile west of the Andes and south of 40° S. The Marine Climate extends well inland in Europe, except in Scandinavia, where it is orographically restricted to western Norway and southern Sweden. Elsewhere in Europe it exists from southern Iceland southward to northern Spain and eastward into Poland, Romania, and Bulgaria. In both Africa and Australia it occurs in portions of the southern coastal regions; it also occupies most of New Zealand. It should also be noted that vast ocean areas in the middle latitudes have a Marine Climate. Indeed, these areas are the source regions for the mild, moist mP air masses that are eventually transported onto the continental margins to the east.

Perhaps the most noteworthy characteristic of the temperatures in Marine Climate areas is their relative constancy. The resistance of the nearby ocean waters to temperature changes causes both daily and annual ranges to be exceptionally small for the latitude. Summertime high temperatures for Marine Climate stations typically average between about 12° and 20°C (the upper 50s and 60s Fahrenheit); nightly lows are commonly in the lower teens Celsius (the 50s Fahrenheit). In the winter, daytime highs generally range between about 5° and 11° C (the 40s or lower 50s Fahrenheit), while nightly lows range from about −4° to +5°C (the upper 20s and 30s Fahrenheit). As might be expected, temperature ranges are lowest along the immediate coasts and increase inland. As a result, coastal regions experience cooler summers and milder winters than do interior locations. For example, compare the climagraphs for Brest, France, and Budapest, Hungary, in Figures 9.28 and 9.29.

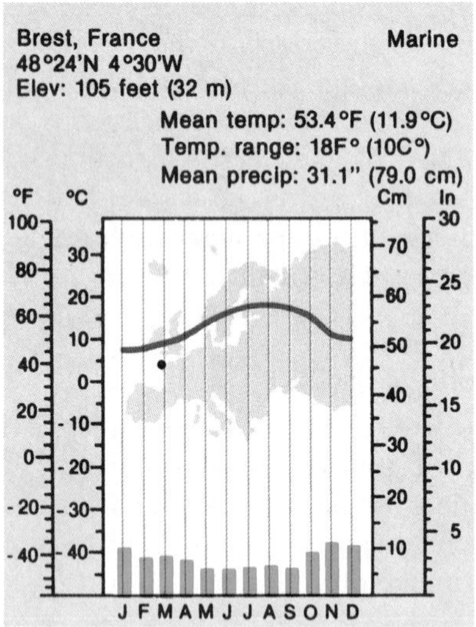

FIGURE 9.28 Climagraph for Brest, France. Brest's coastal location is responsible for its relatively low annual temperature range as well as its winter precipitation maximum.

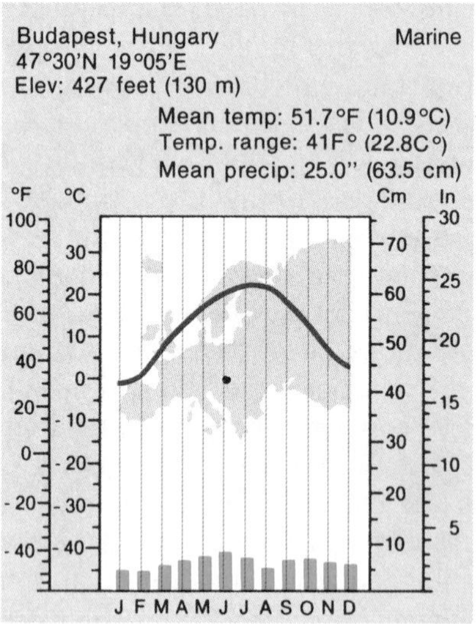

FIGURE 9.29 Climagraph for Budapest, Hungary. Budapest is located in central Europe and has a much larger annual temperature range than does Brest, as well as a summer maximum of precipitation.

FIGURE 9.30 Chilkat Inlet, in Southeast Alaska, has a Marine Climate with heavy orographic precipitation. © Charlie Ott/Photo Researchers)

The Marine Climate is characterized by a large amount of cloudy and rainy or drizzly weather. In fact, the mean annual percentage of cloudiness and total duration of precipitation exceed those of any other climate. Most precipitation is frontal and cyclonic, but in many areas orographic lifting also plays an important role. The storms that affect the west coasts usually are large and often deeply occluded; they tend to produce prolonged periods of leaden skies, light rain or drizzle, and strong winds. Convectional precipitation is minimized by the cool ocean temperatures, which act to stabilize the air.

Precipitation totals vary tremendously, often over rather short distances. This results largely from the highly varying topography of these west coastal areas. The heaviest totals occur along the coasts of Canada, Alaska, Chile, and New Zealand, where lofty north-south mountain ranges lie across the path of the prevailing westerlies (see Figure 9.30). The resulting orographic uplift of the moist oceanic air produces heavy rains and snows, with annual means exceeding 500 centimeters (200 inches) of water in a few localities. These means are greater than those of any other locations outside the tropics and subtropics. Ketchikan, Alaska (see Figure 9.31), is a representative station with a strong orographic influence and excessively heavy precipitation totals.

In contrast, lowland portions of the Marine Climate, notably in Europe, have relatively low annual precipitation means, usually ranging between 50 and 75 centimeters (20 and 30 inches). These modest amounts, however, generally are more than sufficient to maintain adequate surface moisture supplies, because evapotranspiration rates are kept low by the combination of high humidities, cool temperatures, and abundant cloudiness.

Precipitation usually is rather evenly distributed throughout the year, but coastal regions tend to receive a winter maximum because of the frequency of coastal storms triggered by temperature contrasts between the cold land and warmer ocean (see, for example, the climagraph for Seattle, Figure 9.32). Most coastal storms, even in winter, bring only rain and drizzle. Some of the heaviest snowfalls in the world occur in the high mountains backing these coasts, but their elevation causes these areas to have winter temperatures too low to be classified as Marine Climates.

In the inland portions of Europe, the larger annual temperature ranges and reduced frequency and intensity of winter storms (compared with the coasts) typically result in a weak summer maximum of precipitation. This is illustrated by the climagraph for Budapest (see Figure 9.29). The somewhat greater continentality of the inland areas also allows the summer air to become sufficiently heated to trigger occasional convectional showers and thunderstorms. This activity produces a substantial proportion of the summer precipitation for most areas of interior Europe.

Two additional noteworthy weather phenomena within the Marine Climate are strong winds and fog.

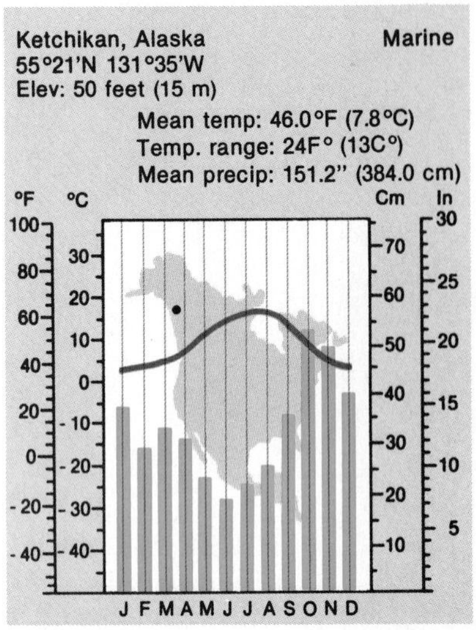

FIGURE 9.31 Climagraph for Ketchikan, Alaska. Ketchikan's heavy precipitation is primarily a result of orographic lifting.

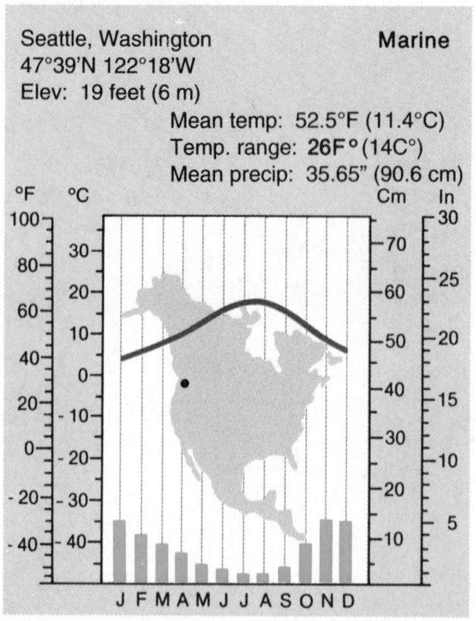

FIGURE 9.32 Climagraph for Seattle, Washington.

High winds commonly are associated with the deep storms that affect the coastal regions in winter. The combination of rugged, rocky coasts and frequent gales has led to many shipwrecks over the centuries. Perhaps the most famous was the destruction of the Spanish Armada off the coast of Scotland in 1588.

The combination of constant dampness and frequent land/water temperature contrasts also favors the formation of dense advection and radiation fogs. Human urban and industrial activities frequently enhance conditions for fog formation by their large-scale production of hygroscopic (water-attracting) condensation nuclei. The infamous smogs of London and the Rhine Valley are largely of human origin.

Humid Continental Climate (Köppen Dfa, Dfb, Dwa, and Dwb)

The interiors and east coasts of continents in the middle latitudes experience some of the most changeable weather on Earth. These regions lie in the latitudinal zone most frequented by the polar front and are in the collision zone between air masses of subtropical and subpolar origins. Because these air masses have not passed over moderating bodies of water, as in the case of the Marine Climate, they retain a large measure of the thermal characteristics of their source regions. As the polar front shifts north and south, these differing air masses alternately occupy a given locality, bringing about sharp changes in weather conditions.

Geographically, the Humid Continental Climate occupies the interiors and eastern (or leeward) coasts of large continents in the middle and upper portions of the middle latitudes. It occurs only in the northern hemisphere, in the continents of North America and Eurasia, because the higher latitudes of the southern hemisphere lack sufficient land to produce continental climate conditions. Specifically, the climate is located in three widely separated areas at latitudes generally ranging from 35° to 55° N. In North America, it extends from the Upper Midwest of the United States north into south-central Canada and east through the Great Lakes region, New England, and the Canadian Maritime Provinces. A second area begins in eastern Europe, extending from southern Scandinavia (where warmth derived from the Atlantic causes the region to extend poleward of 60° N) south into Bulgaria and east into the central Asiatic Soviet Union. The third area, located in eastern Asia, occupies portions of northeastern China, northern Korea, and northern Japan.

The great variability of temperatures on both a seasonal and a short-term basis is an important thermal characteristic of this climate type. Large annual ranges result from major seasonal variations in insolation received by continental land masses that readily heat and cool. They are increased further by the shift of the polar front, which is generally situ-

FIGURE 9.33 A checkerboard pattern of productive farmland occupies a vast area of the midwestern United States within the Humid Continental Climate. The scene here is from southern Wisconsin. (JLM Visuals)

ated to the north of the Humid Continental areas in summer and to their south in winter. Most locations have annual ranges between 22 and 42C° (40 and 75F°), with the greater values at higher latitudes (especially in more interior locations). The climagraphs for Des Moines and International Falls (see Figures 9.34 and 9.35) illustrate the influence of latitude on annual temperature ranges.

During the summer, the weather is almost continuously warm and humid near the southern margin of the Humid Continental Climate, where conditions are similar to those of the Humid Subtropics. Northward, however, temperatures fall to pleasantly cool values near the climate's poleward margins. Midsummer means range from as high as 27° C (80° F) in the central United States to near 16° C (60° F) in south-central Canada. Even the most poleward areas have

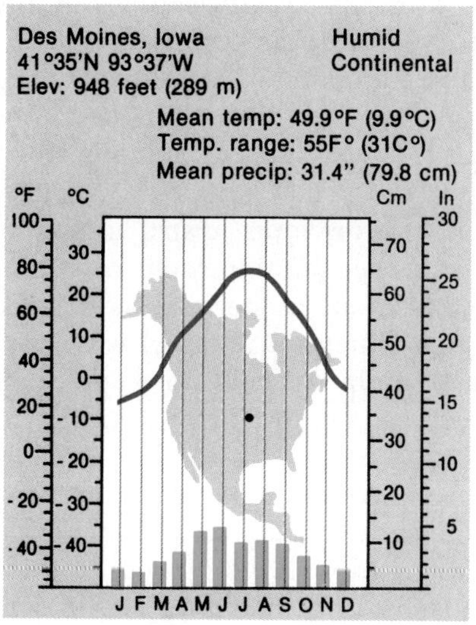

FIGURE 9.34 Climagraph for Des Moines, Iowa. Des Moines has the large annual temperature range and spring precipitation maximum typical of the Corn Belt.

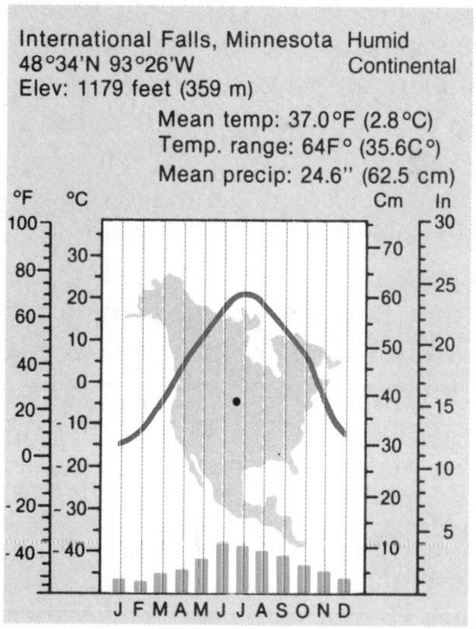

FIGURE 9.35 Climagraph for International Falls, Minnesota.

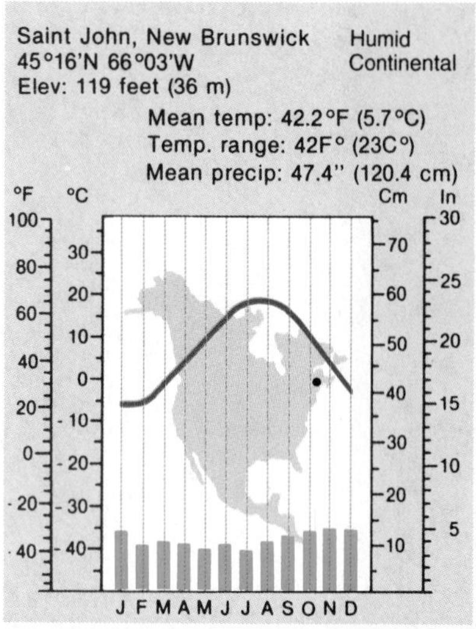

FIGURE 9.36 Climagraph for Saint John, New Brunswick. Saint John is on the North Atlantic storm track and often receives heavy winter snowfalls.

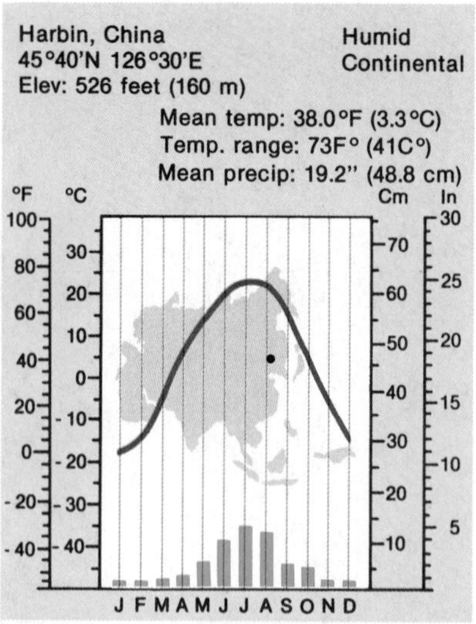

FIGURE 9.37 Climagraph for Harbin, China. Harbin has a very large annual temperature range and a strong monsoonal summer precipitation maximum.

hot spells, with daytime temperatures reaching the middle 30s Celsius (the 90s Fahrenheit). Unlike the southern margins, however, these areas also are within reach of refreshingly cool summer cP air masses that frequently produce highs in the upper teens to low 20s Celsius (the 60s to lower 70s Fahrenheit).

Winters are cold, with the lowest means in poleward and inland locations. Temperatures average well below freezing everywhere, and the ground is frozen and snow-covered for one to five months. Fronts are very active during the winter, so that day-to-day temperatures are subject to large variation, depending on the dominating air mass. The dominant wintertime air mass is cP, which generally is associated with clear, moderately cold weather. On occasion, though, outbreaks of frigid arctic (cA) air bring temperatures below $-18°$ C (0° F) to all areas, and nighttime lows in the coldest regions can plummet to $-40°$ C ($-40°$ F) or lower. At the opposite extreme, midwinter thaws sometimes occur when traveling cyclones draw mT air northward. Variations in the flow of the jet stream may cause extended periods, or even the entire winter season, to be unusually cold or mild.

Because available moisture is limited, precipitation totals within Humid Continental Climate areas generally are modest. The long, cold winters reduce evapotranspiration rates sufficiently to produce an annual moisture surplus, but summer water deficits occur in all areas except near-coastal regions. Precipitation means are greatest along the southern and coastal margins and gradually decline poleward and inland. The largest precipitation totals occur in Japan, one of the few areas within this climate with a strong orographic influence. Annual totals there average 180 centimeters (70 inches) or more in some localities. The New England and Canadian Maritime region has access to Atlantic moisture and also is quite wet, with typical annual totals of 100 to 125 centimeters (40 to 50 inches), as at Saint John (see Figure 9.36). Amounts in both North America and East Asia gradually decline inland, and means near 43 centimeters (17 inches) occur in the northern interiors of all three Humid Continental regions.

Seasonal precipitation distributions are most strongly influenced by the extent of continentality. Along east coasts and in the eastern Great Lakes area of the United States and Canada, winter storms are frequent and substantial cold season precipitation occurs. The climagraph for Saint John illustrates this situation. Progressing inland, the tendency for a summer precipitation maximum increases because the much higher temperatures of summer greatly increase the water vapor holding capacity of the air and reduce its stability. In addition, a weak monsoonal tendency, the poleward shift of the oceanic subtropical highs to the east, and the northward displacement of the polar front all combine to produce an inflow of moist mT air. The tendency for a summer maximum is strongest in continental East Asia, as in Harbin (Figure 9.37) where the monsoonal tendency produces a winter dry season. Elsewhere, however, winter evapotranspiration rates are so low that a winter moisture surplus exists despite the low precipitation totals at this time of year.

Precipitation during the fall, winter, and spring is almost entirely cyclonic and frontal and normally occurs at intervals of two to four days. Most winter precipitation is in the form of snow. Winter snowstorms sometimes are heavy and, if accompanied by strong winds, can produce blizzard conditions capable of bringing outdoor activities to a standstill.

In summer, cyclonic and frontal activity weakens and shifts northward, and most areas come under the dominance of mT air. The weather is much clearer at this time of year than during the cold season, but strong surface heating frequently triggers convectional showers and thunderstorms. In contrast to the general rains and snows of winter, the localized nature of these storms may produce greatly varying amounts and leave localized pockets of drought. More general droughts occur in some years when shifts in the pressure and wind patterns draw in cT, rather than mT, air masses. The cooler northern fringes of the Humid Continental regions receive less summertime convectional precipitation, but this is largely offset by the combination of lower evapotranspiration rates and increased frontal precipitation.

THE HIGH LATITUDE CLIMATES

The high latitude climates are dominated by cold weather during most or all of the year. Low temperatures produce difficult living conditions and, along with perennially frozen subsoil conditions, severely restrict agricultural activities. As a consequence, most areas in the high latitudes are lightly populated or uninhabited. During the brief summer, many inland areas are surprisingly mild, but coastal regions as well as the ice plateaus of Antarctica and interior Greenland remain cold. During the winter, cold and nearly constant darkness reign everywhere. Low temperatures limit the water vapor content of the air, so that precipitation totals are generally low despite the extended periods of rain or snow that frequently accompany high latitude storms. Evapotranspiration rates also are greatly reduced by the cold air, so that an annual moisture surplus exists nearly everywhere.

Subarctic Climate (Köppen Dfc, Dfd, Dwc, and Dwd)

The Subarctic Climate exists only in the northern interiors of North America and Eurasia, where it is found in a broad zone situated mostly between 50° and 70° N. As is true of the Humid Continental Climate, no Subarctic Climate regions exist within the southern hemisphere because of the absence of sufficient land in the proper latitudes. West winds transport in enough oceanic warmth to extend the

FIGURE 9.38 The peaks of the Alaska Range tower above a forest of spruce in southern Alaska, within the Subarctic Climate. (© Charlie Ott/Photo Researchers)

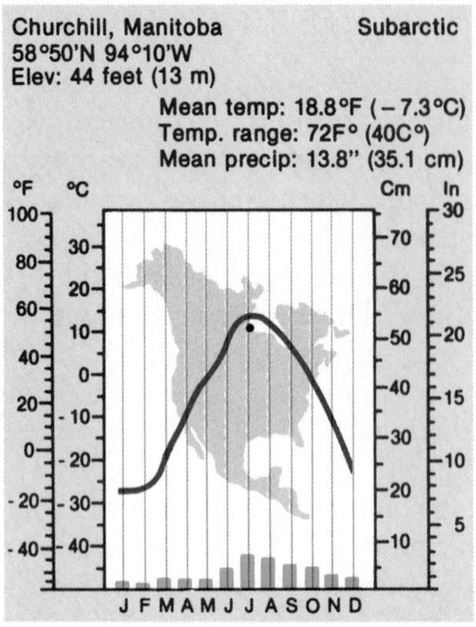

FIGURE 9.39 Climagraph for Churchill, Manitoba.

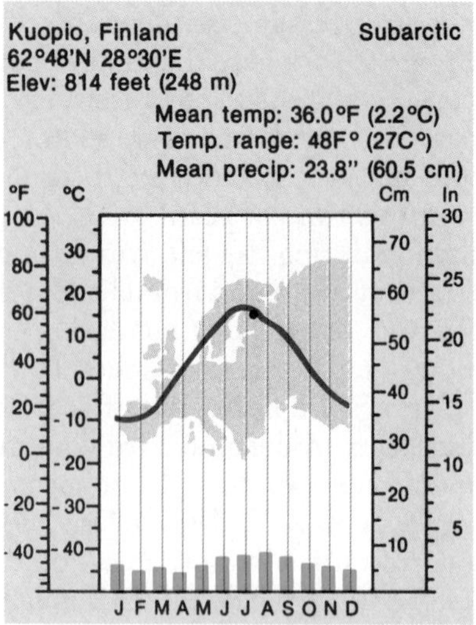

FIGURE 9.40 Climagraph for Kuopio, Finland.

climate poleward on the western sides of the continents, and latitudes gradually decline eastward. In North America, the climate extends from central Alaska east through most of interior Canada to Newfoundland. In Eurasia, it exists from northern and eastern Scandinavia east in a widening swath through the northern interior of the Soviet Union to the Kamchatka Peninsula.

According to Köppen's definition, the coldest month in the Subarctic Climate averages below −3° C (26.6° F), but nearly all areas are far colder. A more critical thermal restriction is that at least one but no more than three months have means exceeding 10° C (50° F). Summers therefore are mild but very short, with a growing season of three months or less. Precipitation exceeds potential evapotranspiration, which is very low in a climate this cold.

One of the most important thermal characteristics of the Subarctic Climate is its extreme seasonal temperature variations. Portions of this climate experience the world's largest annual temperature ranges. In North America, differences between the summer and winter months are generally 33 and 44 C° (60 and 80F°). Ranges are much more diverse in Eurasia, varying from less than 17 C° (30 F°) in coastal northern Norway, where strongly maritime conditions prevail, to well over 56 C° (100 F°) in interior eastern Siberia. The climagraphs of three representative Subarctic Climate stations (see Figures 9.39–9.41) illustrate the diversity of ranges.

Winter, the dominant season in the Subarctic Climate, is long, dark, and extremely cold. The cold winter temperatures, which are lower than anywhere else in the world except for the ice plateaus of Antarctica and Greenland, result from a combination of several factors. One is the lack of insolation. Most areas are near or above the Arctic Circle and receive at most only a few hours of low-intensity heating from a Sun barely above the horizon. A second factor

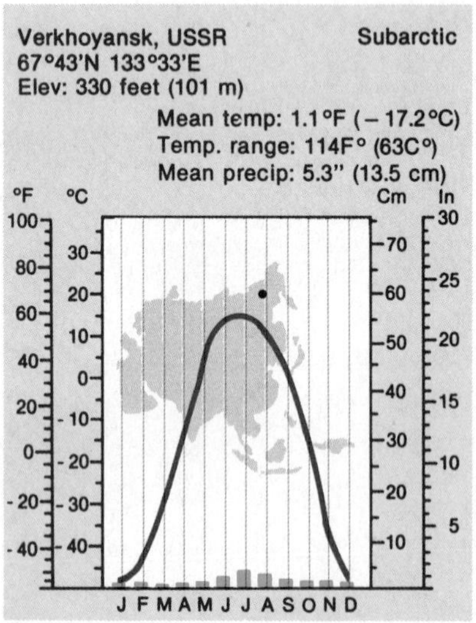

FIGURE 9.41 Climagraph for Verkhoyansk, USSR. Verkhoyansk has one of the world's largest annual temperature ranges.

is the continentality produced by the location of most areas in the interiors of large, snow-covered land masses. Finally, the thermal compaction of the cold air results in the formation of high pressure centers over Canada and Siberia. These high pressure systems generally are associated with clear skies and light winds that favor surface radiation.

Winter readings in most locations average below −18° C (0° F), and subfreezing mean temperatures last anywhere from five to eight months. The lowest temperatures of all occur in eastern Siberia; in the Soviet city of Verkhoyansk, for example, the January mean temperature of −50° C (−58° F) makes it one of the coldest permanently inhabited locations in the world (see Figure 9.41). In western Canada and Alaska, northern Europe, and the east coasts of Canada and Siberia, the proximity to water raises winter means (see the climagraph for Kuopio in Figure 9.40), but the reduction in cold is partially offset by an increased prevalence of clouds, winds, and heavy snows.

Summers are brief but very mild in contrast to the frigidity of the winters. At this time of year the midday Sun is fairly high in the sky and daylight is almost continuous. As a result, total daily insolation receipts are temporarily equivalent to those of the tropics, and only the thermal inertia of the atmosphere and surface keeps temperatures from rising to tropical levels. Means are in the teens Celsius (the 50s and 60s Fahrenheit), with the warmest readings in the southern interiors.

Fronts are active throughout the year so that day-to-day temperature variations are substantial. Under the influence of warm winds from the south, afternoon summertime highs sometimes reach the upper 20s to mid 30s Celsius (the 80s to lower 90s Fahrenheit). Fairbanks, in interior Alaska, recorded that state's all-time high temperature of 38° C (100° F) under such conditions. In contrast, cold air from the ice-covered Arctic Ocean, the subpolar North Atlantic, or Canada's Hudson Bay can keep highs in the single digits Celsius (the 40s Fahrenheit), and frosts and light freezes can occur at night in most places throughout the summer.

The combination of cold temperatures and continentality results in generally light precipitation. Heaviest totals occur along the coasts, where annual mean temperatures are warmest, water sources are present, and winter storms are frequent. Annual means of up to 100 centimeters (40 inches) occur in eastern Newfoundland and northern Norway. Precipitation declines rapidly inland and, in the northern interiors, amounts less than 25 centimeters (10 inches) are the rule.

Most precipitation is frontal and cyclonic. Despite the low totals, storms with limited moisture supplies frequently pass through the subpolar regions, producing a large number of days with cloudy skies and light rain or snow. Except for coastal areas, a summer precipitation maximum occurs almost everywhere. Winter snows generally are light, but are frequently associated with deep low pressure systems that produce strong winds and blizzard conditions. Coastal areas sometimes receive heavy snowfalls from storms with access to oceanic moisture. The coasts in general are cloudier, foggier, and damper year-round than the interiors.

Tundra Climate (Köppen ET)

The last two climates in our world survey are differentiated only by degrees of coldness. Located in the subpolar and polar regions, they receive smaller quantities of insolation than any other places on Earth. As noted earlier, the high latitudes have a net radiation deficit. A substantial proportion of their thermal energy is supplied by the poleward advection of heat from the lower latitudes by winds and ocean currents.

Within the Tundra Climate, the mean temperature of the warmest month is between 0° and 10° C (32° and 50° F). As in the case of the previous two climates, the great majority of land areas experiencing a Tundra Climate are in the northern hemisphere. Most form a coastal rim around the Arctic Ocean. Included are northern Canada and Alaska, coastal Greenland, northern Iceland, and the north coast of Eurasia from Norway eastward to the Bering Strait. Land areas within the southern hemisphere are limited to the southern tip of South America, the Antarctic Peninsula, and a number of small islands. Most of these areas lie poleward of 60° N and S, although the climate extends to about 53° N and S in coastal Labrador and South America. Most high latitude ocean areas also have a Tundra Climate. Finally, this climate occurs in high mountains and plateaus in various parts of the world, notably on the Tibetan Plateau. Its characteristics within these areas are discussed more fully in the Highland Climates section of this chapter.

FIGURE 9.42 A few dwarf spruce trees dot the tundra of central Alaska, near the southern margin of the Tundra Climate. (Visuals Unlimited/© C.P. Hickman)

The Tundra Climate can be described as a high latitude maritime climate. Few areas are more than 160 kilometers (100 miles) from the ocean. The proximity to water reduces annual temperature ranges from the extremes encountered in the Subarctic Climate, causing summers to be colder and winters to be less frigid than in the former climate, despite the higher latitude. The major seasonal temperature control (and one of the most important aspects of the climate) is the duration of sunlight. Most areas are poleward of the Arctic Circle; as a result, sunlight varies from 24 hours a day near the time of the summer solstice to 0 near the winter solstice. The Sun's motion is much more horizontal (or circular) than vertical, so little diurnal variation occurs. Temperatures trend irregularly upward in spring and downward in fall, with the irregularities caused largely by frontal passages.

During the summer most places average between 4° and 10° C (the 40s Fahrenheit). The Sun, though low in the sky, is out all or nearly all day. As a result, insolation receipts in late spring and early summer can reach daily values equivalent to those of the tropics. A south wind in mainland North America and in Eurasia will bring in warm continental air and occasionally can cause temperatures to reach the mid 20s to low 30s Celsius (the 70s and 80s Fahrenheit). The ice-covered Arctic Ocean is always nearby, though, and a shift to northerly winds can quickly drop readings to near freezing, with the likelihood of frost. Coastal fogs frequently form in summer when warm air from the land encounters cold water or the Arctic ice pack.

Winter temperatures are regionally much more variable than those of summer. For the considerable majority of the tundra lands of North America, Eurasia, and northern Greenland, winter means are well below −20° C (0° F). The climagraph for Chesterfield Inlet, on the northwest shore of Hudson Bay, serves as an example (see Figure 9.43). In these areas, mean temperatures are below freezing eight or nine months of the year, and readings can fall as low as −46° to −50° C (−50° to −60° F). Some areas, however, are near water bodies that remain partially or entirely unfrozen throughout the winter. Temperatures here are considerably less cold. One large area that experiences such conditions is located where the North Atlantic merges with the Arctic Ocean. Here the last remnants of the Gulf Stream keep water temperatures above freezing and air temperatures near or above −18° C (0° F) on the adjacent land. The climagraph for Svalbard, an island in the Arctic Ocean north of Norway, provides an example (see Figure 9.44). Extreme examples of the maritime influence occur in the southern hemisphere. Macquarie Island, located southwest of New Zealand (see

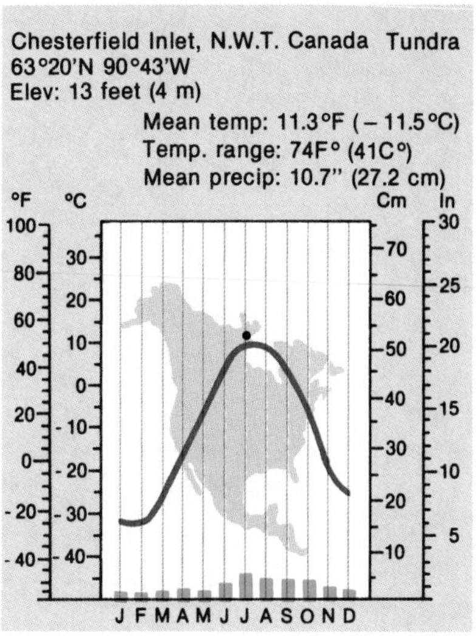

FIGURE 9.43 Climagraph for Chesterfield Inlet, Northwest Territories, Canada.

Figure 9.45), has a warm monthly mean of 6.4°C (43.5°F), but its coldest month averages 2.8°C (37.0°F), producing an annual range of only 3.6C° (6.5F°).

Cold temperatures help reduce annual precipitation means in most places to below 38 centimeters (15 inches). The exceptions are mountainous windward coasts such as southern Greenland, northern Norway, parts of western Alaska, and the southern hemisphere islands. Macquarie Island, for example, receives a mean of 106.2 centimeters (41.8 inches).

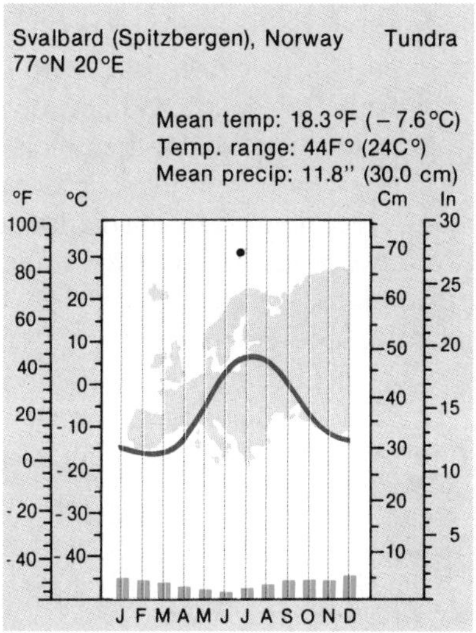

FIGURE 9.44 Climagraph for Svalbard, Norway.

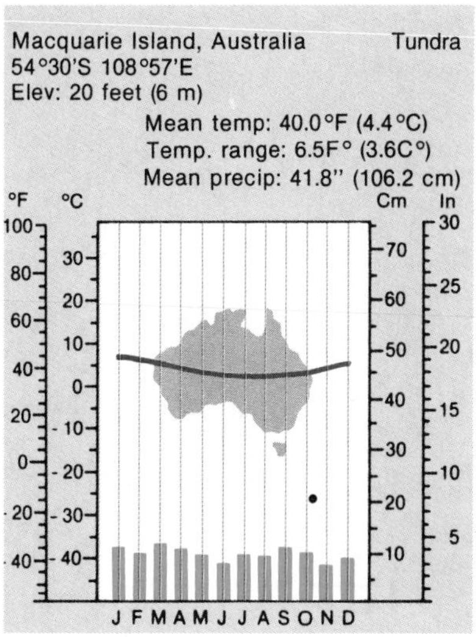

FIGURE 9.45 Climagraph for Macquarie Island, Australia. Macquarie Island's extreme maritime location causes it to have a very small annual temperature range despite its high latitude.

These same coastal areas generally have a winter maximum caused by storms associated with the subpolar low and receive heavy winter snowfalls. Most areas, though, have a strong summer precipitation maximum (as in the case of Chesterfield Inlet) because the winter cold causes the Arctic Ocean to freeze over, removing the primary moisture supply. Summer precipitation is predominantly rain, but some falls as wet snow. Virtually all precipitation is frontal and cyclonic. During the fall, winter, and spring, cyclonic storms can be very strongly developed and generate blizzard conditions because the flat, treeless tundra offers little frictional resistance.

Polar Climate (Köppen EF)

The Polar Climate is the most poleward and the coldest of the world's climate types. It also is the only one with no permanent human inhabitants. Monthly mean temperatures are below freezing all year, and the landscape is covered almost everywhere by a deep layer of ice. Two widely separated areas—Antarctica and the interior of Greenland—have a polar climate; together they comprise about 10 percent of the Earth's land surface.

The extreme coldness of this climate is only partially caused by low Sun altitudes. It also is produced by the high reflectivity of the ice and snow cover and

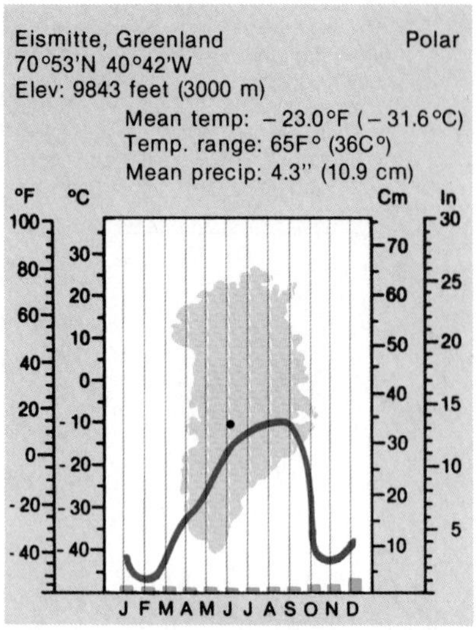

FIGURE 9.46 Climagraph for Eismette, in interior Greenland.

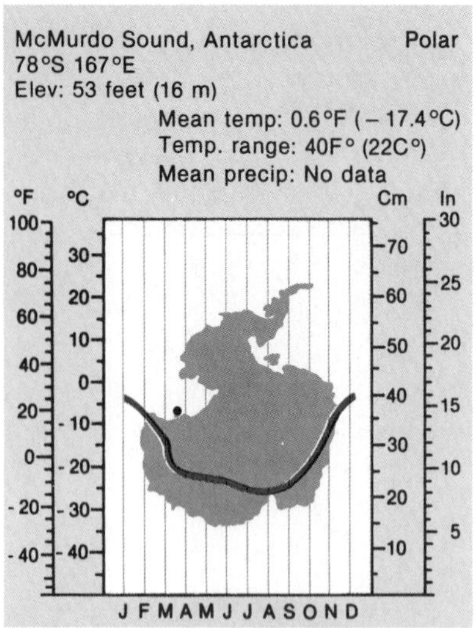

FIGURE 9.47 Climagraph for McMurdo Sound, on the coast of Antarctica.

by the high elevation of the surface. Most areas are situated between 1.5 and 2 kilometers (1 and 2 miles) above sea level.[2]

Although the unrelenting cold is by far the dominant overall characteristic of this climate, the extreme variation in the duration of daylight is its major seasonal attribute. Except for southern Greenland, all areas lie poleward of the Arctic or Antarctic Circles and experience the full 24-hour range in the duration of sunlight between the times of the solstices.

Summer temperatures average below freezing on a monthly basis, but in some parts of coastal Antarctica and the margins of the Greenland ice sheet mild spells occasionally occur, with above-freezing conditions lasting for hours or even several days. The high-elevation interiors of both areas remain well below freezing, as the climagraph for Eismitte, in central Greenland, illustrates (see Figure 9.46). In interior Antarctica, temperatures remain below −18° C (0° F) throughout the year. Antarctica as a whole is colder than Greenland because of its higher latitude, somewhat higher average elevation, and greater degree of continentality.

Winter temperatures, especially in Antarctica, are the lowest on Earth. The coldest readings of all occur not at the South Pole but at the so-called "Pole of Cold" located in the center of the continent near 78° S 107° E. A research station named Vostok established at this location in the 1950s by the Soviets has recorded the lowest temperatures in the world. The current record is −89.2°C (−128.6° F), set on July 21, 1983.

Although the Sun remains below the horizon for several months in the late fall and winter in interior Antarctica, it is near enough to the horizon during much of this period to produce some twilight. Additional light often is provided by the southern lights or **aurora australis** (the southern hemisphere counterpart of the **aurora borealis** of the northern hemisphere).[3] Because of these light sources, the polar regions, though receiving less insolation than any other part of the world, also have less total darkness than any other place.

The Polar Climate is not only the coldest, but also the driest of the world climate types in terms of precipitation amounts received. Evapotranspiration rates are so low, however, that a major surface mois-

[2] Because of their elevation, these areas could well be included within the Highland Climates classification, discussed in the next section, as are the glaciated areas of the lower latitudes. By convention, however, because of their location and large size, they are treated as a distinct climate type.

[3] Auroral displays are produced when electrons streaming outward from the Sun as part of the solar wind collide with and ionize atoms in the Earth's atmosphere in the vicinty of the magnetic poles. Light is emitted when these atoms reconstitute and release the energy they absorbed during the collisions.

FIGURE 9.48 Nearly all Polar Climate areas have a surface cover of ice and snow. Here, scientists in Antarctica conduct research in front of a wall of glacial ice, while volcanic Mt. Erebus steams in the distance. (Visuals Unlimited/© Jeanette Thomas)

ture surplus exists. The answer to the poetic question "Where are the snows of yesteryear?" is obvious in Greenland and Antarctica (see Figure 9.48). They bury the surface to a depth of up to 4 kilometers (2.5 miles)! These areas provide an excellent example of the inadequacy of measuring aridity solely in terms of total precipitation, without taking temperatures and evapotranspiration rates into account.

Precipitation is cyclonic and frontal in nature and, of course, virtually all is frozen. The combination of cold, continentality, and thermal high pressure blocks most storms from the interior areas, especially in Antarctica. As a result, precipitation is heaviest near the coasts and decreases rapidly inland. Amounts received are uncertain due to a lack of reporting stations and the difficulty involved in making accurate measurements of wind-blown snowfalls. It is believed that precipitation totals in the interiors are mostly 5 to 13 centimeters (2 to 5 inches) per year in terms of water equivalent. Much of the precipitation in the interiors occurs in the form of ice needles, formed by the sublimation of atmospheric water vapor at temperatures below −39° C (−38° F), that drift down from a clear sky. (This same fall of ice needles from cirrus clouds in the upper atmosphere of the lower latitudes produces their wispy appearance.) Precipitation totals increase to means of perhaps 50 to 65 centimeters (20 to 25 inches) near the Antarctic coast. Heavy amounts in excess of 125 centimeters (50 inches) occur near the southern edge of the Greenland icecap because it offers an orographic barrier to storms moving along the North Atlantic winter storm track.

Winter storms frequently produce gale- and even hurricane-force winds, largely because snow and ice offer less frictional resistance than any other surface on Earth, including water. Coastal storms also are numerous and often very strongly developed; some coastal stations in Antarctica report gale-force winds nine days out of ten. Under such conditions the air becomes filled with fine, horizontally driven snow and ice crystals, which sometimes produce whiteouts that reduce visibilities to zero.

HIGHLAND CLIMATES

It is widely known that mountain regions have distinctly different climatic conditions than adjacent

FIGURE 9.49 Major climatic variations over short distances exist within Wyoming's Grand Teton National Park. (T. Algire/H. Armstrong Roberts)

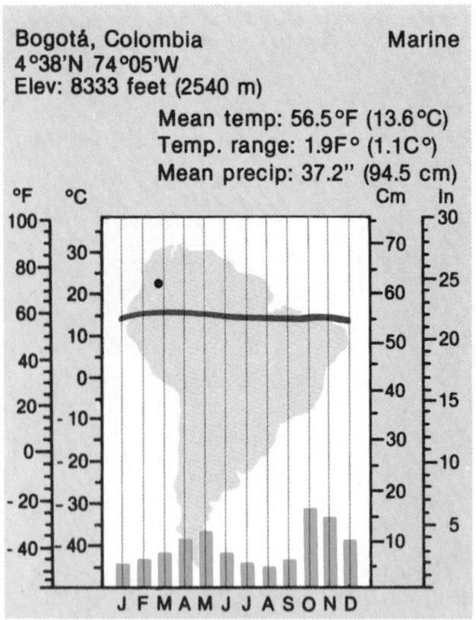

FIGURE 9.50 Climagraph for Bogotá, Colombia. Bogotá's location near the equator accounts for its double precipitation maximum (caused by passages of the ITCZ) as well as its very small annual temperature range.

lowlands (see Figure 9.49). Mountains are popular summer vacation destinations for people seeking to escape the heat of lowland areas, and many people also visit mountainous areas in winter to take advantage of winter sports opportunities. In this section the climatic differences between highland and lowland regions are examined, and the reasons for these differences are analyzed.

The highland climates as a group differ in one basic respect from the world climate types already discussed. Instead of consisting of a single large, relatively homogeneous region, they form a complex mosaic of diverse climate types located in close proximity to one another. In fact, it is their diversity, rather than their homogeneity, that causes them to be grouped together here. Climatic conditions in the world's highland areas vary so greatly and change so rapidly with distance that their total pattern is too complicated to depict on a world map or to analyze fully in a section of a chapter. It should be stressed that the highland areas do not contain any new climate types that have not been examined. We already have covered the entire range of world types, from those that are hot and wet to those that are cold and dry.

Distribution

Description of the global distribution of highland climate types is complicated by a lack of agreement on how greatly the climatic conditions of an area must be topographically modified to warrant its inclusion as a highland climate region. In general, the greater the topographic relief of an area, the greater the diversity of its climatic conditions.

In North America, the Rockies, Sierra Nevadas, Cascades, and the mountains and interior plateaus of Mexico are climatically distinct from the surrounding lowlands. In South America, the Andes produce a continuous wide band of highland climates almost 8,000 kilometers (5,000 miles) long. The world's greatest concentration of highland climates occurs in southern Eurasia, extending from western China to northern Spain, and includes the Himalayas and its associated ranges, the Tibetan Plateau, the Zagros and Elburz Mountains of Iran, the Caucasus of the Soviet Union, and the European Carpathians, Alps, and Pyrenees. In Africa, important areas include the Atlas Mountains and Ethiopian Highlands.

Lower mountain and upland areas that modify their climates to a lesser degree also are extensive. Some of the most notable are the Appalachians, the highlands of southeast Brazil and southern Africa, the Soviet Urals, the Italian Apennines, the mountains of Scandinavia, and the Dividing Range of eastern Australia. Smaller highland areas, often located on mountainous islands, are too numerous to list fully, but include portions of Japan, Madagascar, the East and West Indies, and New Zealand. It therefore is evident that regions with topographically modified climates are extensive and widely distributed.

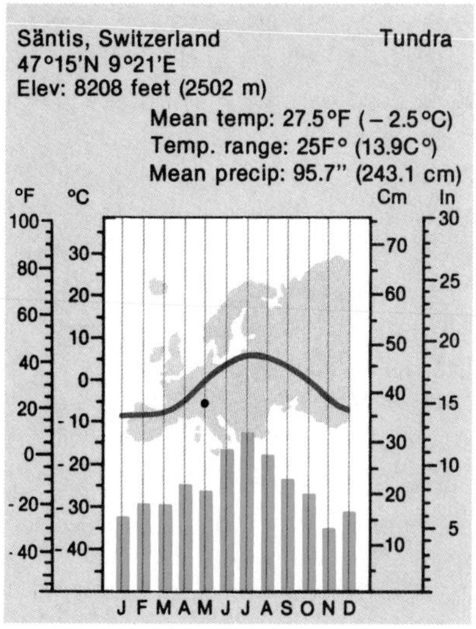

FIGURE 9.52 Climagraph for Säntis, Switzerland. Säntis receives heavy orographic precipitation totals.

Climatic Controls

The climate of any specific highland location is controlled by the combined influences of elevation, latitude, orographic effect, and local topography. The most important of these is elevation. Its significance stems from the fact that, in the troposphere, the temperature declines at an average rate of 6.4C° per 1,000 meters (3.5F° per 1,000 feet). As a result, temperature means in highland areas are substantially lower than those of adjacent lowlands. An example is provided by the climagraph for Bogotá, Colombia (see Figure 9.50). Most sea level stations located at Bogotá's latitude of $4\frac{1}{2}°$ N experience annual means near 27° C (80° F). Bogotá's elevation of more than 2,500 meters (8,000 feet), however, causes its mean temperatures to remain in the low to mid-teens Celsius (mid- to high 50s Fahrenheit) throughout the year. High mountains can cause temperatures to be reduced sufficiently to produce polar climatic conditions, along with permanent ice and snow, even at the equator.

Latitude, a second important climatic control, affects highland areas in much the same way as it does lowlands. As latitude increases, decreased insolation produces lower temperatures. As a result, the higher the latitude of a place, the lower the elevation at which a given climate may be expected to occur (see Figure 9.51). An increase in latitude also is generally associated with larger annual temperature ranges. For example, Bogotá, despite its cool temperatures, has the very small annual temperature range that one would expect of a city at its low latitude. Säntis, in the Swiss Alps (see Figure 9.52), is at nearly the same elevation as Bogotá, but its latitudinal position causes

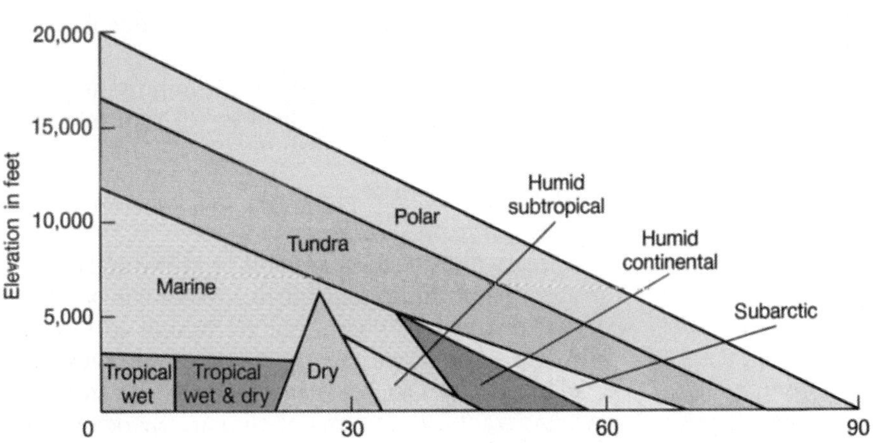

FIGURE 9.51 Idealized relationship of climate type to latitude and elevation.

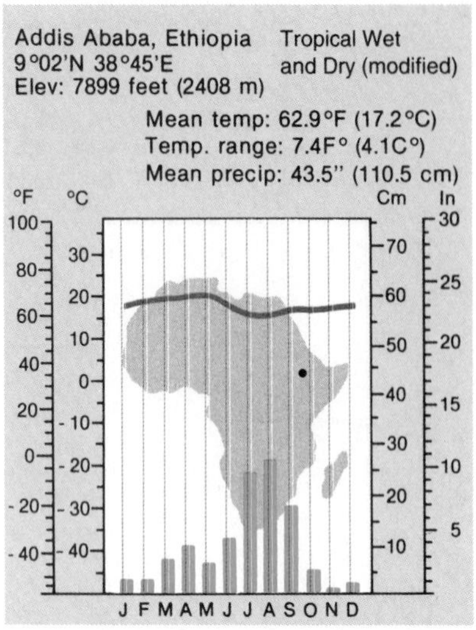

FIGURE 9.53 Climagraph for Addis Ababa, Ethiopia.

it to have a much lower annual mean and a much higher annual range of temperature.

The orographic effect is the dominant control on precipitation amounts and distributions within a highland area. In general, the windward sides of mountains tend to be wet, the leeward sides dry, and large differences in amounts typically occur over small distances. The combination of orographic uplift and reduced evapotranspiration rates resulting from cooler temperatures generally causes highlands to be much more humid than nearby lowlands. Most of the world's major river systems, in fact, have their headwaters in highland regions. The Ethiopian Highlands in which Addis Ababa (see Figure 9.53) is located, for instance, are largely surrounded by arid lands, yet they provide much of the water for the Nile. Säntis receives about four times as much precipitation as do lowland European stations at the same latitude. On the other hand, high interior valleys and plateaus may be relatively dry if surrounded by higher mountains that block moisture-bearing winds. This is the reason for Bogotá's modest precipitation totals.

Two other climatic characteristics of highland areas also are associated with orographic uplift. Fog is common in mountain areas, especially in humid regions. On many windward slopes the continuous generation of fog can obscure mountain peaks for days at a time. The combination of orographic barriers and unstable air also can lead to dangerously changeable weather. Sudden violent thunderstorms and, at high elevations, local blizzards frequent many mountainous areas.

The last of the major climatic controls is the local topography. Topographic variation obviously is much greater in mountainous areas than in flat areas, and the variety and complexity of localized climates are correspondingly increased. Two of the most important topographic factors were discussed in Chapter 5. These are (1) slope aspect, which causes slopes facing the Sun to be warmer and thus generally drier than more shaded slopes; and (2) the nature of the surface, which can vary greatly in material constituency and vegetative cover. For example, a bare rocky slope will likely be warmer, drier, and windier during the day than a forested slope. A third topographically related factor is a site's degree of exposure. This especially influences wind speeds and directions. A sheltered valley nestled beneath high mountains, for instance, is protected from strong winds, but an isolated peak is exposed to the full force of the winds, unreduced by surface friction. A good example of the effects of exposure has been provided by a weather station located atop Mt. Washington, New England's highest peak, which recorded the world's highest wind speed of 90 meters per second (231 miles per hour).

SUMMARY

This chapter has explored the global pattern of climates. World climatic characteristics and resulting climate regions are distributed in a systematic geographic pattern. This pattern in turn results largely from the controlling influences of spatial variations in insolation, continentality, air pressure, wind patterns, and, in some areas, orographic influences and ocean currents.

In order to summarize the most important information presented in the chapter, we return to the Focus Questions:

1. *What are the chief advantages and shortcomings of climate classification systems?*

Climate classification systems are devised largely as an organizational tool for the systematic analysis of world climates. Their basic purpose is to simplify the study of climate by reducing the number of climatic types to a manageable total. By so doing, geographical patterns of climatic characteristics become evident, and a greater understanding of global climate controls can be gained. A basic shortcoming of all climatic classification systems is their artificiality. The world's climates, which in reality display tremendous variety, are grouped into a limited number of more or less broadly defined types, with each type occupying a certain geographical region. This setup may give a misleading impression of regional homogeneity, with sudden climatic changes occurring as boundary lines are crossed. This is not the case. In reality, climatic conditions vary gradually with distance within climatic regions and do not generally change in any dramatic respect when boundary lines are crossed.

2. *What are the major climatic characteristics of each of the global climate regions, and why do these characteristics occur?*

In this chapter, world climates were grouped into low latitude, middle latitude, and high latitude types, with highland climates treated as a separate category. Three low latitude climate types exist. The Tropical Wet Climate, under the domination of the ITCZ, is characterized by continuous warmth and by abundant precipitation of a predominantly convectional nature. The Tropical Wet and Dry Climate also is warm throughout the year, but displays a distinctive pattern of rainy summers and dry winters. The Low Latitude Dry Climate, dominated by the subtropical highs, is characterized by mild winters, very hot summers, and year-round dryness.

The middle latitudes contain five climate types. The Dry Summer Subtropical Climate has a pattern of dry, warm summers dominated by the subtropical high and mild, humid winters under the influence of the westerlies, with their traveling cyclones and anticyclones. The Humid Subtropical Climate has hot, humid summers with scattered convectional precipitation, and cool to mild winters with alternating periods of fair weather and frontal and cyclonic precipitation. The Mid Latitude Dry Climate is cut off from major moisture sources by orographic barriers and therefore is dominated by dry weather with large annual temperature ranges. The Marine Climate, in contrast, is characterized by adequate to abundant precipitation from cyclonic storms and by small annual variations in temperature. The most poleward of the middle latitude climates is the Humid Continental Climate, which has cold winters, mild to warm summers, and modest but generally adequate precipitation that is predominantly frontal and cyclonic in origin.

Three high latitude climates were discussed. All are characterized by varying degrees of coldness and by rather low precipitation amounts from cyclonic storms and their associated fronts. The Subarctic Climate has severely cold winters and cool to mild summers of rather short duration. The Tundra Climate experiences above-freezing mean temperatures for only a few months in summer. Finally, the Polar Climate is characterized by subfreezing temperatures and a surface ice and snow cover throughout the year.

The climates of the high mountain regions of the world display great variability, often over small distances. This variability exists largely because of differences in elevation, orographic influences on precipitation, and local topographic conditions such as slope aspect and exposure.

Selected References

Barry, Roger G. *Mountain Weather and Climate.* New York: Methuen Press, 1981.

Lockwood, John G. *Causes of Climate.* New York: Halsted Press, 1979. Discusses not only the causes of climate, but also the climates of the past and future.

Lydolph, Paul E. *The Climate of the Earth.* Totowa, N.J.: Rowan and Allenheld, 1985. A systematic discussion of global weather and climate. Climate coverage is in a regional context by continents.

Oliver, John E. *Climate and Man's Environment.* New York: John Wiley & Sons, Inc, 1973. Discusses the interaction of climate with other aspects of the physical environment, its effects on human activities and lifestyles, and the nature and causes of climatic change.

Key Terms

climagraph
arid
semiarid

Review Questions

1. What are the chief advantages of using a regional climate classification system to study world climate characteristics? What are the shortcomings of such systems?

2. What are climagraphs? What information do they depict, and what are their advantages?

3. Why could it be said that the Tropical Wet Climate is the world's most monotonous climate? What are the controlling factors for this climate's temperature and precipitation characteristics?

4. Explain how the location of the Tropical Wet and Dry Climate controls its annual precipitation distribution pattern.

5. What factors determine the dryness of an area? Why are there no arid and semiarid climates in the high latitudes?

6. The Low Latitude Dry Climate is located between approximately 18° and 33° N and S. Describe the changes in both amounts and seasonal distributions of precipitation you would encounter if you traveled from the climate's equatorward margin to its poleward margin, making a one-year stop at every 7° of latitude.

7. Why are the world's hottest temperatures recorded in the Low Latitude Dry Climate, rather than nearer the equator? Are winter temperatures in this climate also hot? Why or why not?

8. Describe the annual precipitation distribution pattern of the Dry Summer Subtropical Climate. What causes this pattern?

9. How and why do annual temperature characteristics differ between coastal and interior California?

10. How and why does the annual precipitation distribution pattern of the Humid Subtropics differ from that of the Dry Summer Subtropics?

11. How do the causes of the dryness of the Low Latitude Dry Climate and the Mid Latitude Dry Climate differ? In which climate are the annual temperature ranges larger? Why?

12. Why is the only extensive land area with a Marine Climate located in Europe? What changes in climatic conditions occur as one travels eastward from the west coast of France to the western border of the Soviet Union? Why do these changes occur?

13. What are the reasons for the great seasonal temperature contrasts within the Humid Continental Climate? What are the causes of the large day-to-day temperature contrasts that sometimes occur?

14. Why do the Humid Continental and Subarctic Climates occur only in the northern hemisphere?

15. Why do the world's largest annual temperature ranges occur within the Subarctic Climate? How do these large ranges influence the seasonal distribution of precipitation?

16. Why do Tundra Climate areas remain cold during the summer, despite receiving large amounts of insolation? Why do the Polar Climates remain so cold in summer?

17. In what basic way do the highland climates as a group differ from the other world regional climate types?

18. In what respects would your own climate be different from what it is now, if your elevation were increased by 1,500 meters (5,000 feet)?

19. List several reasons why climatic characteristics often vary substantially over small distances in mountainous areas.

20. In what important respects do the climates of cities differ from those of the adjacent countryside?

21. What causes urban heat islands?

CASE STUDY

Urban Climates

The Case Study at the end of Chapter 5 dealt with possible global temperature changes resulting from human activities, notably the addition of dust and carbon dioxide to the atmosphere. These concerns, although serious, are still speculations because apparently no readily measurable worldwide changes in temperature have yet been produced by human actions. However, human activities already have resulted in the formation of distinct climatic conditions in urban areas, where a large and rapidly growing segment of the world's population lives.

The most significant impact that an urban area has on its climate is to increase local temperatures by producing what has been termed an **urban heat island**. A combination of several factors is responsible for this occurrence. First, and probably most important, large quantities of energy are produced by human activities and eventually released to the atmosphere. Most of these activities, such as heating homes and businesses, running factories and power plants, and operating vehicles, involve the burning of fossil fuels. Heat-producing operations have an especially large impact in the winter, when heating needs are greater and proportionally less solar energy is present. Studies have shown that the production of artificial heat in cities in northern Europe and North America during the midwinter months may rival or exceed that provided by insolation.

A second factor contributing to the formation of a heat island is the abundance of materials such as brick, concrete, and asphalt. These materials readily absorb heat and reradiate it for long periods. A third factor is the buildup of heat-absorbing pollutants over the city. The major product of the combustion of fossil fuels is carbon dioxide—the chief gas responsible for the greenhouse effect. A final important factor is the surface dryness of the city. The materials from which a modern city is constructed are relatively impervious to moisture, and most rain water quickly flows into storm sewers. As a result the surface soon dries, and subsequent insolation that otherwise would be used to evaporate water instead is converted to sensible heat. In contrast, the evapotranspiration of water from soil and vegetation in rural areas uses a large proportion of the Sun's energy—often more than is converted directly to heat.

The urban heat island typically is best developed at night, largely because energy absorbed by city materials during the day continues to be released at this time. The difference in nocturnal temperatures between adjacent urban and rural areas sometimes is large enough that forecasters give separate predicted lows for each.

Cool temperatures, clear skies, and light winds aid in heat island development. Cool temperatures result in a greater production of heat, clear skies allow the day's heat in rural areas to radiate quickly into space, and light winds keep the warmer city air from being blown away.

The heat island patterns of a number of cities have been mapped (see Figure 9.54). Most patterns consist of roughly concentric circles. The highest readings occur in the center of the city, where the heat-producing factors are most concentrated, and progressively lower values are found toward the suburbs. In detail, the patterns are complicated by the presence of especially warm temperatures over industrial or other areas with high concentrations of activity, and by cool pockets, often associated with open areas such as parks and cemeteries.

Urban climates also differ from those of rural areas in a number of aspects involving atmospheric moisture. Relative humidities are somewhat reduced because the warmer city temperatures increase the air's water vapor holding capacity. On the other hand, specific humidity values are at least as high as those of surrounding rural areas because combustion releases large amounts of water vapor. Mean amounts of cloudiness, fog, and precipitation all are higher in urban areas than in rural areas. One of the factors encouraging the formation of these condensation products is the large quantity of dust released by human activities. Dust particles sometimes are present within large cities in sufficient quantities to restrict visibilities greatly and form what has been called an "urban dust dome." The dust particles reduce the intensity of sunlight, and many are also **hygroscopic** (water attracting) so that they serve as condensation nuclei.

continued on next page

City heat also encourages convectional lifting, especially in summer, resulting in the development of cumuliform clouds and sometimes of convectional precipitation. A number of studies have determined that urban heat production is responsible for a 5 to 10 percent increase in mean annual precipitation totals in and immediately downwind of large North American cities.

Wind patterns in a city also are considerably altered from those of the surrounding countryside. A large city has an extremely irregular surface profile, with the vertical walls of buildings acting to block and redirect the wind. The large amount of surface friction reduces average wind speeds, often by a factor of 20 or 30 percent. On the other hand, the wind often is redirected down streets, which are literally "avenues of least resistance" to the wind. Most people who are outdoors are on the streets, so that urban wind speeds in areas frequented by pedestrians actually may be higher than in rural areas.

The combination of explosive world population growth, major urban immigration, and a continued rise in per-capita energy consumption will all further increase the extent and significance of urban climatic modification.

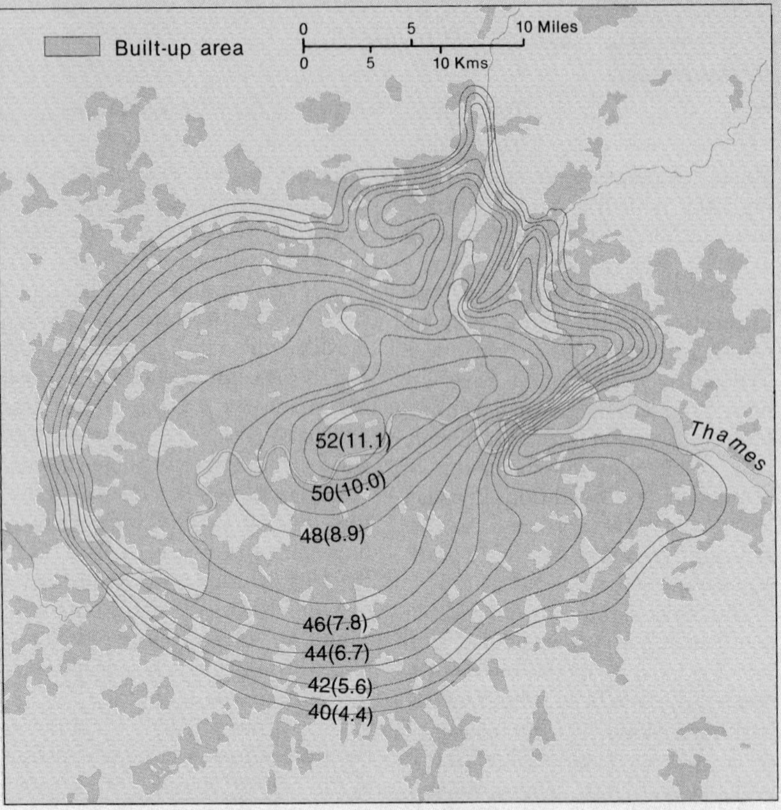

FIGURE 9.54 Map of the London urban heat island as it developed on a clear night. Temperatures are given in F° (C°).

CHAPTER 10

WATER ON AND BENEATH THE EARTH'S SURFACE

OUTLINE

FOCUS QUESTIONS

1. What factors control the movement and storage of subsurface water?
2. How do lakes form and disappear?
3. What factors determine geographic and temporal variations in stream discharge?
4. What are the chief physical and chemical characteristics of ocean water?
5. What are the causes of the three oceanic motions of currents, waves, and tides?

Water in the atmosphere has been discussed in detail in the last several chapters. At any given time, however, only 0.001 percent of our planet's near-surface water is located in the atmosphere; the overwhelming majority is to be found on and beneath the Earth's surface (see Table 7.1). In this chapter, we begin the study of the terrestrial portion of the hydrologic cycle by examining the location, characteristics, and movement of the Earth's surface and subsurface water supplies. Succeeding chapters examine the crucial roles this water plays in influencing the planetary distributions of vegetation, soils, and landforms.

Most precipitation falling on land does not flow immediately into rivers; instead, it enters the soil and sometimes the underlying rock layers, where it may stay for an extended period before eventually returning to the surface. The first portion of the chapter is therefore devoted to the Earth's subsurface waters. We next examine lakes and rivers, which derive their water from a combination of surface and subsurface sources. The final topic is the ocean, which is the ultimate destination of runoff from the land and is the feature that dominates the surface geography of our planet.

SUBSURFACE WATER

The hydrologic cycle is a vast distillation mechanism for water. Salts and other dissolved substances are removed through the process of evaporation, and the water subsequently is returned to the surface as precipitation. Fresh water, both on and beneath the Earth's surface, therefore has not had the opportunity to gain large quantities of dissolved substances since it was precipitated as rain or snow.

Although large quantities of fresh surface water exist in lakes and rivers, the great majority of the Earth's liquid fresh water is beneath the surface. The total quantity of this water has not been accurately determined, but its volume is estimated at some 10^7 cubic kilometers (2,400,000 mi^3). This is approximately 77 times the water in all of the Earth's surface rivers and lakes and is about one-third of the quantity locked up in the Antarctic and Greenland ice caps. It is enough water, if evenly spread over the Earth's surface, continents and oceans alike, to cover it to a depth of about 10.5 meters (35 feet). Nearly half of this water is within 0.8 kilometers (0.5 miles) of the surface and is considered to be accessible for extraction.

It is convenient to separate the Earth's subsurface water into two categories based on depth, movement, and origin. Nearest the surface is **soil water,** which occupies the pore spaces between individual soil particles. Some soil water adheres in a molecular film to the soil particle surfaces, while the rest moves slowly down through the soil under the influence of gravity. Soil water occurs in the **zone of aeration,** where both water and air occupy the pore spaces between the individual soil particles in constantly varying proportions. Below this is the **zone of saturation.** Here, all pore spaces or other openings in the soil or rock are completely filled with **ground water** (see Figure 10.1). The zone of saturation, which extends to about 3 kilometers (2 mi) beneath the surface, contains most of the Earth's fresh subsurface water. Soil water and ground water both are derived primarily from precipitation and are in constant slow movement. Most of this water eventually returns to the surface to continue its participation in the hydrologic cycle.

Soil Water

Although nearly all water in the soil is derived from precipitation, much precipitation on land does not fall directly onto the soil. Instead, it is **intercepted** by the surface cover of vegetation and vegetative litter (see Figure 10.2). These overlying materials must be thoroughly wet before an excess is produced that drips or flows down to the soil level. The interception process is especially effective in reducing available soil moisture at the beginning of a rainfall, and a light shower falling in a heavily vegetated area may be almost entirely intercepted and the moisture subsequently lost to evaporation.

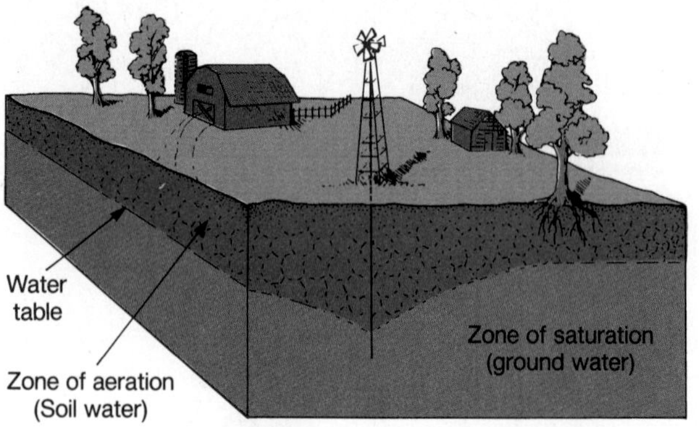

FIGURE 10.1 Subsurface water zones and their relationship to the water table.

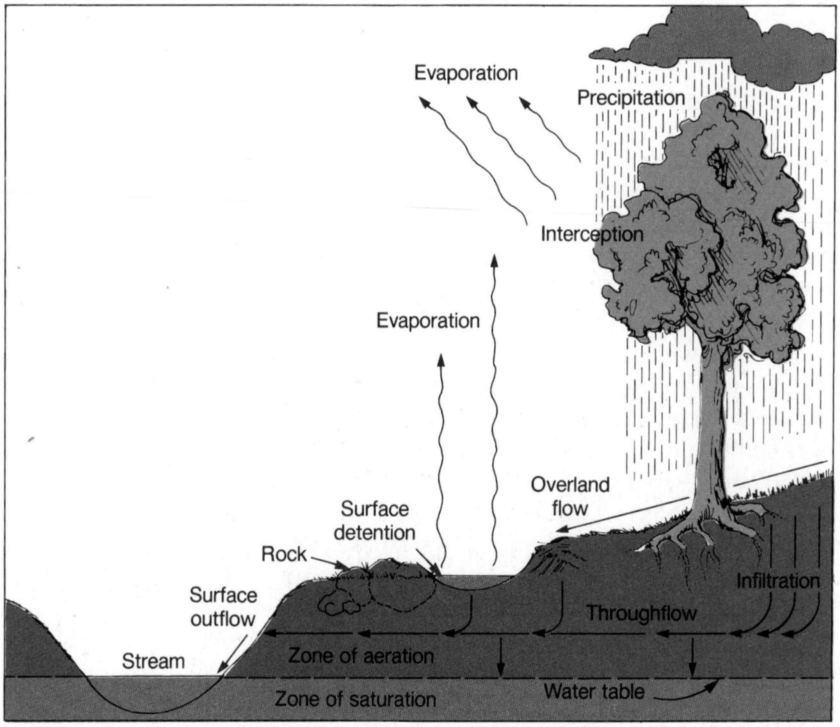

FIGURE 10.2 Water may be intercepted or may be stored on the surface before infiltrating the soil. During rainy periods, considerable subsurface water travels laterally near the surface as throughflow, rather than descending to the water table to become ground water. Most throughflow soon returns to the surface.

When the vegetation cover is fully wetted, additional moisture reaches the soil surface. At this point, the water may collect on the surface and eventually evaporate, it may flow directly into rivers or lakes, or it may be absorbed by the soil. The maximum rate at which the soil can absorb water at any given time depends largely on its porosity and permeability. The **porosity** of the soil is defined as the volume of soil pore spaces divided by the total volume of the soil itself. The more porous a soil, the more void space it contains that is capable of retaining water. Undisturbed soils in humid, well-vegetated environments typically are deep and relatively porous. As a result, only exceptionally intense storms have rainfall rates that exceed the soil infiltration rate and produce surface flow. The thinner soils of arid and semiarid regions, however, more commonly experience surface runoff as a result of heavy rainfalls. The **permeability** of a soil refers to its ability to transmit flow. It is determined by the number and size of the pore spaces, and especially by the extent to which they are interconnected.

Both the porosity and permeability of the soil can be greatly reduced by surface disturbances such as compaction by machinery or livestock. Even the removal of vegetation allows the splash of raindrops on the bare soil to dislodge fine particles and wash them into surface pore spaces, clogging them up and considerably reducing the soil's permeability. Soil disturbances related to agricultural activities have significantly lowered total world soil infiltration rates, producing increased surface runoff and greatly increased erosion rates.

Water that has infiltrated the soil is drawn slowly downward by gravity. The water moves in very circuitous routes around the soil particles, with some moisture adhering to the surfaces of the soil grains. With depth, soils become increasingly compacted by the weight of the overlying material, resulting in a progressive decline in porosity and permeability.

During a prolonged rainstorm, the continued downward percolation of soil water may produce a temporary condition of saturation in the lower soil. Since this water can no longer move downward rapidly enough to offset the addition from above, it is diverted laterally so that it flows parallel to the surface as **throughflow.** Throughflow is probably the major type of hillside flow in humid climates. This flow may intersect the surface along a stream bank or other steep slope and seep out to continue downhill as surface flow (see Figure 10.2).

Soil water that does not seep out at the surface or adhere to the soil particles continues to sink slowly down until it reaches the **water table,** which is the upper boundary of the zone of permanently saturated soil or rock. In most cases, water reaches the water table within 48 hours of the time it initially entered the soil. At that point the soil is left containing only water that adheres to the soil particles and can therefore be held indefinitely against the pull of gravity. This "held" water is considered to be an integral constituent of the soil. Although the amount of soil water varies because of evaporative losses and extraction by plant roots, it always is present to some extent, even in the driest areas. Its characteristics and behavior are further considered in Chapter 12.

Ground Water

Ground water saturates the soil and rock layers below the water table. In general, it is most abundant in areas underlain by extensive layers of porous rock such as sandstone. The quality and quantity of water present as well as the location and extent of the zone of saturation, however, vary greatly in different localities. In some places, the upper surface of the zone of saturation intersects the Earth's surface, producing an outflow of ground water. In other places, it exists hundreds of meters beneath the surface. Any hole, such as a well, that is dug into the zone of saturation will fill with water up to the level of the water table.

In arid areas little or no water is being added to existing ground water supplies. Ground water in such areas generally is a legacy of more humid climatic conditions in the past and is sometimes referred to as **fossil water.** Like oil, it is a nonrenewable resource because it cannot be replaced under the existing climate. The extraction of this water permanently lowers the water table and reduces ground water supplies.

Even in semiarid and subhumid regions, such as the southern Great Plains of the United States, the large-scale extraction of water for irrigation is producing rapidly declining water tables and the threat of eventual ground water exhaustion. Shallow wells in such regions create surrounding **cones of depression** in the water table as the water is withdrawn more rapidly than it can be replaced (see Figure 10.6). If wells are located too close to one another, adjacent cones of depression can coalesce to produce a regional lowering of the water table.

In humid areas, the water table is subject to continual fluctuations in level. During wet periods, the addition of water from precipitation causes it to rise until it reaches the surface in valleys or other low-lying areas, sometimes producing an outflow of ground water known as a **spring.** If the outflow occurs in a valley that continues downhill, the spring will form or add water to a river or stream (see Figure 10.3). If it occurs in an enclosed depression, a lake or swamp will be formed. In this fashion, ground water returns to the surface to continue its movement through the hydrologic cycle.

Most streams in humid areas are at the water table and have a substantial portion of their total volumes of flow provided by the outflow of ground water. This provides a dependable supply of water, causing larger streams to be **perennial streams,** which continue to flow even during extended droughts. The much higher flow volumes during wet periods are derived primarily from the temporary additions of surface runoff and throughflow. In contrast, the water table frequently is far below the level of the river valleys in arid regions, causing such regions to contain intermittent or **ephemeral streams,** which flow only during wet periods when they are supplied by surface or soil water sources. Small streams in humid environments also are frequently intermittent or ephemeral. For the world's land areas as a whole, it is

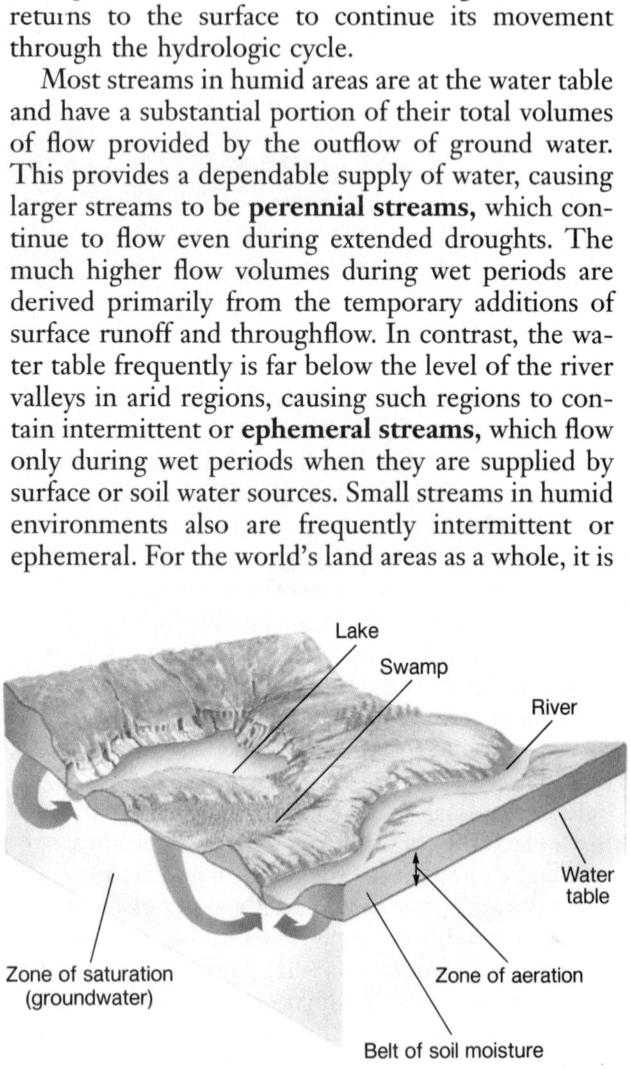

FIGURE 10.3 Relationship between the surface topography and the movement of ground water.

estimated that ground water outflow supplies 30 percent of total stream flow. Total land runoff is believed to average approximately 26.7 vertical centimeters (10.5 inches) of water per year, so that the annual ground water contribution to this flow averages about 8.0 centimeters (3.2 in).

Ground water flow rates usually are imperceptible, ranging from a meter or two per day to a few meters a year. The speed of flow depends on both the pressure head that develops as a result of local differences in the water table level and on the frictional resistance that the water encounters to its movement. The greater the friction (i.e., the lower the permeability of the strata through which the water is flowing), the larger and steeper the pressure head that can be maintained.

As already noted, different Earth materials vary tremendously in their ability to store and transmit ground water. A porous and permeable rock layer that contains large quantities of ground water and permits its ready movement is called an **aquifer.** In many parts of the world large aquifers, often consisting of deep layers of sandstone or limestone, act as subsurface reservoirs that hold tremendous volumes of fresh water. Wells dug or drilled into these strata may be expected to yield large and dependable water supplies.

The world's largest known aquifer, the Nubian Sandstone, underlies much of the western Sahara. It contains an estimated 600,000 cubic kilometers (144,000 mi^3) of water. Although its present recharge rate is almost zero, this vast underground reservoir is still virtually untapped and constitutes a natural resource with enormous potential for the countries of northwestern Africa.

The Ogallala aquifer, which underlies much of the southern and central Great Plains, is the most economically important aquifer in the United States. In contrast to the Nubian Sandstone, this aquifer has in recent decades been greatly lowered in many areas by large-scale withdrawals of water for irrigation agriculture (see the Focus Box on pages 226–227).

Aquifers may be underlain or surrounded by relatively impermeable rock layers known as **aquicludes.** These strata prevent the aquifer from losing its water by acting as seals to its movement. Aquifers often serve as storage tanks, holding their water supplies in one area for extended periods of time. In some cases, however, aquifers can serve as conduits that transport water great distances. It was noted in Chapter 9 that highlands often have substantial precipitation surpluses because of orographic lifting factors and cool temperatures. Aquifers extending from humid highlands to adjacent dry lowland areas therefore may serve as important water suppliers to the lowlands.

Human Impact on Ground Water

Ground water has been extracted by humans since the beginning of civilization, primarily for the irrigation of crops. Until less than a century ago, only near-surface water supplies were tapped through the digging of shallow wells. In recent decades, however, water extraction technology has advanced rapidly. The development of deep drilling methods and powerful pumps has made large quantities of ground water available. It also has unfortunately substantially lowered the water table in virtually all areas where ground water is extracted on a large scale.

Ground water supplies often are most extensively developed in arid and semiarid areas where surface water is inadequate or nonexistent. This creates a problem in maintaining reserves, since the ground water in these areas generally is fossil water. The overpumping of water in numerous drier areas of the world is producing rapidly declining water tables, increasing pumping costs, and decreasing yields. Three areas within the United States where these problems are especially acute are southern Arizona, the Central Valley of California, and the southern Great Plains.

Other problems resulting from the overuse of ground water are saline water intrusion, land subsidence, and waterlogging. **Saline water intrusion** occurs in coastal regions where so much fresh water has been removed that the normal seaward flow of fresh water is reversed, and saline water from the ocean is instead drawn in (see Figure 10.6).

When ground water saturating certain fine-grained sediments is withdrawn, an irreversible dewatering and compaction of the sediments occurs, resulting in the sinking, or **subsidence,** of the surface. Surface subsidence has produced serious problems in places such as Mexico City, Bangkok, Venice, southeastern Texas, and the Central Valley of California. In parts of Mexico City, for example, subsidence has lowered the surface by more than 8 meters (25 feet), and the process continues at a rate of about 2.5 centimeters (1 inch) a year. Subsidence in Venice, Italy, is especially critical because much of the city is situated at

THE DEPLETION OF THE OGALLALA

The vast short grasslands of the American High Plains were first settled by farmers and ranchers in the 1880s. This region, lying in the rainshadow of the Rocky Mountains, has a semiarid climate, and for 50 years after its initial settlement, it supported a low-intensity agricultural economy of cattle ranching and wheat farming. In the early twentieth century, however, it was discovered that much of the High Plains were underlain by a huge aquifer. This aquifer was named the Ogallala Aquifer after the Ogallala Sioux Indians that once inhabited the region.

The Ogallala Aquifer is a calcareous sandstone formation that underlies some 583,000 square kilometers (225,000 square miles) of land extending from northwestern Texas to southern South Dakota (see Figure 10.4). Although it is thin near its southern terminus, it gradually thickens northward, attaining a maximum thickness of over 300 meters (1,000 ft) in Nebraska. Water from rains and melting snows has been accumulating in the Ogallala for the past 30,000 years, and it is estimated that the aquifer contains enough water to fill Lake Huron. Unfortunately, under the semiarid climatic conditions that presently exist in the region, current rates of addition to the aquifer are minimal, amounting to about half a centimeter (0.2 in) a year.

The first wells were drilled into the Ogallala during the drought years of the early 1930s. The ensuing rapid expansion of irrigation agriculture, especially from the 1950s onward, transformed the economy of the region. More than 100,000 wells now tap the Ogallala. Modern center-pivot irrigation devices, each capable of spraying 4.5 million liters (1.2 million gallons) of water a day, have produced a landscape dominated by geometric patterns of circular green islands of crops (see Figure 10.5). Ogallala water has enabled the High Plains region to supply 25 percent of the cotton, 38 percent of the grain sorghum, 16 percent of the wheat, and 13 per-

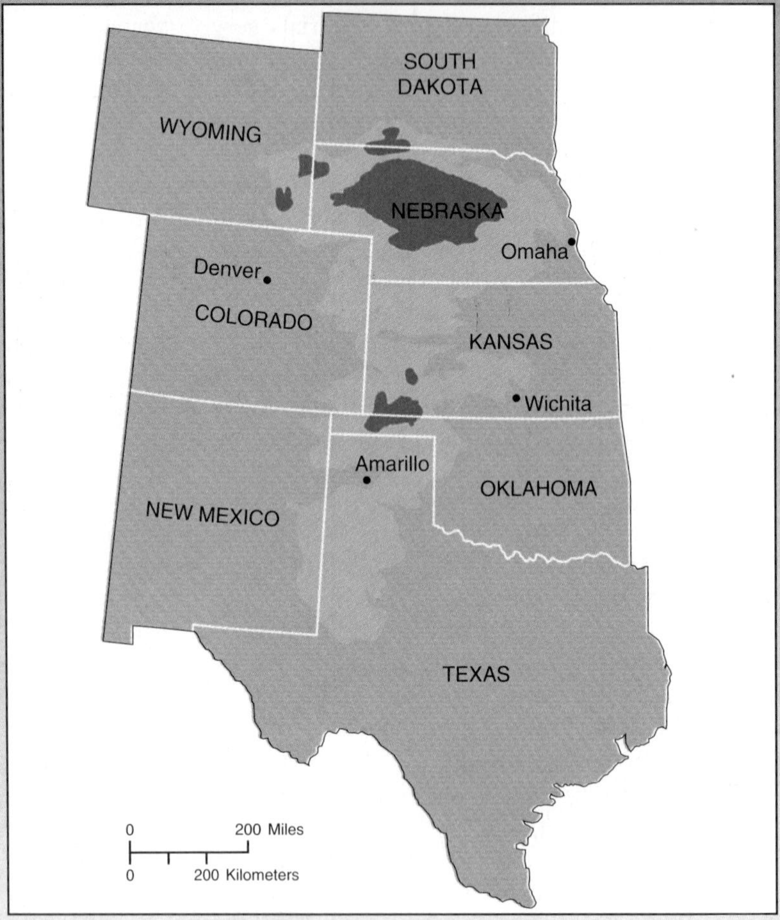

FIGURE 10.4 Location of the Ogallala aquifer.

FIGURE 10.5 Tremendous quantities of water from the Ogallala aquifer are used by center pivot irrigation systems. The resulting circular crop patterns form a distinctive landscape when viewed from the air. (Science VU/Visuals Unlimited)

cent of the corn grown in the United States. In addition, 40 percent of American grain-fed beef cattle are fattened here.

This unprecedented development of a finite groundwater resource with an almost negligible natural recharge rate unfortunately has caused water tables in the region to fall drastically. In the 1930s, wells encountered plentiful water at a depth of about 15 meters (50 feet); currently, they must be dug to depths of 45 to 60 meters (150 to 200 feet) or more. In places, the water table is declining at a rate of a meter (3 feet) a year, necessitating the periodic deepening of wells and the use of ever more powerful pumps. It is estimated that, at current withdrawal rates, much of the aquifer will run dry within 40 years. The situation is most critical in Texas, where the climate is driest, the greatest amount of water is being pumped, and the aquifer contains the least water. More than 25 percent of the Texas portion of the Ogallala has now run dry, and pumping costs for the remainder have increased by 400 percent between the early 1970s and the mid-1980s. It is projected that the remaining Ogallala water will, by the year 2030, support only 35 to 40 percent of the irrigated acreage in Texas that it supported in 1980.

The reaction of farmers to the inevitable depletion of the Ogallala varies. Many have been attempting to conserve water by irrigating less frequently or by switching to crops that require less water. Others, however, have adopted the philosophy that it is best to use the water while it is still economically profitable to do so, and to concentrate on high value crops such as cotton. The incentive of the farmers who wish to conserve water is reduced by their knowledge that many of their neighbors are profiting by using great amounts of water, and in the process are drawing down the entire region's water supplies.

In the face of the upcoming water supply crisis, a number of grandiose schemes have been developed to transport vast quantities of water by canal or pipeline from the Mississippi, the Missouri, or the Arkansas Rivers. Unfortunately, the cost of water obtained through any of these schemes would increase pumping costs at least tenfold, making the cost of irrigated agricultural products from the region uncompetitive on the national and international markets. Somewhat more promising have been recent experiments for releasing capillary water above the water table by injecting compressed air into the ground. Even if this process proves successful, however, it would almost triple water costs. Genetic engineering also may provide a partial solution, as new strains of drought-resistant crops continue to be developed. Whatever the final answer to the water crisis may be, it is evident that, within the High Plains, irrigation water will never again be the abundant, inexpensive resource that it was during the agricultural boom years of the mid-twentieth century.

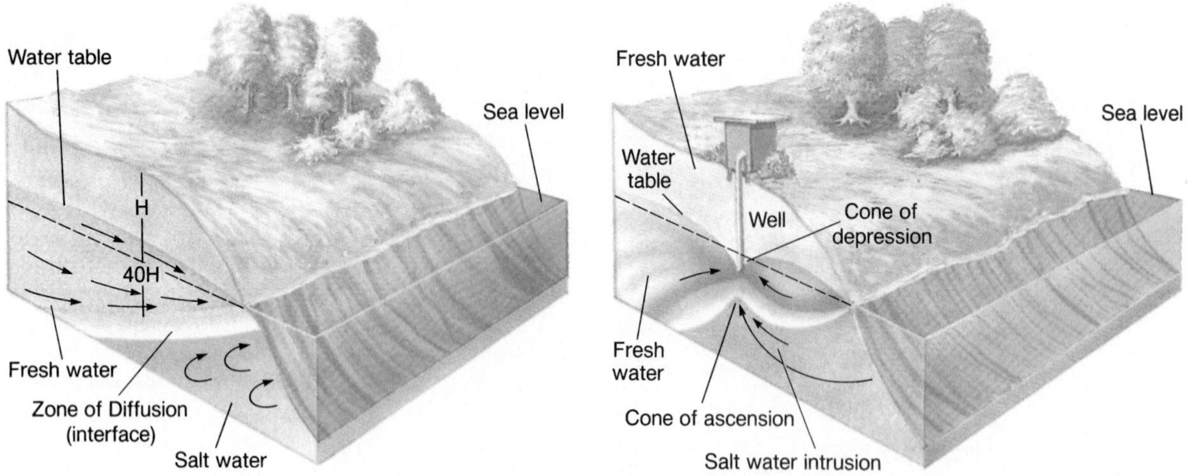

FIGURE 10.6 Saline water intrusion usually results from the excessive withdrawal of fresh ground water from a near-coastal site.

or near sea level; as a result, the famed city of canals has been partly inundated by the Adriatic Sea.

Waterlogging of the soil typically occurs in low areas in dry climates where the application of irrigation water over extended periods has raised the water table to the surface. This process often results in the dissolution and surface deposition of salts, rendering the land unfit for cultivation (see the Focus Box on pages 304–305 in Chapter 12). Waterlogging and salt deposition is a problem of increasing severity in parts of the western United States and has long existed in the Middle East. Large portions of the once fertile Tigris-Euphrates lowland of Iraq, for example, have been converted to a semidesert by this process.

An additional ground water problem resulting from human activities is pollution. An increasing variety of liquid contaminants is finding its way into ground water supplies, especially in urban and industrialized areas with humid climates such as the eastern United States and western Europe. The chief sources of pollutants are waste disposal sites such as landfills, lagoons, and disposal pits. Nonpoint sources such as urban runoff and water from agricultural land treated with fertilizers and pesticides also produce pollution problems.

LAKES

A **lake** is a landlocked body of water with a horizontal surface water level. Lakes are short-lived landscape features, usually lasting only a few centuries or millennia. As long as this may seem in human terms, it is an extremely short timespan from a geological standpoint. Despite the relative brevity of their existence, however, lakes currently contain more than 95 percent of our planet's surface supply of fresh liquid water.

Geographic Significance

Freshwater lakes cover an estimated total of 825,000 square kilometers (320,000 sq mi), or about 1.5 percent of the Earth's land surface. They contain more than 125,000 cubic kilometers (30,000 mi^3) of water. Although probably more than a million lakes are in existence, roughly four-fifths of the total volume of the Earth's fresh surface water is found in the 40 largest lakes.

The world distribution of lakes, like that of ground water and the ocean, but unlike that of rivers, is related more closely to the presence of basins suitable for water accumulation than it is to climate. Most of the world's freshwater lakes exist on three continents—North America, Africa, and Asia (see Table 10.1 and Figure 10.7). The large lakes of North America contain approximately 32,000 cubic kilometers (7,700 mi^3) of water, of which over three-fourths is in the five Great Lakes. Large lakes in East Africa contain some 36,000 cubic kilometers (8,640 mi^3) of fresh water. Lake Baikal in the southeastern

TABLE 10.1 The World's Largest Lakes

NAME	BASIN ORIGIN	SURFACE AREA		VOLUME		MEAN DEPTH	
		Km^2	Mi^2	Km^3	Mi^3	M	Ft
1 *Caspian Sea	Uplift (cut off from sea)	371,000	148,000	79,340	19,035	182	597
2 Superior	Glacial erosion	82,100	31,700	12,150	2,916	149	489
3 Victoria	Uplift of margins	68,780	26,560	2,660	637	40	131
4 *Aral Sea**	Uplift (cut off from sea)	61,980	23,930	900	215	15	51
5 Huron	Glacial erosion	59,570	23,000	3,450	827	59	195
6 Michigan	Glacial erosion	57,750	22,300	4,840	1,161	85	279
7 Tanganyika	Faulting (graben)	33,990	13,120	19,420	4,659	572	1,876
8 Baikal	Faulting (graben)	31,490	12,160	23,260	5,581	740	2,427
9 Malawi	Faulting (graben)	30,790	11,890	8,370	2,009	273	895
10 Great Bear	Glacial erosion	30,400	11,740	2,200	529	72	238
11 Erie	Glacial erosion	25,670	9,910	480	116	19	62
12 Great Slave	Glacial erosion	25,390	9,800	1,550	373	62	204
13 Winnipeg	Glacial erosion	24,520	9,470	320	76	13	43
14 Ontario	Glacial erosion	19,010	7,340	1,640	393	86	283
15 Ladoga	Glacial erosion and uplift	18,731	7,230	917	220	91	300

SOURCE: Academic American Encyclopedia
*Saline Lakes.
**Because of the Soviet diversion of the two major rivers entering the Aral Sea to supply water for irrigation agriculture, this water body has been shrinking rapidly. Between 1957 and 1990, the Aral Sea decreased 35 percent in area and 67 percent in volume. In 1990, its area was estimated at 40,000 km^2 (15,440 mi^2), and its volume at 350 km^3 (84 mi^3).

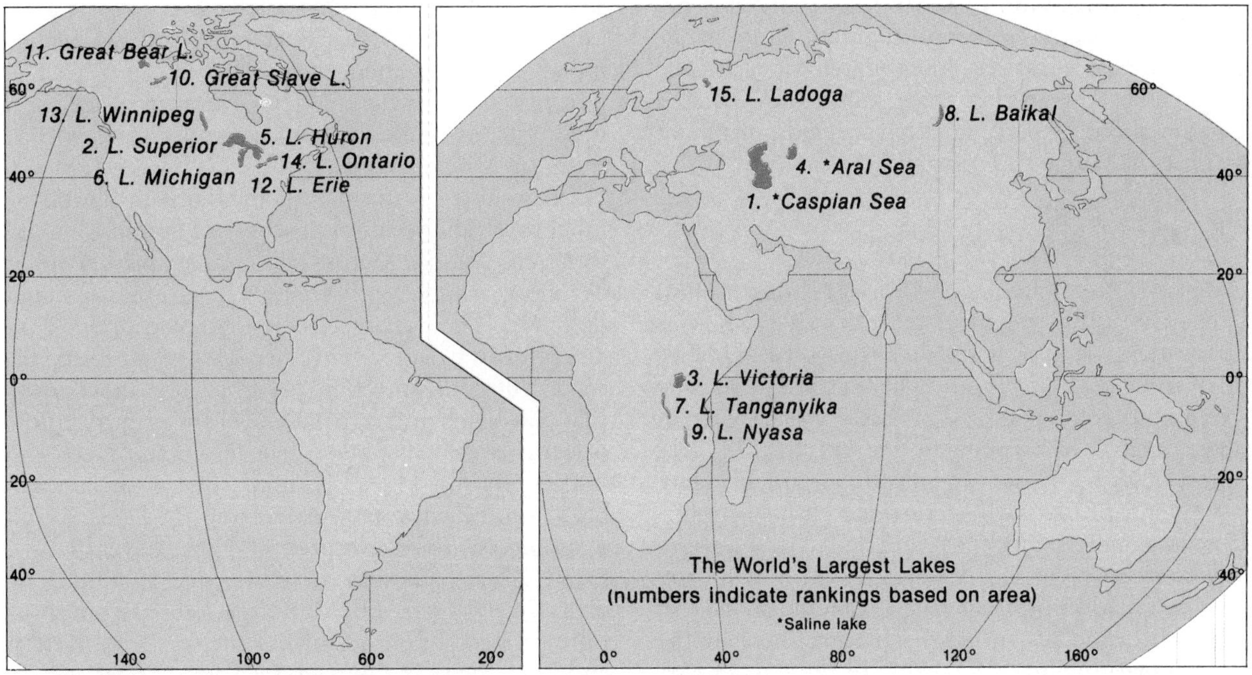

FIGURE 10.7 Locations of the world's fifteen largest lakes.

Soviet Union is the world's deepest and most voluminous body of fresh water; it contains 23,000 cubic kilometers (5,500 mi^3) of water— 18 percent of the world total. In contrast, lakes in South America, Europe, and Australia collectively contain only about 3,000 cubic kilometers (720 mi^3) of water, or about 2 percent of the total.

Although fresh water lakes are much more numerous and important for human activities than are those containing salt water, some of the world's largest lakes are saline. Saline lakes cover an estimated total area of 700,000 square kilometers (270,000 sq mi) and contain approximately 105,000 cubic kilometers (25,000 mi^3) of water. Both the area and volume of all saline lakes combined therefore are about 85 percent of the respective totals for freshwater lakes. Saline lakes are located mostly in Asia and are dominated by the Caspian Sea, which, despite its name, is considered the world's largest lake. By itself, the Caspian Sea contains 76 percent of the total volume of all the world's saline lakes.

Lakes with outlets nearly always contain fresh water, while those with no outlets usually are saline. In some cases, as with the Caspian and Aral Seas, lakes are salty because they are cutoff arms of the ocean. In most cases, however, the salt has been supplied gradually by inflowing rivers and has accumulated because the evaporation of water has left the salt behind. Some lakes are only slightly saline, while others, like the Dead Sea, are saturated with salt. The Dead Sea is the world's most saline large lake. It contains about 24 percent dissolved solids, making it six times as salty as the ocean.

Origins of Lakes

Only two criteria are necessary for the formation of a lake. One is the presence of a relatively nonporous surface depression or basin; the other is the availability of sufficient water from either surface or subsurface sources to fill it at least partially. Surface depressions capable of trapping water are formed in a variety of ways, but the two most important by far are movements of the Earth's crust and glacial action.

Crustal movements, the origins and characteristics of which are discussed in Chapters 14 and 15, are notable for forming large lakes, sometimes of great depth. In some instances, the motion involves the gradual broad uplift of a portion of the Earth's surface that cuts off an arm of the sea to form a lake. The Caspian and Aral Seas formed in this fashion when crustal uplift gradually severed their connection with the Mediterranean. Tectonic activity also can result in a narrow wedge of land (a **graben**) being dropped sharply downward between a pair of faults. Lakes resulting from this type of movement tend to be long, narrow, and exceptionally deep. Lake Baikal was formed this way. The world's greatest group of fault-produced lakes is located along the Great Rift Valley, which extends through Asia and Africa from Syria to Mozambique. Major lakes within this valley include the Dead Sea, Lake Albert, Lake Tanganyika, and Lake Malawi.

Lakes resulting from glacial erosion and deposition are more numerous than those from all other origins combined (see Figure 10.8). They vary greatly in size, and most are located in northern portions of North America and Eurasia, which were covered by continental glaciers between 10,000 and 20,000 years ago (see Chapter 18). Most glacial lake basins were produced by the scooping action of ice either in areas of soft bedrock or in loose ice-deposited sediments. Others resulted from the blocking of surface drainage by glacial deposits. Literally hundreds of thousands of glacial lakes, large and small, are found within North America north of the Ohio and Missouri Rivers. Included are some of the world's largest lakes, such as the five Great Lakes, Great Slave Lake, Great Bear Lake, and Lake Winnipeg. Large numbers of glacial lakes also exist in northern Europe, especially in Scandinavia, although they are less numerous and generally smaller than those of North America.

Lakes also can form in depressions produced by numerous other natural processes. Foremost among these are solution depressions in regions underlain by highly soluble rock types such as limestone and dolomite. Their characteristics and distributions are discussed in Chapter 16. Other lakes result from volcanic eruptions, landslides, or other mass movements that may block drainage outlets; changes in river flow patterns; wind erosion or deposition; and meteorite impacts. It should also be noted that artificial lakes have become important features on the landscape in many areas. Some are produced inadvertently as a result of earth-moving activities such as mining or quarrying. Others, often of large size, are purposefully created. Large reservoirs such as Lake Powell and Lake Mead in the western United States serve a variety of purposes, including flood control, water

FIGURE 10.8 Peyto Lake, in Banff National Park, Alberta, is one of hundreds of thousands of Canadian lakes formed by glacial erosion (© Pat and Tom Lesson/Photo Researchers)

supply, hydroelectricity generation, transportation, and recreation.

Disappearance of Lakes

The basic reason for the brevity of the existence of lakes is that physical and biological factors work against the maintenance of enclosed basins filled, or partially filled, with standing water. In many cases, the inflow of water exceeds a basin's capacity to hold it and the ability of evaporation to remove it. If this occurs, a surface outflow of water results that gradually erodes the lip of the basin and reduces its ability to contain water. Continued erosion will eventually destroy the basin's ability to hold standing water.

The relative lack of motion of lake water also makes the basin a sediment trap. Sediments carried by inflowing streams are deposited on the floor of the lake, gradually filling it. This process has shortened the expected useful lifespans of even large artificial lakes in various parts of the world to only a century or two. The process of infilling often is hastened by the growth of aquatic plants and the consequent deposition of organic matter. Most shallow lakes gradually become **swamps** when their water depths decline sufficiently to allow water-tolerant vegetation to cover the entire lake surface, and with continued infilling they eventually become dry land.

STREAMS

Rivers and **streams** are channeled bodies of flowing water. They engage in the gravitational transport of excess surface moisture derived from precipitation. The water is ultimately transported either to interior basins on land or, more commonly, to the sea. Although the two terms are often used interchangeably, a stream has no particular size connotation, whereas a river generally is considered to be large—often the master stream in a drainage system. Stream flow constitutes the second most dynamic, or short-term, phase of the hydrologic cycle, after water vapor transport by wind. It forms a vital final link in the cycle by returning vast quantities of water to the oceans.

In addition to carrying water, streams also transport sediments derived from the land. In doing so, they serve as the major agent responsible for the gradual erosion of the Earth's land surface. Stream-produced landscapes are so important that an entire chapter (Chapter 17) is devoted to it. In this section we concentrate on the hydrologic aspects of streams.

Geographic Distribution and Significance

Streams vary greatly both in number and in volume of flow in different parts of the world. They are generally absent only from the 10 percent of the Earth's

THE GREAT LAKES BECOME GREATER

Following an extended period of cool, wet years, the Great Lakes in the mid 1980s rose to their highest levels in 125 years of record-keeping. In late 1986, their surface water levels averaged about two meters (6 ft) higher than at low points reached in 1964. In October 1986, the surface of Lake Michigan was at an average elevation of 177.3 meters (581.6 ft), more than 0.3 meters (1 ft) higher than a year earlier. Drier conditions in 1987 and 1988 lowered lake levels somewhat from the record highs of 1986, but this decline may only be temporary.

The recent rise in water levels has occurred because the Great Lakes, like many lakes, respond in a highly sensitive fashion to the balance between water gains, resulting from precipitation and water inflow, and losses, resulting from evaporation and outflow. Evaporation is responsible for an estimated 40 percent of water losses from the Great Lakes, and outflow is responsible for the remaining 60 percent. For 15 of the 18 years from 1969 through 1986, precipitation in the Great Lakes area was above normal, while cooler than normal temperatures reduced evaporational losses.

Although the lakes are at record levels for the historical period of record-keeping, they may be only just approaching mean levels for the past 1,500 years. There is evidence of a cyclical rise and fall of lake levels of 3 to 5 meters (10 to 15 ft), with a periodicity of about 500 years.

The rising water levels pose serious problems for the millions of people who live along the shores of the lakes in both the United States and Canada. Lakefront property is heavily built up right to the water's edge in many urban areas such as Chicago, Milwaukee, and Cleveland, so that the effect of any change in water level is felt immediately. During the winter storms that are common in this area, giant waves destroy seawalls and toss sheets of water into streets and buildings, breaking windows and flooding basements. Chicago's famed Lake Shore Drive has had to be closed on occasion because of high water. Along less developed stretches, storm waves have resulted in major property losses when they have undermined lakeshore bluffs (see Figure 10.9). The problems are worst when strong winds blow onshore, piling up the water even higher than normal.

It may be wondered why water levels are able to rise substantially,

land surface that is perennially covered by ice. Even those areas, located mostly within Antarctica and interior Greenland, have excess precipitation removed by rivers; the only difference is that they are rivers of frozen water, or glaciers. In addition, meltwater streams drain many glaciated areas during the summer. Elsewhere, as might be expected, the extent of river development is determined largely by the balance between mean precipitation amounts and evapotranspiration rates. Even the driest deserts, however, have stream channels that occasionally carry water.

The volume of water contained in the world's streams at any one time is very much smaller than the amount held in lakes and as ground water. It amounts to only about 0.025 percent, or 1/4,000th, of the Earth's total near-surface supply. The practical significance of this water, though, is much greater than its volume implies, both because of its ready accessibility and because of its speed and power of flow. While water in lakes and beneath the surface can be considered to be in at least temporary storage, the water in streams is in active movement so that, in time, large quantities are involved. In one year, the total volume of water discharged into the ocean by streams is estimated at 33,000 cubic kilometers (8,000 mi^3). This is approximately one-third of the total volume of the Earth's freshwater lakes.

Although virtually all stream flow ultimately is derived from precipitation, the water in a direct sense comes from several different sources. The most obvious of these is surface runoff from storms, which in some areas supplies the majority of the water entering stream systems. It currently is believed, however, that in most humid tropical and temperate regions, the largest volume of water in streams is supplied by the outflow of ground water and by throughflow.

FIGURE 10.9 Active erosion along the north shore of Lake Ontario near Toronto is responsible for the formation and recession of these bluffs. (B.F. Molnia)

since the Great Lakes drain into the Atlantic by way of the St. Lawrence River and into the Mississippi River system through the Chicago River. These outlets, however, are simply far too small to carry off the water at the rate at which it has been accumulating. One expert likened their potential for draining the lakes to "pinholes in a bathtub." For example, a recent proposal to triple the outflow of Lake Michigan water through the Chicago River would not only produce flooding along the river, but also would remove only enough water to drop the lake's water level by 6 centimeters (2.5 in) in a fifteen-year period.

Scientists agree that no quick solution to the problem is in sight. Lake levels should remain high for the foreseeable future and may even rise further, resulting in extensive flooding along the south shores of all five lakes.

This is especially true in areas of porous and permeable rock and of deep, relatively undisturbed soils. Some streams also are fed by glacial meltwater, and a small quantity of additional water is added directly by precipitation falling into the streams.

A stream's **discharge,** or volume of flow past a given point in a specified time interval, normally increases downstream because of the increase in the total area drained. Losses occur to seepage and especially to evaporation, however, and streams flowing through dry climate regions actually decrease in discharge in a downstream direction, sometimes even drying up completely. Two well-known examples of major rivers that decrease naturally in flow because their courses traverse arid regions are the Nile and the Colorado. (In recent times, the volumes of flow of both the Nile and Colorado have further diminished because of water withdrawals for human use. At present, in fact, no water from the Colorado River reaches its mouth at the head of the Gulf of California except during times of exceptionally high flow!)

The total global discharge of streams into the ocean is believed to be slightly less than 30 percent of all the precipitation that falls on the land areas of the Earth. Most of the remainder is evaporated, making evaporation the leading process by which moisture leaves the land. A relatively small amount of water also enters the ocean as subsurface flow.

Although many thousands of streams of all sizes flow into the ocean, total global discharge is dominated by a small number of large rivers (see Table 10.2 and Figure 10.10). The Earth's major river system in all respects except length is the Amazon (see Figure 10.11). By itself, it drains 4.8 percent of the

TABLE 10.2 The World's Major Rivers

NAME	DRAINAGE AREA (x 1000)		PERCENTAGE OF WORLD'S LAND AREA DRAINED	LENGTH		MEAN DISCHARGE (x 1000)		PERCENTAGE OF WORLD'S TOTAL DISCHARGE
	Sq mi	Sq km		mi	km	F^3/sec	M^3/sec	
Amazon	2,722	7,052	4.8	4,000	6,440	6,350	180	19.2
Rio de la Plata/Paraná	1,600	4,145	2.8	2,485	4,000	777	22	2.3
Congo	1,314	3,404	2.3	2,914	4,692	1,458	41	4.4
Nile	1,293	3,350	2.3	4,132	6,653	110	3	0.3
Mississippi/Missouri	1,244	3,223	2.2	3,741	6,020	650	18	2.0
Ob/Irtysh	1,149	2,977	2.0	3,362	5,413	558	16	1.7
Yenisey	996	2,580	1.7	3,332	5,365	671	19	2.0
Lena	961	2,490	1.7	2,734	4,402	575	16	1.7
Yangtze	756	1,959	1.3	3,434	5,526	1,200	34	3.6
Niger	730	1,891	1.3	2,600	4,186	215	6	0.7
Amur	716	1,855	1.3	1,755	2,826	438	12	1.3
Mackenzie	711	1,842	1.2	2,635	4,242	400	11	1.2
Ganges/Brahmaputra	626	1,622	1.1	1,800	2,898	1,360	39	4.1
St. Lawrence/Great Lakes	565	1,464	1.0	2,500	4,025	360	10	1.1
Volga	525	1,360	0.9	2,293	3,692	282	8	0.9

SOURCE: Encyclopaedia Britannica.

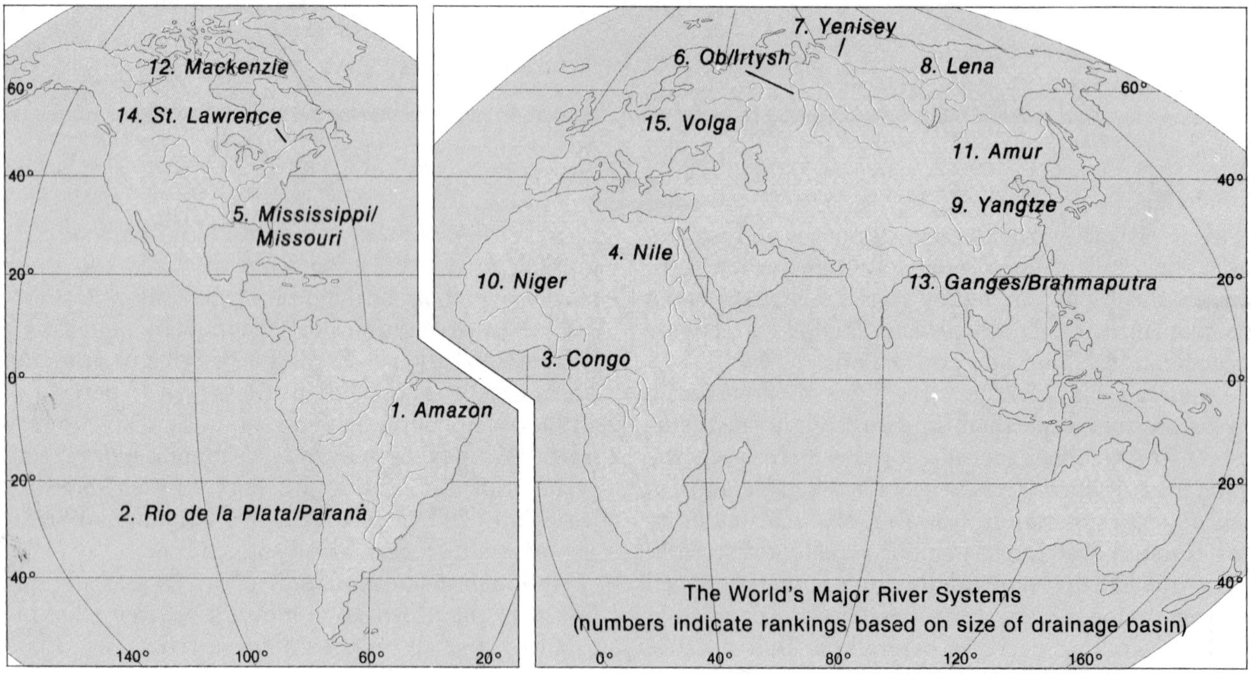

FIGURE 10.10 Locations of the world's fifteen largest rivers.

FIGURE 10.11 The drainage basin of the Amazon River (depicted in green) is the world's largest.

Earth's land area and contributes nearly one-fifth of the total world discharge, a figure more than four times that of the second largest river, the Congo. The Mississippi, North America's dominant river, ranks only seventh in total discharge among the rivers of the world. The Nile, with a length of 6,650 kilometers (4,132 mi), has the distinction of being the world's longest river, and it also has the fourth largest **drainage basin,** or area drained. Because of the arid nature of much of its course, however, it is not among the top twenty rivers in discharge.

The areal distribution of stream discharge volumes is strongly correlated with climate, especially with precipitation. As might be expected, regions with high precipitation generally have streams with large discharge volumes. Temperature and humidity also are important criteria, since they largely determine evapotranspiration rates. The geographic influence of these two factors is to increase discharge at any given level of precipitation in areas at high latitudes and elevations, as these areas tend to have cooler climates.

Several nonclimatic variables also significantly influence discharge rates. One of the most important is the amount and type of vegetation growing in the drainage basin. In general, a high vegetation density results in decreased water entering nearby stream systems because of the role of vegetation in intercepting precipitation, increasing friction for runoff (giving it more time to be absorbed or evaporated), and especially in transpiring large quantities of moisture. For these reasons, lumbering or the removal of vegetation for agricultural purposes typically increases runoff and erosion.

The nature of the surface also strongly influences local discharge volumes. Irregular surfaces, and particularly the presence of surface depressions, reduce runoff by promoting the storage of water in puddles, ponds, or lakes. Areas of steep slopes tend to experience greater stream discharge because rainwater reaches stream systems rapidly, giving it less time to be absorbed or to evaporate. Even water that immediately infiltrates the soil in areas of steep slopes travels much more rapidly than subsurface water in more level areas. The porosity and permeability of the soil and rock also play important roles in controlling rates of water absorption and movement because the pores provide conduits for subsurface flow. A final factor that is a crucial determinant of stream discharge is the size of its drainage basin. Within a given region, where other controlling factors are relatively constant, the volume of discharge of each stream is strongly correlated with the drainage basin area.

Temporal Variations in Stream Flow

Although the size of a stream can be described in a general way by citing its mean volume of flow, the discharge of most streams varies considerably on both a seasonal and a short-term basis. The causes of these variations are mostly climatic, because the great majority of the Earth's land surface has distinctive seasonal fluctuations in precipitation, temperature, or both. Variations in the quantities of water transpired by plants also are significant in areas of seasonally active vegetation.

Near the equator, large rainfall totals produce high runoff rates throughout the year. Somewhat farther poleward, in the seasonally wet and dry tropics, streams tend to have a single well-developed summer discharge maximum. Many smaller streams here are intermittent, with no flow at all near the end of the dry season. Within the middle and higher latitudes, flow patterns vary considerably, depending on precipitation distribution, but streams most commonly exhibit winter and spring discharge maxima and late summer and autumn minima. High discharges in spring are associated with the combination of spring rains, snowmelt, the possible presence of impermeable frozen subsoils, and relatively low evapotranspiration rates. Peak flows tend to occur progressively later in spring with increasing latitude as warm moist air and melting conditions advance poleward with the Sun's rays.

In dry climates, especially within the subtropics, only larger rivers exhibit year-round flow. These rivers are **exotic;** their water is derived from runoff from more humid areas at much higher elevations or in differing latitudes (the Nile and Colorado Rivers again serve as examples). Smaller local streams in these areas usually are situated well above the water table and are ephemeral, flowing only for short periods during and immediately following storms. In most cases their water either evaporates, is absorbed by the ground, or flows into saline lakes occupying basins of interior drainage (see Chapter 19).

Short-term variability in flow is characteristic of virtually all streams and is especially significant for those fed principally by surface runoff. Changing stream-flow volumes result from variations in the quantity of water supplied to the stream systems primarily by storms and sometimes by snowmelt. The irregularity in the timing and quantity of these water influxes makes it much more difficult to predict short-term stream flow than seasonal variations in flow.

The surface runoff from an intense storm is relatively rapid and usually does not travel far before it enters a tributary of the local drainage system. Small streams respond quickly and generally **crest,** or reach peak flow, shortly after the period of most intense rainfall. Crests normally are reached after progressively longer time intervals in streams draining larger basins, since most water must travel greater distances, and the rise in water levels is more gradual. Stream levels within drainage basins of all sizes tend to fall more gradually than they rose following a precipitation event. The gradual decline occurs because the much slower movement of throughflow and ground water outflow normally causes water inputs from these sources to peak after the surface flow has begun to decline. For perennial streams draining humid areas, the flow eventually subsides to a stable base flow fed almost entirely by ground water.

A variety of government agencies continuously monitor stream water levels at thousands of sites in North America (see the Focus Box on page 238). One of the most practically important reasons for carefully monitoring stream discharge rates is the danger of flooding. A **flood** occurs when the volume of flow of a stream exceeds the carrying capacity of its channel, and excess water spills over the banks to cover adjacent areas. Occasional flooding is characteristic of nearly all rivers and for some is an annual event.

Major river floods rank with earthquakes and hurricanes as one of the world's foremost categories of natural hazards (see Figure 10.12). Between 1947 and 1967, for example, floods claimed an estimated 173,000 human lives. In the United States they are responsible for average annual property losses of hundreds of millions of dollars.

The frequency of river flooding contributes to the fact that many rivers flow within **floodplains.** These are broad belts of low, flat land adjacent to rivers that are covered periodically by water during times of flooding. Floodplains, with their level surfaces, generally fertile soils, and proximity to river water for irrigation and transportation, are favorite settlement sites, so that occasional floods are a hazard of floodplain living.

FIGURE 10.12 A flood in Wisconsin inundates a small town. (Wisconsin Dept. of Natural Resources/JLM Visuals)

The flooding potential of streams generally is increased by human activities associated with agriculture and especially with urbanization. The removal of vegetation and the compaction of the surface by the passage of humans, livestock, and vehicles reduce the soil's infiltration capacity and increase surface runoff. In suburban and urban areas, extensive areas are surfaced with impervious materials such as asphalt and concrete. This greatly increases surface runoff, which then enters storm drains designed to conduct the water as quickly as possible to stream systems. As a result, many urban streams that in the past rarely overtopped their banks now do so frequently, adversely affecting large numbers of people.

THE OCEAN

The **ocean** is the interconnected body of salt water that dominates the surface of our planet. The Earth is much more a world of water than of land, since the ocean covers nearly 71 percent of the surface and an additional 2 percent is covered by rivers and lakes. Even the quarter of the Earth's surface that consists of dry land owes some of its most basic characteristics to the ocean. Life on Earth, which currently abounds in water and on land alike, is believed to have begun in the ocean. In addition, as we have seen, the ocean waters exert a tremendous influence on global climates.

The ocean is commonly considered to be both the beginning and the end of the hydrologic cycle. It provides most of the water that evaporates to enter the atmospheric portion of the cycle, and runoff from the land returns to it to complete the cycle. The ocean is sometimes referred to as the "hydrologic sink" because some 96.7 percent of all water at or near the surface of the Earth is located there at any one time, and because millennia may pass between the time a given water molecule enters the ocean as precipitation or runoff and the time it evaporates to begin another journey through the hydrologic cycle.

Origin, Distribution, and Extent

Ocean water originated in the same way as did the surface and subsurface water of the land. It was released from its original solution in rock deep within the Earth as the rock moved surfaceward as magma; subsequently, it entered the atmosphere in the form of water vapor during volcanic eruptions. Cooling and condensation of the water vapor caused it to precipitate to the surface, where it accumulated in enclosed basins.

The present geographical distribution of the ocean is closely related to the distribution of the types of rocks composing the Earth's crust. The entire crust "floats" on the semisolid material of the underlying mantle. Much of the crust consists of dense basaltic rock that floats at a relatively low level, producing the basins that contain the ocean water. The less dense granitic rock from which the continents are constructed is sufficiently buoyant and high-standing to remain above the ocean level (see Chapter 14). Put another way, the ocean has simply covered the lowest 71 percent of the Earth's surface. This includes approximately 60.6 percent of the northern hemisphere and 81.0 percent of the southern hemisphere (see Figure 10.13).

Several major subdivisions of the ocean exist; they also are termed oceans, and their areas and mean depths are listed in Table 10.3. As the table indicates, the mean depths of the water in the various oceans, with the exception of the comparatively shallow Arctic Ocean, are rather consistent. Depths range from 0 along the coasts to a maximum known value of 11,033 meters (36,198 feet) in the Marianas Trench

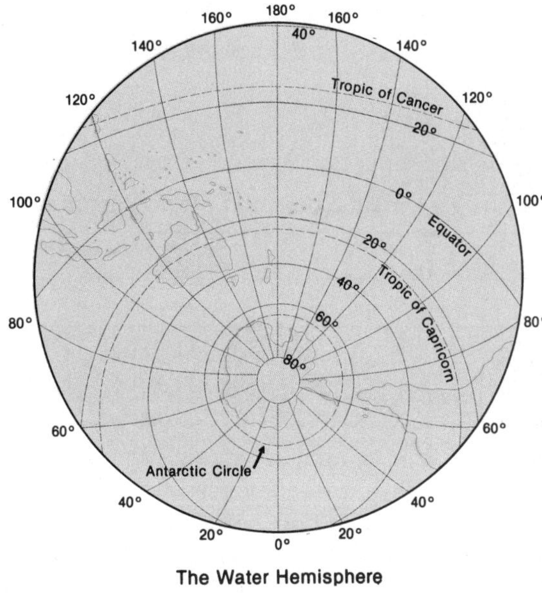

FIGURE 10.13 The "ocean hemisphere." This orientation of the Earth stresses the areal predominance of the oceans over the continents.

MEASURING STREAM DISCHARGE

Streams vary greatly in volume of flow over time as a result of variations in the amount, duration, and intensity of precipitation events as well as in preexisting soil water and ground water conditions. Each stream has a unique hydrologic "personality," which should be familiar to those involved in the design of dams and other river-modification projects, in water-supply planning, and in flood forecasting.

The U.S. Geological Survey (USGS) is the federal governmental agency chiefly responsible for monitoring the discharge of the more important rivers within the United States. It maintains over 11,000 stream-gauging stations that regularly measure flow volumes. In addition, many individual states have hydrologic agencies that operate stream-gauging stations. Streamflow variations over days or weeks often are displayed pictorially on **hydrographs** (see Figure 10.14). Analyses of these graphs help hydrologists to better understand the hydrologic characteristics of individual streams and to determine the contributions of surface flow, throughflow, and ground water flow to their total discharge.

Stream discharge is measured in cubic feet per second (cfs) or cubic meters per second (cms). The formula employed is $Q = wdv$, where "Q" is stream discharge, "w" is stream width, "d" is stream depth, and "v" is mean current velocity. Because a given stream cross-section typically has an irregular shape and differs in its speed of flow at different points, it is normally divided into a number of evenly spaced segments for purposes of measurement. The discharge is determined separately for each segment, and the total discharge value is then derived by summing the segmental values.

Numerous stream-flow velocity readings must be taken at different points within a stream in order to obtain an accurate overall velocity figure. A variety of techniques have been developed for this purpose. The width, depth, and velocity of a stream change almost continually

TABLE 10.3 Areal Extent and Mean Depths of the Ocean Basins

OCEAN	AREA		MEAN DEPTH	
	Sq Km	Sq Mi	M	Ft
Pacific	165,427,000	63,855,000	4,070	13,350
Atlantic	82,328,000	31,744,000	3,642	11,950
Indian	73,500,000	28,371,000	3,892	12,770
Arctic	14,060,000	5,427,000	1,204	3,950
"Antarctic"*	14,847,000	5,731,000	3,730	12,240
Other**	10,339,000	3,991,000		
Total***	360,412,000	139,119,000	3,865	12,680

*This refers to the ocean water surrounding Antarctica.
**Includes restricted bodies of ocean water such as the Mediterranean Sea, Black Sea, and Red Sea. All figures are approximate because boundaries between adjacent ocean basins are arbitrary.
***Estimated total volume of water is 1,192,625,000 cubic kilometers (286,230,000 mi^3).

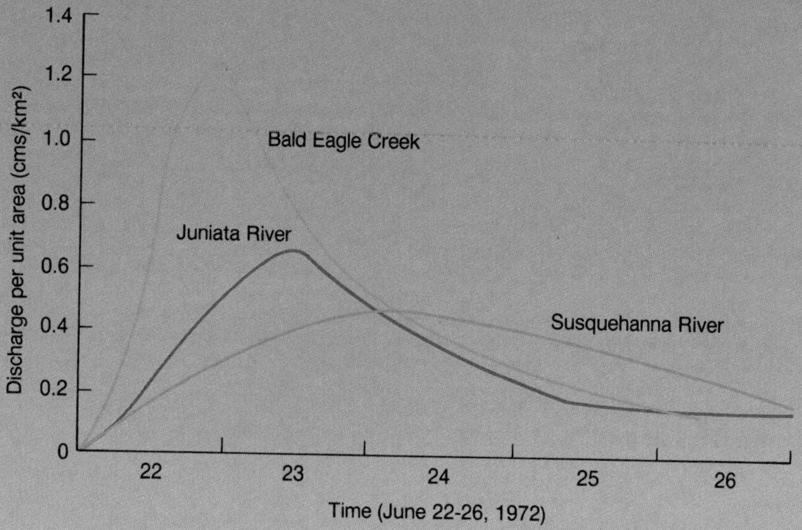

FIGURE 10.14 Hydrographs plot stream discharge over periods of time. These three hydrographs show the response of streams of differing size to the heavy rains associated with Hurricane Agnes in June, 1972. Bald Eagle Creek, the smallest of the three, had the "flashiest" response, rising and then falling relatively quickly after the rainfall event. The largest of the three streams, the Susquehanna River, rose and fell much more slowly.

with time, so that frequent measurements are required. Fortunately, all three of these parameters for a given stream tend to vary systematically with the stream's water level, or **stage.** Stream stage is easily determined by observing the height of the water on a measuring staff at the gauging station. Once the relationships between stage and discharge have been established, the discharge of a stream usually can be accurately estimated from its stage.

in the western Pacific. The estimated mean oceanic depth of 3,865 meters (12,680 feet) is almost five times the average elevation of the world's land areas, which is estimated at 841 meters (2,760 ft). The vertical dimensions of the ocean therefore exceed those of the continents even more than do its horizontal dimensions (see Figure 10.15).

Physical and Chemical Characteristics

Ocean water displays substantial spatial variation in a number of physical and chemical characteristics. Among the most important are chemical constituency, temperature, and the presence or absence of ice.

Constituency

Seawater contains an average of about 96.5 percent water and 3.5 percent dissolved solids by weight. Nearly every known substance is soluble to at least some extent in water, and about two-thirds of the 92 naturally occurring chemical elements have been found in seawater. The great majority of these solids are ions of salts.[1] More than 99 percent by weight of the dissolved solids consist of the six elements chlorine, sodium, magnesium, sulfur (as sulfate), calcium, and potassium. Chlorine and sodium are the two dominant elements by far and, upon evaporation of sufficient water, will combine to form common table salt, or sodium chloride (NaCl).

The origin of the salinity of the ocean is similar to that of saline lakes. Several billion years of runoff from the land has carried to the sea tremendous quantities of dissolved salts. Because there is no natural mech-

[1] An ion is an electrically charged atom or group of atoms. The dissolution of minerals in seawater causes them to dissociate into their constituent atoms, which will recombine if the water is evaporated. A salt is formed when the hydrogen atoms of an acid are chemically replaced by atoms of a metal. Salinity causes sea water to conduct electricity better than does fresh water, and electrical conductivity is normally used to measure oceanic salinity.

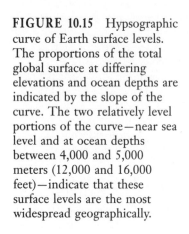

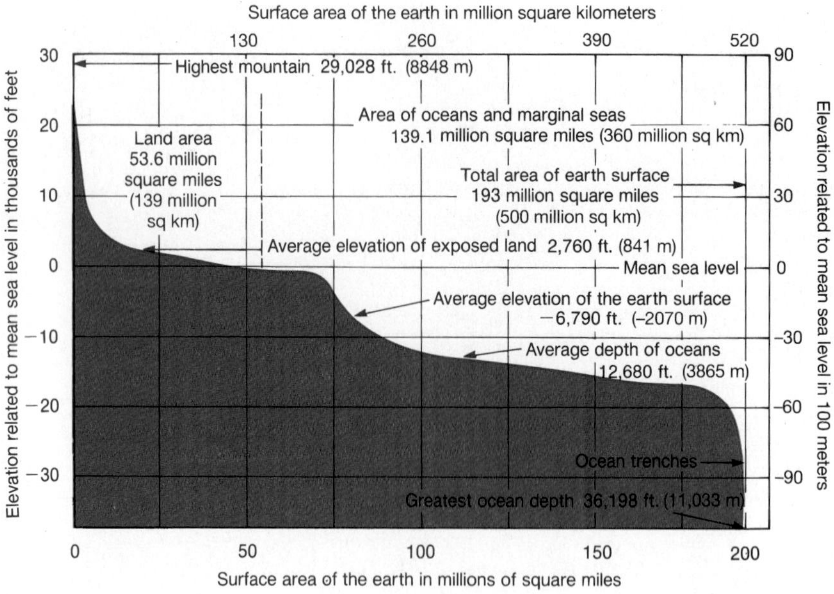

FIGURE 10.15 Hypsographic curve of Earth surface levels. The proportions of the total global surface at differing elevations and ocean depths are indicated by the slope of the curve. The two relatively level portions of the curve—near sea level and at ocean depths between 4,000 and 5,000 meters (12,000 and 16,000 feet)—indicate that these surface levels are the most widespread geographically.

anism for removing these salts until saturation occurs (and the oceans are still well below the salt saturation level), salinity continues to gradually increase. The quantity of salt in the ocean is now so great that, if removed, it would exceed in volume the portion of the continent of Africa that is above sea level.

Near-surface salinity levels vary somewhat over the open ocean because of losses of water to evaporation and gains by the influx of fresh water from rivers or precipitation. In general, salinity levels are highest in the subtropics, where large net evaporative water losses concentrate the salts (see Figure 10.16). Large variations in salinity occur only in restricted seas where communication with the major ocean basins is limited. For example, the Black Sea, which receives large quantities of fresh water from the Danube and other rivers, has a salinity of only 1.6 percent. The shallow Baltic Sea in the spring snowmelt season sometimes has salinities as low as 0.1 percent. On the other hand, the eastern Mediterranean and Red Seas are located in arid climates and have summer salinity levels of 4 percent or more.

Temperature

The temperature of the ocean surface varies considerably in different parts of the world (see Figure 10.17). As might be expected, the temperature distribution pattern generally is similar to that of the atmosphere. Extensive areas of tropical ocean water have surface temperatures of between 26.5° and 29.5° C (80° and 85° F), and values generally fall with increasing latitude to readings near or slightly below 0° C (32° F) in the subpolar and polar regions. The mean temperature of the ocean surface water as a whole is about 21° C (70° F).

A basic difference in the thermal behavior of the continents and ocean is that the latter experiences smaller temperature variations, especially on a seasonal and daily basis (see Chapter 5). Annual temperature variations of the open ocean waters reach a maximum of 8 to 10 C° (15 to 18 F°) in the middle latitudes of the northern hemisphere and 5 to 6 C° (9 to 10 F°) in the middle latitudes of the more highly maritime southern hemisphere. Variations of at most 1.5 to 2 C° (3 to 4 F°) are typical of the tropics. Seasonal temperature variations decrease rapidly with depth and, in most places, do not occur at all at depths greater than 180 meters (600 ft). Daily temperatures over the open ocean vary less than 0.5 C° (1 F°).

Despite the transparency of the ocean, most solar energy is absorbed in the top 1.5 meters (5 feet) of water. As a consequence, outside the polar regions, water temperatures decline everywhere with depth. The warm surface water is lighter and more buoyant than the underlying water and remains at the surface.

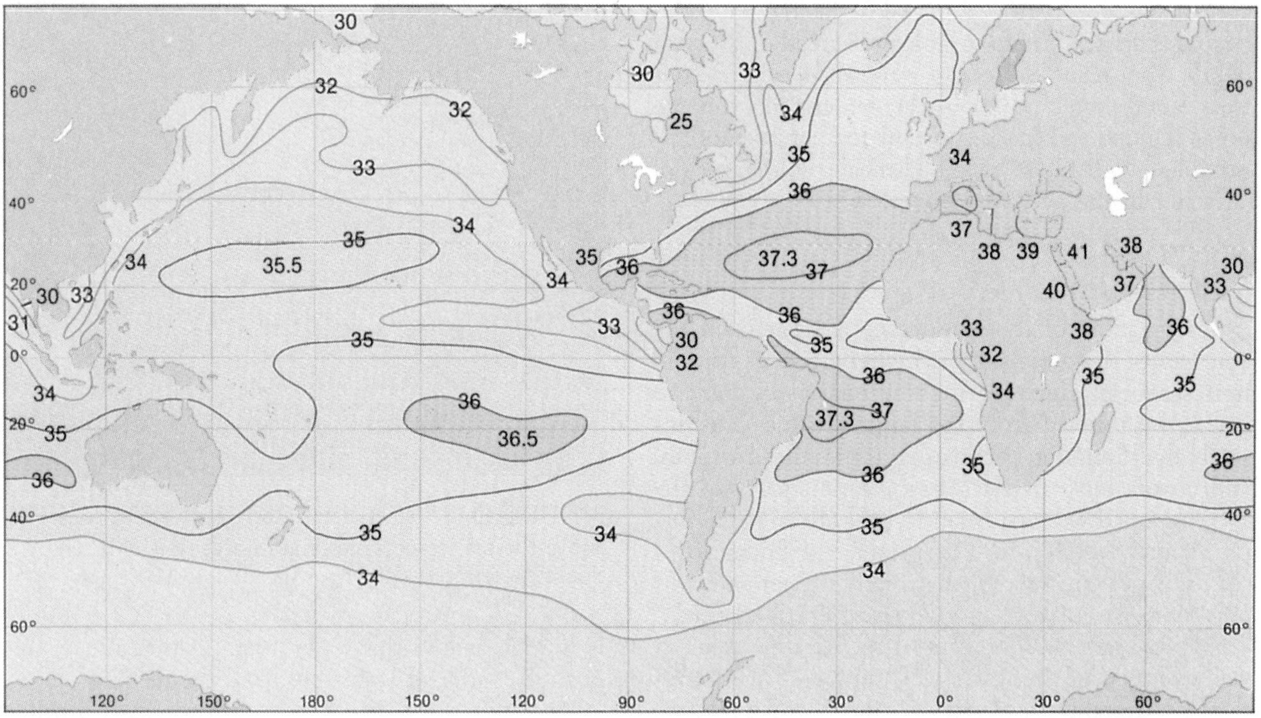

FIGURE 10.16 Map of mean oceanic surface water salinity levels in parts per thousand (‰) in August. Green shaded areas have salinity levels exceeding 36 ‰, yellow areas have levels between 34 and 36 ‰, and light gray areas have salinity levels of under 34 ‰.

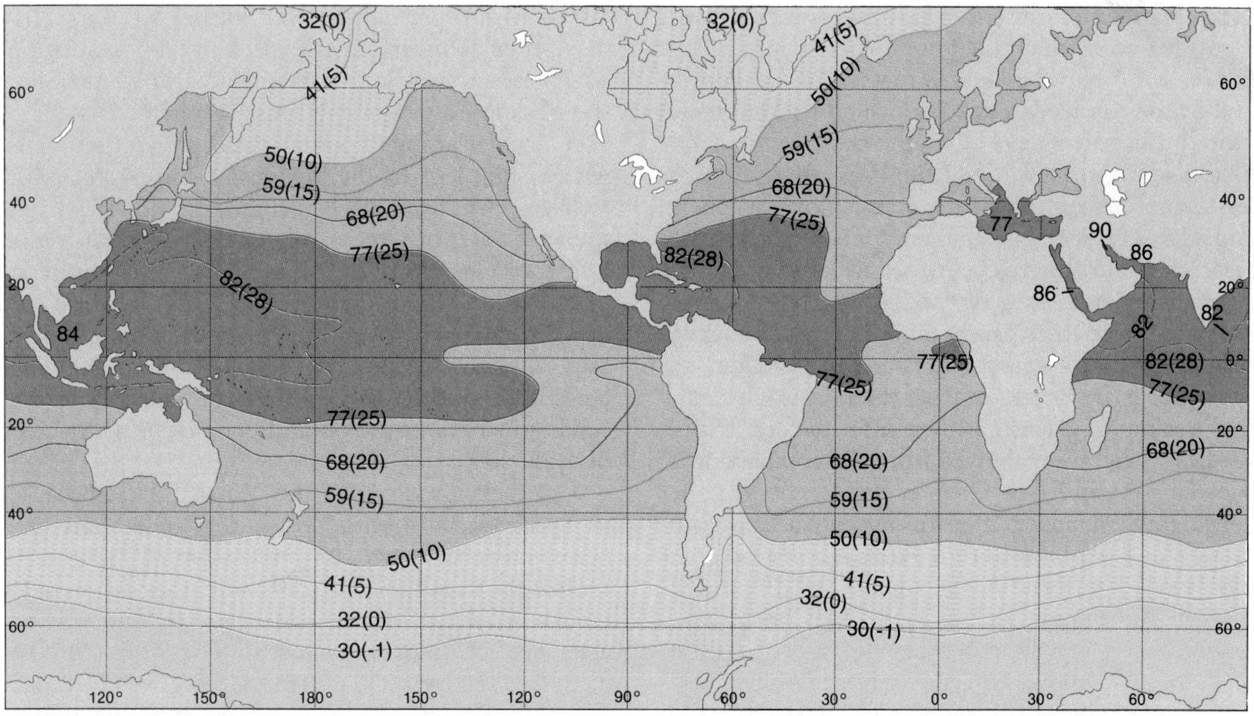

FIGURE 10.17 Mean ocean water surface temperatures. Readings are in degrees Celsius, with Fahrenheit equivalents in parentheses.

Conversely, cold water produced in the high latitudes becomes dense enough to sink and spread outward at depth. As a result, all ocean water at depths greater than 3,000 meters (10,000 feet) is derived from the polar regions and has a constant temperature of approximately 2° C (35° F). The mean temperature of all water in the ocean is about 4° C (39° F).

Oceanic Ice

Although most of the world's ice is located on the continents, the oceans of the higher latitudes contain both sea ice produced by the freezing of seawater and glacial ice derived from the land. **Sea ice** is extensively distributed in the oceans of the high latitudes, and covers most of the Arctic Ocean, parts of the extreme northwestern Atlantic, and much of the water around Antarctica. The extent and thickness of sea ice varies considerably between summer and winter, but most is less than 3.5 meters (11 feet) thick.

Approximately 70 percent of the Arctic Ocean, as well as much of the water near the coast of Antarctica, is covered by a type of sea ice called **cap ice** (see Figure 10.18). This ice is not smooth, because the expansive force of freezing causes pressure ridges up to several meters high to form. In addition, wave action sometimes causes the ice to separate into large plates between which open water may appear. Most cap ice is 2.5 to 3.5 meters (8 to 11 ft) thick. Peripheral to the cap ice is **pack ice,** which is thinner and more broken. It occurs along the coast of the Arctic Ocean south to Newfoundland. Pack ice generally is less than 2 meters (6 feet) thick. It is not permanent, and most of it melts in summer. Outside the areas of pack ice are drifting pieces of sea ice of varying size. Large, tabular masses of sea ice that have become separated from their parent bodies of cap or pack ice may range up to several kilometers in diameter.

Icebergs are masses of freshwater ice produced by the breakup of glaciers as they enter the sea. Unlike sea ice, glacial ice is derived from the compaction of snowfall on land (see Chapter 18). Icebergs differ considerably in appearance from sea ice because they do not form continuous thin sheets but instead commonly take the form of individual rugged islandlike masses (see Figure 10.19). Sizes vary greatly, but large icebergs may be more than 100 meters (300 ft) high and have above-water lengths exceeding 1.6 kilometers (1 mile). Between 80 and 85 percent of an iceberg (depending on the amount of entrapped air) is hidden below the water line, so that the visible portion is indeed "just the tip of the iceberg."

FIGURE 10.18 A tabular fresh water iceberg rises above the sea ice-covered ocean north of McMurdo Sound, Antarctica (Rodman Snead/© JLM Visuals)

Although sea ice movements are largely controlled by the wind, icebergs are carried predominantly by ocean currents. The largest travel longest and farthest; large icebergs may move 16 to 24 kilometers (10 to 14 mi) per day and last up to two years. (Indeed, their longevity has resulted in the formulation of schemes to tow large icebergs to arid countries in the Middle East to supply fresh water!)

The chief problem posed by icebergs is their danger to shipping. In the northern hemisphere, most icebergs form from glaciers entering Baffin Bay from the west coast of Greenland. They drift south in the Labrador Current to the Grand Banks area southeast of Labrador, where they are picked up by the Gulf Stream, carried eastward, and eventually melt. In this area, they are within the North Atlantic shipping lanes. One of these icebergs was responsible for the sinking in 1912 of the ocean liner *Titanic* with a loss of 1,517 lives.

Few icebergs occur in the North Pacific because no major glacial sources exist in that area. Many icebergs, however, form along the coast of Antarctica. Unlike those of the North Atlantic, they are usually tabular in shape because they have broken off the extensive ice shelves located along portions of the coast (see Figure 10.18). These icebergs can be huge. For example, an iceberg 153 kilometers (95 miles) long and 35 kilometers (22 miles) wide broke off the

FIGURE 10.19 A large iceberg that has grounded at Godhavn, on the West Greenland coast, dwarfs the houses in front of it. (P. J. Bryant, Univ. of California–Irvine/BPS)

Antarctic ice shelf in October 1987. Such icebergs have been observed as far equatorward as 30° S.

Motions of the Ocean Waters

The ocean is never completely still. It is influenced by a variety of forces originating on, above, and below the surface that keep it in a state of constant motion. The water movements are commonly divided into the three categories of currents, waves, and tides.

Currents

Ocean **currents** are continuous movements of water streams of similar temperature and density. As noted in Chapters 5 and 7, these currents aid in the transport of excess energy from the tropics to the higher latitudes. The presence of ocean currents of differing temperatures has a profound influence on global patterns of both temperature and precipitation.

The most important ocean currents are those at and near the surface because they interact with the atmosphere and coast and affect shipping. The energy that sets these currents in motion is the frictional drag of the wind on the water surface. As a result, both the speed and the direction of the surface currents are determined largely by the global wind belts, and there is considerable similarity in the planetary patterns of winds and ocean currents.

The vast inertia of the ocean water causes the ocean currents to be more consistent in their position, speed, and direction of flow than are the wind currents. Satellite imagery and surface observations have revealed, however, that the ocean currents develop constantly changing patterns of swirls and eddies along their routes, and that extensive eddy fields generated by the flow of the currents involve much of the ocean surface waters in slow swirling motions. The ocean currents flow at a much slower average speed than the wind; flow rates average only 1.5 to 3.0 kilometers per hour (1 to 2 mph) and do not exceed 9 kilometers per hour (5.5 mph).

Two factors other than the wind also play important roles in determining the overall pattern of ocean current flow. One is the Coriolis effect produced by the Earth's rotation, which deflects currents to the right of the prevailing wind flow in the northern hemisphere, and to the left in the southern hemisphere. The net influence of the Coriolis effect is to produce an average 45° deflection of the surface currents with respect to the winds.

The other major influence on the global ocean current pattern is the configuration of the ocean basins and the continental margins. Unlike the wind, which is free to pass over land and water alike, the ocean currents are, of course, restricted to the ocean basins. As they approach land, they are redirected and often split into complex patterns that may bear little relation to the prevailing wind pattern.

If the three largest oceans are divided by the equator, a total of five extensive ocean basins are produced: the North and South Pacific, the North and South Atlantic, and the South Indian Oceans. (The North Indian Ocean is too small in size to be considered an additional member.) The basic pattern of currents for each basin is essentially similar; it consists of a giant elliptical loop, called a **gyre,** elongated in an east-west direction (see Figure 10.20). The oceanic gyres are centered at roughly 30° N and S somewhat to the west of the midpoints of their respective basins. They correspond closely in position to the oceanic centers of the subtropical highs and are, in fact, produced largely by the wind circulations around these highs. Like the wind, the currents flow clockwise around their centers in the northern hemisphere and counterclockwise in the southern hemisphere.

Currents largely unrelated to those at the surface exist at depth in the oceans (see Figure 10.21). Unlike surface currents, the oceanic **deep currents** are produced by density differences resulting from variations in temperature and salinity. Deep currents have significant vertical as well as horizontal components of motion. A slow circulation occurs between the surface water and the deep water. Surface water descends to great depths in zones of convergence, mostly located where cold water from the high latitudes encounters and plunges beneath warmer water from the middle latitudes. Extensive convergence occurs in the waters around Antarctica and in the North Atlantic south of Greenland. The zones of convergence are the sources of the cold deep water that exists throughout the ocean. Deep water eventually returns to the surface in zones of **upwelling.** Most of these zones are situated near the west coasts of continents in the subtropics, where equatorward-flowing water is redirected westward by the trade winds. The seaward push of this water causes cold water from great depths to well up to replace it.

A period of several centuries may be required for one complete cycle of convergence, descent, and upwelling. This period is estimated at about 700 years

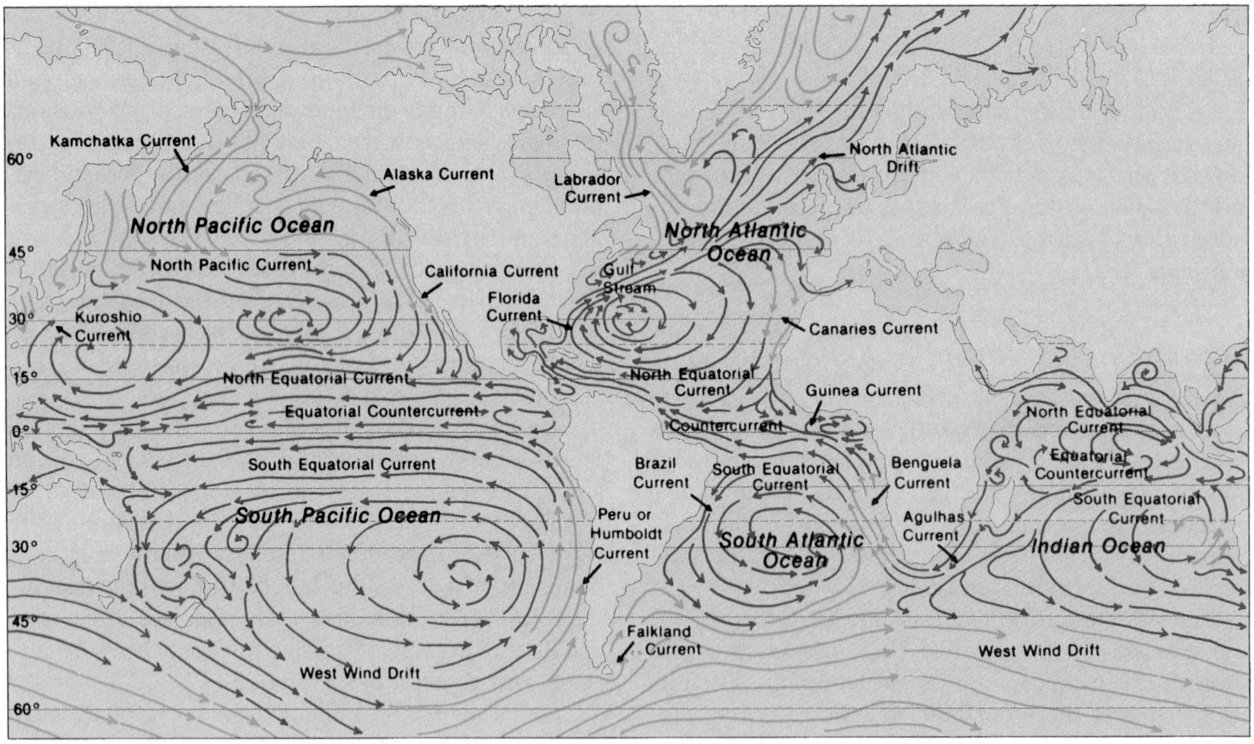

FIGURE 10.20 Global surface ocean currents. Warm currents are shown in red; cold currents in blue. The centers of the major current circulation patterns, or gyres, roughly correspond to the oceanic centers of the subtropical highs.

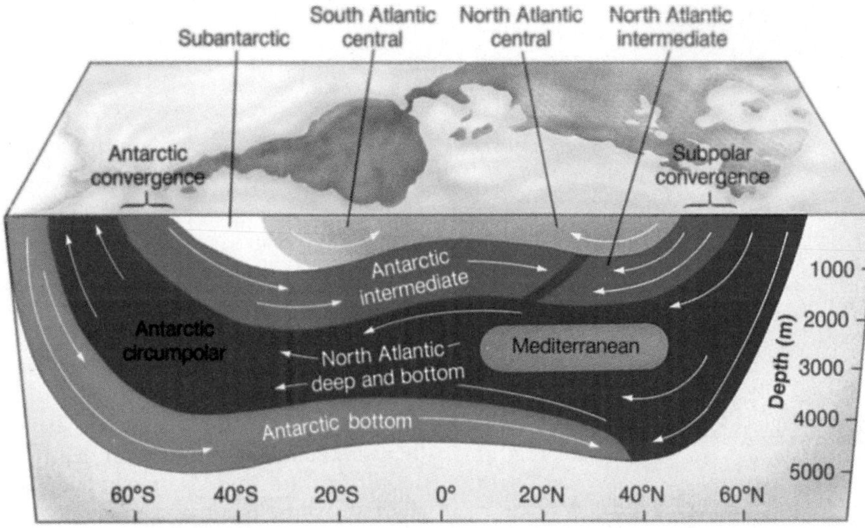

FIGURE 10.21 A profile view of the Atlantic, showing the major deep ocean currents. Most of the oceanic deep water is derived from the two zones of convergence located in the subpolar regions, because only there is the water sufficiently cold to become dense enough to sink.

for water of the Atlantic and 1,500 years for that of the Pacific. Slow as it is, this vast vertical circulation system is crucial to sea life. The descent of surface water in areas of convergence carries dissolved carbon dioxide and oxygen needed by marine plant and animal life into the depths. Conversely, the rise of bottom water in zones of upwelling transports nutrients from the sea floor to the surface. As a result, zones of upwelling are the sites of some of the world's most productive fishing grounds (see the Case Study at the conclusion of this chapter). Water in the vast mid-ocean basins is largely cut off from this circulation. As a consequence, it is low in dissolved gases and nutrients and nearly devoid of life.

Waves

Waves are the most familiar of the ocean movements and have the greatest influence on the characteristics of the coastline (see Chapter 20). Like currents, they are generated primarily by the wind. Unlike currents, though, waves do not produce a significant linear transport of water, although they are efficient transporters of energy over great distances.

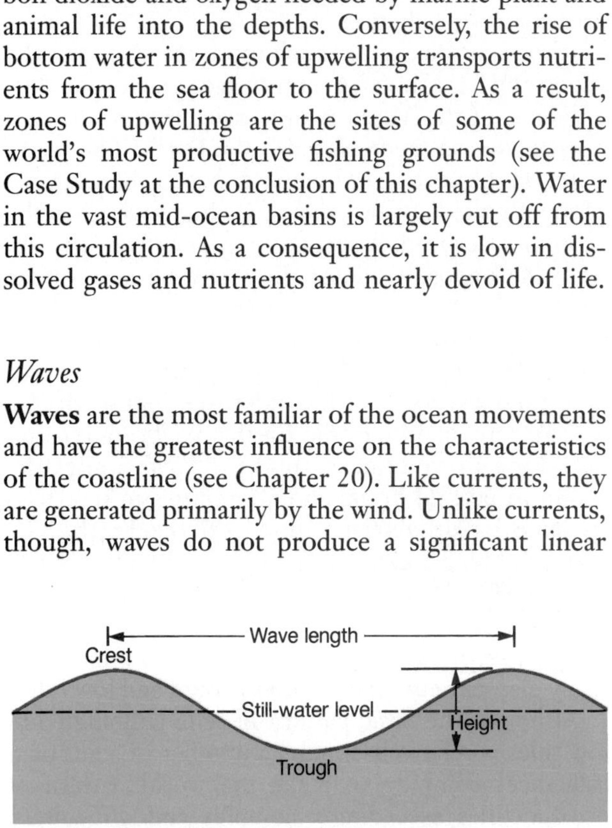

FIGURE 10.22 Ocean wave terminology

Waves are produced by the frictional drag of the wind blowing over the water surface (see Figure 10.22). Once started, the size and energy of waves is increased by the push of the wind against their sides. Three factors determine wave sizes. These are wind speed, **fetch** (the distance over which the wind blows), and the length of time the wind has been blowing from the same approximate direction. An increase in any of these factors increases wave dimensions. The largest wind-generated waves ever reliably reported were 17 meters (55 feet) high. One combination of conditions that would produce waves of this size is a 110 kilometer per hour (70 mph) wind blowing for 24 hours over a fetch exceeding 800 kilometers (500 miles). The largest storm waves are capable of damaging or sinking even large ships and of producing tremendous erosional and structural damage along coasts (see Figure 10.23).

The waves that reach our coastlines are at any given time generated only in certain restricted areas, usually located within storms. They are able to travel hundreds or even thousands of kilometers before they eventually reach distant shores. Large waves along the California coast, for example, are commonly produced by storms near Alaska. On occasion, waves reaching Hawaii are generated in storms off the coast of Antarctica.

Waves lose little energy as they travel through deep water, but this situation begins to change as they approach the shore and move into shallow wa-

FIGURE 10.23 Large waves such as these are generated in storms that may be centered thousands of kilometers from shore. (H. Armstrong Roberts)

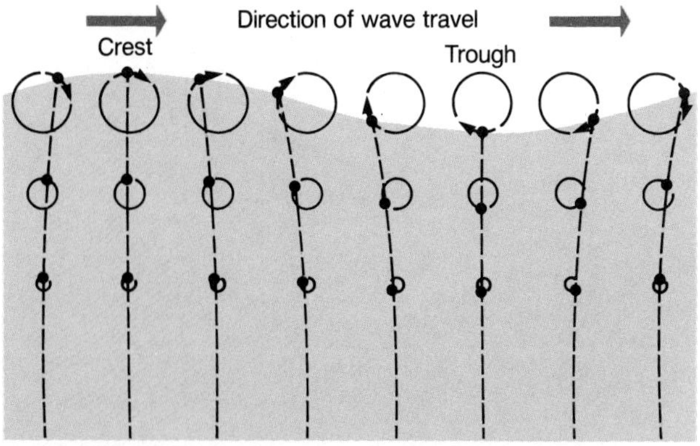

FIGURE 10.24 The water in waves moves in circular paths that gradually decrease in magnitude with depth. Waves begin to rise, tilt forward, and eventually break as they enter areas where the bottom is too shallow to permit the orbital motion to continue unimpeded.

ter. When the water depth decreases to less than half the wave length, the wave starts to "feel bottom" as its forward progress begins to be slowed by friction. The restraining influence and upward push of the bottom causes the wave to rise gradually and tilt forward. When the water depth has been reduced to about 1.3 times the wave height, the upper portion of the wave pitches forward and breaks. (The process is much like that of a person who falls forward because he or she has been tripped.) The water now becomes foam-filled surf, which for the first time experiences significant net horizontal motion as its forward momentum causes it to rush up the beach as **swash** and then, its energy spent, to return as **backwash.** During this process, the energy imparted to the wave days earlier in a distant storm finally is expended on the coastline.

Tides

The ocean **tides** are the rhythmic rise and fall of the ocean produced by the gravitational influence of the Moon and Sun. Unlike waves and surface currents, the tides affect the ocean throughout its entire vertical extent. Gravity-induced tides are not restricted solely to the ocean. They also occur in large lakes, on the continents (with displacements of up to 30 centimeters or 12 inches), and in the atmosphere (with displacements of many kilometers). Their practical significance with respect to human activities, however, is greatest by far in the ocean. The ebb and flow of the tides has been observed, pondered, and put to practical use by seafarers for thousands of years. Only in recent times, however, have sufficiently detailed records been compiled in most areas to allow for precise tidal predictions.

Although the Sun is much more massive than the Moon and has a much greater surface gravity, the tidal influence of the Moon is over twice that of the Sun because it is so much closer to the Earth. The ocean water on the portion of the Earth directly beneath the Moon is pulled upward, creating a high tide when the Moon is overhead. A second high tide occurs on exactly the opposite side of the Earth because of the rotational inertia of the Earth-Moon system and because that area is farthest from the Moon, making its pull weakest there. Midway between the two areas of high tides are zones of low tides (see Figure 10.25).

The combination of the Earth's axial rotation and the Moon's revolution around the Earth causes the Moon to take 24 hours and 50 minutes to return to its same longitudinal position over the Earth. The pattern of high and low tides rotates with the Moon's changing position, so that most places experience two high tides and two low tides during this period. A timespan of 6 hours and 12 minutes therefore normally elapses between subsequent high and low tides.

Although the basic pattern of semidaily high and low tides is relatively simple, a number of additional influences combine to make the world pattern of oceanic tides exceedingly complex and difficult to predict. One of the most important is the gravita-

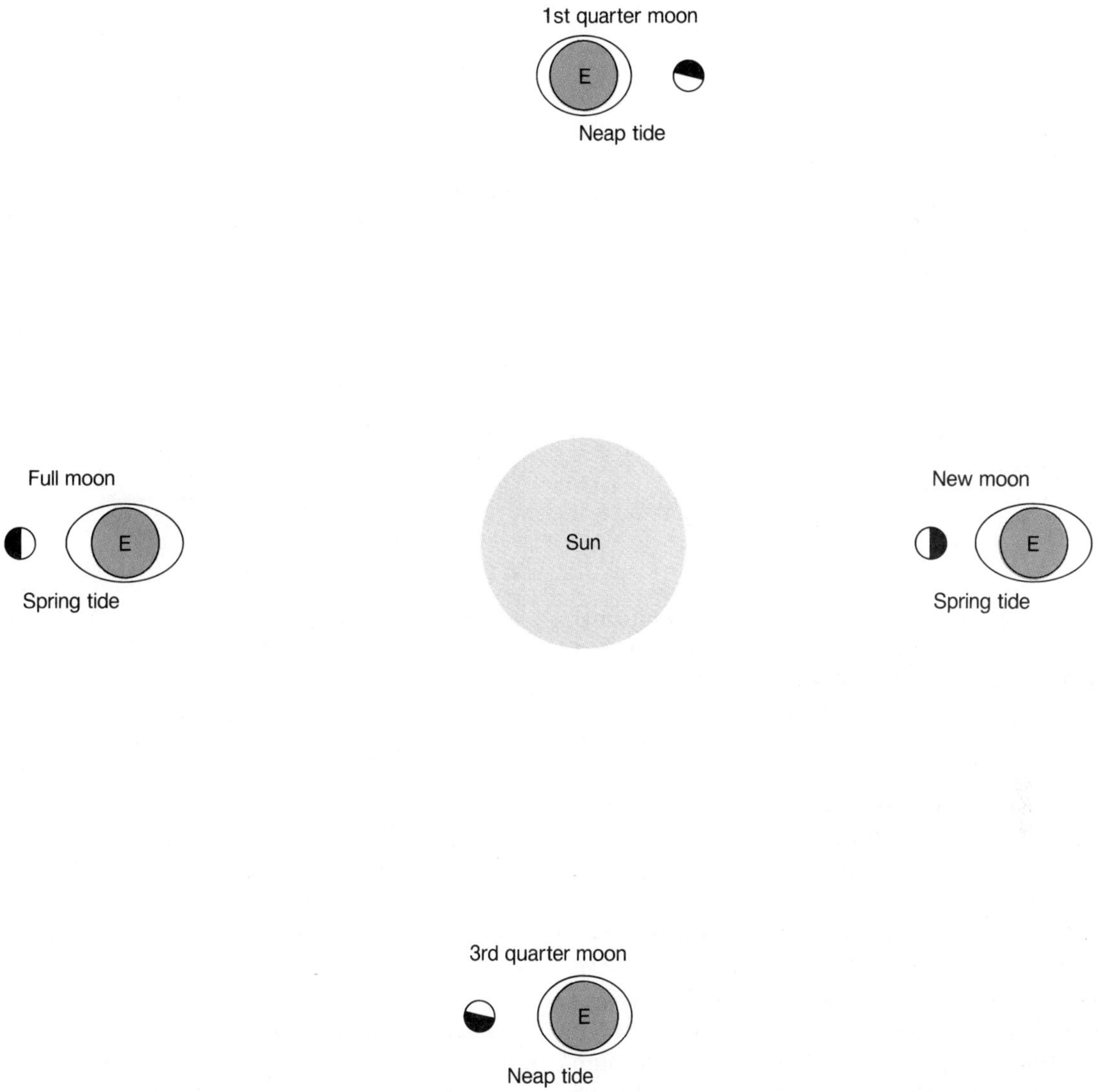

FIGURE 10.25 The ocean tides are caused by the gravitational pull of the Moon and Sun. The diagram shows the relative positions of these bodies with respect to the Earth at the times of the spring and neap tides. The more pronounced bulges around the Earth at the time of the spring tides indicate that tidal ranges are highest then because the gravitational attractions of the Moon and Sun are working in concert.

tional influence of the Sun. Like the Moon, the Sun acts to produce a pair of high tides in those areas facing directly toward and away from it, with an intervening pair of low tides. The tides actually experienced at any one place therefore result from the combined influence of the Moon and Sun. When these two bodies are aligned either on the same side of the Earth (at the time of the new moon), or on opposite sides (at the time of the full moon), their gravitational effects work in concert, and tidal ranges are largest. This produces the so-called **spring tides,** whose occurrences are actually unrelated to the season of the year. When the Sun and Moon are situated at right angles with respect to the Earth (at the times of the half moon), their gravitational influences are opposed to each other, and tidal ranges are smallest. These are called **neap tides.**

Tidal ranges also are affected by the varying distances of the Sun and Moon from the Earth. The relative variations in these distances are rather small,

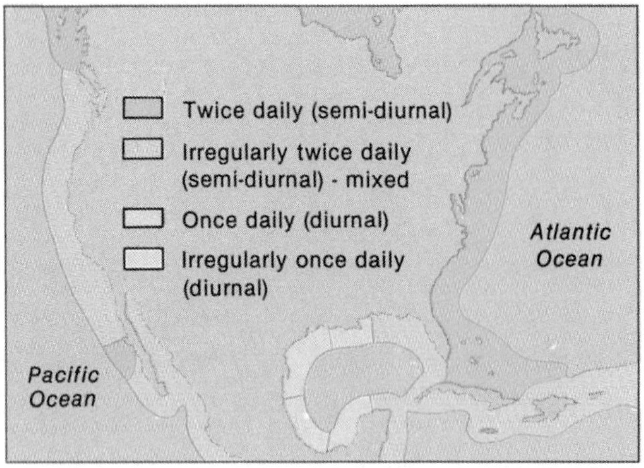

FIGURE 10.26 Tidal patterns along the North American coast. Most areas have two sets of high and low tides each day.

but their influence is significant because gravitational attraction, and therefore tide-generating power, is inversely proportional to the cube of the distance between two bodies.

Probably the most important and certainly the most complex of the tidal influences is the geographic pattern of the ocean basins and coastal margins. Their irregular shapes cause the tides to be redirected or reflected in countless ways and often to be augmented, diminished, or even canceled altogether in some places. Highly uneven coastlines frequently experience tides that vary greatly both in range and in time of occurrence over small distances. In broad water bodies, including the major ocean basins, the tides display a circular pattern, with high and low tides rotating around one or more nodal points or lines that experience no tides at all (see Figure 10.27).

Tidal ranges of 1 to 2 meters (3 to 6 ft) are most common for the world as a whole. The highest tides occur in funnel-shaped bays or estuaries, where the tidal influence is focused by the converging shorelines. Nova Scotia's Bay of Fundy has the highest tides of all, with one site experiencing a mean spring tidal range of 14.5 meters (47.5 feet) and an extreme range of 16.5 meters (53.5 feet).

Tides in restricted bodies of water may be so concentrated that strong **tidal currents** are formed. These consist of currents that flow landward during the period of the rising tide and seaward during the period of the falling tide. If a current is forced to

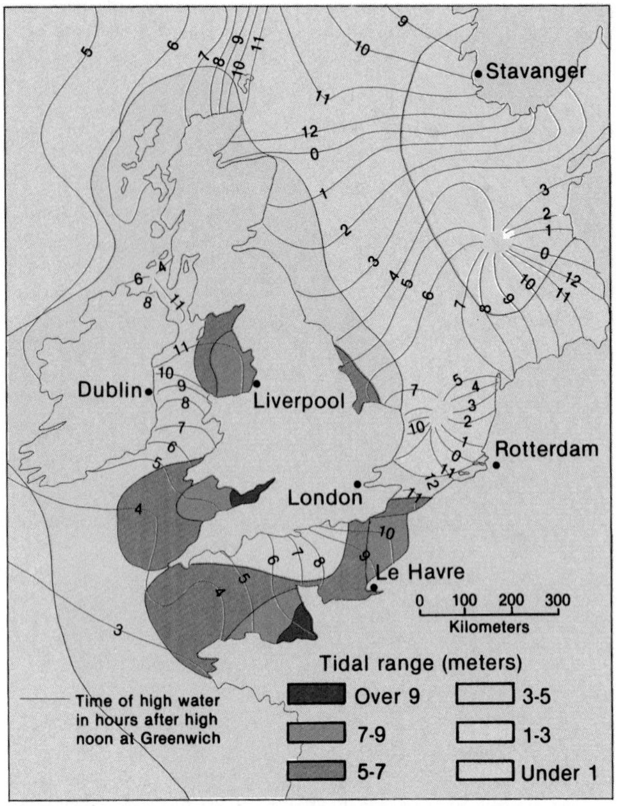

FIGURE 10.27 Tidal pattern in the North Sea region. The tidal range is given in meters, and the time of the high tide is in hours after Greenwich noon. Note the rotation of the tides around the two nodal points.

advance upstream against a strong downstream flow of river water, its progress is slowed and water may be piled up at its forefront to produce a **tidal bore.** In the lower Amazon, such a tidal bore advances upriver as a wave for a distance of 480 kilometers (300 miles). In places it forms a wall of water 5 meters (16 feet) high that travels at a rate of 22 kilometers per hour (14 mph) and produces a roar that can be heard for great distances.

The net result of the combination of tidal influences is that most localities experience alternating higher and lower high and low tides. Even a thorough understanding of all the tidal influences does not make it possible to forecast the height of the tides precisely. This is because ocean levels also are strongly influenced by constantly changing patterns of wind direction, fetch, and speed and, to a lesser extent, by variations in air pressure.

SUMMARY

This chapter has examined the characteristics and geographical distribution of the Earth's surface and near-surface water supplies. Water occurs in quantity in several different sites. It exists in a dispersed state in virtually all soils, and substantial volumes of deeper ground water occur within porous rock and sediment layers in many regions. Surface water bodies include lakes, rivers, and the ocean. With the exception of some of the deeper ground water supplies, all of these sites contain water that participates in the hydrologic cycle, and all continually experience both additions and losses of water.

We now return to the Focus Questions posed at the beginning of the chapter:

1. *What factors control the movement and storage of subsurface water?*

Subsurface water enters the ground primarily through the soil pore spaces. Large quantities of water adhere to the soil particles, while the remainder makes its way to the water table to become ground water. Ground water normally moves slowly downhill under the influence of gravity and hydrostatic pressure, with the speed of movement controlled largely by the permeability of the subsurface rock or sediment layers through which it passes. Large volumes of ground water are confined more or less permanently within aquifers.

2. *How do lakes form and disappear?*

Lakes occupy enclosed basins that are supplied with enough water to fill them at least partially. These basins form in a number of ways, but the two most important are by glacial erosion and deposition and by crustal movements, especially faulting. Lakes have relatively brief spans of existence because of the action of two processes that eventually destroy the water-holding capacity of the basin. The first is erosion of the basin rim, which may cause the lake to drain. The second is the filling of the basin by sediments and organic matter, converting it to a swamp.

3. *What factors determine geographic and temporal variations in stream discharge?*

Streams transport excess precipitation to enclosed interior basins or, more commonly, to the ocean. The discharge of a stream is determined primarily by climate and by the size of its drainage basin. Streams with drainage basins of similar size tend to have greater discharges in areas with heavy precipitation, cool temperatures, steep slopes, and limited vegetation. Nearly all streams display large short-term variations in flow as a result of precipitation, and many streams experience major seasonal variations in discharge. These variations are correlated with seasonal fluctuations in precipitation, temperature, and vegetative cover.

4. *What are the chief physical and chemical characteristics of ocean water?*

The ocean is the interconnected body of saline water that covers the lowest 71 percent of the Earth's surface. It consists of approximately 96.5 percent water and 3.5 percent dissolved solids, mostly chlorine and sodium. Variations in salinity are minor, except in the vicinity of the mouths of large rivers and in basins with restricted access to the open ocean. Ocean surface temperatures are high in the tropics and decline with latitude, but the deep ocean water everywhere has a temperature near 2° C (35° F). The Arctic Ocean and much of the water around Antarctica is covered with sea ice or contains ice shelves or icebergs derived from coastal glaciers.

5. *What are the causes of the three oceanic motions of currents, waves, and tides?*

Currents are continuous movements of water streams of similar temperature and density. Surface currents are set in motion by the wind and display a general similarity in flow pattern to that of the global wind currents. Deep currents are produced largely by water density variations resulting from differences in temperature and salinity. Waves are oscillations of the surface water generated by strong winds associated with storms. The three factors of wind speed, fetch, and wind duration determine the size of waves. The tides are produced by the gravitational pull of the Moon and, to a lesser extent, of the Sun. They result in a rhythmic rise and fall of the ocean surface.

Review Questions

1. What are the two categories of subsurface water? Describe the location and characteristics of each.

2. What factors determine the soil's infiltration rate? What influence does the presence of vegetation exert on the infiltration rate?

3. How and why does ground water return to the surface? Where is it most likely to go?

4. Portions of the New York City area periodically experience problems with saltwater intrusion. What factors might cause this area to be especially susceptible to this problem? What factors are likely to be responsible for Phoenix, Arizona's problem of rapidly declining ground water supplies?

5. What determines whether a lake will contain fresh water or salt water?

6. What are the two chief methods by which lake basins are formed? Give some examples of large lakes produced by each method. How else may lake basins be formed?

7. Why do lakes have relatively short spans of existence?

8. What basic factor controls the geographic distribution of rivers? Is this same factor the major distributional control for lakes, ground water, and the ocean? Explain why or why not for each.

9. What climatic factors determine the proportion of a region's total precipitation that will be discharged by streams? How much of the world's land precipitation is transported to the ocean by rivers? How is the remainder removed from the surface?

10. What is a flood? Explain how human landscape modifications in agricultural and urban areas can increase the likelihood of flooding.

11. What factors influence the salinity of the ocean? In what general geographical areas is ocean water most saline and least saline?

12. In what ways are ocean temperature distributions, both vertically and horizontally, similar to those of the troposphere? In what ways are they different? Why are the deep oceanic waters cold everywhere?

13. Where are the chief areas of sea ice? Explain why they are located where they are. How do sea ice and icebergs differ in origin and characteristics?

14. Describe, in general terms, the pattern of surface ocean currents within the five major ocean basins. What factors are responsible for this pattern?

15. What causes deep ocean currents? Why are these currents vital to sea life?

16. How are ocean waves formed? In what parts of the ocean are large waves produced? What factors determine the size of waves?

17. Explain the cause of the tides. Why do they vary in height in most places at regular intervals through the month? What factors complicate the global pattern of tides?

18. What is El Niño? What do the climatic consequences of recent El Niños indicate about the relationship between the ocean surface temperature and the atmosphere?

Key Terms

soil water
ground water
interception
throughflow
water table
spring
aquifer
stream discharge
drainage basin
flood
sea ice
iceberg
currents
waves
tides

CASE STUDY

EL NIÑO—AN OCEAN CURRENT WITH WORLDWIDE ENVIRONMENTAL IMPLICATIONS

Off the west coast of South America lies a vast sweep of northward-drifting cold water known as the Humboldt Current. The best-developed and most extensive of the world's cold ocean currents, it provides the coast from southern Ecuador to northern Chile with a cool, misty, virtually rain-free climate (see Figure 10.28). Water temperatures off the coast of Peru range from as low as 10° C (50° F) in the south to about 22° C (71° F) in the north. These readings are at least 5.5 C° (10 F°) colder than the means for their latitudes.

In normal years the flow of the Humboldt Current is strongly maintained by the counterclockwise wind flow around the South Pacific center of the subtropical high, located some 3,200 kilometers (2,000 mi) west of South America. These winds blow northward along the coast, inducing an equatorward flow of cold water originating off the coast of Antarctica. As they reach the latitude of the Peruvian coast, the winds are gradually redirected westward as they are incorporated into the southeast trade winds. This change in wind direction, coupled with the influence of the Earth's rotation, slowly swings the northward-flowing water toward the west. The westward movement of the surface water, in turn, causes large volumes of cold water to well up from great depths. As this deep water slowly rises, it carries to the surface a great quantity of nutrients, mostly phosphates and nitrates, derived from the sea floor. Microscopic plants and animals abound in these rich waters and provide sustenance for larger forms of sea life, making this area, until recently, one of the most productive fishing grounds in the world. In the early 1970s, more than 20 percent of the world's total fish catch was taken from these waters, and Peru was the leading fishing nation.

As the trade winds off the South American coast blow seaward, their

continued on next page

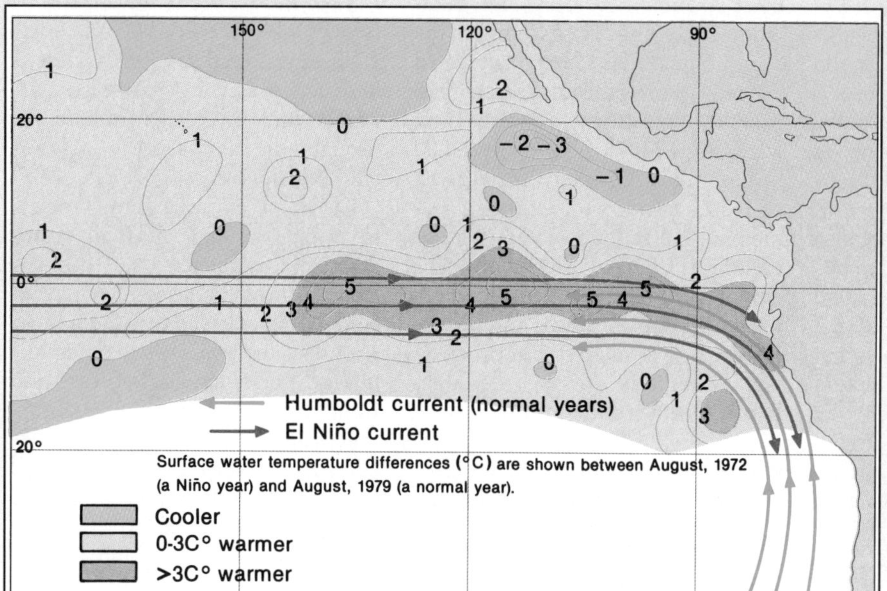

FIGURE 10.28 The effects of the 1972 El Niño on surface water temperatures off the South American coast.

westward transport of cool water slightly lowers the ocean level. It is this westward movement of cool water that induces the bottom water to well up to take its place. At the same time, the trade winds produce a pileup of tropical water in the western Pacific as the water is pushed toward the land. The water here is too warm to sink, and mean sea levels are raised about 40 centimeters (16 inches) above normal.

Once every five years or so, for reasons still unknown, air pressure patterns over the South Pacific undergo a major alteration, and the southeast trade winds weaken and die, often to be replaced by light winds from the west. The warm water that has accumulated in the western Pacific no longer is pressed against Asia and the East Indies and surges eastward. It crosses the Pacific in about two months, overrunning the denser, cold water of the Humboldt Current to produce a southward flow of tropical water down the South American coast. Sea levels may be raised by up to 30 centimeters (1 ft), water temperatures rise as much as 8 C° (14.5 F°) above normal, and the warm water may advance down the coast as far as 1,000 kilometers (600 mi) south of the equator. This phenomenon often begins during the Christmas season and has been named **El Niño,** the Spanish name for the infant Jesus.

The most recent major occurrences of El Niño have been in 1957, 1965, 1972, 1976–77, 1982–83, and 1987–88. Although they have differed in detail, the consequences of each have been widespread and locally disastrous. The effects generally are felt most strongly in the immediate vicinity of the South American coast. The overspreading of the ocean surface by tropical water keeps upwelling nutrients from reaching the surface and reduces the water's dissolved oxygen supply. This results in the death or departure of the commercial fish and spells economic disaster for the Peruvian fishing industry.

The warm water also moistens and destabilizes the air, which, when it penetrates inland and is lifted by the Andes, often unleashes torrential rains and severe windstorms. The resultant flooding is especially devastating because the desert surface that is affected lacks stabilizing vegetation and a well-developed system of river channels to transport the runoff.

The El Niño of 1982–83 was the most strongly developed and disastrous on record. At its height, the South Equatorial Current reversed direction across the entire Pacific to produce an eastward flow of tropical water some 13,000 kilometers (8,000 mi) long. Warm water off the coast of Peru extended to depths of 140 meters (450 ft). Associated weather calamities affected every continent except Antarctica and Europe, causing 1,500 human deaths and inflicting damage estimated at between $2 and $8 billion. Among the areas hardest hit were the following:

- Peru and Ecuador, where the fishing industry that was just beginning to recover from the 1972 El Niño was destroyed, and where tremendous floods triggered by El Niño left many dead and homeless.
- Australia, where a record drought caused more than $2 billion worth of destruction to farmland, crops, and livestock, and where 75 people were killed by brush fires. Severe drought also affected Indonesia, the Philippines, India, Mexico, Central America, and southern Africa.
- French Polynesia, which was visited by six tropical storms and hurricanes; this followed a period of 75 years in which it had experienced none. The Hawaiian Islands also experienced a rare hurricane.
- The U.S. West Coast, which received a spate of severe winter storms, resulting in much coastal flooding and erosion, numerous landslides, and a record mountain snowpack. The Gulf coastal states also experienced torrential rainfalls and severe flooding.

The most recent advent of El Niño took place in 1987 and early 1988. It was neither as strong nor as destructive as the 1982–83 occurrence, but was linked to severe droughts in India and the U.S. Pacific Northwest.

El Niño is considered by some experts to be the single most important disruptive influence on world climate patterns and has recently become a major topic for research by scientists in several fields. An improved understanding of El Niño would help us predict future occurrences and better prepare for future climatic disasters. The nearly global impact of El Niño provides a graphic illustration of the fact that all terrestrial phenomena are interrelated to form one single complex system, and that a significant change in any one component can have far-reaching effects on the others.

CHAPTER 11

BIOGEOGRAPHY

OUTLINE

FOCUS QUESTIONS

1. What are the chief environmental controls on the global distribution of plants and animals?
2. What are the major plant associations and subassociations, and what are their basic botanical characteristics?

The Earth is home to a great variety of life forms. It is estimated that more than 5 million species of plants and animals inhabit our planet. About 1,500,000 species have been classified, and many more—particularly small and microscopic life forms—await discovery.

The systematic study of plants and animals forms the subject matter of biology; the branch of physical geography devoted to this subject is known as **biogeography.** Biogeographers strive to develop a better understanding of the global distributions of associated groups of plants and animals and to better comprehend the interactions of these groups with their environment.

In most parts of the Earth, plant life is much more abundant, and evident, than is animal life. Vegetation covers approximately 80 percent of the Earth's land surface and is remarkable in its diversity. Plants range in size from microscopic algae, little changed from the earliest life forms to evolve on our planet, to the giant sequoia trees that are the largest living organisms on Earth. Plants grow from the margins of glaciers to the heat of the equator, from areas submerged in water to almost rain-free deserts. They not only comprise the vast bulk of life on Earth, they also have made possible the evolution of animals, including human beings, by providing both food and a breathable atmosphere.

Most plants can produce the bulk of their solid matter directly from water and carbon dioxide in the presence of light. The process by which they do so, termed **photosynthesis,** is accomplished by chlorophyll, a green-colored organic chemical. The Sun's energy powers plant growth, but only about 0.1 percent of the total insolation is captured by growing plants during photosynthesis. Somewhat over half of this energy is used directly by the plant in its own metabolic processes, while the remainder is stored as carbohydrates and is later released as the plant is consumed by animals or decomposed by microorganisms. These organisms convert the carbohydrates once again to carbon dioxide, which is released back into the atmosphere for recycling (see Figure 1.3).

The photosynthetic process releases free oxygen (O_2) into the air. This gas, which is essential to all animal life, apparently did not exist in appreciable quantities in the Earth's early atmosphere. Plants therefore must have evolved well before animals, which appeared only after vegetation had had sufficient time to produce a breathable supply of atmospheric oxygen.

Animals, unlike plants, are incapable of manufacturing their own food by photosynthesis. Instead, they depend on plants or on other animals for sustenance. Animals also are capable of independent movement, while most plants are not. In addition, most nonmicroscopic animals are aware of their surroundings and possess at least a rudimentary intelligence. These characteristics allow them to respond rapidly to stimuli, giving them a measure of self-protection from threats to their well-being.

The great majority of animal species are microscopic. The larger animals—the ones with which biogeographers are most concerned—may be classified as either invertebrates or vertebrates. **Invertebrates,** animals without backbones, comprise over 90 percent of all nonmicroscopic animal species. Important invertebrate groups include worms, mollusks, and especially insects. **Vertebrates,** which have backbones, are divided into five principal groups: fishes, amphibians (including frogs and salamanders), reptiles (especially snakes and lizards), birds, and mammals. The vertebrates are, as a group, more highly advanced on the evolutionary scale than are the invertebrates.

In general, the distributions of animals are less studied by biogeographers than are those of plants. This is chiefly because animals are much less prominent on the landscape than are plants. In addition, the mobility of animals often makes it difficult to analyze the interrelationships that exist between them and the environment.

As far as we know, Earth is the only planet in our solar system (and indeed, in the universe) on which life has developed. The first living organisms evolved very early in our planet's history—probably at least 3.5 billion years ago. Between that time and the present, evolution has produced all the plants and animals currently occupying the Earth's surface. The two keys to the evolutionary process are (1) biological changes that occur through mutations, and (2) the ability of individuals that are better adapted to their environment to survive while those not so well adapted perish.

The **mutations** that have produced the world's plant and animal species involve molecular rearrangements in their genetic material that create changes in their physical characteristics. Most mutations affect life forms adversely, making them less able to survive. A very small proportion of all mutations, however, produce changes that enable a given individual to compete more successfully than before, resulting in its multiplication and areal expansion. During the long biological history of the Earth, literally hundreds of millions of successful mutations have produced new species of gradually increasing variety and complexity. This process has culminated in the Earth's present **biota,** or population of plants and animals.

The global distribution of life forms currently is being modified profoundly by the recent emergence of a single dominant species—Homo sapiens, or human beings. The extent of this modification is illustrated by the fact that the great majority of the Earth's land surface no longer has an undisturbed natural vegetation cover. In addition, many species of wild animals currently are being decimated at an alarming rate. As a result, many of the plant and animal assemblages discussed in this chapter no longer dominate extensive portions of the Earth's surface, as they did before their alteration or removal by humans over the past few centuries or millennia. A detailed examination of human modifications of the environment lies beyond the scope of this book, but a prime example of the human impact on the natural environment is provided in the Case Study at the end of the chapter. It examines the causes and implications of the ongoing destruction of a major biological assemblage—the world's tropical rainforests.

The remainder of this chapter is divided into three sections. We will first discuss the development and function of biological communities on a local level. Next, the important environmental factors controlling the geographic distributions of large-scale biological assemblages are identified. Finally, the characteristics and global distributions of each of these assemblages are examined.

LOCALIZED BIOLOGICAL COMMUNITIES

The global distribution of plants and animals is not random. Each species has specific environmental requirements, so that different biological groupings or assemblages exist in different settings. A specific site that has environmental conditions favoring its habitation by a given species or group of species is termed a **habitat.** Habitats can be identified at differing scales. **Macrohabitats** generally are climatically delineated and usually are differentiated on the basis of their vegetation. **Microhabitats** exist within macrohabitats. They may be very small, such as around the edge of a pond or under a rotting log. Microhabitats are especially important to small plants and animals, but even large animals often confine themselves to only limited portions of macrohabitats.

An assemblage of plants and animals whose members occupy a particular environmental setting of restricted size is referred to as a **biological community.** The members of such a community share their environment, and the various resources within it are divided so as to use them as completely and efficiently as possible. Biological communities of all sizes exist and are determined to a great degree by climatic characteristics at both the macro and micro levels.

Carrying the idea of the biological organization of the local environment farther, we have the concept of the ecosystem. An **ecosystem** is a functioning entity consisting of all the organisms in a biological community, as well as the environment in which and with which they interact. A balanced ecosystem is largely self-sustaining in its resources and is relatively stable in its assemblage of life forms. An essential characteristic of a healthy ecosystem is the maintenance of flow cycles of energy and of nutrients (see Figure 11.1). As we have seen, energy is derived from insolation and is fixed or "captured" by plants during photosynthesis. It may then pass to **herbivores** (plant-eating animals), to **carnivores** (meat-eating animals), and, upon their deaths, to bacteria and the soil. Mineral nutrients originally derived from the weathering of bedrock (see Chapters 12 and 16) also are cycled through vegetation, animals, the soil, and back again to vegetation. This often occurs with little loss, and what losses do occur are replaced by minerals released by subsequent weathering or animal input.

Ecological communities generally are treated as discrete entities that differ significantly from neighboring communities. In reality, however, they often are separated by transitional communities called **ecotones.** Ecotones share attributes of the ecological communities they border and therefore tend to exhibit relatively complex biological characteristics.

Ideally, a biological community is relatively stable. If environmental conditions are altered significantly, however, the community may undergo drastic changes. Disturbances such as high winds, floods, droughts, fires, insect invasions, or human interference can remove some species and encourage others, thereby greatly modifying the makeup of the community. If the disturbance is minor, and environmental conditions return to those that originally existed, the community will tend to revert gradually to its

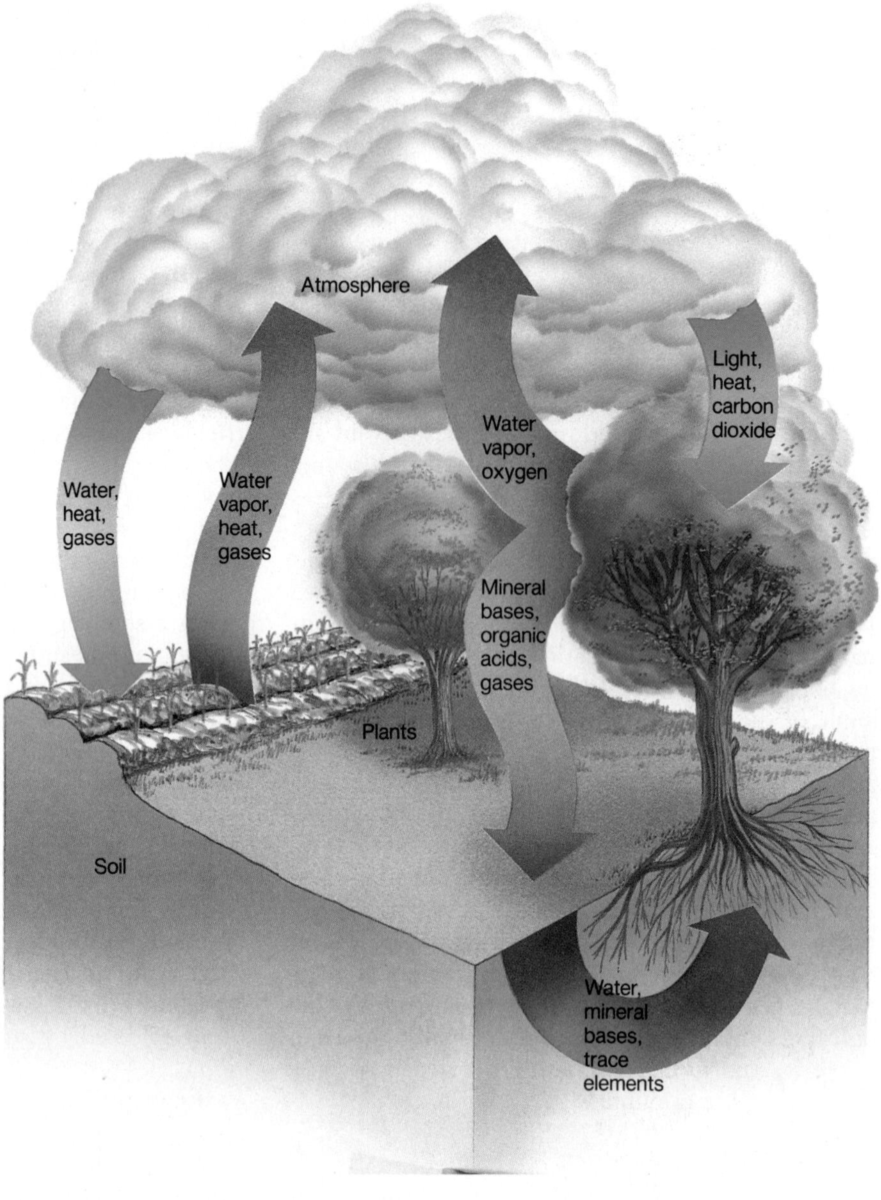

FIGURE 11.1 Diagrammatic depiction of the flow of energy and materials between the atmosphere, soil, and vegetation.

original composition. Reversion occurs because the original assemblage of plants and animals formed the most stable and efficient community under the existing environmental conditions, and a return of those same conditions should produce the same biological response.

The gradual change in the composition of a plant community in response to changing environmental conditions following a disturbance is referred to as **vegetative succession** (see Figure 11.2). Succession normally proceeds at a relatively rapid rate when a soil cover is already present and seeds and some vegetation exist within the disturbed area. In some instances, however, physical processes can produce a new surface that has no soil or vegetation cover. Examples of such processes include volcanic eruptions, landslides, glacial erosion, and the drainage of water bodies. In such cases a **primary succession** of vegetation occurs. Primary succession takes place relatively slowly because its progress is largely dependent upon the development of a soil cover.

Greatly differing successional sequences occur under dissimilar environmental conditions. Eventually, the successional process produces the biological

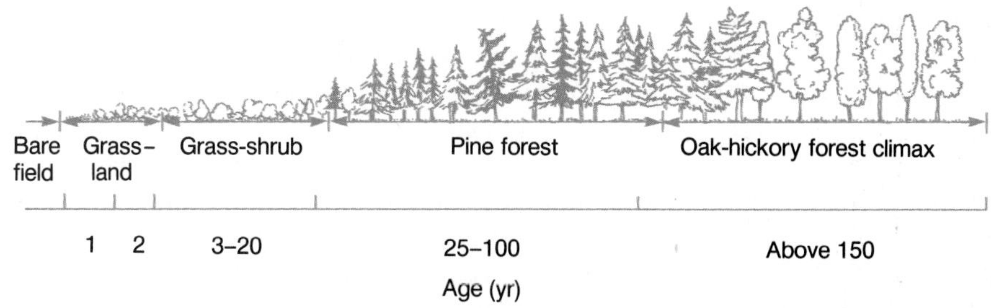

FIGURE 11.2 An idealized vegetative successional sequence in the southeastern United States following the abandonment of farmland. Each successional stage alters the environment in such a way as to allow the next stage to develop.

community best suited to survive under existing environmental conditions. This **climax community,** as it is called, theoretically continues to occupy the area until the next disturbance or environmental alteration occurs.

A climax community has several important attributes. It usually contains the greatest **biomass,** or mass of living organisms, that the area is capable of supporting. A high degree of species diversity typically exists, so that all available habitats are utilized effectively. The composition and relative distribution of species tend to be stable, and birth and death rates are balanced. Finally, most energy and nutrient needs are supplied and recycled by the organisms within the community, so that losses are minimized.

Most areas of the world do not presently have climax biological communities. This is primarily the result of widespread human disturbance and, to a lesser extent, of geologically recent worldwide climatic changes (see Chapter 18). Nevertheless, it is possible to divide the world's plants and animals into several major global-scale communities, or **biomes,** that exist in relative equilibrium with their environment. We will describe these communities shortly. Before doing so, however, we will examine the basic environmental controls that are responsible for their geographic distribution.

BIOGEOGRAPHIC DISTRIBUTION CONTROLS

Plants and animals display a tremendous range of physiological and behavioral characteristics that enable them to cope with the multiplicity of terrestrial environments. Some environments, of course, can support a greater abundance and variety of life forms than can others.

In order for a given plant or animal species to occupy a given geographic area, it must have had access to that area. Most species originally evolved in a restricted area and their current ranges are a result of reproductive success and migration to other areas. Plant migration occurs primarily during the seed stage and is accomplished largely through the mechanisms of wind, water, and animal transport. Animals, unlike plants, can purposefully change location to reduce stressful conditions; they are therefore less tied to the local environment. Most animals possess mechanisms for controlled movement such as wings, legs, and fins. Many animals migrate seasonally in order to follow food sources or to avoid weather extremes (see Figure 11.3).

The local abundance of an organism usually is related closely to the availability of its preferred habitat. Unfavorable areas between habitats can become **barriers** to dispersal, and they must be crossed in order for the organism to reach other favorable potential habitats. Few barriers are absolute for all species, although they vary greatly in effectiveness for different plants and animals. In this respect, barriers

FIGURE 11.3 A migrating herd of bison in Custer State Park, South Dakota. (South Dakota Dept. of Tourism)

may act as filters, allowing some species to pass through, while keeping others out. On a global scale, the three most important categories of barriers are oceans, mountain ranges, and deserts. Collectively, these three types of barriers are largely responsible for the existing geographical patterns of most terrestrial life forms.

Despite the successful adaptations of terrestrial biota to diverse and often adverse environments, large portions of our planet are largely or entirely devoid of life. For the Earth as a whole, the leading barrier by far in restricting the areal distribution of land biota is the presence of surface water. The oceans, of course, comprise a complex biological realm that is both rich in variety and distinctively different from the land. Because of space limitations in this book, however, the subsequent discussion is limited to land biota.

Approximately 80 percent of the world's land surface contains significant plant and animal life. The remaining 20 percent is subject to environmental extremes that are prohibitive to the survival of plants and of most animals (see Figure 11.4). About half of this area is too cold for plants to carry on their life processes. These regions, primarily consisting of Antarctica and interior Greenland, are mostly covered by snow and ice throughout the year, so that they lack soil or even a bedrock surface from which plants could extract nutrients. Most of the remaining surface area devoid of vegetation is too dry and consists of almost lifeless desert. Most notable here are portions of the central Sahara, the Arabian Peninsula, and interior Asia. Perhaps an additional 1 or 2 percent of the land surface is lacking in natural vegetation because it is too steep, contains toxic surface materials such as excessive quantities of salt, or has been subject to recent natural disturbances that have temporarily removed the vegetation cover. Where vegetation is lacking, animal life is eliminated or severely restricted by harsh environmental conditions and a lack of food.

From the preceding paragraph, it can be deduced that moisture, temperature, slope, and soil are among the factors that most strongly influence the world distribution of life forms. These factors can be grouped into five major categories of biological controls: climate, topography, migrational access, soil, and organisms. We now examine, in turn, the influence of each on the global distributions and types of plants and animals.

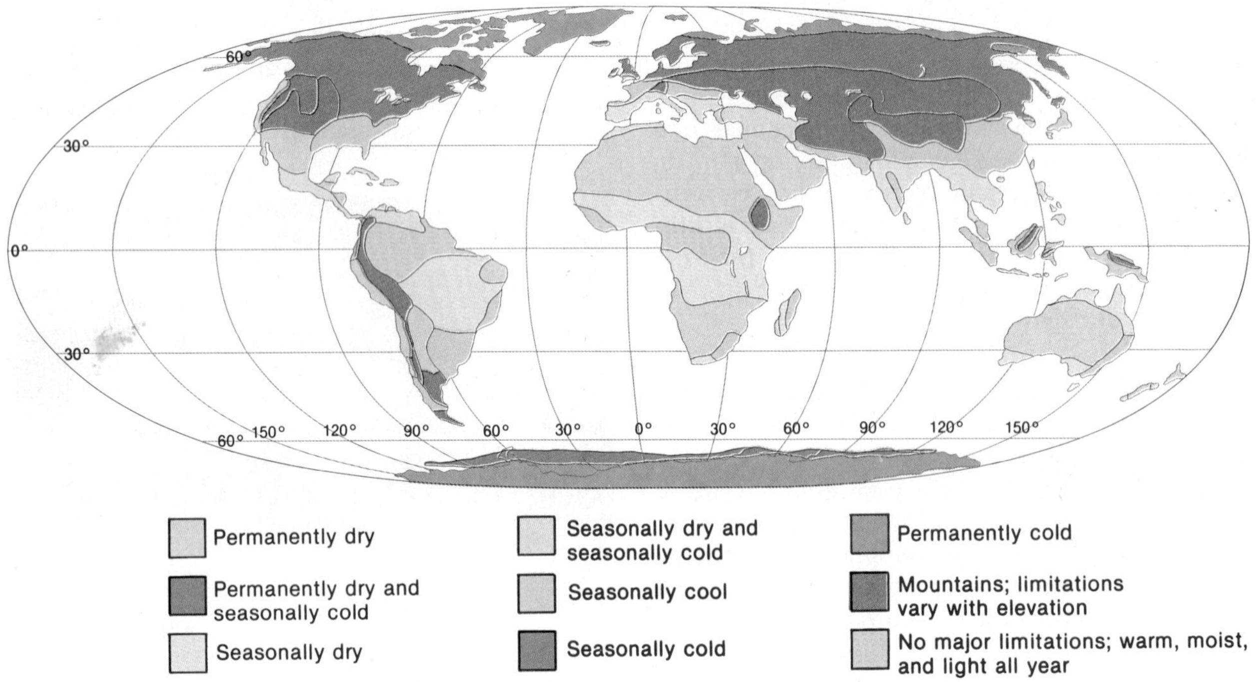

FIGURE 11.4 Global distribution of major climatic stress factors on natural vegetation.

Climatic Influences

Climatic characteristics, especially moisture and temperature, are the most important biogeographical controls. Climate particularly controls vegetation patterns, and vegetation, in turn, largely controls animal populations. Natural vegetation depends so strongly upon climate that considerable similarity exists between the geographic patterns of climate and vegetation. Natural vegetation tends to be most abundant and diverse where there are no major climatic stresses. These stresses, the most crucial of which are the occurrence of cold or dry seasons, greatly reduce the variety of species of both plants and animals. As a result, both the variety and quantity of life forms generally decline as one proceeds from wet to dry and from warm to cold regions of the Earth.

Moisture

The availability of moisture is the single most important criterion controlling the large-scale pattern of natural vegetation. Moisture availability is considered here as a climatic factor because it is determined largely by the balance between precipitation amounts and evapotranspiration rates, but it also is influenced significantly by soil and topographic factors, and will be discussed again under those headings.

Water is essential to plants in a number of ways. Vital soil nutrients must be absorbed by plant roots and distributed throughout the plant in a dissolved state, and water is the necessary solvent. Water also is an essential reactant in the synthesis of the organic compounds (starches and sugars) manufactured by plants. In addition, the rapid transpiration of water during hot weather helps to reduce vegetative surface temperatures to tolerable levels.

The amount of soil water available is a vital factor in controlling the quantity of natural vegetation that a locality can support. Being literally "rooted to the spot," plants must rely on moisture that comes within reach of their roots. The total biomass within an area normally increases with available moisture until a point of excess moisture is reached, after which the quantity of vegetation again diminishes. Perhaps the most striking characteristic of a desert is its relative lack of vegetation, while the density and lushness of a well-watered tropical forest are equally obvious and impressive (see Figure 11.5).

Plants are sometimes categorized on the basis of their adaptations for coping with available moisture.[1] **Xerophytes** (from the Greek *xeros*, "dry" and *phytos*, "plant") are plants especially adapted to regions deficient in water. These regions are mostly arid deserts, but also include areas such as beaches or steep slopes, where the nature of the soil, rather than the climate, limits moisture availability. Some of the water-conserving characteristics commonly displayed by xerophytes include extensive root systems, water storage ability, waxy leaves, thickened epidermal layers designed to reduce transpiration, small leaves or none at all (as in the case of some cacti), and the ability to minimize insolation receipts by turning leaves edgewise to the Sun. In addition, xerophytes tend to be small and to grow slowly.

Hygrophytes (*hygros*, "wet") are plants especially adapted to excessively moist environments. Because their environmental stress factor is the opposite of that for xerophytes, many of their physical characteristics are also directly opposed to those exhibited by xerophytes. For example, hygrophytes tend to have shallow root systems, large leaves with numerous stomata (pore spaces), thin epidermal layers, and the ability to turn leaves directly toward the Sun. In addition, they usually grow rapidly and may attain a large size when mature. Large size helps these plants compete for sunlight—an important consideration because vegetation is usually dense in areas where hygrophytes grow.

A third category of plants, the **mesophytes** (*meso*, "middle"), have physical characteristics intermediate between those of xerophytes and hygrophytes. They typically occupy well-drained habitats in areas of the low and middle latitudes with ample year-round precipitation and without excessively low temperatures.

A fourth category, the **tropophytes** (*tropos*, "to turn"), is defined somewhat more broadly than the other three. Tropophytes are adapted to alternating favorable and unfavorable climatic conditions for growth. These conditions generally consist of seasonally varying warm and cold periods, such as occur in the middle and high latitudes, or wet and dry pe-

[1] It should be noted, though, that the definitions of these groups are not exact, that they permit considerable overlap, and that experts are not in complete agreement as to the geographic boundaries of the areas dominated by each group or even of their constituent plant members.

FIGURE 11.5 Contrasting vegetation assemblages growing in response to differing moisture conditions: (*left*) temperate rainforest in Olympic National Park, Washington; (*right*) semi-desert scene in the Flinders Ranges, South Australia. (*left*: G. Ahrens/H. Armstrong Roberts; *right*: © Bill Bachman/Photo Researchers)

riods, as experienced, for example, in the Tropical Wet and Dry and the Dry Summer Subtropical Climates. Because most areas experience at least one of these two types of seasonal alternations, the tropophytes are distributed more extensively than are the other three plant groups.

Most tropophytes are large, leafy plants commonly categorized as trees or shrubs. A **bush** (or **shrub**) is a woody plant with several permanent stems rising from the ground. A **tree,** on the other hand, normally is larger and contains a single primary stem or trunk that develops many branches. Many tropophytic shrubs and trees, especially those with broad leaves, are **deciduous;** that is, they lose all their foliage at the beginning of the unfavorable season. Conversely, most needleleaf bushes and trees are **evergreen;** they retain their foliage throughout the year. Tropophytes, whether deciduous or evergreen, are more versatile than xerophytes, hygrophytes, and mesophytes in that they are able to grow vigorously during favorable seasons and to protect themselves by becoming dormant during unfavorable seasons.

Temperature

The second dominant climatic control on the distribution of the Earth's biota is temperature (see Figure 11.6). Plants vary greatly in thermal tolerance, so that only the coldest areas do not support vegetation. Some plant species are adapted to a broad range of thermal environments; others are restricted to a narrow range. Each plant species, though, has a maximum and minimum thermal limit beyond which it cannot survive, as well as an optimum temperature range within which it grows most vigorously.

FIGURE 11.6 Contrasting vegetation assemblages growing in response to differing thermal conditions: (*left*) palm forest on a beach in Sri Lanka; (*right*) dwarf spruce and alpine tundra in the Olympic Mountains of western Washington. (*left:* M. Thonig/H. Armstrong Roberts; *right:* B. J. O'Donnell/BPS)

Animals also are greatly affected by temperatures. Unlike plants, however, most have the ability to seek shelter, especially underground, during extreme conditions. Desert animals often are nocturnal, thus avoiding the heat and dryness of the day. Birds and mammals are warm-blooded and most have a body covering of feathers or fur to help protect them from excessive cold.

Many plants outside the tropics require varying temperatures to initiate different crucial processes such as germination, fertilization, flowering, fruiting, and foliage growth. Because most plant growth stops when temperatures fall below 6° C (42° F), little vegetation exists in areas where temperatures never rise above this level. As temperatures increase (assuming other vegetative controls present no hindrance), growth rates and biomass concentrations increase to a maximum at about 30° C (86° F).

Of the two thermal extremes, cold has the most detrimental impact on vegetation. Tropical plants often have cold temperature limits well above the freezing point. Many plant species of the higher latitudes, conversely, can survive extended periods with temperatures far below −18° C (0° F). The most critical temperature for the survival of many plant species, however, is the freezing point. Numerous plants, including most flowers and garden vegetables that are familiar and important to residents of the middle latitudes and subtropics, are killed annually by the first autumn freeze. In contrast, excessive heat rarely kills plants directly, and even the hottest surface locations on Earth can support vegetation. High temperatures do, however, greatly increase transpirational losses. This frequently causes plants to require more water than is available, resulting in wilting and possible eventual death.

In regions that experience the climatic stress of a cold season with subfreezing temperatures, plants must adjust their life cycles in order to survive. One group of plants, the **annuals,** do not have mechanisms that enable individual plants to survive the rigors of the cold season. Instead, they have a life cycle that is sufficiently short to enable them to mature in one warm season and produce the seeds that ensure the growth of the next generation of plants the following year. The amount of time they have to accomplish this task is termed the **growing season.** In the United States, this period generally is defined as the number of days between the last killing frost of spring and the first killing frost of fall. The length of the growing season tends to decrease with latitude and also with increasing elevation and continentality. As a result, the variety of annuals that can grow is increasingly restricted in these directions to types with short growing seasons. Most of our food crops (including grains and vegetables but excluding tree crops) are annuals, and must be replanted each year.

A second group of plants, the **perennials,** have multiple-year lifespans. In areas with a climatic stress, they have developed protective mechanisms

THE ENVIRONMENTAL CONSEQUENCES OF ACID RAIN

The phenomenon of acid rain has become a major environmental concern in recent years. It is a complex problem that is far from completely understood, but seems to be caused largely by the combustion of fossil fuels. Natural factors, including soil and bedrock characteristics and wind and precipitation patterns, also influence the severity of this problem in some regions.

Although natural rainwater is mildly acidic, the presence in the atmosphere of industrially derived acids can increase the acidity of precipitation to the point that it becomes a corrosive substance that can be toxic to plant and animal life. The release of sulfur dioxide gas from coal-fired electric power plants is believed by most experts to be the major single factor contributing to the formation of acid rain. Within the atmosphere, this gas is converted to liquid sulfuric acid droplets that are carried to the surface by precipitation.

In many areas, the effects of acid precipitation on the soil is not a concern. This is because the surface is naturally buffered by the presence of lime or other alkaline substances in the rock or soil. In other areas such as the northeastern United States, southeastern Canada, and much of Europe, however, the presence of numerous coal-burning power plants, coupled with naturally acidic rock and soil types, has resulted in a serious acid rain problem. Globally, an estimated 6.4×10^{10} kg (70 million tons) of sulfur is released each year into the atmosphere by human activities.

Several major adverse environmental consequences of acid rain have been widely reported. In large portions of Scandinavia and Germany, needleleaf forests have been slowly dying. Needles near the tops of the trees turn yellow and fall off, and eventually the entire tree may be killed. In Germany, this phenomenon was first noticed in the Black Forest in the early 1970s, and by 1990 more than half of the nation's trees were affected. In eastern Europe, entire forests are reported to have been destroyed. Forest dieback also has taken place recently in portions of the Appalachian Mountains of the eastern United States (see Figure 11.7). Although the exact mechanism by which the trees are killed by acid precipitation remains uncertain, it is believed that the acidification of the soil kills organisms that participate in the soil-plant nutrient cycle and also mobilizes potentially toxic aluminum and heavy metal ions in the soil that subsequently are absorbed by the trees.

Acid rain also has long been blamed for the destruction of fish and other aquatic life in New England, upstate New York, eastern Canada, and Scandinavia. Many lakes within these areas, which previously supported large fish populations, have been virtually sterilized.

that enable them to survive as individuals. Often this involves a period of dormancy. Most broadleaf trees and shrubs, for example, lose their leaves and become dormant during autumn in the middle and higher latitudes, resuming their growth and producing new foliage when warm weather returns the following spring. Dormancy and a deciduous habit therefore protect plants from the stresses of a dry or cold season. Perennials have a tremendous advantage over annuals in terms of size, and therefore in spatial dominance, because they can add to their growth over an extended period. As a result, most natural vegetation groups are dominated by perennials; all the world's larger plant species are of this type.

Light

Light is a third climatic necessity for most life forms. Plants use light as the energy source for photosynthesis[2], in which carbon dioxide and water are converted to glucose (sugar), with oxygen being released

[2] The fungi, which can live without light because they obtain their food by absorbing dissolved organic compounds, are an exception. They recycle existing organic compounds and can live only in the presence of living or dead organic material. The fungi, which include yeasts and mushrooms, are so unlike other plants that botanists increasingly consider them not as plants at all, but as a third biological kingdom. Along with bacteria, they serve as decomposers of organic matter.

FIGURE 11.7 Forest destruction caused by acid fog and rain on Clingmans Dome in the Great Smoky Mountains of North Carolina. (Breck P. Kent, © JLM Visuals)

A cultural concern resulting from acid rain has been the damage suffered by numerous priceless works of art in Italy, Greece, and several other countries. Although acid rain can corrode metals, fade paints, and even etch glass, marble statues are especially affected because the acid chemically converts the marble to calcium sulfate, which is soft and soluble. This process, when combined with the effects of freezing water, causes the outer portions of the marble to be shed in layers, allowing progressively further penetration of acidic water.

The release of sulfur dioxide by power plants in the American Midwest has been much criticized by residents of the Northeast and adjacent southeastern Canada, who have experienced severe acid rain problems in the last few decades. The Midwest contains a large concentration of coal-burning power plants. Midwestern coal is plentiful and easy to extract, but it unfortunately has a high sulfur content. In an effort to meet U.S. Environmental Protection Agency regulations, these plants are in the process of installing expensive scrubbing systems that reduce the sulfur content of the coal before it is burned and are also increasingly importing low-sulfur coal from the western United States. The complete abandonment of the productive, high-sulfur coal deposits, however, would have a devastating impact on the economy of many regions. Further reductions in the use of this coal are being vigorously resisted as long as any doubt exists about the degree to which its combustion affects acid rain contamination farther to the east.

The acid rain controversy illustrates an important environmental dilemma that societies face to some extent in the pursuit of nearly all major economic activities. It involves the "acceptable" degrees to which we should expect the natural environment to be degraded as a consequence of our actions. Some degradation is inevitable, no matter how careful we are. The minimization of the environmental consequences of human activities, however, although frequently viewed as highly desirable, can be tremendously expensive both monetarily and in terms of human livelihood.

to the atmosphere as a by-product. The generalized equation for photosynthesis is:

$$\text{carbon dioxide} + \text{water} + \text{light energy} = \text{glucose} + \text{oxygen}$$

or

$$6CO_2 + 6H_2O + 686 \text{ kilocalories} = C_6H_{12}O_6 + 6O_2$$

The competition for light provides the stimulus for vertical growth in plants. In densely vegetated areas, the heights of the various plant species generally are associated closely with shade tolerance. Plant species frequently are excluded from a given habitat solely because of their inability to secure the required amount of sunlight. The changing availability of light is largely responsible for vegetative succession in an area recovering from some disturbance, such as a fire, that has removed the vegetation cover.

Although the amount of light available beneath a dense cover of vegetation is inadequate for most plant species, the intensity of light in an open setting is less important than its duration. The length of daylight varies significantly during the year outside the tropics. Like temperature, seasonal differences in daylight trigger biological activities such as budding, flowering, and leaf growth or shedding in plants. They also control mating and migration patterns in many animal species.

Other Climatic Influences

The relative humidity and the amount of cloudiness and fog all affect evapotranspiration rates, both on a short-term and a long-term basis. The presence of clear skies and dry air during the summer causes especially rapid water losses because it is combined with warm temperatures and occurs when plant and animal water requirements are at their peak.

Wind is another climatic element with both favorable and unfavorable effects on organisms. On the positive side, it increases the carbon dioxide intake of plants and is essential for pollen and seed dispersal for many species. It also aids in the movement and dispersal of many lightweight or flying animals. Conversely, the persistent high winds experienced along some exposed high latitude coasts or in mountainous areas can lead to excessive transpirational losses. Under these conditions, trees may not grow or they may be stunted or deformed (see Figure 11.8). Strong winds in storms may blow migrating flocks of birds off course or blow insects out to sea.

Topographic Influences

Topographic influences on the biota range from the large-scale to the extremely localized. The vegetation of the American Great Plains, for example, differs greatly from that of the adjacent Rocky Mountains largely because of the topographic diversity of these two large regions. On a much smaller scale, differing ecosystems frequently are encountered on different portions of a hill, or between an area of sloping land and a nearby flat stream valley. The primary effect of the topography is to produce a variety of differing microhabitats for plants and animals. As a result, biogeographic patterns tend to be much more diverse in rugged areas than in areas of low relief.

FIGURE 11.8 Stunted and wind-deformed trees at the tree-line in the Colorado Rockies. Branches grow only on the downwind (east-facing) side of the trees. (Ralph Scott)

The biota is influenced by a variety of topographically related factors. These include soil characteristics, insolation receipts, surface temperatures, and several factors relating to water availability such as precipitation amounts, evapotranspiration, and runoff rates. Large-scale topographic influences are mostly climatic and are associated with the cooler and moister conditions that normally exist in mountain regions as compared with adjacent lowlands. The biological consequences of these climatic differences are discussed later in this chapter. Probably the two most important localized topographic influences on vegetation are slope steepness and aspect.

Slope Steepness

Slope steepness affects vegetation primarily through its influence on water drainage and soil erosion. As slope angles increase, gravity becomes increasingly effective in causing excess water to drain rapidly downhill. As a result, steep slopes may experience such rapid runoff that little water remains to enter the soil and become available to plants. The rapid runoff also produces a high erosion potential for the slope, so that an inverse correlation exists between slope steepness and soil depth. The net effect on the vegetation of steeply sloping terrain is a reduction in plant size and diversity, as well as a significant lowering of the total biomass. Steep mountain slopes with little soil are likely to be devoid of vegetation or to contain a sparse cover of xerophytes.

Excessively flat topography, especially surface depressions, also can reduce the favorability of a habitat for vegetation. In this case, poor drainage and waterlogging is likely. An excess of surface water leads to a lack of aeration, which hinders soil development and interferes with plant respiration. In dry climates, it also may lead to the surface accumulation of salts.

The optimum condition for plant growth generally is considered to be a very gentle slope with an inclination of about 2 or 3 degrees. This allows excess water to drain from the surface, but does not permit drainage to proceed so rapidly that it results in significant erosion.

Slope Aspect

The directional orientation or **aspect** of a slope influences its biota by controlling the duration and intensity of sunlight. Slopes inclined toward the Sun are substantially warmer during the day than are slopes inclined away from the Sun. Sunward slopes also have drier surfaces because evapotranspiration is greatly increased by the warmer temperatures and strong insolation. The presence of these contrasting microclimatic conditions may produce major differences in the nature of the habitat on different sides of the same hill (see Figure 11.9).

Slope aspect has only minor significance near the equator because the Sun is high in the sky at midday and because its path carries it into both the northern and southern sky at different times of the year. The significance of slope aspect increases with latitude as the angle of incidence of the Sun's rays becomes progressively lower and more concentrated on the equatorward-facing slopes. In the northern hemisphere, south-facing slopes have the warmer and drier conditions, while in the southern hemisphere this situation is reversed. East- and west-facing slopes have intermediate conditions, with west-facing slopes somewhat warmer and drier in both hemispheres because daytime temperatures peak in the afternoon hours.

FIGURE 11.9 Vegetation differences resulting from variations in slope aspect in southwestern Colorado. The north-facing slopes on the left are the most densely vegetated. (J.K. Nakata: Sight and Sound Productions © 1990)

Accessibility Influences

Plants and animals that migrate, or are otherwise transported, to isolated locations usually will evolve within their restricted habitats. This causes these populations to become progressively more distinct from others of their kind, until they eventually form new subspecies and species. Thus California, largely separated from the rest of North America by mountains and deserts, has its own distinctive flora and fauna; similarly, the biological communities that have developed on many isolated islands or island groups are unique. The number of species occupying an isolated land mass depends on several factors, including the size of the land mass, its diversity of habitats, its accessibility to sources of plant and animal colonists, and the biological richness of these sources.

The Hawaiian Islands serve as a prime example of a large and extremely isolated island group that has developed a restricted and unique flora and fauna. These islands contain no true freshwater fish, no amphibians, reptiles, or mammals (except for human beings and one species of bat). Seed plants apparently were introduced largely via the digestive systems of birds; a few seeds, such as coconuts, arrived by water. Of the 1,729 known species and varieties of Hawaiian seed plants, 40 percent are of Indo-Malaysian origin, 18 percent of American origin, and the rest are of local or of uncertain origin.

Australia also has long been isolated by water bodies from the mainstream of evolutionary development and thus has its own unique, and relatively primitive, assemblage of plants and animals. More than 90 percent of the continent's native trees are varieties of a single genus, Eucalyptus. Likewise, most Australian mammals are marsupials, such as kangaroos and koalas, which carry their young in an abdominal pouch (see Figure 11.10).

Many other isolated islands have developed distinctive biological assemblages. These include Madagascar, with its lemurs, and the Galápagos Islands off the coast of Ecuador, with their iguanas and giant tortoises.

Soil Influences

Soil characteristics affect the biota on a variety of scales, the smallest of which often are very localized. The influence of soil on the distribution of plants is closely related to the effects of climate and topography. Some of the numerous attributes of the soil that

FIGURE 11.10 Marsupials such as kangaroos carry their young in abdominal pouches. Shown here is a gray kangaroo and her "joey." (© Mike James/Photo Researchers)

can affect vegetation include the size of the soil particles, the tendency of these particles to form larger aggregates, the compactness of the soil, its mineral and organic matter content, its water-retention ability, and its acidity or alkalinity. These factors are discussed more fully in the next chapter; here we briefly discuss the vegetative influence of the soil's mineral content and the effect of soil water on the availability of soil nutrients to plants.

Plant Nutrient Availability

Plants obtain the water and nutrients they require from the soil. The nutrients are absorbed through the plant roots after being dissolved in soil water. It therefore is essential to vegetation that the soil not only contains the needed minerals, but that it contains adequate, but not excessive, quantities of water.

Minerals required by plants are derived from the two solid soil-forming constituents—bedrock and organic matter. Plant species differ in their mineral requirements, so that soils formed in areas with different types of bedrock may encourage the growth of different vegetative assemblages. This, in turn, will help determine the composition of the local animal population. The **fertility** of the soil, or its productivity with respect to vegetation, is a somewhat relative term, because a given soil may contain adequate nutrients for one plant species but not for another. Some types of rock, however, are noted for forming soils of generally higher fertility than are others.

Assuming that the important plant nutrients are present in the soil, the ability to support vegetation is still critically dependent upon the presence of sufficient water to dissolve the minerals. Soils in arid environments are often fertile, but they receive insufficient moisture to dissolve the available nutrients and hence are highly limited in their ability to support vegetation. Conversely, soils in excessively wet climates frequently are infertile because most soluble minerals have been removed by the downward percolation of soil water.

Biotic Influences

The joint occupation of most areas by a diversity of plant and animal species modifies the environment in a variety of ways. The presence and activities of other life forms increase the favorability of the habitat for certain plant species. For example, the shade provided by a forest helps conserve soil moisture and affords protection from sunlight for many shade-loving plants of the forest floor. In addition, various plants utilize the mobility of animals for pollination and seed dispersal. Organic matter produced by the decomposition of vegetation also releases nutrients to the soil for use by new generations of plants.

On the other hand, the presence of numerous plant and animal species within a region of limited size and resources necessarily results in competition for survival. Even if the climate, topography, and soil type are all favorable for a given species, it cannot survive within a given area unless it can successfully occupy a habitat despite the likely presence of competitors. Of all the species with access to a given habitat within a biological community, only those best adapted to the environment will be successful in occupying it. If a species is out-competed by other plants or animals within an area, it will be eliminated from that area. If a species is out-competed everywhere, it will become extinct.

Because of its absolute finality, the extinction of plant and animal species is of concern to many individuals, especially when it is carelessly brought about by human activities, as so many extinctions are. Natural extinction (that is, extinction not brought about by human actions), however, is a necessary biological process. There simply is not enough room on Earth for all the species of life forms that ever were, are, or will be. The removal of organisms less able to compete makes room for species that are better suited to survive within a given environment; it is the means by which biological diversity and ultimate advancement occurs.

WORLD DISTRIBUTION OF NATURAL VEGETATION AND WILDLIFE

As noted earlier, plants and animals can be divided geographically into global-scale communities called biomes. Biomes usually are delineated on the basis of their dominant forms of natural vegetation; they therefore are synonymous geographically with the large-scale groupings of natural vegetation known as **plant associations.** Four plant associations are commonly recognized: forest, grassland, desert, and tundra. Major vegetative differences exist within associations. The associations therefore can be divided into subassociations that exhibit a greater degree of structural similarity in their vegetative characteristics (see Table 11.1).

In this section, we shall examine each of the four plant associations, giving primary emphasis to their vegetative characteristics, but noting their faunal assemblages as well. (As we have seen, faunal distributions are much less clear-cut than floral, largely because of the greater mobility of animals). The various subassociations of the forest and grassland associations are sufficiently large and important to merit individual treatment.

The global distribution of plants and animals is complex and varied when examined in detail (see Figure 11.11). Such complexity makes any limited classification system such as the one used in this book merely a convenient abstraction of reality. In actuality, there is no set number of vegetation groupings, and the character of the vegetation normally does not suddenly change when vegetative "boundaries" are crossed. It can be argued that biogeographical regions are devised more to fit our intellectual limitations than they are to delimit natural boundaries.

The formulation of natural vegetation groupings, however inadequate they may be in reflecting the complexity that actually exists, does serve useful purposes. It indicates that vegetation differs greatly on a regional basis and facilitates a generalized discussion of these differences. In addition, a geographic analysis of vegetative patterns emphasizes the crucial fact that the distribution of natural vegetation is not random, but instead is closely related to other environmental parameters, especially global patterns of temperature and available moisture.

In the following discussion of vegetation associations, we attempt to provide an element of geograph-

TABLE 11.1 Natural Vegetation Associations and Subassociations[1]

I. Forest
  A. Tropical Rainforest
  B. Tropical Semideciduous Forest
  C. Tropical Scrub Woodland
  D. Subtropical Broadleaf Evergreen Forest
  E. Mediterranean Woodland and Scrub
  F. Mid Latitude Deciduous and Mixed Forest
  G. Mid Latitude Evergreen Forest
    1. West Coast Coniferous Forest
    2. Southern Pine Forest
  H. Northern Coniferous Forest
II. Grassland
  A. Savanna
  B. Steppe
    1. Low Latitude Steppe
    2. Mid Latitude Steppe
  C. Prairie
III. Desert
IV. Tundra

[1]Roman numerals precede natural vegetation associations; letters precede subassociations. Because of its diverse nature, mountain vegetation does not fit into this classification.

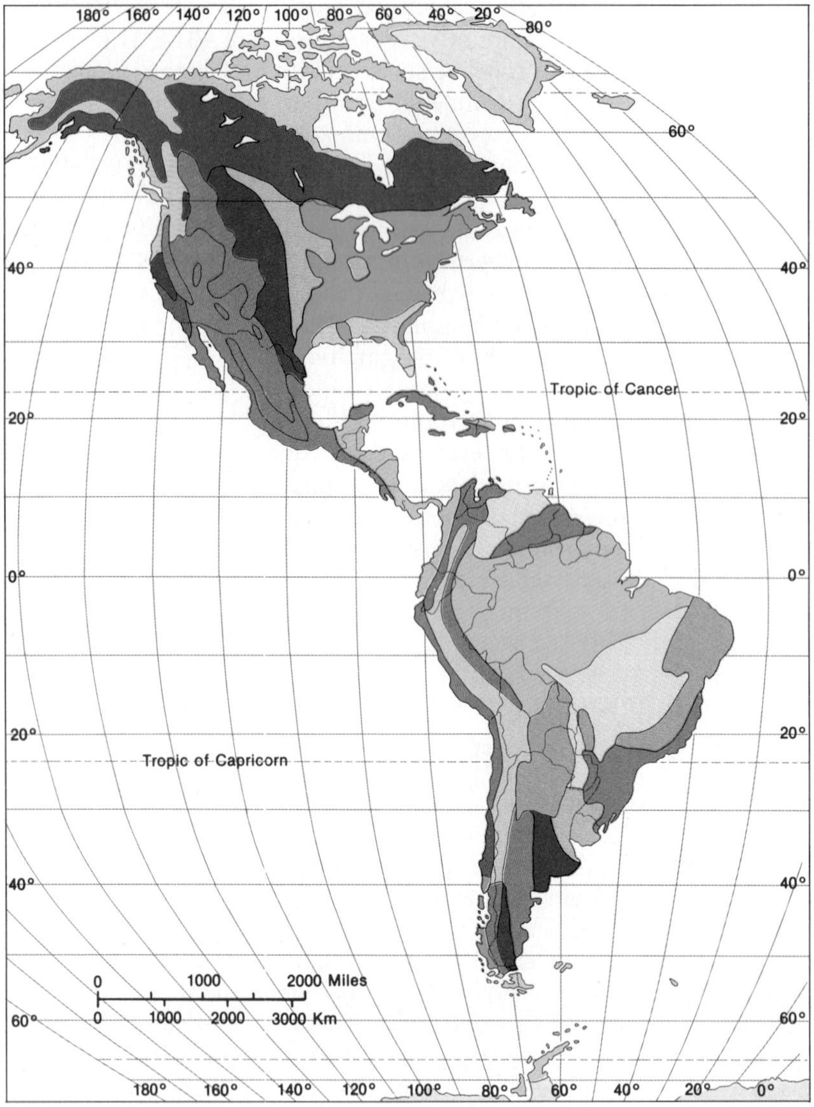

FIGURE 11.11 Global distribution of natural vegetation regions.

ical continuity by proceeding in a general poleward direction and by stressing the factors responsible for the basic changes in characteristics from one grouping to the next. We begin with the forest association, which not only dominates the equatorial regions, but is the most areally extensive and important association from both a biological and a human occupancy standpoint.

Forest Association

A **forest** is an assemblage of trees that grow sufficiently close together for their foliage to overlap and largely shade the ground for at least a portion of the year. Trees are not the only type of vegetation in a forest, but they are the dominant form.

Trees far exceed all other terrestrial life forms combined in total biomass. The key to their success in the competition with other forms of vegetation is their size, which has enabled them to secure the sunlight they need and to shade out rivals. In essence, forests dominate wherever the soil, the topography, and especially the climate allow their growth. They are, however, subject to two important climatic limitations that exclude them from large areas and thereby allow other forms of vegetation to dominate.

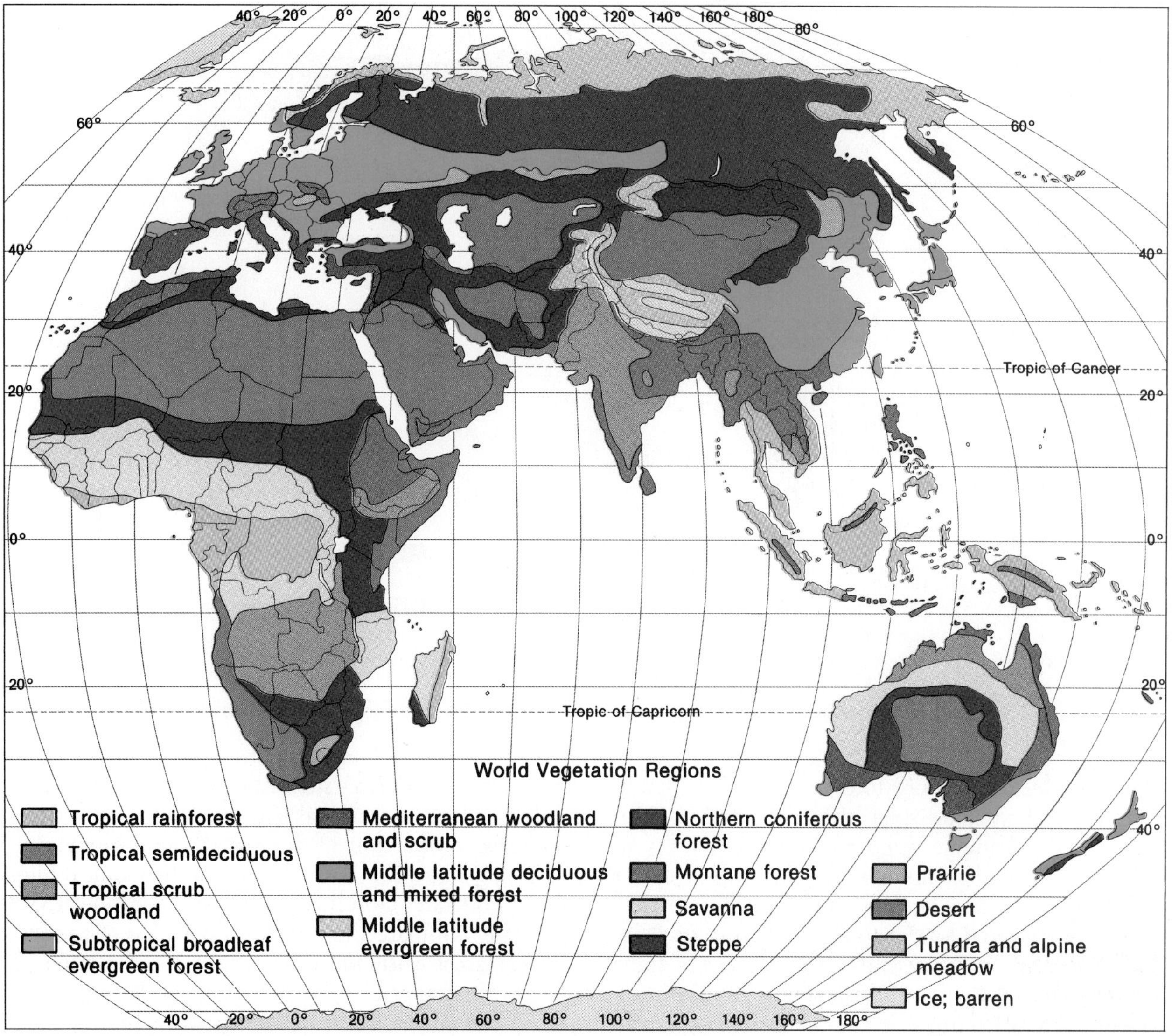

The first is that trees need a warm growing season of reasonable length. Even the most cold-tolerant forests cannot grow where the warmest month has a mean temperature below 10° C (50° F). The second and more geographically important limitation is the large water requirement of forests. Because they are large perennials, trees need substantial moisture during the growing season and at least some water even when they are dormant. This restriction excludes forests from extensive areas with perennial or seasonal soil moisture deficiencies, including most of the Low Latitude and Mid Latitude Dry Climates, the Tropical Wet and Dry Climate, and portions of the Dry Summer Subtropical Climate (see Figure 11.12.)

Trees commonly are classified on the basis of leaf characteristics. **Broadleaf** trees, as the term implies, have leaves of considerable width, while **needleleaf** trees have leaves that are very narrow and sometimes pointed. Many species of needleleaf trees bear seeds in cones and therefore are described as **conifers.** Familiar middle latitude examples of broadleaf trees include oak, maple, and hickory, and well-known needleleaf trees include pine, spruce, and fir. The broadleaf trees, as a group, are more complex and biologically more advanced than the needleleaf trees and, under favorable environmental conditions, hold the competitive edge because of their higher shade tolerance. As a result, they are the more widely distributed of the two groups. On the other hand, the

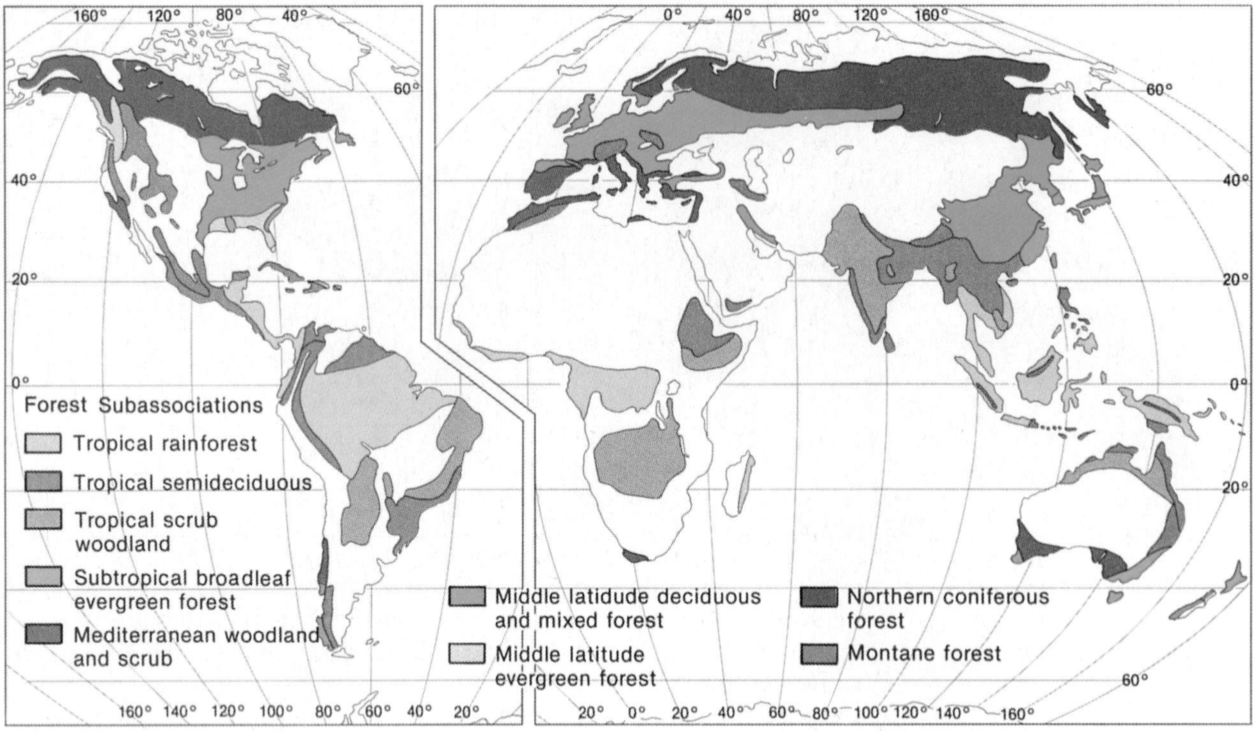

FIGURE 11.12 World distribution of forest subassociations. Unshaded areas did not have forest or woodland as their natural vegetation.

needleleaf trees are more tolerant of cold climates, drought, and infertile soils; therefore, they gain a competitive advantage when any of these conditions prevail. Trees and shrubs also are classified as either deciduous or evergreen, depending on whether they undergo a seasonal loss of foliage. The use of these two pairs of antonyms permits trees and shrubs to be separated into four groups in a simple classification system. These groups are broadleaf evergreen, broadleaf deciduous, needleleaf evergreen, and needleleaf deciduous.

For the purposes of this book, we divide forests into eight subassociations based on their structure and leaf characteristics. Most of the subassociations exist in several widely separated areas of the world that have similar climates. This underscores the dominance of climate as the major natural vegetation distribution control. The following vegetative descriptions are of climax or near-climax plant assemblages in areas undisturbed by human activities.

Tropical Rainforest

The tropical rainforest is the most equatorward and the most extensive forest subassociation. It covers 6 percent of the Earth's surface and contains up to four million plant and animal species, which collectively comprise 80 percent of the world's total biomass.

The geographical distribution of the tropical rainforest coincides closely with that of the Tropical Wet Climate (compare Figures 9.1 and 11.11). The three largest areas with this forest type are the Amazon Basin of northern South America, the Congo Basin of central Africa, and most of the East Indies.

A highly favorable climate for vegetative growth is responsible for the lushness of the rainforest (see Figure 11.13). The absence of both a cold season and a lengthy period of drought means that no major climatic stresses are present. As a result, a twelve-month old growing season exists, vegetation is dense, and the structure of the forest is determined primarily by the competition for light.

The tropical rainforest is characterized by an amazing variety of broadleaf evergreen trees. While an acre of land within the forests of the middle latitudes may contain a dozen or so tree species, an area of similar size within the rainforest may contain a hundred or more species. Some of the better known varieties include mahogany, rosewood, and ebony, all prized sources of wood for furniture; ironwood, which has wood so dense that it cannot float; balsa, the lightest wood in commercial use and familiar to makers of model planes; and cinchona, the source of quinine used to treat malaria. The numerous tree species are well integrated, so that "stands" of a single type do not occur. This characteristic has made human exploitation of the forest for specific tree species difficult and has been a factor in the preservation of extensive tracts of virgin rainforest.

Despite the great diversity of types of wood produced by the trees of the rainforest, they are surprisingly similar in structure. The forest is tall and very dense. Trees are straight and relatively slender, branching only near their crowns and typically producing an interlacing canopy of leaves and branches sufficiently dense to shade the ground completely. Larger trees often have developed buttress roots for support in the soft, damp soil. The root systems of these hygrophytic trees also are shallow because of the ready availability of near-surface moisture and because most of the limited supply of soil nutrients is concentrated very near the surface.

Because of the density of the vegetation, success in the competition for sufficient sunlight to carry on photosynthesis becomes the key to survival for the plants of the rainforest. The forest typically exhibits a layered appearance caused by the tendency of large numbers of trees to attain similar heights. This produces canopies of leaves and branches at two, three, and sometimes four distinct levels. This vertical zonation of tree heights results not so much from age differences of the trees as from variations in shade tolerance between different groups of species. A typical three-tiered rainforest contains an upper story of solitary giant trees, consisting of species that require the most sunlight, and averaging perhaps 55 meters (180 ft) in height. An intermediate story forms a continuous dense canopy 9 to 18 meters (30 to 60 ft) above the ground. A third discontinuous layer of highly shade-tolerant trees such as palms and tree ferns constitutes the lowest story (see Figure 11.14).

The floor of the tropical rainforest is surprisingly open, allowing for reasonably easy passage on foot.

FIGURE 11.13 View of the canopy of virgin tropical rainforest in the upper Amazon Basin near Iquitos, Peru. (J. L. Medeiros)

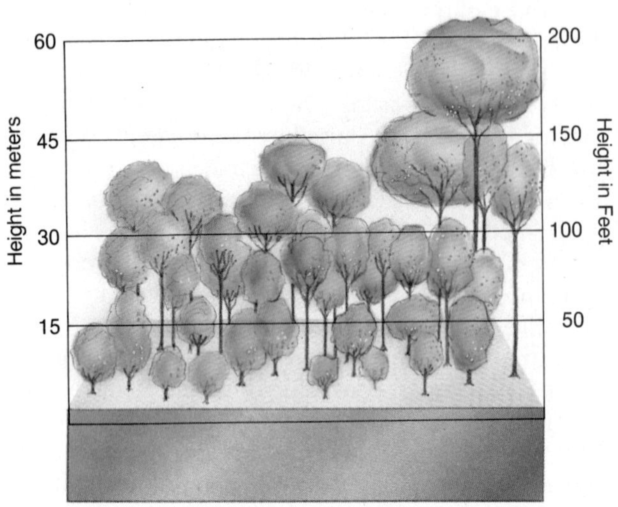

FIGURE 11.14 Vertical zonation of tree heights in a tropical rainforest. The tropical rainforest typically contains 3 or 4 distinctive layers of trees differentiated on the basis of shade tolerance. Each layer contains its own assemblage of smaller plant and animal species.

This openness results from the density of the vegetative canopy, which normally allows less than 1 percent of the available light to penetrate to the surface. The forest floor is a still, damp, and dim world, permeated with the moldy odor of decomposing vegetation and containing a scattering of only the most shade-tolerant plants. Light, life, and growth are concentrated in the treetops. Here reside a tremendous assortment of plants and animals that greatly exceed, in the number of species represented, the biota associated with any other vegetative assemblage. Many thousands of the plant and animal denizens of the tropical rainforest have not as yet even been biologically classified and cataloged. Among the plants are giant woody vines called **lianas,** which festoon the trees and whose trunks can reach diameters of up to 20 centimeters (8 in) at ground level. **Epiphytes,** or "air plants," are very common. These small plants typically grow in hollows or on the upper surfaces of horizontal branches and derive their moisture and nutrients from the air. Included in this category are mosses, bromeliads, ferns, and thousands of species of orchids whose flowers add color that helps relieve the monotony of the sea of dark green rainforest vegetation.

The tropical rainforest contains the greatest variety of animal life of any biome because of the richness of its food resources and the relative constancy of environmental conditions throughout the year. Most animals live in the treetops. Large mammals are relatively rare, but birds and monkeys fill the air with their raucous chatter, snakes slither silently along the branches, and a wide variety of other small mammals, reptiles, and amphibians are to be found. While many of the mammals are arboreal, there also are many ground-dwelling forms such as deer, wild pigs, and a variety of rodents. Most numerous of all, however, are the insects in their countless millions, exhibiting tremendous variety and far outnumbering all other species of animals combined.

As already noted, most of the tropical rainforest is free of dense undergrowth because of insufficient light. Where breaks in the canopy permit sunlight to reach the surface, though, nearly impenetrable thickets of leafy vegetation, commonly known as **jungle,** are likely to occur. Often patches of jungle result from human activities. Subsistence farmers in many rainforest areas clear small plots of land, raise crops on them for several years, and then abandon them, allowing them to turn to jungle. Settlements often are surrounded by extensive tracts of jungle that were previously farmed. Strips of dense jungle also are common along the banks of the numerous rivers that flow through the rainforest. They form because the rivers create breaks in the forest that allow sunlight to reach each bank for at least a portion of the day. Riverside strips of jungle are readily apparent to travelers on river craft—still the predominant means of transportation in many rainforest regions—and have helped foster the erroneous impression that impenetrable undergrowth is widespread throughout the rainforest.

Low-lying coasts in the tropics frequently have forests of mangrove trees, which have a tolerance for brackish water. Their stiltlike prop roots trap sediment and gradually extend the land in a seaward direction. Higher and drier coasts with sandy beaches are typically backed by forests of palm trees, which are tolerant of salty air and occasional coastal storms.

Tropical Semideciduous Forest

Progressing poleward from the equatorial regions dominated by rainforest vegetation, a climatic stress begins to appear in the form of a winter dry season of increasing length. Trees have sufficiently extensive roots to withstand relatively short periods of dry weather by tapping soil moisture at lower levels; as a consequence, rainforest vegetation extends into areas with as much as two months without rainfall. Where the dry period increases to between two and five months, however, the vegetation begins to exhibit important differences in its characteristics and becomes tropical semideciduous forest.

This forest represents what might be considered a deterioration in vegetative quantity and quality from that of the rainforest as a result of less favorable moisture conditions. The mean size of the trees diminishes, with maximum tree heights generally of 15 to 30 meters (50 to 100 ft). The variety of plants also is reduced, with perhaps 30 or 40 tree species per acre and an increasing tendency for large stands of a single type to occupy some localities. As the dry season increases in length in a poleward direction, the forest undergoes a gradual transition from broadleaf evergreen to broadleaf deciduous, as progressively more trees lose their foliage and become dormant in order to conserve moisture. The term **semideciduous** indicates that this forest contains a mixture of both evergreen and deciduous trees.

One of the most important differences between the tropical rainforest and the tropical semideciduous forest is that the latter has a significantly reduced vegetative density and a much lower biomass. Trees are not crowded so closely and, coupled with the smaller average size of the trees, the canopy of leaves and branches is much less dense and continuous (see Figure 11.15). Branches tend to start at relatively low levels, and tree crowns are much more rounded than those of the rainforest. A vertical zonation of tree heights is often still apparent, but usually only two foliage layers are present. The larger trees are most frequently deciduous because of their greater transpirational demands, while the smaller trees are predominantly evergreen. The decreased density of the forest, coupled with the deciduous habit of many of the tree species, enables a considerable amount of sunlight to reach the forest floor, especially during the dry season. This in turn supports substantial amounts of undergrowth.

Animal life within the tropical semideciduous forest is diverse, and generally is similar to that of the Tropical Rainforest. It is rich in birds, insects, and in arboreal reptiles and mammals, and many areas also contain a considerable variety of ground-dwelling vertebrate species, including large cats such as jaguars and panthers.

FIGURE 11.15 The ruins of the Mayan city of Tikal, in Guatemala, have been invaded by a mixture of broadleaf evergreen and broadleaf deciduous trees typical of the tropical semideciduous forest. (J. K. Nakata: Sight and Sound Productions © 1990)

Tropical semideciduous forests are scattered in a number of widely separated regions of the tropics. In theory this vegetation type should be found in continuous strips on both poleward margins of the tropical rainforests. The impact of dry season fires, both of natural and increasingly of human origin, however, has considerably reduced the extent of this vegetation type. Where fires have been numerous, the trees have largely been destroyed and the growth of grass encouraged. In many areas, therefore, the rainforest is bordered by the savanna grasslands discussed later in this chapter. Currently, large expanses of tropical semideciduous forest are located in Southeast Asia, India, the southern East Indies, portions of northern South America and Central America, and in southern Brazil.

Tropical Scrub Woodland

Progressing into tropical environments with a still longer dry season, the natural vegetation is further modified. Many areas with a dry season four to seven months long are vegetated with a mixture of small trees, bushes, grasses and other herbaceous (nonwoody) plants; these areas are termed tropical scrub woodland (see Figure 11.16). The term **woodland** indicates that, although woody trees and shrubs are the dominant forms of vegetation, no continuous

FIGURE 11.16 A tropical scrub woodland/savanna transition area in the Lubombo Mountains of Swaziland. (Sinclair Stammers/Science Photo Library, Photo Researchers)

tree canopy exists, and it would therefore be improper to characterize this plant assemblage as a forest. The term **scrub** refers to small, stunted trees and shrubs growing thickly together.

The vegetation within this subassociation varies greatly in composition and appearance in different areas. In some localities it consists of small individual trees 6 to 12 meters (20 – 40 ft) in height surrounded by tall grasses and other low-growing plants. Elsewhere it takes the form of dense thickets of thorny shrubs separated by grassy openings. In still other instances it consists of a variable mixture of small trees, shrubs, and grasses. In all cases, the vegetation consists predominantly of tropophytes that grow actively during the rainy season and become dormant during the long, often hot, dry season. Because of the striking alteration in appearance between the lush greens of summer and the sere buffs and browns of winter, these areas often are described as having "rain-green vegetation" (see Figure 9.9 on page 189).

The geographic distribution of the tropical scrub woodland subassociation also has been complicated and generally reduced by the destructive effects of fire so that several large but detached areas currently exist. One of the largest is in portions of southern and Southeast Asia, where thickets of bamboo often are encountered. In northern Australia, species of eucalyptus trees dominate the landscape. A large section of southern and eastern Africa is dominated by species of acacia trees. In portions of eastern Brazil, Paraguay, and northern Argentina, a mixture of thorny bushes, cacti, and trees such as the quibracho make some areas almost impassable.

Most fauna in the tropical scrub woodland is ground-dwelling. The extensive grasses and low-growing trees encourage the presence of groups of large herbivores such as zebras, elephants, and antelope in Africa, and kangaroos in Australia. These animals are in turn preyed upon by carnivores such as lions and scavengers such as hyenas.

Subtropical Broadleaf Evergreen Forest

The occurrence of a dry season of increasing length, as the subtropical high pressure belt is approached, is responsible for the transition (at least ideally) of tropical rainforest to tropical semideciduous forest and then to tropical scrub woodland. Progressing still farther poleward, the climate eventually becomes so dry that in most places trees cannot grow. A major climatically imposed gap thus divides the world's forests into tropical and nontropical segments (see Figure 11.12). The subtropical high, however, is not strong enough everywhere to produce this vegetative transition. Especially along the east coasts of continents paralleled by warm ocean currents, enough precipitation falls that trees are able to continue to flourish, resulting in a link between the tropical and nontropical forest subassociations. This subtropical forest, the subtropical broadleaf evergreen forest, differs in several important respects from those found at either higher or lower latitudes.

Within the areas occupied by this forest type, precipitation is moderate to abundant and is distributed evenly enough throughout the year that no period of severe soil moisture depletion occurs. A moderate climatic stress is produced by occasional subfreezing winter nights in most areas, but most trees are sufficiently cold tolerant to retain their foliage. In contrast to the broadleaf evergreen forests nearer the equator, the subtropical broadleaf evergreen forest contains a limited number of tree species within any given locality, and large stands of a single species are common. The forest is not particularly dense, and average tree sizes are modest. Leaves are smaller and more leathery than those of the rainforest. Some vertical zonation is normally apparent, with an upper story of larger trees that require substantial sunlight and a lower story of small, shade-tolerant trees, shrubs, and sometimes grasses. Although the exact vegetative composition varies considerably from place to place, the dominant tree species in many areas include live oaks, magnolia, and laurel. The trees often are draped with lianas and epiphytes. One epiphyte familiar to residents of the southeastern United States is Spanish moss (see Figure 11.17).

The most extensive subtropical broadleaf evergreen forests grow in eastern Asia, particularly in eastern China, southern Korea, and southern Japan. Other areas of significance include portions of the extreme southeastern United States, New Zealand, and southeastern Australia.

Mediterranean Woodland and Scrub

On the poleward sides of the subtropical highs, the climate again becomes moist enough to support the growth of trees, and forests return as the dominant vegetation type. Once again, the transition from dryland vegetation to humid forest is not abrupt, but proceeds through a series of intermediate stages. These stages are well illustrated within the Mediterranean woodland and scrub subassociation.

This subassociation occurs mostly on the west sides of continents and is closely associated with areas having a Dry Summer Subtropical Climate. Included are southern and central California, the Mediterranean Basin region, central Chile, the southern tip of Africa, and much of southern Australia. These areas have dry, warm summers dominated by the poleward-shifting subtropical high and mild, humid winters influenced by traveling cyclonic storms within the westerly wind belt. The dry summer season, usually lasting from four to seven months, is the major climatic stress factor. It is largely responsible for producing a vegetative subassociation dominated by shrubs and small trees that somewhat resembles the tropical scrub woodland.

FIGURE 11.17 Live oak trees draped with Spanish moss are a common sight in the subtropical broadleaf evergreen forest of the southeastern United States. (Visuals Unlimited/© John D. Cunningham)

FIGURE 11.18 The chaparral vegetation on this California hillside includes manzanita, digger pine, and chamise. (J. L. Medeiros)

The composition of the constituent plant species varies greatly between the widely separated regions of the Mediterranean woodland and scrub subassociation. Despite their biological differences, however, a remarkable physical similarity in vegetative forms and community organizations is exhibited in response to similar environmental conditions. Most trees and shrubs are classified as **sclerophylls** (from the Greek *sclero*, or "hard," and *phyllos*, or "leaf"). They have small, hard, thick, leathery leaves that minimize transpirational losses. Other common xerophytic adaptations include deep root systems, viscous fluids, waxy leaf coatings, and thick bark. Many plants also have thorns as a protection from grazing animals.

Better vegetated regions typically display one of two vegetation assemblages. **Mediterranean Woodland** consists of a parklike combination of scattered small- to medium-sized oak trees surrounded by grasses and flowering herbaceous plants. An assemblage known as **Chaparral** in California and **Maquis** in the Mediterranean region is somewhat more common (see Figure 11.18). It consists of an often dense cover of woody shrubs and small trees ranging up to 6 meters (20 ft) in height. Species of oak most frequently predominate, although in Australia eucalyptus and acacia are most frequent and in California manzanita shrubs are abundant.

Climax vegetation is extremely rare within this subassociation because of a long history of human occupancy and exploitation. The removal of trees for lumber and firewood and overgrazing by sheep and goats, especially in extensive areas within the Mediterranean Basin, have produced a serious decline in both the quality and quantity of the vegetation. Fire, too, has played a major destructive role. Brush fires in the tinder-dry vegetation are an annual late summer occurrence in many regions, but are probably most notorious in California and Australia. Today, those areas most adversely affected by overgrazing and fire are covered by a dense growth of thorny shrubs commonly less than a meter (3 ft) in height.

The Mediterranean woodland and scrub subassociation at one time supported a varied fauna, with numerous herbivores such as ground squirrels, deer, and elk, and with wolves and mountain lions as predators. The larger animals, however, have largely disappeared as a result of a long history of human occupancy. Smaller animals that coexist well with humans, such as birds and rodents, are now dominant.

Mid Latitude Deciduous and Mixed Forest

Progressing poleward from the subtropics into the middle latitudes, the dryness produced by the subtropical highs is left behind, and conditions become more favorable for forest growth. At the same time, however, a new climatic stress begins to appear in the form of a cold season of increasing length and severity. The result is an increase in cold-tolerant needleleaf evergreen trees and the development of a period of vegetative dormancy, during which the cold-sensitive broadleaf trees that dominate the forest must reduce their moisture content and shed their leaves to survive.

The mid latitude deciduous and mixed forest, as this subassociation is often termed, dominates areas with warm summers, subfreezing periods in winter, and moderate to abundant precipitation well distributed throughout the year. It is the most extensive subassociation in the middle latitudes. The greatest expanses occur within the northern hemisphere, which, unlike the southern hemisphere, has large land masses with substantial annual temperature ranges in the middle latitudes. Three large regions of this forest type within the northern hemisphere are the eastern United States, much of Europe and a portion of the Asiatic Soviet Union, and east-central mainland Asia. In the southern hemisphere, its only significant area of occurrence is in the southern Andes of Chile and Argentina.

During the summer season, the closed canopy of broadleaf trees bears some similarity to that of the tropical rainforest (see Figure 11.19). Important differences, though, exist between the characteristics of the two forest types. One difference is the much lower vegetative biomass of the middle latitude forests. Mature trees reach only intermediate heights of perhaps 30 to 45 meters (100 to 150 ft). Individual trees are quite variable in size, and size is related more to age than to species type. No clear-cut vertical zonation of heights is normally observable, although there frequently is an understory of shade-tolerant trees and shrubs. The variety of tree species is rather limited in any one area, and often only three or four are locally dominant. Likewise, fewer vines and epiphytes use the trees for support, although certain climbing plants such as the honeysuckle, poison ivy, and Virginia creeper of the eastern United States may be locally abundant. Although the forest is dense enough to produce a closed canopy, even in summer patches of sunlight penetrate to the forest floor, which is not dark and gloomy like that of the rainforest. The intensity of shade does, however, inhibit the growth of near-surface vegetation.

Probably the major difference between the tropical rainforest and the mid latitude deciduous and mixed forest is that the former is evergreen and the latter experiences a seasonal loss of most foliage. While the rainforest is constant and rather timeless in appearance, the deciduous forest undergoes a seasonal transformation, in visual terms, between life and death. Because the trees spend much time in a dormant state, growth is much slower than in the tropics.

FIGURE 11.19 Autumn begins to color the trees in a rural portion of the mid latitude deciduous and mixed forest in the eastern United States. (Richard Jacobs/© JLM Visuals)

THE GYPSY MOTH INVASION OF NORTH AMERICA[1]

The gypsy moth, *Lymantria dispar* (L.), was introduced to North America when M. Henri Trouvelot accidentally released the insects in Medford, Massachusetts, sometime in 1868 or 1869. Trouvelot, an amateur entomologist experimenting with silk production, was aware of the dangers inherent with nonnative species but he could not have fully anticipated the scale of the environmental, economic and sociological consequences that followed his mishap. The gypsy moth is a destructive pest of temperate mixed forests with a strong preference for oak leaves as a food source. Perhaps because of a lack of predatory controls, the insect's population waxes and wanes in boom and bust cycles. Boom years, such as 1981, see the outbreak of billions of caterpillars, the destruction of thousands of hectares of forest, and inconvenience and losses suffered by many people.

Although insect life stages have been recorded across the continent, it is in the Northeast where populations currently reach destructive proportions (see Figure 11.20). Two agents of diffusion can be identified as contributing to the expansion of its range: the insect itself and human beings. As North American gypsy moth females are flightless, spread occurs primarily in the first caterpillar stage when it floats on a silk thread in search of suitable food. This method of transport usually restricts movement to a few kilometers at most. Long distance movement occurs when fertile females or egg masses are inadvertently transported on vehicles by human beings.

Available historical data have revealed irregular pulses in the invasion process: a high rate of expansion from 1900–1915, a period of slow expansion from 1916–1965,

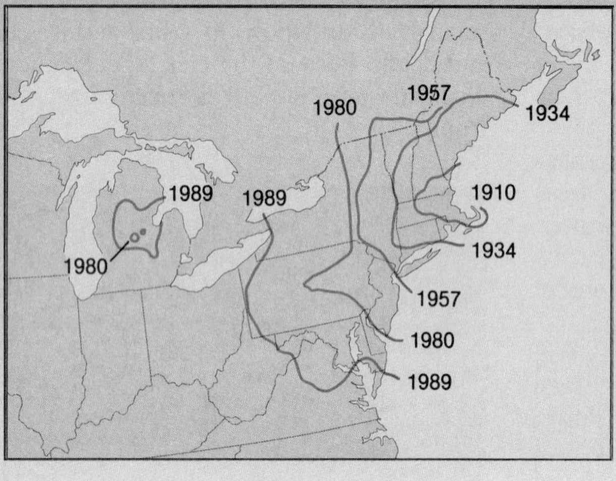

FIGURE 11.20 The expansion in the North American range of the gypsy moth.

Diversity in the mix of tree species is sufficient to give a distinctive character to the forests of different areas. In the southeastern United States, oak, hickory, and poplar are among the dominant tree types. Farther north, maple, birch, and beech are dominant. Many of these trees are noted for their brilliant autumn colors and are an important tourist attraction in New England, the Canadian Maritimes and the Great Lakes region. In Eurasia, species of oak, beech, and ash are most common, and in southern Chile, beech trees are dominant.

Coniferous needleleaf evergreen trees are scattered throughout many portions of the mid latitude deciduous and mixed forest. These trees generally

FIGURE 11.21 The effectiveness of aerial spraying operations is evident in this photo showing both treated and untreated forest plots in Pennsylvania following a gypsy moth outbreak. (Noel F. Schneeberger)

and a greatly accelerated rate since 1966. Explanations for the different rates may be related to geographical and temporal changes in mean minimum temperature, the varying distribution of forest species composition, and differing levels of human intervention. Increased recreational use of forests may account for the increased rate of expansion since 1965.

Suppression and control have been attempted since the time of the initial release. Methods have varied from the manual removal of caterpillars and egg masses to widespread aerial spraying with broad-spectrum insecticides. In the 1940s an attempt was made to halt the spread at the Hudson River. The intensive eradication campaign used the pesticide DDT and may have slowed, but eventually did not prevent, the expansion of the insect's range into Pennsylvania, New Jersey, and Maryland. Today, specific pesticides are applied to carefully selected target areas (see Figure 11.21). The development and use of geographic information systems has been beneficial in monitoring insect populations, selecting treatment areas, matching treatments to environmental conditions, and assessing the effectiveness of alternative suppression methods.

The gypsy moth currently infests only one-fifth of its potential range, and there is no evidence that suppression activities have reduced the rate of expansion. As the insect moves south it encounters more favorable climatic conditions and prime habitat, much of it of great economic value. Humans will continue to have to adapt to the dynamics of the insect rather than attempt to remove the threat from our environment.

[1]This focus box was written by Gregory Elmes, Associate Professor of Geography at West Virginia University.

are out-competed by the broadleafs, but are more tolerant of adverse conditions such as cold temperatures and rocky, sandy, or highly acidic soils. They therefore tend to be more numerous and even locally dominant in mountainous areas within the higher Appalachians and the Alps and in sandy areas such as the Pine Barrens of southern New Jersey and the Landes district of southwestern France. Needleleaf trees also become increasingly abundant near the northern border of the forests in North America and Eurasia, as these areas are transitional to the adjacent northern coniferous forest.

The mid latitude deciduous and mixed forest has been greatly altered and in many cases completely

FIGURE 11.22 *(left)* Avenue of the Giants in Redwood National Park, California. *(right)* An open forest of pine trees grows in sandy soil in Florida. *(left:* Redwood Empire Association; *right:* JLM Visuals)

removed by humans. Most regions containing this forest subassociation have dense human populations and high agricultural and industrial productivities, conditions that are incompatible with extensive forest coverage. Large areas now consist of farmland, roadways, and urban sprawl. The removal of the forests is most complete in eastern Asia, where only the most rugged and inaccessible regions contain substantial numbers of trees. Only in the United States is the majority of the surface still forested, and nearly all is in secondary growth.

Animal life in the mid latitude deciduous and mixed forest also has been influenced greatly by humans. Animals well suited for coexistence with humans, such as deer, squirrels, rodents, and a great variety of birds, are still abundant in many areas. Other animals that once inhabited the forests, such as bears, wild boar, wolves, and several species of wild cats, are now rare or absent in most regions.

Mid Latitude Evergreen Forest

Broadleaf trees predominate over needleleaf trees in most of the middle latitudes. Two large areas, both in North America and primarily within the United States, however, contain needleleaf evergreen forests. The reasons for the existence of these forests are unclear, because they do not represent the climax assemblages that would be expected within their respective areas based on climatic conditions. It is believed that fire may contribute to their maintenance.

One of these two forests is the **west coast coniferous forest** (see Figure 11.22). The only fully developed forest of this type extends along the west coast of North America from northern California to southern Alaska. In other mountainous Marine Climate locations such as southern Chile and western Norway, mixed broadleaf and needleleaf forests exist that appear to represent a transitional phase between this and the mid latitude deciduous and mixed forest.

The west coast coniferous forest contains the world's most impressive assemblage of needleleaf evergreen trees. The forest has a high biomass, probably second only to the tropical rainforest. Trees are very large, with tall, straight trunks. Heights can reach well over 60 meters (200 ft), with some coastal redwoods more than 90 meters (300 ft) in height and 4.5 meters (14 ft) in diameter. The size of these trees exceeds that of all other living organisms on Earth.

The largest and most dense stands occur on the moist western (windward) sides of the Cascades and Coast Mountains of British Columbia. In these areas Douglas fir, western hemlock, and Sitka spruce are among the dominant species. In the drier lands to the east of the mountains are less dense forests of smaller Douglas fir, Ponderosa pine, and white pine. Near the coasts where hard freezes are rare, there is some admixture of broadleaf deciduous and broadleaf evergreen trees.

An extensive portion of the southeastern United States stretching from eastern Virginia south to Florida and west to Texas is covered by another forest composed mostly of species of pines (see Figure 11.22). Known as the **southern pine forest,** it is closely associated with the sandy, drought-prone, and rather infertile soils of the Atlantic and Gulf coastal plain. Its existence probably can be attributed to the resistance of pine trees to droughty soil conditions, their low soil nutrient demands, and their ability to withstand ground fires.

The pine forests occupy well-drained sandy surfaces that are situated between stream valleys. The most common tree species include longleaf, shortleaf, pitch, and loblolly pines. These trees are valuable sources of lumber, pulpwood, and naval stores (turpentine and resin). Within the swampy stream valleys, hardwoods such as live oak, bald cypress, tupelo, and red gum predominate.

Northern Coniferous Forest

The most poleward forest subassociation is the **northern coniferous forest,** also known by its Russian name **Taiga.** It is second only to the tropical rainforest in total areal extent. The northern coniferous forest primarily occupies regions having a Subarctic Climate with short mild summers and long frigid winters, but the forest's southernmost margins enter Humid Continental Climate regions. It exists within two giant areas, both in the northern hemisphere between approximately 47° N and 70° N. These are the northern interior of North America, extending from Alaska to Labrador, and the even larger interior of northern Eurasia, reaching from Scandinavia to the Pacific Ocean. Logging activities have removed or modified the forest over portions of its southern range. The central and northern portions of the forest, however, have been influenced relatively little by human activities so that this forest subassociation as a whole remains more nearly in a natural state than any other.

The northern coniferous forest is dominated by cold-tolerant needleleaf evergreen trees (see Figure 11.23). These trees have an advantage over deciduous varieties because they need not grow new foliage each spring, but can instead begin adding to their growth as soon as warm weather arrives. The needleleaf trees also are sufficiently drought resistant to withstand the long winter period of physiological dryness imposed by the freezing of soil moisture.

Only a small number of tree species have developed the ability to cope with the severe climatic stresses imposed by the short growing season and extremely low winter temperatures of these areas. As

FIGURE 11.23 Taiga forest with muskeg in the foreground. Gore Range, Colorado. (John S. Shelton)

a result, the northern coniferous forest is biologically the least complex forest subassociation. It is common for only one or two tree species to dominate over large areas, and the forest usually is locally uniform in height and density. Spruce and fir trees are most abundant for the region as a whole. In the northern United States, these trees are mixed liberally with pine trees, especially white pine, red pine, and jack pine. In western Eurasia, Scotch pine and stone pine are locally abundant. Eastern Siberia, the coldest region of all, is unique in being dominated by two species of needleleaf deciduous larch trees. Apparently winter temperatures here are too extreme even for most cold-tolerant needleleaf evergreens, and the larch becomes the most successful competitor.

The trees of the northern coniferous forest are moderate in size near their southern margins, but become progressively smaller poleward. Most attain heights of 15 meters (50 ft) or less. Near their northern limits, they are dwarfed by the shortness of the growing season, and mature specimens as much as a century old may be only a meter or two tall. The density of the forest also tends to diminish in a poleward direction. The forest floor typically is covered with a mat of lichens and mosses. Especially dominant in northern Canada is a pale-colored species of lichen referred to as "caribou moss."

During dry periods in summer, the forest can become highly susceptible to lightning-set forest fires. Forest openings produced by fires normally are occupied by certain broadleaf deciduous trees, including birch, aspen, alder, and poplar. These deciduous patches are conspicuous from even great distances because of their distinctively lighter color. They eventually are replaced by the climax needleleaf trees, but the pace of change is very slow in this cold environment, and the transition can take centuries.

In addition to forest fires, breaks in the continuity of the forest have resulted from the lasting effects of the continental glaciers that covered parts of the area as little as 5,000 years ago (see Chapter 18). Their legacy is a landscape dotted by myriad lakes of all sizes. Lakes are relatively ephemeral features on the landscape; as they gradually fill with sediment, they are occupied by a succession of hygrophytic vegetation assemblages. Many small lakes of glacial origin within this region already have been converted to **muskegs**—swampy areas covered with mosses, swamp grasses, shrubs, and scattered water-tolerant trees such as larches. In Canada and Scandinavia, some severely glaciated areas still have patches of exposed bedrock on which little vegetation grows.

The diversity of fauna in the northern coniferous forest is limited by severe winters, a lack of food, and a limited variety of habitats. Most evident are large herbivores, especially species of deer, which are often found in large herds. Carnivores include wolves, weasels, mink, and bears. Birds are most numerous in the summer, when the swarms of insects that breed in the numerous lakes and marshes provide a plentiful supply of food. Most bird species migrate south for the winter.

Grassland Association

The **grasses** are plants with bladelike, opposing leaves connected to a jointed stem by a sheathlike attachment. A great number of species of grasses exist, including many that are commercially important. Both directly and indirectly, the members of the grass family are largely responsible for producing the basic foods on which humanity depends. All of the cereal grains such as rice, corn, wheat, oats, and barley are grasses. In addition, most animals raised for meat and dairy products feed primarily or exclusively on grasses.

Grasses grow on virtually all vegetated portions of the Earth's surface and are the dominant form of natural vegetation on more than a third of the world's land area (see Figure 11.24). Their widespread geographical distribution results both from their hardiness and their diversity. Grasses are abundant from the equator to the tundra and from swamps to deserts. A key to their ability to survive in environments with severe periodic climatic stresses is the fact that many grasses are annuals and can simply regrow from seeds during favorable times. Numerous other grass species, unlike trees, can grow new shoots from their subsurface stem sections after the above-ground portions have been killed by fire or adverse weather conditions, or have been removed by grazing.

The chief competitive disadvantage of grasses is, of course, their small size. Most grasses attain heights of only a meter or two. Grasses are not sufficiently xerophytic to dominate the vegetation in most arid regions, although they are well represented in many semidesert areas. As a result, most of the world's grasslands are situated on the margins of forests—in those areas either too wet or too dry for trees. This generally includes regions with subhumid, semiarid,

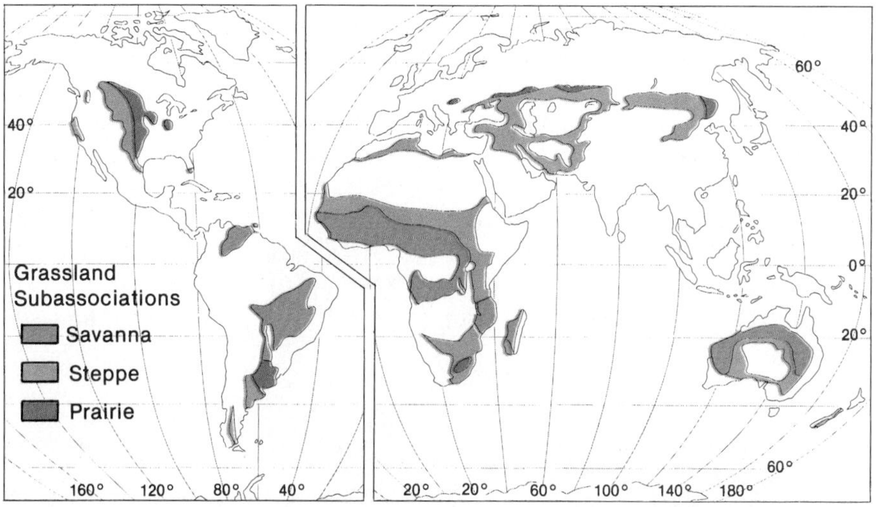

FIGURE 11.24 Global distribution of grassland subassociations.

and seasonally dry climates, as well as more restricted areas that experience poor drainage or periodic flooding.

Over the last several millennia, the world's grasslands have been expanding gradually in area as they have encroached upon their forest margins. This trend may be partially the result of post-Pleistocene climatic changes, but is believed to be caused primarily by human activities such as the gathering of firewood, overgrazing by domesticated animals, and especially the use of fire. Expansion has been occurring for such a long time that researchers are unsure about the original extent of the world's "natural" grasslands. It is certain that, especially in the tropics and subtropics, both grasslands and deserts will continue to expand at the expense of forests in the near future.

The grassland vegetation association is divided into three subassociations: the savanna, steppe, and prairie. The savanna grasslands occur in the low latitudes, mostly within drier portions of the Tropical Wet and Dry Climate. They will be discussed first. Next to be examined are the steppe grasslands, which are associated with semiarid climates in both the low and middle latitudes. The prairie, which is restricted largely to subhumid portions of the middle latitudes, is covered last.

Savanna

The savanna is unique among the plant groupings discussed in this chapter in that it contains, by definition, a mixture of two greatly different types of vegetation. It may be described as a tropical grassland that contains scattered trees or shrubs (see Figure 11.25). In theory, the savannas are situated on the dry (normally poleward) margins of the tropical scrub woodland subassociation, with the boundary between the two being located where grass becomes areally dominant over trees and shrubs. They generally occupy the drier portions of the Tropical Wet and Dry Climate, where the winter dry season lasts between six and eight months. The actual boundary between the tropical scrub woodland and the savanna is, of course, very difficult to define precisely because it is subject to individual interpretation and is determined not only by the length of the dry season but also by the effects of fire, grazing animals, and human activities. The dry margins of the savannas occur where the climate becomes semiarid and the mixture of tall grasses and woody plants is replaced by short tropical steppe grasses.

The most extensive savannas are located on the continents of Africa, South America, and Australia. The African savannas are the largest and best known. Most occur in a broad arc surrounding the tropical rainforest of the Congo Basin to the north, east, and south. These savannas, especially those of eastern Africa, are frequently seen on "big game" movies or television documentaries because of their great herds of large animals. In South America, extensive savannas are located in Venezuela and Colombia as well as in southern Brazil and Paraguay. Large portions of northern, eastern, and western Australia also are of-

FIGURE 11.25 The East African savanna during the dry season. Scattered acacia trees and a solitary giant baobob stand above a ground cover of dry bunch grasses.

ten considered to be savanna, although the mix of plants here is somewhat different. Other smaller but noteworthy savannas exist in much of western India, western Madagascar, and the Everglades of southern Florida.

Savanna grasses thrive because they consist of species that become dormant during the dry season. The above-ground portions of the plants die, but new shoots grow with the onset of the summer rains to turn the landscape lush and green. Great variations in grass heights exist in different areas, ranging from 0.3 meters (1 ft) to more than 4 meters (12 ft). Most savannas are dominated by coarse, short-bladed bunch grasses that grow in dense tufts, with intervening patches of bare earth. They have well-developed root systems that support rapid growth during the rainy season.

Trees may be scattered throughout the savanna as individuals to produce a parklike appearance, or they may exist in groups. They consist of a mixture of stunted broadleaf deciduous and broadleaf evergreen species, which, with their thick bark, lack of lower branches, and flattened or umbrella-shaped crowns, are highly similar to those of the tropical scrub woodlands. Trees do not dominate because the seasonal rainfall distribution is too uneven to readily support large perennials with year-round moisture requirements. In general, the number of trees declines toward drier areas, as does grass height.

Dense strips of trees commonly occur along the banks of streams. Their presence indicates that when water is available, trees will assert dominance over grasses. Often, the trees along either bank are sufficiently large that their foliage intermeshes, enclosing the stream within a tunnel of vegetation. These strips of forest are termed **galeria,** the Italian word for tunnel.

In some savanna areas, the woody vegetation is in the form of shrubs rather than trees. The shrubs are fire-resistant and often thorny, and tend to grow in

FIGURE 11.26 Zebras and giraffes are two of the large herbivores native to the East African savanna. (© Gregory Dimijian/Photo Researchers)

dense thickets. Most shrubs are xerophytic broadleaf evergreens with small, hard leaves.

The dominant fauna of the savanna are the large herbivores. These animals are most numerous and best known in Africa, where they include elephants, zebras, wildebeests, antelope, giraffes, and many others (see Figure 11.26). Large carnivores, of which the lion is best known, also thrive here. A diversity of smaller animals, especially birds and rodents, occupy the savanna as well.

Steppe

The steppe comprises the short grasslands of the low and middle latitudes. It is closely associated with semiarid climates. Many geographers use the term *steppe* to refer only to the short grasslands of the middle latitudes and, indeed, it is there that they attain their greatest degree of distinctiveness as a separate vegetation subassociation. Large areas of the tropics, however, also are covered primarily by short grasses because of similarly marginal amounts of rainfall; for this reason, it seems justifiable to include them in this category.

The low latitude steppes are located in regions with a dry, hot climate typically punctuated by a brief, often intense, rainy season lasting two to four months. Ideally, they occupy the semiarid margins of the subtropical deserts. Larger areas include the borders of the Sahara and Kalahari Deserts in Africa and portions of north-central Australia. Smaller tropical steppes occur in several localities, including northwestern Argentina and northern Venezuela.

The vegetation of these regions consists predominantly of a short grass cover that lies dormant for much of the year. The grass is similar to that of the savanna in that it is tufted, with patches of bare earth between. Varying amounts of often thorny broadleaf deciduous shrubs are interspersed with the grasses and may be locally dominant.

The steppes of the middle latitudes, like their low latitude counterparts, generally occupy the semiarid margins of deserts. On their humid sides, they grade into either tall-grass prairies or forests. Two principal areas exist; one occupies the Great Plains of the United States and south-central Canada, the other extends more than 7,200 kilometers (4,500 mi) from the north shore of the Black Sea east through the southern Soviet Union to northern China. Additional steppe areas occur in interior portions of the Middle East and southwestern Asia. Only limited expanses of steppe exist in the southern hemisphere; these include eastern New Zealand and the South African veldt, both of which are transitional to prairie.

Most steppe lands of the middle latitudes consist of vast expanses of flat or gently rolling terrain covered almost exclusively with short grasses less than half a meter (2 ft) high (see Figure 11.27). In some areas the grass is dense and continuous, and the combination of soil and matted grass roots forms a rough

FIGURE 11.27 Short steppe grasses and unrestricted visibilities are typical of the North American Great Plains. (C. Donald Ahrens)

sod that protects the surface from erosion. Important sod-forming species in the American Great Plains include buffalo grass and bluestem wheatgrass. In other areas, especially toward the desert margins, the vegetative cover consists predominantly of bunch grasses such as feather grass. Most grasses are perennials that lie dormant during the winter and sprout vigorously in the spring. During the summer, high temperatures and evapotranspiration rates produce severe soil moisture deficits, and grass growth is greatly reduced.

Along with the grasses are various flowering annuals as well as some shrubs. Shrubs are especially common on rocky hillsides where the decreased density of grasses reduces the risk of destructive fires. An increase in xerophytic shrubs also is evident toward the dry margins of the steppes and in regions overgrazed by domesticated animals. In most cases, trees are lacking in the middle latitude steppes except along the banks of streams, where such water-tolerant species as cottonwoods and willows are found.

Prairie

The prairie is one of the smallest and most distinctive of the world's major vegetation groups. It has also been more nearly removed by human activities than any other plant subassociation. The underlying soils have proven so fertile that virtually all prairie areas have been plowed up and converted to cropland. Remaining small patches of prairie along roadsides, the edges of fields, and on steeper slopes can provide an idea of the original appearance of the vegetation, but they certainly cannot convey an adequate impression of the original immensities of shoulder-high grasses extending to the horizon (see Figure 11.28).

The prairie regions are located in the middle latitudes, mostly near the dry (subhumid) margins of the Humid Continental Climate. The largest area is centered in the midwestern United States, extending in a rough triangle from Indiana northwestward to southern Alberta and southward to eastern Texas.

FIGURE 11.28 Tucker Prairie, a protected tall grass prairie in central Missouri. (Marbut Memorial Slide Collection/American Society of Agronomy)

Smaller patches within the United States include the "Black Belt" of western Alabama and the Palouse country of eastern Washington, which is transitional to steppe. A second extensive area of prairie is the pampa of South America, located within northeastern Argentina, Uruguay, and southern Paraguay and Brazil. A small but important prairie region also occupies the Danube plain in Hungary, and vegetation transitional to prairie is found along the humid northern margin of the steppe lands of central Eurasia in a narrow discontinuous band stretching from the Ukraine to Manchuria in northern China.

The predominant vegetation of the prairie was a mixture of tall grasses that were sufficiently dense to form a continuous sod cover over large areas. The grasses were mostly 0.6 to 1.2 meters (2 to 4 ft) high, but were reputedly as tall as 3 meters (10 ft) near the eastern margin of the prairie in Indiana and Illinois. They diminished in both height and density toward the drier steppe margins. A variety of annual and perennial species contributed to the grass cover, and in the American Midwest, species of bluestem were predominant. Interspersed with the grasses were a large number and variety of herbaceous flowering plants that produced a colorful landscape, especially in late spring. Particularly common in the American Midwest were the sunflower and black-eyed Susan.

On the original prairie, trees existed only in scattered patches near the forest margins and, somewhat in galeria fashion, in strips along streams. With settlement, large numbers of trees were planted for decoration, shade, and as windbreaks; these generally have thrived without special care and often have attained large dimensions. The reason for the original lack of trees therefore poses a rather perplexing problem. Most theories concerning the absence of trees revolve around the following four ideas:

1. Fires of both natural and human origin may have been sufficiently frequent to destroy tree seedlings while doing little harm to the herbaceous vegetation.
2. The high lime content of most prairie soils may have encouraged the growth of grasses while suppressing tree growth. This factor is known to have been instrumental in the dominance of grasses in some areas, such as the Alabama Black Belt.
3. The large herds of bison and other grazing animals that ranged across these areas may have trampled and eaten vulnerable tree seedlings.
4. The climate may have been significantly drier during the Pleistocene epoch, becoming humid enough to support trees so recently that a relict grassland cover still persisted. In support of this idea, it is known that forests were gradually encroaching on the eastern margins of the American prairie at the time of settlement.

It is, of course, quite possible that the presence of a grass cover resulted from a combination of these factors.

At present, the prairies consist largely of a checkerboard of fields of crops. It should not be surprising that cereal grains such as wheat and corn are currently most important, because these domesticated tall grasses are biologically quite similar to the wild tall grasses that once grew so profusely in the prairie.

The dominant fauna of both the steppe and the prairie are large migratory grazing animals. In North America, vast herds of bison once roamed these grasslands, and deer and pronghorn antelope also were numerous. In Eurasia, saiga antelope and wild horses were among the dominant animals, and in Australia, kangaroos were numerous. These herbivores have now been largely replaced by domesticated animals such as cattle and sheep.

Desert Association

Desert vegetation is closely associated with the world's arid climate regions. These regions are located mostly in the subtropics, but also occur in portions of the middle latitudes (see Figure 11.29). The largest desert area by far extends across the Sahara of North Africa, through the Arabian Peninsula, and well into southwestern Asia. Other important subtropical deserts include the Kalahari of southwestern Africa, the Mojave and Sonoran Deserts of the southwestern United States and adjacent Mexico, the Atacama of western South America, and the Australian Desert. Within the middle latitudes, deserts occupy drier portions of the American Great Basin, southern and western Argentina, and much of Central Asia.

Despite their aridity, most deserts contain a variety of vegetation (see Figure 11.30). The one-quarter, or less, of all deserts that are essentially devoid of plants occur mostly in areas that are extremely dry, have shifting sands, or have steep, rocky slopes. Desert vegetation typically consists of a rather diverse mixture of small, predominantly xerophytic plants that are especially adapted to survive in

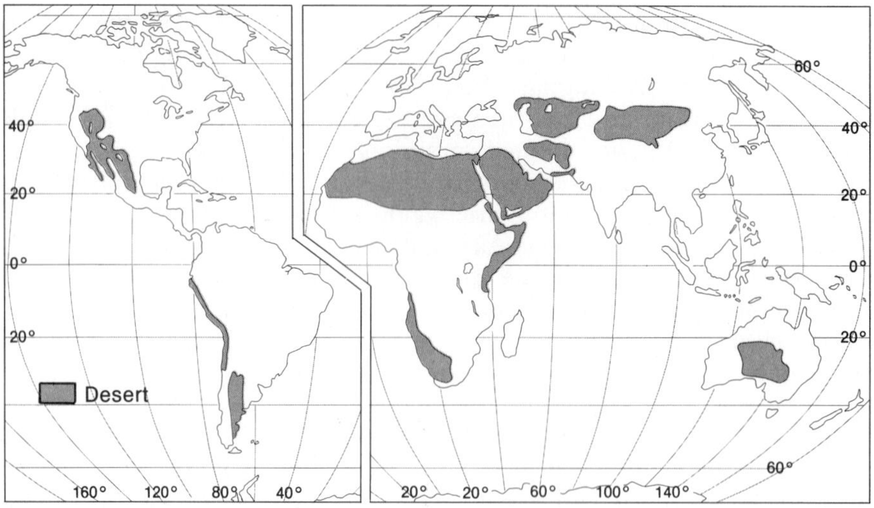

FIGURE 11.29 Global distribution of desert regions.

an arid environment. Desert plants also must be tolerant of extremes of both heat and cold. Species vary greatly in different areas, but the environmental similarity of these areas leads to the development of plant groups that are similar in general appearance and characteristics.

FIGURE 11.30 A variety of diverse forms of vegetation grows in the desert of southern Arizona. (David Muench/H. Armstrong Roberts)

Perhaps the most basic attribute of desert vegetation is its wide spacing. Most other plant assemblages produce a more or less continuous ground cover. The lack of surface moisture does not permit this in deserts, despite the xerophytic nature of the plants. In general, the spacing between plants increases with aridity, but even in better watered areas, most of the surface is bare. Two other basic characteristics of desert plants are small size and slow growth. Most plants attain heights of only a meter (3 ft) or less, and many take years to reach maturity. The vegetation of the desert can almost be viewed as progressing through life in slow motion, so that the desert normally projects a somewhat timeless image.

Desert plants can be divided into several botanical groups. Most widespread and important are shrubs of both evergreen and deciduous varieties. These are very common in the southwestern United States, where familiar species include sagebrush and mesquite.

The best-known desert plants are the leafless evergreen herbs, which include most cacti. It is commonly believed that cacti are the dominant plants in deserts. Actually, they are native only to the Americas and are abundant in only a relatively few localities. Their striking appearance and often rather large size, however, cause them to stand out from the more mundane desert plants, and this probably accounts for their wide renown.

A third group of desert plants are the **sclerophylls,** broadleaf evergreen trees or shrubs with

THE MESQUITE INVASION OF THE AMERICAN SOUTHWEST

When the early explorers and settlers entered the southwestern United States during the sixteenth through the nineteenth centuries, they encountered a region covered largely by native grasses. These grasses provided sustenance for a variety of animals, including deer, antelope, and bison. The early farmers and ranchers also used these grasses as forage for their livestock. Woody plants were present, but in most places they were widely scattered and were often concentrated in stream valleys.

During the past century, however, there has been an almost explosive increase of trees and shrubs throughout the Southwest. This has not only considerably reduced the grazing potential of most areas, but has altered the basic vegetative composition of the entire region.

The plant that has most greatly increased its range during the past century is mesquite. Mesquite is a spiny, deciduous tree or shrub with an extensive root system that makes it highly drought resistant (see Figure 11.31). Originally confined largely to draws and washes, it has invaded extensive areas of flatter surfaces from Texas and Oklahoma west to central Nevada. Today, a low, scrubby covering of mesquite occupies some 17 million hectares (70 million acres) of rangeland in the southwestern United States.

A number of researchers have attempted to determine the cause of mesquite's rapid expansion in range.

FIGURE 11.31 A mesquite tree near Phoenix, Arizona. (Ralph Scott)

They have concluded that it is largely a culturally induced phenomenon resulting from changes in land-use patterns following European and American settlement of the region. Two factors are believed to be crucial:

1. *Overgrazing and trampling.* The wild herbivores of the Southwest did not overly tax the recuperative ability of the grasses, because their numbers were kept in check by natural predators and Native American hunters. The grasses therefore were able to hold their own against the encroachment of mesquite or other woody plants. The cattle and sheep ranchers of the nineteenth and early twentieth centuries, however, fenced in their lands and often overstocked them, so that the grass cover was reduced greatly by overgrazing and trampling. This opened up large areas to colonization by mesquite and other unpalatable shrubs such as sagebrush and creosote bush.
2. *Suppression of fires.* Summer fires set by lightning were once common on the dry grasslands of the Southwest. The Native American inhabitants also set fires both to promote fresh grass growth and to drive game during hunts. These fires did little permanent damage to the grasses, which quickly sprouted new shoots from their undamaged root systems, but they were deadly to small trees and shrubs. The European and American farmers and ranchers, however, viewed fire as their enemy and did their best to suppress it. In addition, overgrazing in many areas reduced the supply of dry vegetative matter on which wildfires fed. The reduction in the frequency of fires generally enables mesquite and other woody plants to reach maturity.

The expansion of mesquite shrubland at the expense of grassland is one of the major problems facing ranchers in many parts of the Southwest today. Vigorous efforts are underway in many areas to eradicate or control mesquite growth. To what extent these efforts will prove successful, only time will tell.

small, hard, leathery leaves. These plants are most closely associated with the Mediterranean woodland and scrub forest subassociation, and they are most frequently found near the poleward margins of the subtropical deserts that border the Mediterranean Climate. The dominant sclerophyll of the American West is the creosote bush.

A fourth group, the **halophytes** (from the Greek *halos*, or "salt"), are salt-tolerant plants that normally occupy lowland basins of interior drainage where the soils are highly saline. Because these plants have developed an ability to cope with highly saline surface conditions, they are able to grow where no other plants can survive and are therefore free from competitors. Important examples of halophytes in the western United States are salt bush and shadscale scrub.

A fifth group of desert plants are the **ephemerals.** They are small, short-lived plants that go through their life cycles quickly following a substantial rainfall. The ephemerals take advantage of those brief, rare periods when the desert is not dry, and they therefore cannot be classified as xerophytes. They include a wide variety of plants; most are grasses or flowering herbs whose seeds or bulbs have lain dormant for months or even years. The abundance of brightly colored flowers among the ephemerals has led to the frequent observation that the desert "blooms" after a rainstorm. Within a month or so, however, the moisture is gone, the flowers have produced their seeds and died, and the desert reverts to its normal sere appearance.

Most desert localities, of course, contain a mixture of several of these plant groups in varying proportions. The somewhat better watered semidesert regions usually are dominated by scattered shrubs with a considerable number of smaller herbaceous plants such as cacti and grasses interspersed. Within the more sparsely vegetated drier deserts, widely spaced individual shrubs, cacti, and bunch grasses are common.

Trees are not found in most desert areas because their moisture requirements are too high. On occasion, though, scattered xerophytic trees do occur, especially near the desert margins. This is especially true in Australia, where occasional drought-resistant eucalyptus trees are found in many desert areas. Other localities where trees can secure enough moisture for growth are **oases,** where the water table intersects the surface to produce a spring, and riverbanks, where dense **riparian** (riverbank) vegetation often flourishes. Riverbanks in parts of the southwestern United States have been invaded recently by the tamarisk, an imported tree whose dense foliage has choked out the native vegetation.

The desert fauna consists mostly of small, nocturnal animals that remain hidden from the often intense heat of the day under stones or in burrows so as to conserve water. Rodents are common, as are reptiles such as snakes and lizards. Arachnids such as scorpions and tarantulas also are well represented.

Tundra Association

The **tundra** is the most cold-tolerant vegetation association and consequently is found at higher latitudes and elevations than any other. Because the tundra has been influenced relatively little by human activities, it is the least modified of all the major plant groups.

Most tundra areas are located within the northern hemisphere (see Figure 11.32). The three largest regions are northern North America from Alaska to Labrador, the Arctic coast of Eurasia from Norway to the Bering Sea, and the coastal rim of Greenland. These areas all have a Tundra Climate characterized by long cold winters, summers with monthly means below 10° C (50° F) and occasional frosts, and a growing season of only about two months. Only during the summer, extending from late June or early July into early September, does the snow melt and the vegetation appear.

Tundra vegetation consists of a mixture of small plants, most of which are herbaceous perennials (see Figure 11.33). Two layers of plants normally cover most of the surface. The upper layer of grasses, grasslike sedges, flowers, and sometimes small shrubs often produces a meadowlike appearance. Beneath the upper layer exists a well-developed surface layer of lichens and mosses.

Probably the two most important plant groups are the grasses and the lichens. Grasses are so common that some biogeographers consider the tundra as a grassland subassociation rather than as a separate vegetation association. Several species are common, but perhaps the best known is cotton grass, a tall grass that grows in poorly drained areas and is surmounted by a cottony tuft. Lichens consist of algae and fungi growing together in a symbiotic relationship to produce a plant form unlike either one alone.

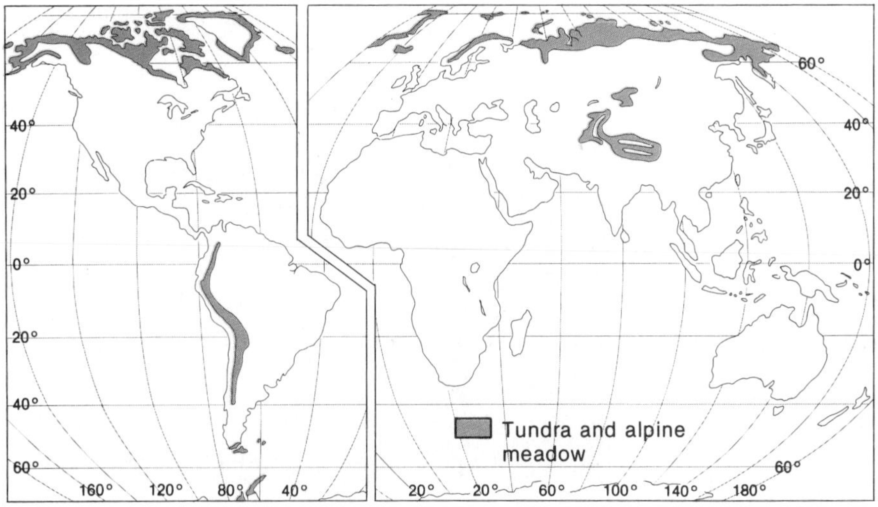

FIGURE 11.32 Global distribution of tundra vegetation.

The best-known example is the so-called "reindeer moss," which grows to heights of 5 to 10 centimeters (2 to 4 in).

Near the southern margin of the tundra, dwarf trees and shrubs are found on better drained sites. These trees are of the same species as those of the neighboring northern coniferous forest and include spruce, willow, birch, and alder. They grow extremely slowly and at maturity may attain heights of only a meter or two. On occasion, they display a tendency to grow along the ground more like vines than trees. The very small size of the trees and shrubs is believed to result largely from the alternate freezing and thawing of the surface, which causes the ground to heave, thus tearing and crushing the roots. In addition, an underlying layer of **permafrost,** or perennially frozen soil, acts as a barrier to root development. In better drained river valleys, the depth to permafrost is greater, and larger trees are able to grow. As a result, narrow strips of coniferous forest extend up river valleys well into the tundra.

FIGURE 11.33 Tundra or alpine meadow vegetation above the tree-line in the Colorado Rockies. (Ralph Scott)

Localized topographic variations exert perhaps a greater influence on the vegetation of the tundra than on any other plant group. Permafrost and past glaciation have caused large areas to be poorly drained, and lakes and marshes abound. The marshes are sites where hygrophytic grasses grow. Slightly higher and drier sites a short distance away are likely to be covered with completely different types of vegetation, including patches of colorful flowers such as buttercups, anemones, and poppies. With the low Sun angles that always exist in the high latitude tundra, slope orientation also is crucial; south-facing hillsides are sunnier, warmer, and better vegetated.

Proceeding poleward through the tundra, the density, size, and biological sophistication of the plants gradually decline. At its poleward edge, the tundra surface consists largely of bare rock, with only widely scattered patches of moss and lichen.

Despite its harsh climate, the tundra is home to several species of large migratory herbivores, including caribou, reindeer, and musk ox. Smaller mammals include snowshoe hares and lemmings, while carnivores are represented by wolves, Arctic fox, owls, and hawks. Numerous birds summer in the Arctic, feeding on the huge insect population, and migrate south to warmer climates for the winter.

Mountain Biogeography

Within mountainous regions, all the factors that control the distribution of plants and animals are subject to large variations over small distances. The climate rapidly becomes colder and moister with increasing elevation and, in addition, large differences in precipitation usually exist between windward and leeward slopes. Rugged topography is the hallmark of a mountainous region; slope steepness and orientation vary greatly. Soils may be deep and well developed in valleys but completely absent on the steepest slopes; because of the complex geology typical of many mountainous areas, soils frequently have formed from a variety of parent materials.

As a result, great differences in life forms are encountered within small distances, especially in higher mountains. Because of their isolated nature, even mountainous regions with similar environmental conditions do not necessarily contain the same biological assemblages. Distances often are too large for plants and animals to migrate between such regions, resulting in the evolution of species unique to a given highland area or even a single mountain peak.

The resulting biogeographical patterns are too complex to discuss in detail except at the local level. At any specific site, however, within the constraints imposed by topography and soils, the flora and fauna may be expected to be associated closely with the local climate. This correlation again indicates that climate is the most powerful of the biological distribution controls.

Local climatic conditions may cause the vegetation within a highland area to be completely unlike that of the surrounding lowlands. Areas of montane forest therefore can be found in the midst of deserts, and mountain tundra (or **alpine meadow**) can exist at the equator. Patterns of mountain vegetation are by no means random, however, especially when examined at the broader scale. Because climatic influences are dominant, and elevation largely determines mountain climate, a vertical zonation of plant assemblages that represents a compression of latitudinal vegetation gradients is normally well developed (see Figures 9.49 and 11.34). A very important natural vegetation boundary in mountain areas is the **treeline,** or uppermost limit of tree growth. It is located approximately at the elevation at which the warmest month of the year averages 10° C, or 50° F (see Figure 11.35). In general, the biological assemblages of mountain areas are those that thrive under cooler and moister conditions than those of the surrounding lowlands.

FIGURE 11.34 A zonation of vegetation types based on elevation is apparent in this view of Wyoming's Grand Teton Mountains. Prairie vegetation in the foreground gives way to coniferous forest, which in turn merges with tundra and, finally, with ice and bare rock. (Richard Jacobs/© JLM Visuals)

SUMMARY

Plants and animals occupy about 80 percent of the Earth's land surface. Through the processes of mutation and natural selection, a great variety of life forms adapted to a diversity of environmental conditions have evolved. Given a stable environmental setting, a given habitat will develop a biological community that will utilize the available space and resources to its maximum advantage. When a community is disturbed or removed, life forms generally will return to the site in an organized successional sequence, eventually reaching ecological stability when a climax assemblage is achieved. The combination of human activities and recent global-scale climatic changes has prevented stable climax communities from currently existing in many areas. Geographers

FIGURE 11.35 The tree-line is located at about the 3,500 meter (11,000 foot) level in the Colorado Rockies. (© Joseph T. Collins/Photo Researchers)

are concerned primarily with the world pattern of flora and fauna and with the reasons for this pattern. In order to summarize this information, we return to the Focus Questions posed at the beginning of the chapter.

1. *What are the chief environmental controls on the global distribution of plants and animals?*

The five chief environmental controls on the distribution of the biota are climate, topography, migrational access, soil, and organisms. Climate is the most important, especially because it influences moisture availability and temperature. Moisture availability influences such vegetative characteristics as plant size, root development, transpiration rates, leaf number and size, growth rates, and regional biomass. Temperature is especially important for its influence on the length of the growing and mating seasons, vegetative growth rates, and leaf shedding. Topographic influences often are localized. In general, areas of steep slopes support biological assemblages of reduced complexity, and poleward-facing slopes support life forms adapted to cooler and moister surface conditions. Geographical barriers such as oceans, deserts, and mountain ranges determine the degree of migrational access to a region. Isolated habitats such as islands tend to have biological communities that are more primitive and limited in variety than those with good access to other regions. A number of soil characteristics are important to plants and to the animals that feed on them. Crucial are soil depth, mineral content, and water retention ability. Organisms must successfully coexist within a given habitat, or ecosystem alteration will occur. The interaction between the life forms inhabiting an ecosystem is highly complex, but an overriding factor is the competition for available resources.

2. *What are the major plant associations, and what are their basic botanical characteristics?*

The four major plant groupings, or associations, are forest, grassland, desert, and tundra. Each is associated with a different range of environmental conditions. Forests cover the largest geographical area of all the associations and comprise the great majority of the total biomass of land areas. They generally dominate in areas with at least a five-month period of annual soil moisture surplus and with a warmest monthly mean temperature of 10° C (50° F) or higher. Most of the tropics, middle latitudes, and subpolar regions meet these criteria and are forested. Broadleaf evergreen trees dominate near the equator, while broadleaf deciduous trees are dominant in forested portions of the subtropics and middle latitudes. Needleleaf trees are most common in subpolar and alpine environments. Grassland regions typically are encountered on the dry margins of forests. Savanna grasslands, which are tropical grasslands with scattered trees, occupy large portions of the low latitudes with dry seasons of from five to eight months. Short steppe grasslands are found in semiarid regions of both the low and middle latitudes. Tall prairie grasslands formerly occupied subhumid areas within the middle latitudes, but have largely been removed for agriculture. The driest deserts are unvegetated, but most deserts are covered by a variety of xerophytic herbs and shrubs. Deserts are most widespread in the subtropics, but also exist in portions of the middle latitudes. Tundra vegetation consists of a mixture of small, cold-tolerant herbaceous perennials, which often present a meadowlike appearance. The plants of the tundra are only visible, and biologically active, for approximately three months in summer. They dominate subarctic regions with warm season monthly means between 0° C and 10° C (32° F and 50° F).

Review Questions

1. Explain how each portion of the world happens to be occupied by life forms that are adapted to the local environmental conditions.

2. What is a climax biological community? How does the concept of vegetative succession relate to this concept?

3. List the five major biological distribution controls. Which one has the greatest effect on world plant and animal distributions?

4. What environmental factors other than the amount of precipitation influence the quantity of moisture available to plants?

5. List several structural differences between the characteristics of xerophytes and hygrophytes.

6. Would you expect to encounter greater diversity in the flora and fauna of a level region or of a hilly region? Explain several reasons for this.

7. What competitive advantage do trees have over other types of vegetation. Why don't trees dominate the vegetation of all areas?

8. Discuss the basic vegetation differences between the tropical rainforest and the tropical semideciduous forest. What factor causes these differences?

9. Why are the world's forest areas separated into tropical and nontropical segments? Are these two segments connected? Where and why?

10. Through what natural vegetation associations and subassociations would you pass on a trip from New York City to San Francisco? Explain briefly the environmental factors responsible for each of the changes in vegetation.

11. What basic similarities and differences exist between the vegetation of the tropical rainforest and the mid latitude deciduous and mixed forest?

12. Name two regions of the coterminous United States where needleleaf evergreen trees dominate. Explain why they exist in these areas.

13. In what respects is the northern coniferous forest different from all the other forest subassociations?

14. Describe the basic environmental characteristics that favor each of the four natural vegetation associations.

15. Describe the typical appearance of the savanna grasslands. In what ways do they differ from the grasslands of the middle latitudes?

16. Discuss the controversy over the origin of the prairie. What factors may have contributed to its formation?

17. What are some common misconceptions regarding the nature of desert vegetation? Which of the major groups of desert plants is most widespread and important?

18. Discuss the major environmental limitations of the tundra, and explain how they have influenced the tundra flora and fauna.

19. Imagine a supercontinent extending from the equator to the North Pole. Starting with the tropical rainforest, list in order the hypothetical sequence of vegetation groups that a person traveling poleward would encounter.

20. In what natural vegetation association or subassociation is your own area located? Explain how local environmental conditions have resulted in the dominance of this type of vegetation.

Key Terms

photosynthesis	**mesophyte**
biogeography	**tropophyte**
habitat	**deciduous**
biological community	**evergreen**
ecosystem	**annual**
vegetative succession	**perennial**
climax community	**broadleaf**
biome	**needleleaf**
xerophyte	**plant association**
hygrophyte	**jungle**

CASE STUDY

THE DESTRUCTION OF THE WORLD'S TROPICAL RAINFORESTS

Reference was frequently made in this chapter to the impact of human activities on global vegetation patterns. In some areas, this impact has taken place over many centuries, and little now remains of the original surface cover. In other areas, only minor modifications have occurred, and the original vegetation remains largely intact. In at least one instance—that of the tropical rainforests—humans are currently destroying a major vegetation assemblage. This destruction is proceeding with unprecedented speed and is likely to produce major adverse consequences for the tropics and perhaps even for the entire planet.

CAUSES AND EXTENT OF DEFORESTATION

Until a few decades ago, most portions of the tropical rainforest were untouched and largely unexplored. Little lumbering took place because of the lack of roads, the long distances to markets, and the great diversity of tree species. Beginning in the 1950s and 1960s, however, a number of factors combined to alter the economic conditions that previously had protected the rainforests. This initiated the large-scale forest destruction that continues at an accelerating pace to the present. One of the chief factors in this turnabout was population pressure caused by the tremendous growth rates of the developing countries that contain tropical rainforests. Many landless peasants in Asia, Africa, and South America are willing to settle and farm rainforest areas despite the land's low fertility and are burning rainforest to clear land for agriculture.

Commercial lumbering is increasingly profitable, despite the inherent difficulties already mentioned (see Figure 11.36). Modern machinery has greatly increased the efficiency of operations, and uses have been found for more and more tree species. Clear-cutting methods are often employed, and even if they are not, uncut trees frequently are damaged and later die from insect infestation and disease. Settlers also need wood for lumber and fuel. Tropical rainforest lumber sales, largely to Japan, total several billion dollars a year. This income is desperately needed by the developing countries, most of which are burdened with huge foreign debts. Other important causes of rainforest destruction include cattle ranching, mining, and the flooding of vast areas behind dams built for hydroelectric power generation.

In 1990, the remaining area of tropical rainforest totaled some 7.8 million square kilometers (3 million mi^2), distributed among 70 countries. This is an area nearly the size of Europe. Each year, approximately 113,000 square kilometers (44,000 mi^2) of rainforest are destroyed by land clearing for crop production, fuelwood gathering, and cattle ranching, and an additional 45,000 square kilometers (17,000 mi^2) are damaged or destroyed by commercial lumbering operations. A study released in June 1990 estimated the rate of tropical

FIGURE 11.36 Appearance of the surface following clear-cutting operations in the Brazilian Amazon rainforest. (© Ulrike Welsch/Photo Researchers)

continued on next page

rainforest destruction to be 0.6 hectares (1.5 acres) each *second.* In total, an area slightly larger than New York and Vermont combined is lost each year. Within a decade or two, only two large rainforest areas will remain; these will be in the western Amazon Basin and in the central Congo Basin.

The destruction of the tropical rainforests after some 60 million years of continuous existence is a major event in the Earth's biological history, and will considerably alter the face of our planet, even as viewed from far out in space. The scope of the clearing operations is larger in magnitude than the earlier removals of the middle latitude forests of North America and Europe, and is being accomplished in a much shorter time span.

CONSEQUENCES OF DEFORESTATION

The removal of the rainforests is likely to produce a number of adverse consequences, the exact nature and magnitude of which are difficult to forecast. One is the permanent deterioration of the vegetative cover. The removal of the trees will not, of course, produce a bare surface for long. Replacement by a dense secondary growth of weeds, shrubs, vines, and saplings is rapid. If undisturbed, the area eventually may return, after several centuries, to its original state. In most cases, though, cutover areas are settled or grazed, thus inhibiting reforestation. In addition, if cutting is too extensive, few seeds are locally avail gradually to my clear consciousness" ("Words of Professor Royce at the Walton Hotel at Philadelphia, December 29, 1915," *Philosophical Review* 25:3 [May 1916]: 510).

A second important concern is soil deterioration. Soils in the humid tropics are naturally low in fertility because of the leaching of soluble plant nutrients by downward-moving soil water from heavy rains. (This process is discussed further in the next chapter.) The rainforest trees, however, are highly efficient at recycling nutrients by absorbing them into their root systems before they can be leached away. At any one time, a large proportion of the total nutrients are locked up in the trees themselves, to be released gradually as they die and decompose. Lumbering activities therefore directly remove many of these nutrients and also destroy the mechanism for protecting the rest from the leaching process. In addition, erosion is greatly accelerated in these rainy regions by the absence of both a protective canopy of leaves and branches and the soil-binding influence of tree roots. Flooding also becomes more common when the trees, with their large water-absorbing capabilities, are gone and the soils become more compact. Finally, some tropical soils tend to bake brick-hard upon exposure to sunlight.

Concern has also been expressed that rainforest destruction may increase global temperatures because the rainforests currently are responsible for the absorption of more carbon dioxide than the rest of the world's vegetation combined. Without this continuing CO_2 removal and storage in the form of trees, the greenhouse effect may intensify. The rainforest areas also may become considerably drier because vegetative transpiration provides a substantial proportion of the atmospheric water vapor content for rainfall received in rainforest areas.

The greatest tragedy from a biological standpoint is the impending extinction of countless plant and animal species. The tropical rainforests contain literally millions of species of plants and animals that exist nowhere else on Earth and that would not survive if their habitat were destroyed. Many of the denizens of the rainforests have not yet been studied scientifically; extinction would eliminate them before we have a chance to learn about them or, in some cases, even to become aware of their existence. In addition, these varied life forms represent a great storehouse of genetic diversity. Their destruction would considerably diminish the planetary pool of genetic resources, probably altering future evolutionary history.

PARTIAL SOLUTIONS

The fate of the tropical rainforests as vast expanses of virgin forest appears to be sealed. The human transformation of the face of the Earth, which is an inevitable consequence of the scientific revolution, is well underway. Sadly, the maintenance of large uninhabited and unutilized wilderness areas such as the rainforests appears to be incompatible with this transformation.

If the rainforests themselves cannot survive, at least some of the adverse consequences attending their removal can be minimized. For instance, trees can be replanted immediately after logging operations within an area are completed. Replanting would protect the soil and would likely reduce the climatic consequences of deforestation. Little replanting is now taking place. A second action, currently being undertaken by several countries including Brazil, is the establishment of large natural rainforest reserves. These reserves should save many plant and animal species from extinction and preserve for future generations at least some segments of a unique ecological system.

CHAPTER 12

SOILS

OUTLINE

FOCUS QUESTIONS

1. What physical and chemical characteristics of the soil most affect its ability to support vegetation?
2. What are the five factors in soil development, and what influence does each have on soil characteristics?
3. What are the major world soil types, and what are the chief attributes of each?

The soil is one of our most essential natural resources, on which we depend directly or indirectly for nearly all our food, and without which we, and all other advanced life forms, could not exist.

Unlike vegetation, soil is not a living entity, but it is intimately associated with life in several major respects. In the first place, the soil contains a great deal of life. Soil organisms include a tremendous variety of both plants and animals. Much of this life is microscopic, but some, like the root systems of large trees, dwarfs all but the largest animals living on the surface. Second, the soil supports nearly all plant life, and this in turn supports animal life. Even aquatic plants and animals largely depend upon minerals derived from soil particles eroded from the land by water or wind. A third association with life is the soil's life-like ability to adjust to its environment. The soil is sensitive, to some degree, to almost every aspect of its natural environment. Soil characteristics at any given site therefore are determined by the interplay of past and present environmental factors at that site. The soil's response to its environment is dynamic. Not only does the soil develop characteristics that are in equilibrium with the environment in which it forms, it also responds to environmental changes such as variations in climate or vegetation. Any such change produces a slow alteration of soil characteristics until a new equilibrium is achieved. The geographic distribution of soil types, then, is not random, but is closely related to other components of the environment.

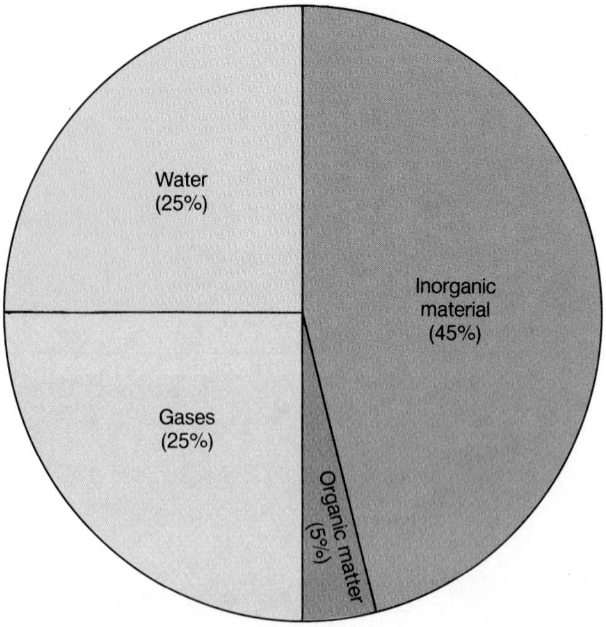

FIGURE 12.1 Relative proportions of the four components of a typical soil.

SOIL ORIGIN AND CHARACTERISTICS

The soil is so complex and variable in composition, and means so many different things to different people, that it is very difficult to define. In general, though, **soil** can be described as a loose mixture of weathered rock material, organic matter, water, and air that can support plant growth. Over extended periods, soils tend to develop distinctive layers that differ in their physical, chemical, and biological characteristics. By volume, a typical soil might contain 45 percent inorganic material, 5 percent organic matter, 25 percent water, and 25 percent gases (see Figure 12.1). These last two components fill the voids or **pore spaces** between the solid soil particles. The proportions of the four components vary greatly among soils, but nearly all soils contain at least a small quantity of each. Each component plays a vital role in the ability of the soil to perform its most important function—the support of plant life.

The chief constituent of most soils is weathered rock material. This material is a major source of the minerals required by plants and animals living in the soil. Weathering, which is discussed in more detail in Chapter 16, refers to the action of physical and chemical processes that break down rock and may eventually convert it to soil-forming materials. Weathering alone does not produce soil, although it certainly is a major and necessary preliminary step. It does, however, form a **regolith** (weathered rock) layer above the unaltered bedrock. The regolith is transformed gradually into soil through physical, chemical, and biological processes. These processes, discussed in subsequent sections of this chapter, involve the movement of some materials, the formation of new materials, the addition of organic matter, and the development of distinctive layers within the soil.

The organic components of the soil consist of the remains of plants and animals. The great majority are derived from vegetation such as leaves, grass, roots, and branches. The organic matter undergoes gradual decomposition, which converts it to a dark substance called **humus** and slowly releases its mineral content.

Many of these organically derived minerals are required by the soil's living population of plants and animals. Organic matter therefore is an important source of plant nutrients. In addition, chemical reactions involving organic material release nutrients bonded to inorganic mineral particles so that they become available for plant use.

Soil water and air both are derived primarily from the atmosphere. The proportions of these two components are to some extent inversely related, since both share the soil pore spaces. Following a rainstorm, water floods the pore spaces, expelling much of the air. As the water gradually drains away, evaporates, or is used by plants, air refills the pores.

We now examine several important soil characteristics.

Soil Color

Soils exist in a great variety of colors (see Figure 12.2). One group varies from white through all shades of gray to black. Other soils range through yellows, reds, and browns. Even greenish and bluish soils are sometimes formed, and colors often differ considerably with depth at a single site. Color is not important for its own sake; its significance derives from the information it provides about the mineral content and the physical and chemical condition of the soil. Because color is the most easily observable soil characteristic, it usually is the first noted during an examination. Soil color charts, such as the Munsell chart that contains 175 gradations of color organized by hue, value, and chroma, are used for soil identification and description.

FIGURE 12.2 Soils with contrasting colors. The soil on the left is a dark, humus-rich South Dakota mollisol that contains a white layer of calcium carbonate ($CaCO_3$) within the E horizon. The soil on the right is an alfisol from California that has been colored orange by iron and aluminum oxides. (*left:* Marbut Memorial Slide Collection/American Society of Agronomy; *right:* Soil Taxonomy AH 436, Soil Conservation Service)

Most soil color is imparted by just two substances—humus and iron. As the humus content increases, soils tend to darken through shades of brown to nearly black. Oxides of iron usually are responsible for imparting a reddish or yellowish color. These two colors are especially typical of the soils of humid tropical and subtropical areas, because iron tends to form relatively insoluble compounds with oxygen under warm, moist conditions. With good drainage and aeration, oxidized iron compounds produce a bright rust red color. If moderately poor drainage conditions prevail, yellow colors are produced by hydrated iron compounds. Under very poor drainage conditions, reduced iron compounds can give the soil a greenish or gray-blue color.

In arid environments, the soil may contain whitish flecks or layers due to the presence of calcium carbonate ($CaCO_3$) or salts. These substances are soluble in water and do not collect in the soils of humid climates. In humid climates, nearly white subsoils exist where iron compounds and other coloring agents have been removed by downward-percolating water. Sandy soils composed largely of quartz (SiO_2) also are typically white or buff in color.

Soil Texture

In contrast to soil color, which is important largely for the information it provides about the composition and condition of the soil, the **soil texture,** or size of the individual soil particles, has important direct effects on the soil. Excluding gravel and larger rocks, soils consist of three basic sizes or "grades" of particles. The largest or coarsest grade is **sand,** while **silt** particles have an intermediate texture, and **clay** particles are the smallest or finest. Each texture grade has been assigned a specific range of particle diameters, as indicated in Table 12.1. Only inorganic materials are included when texture measurements are made, organic matter being removed beforehand. Many soils contain gravel and stones exceeding the maximum sand-size diameter of 2 millimeters (0.08 inches).

TABLE 12.1 Soil Texture Grades

SIZE GRADE	DIAMETER (MM)	DIAMETER (IN)
Gravel	greater than 2.0	greater than 0.08
Very coarse sand	1.0–2.0	0.04–0.08
Coarse sand	0.5–1.0	0.02–0.04
Medium sand	0.25–0.50	0.01–0.02
Fine sand	0.10–0.25	0.004–0.01
Very fine sand	0.05–0.10	0.002–0.004
Silt	0.002–0.05	0.00008–0.002
Clay*	below 0.002	below 0.00008

* The term "clay" denotes the grade, or size, of the soil particles and does not mean that they consist of clay minerals.

A factor complicating the description of soil textures is that nearly all soils contain a mixture of soil particles of differing sizes. Various proportions of particles of different sizes thus can produce a variety of texture classes. The classes adopted by the U.S. Department of Agriculture, which are in general use in the United States and many other countries, are shown in Figure 12.3 and are explained in the accompanying legend. In general, a soil containing 50 percent or more silt-sized particles has "silt" or "silty" in its name. Clay-sized particles exert a strong influence on the soil, so soils containing more than about 27 percent clay have this term in their names. Soils that are significantly influenced by all three particle grades, and therefore have an intermediate texture, are referred to as **loam** soils.

The soil texture in any given area is controlled largely by the climate and by the chemical stability of the material from which the soil was derived. Regions with warm, moist climates typically have deep, fine-textured soils because chemical weathering processes are exceptionally active under these conditions. Conversely, cold, dry areas usually have thin, coarse-textured soils, if a soil cover has developed at all. Likewise, soils developing from chemically stable rock or sediment, especially if it has a high quartz (SiO_2) content, tend to be coarse while those developing from parent material with unstable mineral assemblages weather rapidly to produce deeper, finer textured soils.

Both sand- and silt-sized particles are granular and consist primarily of quartz in most cases. Clay-sized particles usually have been highly altered chemically from their original state; many are sheetlike in shape.

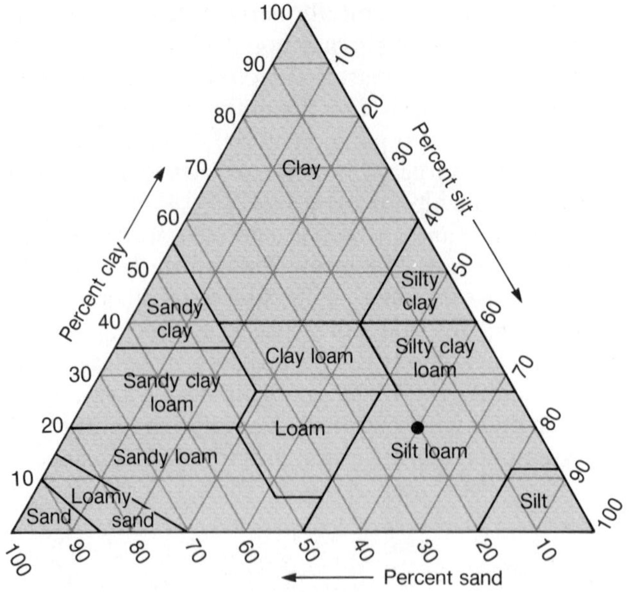

FIGURE 12.3 The soil texture triangle is used to determine the textural class of the soil, based on the proportions of sand, silt, and clay it contains. Each side of the triangle indicates the percentage content of one of the three texture grades. The three lines that represent the proper proportions of sand, silt, and clay in a soil sample are traced inward to their point of juncture. For example, a soil consisting of 20 percent sand, 60 percent silt, and 20 percent clay is classified as a silt loam. (Note the location of the dot in the diagram for this soil type.)

They vary considerably in their mineral composition. The finest clay particles are so small that they can remain suspended in water indefinitely and are soft and gelatinous. They are termed **colloids** and, as will be seen, are essential to the biochemical reactions that allow plants to extract nutrients from the soil.

The texture of the soil is of practical significance in two major ways. First, it affects the looseness and "workability" of the soil from an agricultural standpoint. Coarse-textured, sandy soils are "light" and easily worked. Plant roots penetrate readily, but the soil may lack the firmness to keep plants from being bent or uprooted during floods or windstorms. Fine-textured soils, in contrast, tend to be dense, sticky when wet, and "heavy." They may be difficult to till, but provide strong physical support for plants.

Second, the soil's ability to absorb and retain water is greatly influenced by soil texture. Coarse textured soils have relatively large intergranular pore spaces because sand-sized particles generally do not fit together as snugly as do silt and clay particles. As a result, sandy soils have high water infiltration rates but low water retention capacities because water passes readily through the pore spaces and relatively little adheres to the soil particles. The situation is reversed for fine-textured soils, which have large water retention capacities, but absorb water very slowly. This may lead to poor drainage and excessive runoff, resulting in rapid erosion and in sediment pollution of streams near cultivated fields. Loamy textures generally are considered most favorable for agriculture because they provide a balance between the problems resulting from excessively fine- and coarse-textured soils.

Soil Structure

Individual soil particles normally do not remain detached from one another. If they did, fine-textured soils, when dry, would behave like the finest dust, to be easily washed away by rainwater or blown away by the slightest wind. Instead, soil particles tend to adhere to one another, forming larger aggregates called **peds.** The shape, size, and organization of these peds constitute the soil's **structure.**

Peds are held together largely by the clay-sized particles and by the soil colloids of organic origin. These fine particles develop strong electromagnetic charges that bind the soil particles together. As a result, coarse textured soils, with little colloidal material, tend to have weakly developed structures or even a complete lack of structure. As a general rule, the coarser the texture of a soil, the weaker its structure. The type and strength of the structure also are influenced by the chemical nature of the clay mineral particles, the type and abundance of soil microorganisms, the frequency and extent of wetting/drying and freezing/thawing cycles, and cultivation practices.

Although a wide variety of specific structural forms can develop, soil structures are divided into four primary groups (see Figure 12.4). **Blocky** structures contain peds that are compact and relatively flat-sided. Soils with **platy** structures consist of a large number of thin platelet-shaped peds paralleling the surface. **Prismatic** or **columnar** structures sometimes are associated with excessively drained soils, because the pore spaces between the peds offer water a direct path downward through the soil. They are the most massive of the four types. **Granular** or **crumb** structures are common in soils with a moderately high humus content. The degree of structural development and even the type of structure may vary with soil depth.

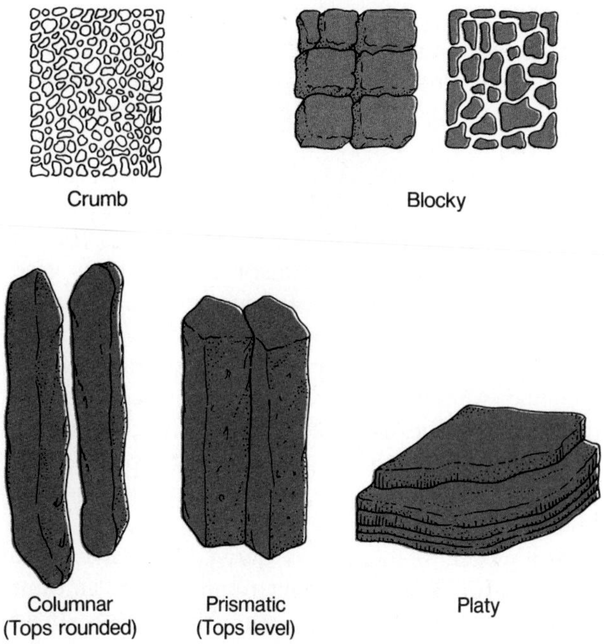

FIGURE 12.4 The primary soil structural groups.

One of the most important practical effects of a soil's structure is the influence it exerts on the ease of absorption and movement of soil air and water. A soil with a well-developed structure typically is less compact and has greater permeability and porosity than does a structureless soil. Some types of structure, however, may restrict the absorption of water. For example, a soil with a strongly developed platy structure may be poorly drained because each "plate" acts as a barrier to water infiltration.

The strength with which the individual peds are internally bound influences both the soil's resistance to erosion and the ease of cultivation. A strong structure holds the soil together and causes it to resist erosion. This same characteristic, though, makes the soil more difficult to plow. Plowing tends to alter and usually to weaken soil structure, and the passage of farm machinery leads to soil compaction. As a result, the soils of agricultural areas typically have poorly developed structures and lower water-infiltration capacities than do nearby unfarmed areas.

Soil Air and Water

Although soil water and air come mostly from the atmosphere, their chemical constituencies are modified by their residence in the soil. These chemical changes have a significant effect on the soil biota.

The soil atmosphere is critical to soil organisms because it supplies the oxygen needed for animal and plant respiration and the carbon dioxide required for plant photosynthesis. A lack of oxygen in a saturated soil can suffocate both plants and animals. In addition to air from the atmosphere that enters the soil through its pore spaces, the soil also contains gases released from chemical reactions and biological activities. As a result of evaporation and plant respiration, soil air contains slightly less oxygen and nitrogen and somewhat more water vapor and carbon dioxide than does the atmosphere. The constant exchange of soil air with air from the atmosphere, however, keeps the differences from becoming too pronounced.

Soil water, of course, is derived primarily from precipitation. During a heavy rainfall, large amounts of water typically enter the soil. Most passes downward through the pore spaces toward the water table in response to gravity. If the soil is freely drained, the excess water takes about two days to pass completely through. The soil at this point retains its **field capacity** for water, which is the maximum amount of water the soil is capable of holding against the pull of gravity. This water is vital to the soil's plant and animal population because it is retained until used or evaporated.

Soil water can be held against the pull of gravity because it is attracted to the surfaces of the soil grains by **capillary tension.** This property results from the electromagnetic charges developed by the soil particles. Because fine-textured soils have a large total soil surface area per unit of volume, they have higher water-retention abilities than do medium- or coarse-textured soils. Plants are able to withdraw much of the soil water for their own use, but finally, at the **wilting point,** the remaining water is bound so tightly to the soil particles that plants are unable to extract it.

Water in the soil dissolves soil minerals, making them available to plants as they absorb water through their root systems. As water passes through the soil, dissolved minerals and fine-textured particles are carried, or **translocated,** from one place to another. These materials often are deposited a considerable distance from their points of origin.

In humid climates, an excess of downward-percolating water removes minerals from the upper soil in a process called **leaching.** Leaching involves the removal or **eluviation** of dissolved and fine-textured solid materials from one portion of the soil, followed by the translocation and deposition (or **illuviation**) of these materials in another, usually deeper, layer of the soil. Leaching, where strongly developed, often results in soluble plant nutrients being deposited so deeply in the soil that they become inaccessible to most vegetation; under very wet conditions, many of these materials are carried completely out of the soil. The leaching process is the primary reason for the general lack of fertility of soils in humid climates. Fertilizers may be used to replace leached or otherwise unavailable plant nutrients temporarily, but these, too, are susceptible to leaching so that reapplications on a periodic basis are necessary.

In arid climates, the drying of the surface soil causes capillary water to be drawn toward the surface. This water transports minerals dissolved from the parent material or lower soil upward to be deposited near the surface in a sort of reverse leaching process. Although the upper soil's supply of plant nutrients may be increased by this process, some of the substances brought up are toxic to most plants. A prime example is provided by salts of various types. In many arid areas where water is available for agriculture, the high salinity of the soil makes the raising of crops impossible or impractical. The use of irrigation water for dry-land agriculture has in many cases intensified the problem by dissolving large quantities of salts from the lower layers and carrying them to the surface. The problem is worsened if the irrigation water itself has a significant dissolved salt content. Approximately one-third of the irrigated farmland in the western United States has been damaged by increasing salinity, and some once-productive land has had to be abandoned. The problem of soil salinization resulting from irrigation is discussed in the Focus Box on pages 304–305.

The surfaceward movement of water also may deposit substances that form a resistant **duricrust layer** on or just beneath the surface (see Figure 12.5). When well developed, a duricrust layer may resemble a buried layer of concrete, and under extreme conditions dynamite has been employed to break it up.

As a general rule, the most naturally fertile soils exist in subhumid or semiarid climates, where precipitation approximately equals evaporation. Under these conditions, vertical translocations of soil water and dissolved minerals in either direction are minimized.

FIGURE 12.5 A duricrust layer appears between the 1- and 2-foot levels in this aridisol from southern New Mexico. The horizontal structure of this layer contrasts markedly with the vertical prismatic structure of the overlying A horizon. (Marbut Memorial Slide Collection/American Society of Agronomy)

Organic Matter in the Soil

Organic matter is incorporated into most soils in the form of humus. **Humus** is finely divided, decomposed organic matter with a brownish black color. The remains of both plants and animals serve as sources of humus, but most humus has a vegetative origin, largely from leaves, branches, grass, and roots. Humus forms from the slow oxidation of organic litter by bacteria. It is a relatively stable residue remaining after the bulk of the decomposition has already occurred. Further slow oxidation of the humus itself gradually releases nutrients to the soil and its biota.

The amount of humus present in the soil is determined by the balance between the rates of humus formation and consumption. Because humus forms primarily from plant matter, substantial quantities exist only in places where vegetation is abundant. In most cases, these are humid or subhumid forest or grassland environments. Most plants and animals from which humus is derived live near or above the soil surface, so the humus content of most soils is greatest in the near-surface layers. Under some conditions, though, translocation results in the illuviation of humus into the subsoil. Although humus is formed by the bacterial decay of organic litter, excessive bacterial decay rates can cause humus to be consumed as rapidly as it is formed. Bacterial activity is largely related to temperature; as the temperature increases, the quantity and activity of bacteria also increase, and net humus production declines (see Figure 12.6). For example, a forest soil in an area with an annual mean temperature of 5° C (41° F) might be expected to contain about 8 percent humus, by volume. Under similar vegetative conditions and an annual mean temperature of 20° C (68° F), the humus content is likely to be only about 1 percent.

Humus has profound physical, chemical, and biological effects on the soil. Perhaps most important, it is a major source of plant nutrients and therefore of soil fertility. Because humus is derived from the remains of previously living plants, it contains many of the nutrients required by currently living plants. It is an especially important source of nitrogen, sulfur, and phosphorous.

The spongy consistency of humus increases the capacity of the soil to hold water, along with its dissolved minerals. This characteristic is especially valuable in coarse-textured soils, which have limited water-retention abilities. As discussed in the next section, humus also plays an essential role in the process

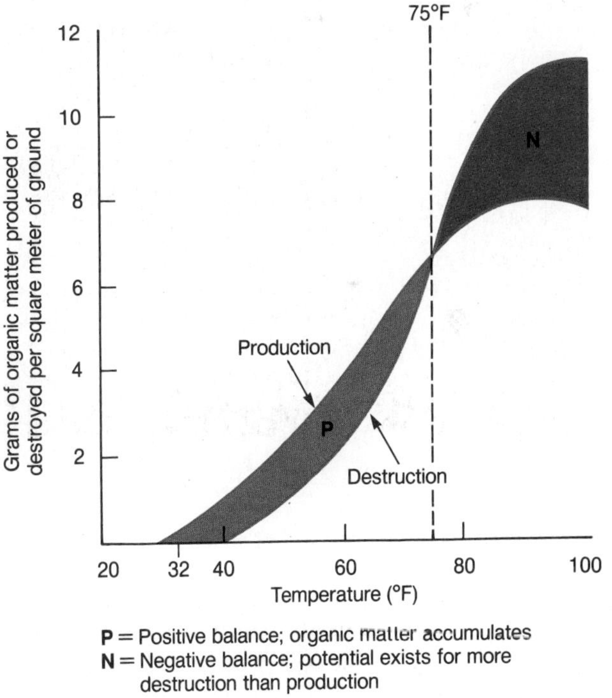

FIGURE 12.6 The relationship between temperature and the rates of humus production and destruction.

SOIL SALINIZATION–A GROWING PROBLEM IN ARID LANDS

The soils of many desert regions are inherently fertile, and the use of irrigation water can make these regions "bloom," creating some of the most productive agricultural lands on Earth. Irrigated cropland, which constitutes about 15 percent of all cultivated land, produces up to 30 percent of the world's food supplies. The extensive irrigation of California's Imperial Valley by water from the Colorado River, for example, has helped make that state the nation's leader in value of agricultural production. Unfortunately, the irrigation of arid lands often causes a gradual increase in soil salinity. This problem, unless controlled, eventually can ruin even the most productive agricultural lands. Approximately one-fourth of the world's irrigated land suffers from salinity to some extent, and that figure is increasing.

Salinization of soils caused by irrigation is probably the most ancient of all factors promoting desertification (see the Case Study at the end of Chapter 19). In the fourth millennium B.C., the Sumerians occupied the southern Tigris-Euphrates Valley and built a prosperous civilization based on irrigation agriculture. By the second millennium B.C., this region, often referred to as one of the "cradles of civilization" in world history texts, was so badly salinized that agriculture was largely abandoned. Even today, many parts of this valley are semidesert wastelands.

The rapid spread of irrigation agriculture has greatly increased the geographical scope of the problem during the past century. For example, extensive areas of Pakistan have been affected, as have large portions of the cotton-producing lands of the southwestern Soviet Union. Salinization is becoming a problem in Egypt's Nile Valley now that the Aswan High Dam has ended the annual flood that once flushed the salts from the soil. Extensive portions of the American West also suffer from this problem, and some areas have already been abandoned.

THE SALINIZATION PROCESS

Soil salinization is a problem because salts in large quantities are toxic to most crops. Common table salt (NaCl) is the worst offender. As it dissolves, the sodium cations increase the pH of the already alkaline soil to 9 or 10. Sodium attacks the soil structure, disintegrating the structural peds that provide channels for the percolation of moisture. Precipitated salts clog the soil pores, making the soil impervious to air and water and producing a white crustal layer.

Most irrigation systems use canals to transport water from a river or well to the fields. These fields are periodically flooded, either wholesale or along furrows. The fields often are not leveled, so that excess water is needed to flood the higher portions. Frequently, two or three times the amount of water actually needed by the crops is used. This excess water percolates down to the water table. If natural ground water drainage is poor and no artificial drainage system is provided, the water table rises. The rising water carries with it dissolved salts from both the lower soil and the irrigation water (which in arid regions often is slightly saline). As capillary action causes the water to approach the surface, the water evaporates and the salts are left behind. Salinization and waterlogging increase in severity until further farming becomes uneconomical and the land is abandoned.

POSSIBLE SOLUTIONS

There are a number of existing and potential solutions to the soil salinization problem that provide a measure of optimism for the future. In areas where fresh water is abundant and the water table is not too high,

of cation exchange, whereby plant nutrients are transferred to plants from the soil. The presence of humus aids in developing and maintaining a soil structure favorable for plant growth and for cultivation. A high humus content also gives the soil a looseness and high aeration capacity that favor vegetative growth.

As is true with almost anything that is beneficial in moderate amounts, it is possible for soils to develop,

FIGURE 12.7 This abandoned, salt-encrusted farmland in California's San Joaquin Valley has been ruined by salinization resulting from the surfaceward movement of irrigation water. (Howard Wilshire)

the application of large quantities of water can flush away the accumulated salts. If the irrigation water itself contains a large quantity of dissolved salts, it can be desalinated before use. Unfortunately, this is a costly process because of the large amounts of energy required.

Irrigated fields in some regions now employ "tile" drainage—a system of perforated plastic pipes under the surface that conducts away excess irrigation water and dissolved salts. This system is expensive to install, but is widely used in California's Imperial Valley and in the Uzbekistan region of the southwestern Soviet Union.

The process of drip irrigation, pioneered by the Israelis in the 1960s, currently is considered one of the most promising methods of conserving water and combating salinity in arid lands. It employs a system of above-ground plastic pipes containing perforations that drip water into the soil directly beside each plant.

Potentially the most promising solution of all involves the genetic engineering of crops to produce salt-tolerant varieties. Irrigation water containing up to 3,000 ppm (parts per million) of dissolved salts is now being routinely used in some areas for a wide variety of salt-resistant crops, including wheat, cotton, potatoes, and sugar beets. Some crops are now being irrigated with water containing up to 10,000 ppm dissolved salts—close to one-third the salinity of seawater. The dream of researchers is to develop strains of crops that can be irrigated with seawater itself, thereby assuring humanity of an unlimited supply of irrigation water. This dream may someday become a reality; already salt-resistant varieties of barley have been grown with seawater irrigation under laboratory conditions.

and to be unfavorably affected by, an excessive organic matter content. This usually occurs in locations with dense vegetation, poor drainage, and a cool climate. Under this set of conditions, organic decomposition does not reach the humus production stage. Carbon dioxide from decomposing vegetation reacts with water to form an excess of carbonic acid (H_2CO_3). This produces infertile, strongly acidic

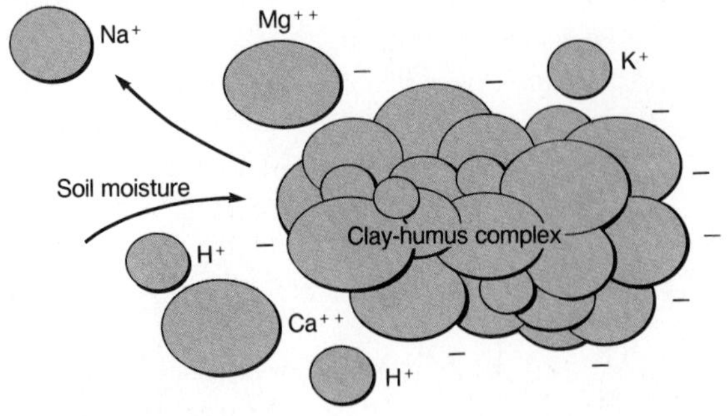

FIGURE 12.8 The cation exchange process. Plants gain needed base cations from the clay-humus colloids in the soil by exchanging other cations such as hydrogen and aluminum for them.

soils with a bluish black color resulting from reduced iron compounds. The continued accumulation of organic matter can result in the formation of peat layers.

Soil Chemistry

The fertility of a soil is determined by its plant nutrient content and by the availability of these nutrients to plants. The nutrients are derived almost entirely from the weathering of parent material and from the decomposition of humus. By what mechanisms, though, are nutrients transferred from the soil to plants? The answer to this question lies in the properties and activities of the soil **clay-humus complex.**

This complex consists of colloidal combinations of tightly bound clay and humus particles. Soil colloids develop negative electrical charges and therefore become capable of attracting and holding positively charged ions (cations). The cations are formed by the separation or dissociation of molecules when they are dissolved in water—in this case, soil water. The substances ionized in this fashion are mostly bases, such as calcium, magnesium, phosphorus, and potassium, which plants require and must be able to withdraw from the soil. The cations are electrically attracted to the negatively charged clay-humus colloids and therefore are kept from leaching away. Plants gain needed cations by exchanging other cations for them; the most commonly exchanged cations are aluminum and especially hydrogen, which are more strongly attracted to the clay-humus colloids (see Figure 12.8).

The ability of a soil to exchange cations is called its **cation exchange capacity (CEC).** The CEC of the soil is determined not only by its supply of exchangeable base cations, but also by the quantity of hydrogen (H^+) ions that can be traded for the base cations. The quantity of available H^+ ions is measured by the pH scale, which indicates the logarithmic concentration of H^+ ions (see Figure 12.9). The pH scale is numbered so that a pH of 7, which means that each 1,000 cubic centimeters of solution contains 10^{-7} grams of hydrogen ions, is chemically neutral. Each whole number below 7 means that there are 10 times as many H^+ ions, making the soil increasingly **acidic.** Each number above 7 means that there are only one-tenth as many $H+$ ions as for the previous number, making the soil increasingly basic or **alkaline.**

Soils range in pH between approximately 3 and 10. In general, a soil pH between 5 and 7 is best for most agricultural crops, although different soil flora and fauna have developed greatly different pH tolerances. A highly acidic soil contains an excess of H^+ ions and probably does not have enough exchange-

FIGURE 12.9 The pH scale is a logarithmic scale for measuring the acidity or alkalinity of a substance.

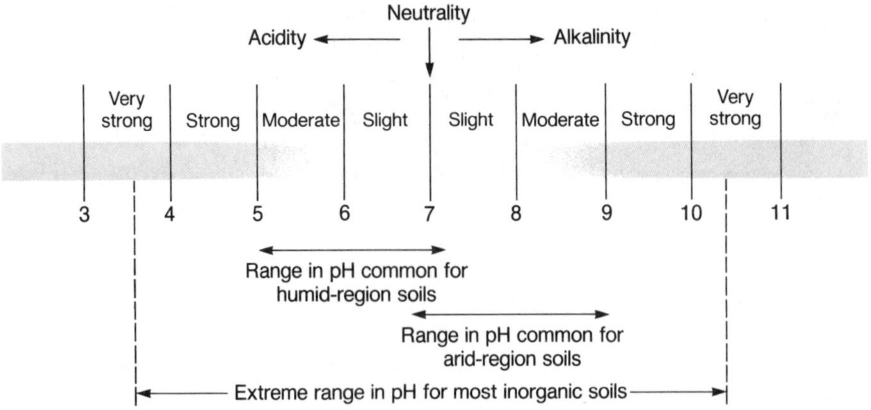

able base cations. Even if the cations are available, they are so soluble that the clay-humus colloids cannot hold them in the soil, and they readily leach away. On the other hand, a highly alkaline soil is generally infertile because it lacks H^+ ions to exchange for bases and also possibly contains toxic concentrations of sodium.

Soils of humid climates tend to be acidic due to an abundance of H^+ ions produced when water (H_2O) molecules dissociate, or break apart. In addition, rainwater is slightly acidic and has been increasing in acidity recently because of the widespread burning of fossil fuels by humans. Arid and semiarid climates, conversely, generally have alkaline soils. Because most forests grow in humid environments, forest soils usually are somewhat acidic. The soils of grassland and especially of desert environments tend to be alkaline.

Soil Horizons

Although soils vary greatly in their characteristics in different areas, one may have to travel a considerable horizontal distance before major changes in soil type are encountered. In a vertical direction, however, soil characteristics may vary greatly over distances of only a few centimeters or, at most, a meter or two. In this respect, the soil is similar to the atmosphere, which displays only a gradual variation in temperature, moisture content, pressure, an so forth horizontally, but which changes rapidly in all these characteristics vertically. In the soil, as in the atmosphere, distinctive horizontal layers tend to form as a result of different environmental controls operating at differing levels. In the soil these layers are termed horizons.

Soil horizons are layers, roughly paralleling the surface, that differ in their physical, chemical, and biological characteristics. Any of the soil characteristics, including color, texture, structure, consistency, physical and chemical maturity, numbers and types of organisms, humus content, and pH, may vary between horizons.

Soils develop horizons for two basic reasons. First, as we have seen, the soil is sensitive to its environment, and nearly all **pedogenic** (soil-forming) environmental controls vary rapidly in a vertical direction. Second, vertical translocations of materials occur in all soils because of the movements of soil water. As a result, selected materials are eluviated from some portions of the soil and illuviated into others.

The soil horizons are identified by a lettering system, as shown in Figure 12.10. From the surface downward, five major horizons, designated the O, A, E, B, and C horizons, are recognized. In order to provide greater precision, transitional horizons often are identified. For example, the subhorizons EB and BE are sometimes recognized between the E and B horizons. The characteristics and extent of development of horizons vary greatly in differing pedogenic environments, and some of the subhorizons, or even horizons, may not be present in a given soil sample.

The uppermost horizon, referred to as the **O horizon,** consists of fresh or partially decomposed organic matter. This horizon develops only where there is a substantial cover of natural vegetation; it therefore is absent in arid and some tropical regions.

The second, or **A horizon,** often is referred to as the **topsoil.** It contains a mixture of inorganic and organic matter (much of the latter in the form of humus) and typically is low in density and dark in color. The topsoil generally is the most fertile soil horizon and is the horizon in which most plants germinate.

Below the A horizon is the **E,** or **eluvial horizon.** It is a mineral horizon that is lighter in color than the overlying horizons, and is best developed in high rainfall areas, especially under a forest cover. Clay-sized particles and minute organic substances have been washed or eluviated from it by downward percolating water.

The underlying **B horizon,** sometimes termed the **subsoil,** often is a zone of illuviation. It typically is denser than the A and E horizons and may develop zones of clay accumulation. In some cases, iron oxides give the B horizon a reddish color. Unless humus has been translocated downward from the A and E horizons, the B horizon contains little organic material. The A, E, and B horizons together comprise what is sometimes referred to as the "true soil," or **solum.**

The **C horizon** consists of weathered rock material that as yet has been little affected by pedogenic processes. This material is undergoing changes that eventually will convert it to soil, but has not yet crossed the threshold that separates soil from weathered regolith. Rapid weathering in the tropics can form a C horizon more than 50 meters (150 ft) thick. In the middle latitudes, thicknesses of 1 to 3 meters (3-10 ft) are typical, but much greater thicknesses occur in some places. Below the C horizon lies unweathered bedrock or other parent material, sometimes labeled the **R horizon,** that represents the raw material for future soil.

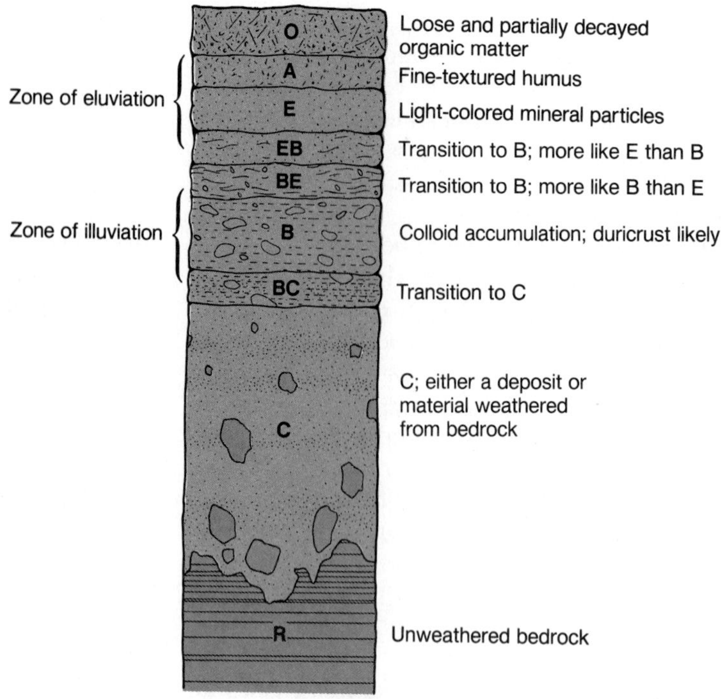

FIGURE 12.10 Idealized soil profile from a humid climate region in the middle latitudes. The relative positions of the various horizons are indicated.

Considerable time is required for soil horizons to develop. For this reason, the presence of well-developed horizons indicates that the soil is mineralogically mature and that environmental conditions have remained relatively constant over a prolonged period.

Pedologists study soils by examining **soil profiles,** which are vertical slices of the soil in which the various horizons are exposed. Each soil type exhibits a unique profile that can be used to identify it. In this respect, a soil profile is much like a fingerprint, no two of which are exactly alike. A person's fingerprint, though, tells little or nothing about the characteristics of its owner, while a soil profile can provide a great deal of information about a soil's physical and chemical characteristics. In many places soil profiles have been destroyed by human activities such as plowing and the construction of buildings and highways.

A mature soil constitutes a good example of a system in **dynamic equilibrium** with its environment. The term "dynamic equilibrium" may seem paradoxical, since dynamic implies constant change and equilibrium implies balance or changelessness. The two terms, however, each refer to a different attribute of the soil. The soil is dynamic because it undergoes a constant progression of soil-forming activities. Each soil mineral particle passes, in turn, through a sequence of horizons, beginning as unweathered R horizon material and eventually being eroded at the surface as A horizon material. On the other hand, the soil can achieve and maintain equilibrium with its environment. If the environmental controls in a given area remain constant long enough, the soil type will remain the same even though different material is constantly passing through the soil "assembly line" from initial weathering to erosion.

FACTORS IN SOIL DEVELOPMENT

To this point, our discussion has centered on the physical and chemical characteristics of soils. Although we have briefly alluded to the reasons, or controls, for some of these characteristics, we have not covered this subject in any systematic fashion. It is the geographic distribution of the pedogenic controls, however, that has determined the global pattern of soil types. Before proceeding to a discussion of world soil types, then, we need to examine the five major factors controlling the type of soil that will develop in any given locality. These five factors can be summarized by the so-called "clorpt equation," written as follows:

$$\text{Soil} = f(CL, O, R, P, T)$$

This equation indicates that the soil at any given site is a function of climate, organisms, relief, parent material, and time.

Parent Material

The first factor to influence a newly developing soil is its **parent material,** the material from which the soil develops. The parent material is the "raw material" of the soil, which undergoes progressive physical and chemical alterations as the soil forms. As might be expected, the parent material has the greatest influence on recently formed soils. Its influence diminishes, but does not entirely disappear, as the soil matures. Pedogenic processes operating in differing environments can cause greatly differing soils to form from the same parent material.

Nearly all parent materials formed from combinations of the most abundant elements in the Earth's crust. Table 12.2 lists the eight elements that individually comprise 2 percent or more, by weight, of the crust. Differing parent materials vary greatly in the numbers and proportions of the elements they contain. In addition, some of these elements are essential for plants, and some are not. As a result, the type of parent material is an important factor in determining soil fertility. For example, soils formed from sandstone or quartzite generally are much less fertile than those formed from limestone or basalt because the latter two rock types contain more of the minerals that provide important plant nutrients.

TABLE 12.2 Abundant Elements of the Earth's Crust

ELEMENT	PERCENTAGE OF CRUST BY WEIGHT
Oxygen	46.6
Silicon	27.7
Aluminum	8.1
Iron	5.0
Calcium	3.6
Sodium	2.8
Potassium	2.6
Magnesium	2.1
Total	98.5

The mineral content of the parent material also plays a part in determining how rapidly the soil matures. If chemically unstable minerals predominate, their rapid weathering soon produces a soil bearing little similarity to its parent material. Conversely, a predominance of stable minerals results in slower weathering rates and in less variation of the soil from its parent material.

Areas with complex geologic histories may contain a variety of parent materials for soils to develop upon. Abrupt changes in soil type often result as boundaries between different parent materials are crossed. Conversely, the soils over a large area may be made more similar if the entire area is underlain by the same type of parent material.

From the standpoint of origin, soil parent materials often are designated as either residual or transported. **Residual** parent materials consist of weathered or unweathered bedrock that remains in place before and during the pedogenic process. **Transported** parent materials, which are geographically widespread, have been brought in from other sites through the actions of gravity, water, ice, or wind, before being converted to soils. These transport mechanisms and the chief areas in which they operate are discussed in some detail in Chapters 16 through 20.

Relief

Topographic relief, like parent material, is a pedogenic control that can vary greatly over a very short distance, leading to a variety of localized soil types. As noted in Chapter 11, both slope steepness and aspect influence the soil.

Most differences in soil characteristics relating to slope steepness occur because of its effects on soil erosion and drainage. Because erosional processes increase in effectiveness as slope angles increase, soils on steep slopes tend to be thin, coarse textured, and poorly developed. Soil depths decrease as steepness increases until, at slope angles of about 45° or more, loose material washes, slides, or falls off slopes as soon as it forms, leaving bare rock surfaces. Steep slopes are largely responsible for the stark, rocky surfaces of the mountains of much of the western United States as opposed to the soil-covered and forested Appalachians.

The combination of thin, coarse-textured soils and steep slopes frequently results in excessive drainage. This may cause soils in mountainous areas to be droughty despite precipitation amounts that would normally be considered adequate. On the other hand, flat valley areas, which receive both water and soil from adjacent slopes, tend to have deep and often poorly drained soils. Globally, the best developed and most agriculturally productive soils are located in low relief areas.

Slope aspect affects soil for the same reasons that it affects vegetation. North-facing slopes in the northern hemisphere and south-facing slopes in the southern hemisphere are inclined away from the Sun. As a result, they develop under conditions of lower temperature and greater moisture (resulting from decreased evapotranspiration rates) than do slopes oriented toward the Sun. East- and west-facing slopes, as well as all slopes in the tropics, are not so greatly influenced, because the directional orientation of sunlight is less constant in these settings.

Organisms

Organisms play an essential role in soil formation. The numerous plants and animals living in the soil release minerals from the parent material, supply organic matter, aid in the translocation and aeration of the soil, and help protect the soil from erosion. The types of soil organisms present greatly influence the soil's physical and chemical characteristics. In fact, for mature soils in many parts of the world, the predominant type of natural vegetation is considered the most important direct influence on soil characteristics. For this reason, the description of a soil as a "prairie soil" or "tundra soil" conveys to the pedologist a great deal of information about the soil's attributes.

The quantity and total weight of soil flora generally far exceed that of soil fauna. By far the most numerous and smallest of the plants living in the soil are the **bacteria.** Under favorable conditions, a million or more of these tiny, single-celled plants can inhabit each cubic centimeter of soil. It is the bacteria, more than any other organisms, that enable rock or other parent material to undergo the gradual transformation to soil. Some bacteria produce organic acids that directly attack soil parent material, breaking it down and releasing plant nutrients. Others decompose organic litter to form humus. A third group inhabits the root systems of plants called legumes. Many important agricultural crops, including alfalfa, clover, soybeans, peas, and peanuts, are legumes. The bacteria they host within their root nodules have the unique ability to extract nitrogen from the atmosphere and change it to a form available to plants. This process, termed **nitrogen fixation,** can actually make the soil more fertile. Other microscopic plants also are important in soil development. For example, in highly acidic soils where few bacteria can survive, fungi frequently become the chief decomposers of organic matter.

Higher vegetation plays several vital roles with respect to the soil. Trees, grass, and other large plants supply the bulk of the soil's humus. The minerals released as these plants decompose constitute an important nutrient source for succeeding generations of plants as well as for other soil organisms. In addition, trees in particular are able to reach and recycle nutrients located deep within the soil, returning them to the surface (see Figure 12.11). Finally, plants perform the vital function of slowing runoff and holding the soil in place with their root systems, thus combating erosion. The accelerated erosion that often accompanies agricultural use of sloping land is principally caused by the removal of its protective cover of natural vegetation.

The faunal counterparts of the bacteria are the **protozoa.** These single-celled organisms are the most numerous representatives of the animal kingdom and, like the bacteria, a million or more can sometimes inhabit each cubic centimeter of soil. Protozoa feed on organic matter and hasten its decomposition.

Of the larger animals, the earthworm probably is the most important. Up to a million earthworms, with a total body weight exceeding 450 kilograms (1,000 lbs), may inhabit an acre of soil under exceptionally favorable conditions. Earthworms ingest large quantities of soil, chemically alter it and excrete it as casts. The casts form a high-quality natural fertilizer. In addition, earthworms mix soil both vertically and horizontally, improving aeration and drainage.

Insects such as ants and termites also can be exceedingly numerous under favorable climatic and soil conditions. In addition, mammals such as moles, field mice, gophers, and prairie dogs sometimes are present in sufficient numbers to have a significant impact on the soil. These larger animals primarily work the soil mechanically. The soil is aerated, broken up, fertilized, and brought to the surface, hastening soil development.

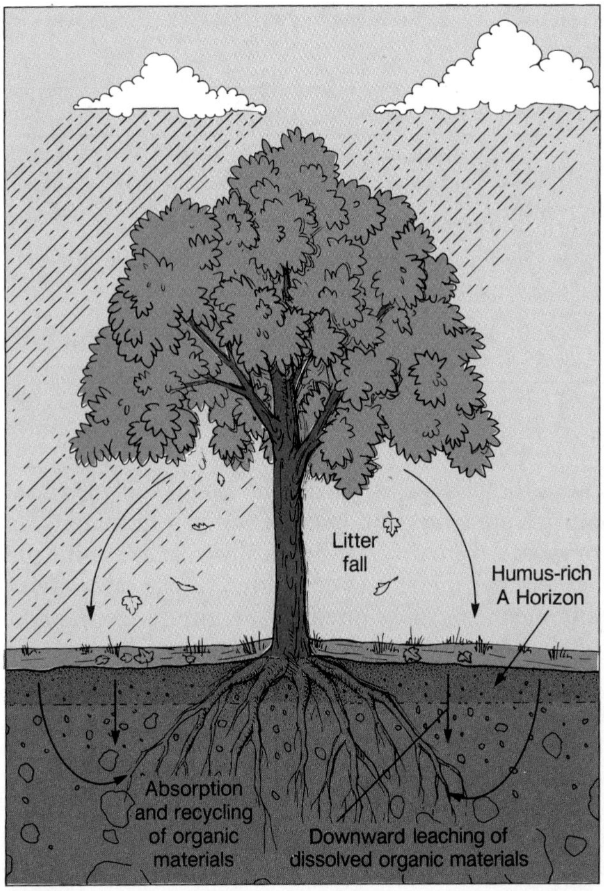

FIGURE 12.11 Trees help to maintain soil fertility by drawing up nutrients from the deeper soil layers through their root systems. These materials are subsequently added to the surface soil through the fall of organic litter.

Climate

While natural vegetation in a direct sense may be considered the most important single control in determining the characteristics of a mature soil, climate certainly is the most important control in an indirect or ultimate sense. This is because climate is the major controlling factor for the world distribution of natural vegetation. Indeed, climate has at least some effect on all of the factors in soil development discussed in this section. Only the nature of the residual parent material (if present) and, in some cases, the amount of time the soil has had to develop are factors independent of at least some climatic influence. The same two climatic elements that most strongly affect human activities—temperature and precipitation—exert the greatest effect on the soil. Extremes of either element have an unfavorable effect on soil characteristics.

Soil formation begins when minerals in the parent material combine chemically with water from precipitation. If this water is not available, as in an arid climate, most physical and chemical weathering reactions cannot occur, and a mature soil will not form. In addition, a lack of precipitation means that little moisture will be available to support biological activity and humus formation. Conversely, an excess of precipitation may lead to conditions of poor drainage, low aeration, leaching, the formation of organic acids, and possible excessive erosion. Many of the soil-forming weathering reactions again will be inhibited, especially by a lack of oxygen, and most plant organisms will not be able to survive. An intermediate amount of precipitation therefore is most favorable for the formation of fertile soils.

The precise amount of precipitation needed for optimal soil development varies greatly from place to place, depending on factors such as evapotranspiration rates, slope steepness, and the nature of the parent material. As a general rule, the most fertile soils are located in areas where precipitation amounts roughly equal evapotranspiration rates. When this balance occurs, no large-scale translocations of soil water and minerals take place, and neither significant leaching nor duricrust formation occurs. Table 12.3 summarizes the differences in soil characteristics between excessively wet and dry environments. It should be kept in mind that the most productive soils develop under moisture conditions approximately midway between those represented in the table.

In the absence of inhibiting factors such as insufficient moisture, the rate of soil formation is positively correlated with the mean temperature of an area. The tropics therefore contain the deepest soils, and soil depths generally decrease poleward. In fact, soil formation rates in the humid tropics average about 9 or 10 times those of the subpolar regions because the rate of chemical weathering is greatly accelerated by warm temperatures and ceases entirely when the soil is frozen. The soils of warm and humid environments also tend to display an increased rate of differentiation from the characteristics of their parent materials. This is because they pass through a greater variety of chemical weathering reactions in a shorter time than do the soils of colder climates.

TABLE 12.3 Differences between Soils in Wet and Dry Environments

WET	DRY
1. Deep (unless formation inhibited by standing water)	1. Shallow
2. Intermediate to fine texture	2. Coarse texture, stony
3. Chemical weathering predominant	3. Physical weathering predominant
4. Leached of soluble minerals	4. Near-surface accumulation of soluble minerals
5. Acidic condition	5. Alkaline condition
6. Variable humus content (depending on temperature and abundance of vegetation)	6. Low humus content due to lack of vegetation

Another important difference between the soils of warm and cold environments is that soil microorganisms are much more numerous and active in warm environments. This factor, in turn, affects the rate of humus formation and consumption. In cold, heavily vegetated regions such as the taiga, humus forms relatively slowly. Once formed, though, it may remain for many years before being consumed by microorganisms. As a result, humus in the high latitudes can accumulate to excessive amounts, leading to highly acidic soil conditions. Within the tropics, conversely, organic materials on or in the soil are rapidly consumed by microorganisms, producing soils deficient in humus. Thus, intermediate temperatures produce soil conditions that generally are most favorable for agriculture. As at least a partial consequence of this fact, the most fertile soils are concentrated chiefly in the middle latitudes. Table 12.4 summarizes the major characteristics of soils developing in warm and cold environments.

Time

Each of the four factors in soil formation just discussed needs time to influence the soil. Therefore, a fifth and final essential factor determining the characteristics of the soil in any locality is the amount of time that has elapsed since soil formation began.

The amount of time needed for soil formation and maturation varies geographically, making it impossible to cite precise figures that will hold true everywhere. A rough approximation of 2.5 centimeters (1 inch) per century as a world average rate of soil formation is sometimes given, but actual rates vary so greatly in different areas that a world average is of limited value. In general, soil development is fastest where the climate is warm and moist, soil microorganisms are abundant, and the soil parent material is composed primarily of chemically unstable minerals that weather quickly. Specifically, average rates of development are most rapid in the humid tropics and slowest in polar and arid environments. Rates of soil formation range from almost zero in extremely cold and arid environments to measured rates of 1 centimeter (0.4 inches) per *year* on recent volcanic ash in the East Indies.

Even at a particular site, rates of soil formation and development do not remain constant over time. A recently formed soil is not in equilibrium with many of the soil-forming factors in its environment. Change therefore is relatively rapid. As the soil ma-

TABLE 12.4 Differences between Soils in Warm and Cold Environments

WARM	COLD
1. Deep	1. Shallow
2. Fine textured	2. Coarse textured
3. Chemical weathering predominant	3. Physical weathering predominant
4. Rapid rate of soil formation	4. Slow rate of soil formation
5. High degree of mineralogic maturity	5. Low degree of mineralogic maturity
6. Soil microorganisms abundant	6. Soil microorganisms sparse
7. Humus lacking	7. Humus abundant
8. Less acidic	8. More acidic

tures, it gradually approaches equilibrium with its environment, and the pace of development slows. Eventually a steady-state condition may be reached in which soil characteristics remain essentially unchanged with time. This condition of equilibrium can be attained only if all factors affecting the soil remain constant for a prolonged period. In the face of major recent geologic, climatic, and especially human influences, few places on Earth currently have the necessary degree of environmental stability for this situation to exist.

SOIL CLASSIFICATION AND WORLD SOIL TYPES

The geographic pattern of world soil types is highly complex. As is true of all natural Earth phenomena, however, the pattern is not random; it displays an organization that is closely related to the distribution of the pedogenic factors just discussed. Because the two most important pedogenic controls are climate and natural vegetation, the pattern of world soil types is more closely correlated with these two factors than with the other pedogenic influences. It should be stressed, though, that each of the five factors in soil development is capable, under certain conditions, of exerting a dominating influence on local soil characteristics, and that each will exert at least some influence on the soil at any given site. Some of these factors, such as parent material and slope, can produce large differences in soil characteristics over very short distances, especially in areas of rugged topography.

Soil Classification Systems

Numerous attempts have been made to classify soils. The problems involved in devising an "ideal" soil classification system, however, are even more imposing than those encountered in developing classification systems for climate and for natural vegetation. Among the major problems are the following:

1. Soils display many different physical, chemical, and biological characteristics. Which ones should be stressed in a classification system? A key factor is the relative importance of the various soil characteristics. As already noted, the importance of each characteristic varies spatially, depending on the soil's environmental setting.
2. Significant variations in soil characteristics exist on a much more localized geographic scale than they do for climate and for natural vegetation. Any detailed soil classification system therefore must be exceedingly complex, and at its most detailed level of classification will contain thousands of soil types. There is, in fact, no set limit to the number of soil types that exist, and any soil classification scheme will, to some extent, be imposing arbitrary divisions on what is actually a continuum of soil types.
3. In many areas soils have been altered greatly by human activities. Should the classification system recognize these modifications, or should it be based only on natural factors?

Differences in soil types have been recognized from antiquity. For example, there is evidence that the Chinese divided soils into different types more than 4,000 years ago. Early soil classification systems were primarily descriptive and differentiated soils on the basis of physical characteristics such as color, structure, and depth. European researchers in the early and middle 1800s assumed that regional soil differences were produced chiefly by variations in parent material and used this factor, along with the texture of the topsoil, as the primary basis of classification.

Russian soil scientists first developed modern concepts of soil formation and classification in the late 1800s. They realized that the soil is not a static medium produced merely by the weathering of parent material, but rather that it evolves through time as the result of energy and material gains and losses from climate and biological activity. Emphasis on soil classification therefore shifted from parent material to the process of soil development and evolution as determined by the analysis of soil profiles. Further work, especially that undertaken by pedologists in the U.S. Soil Survey of the Department of Agriculture, led to the publication in 1938 of a genetic (formation-oriented) soil classification system that came into widespread use around the world. This system, which was modified slightly in 1949, was in official use in the United States for 27 years.

The Soil Taxonomy

During the 1950s growing dissatisfaction with various shortcomings of the Russian-American system led the U.S. Soil Conservation Service to appoint a group of pedologists to devise a completely new soil classification system for official use. This system, which was called the "U.S. Comprehensive Soil Classification System," was first published in 1960

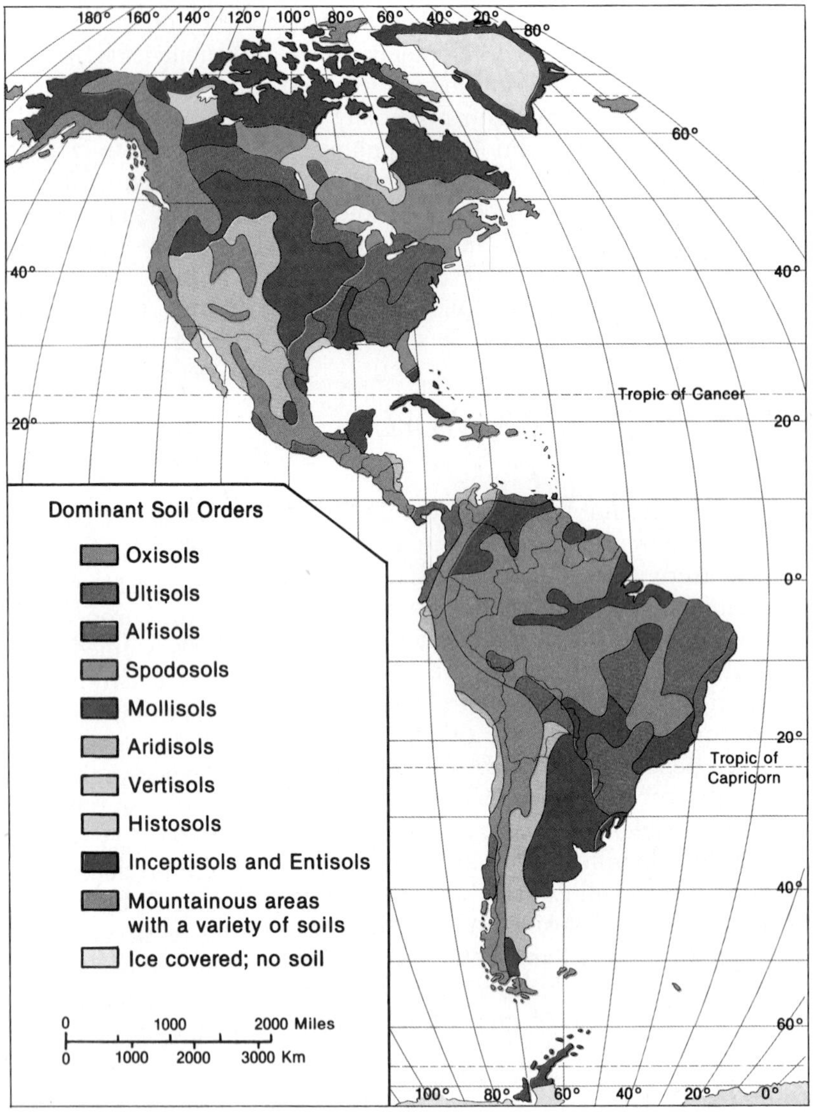

FIGURE 12.12 Global distribution of the ten soil orders of the U.S. Comprehensive Soil Classification System (the Soil Taxonomy).

and was adopted for official use in the United States in 1965. The system, often referred to as the "Soil Taxonomy," has been expanded and refined and remains in use today. It is the system used in the inventory of world soil types appearing in this section. It should be noted that although the Soil Taxonomy is used by a number of countries, others including Canada, have their own national soil classification systems. As a consequence, no one soil classification system has a clear dominance in world usage.

The major characteristics of the Soil Taxonomy are as follows:

1. Soils are classified on the basis of their characteristics rather than their method of formation. It is thus a descriptive, rather than genetic, system.
2. Soils modified by human activities are classified along with "natural" soils. (The earlier 1938 system did not recognize the profound influence of human activities on soil characteristics.)
3. The names of soil types convey exact information concerning their characteristics. Every syllable has a meaning, so that each name describes the soil in almost the manner of a chemical formula.

The system is organized into a hierarchy of six levels of classification, with individual soil types forming a seventh level, as follows:

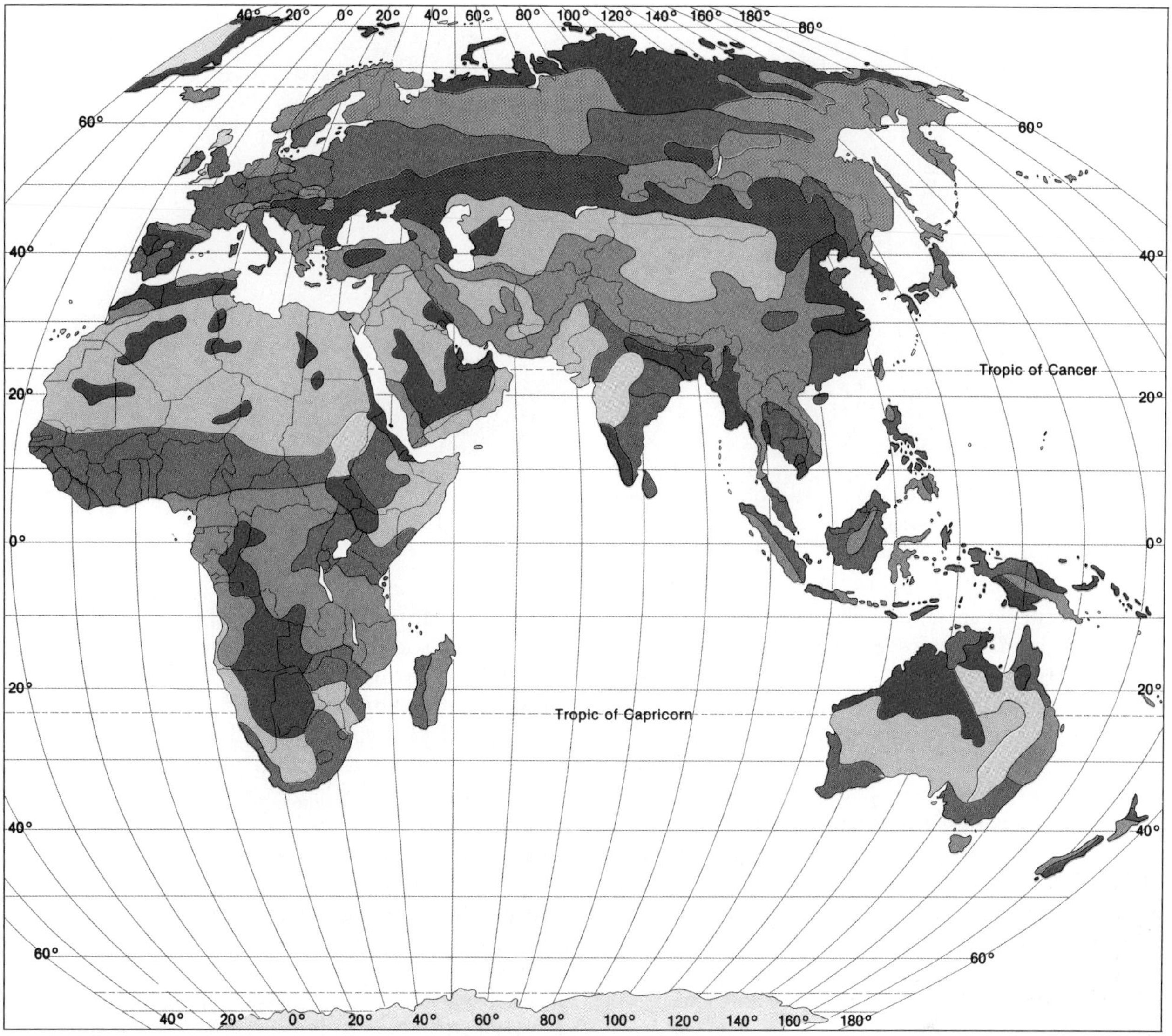

FIGURE 12.12 *continued*

Orders
 Suborders
  Great groups
   Families
    Series
     Individual soil types

Ten soil orders exist in the system. The number of entries increases at each lower level; for example, 14,000 soil series are recognized in the United States alone. (Several times as many series exist for the entire world, but they are as yet incompletely classified.) The chief characteristics of the ten soil orders are described in the next section. Their geographical distribution is mapped in Figure 12.12. As this map shows, some soil orders are much more geographically extensive and important than others.

World Soil Types

Oxisols

The **oxisols** (*oxi* for "oxide") are the deepest and most chemically altered of the soil types. Almost all minerals other than quartz have been weathered to form oxides and hydroxides, and virtually all soluble bases have been leached from the upper soil layers.

FIGURE 12.13a An oxisol developed from weathered basalt on the island of Kauai, Hawaii. (Marbut Collection/Am. Soc. of Agronomy)

FIGURE 12.13b Oxisol exposed on an eroding bank in Sri Lanka. (JLM Visuals)

Oxisols occur on ancient, geologically stable surfaces within the humid tropics, especially in South America and Africa.

The intense leaching that characteristizes these soils results from heavy precipitation, which at least seasonally greatly exceeds evapotranspiration. Not only are dissolved soil constituents transported downward, but even clay-sized particles frequently are removed as they are flushed through the pore spaces separating the larger soil grains. The remaining clay particles often are aggregated into sand-sized peds that allow water to percolate readily through the soil. Much of the leached soil material is carried completely through the soil profile and is lost permanently to the soil. Because there is no soil layer in which any substantial illuviation takes place, soil horizons are rather weakly developed and difficult to recognize (see Figure 12.13).

Intensely leached soils usually are highly acidic because bases are especially susceptible to removal. The acidity of the oxisols, however, is kept in check by their lack of humus and its associated organic acids. An only slightly acidic pH, usually in the 5 to 7 range, results. At this acidity level, silica, which usually is a highly stable and inert soil component, enters into weathering reactions that result in its solution. Large amounts of silica therefore are leached from the soil. Iron and aluminum, conversely, weather to form insoluble oxides that accumulate in the soil. Iron compounds typically impart an orange-red color that is one of the most striking characteristics of the soils of the humid tropics.

The low natural fertility of the oxisols presents imposing problems for agricultural use. Despite the presence of year-round warmth and often abundant moisture, the lack of humus and base nutrients necessitates the use of large amounts of fertilizer for any sustained agricultural use of these soils. The removal of the trees from forested areas often causes further deterioration of soil quality. Trees draw up a modest supply of soluble bases, which are placed on the surface as the plants die and decompose. A nutrient-cycling system is thus established, and a limited amount of soil fertility is maintained. Tree removal causes these bases to be leached, increases the likelihood of erosion, and in some places encourages the formation of a rock-hard subsurface layer of iron oxide called **laterite.** The current large-scale removal of tropical forests, as discussed in the Case Study at the end of Chapter 11, is causing the general deterioration of the soils over large areas.

Ultisols

The **ultisols** (*ult* for "ultimate") are highly chemically weathered soils that have developed under warm, moist climatic conditions. They are somewhat similar to the oxisols, but display less mineral alteration.

Ultisols have a distinctly reddish A horizon that has been colored by iron and aluminum compounds

FIGURE 12.14a An ultisol developed on metamorphosed sandstone in the Ouachita Mountains of western Arkansas. (Marbut Collection/Am. Soc. of Agronomy)

FIGURE 12.14b A typical view of rural farm and forest land in a portion of the North Georgia piedmont containing ultisols. (Marbut Collection/Am. Soc. of Agronomy)

and is low in humus because of rapid bacterial decomposition. The illuvial B horizon has a considerable accumulation of clay minerals; it is composed of highly weathered minerals with a relatively low cation exchange capacity (see Figure 12.14).

Most ultisols have developed under a broadleaf forest cover where the trees maintain some soil fertility by recycling soluble bases from the B horizon to the A horizon. The clearing of the forests in many areas, though, has led to the rapid leaching of these plant nutrients. The fine texture of the soil also inhibits water infiltration and has resulted in severe erosion in many areas that were cleared for agriculture.

The ultisols are especially associated with humid subtropical climate regions, but also occur in portions of the humid tropics. They are the dominant soil type of the southeastern United States, where in many areas they have suffered from nutrient depletion and erosion resulting from excessive cotton cultivation. They also are important in much of China, southern Brazil and Uruguay, the northern Andes, West Africa, northeastern Australia, and parts of the East Indies.

Alfisols

The **alfisols** (*alf* for "aluminum and iron") are fine-textured soils with a yellowish brown A horizon that has been colored by iron and aluminum compounds (hence the name) and is relatively low in humus. They have a clay-rich B horizon with a moderately high cation exchange capacity. Their fine texture often interferes with drainage, and in some areas an impermeable clay layer accumulates in the B horizon (see Figure 12.15).

The alfisols are well-developed soils that are distributed widely throughout the low and middle latitudes. They tend to develop in areas that contain clay-forming parent materials and that have a season in which evapotranspiration exceeds precipitation. They therefore are especially typical of forest/grassland transitional environments. In the United States, they occur mostly in the southern Great Lakes, Ohio Valley, Texas, and Mississippi River Valley regions. Other middle latitude sites of importance include eastern Europe, the central Asiatic Soviet Union along the taiga/steppe transition zone, and southern Australia. Within the tropics, they are especially important in southern and eastern Africa and in the African Sahel, as well as in large areas of India and Pakistan.

Their widespread geographical distribution and moderate to high natural fertility have caused the alfisols to rival the mollisols in worldwide agricultural potential. The wide variety of crops raised in these soils constitute a substantial proportion of total world food production.

Spodosols

The **spodosols** (*spodos* for "wood ash") are acidic soils with a light-colored eluvial E horizon and an

FIGURE 12.15a An alfisol developed under an oak-hickory forest in southern Michigan. (Marbut Collection/Am. Soc. of Agronomy)

FIGURE 12.15b This rolling farmland in the Finger Lakes district of upstate New York contains alfisols that are used for the raising of corn, small grains, and forage crops. (Marbut Collection/Am. Soc. of Agronomy)

illuvial B horizon (see Figure 12.16). Spodosols generally develop from a quartz-rich and often sandy parent material in a cool, humid climate. They usually occur in association with a needleleaf evergreen forest cover; the most extensive areas containing this soil type contain the taiga forests of North America and Eurasia.

The acidity of the spodosols results from the presence of abundant organic matter in the A horizon and surplus moisture. The soil pH is especially low in needleleaf forests because these trees have low nutrient requirements and do not cycle many bases from the subsoil, where soluble bases tend to collect. The leaching of soluble constituents is promoted by the abundant precipitation and by the acidity of the soil. Soil acidity restricts the removal of silica in solution, while promoting the leaching of iron and aluminum compounds. As a result, silica is the primary mineral

FIGURE 12.16a This spodosol, developed on sandy glacial outwash in northern New York, has a well-defined whitish silica-rich E horizon. (Soil Taxonomy Alt 436, Soil Conservation Service)

FIGURE 12.16b Spodosol supporting a spruce forest in Quebec, Canada. (Breck P. Kent/© JLM Visuals)

FIGURE 12.17a This mollisol, developed on glacial drift in central Iowa, has a thick, humus-rich A horizon and a granular structure. (Marbut Collection/Am. Soc. of Agronomy)

FIGURE 12.17b Recently plowed mollisol near Macomb, Illinois. This soil, derived from weathered glacial till and developed under a prairie grass cover, is among the most naturally fertile in the world. (Richard Jacobs/© JLM Visuals)

constituent in the E horizon, giving it a pale gray, ashy appearance. This characteristic led to the name given this soil type.

Dissolved materials and clay-sized particles, as well as some humus, are deposited in the B horizon. The B horizon is therefore an illuvial horizon that is denser, finer in texture, and higher in soluble bases than the E horizon. It generally has a reddish brown coloration resulting from the presence of iron oxides.

The combination of low natural fertility, high acidity, a short growing season, and sometimes a lack of soil depth presents serious drawbacks to agricultural development in regions containing spodosols. These regions are most noted for the production of root crops, such as potatoes, that have modest soil nutrient requirements and a tolerance for acidic conditions.

Mollisols

The **mollisols** (*molli* for "soft") are soils that contain a thick, dark, humus-rich A horizon and have a soft consistency when dry (see Figure 12.17). They exhibit a high CEC and a B horizon accumulation of calcium carbonate. Mollisols are associated chiefly with the prairie and steppe grasslands of the middle latitudes. A variety of suborders are recognized, based largely on moisture conditions. Nearly all mollisols are developed on transported parent material. This material usually consists of alluvium, or of glacial or **loessal** (windblown silt) deposits.

The amount of soil moisture is perhaps the most important influence on the mollisols. These soils form in regions that have a rough balance between precipitation and evapotranspiration. As a result, there is little translocation of soil materials and no significant leaching or accumulation of soluble minerals.

Climatic and vegetative conditions combine to allow the mollisols to maintain a high humus content. Water from winter snowmelt and spring rains promotes rapid early luxuriant grass growth. Drier, hotter conditions in mid and late summer as well as cold winter temperatures greatly reduce vegetative decomposition during these periods, allowing humus to accumulate. The soils of the tall-grass prairies have the highest humus content, which is generally well mixed by the upper soil fauna throughout a thick, black-colored A horizon. In the drier steppe areas, the humus content is reduced by the smaller quantity of available organic matter, and the soils are lighter brown in color. Soil depths also decrease toward drier areas because less water is available for chemical weathering.

Soil pH values are conducive to the growth of a wide variety of vegetation. The better watered areas are almost neutral or even slightly acidic, but pH increases with dryness, producing soils that are moderately alkaline and causing a subsoil concentration of calcium carbonate to develop.

FIGURE 12.18a A stony Utah aridisol is covered by semidesert vegetation dominated by sagebrush. (Marbut Collection/Am. Soc. of Agronomy)

FIGURE 12.18b A sparsely vegetated aridisol in the Atacama Desert of Peru. (Breck P. Kent/© JLM Visuals)

The mollisols have a high degree of natural fertility and are associated with some of the world's most productive agricultural regions. The better watered areas support major grain-producing economies. The American Midwest, South American pampas, Soviet Ukraine, and a part of Chinese Manchuria, which are considered the major "breadbasket" regions of the world, produce huge surpluses of corn, wheat, and oats. The drier steppe areas generally receive too little precipitation to support intense cash crop economies (except on irrigated lands) and are used primarily for raising livestock.

Aridisols

The **aridisols** are soils lacking in available water for extended periods. They occupy large areas of the low and middle latitudes that receive less than about 25 centimeters (10 inches) of liquid precipitation annually. In all, they are estimated to cover 17 percent of the world, largely within the subtropical deserts.

Chemical weathering processes are considerably hampered by a lack of water in arid regions. As a result, the aridisols are shallow, stony, and mineralogically immature, with poorly developed horizons (see Figure 12.18). Soil textures generally are coarse and sandy, leading to a low water-retention ability even when water is available. The humus content is low to completely absent because of the sparseness of the vegetation cover. In addition, organic decomposition is very slow in an arid environment, and much vegetative litter simply blows away before it can be incorporated into the soil.

An important characteristic of the aridisols is their high alkalinity. Evapotranspiration exceeds precipitation, producing a sufaceward movement of ground water and dissolved minerals. Well-drained soils typically experience an accumulation of calcium carbonate and other soluble bases at the site of water evaporation, which is normally several centimeters below the surface. This frequently produces a duricrust layer. In poorly drained depressions, salinization results in toxic conditions for most vegetation.

Despite their shortcomings, aridisols often are potentially fertile and have proven highly productive under certain conditions. The chief problem is a lack of water. Irrigation is essential not only to support plant growth, but often initially to flush excess salts from the soil. Most agricultural activity occurs in enclosed basins where water is available, the surface is flat, and soils are deep enough for farming.

Vertisols

The **vertisols** (*vert* for "invert") contain large amounts of montmorillonite clay that swells when wetted and shrinks when it dries. These soils are located in areas that experience alternating periods (often seasonal) of moisture surplus and deficiency. During dry periods, vertisols shrink and develop cracks that may be as large as 2.5 centimeters (1 inch) wide and 75 centimeters (30 inches) deep. Surface soil crumbles and is blown, or washes, into the cracks. When moistened, the soil swells, closing the

FIGURE 12.19a A vertisol consisting largely of montmorillonite clay in the Lajas Valley of Puerto Rico. (Marbut Collection/Am. Soc. of Agronomy)

FIGURE 12.19b A South Dakota vertisol. This area displays gilgai relief: the ridges, composed of calcium-rich soil, alternate with depressions underlain by soil low in calcium. (Marbut Collection/Am. Soc. of Agronomy)

cracks and covering the soil that has entered the cracks from above. The vertisols thus possess the unique property of mixing themselves and even, to some extent, of turning themselves "inside out." As a result of this natural churning action, they are almost totally lacking in horizons. Vertisols generally develop in semiarid regions under a grass cover, have a moderately high cation exchange capacity and organic matter content, and form fertile and agriculturally productive soils (see Figure 12.19). Unfortunately, their sticky consistency often makes tilling difficult, and they may contain large amounts of sodium.

The vertisols are relatively restricted in geographical extent. Most are located in areas with Tropical Wet and Dry climates in western India, eastern Australia, and parts of northeastern Africa. They occur to a lesser degree in portions of the southern and western United States, notably in Texas and California.

Histosols

The **histosols** (*histos* for "tissue") are dark organic soils of poorly drained areas. They consist primarily of decomposed vegetation, often forming a peat or muck. A lack of oxygen inhibits organic decay, so these soils can attain considerable depth through the continued addition of vegetative matter at their surface. Most histosols are highly acidic and low in fertility (see Figure 12.20).

The histosols are the least extensive and least important of the 10 soil orders. They can occur in al most any nonarid environment where drainage is poor, but are most common in the tundra regions of North America and Eurasia, where the underlying permafrost prevents adequate drainage. They also are found in lowland areas to the south that once were covered by continental glaciers. Glacial erosion and deposition in these areas disrupted the previously existing drainage systems. Significant sites of histosol formation in the coterminous United States include the Florida Everglades and the Mississippi River Delta in Louisiana.

The artificial drainage of histosol areas leads to improved aeration and often to a major temporary improvement in their fertility and agricultural potential. Oxidation of the organic matter, though, leads to eventual compaction and subsidence and to fire danger in dry periods. Smoky fires in the Everglades in recent years, resulting from a decline in the region's water table, have burned over extensive areas and have caused severe air pollution problems for the coastal cities.

Entisols

The **entisols** (*ent* for "recent") are poorly developed soils with little or no profile development (see Figure 12.21). As their name implies, most have formed very recently.

Entisols develop in a number of highly dissimilar and often localized environments; consequently, they

FIGURE 12.20a A histosol occupying a poorly drained depression in southern Michigan. (Marbut Collection/Am. Soc. of Agronomy)

FIGURE 12.20b This swampy area in the Mississippi River delta near Houma, Louisiana, is a prime site for histosol formation. (Richard Jacobs/© JLM Visuals)

vary greatly in their physical and chemical characteristics. Their most widespread areas of occurrence are in sandy deserts and in areas of regolith in dry environments. They also exist in tundra areas, in areas that have been recently glaciated, and in areas recently covered by **alluvium** (deposits of river sediments) or volcanic ash. Soils at construction sites as well as other sites at which the soil has been extensively eroded or disturbed by human activities also are classified as entisols.

Inceptisols

The **inceptisols** (*incept* for "inception or beginning") are immature soils that have some profile development, but contain weatherable minerals. Although recently formed, they generally are somewhat more mature and better developed than the entisols. They show little indication of illuvial layers or the vertical translocation of materials (see Figure 12.22).

FIGURE 12.21a This southwestern Wisconsin entisol developed on a river floodplain and contains many horizons. The development of each horizon was interrupted by an overlying deposit of fresh alluvium during a flood. (Marbut Collection/Am. Soc. of Agronomy)

FIGURE 12.21b Sea oats and beach grass partially cover an entisol formed from wave-deposited sand at Cape Hatteras, North Carolina. (Breck P. Kent/© JLM Visuals)

FIGURE 12.22a A New Zealand inceptisol that has developed on loess deposits. (Soil Taxonomy Alt 436, Soil Conservation Service)

FIGURE 12.22b An inceptisol supports this rocky tundra meadow in Norway. (JLM Visuals)

The inceptisols are found largely in three types of environmental settings, each of which lacks the physical stability for the development of a mature soil. The first is in moderately steep mountain areas, such as the Appalachians, where erosion and landslide activity are relatively common. The second is in tundra areas, where soil development is inhibited by cold temperatures, poor drainage, and permanently frozen subsoils (**permafrost**). The third type of setting, and the most important from a human standpoint, is in floodplains or deltas. These alluvial soils typically have a loamy texture, a high humus content, and a high cation exchange capacity. The alluvial soils of large floodplains, such as those of the Mississippi, Nile, and Ganges Rivers, are intensively farmed and highly productive.

SUMMARY

The soil is the loose mixture of weathered rock material, organic matter, air, and water that mantles most of the Earth's land surface. It is the medium in which most plants and many animals live, and it provides the water and nutrients they need in order to survive. The global distribution of soil types is complex and the type of soil at any given site is determined by the interplay of a variety of environmental controls.

In order to summarize systematically the most important information contained in the chapter, we return to the Focus Questions:

1. *What physical and chemical characteristics of the soil most affect its ability to support vegetation?*

The soil characteristics of greatest importance to vegetation include its texture, structure, organic matter content, and base saturation. The soil texture is determined by the mix of sand-, silt-, and clay-sized particles. Texture especially affects the soil's density and ease of cultivation, as well as its ability to absorb and retain water. The soil structure is produced by the tendency of the individual soil particles to adhere to one another, forming peds of distinctive sizes and shapes in differing environments. The structure notably influences the ability of the soil to absorb water and to resist erosion. Soil organic matter supplies a variety of plant nutrients, plays an essential role in the transfer of the nutrients to plants, and improves the soil's water-retention ability. The base saturation of the soil relates to its ability to exchange base cations that are essential for plants. It therefore is a quantifiable measure of soil fertility. The soil's base cation exchange capacity (CEC) is influenced by a variety of factors, including its humus content and its pH.

2. *What are the five factors in soil development, and what influence does each have on soil characteristics?*

The five factors in soil development collectively determine the type of soil that will develop at a given site. An unlimited number of combinations of these five factors is possible. The first factor to influence the soil is its parent material, which may be either residual or transported, and may consist of bedrock, sediment, organic matter, or some combination of these materials. A second factor is the topographic relief of the land. Steep slopes usually are adversely affected by erosion, and their soil cover typically is shallow, coarse textured, and often relatively infertile. Slope aspect also is significant because slopes that face the Sun tend to be hotter and drier than shaded slopes. A third factor is biological organisms. Microscopic plants and animals, notably bacteria and protozoa, are crucial in decomposing soil organic matter to form humus. Larger plants and animals help to loosen and aerate the soil and add organic matter. The characteristics of mature soils in well-vegetated regions are often more a direct product of the natural vegetation than of any other factor. A fourth factor, the climate, plays both a direct and an indirect role in soil characteristics because of its influence on vegetation. The two climatic elements with the greatest direct influence are moisture and temperature, and moderate conditions with respect to each are generally associated with the most fertile soils. Excessively moist conditions promote poor aeration and leaching, while dry conditions are associated with shallow and immature soils, alkaline conditions, and a lack of humus. Soils developing in high temperature regions often lack organic matter, while cold regions often have shallow, acidic soils. The final factor in soil development is time. All soils need time to develop, and all gradually evolve with time, so that a variety of different soils eventually will form from the same parent material in diverse environments. Time also allows soil horizons to form because of the differential operation of pedogenic processes at varying depths.

3. *What are the major world soil types, and what are the chief attributes of each?*

Pedologists have long been concerned with developing ways of logically and effectively classifying the great variety of existing soils. The Soil Taxonomy, which divides all soils into 10 major soil orders, currently is in official use in the United States and was used in this chapter.

The **oxisols** are the deepest and most chemically weathered of the soil types. They are closely associated with the humid tropics. High bacterial decomposition rates and intense leaching have removed most humus and plant nutrients from these soils and have left concentrations of insoluble oxides of iron and aluminum that give them a distinctive red color. The **ultisols** are reddish soils that are highly chemically weathered, have a high aluminum and iron content and limited natural fertility, and typically are associated with humid subtropical climates. They are common in the southeastern United States. The **alfisols** are mature soils that are fine textured and relatively low in humus; they also have a high aluminum and iron content. They are usually rather fertile, are typical of forest/grassland transitional environments, and are widely distributed in North America. The **spodosols** are acidic soils with a silica-rich eluvial E horizon and an illuvial B horizon. They are relatively low in fertility and typically develop under a needleleaf forest cover in the higher latitudes. The **mollisols** typically develop in middle latitude grassland regions. They have a thick A horizon that is rich in humus and has a high cation exchange capacity. They often are considered the most naturally fertile of the 10 soil orders. The **aridisols** develop in arid regions and usually are shallow, mineralogically immature, and alkaline. The **vertisols** are clay-rich soils that swell when wetted and that shrink and crack as they dry. The resulting churning action allows for little profile development. Vertisols are of limited global extent, but most are associated with grasslands. **Histosols** are composed largely of organic matter that has accumulated in swampy areas. In their natural state, most are highly acidic and infertile, but they can be productive when drained and treated for acidity. The final two soil orders, the **entisols** and **inceptisols,** are immature soils with poorly developed profiles. They have formed in areas where pedogenic processes are disturbed by natural or human-induced erosion or deposition.

Review Questions

1. From what four constituents is the soil formed? What proportion of each comprises a typical soil? What alterations must occur to the soil's two solid constituents before they can be incorporated into the soil?

2. What is soil texture? What practical effects does it have on the soil?

3. What is soil structure? What relationship exists between texture and structure? What practical influences does structure have on the soil?

4. What factors determine the amount of soil water that is available for plants at any given time? What unfavorable long-term effects does an excess or deficiency of water entering the soil have on soil characteristics?

5. What is humus? What favorable effects does humus have on the soil? What environmental conditions maximize humus formation?

6. By what process do plants withdraw nutrients from the soil? What is soil pH, and how does this factor relate to the availability of plant nutrients?

7. Why do most soils develop horizons? In what major respects does the A horizon differ from the B horizon?

8. Which of the soil microorganisms exert the greatest influence on soil characteristics? How does the influence of the larger plants and animals inhabiting the soil differ from that of the soil microorganisms?

9. What are the basic differences in the characteristics of soils that have developed in wet versus dry environments? In warm versus cold environments?

10. Why are soil classification systems devised? What are the chief difficulties in classifying soils?

11. The entisols, inceptisols, aridisols, and histosols all are regarded as poorly or incompletely developed soils. Explain how the environmental settings of each of these soil types contribute to its lack of development.

12. Which of the 10 major soil types is generally considered to have the highest natural fertility? What soil characteristics produce this fertility?

13. Which of the 10 major soil types is the deepest? Why? Explain what physical and chemical characteristics limit its agricultural potential, and what environmental factors have resulted in the development of these characteristics.

14. Which of the ten soil types dominates in your area? Describe its basic characteristics and agricultural potential in a short paragraph.

Key Terms

soil
soil texture
loam
ped
soil structure
translocation
leaching
duricrust
humus
cation exchange capacity
soil horizons
soil profile
parent material
soil taxonomy
oxisol
ultisol
alfisol
spodosol
mollisol
aridisol
vertisol
histosol
inceptisol
entisol

CASE STUDY

AMERICAN AGRICULTURE AND ACCELERATED SOIL EROSION

The United States has been endowed with vast expanses of fertile soils. The presence of these soils, coupled with the enterprise of the American farmer and the availability of modern technology, has allowed American agricultural output to far surpass that of any other country. The successes of American agriculture have not been without cost, however. The greatest problem—one that raises serious questions about our ability to continue to expand our agricultural production—is the alarming rate of soil erosion from American agricultural land (see Figure 12.23).

Soil erosion is a natural and unavoidable phenomenon that becomes detrimental when it occurs more rapidly than new soil can form. Such an unbalanced situation results in a net loss of soil and eventually of agricultural potential. Most human activities that change the landscape disrupt the delicate natural balance between soil formation and removal and greatly increase the rate of erosion.

By far the leading cause of accelerated erosion throughout the world is the clearing of land for agriculture. Removal of the protective vegetation cover exposes the soil to rapid erosion by water in wet periods and by wind in dry periods. Both agents of erosion remove topsoil, which in most cases comprises the most fertile segment of the soil. In the United States, the estimated average rate of soil formation is 2.5 centimeters (1 inch) per century, or approximately 1,360 kilograms (1.5 tons) per acre per year. Estimates of the annual loss of topsoil from agricultural land in the United States range from 5,500 to 12,500 kilograms (6 to 14 tons) per acre per year. Our farmland therefore is being eroded at several times the rate at which it is being formed. As a result, the sediment loads of rivers draining into the Atlantic are estimated to be four or five times greater than before European settlement.

Most intensive land-use activities increase the susceptibility of the soil to erosion. Unfortunately, farming is a highly intensive land-use practice. At present, approximately 20 percent of the land area of the United States is under cultivation; about 75 percent of the 4.5×10^{12} kilograms (5 billion tons) of soil lost annually to water and wind erosion in the 48 contiguous states is derived from agricultural land. During the past two centuries, at least one-third, and perhaps as much as one-half, of the topsoil on U.S. farmland has been lost. For every kilogram of grain produced in the fertile Mississippi Valley, 4 kilograms of soil are eroded from the grain fields. Of the 3.6×10^{12} kilograms (4 billion tons) of topsoil removed annually by water erosion, 0.9×10^{12} kilograms (1 billion tons) is carried to the sea, where it is irretrievably lost. The remaining 2.7×10^{12} kilograms (3 billion tons) settles in lakes, reservoirs, and river floodplains, where it causes serious sedimentation problems.

Sedimentation can be viewed as the "other side of the coin" in the soil erosion problem. In contrast to erosion, where the effects often are gradual and sometimes unnoticeable over the short term, the results

FIGURE 12.23 The erosional effect of a single heavy rainstorm on an Iowa soybean field. (Tim McCabe/Soil Conservation Service)

of greatly accelerated sedimentation rates often are painfully apparent. For example, river channels can be raised so greatly by deposition that they become highly susceptible to flooding. In addition, the rivers frequently develop wide, shallow, and constantly changing braided flow patterns, instead of stable single channel patterns. Reservoirs are filled so rapidly by siltation that their useful lifespans often are reduced to only a few decades. Harbors and river channels must be dredged constantly in order to permit continued shipping.

Many factors affect the severity of soil erosion. The most important is slope steepness, while others include length of slope; soil texture, structure, and organic content; cultivation practices; types of crops grown; and climatic characteristics—especially the frequency of heavy rainfalls and strong windstorms. The U.S. Department of Agriculture estimates that approximately 70 percent of American farmland has experienced moderate to severe erosion, while 30 percent—mostly on very flat surfaces—has experienced slight or no erosion. Figure 12.24 shows the distribution of soil erosion in the United States produced by the combined actions of water and wind. It can be seen that the most severe problems are occurring in the Southeast, the western Midwest, the Great Plains, and in parts of the Southwest. These areas include the most intensively farmed and agriculturally productive areas of the country. Within portions of these areas, considerable farmland has been abandoned because of erosion and gullying. Erosion, coupled with rapid urbanization and highway construction, has removed some 200 million acres of land from agricultural availability. This is more than half the total acreage still being farmed in the United States.

Some reduction of soil erosion rates in the United States has occurred over the past few decades, although erosion still remains unacceptably high. Conservation measures such as contour plowing, minimum tillage, strip cropping, and mulching, as well as government incentives and technical assistance, are all playing a part in reducing the problem. Unfortunately, the implementation of many of these conservation practices results in at least a short-term reduction of income for the farmer, who often finds it difficult to appreciate potential benefits that may be years in coming. Many farmers are currently in a precarious financial situation and simply cannot afford to sacrifice immediate income for long-term gains. It also is difficult to argue with the continued success of American agriculture. Many of the farmers' current economic problems stem from too much rather than too little production, and crop yields continue to increase. This increase is not, however, due to any improvement in the soil. Rather, it is a result of the continuing impact of the "Green Revolution"—the highly successful application of modern technology to agriculture—including such factors as hybrid crops, modern pesticides, increased irrigation, modern marketing facilities, and, above all, the extensive and increasing use of fertilizers. It is tempting, but dangerous, to place faith in the continued ability of science to offset soil deterioration. An increased commitment to do more to provide for the welfare of one of America's most vital natural resources—its soil—is of the utmost importance.

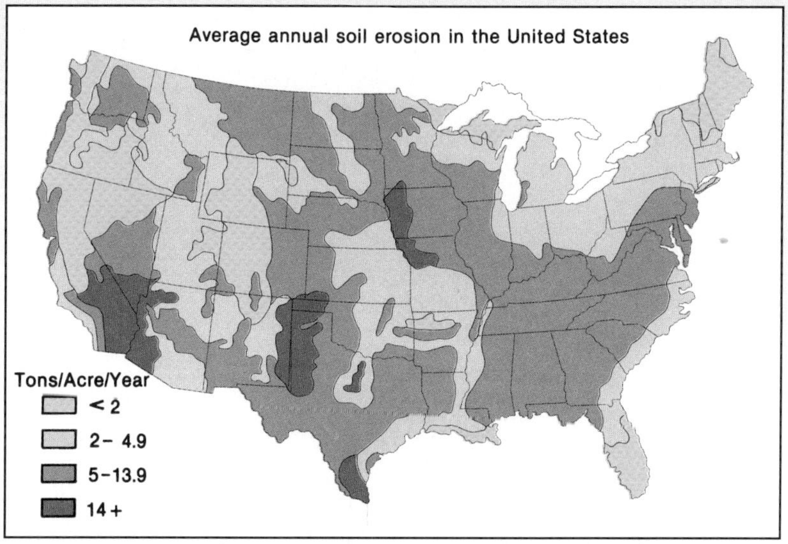

FIGURE 12.24 Soil erosion severity map of the United States. (Information provided courtesy of the U.S. Department of Agriculture.)

C H A P T E R 13

INTRODUCTION TO LANDFORMS

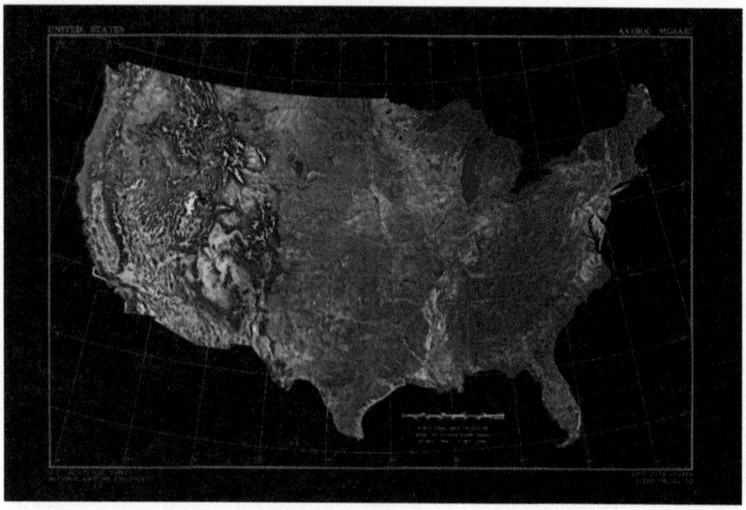

OUTLINE

Basic Terminology and Concepts
Landform Classification
Descriptive Approach to Landform Study
Global Distribution of Landforms
Genetic Approach to Landform Study

FOCUS QUESTIONS

1. What are the important landform-producing forces, and where do they originate?
2. Into what classification systems are landforms commonly organized?

This last section of the book examines the lithosphere—the massive, predominantly solid spheroid of rock and metal that comprises the heart of the Earth system. Because geography is concerned chiefly with the Earth's surface, the focus of this section is primarily on the outermost boundary of the lithosphere. It is here that the lithosphere, atmosphere, hydrosphere, and biosphere all meet and interact, and it is here that humanity resides.

BASIC TERMINOLOGY AND CONCEPTS

Despite the leveling influence of gravity, the Earth's surface is highly irregular and a great variety of surface features exist. Most areas contain a diverse assemblage of features that vary in age and origin, so that an infinite regional variety of surface form is possible.

Individual surface features of the Earth are referred to as *landforms.* A given landform may exist at almost any scale of magnitude, from a small mound to a major mountain range, but it should be recognizable as a distinct surface entity. Geographers have an interest not only in the characteristics of landforms, but also in their surface distribution. The combination of landform characteristics and distributions within a region is referred to as the region's **topography.** Maps are usually employed when landforms are studied, and those displaying both the vertical dimensions of the landforms and the spatial relationships existing among them are termed *topographic maps* (see Figure 3.10)

Geomorphology (from the Greek *geos* "Earth", *morphe* "form", and *logos* "description") is the systematic study of landforms, including their origin, characteristics, and distribution. It is a subject shared by both physical geography and geology, and geomorphologists come from both of these disciplines. Although both geographical and geological geomorphologists stress the origin and evolution of landforms, geographical geomorphologists place primary emphasis on surface characteristics, topographic relationships, and human applications, while geological geomorphologists emphasize the composition, age, and internal structure of landforms.

Like climate, vegetation, and soils, the landforms of an area have multiple controls. A number of these controls are dynamic, causing the topography to change through time. Some changes occur rapidly and can be witnessed readily by human beings. Examples include changes in the course of a river during a flood, beach erosion as the result of a storm, or the eruption of a volcano. Most changes, though, occur very slowly, and centuries or millennia may pass before the landscape is significantly altered.

A region's topography, like its climate, vegetation, and soils, strongly influences the numbers, distribution, and economic characteristics of its human inhabitants. Lowlands, especially those near large rivers or the sea, often support large populations and contain few, if any, uninhabited areas. In contrast, rugged mountainous regions are typified by localized population clusters in favored sites, separated by large areas with few, if any, inhabitants (see Figure 13.1). The relative lack of flat or gently sloping land in such regions usually greatly restricts their agricultural potential, and the difficulties of movement into and out of such regions may hinder their economic development.

Until quite recently, the major terrestrial landform-producing controls were very poorly understood. Well into the nineteenth century, geologists greatly underestimated the age of the Earth,

FIGURE 13.1 In rugged topography like the Swiss Alps, flat valley sites are often densely settled, while the surrounding mountains are unoccupied and unused. This photo shows the resort town of Graechen in the Matter Valley. (Swiss National Tourist Office)

believing it to be only a few thousand years old. One of the chief reasons for this misconception among European and American scientists was the literal interpretation of events recounted in the Old Testament of the Bible, including the creation of the Earth in a period of seven days and the worldwide biblical Flood. One biblical scholar, by adding the lengths of events in the Old Testament, was able to "prove" that the Earth was created in 4004 B.C. This allowed a period of less than 6,000 years for all the geological events that have shaped our planet's surface. Scholars could easily see that such a restricted timespan was insufficient to allow for the production of large landforms such as mountain ranges and river gorges at the very slow rates at which they currently were forming. As a result, they concluded that past cataclysmic events must have produced the Earth's major surface features. This concept has been referred to as the principle of **catastrophism.**

In the 1790s the Scottish geologist James Hutton proposed that the Earth was actually far older than previously believed, and that its surface features had resulted simply from the long continued application of the same geological processes that were currently being observed. Nearly a century passed before this so-called principle of **uniformitarianism,** the basic tenet of which is "the present is the key to the past," was generally accepted by Earth scientists. It is now known that the individual geomorphic events that shape our planet's surface are mostly small in scope and often slow in their occurrence, but the passage of vast amounts of time allows them to have a major cumulative effect.

In recent years it has become increasingly apparent that strict uniformitarianism is too rigid a concept to be fully satisfactory as the guiding philosophical principle in geomorphology. The geomorphic importance of occasional catastrophic events, such as those described in some of the Case Studies in this book, are now more appreciated than before. Such events often are capable of producing rapid changes in landforms because they exceed the existing thresholds of stability of the materials comprising the landforms. The importance to a landscape of major changes through time in the intensity and distribution of geomorphic processes, such as those accompanying the onset of an ice age, also are realized more fully than before.

Figure 13.2 depicts the geological time scale, indicating the major geologic eras, periods, and epochs; representative dates; and major geological and biological events of the past. The Earth is believed to be about 4.6 billion years old, but little detail is available about the geography of our planet's surface features before the beginning of the Paleozoic era, about 600 million years ago. Some of the major geologic events that helped shape North America took place tens or even hundreds of millions of years ago. For example, the Appalachian Mountains began forming in the early to middle Paleozoic Era, and the Rockies, in the late Mesozoic. Most contemporary individual landform features, however, are the comparatively recent products of events in the Quaternary Period, and are less than a million years old.

The geographic distribution of landforms is considerably more complex than the distributions of climate, natural vegetation, and soils. It will be recalled that these three phenomena, despite their variations from place to place, display large-scale patterns that are repeated to some extent on the continents of both hemispheres. The tendency is for the formation of latitudinal zones of similar conditions, with sequential changes occurring in equatorward or poleward directions. Solar energy powers all three systems, and its gradual reduction in intensity with latitude is ultimately responsible for the progression of changes that occur. The Sun also is a basic energy source for the production of landforms. Its influence results in a general latitudinal zonation of those features resulting primarily from solar-powered processes.

There exists, however, a second group of geomorphic processes powered not by solar energy but rather by the atomic fission of radioactive elements in the interior of the Earth. These are the tectonic processes, which, together with their associated landforms, are discussed in the next two chapters. No organized latitudinal zonation exists with regard to these forces and their resulting features. As a consequence, nearly all large regions contain an assemblage of landforms, some produced by solar-powered processes and others by internal Earth processes, while still others are produced initially by one of these groups and subsequently modified by the other. The net result is the production of topographic patterns considerably more complex, at least on the large scale, than the global patterns of climate, vegetation, or soils.

Because surface features are generally slow to change, the existing assemblage of landforms within

Era	Period	Epoch	Yrs. Ago	Significant Events
Cenozoic	Quaternary	Holocene	±10 Thousand	
		Pleistocene		Continental glaciation Europe & N. America
			±2 Million	Cascades & Sierra Nevadas uplifted
	Tertiary	Pliocene		
		Miocene		Evolution of human beings
		Oligocene	±20 Million	
		Eocene		
		Paleocene	±60 Million	Beginning of Rocky Mtns.
Mesozoic	Cretaceous			Extinction of dinosaurs
	Jurassic			First birds
	Triassic			First mammals
			±200 Million	First dinosaurs
Paleozoic	Permian			
	Pennsylvanian			
	Mississippian		±250 Million	Climax of Appalachian Mtn. building
	Devonian			
	Silurian		±350 Million	First land plants & animals, first fishes
	Ordovician			Beginning of Appalachian Mtn. building
	Cambrian			Trilobites predominant
			±600 Million	
Pre-Cambrian				First life
			±5 Billion	

FIGURE 13.2 The geological timescale.

any area is the result of both present and past geomorphic processes. In some instances, the controlling processes have changed greatly, and the landscape is still dominated by **relict** features produced by environmental processes no longer active within the region. An important group of relict landforms is the glacial features that formed during the Pleistocene epoch over much of northern North America and northwestern Europe (see Chapter 18). At any given time, however, the topography is actively evolving as it is modified by existing conditions. The tendency is for relict landforms to be removed gradually and to be replaced by others in equilibrium with the current environment. Large-scale landform changes occur so slowly, and environmental processes that can influence landforms vary so frequently, that topographic changes are occurring nearly everywhere.

LANDFORM CLASSIFICATION

Despite the complexity of landforms in terms of origin, form, and distribution, all terrestrial landform features are produced by understandable natural processes. The Earth's surface can be likened to a wonderfully complicated jigsaw puzzle that has already been assembled. It is the task of Earth scientists, especially geomorphologists, to gain an understanding of how the pieces were constructed and assembled, and why they exist in their current positions. We may be assured that, even though we do not as yet fully understand all the natural laws by which these features were produced, they are at least susceptible to understanding and operate in systematic fashion everywhere.

Physical geographers have long been confronted with a choice of two differing conceptual approaches to the study of landforms. One approach is to categorize landforms on the basis of physical appearance. This is sometimes referred to as the "descriptive approach" to landform study. A shortcoming of this approach is that features of similar general appearance, such as mountain ranges, may have greatly different modes of origin. The alternative approach is to categorize landforms by their method of formation or, more precisely, by the geomorphic processes responsible for their production. This is sometimes referred to as the "genetic approach" to landform study. Difficulties are also inherent in this approach. First, virtually all landforms are polygenetic in origin; that is, they have been produced by not one, but by a number of processes. A second problem is that features that differ greatly in physical form may have been produced by the same geomorphic processes and therefore will be grouped together.

The choice of which of the two approaches to landform study to use rests largely on the priorities that are given to the two basic landform attributes of form and process. A discussion of the origin and history of development of a landform is needed to understand its appearance and distribution. Likewise, the description of a feature's physical appearance is important when discussing its formation so that the totality of the forces acting upon it can be understood. The situation is somewhat like having to choose between two different roads to a given destination; both will reach the same end but one route may be more desirable than the other.

In recent years, most physical geographers have employed the genetic approach to study landforms, and this approach is stressed in the following chapters. When using the genetic approach, the cause-and-effect relationship between process and form is treated in chronological order, rather than being reversed, as it is with the descriptive approach. Before commencing a systematic treatment of landforms based on their processes of formation, however, it seems desirable to discuss briefly the major descriptive categories of landforms and to provide a brief overview of their global distribution.

Descriptive Approach to Landform Study

When the topography is to be divided into categories based on physical characteristics, the chief factor taken into consideration is the extent of local differences in elevation, or the topographic **relief.** Subdivisions are then made according to the origin, distribution, or characteristics of the surface features. Some differences of opinion exist as to the number and identity of the major topographic categories, but a popular division recognizes four: plains, hills, mountains, and plains with areas of high relief.

Plains

Plains are areas of restricted local relief and have a predominance of flat to gently rolling surfaces (see Figure 13.3a). The maximum amount of allowable relief is somewhat arbitrary, but a figure of 60 meters (200 feet) is frequently used. Most plains occur at rather low elevations, but plains in large interior basins of the Andes and Himalayas have elevations as great as 3,650 meters (12,000 feet).

Plains cover extensive portions of the Earth's land surface. As indicated in Table 13.1, the combination of flat plains and rolling and irregular plains comprises about a third of the Earth's land area. Plains also contain most of the world's population, largely because their relatively level surfaces produce fewer impediments for human activities such as agriculture and transportation. Plains are produced by a variety of processes and typically are located in several different portions of any given continent. The most extensive plains are situated in the interiors of continents and are largely or entirely separated from the coasts by intervening ranges of hills and mountains. **Interior plains,** which normally have elevations of less than 300 meters (1,000 feet), result from the long-term stability of the Earth's crust in these inte-

(a)

(b)

(c)

(d)

(e)

FIGURE 13.3 Major descriptive landform types. (a) Plain. Flat farmland southeast of Aberdeen, South Dakota. (b) Hills. (c) Mountains. View in the high Himalayas of Nepal. (d) Plateau with canyon. Canyonlands National Park, Utah. (e) Plain with scattered mountains. View in the Great Basin of Nevada. (a: John S. Shelton; b: Camerique/H. Armstrong Roberts; c: Tony Stone Worldwide; d: JLM Visuals; e: Marbut Memorial Slide Collection/American Society of Agronomy) rior locations. **Coastal plains** have very low elevations and typically are flatter than interior plains. Most result from processes that have raised portions of the ocean floor above sea level in the geologically recent past. As a result, they contain surface layers of weak rocks initially deposited as sediments on the sea floor. This is the case on most of the Atlantic and Gulf coastal plain of the United States.

Hills and Mountains

In contrast to the low relief and gentle slopes that characterize plains, surfaces classified as hilly or mountainous contain moderate to high relief and a predominance of sloping land (see Figures 13.3b and c). Their elevations also are much higher on average than those of plains; mountainous regions contain the most elevated portions of the Earth's surface.

Hills and mountains, although normally classified separately, actually are variations on the same theme and are discussed jointly here. Although it is agreed that hills are smaller than mountains, no generally accepted point of demarcation between the two exists. This sometimes gives rise to confusing or inconsistent terminology, since what are mountains to one group of people may be classified as hills by another. For our purposes, areas with a local relief of 60 to 300 meters (200 to 1,000 feet) are classified as hilly, and areas with a relief exceeding 300 meters (1,000 feet) as mountainous. In Table 13.1, mountainous areas have been separated into areas with low mountains and high mountains. Together, hilly and mountainous lands comprise approximately 35 percent of the Earth's surface.

Hills and mountains both result, directly or indirectly, from the uplift of a portion of the surface. They may be produced directly by internal Earth forces that thrust up individual peaks or ranges of hills or mountains. They also may result from the uplift of an extensive surface area, followed by erosion that leaves hills or mountains as erosional remnants. Both the amount of uplift and the nature and extent of subsequent erosion are critical in determining the relief and ruggedness of the topography. This in turn determines whether the landscape should be classified as hilly or mountainous. The most recent uplift of the Earth's major mountain systems have all been in the geologically recent past; that is, within

TABLE 13.1 Percentage of Land Area Occupied by Major Topographic Categories

	NORTH AMERICA (INCLUDING GREENLAND)	SOUTH AMERICA	EURASIA	AFRICA	AUSTRALIA AND NEW ZEALAND	ANTARCTICA	WORLD
Flat plains	8	15	2	1	4	0	4
Rolling and irregular plains	23	30	30	44	51	0	30
Plateaus	9	10	3	5	1	0	5
Plains with hills or mountains	18	18	11	22	19	0	15
Hills (60-300 m relief)	8	5	10	11	12	0	8
Low mountains (300-900 m relief)	10	11	21	13	12	1	14
High mountains (900+ m relief)	16	11	23	4	1	1	13
Ice caps	8	0	0	0	0	98	11
Percentage of world area	16	12	36	20	6	10	100

SOURCE: Glenn T. Trewartha, Edwin H. Hammond, and Arthur H. Robinson, *Fundamentals of Physical Geography*, 2d ed. (New York: McGraw-Hill, 1968), p. 231.

the last few tens of millions of years. Hilly regions may not have been uplifted as far, or they may represent the erosional remnants of mountain systems that were once much higher.

Plains with Areas of High Relief

In most regions where hills and mountains occur, they occupy a large enough proportion of the land that the majority of the surface is in slope. In plains, conversely, the majority of the surface is relatively level. Large portions of the Earth's surface, however, while predominantly level, contain localized areas of high relief. In these instances, at least two different groups of geomorphic processes are reflected in the landscape.

Plains with areas of high relief are divided into two categories on the basis of their surface characteristics. The first category may be described as *plains with scattered hills or mountains* (see Figure 13.3e). In these areas most of the surface exists near the low end of the elevation range, and the upland surfaces are limited in area. One of several processes may be responsible for the areas of high relief. If they are isolated, roughly conical hills or mountains, a volcanic origin is likely. If they are linear ranges, they probably were produced by folding or faulting, as in the Basin and Range region of Nevada and western Utah (see Chapter 15). If they are widely separated blocks of variable form with little linear organization, they probably represent the erosional remnants of a previously more extensive upland surface. Portions of the Piedmont of the eastern United States are an example.

The second category may be described as *plateaus with canyons or marginal escarpments* (see Figure 13.3d). A **plateau** is a relatively level upland surface that frequently is bounded on at least one side by an abrupt descent to lower elevations. In this case most of the surface is at the high end of the range of elevations, and the low elevation surfaces are limited in area.

Plateaus are most commonly created by the geologically recent uplift of a large area with an initial flat surface. One example is the Colorado Plateau, in the southwestern United States, whose level surface originally formed on the ocean floor. In some instances, however, plateaus are produced by outpourings of tremendous volumes of lava, as in the cases of the Columbia Plateau of the northwestern United States and the Deccan Plateau of India. The high elevations of plateaus cause streams draining them to have a large amount of gravitational energy, frequently enabling them to carve deep, steep-walled **canyons** or **gorges** into the plateau surfaces. The Grand Canyon, which has been eroded by the Colorado River into the rock surface of the Colorado Plateau to a mean depth of 1.6 kilometers (1 mile), is undoubtedly the world's best-known example.

Global Distribution of Landforms

The great majority of the Earth's land surface is contained within six large blocks of land termed **continents.** They consist of Eurasia (comprising both Europe and Asia, which geographically form a single continent), Africa, North America, South America, Antarctica, and Australia. Thousands of smaller islands are scattered throughout the oceans, mostly near the margins of the continents. Slightly over two-thirds of the total land surface is located in the northern hemisphere.

The global distribution of landforms is highly complex, and each continent has a unique assemblage of topographic features and patterns. It has already been noted that there is no systematic global arrangement of landforms comparable to the broad latitudinal belts of climates, natural vegetation, and soils. The distribution of the major landform units, however, is not completely random.

What might be considered the "backbones" of the continents are the major **cordilleran belts**. These consist of the rugged mountain systems, with their associated **basins** (enclosed lowlands) and plateaus, that form the world's most prominent landforms. They are the sites of the greatest current geomorphic activity. The cordilleran systems have a distinctive linear pattern, extending in continuous belts for thousands of kilometers. Most of the smaller landform features within these belts, such as individual mountain ranges and valleys, are aligned parallel to the major cordilleran axes.

For descriptive purposes, the cordilleran systems can be divided into two great belts with a collective length of some 48,000 kilometers (20,000 miles) and mean widths of several hundred kilometers; these belts pass through every continent except Australia (see Figure 13.4). One cordilleran system takes the form of a giant horseshoe that largely encircles the Pacific. It is sometimes referred to as the "Pacific

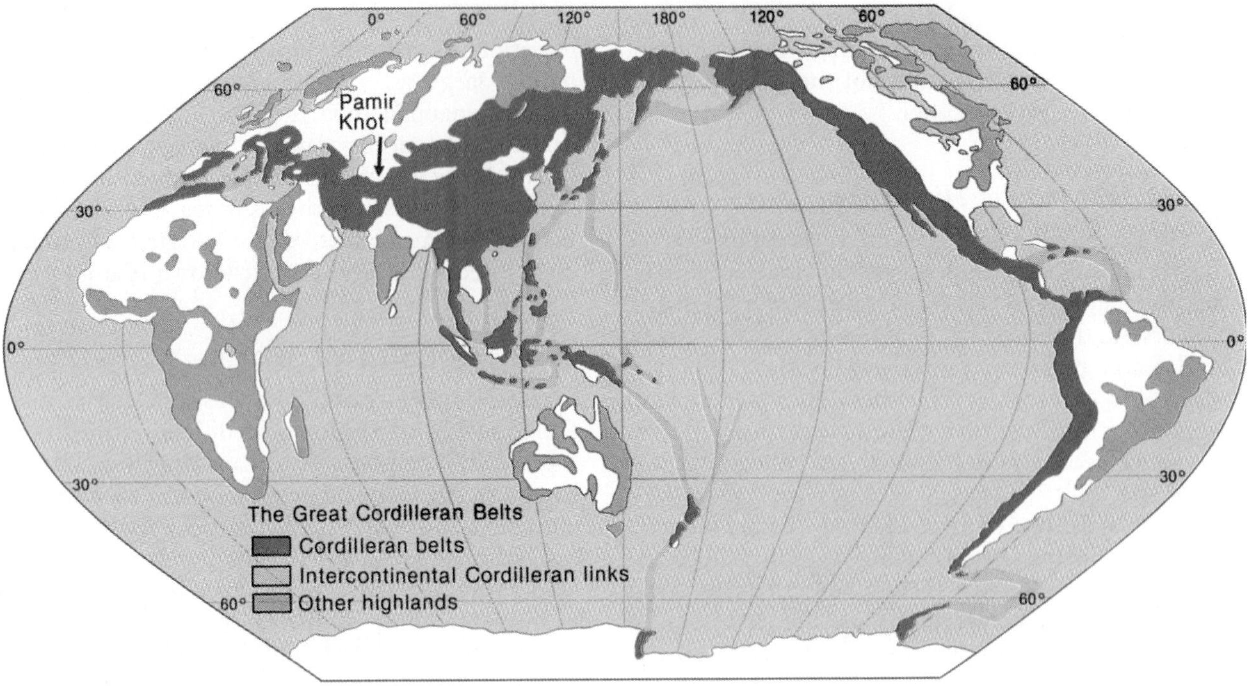

FIGURE 13.4 The global cordilleran belts ring the Pacific and extend across southern Eurasia.

Ring of Fire" because it contains most of the world's active volcanoes. With one end in the vicinity of the Antarctic Peninsula, it extends north through the Andes of South America and through Central America; it then gradually broadens to include the Rockies, the Sierra Nevada, and the Cascade Mountains of the United States. It continues through the Canadian Rockies, southern Alaska, and across the rim of the North Pacific into Siberia. From there it extends southward through the complex mountain systems of East Asia, then east through the East Indies, and south through New Zealand to reach Antarctica once again.

The second cordilleran belt is much more linear than the first and contains mountain ranges that are more diverse in origin and physical characteristics. From a connection with the first belt in eastern China, it extends westward through the Himalayas, the Hindu Kush of Afghanistan, the Zagros and Elburz mountains of Iran, the Caucasus of the Soviet Union, and the Pontus and Taurus mountains of Turkey. It continues westward to include the Balkans, Carpathians, and Apennines of southern and eastern Europe, and the Alps and Pyrenees of southwestern Europe, as well as the Atlas Mountains of Northwest Africa.

Older, lower mountain systems tend to be located on opposite sides of the continents from the cordilleran belts. These systems include the Appalachians of eastern North America and the Brazilian and Guiana Highlands of South America. In Eurasia, the Urals of the Soviet Union and the mountains of Scandinavia and the British Isles are also included. Between these peripheral mountain systems, deep within the continents, are the great interior plains that comprise the most extensive portions of the Earth's land surface. Figure 13.7 depicts the global distribution of the Earth's major topographic provinces.

Genetic Approach to Landform Study

The genetic approach to the study of landforms emphasizes the causal factors in the development of the topography rather than its physical attributes. The chief subdivisions within this system are the various landform-producing forces outlined in Table 13.2. As indicated by this table and the earlier text, two basic groups of geomorphic forces exist (see Figures 13.5 and 13.6). Each is powered by a different energy source; one influences the Earth's surface from below and the other from above.

TABLE 13.2 Landform-Producing Forces

I. Tectonic forces
  A. Diastrophism
    1. Folding
    2. Faulting
  B. Volcanism
    1. Intrusive
    2. Extrusive
II. Gradational forces
  A. Flowing water
  B. Ice
    1. Expansionary forces
    2. Glacier movement
  C. Wind
  D. Ocean waves and currents
  E. Gravity

FIGURE 13.5 An example of the effect of tectonic forces: intense folding of sedimentary rock strata exposed in a California roadcut. (© JLM Visuals)

FIGURE 13.6 An example of the effect of gradational forces: the gorge of the Yellowstone River in Yellowstone National Park, Wyoming. (© JLM Visuals)

The first group of forces, **tectonic forces,** originate within the Earth. They are powered primarily by the nuclear decay of unstable isotopes of the elements uranium and thorium (see Chapter 14). The general tendency of the tectonic forces is to increase the elevation and relief of the surface. Consequently, they are often described as being constructive forces in landform development. From a global standpoint, the tectonic forces are largely responsible for raising land above sea level and for initially producing mountains and plateaus.

The tectonic forces are, in turn, divided into two categories. **Diastrophism** involves solid-state movements of the Earth's crust. This displacement may occur either through the **folding** (bending) of the rock material or through **faulting,** which involves rock fracturing followed by the movement of the two sides of the fracture relative to one another. **Volcanism** involves the transfer of molten rock material, either from one point to another beneath the surface or, more rarely but spectacularly, its expulsion onto the surface. The mechanisms by which tectonic forces are generated and the nature and distribution of the resulting diastrophic and volcanic landforms are examined in the next two chapters.

The second group of landform-producing forces, the **gradational forces**—originate above the Earth's surface. They are powered by solar energy and are given a downward directional component by gravity.

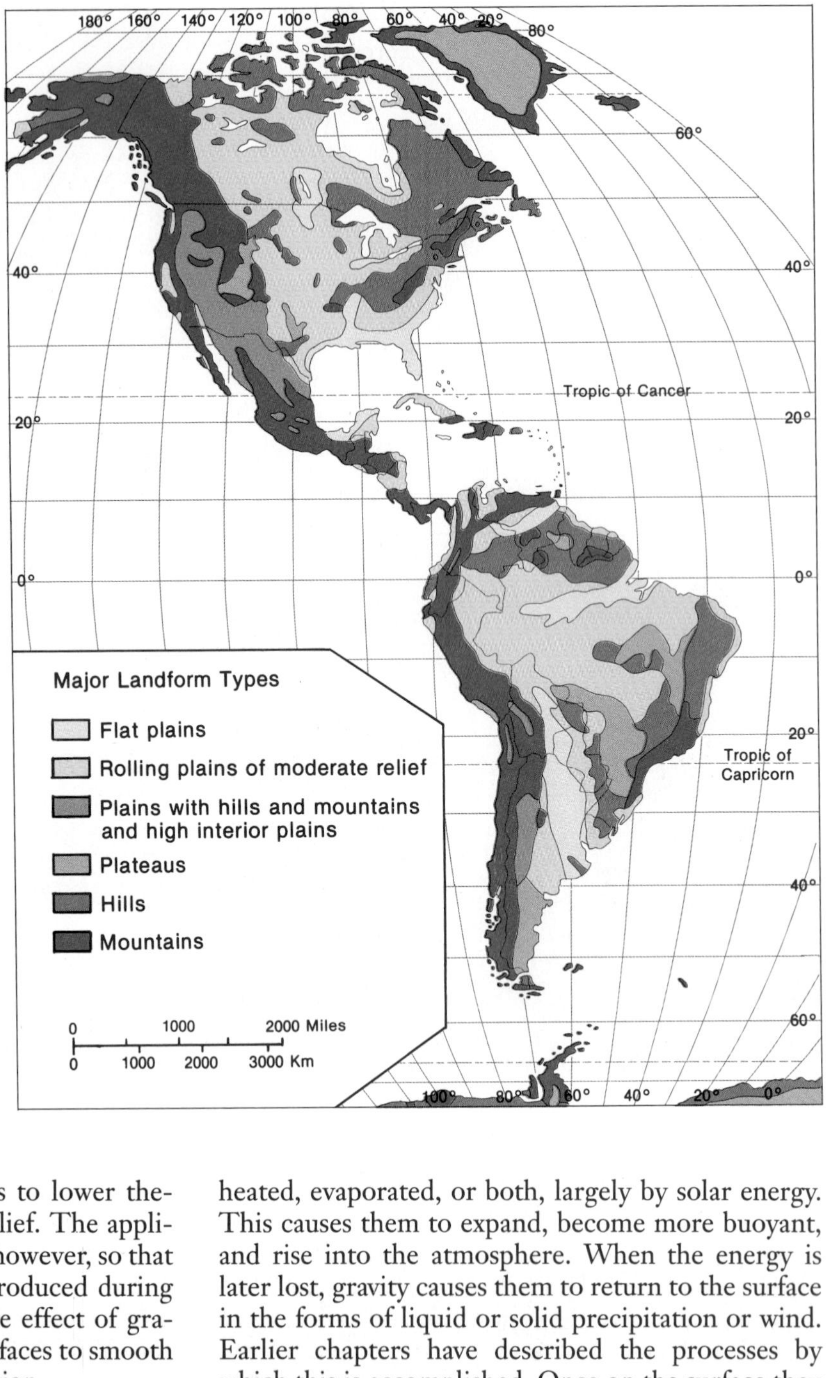

FIGURE 13.7 Global distribution of major landform types.

The overall tendency of gradation is to lower the surface and to reduce topographic relief. The application of these forces is very uneven, however, so that a sequence of changing features is produced during the gradational process. The ultimate effect of gradation would be to reduce all land surfaces to smooth plains approaching sea level in elevation.

Three different substances accomplish most of the actual gradational reduction of the surface. These are water, ice, and air (as wind), often collectively referred to as the "three tools of gradation." It is noteworthy that one is a liquid, one a solid, and one a gas. These substances accomplish their work by being heated, evaporated, or both, largely by solar energy. This causes them to expand, become more buoyant, and rise into the atmosphere. When the energy is later lost, gravity causes them to return to the surface in the forms of liquid or solid precipitation or wind. Earlier chapters have described the processes by which this is accomplished. Once on the surface they tend to flow downhill, or, in the case of air, to areas of lower pressure. The frictional drag exerted on the surface by this movement accomplishes the work of gradation. The force of gravity acting alone can also transport materials downhill in areas with loose or weak surface materials.

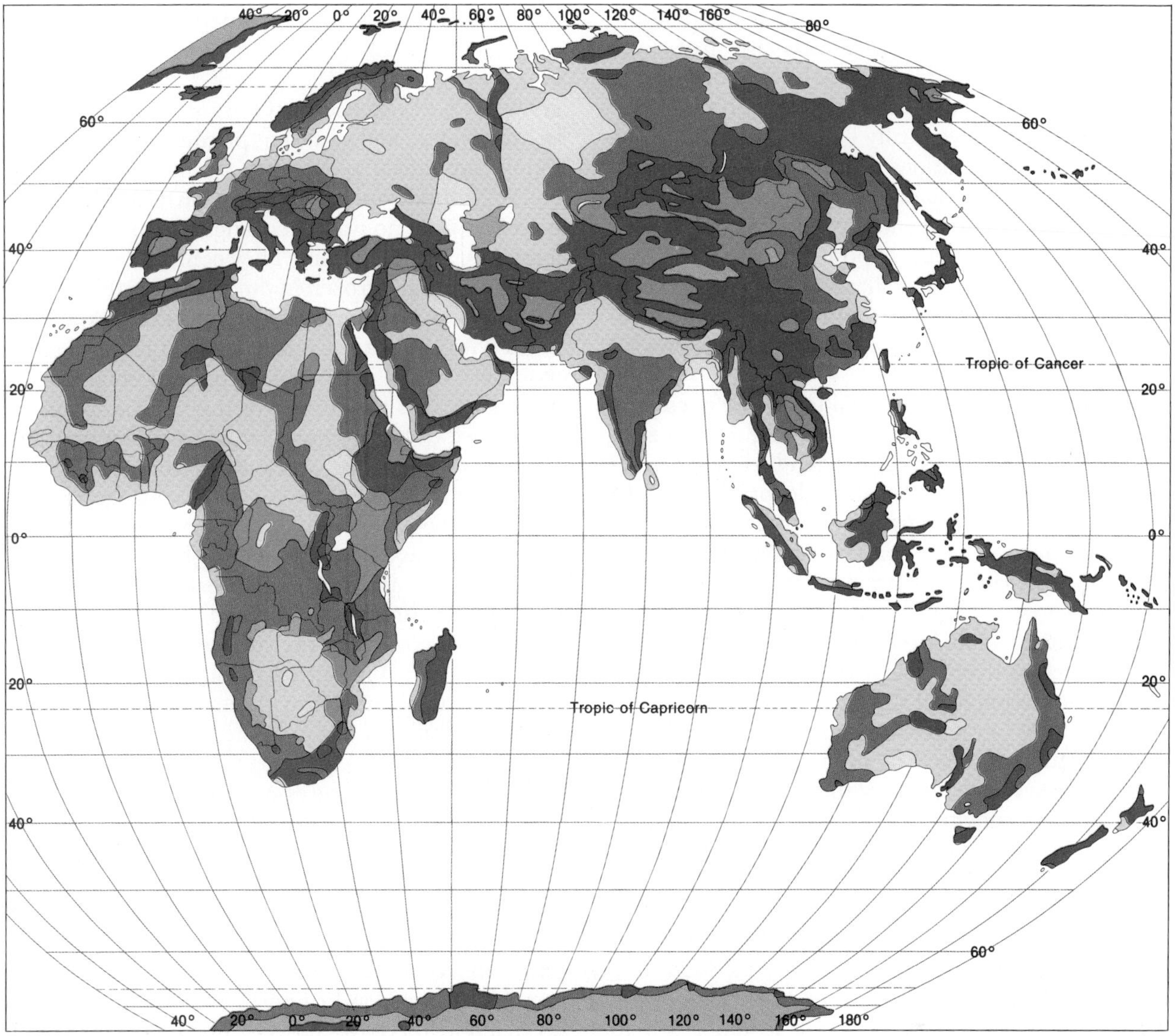

Examined in a temporal sequence, there are four steps in the gradational process, regardless of which "tool" is involved. The first step is **weathering.** Weathering actually is a preliminary process that makes gradation possible. It involves the breakup of solid rock or large rock fragments into pieces small enough to be moved by the three tools of gradation. This is necessary because the forces of gradation are not as powerful as the tectonic forces. For example, although a mountain range may be raised tectonically as a single unit, it must be carried off by water, ice, or wind piece by piece.

Once weathering (discussed further in Chapter 16) has prepared the surface, the weathered materials are in relatively quick succession picked up or **eroded,** transported, and finally deposited by one or more of the three tools of gradation. Because erosion involves the removal of material, it lowers the surface. Deposition, conversely, increases surface elevations. Erosion occurs where the tools of gradation have excess energy and are able to perform the work involved in picking up and transporting weathered material. Deposition usually occurs at relatively low elevations and in areas of low relief. It takes place where the gradational tools lack sufficient energy to continue to transport the materials they have been carrying. For the world as a whole, erosion must in the long run equal deposition, since everything that is picked up must eventually be put back down. The geographical distributions of the sites in which erosion and deposition dominate, though, differ greatly, because the environmental conditions that favor these processes directly oppose one another. As a basic geographical generalization, erosion exceeds deposition on land areas, resulting in their net gradational reduction, while deposition exceeds erosion on the ocean floor.

The tectonic and gradational forces are in essence diametrically opposed to one another in terms of their effects on the surface, and can be envisioned as being engaged in a constant struggle for a controlling influence on the topography. As soon as tectonic forces produce surface uplift, forming hills, mountains, and plateaus, gradational forces attack these uplands, carrying them bit by bit to lowlands or to the sea. The tectonic forces are much more powerful and fast-acting than the gradational forces; they are capable, for example, of thrusting up a volcanic peak in only a few years—an amount of time so short that the gradational effect is minimal. The tectonic forces, however, operate very sporadically, with periods of intense activity followed by long interludes of quiescence. A good example was provided by the violent eruption in 1980 of Mount St. Helens after 123 years of dormancy. The gradational sequence, conversely, is generally slower in its speed of operation, but it is continuous and inexorable. As soon as a tectonic feature is produced—even as it is being uplifted—gradational forces attack it and begin to wear it back down.

Insofar as astronomers and geomorphologists are able to tell, the Earth's topography is unique among the planets and satellites of our solar system. Detailed surface observations of many of these astronomical bodies have been made in recent years, and none has been found to have a landform assemblage even remotely similar to that of the Earth's. The uniqueness of the Earth's topography results more from the gradational forces than from the tectonic forces. All the planets and larger satellites of the solar system apparently either currently have, or at one time had, tectonic forces roughly comparable to those of the Earth. (Some bodies, such as the Moon, seem at present to be tectonically inactive, apparently due to the depletion of their internal radioactive fuel supplies). Not all these astronomical bodies, however, have gradational forces other than the force of gravity. The existence of these forces requires both an atmosphere and sufficient proximity to the Sun to supply an effective quantity of insolation. Those bodies without an atmosphere, such as the Moon, therefore have no way of erasing tectonic features or meteorite impact craters. They contain ancient landscapes essentially frozen in time. The Earth's surface, in contrast, is dynamic and is unique even when compared with the planets and satellites that have an atmosphere and a reasonable supply of solar energy. This is because the surface of the Earth contains an abundance of life forms, is largely covered by water, and has an overlying atmosphere containing water vapor that is engaged in a constant exchange with the water at the surface. The presence of life forms makes possible a large number of biochemical reactions that greatly accelerate weathering processes, while the cycling of surface and atmospheric water permits the erosion of the surface by streams, ocean water, and ice. Without a hydrologic cycle, only the wind, the weakest of the three tools of gradation, would be available to erode the surface. This would greatly reduce both the speed of gradation and the variety of surface features produced.

If one looks at the workings of the tectonic and gradational forces from a long-term global standpoint, a complex geological cycle can be discerned (see Figure 13.8). The same sequence of activities occurs over and over again, and the same Earth materials are continually used and reused. The sequence can be envisioned as beginning with the tectonic uplift of a portion of the surface. The uplifted region is then subjected in turn to weathering, erosion, transport to the ocean floor or some other basin, and deposition. The deposited sediment is eventually buried, compacted by the weight of the overlying materials, and gradually reconstituted as rock. At some later time this rock is once more uplifted by another episode of tectonic activity, and the sequence begins anew.

SUMMARY

The Earth's surface contains a great diversity of landform features, produced by a variety of forces acting on numerous types of materials. The branch of Earth science devoted to the systematic study of

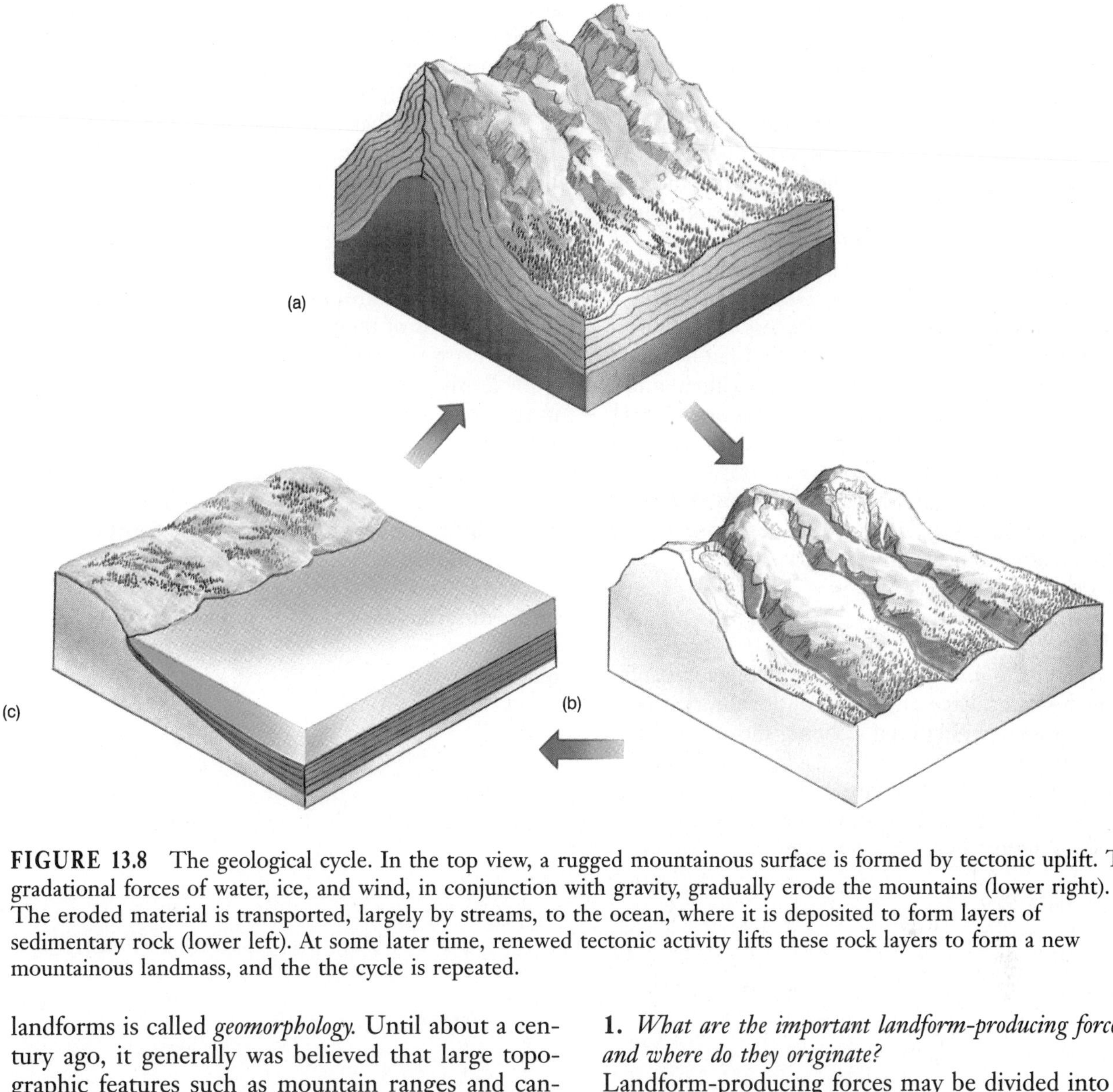

FIGURE 13.8 The geological cycle. In the top view, a rugged mountainous surface is formed by tectonic uplift. The gradational forces of water, ice, and wind, in conjunction with gravity, gradually erode the mountains (lower right). The eroded material is transported, largely by streams, to the ocean, where it is deposited to form layers of sedimentary rock (lower left). At some later time, renewed tectonic activity lifts these rock layers to form a new mountainous landmass, and the the cycle is repeated.

landforms is called *geomorphology.* Until about a century ago, it generally was believed that large topographic features such as mountain ranges and canyons were created suddenly as a result of cataclysmic events. It is now known that, although catastrophes such as earthquakes and floods can quickly produce small-scale landscape modifications, large-scale surface features form and evolve very slowly. As a result, most landscapes exhibit features of widely varying age and often contain landforms produced by forces no longer regionally active.

We turn now to the Focus Questions posed at the beginning of the chapter:

1. *What are the important landform-producing forces, and where do they originate?*

Landform-producing forces may be divided into two major groups. The first group, the tectonic forces, originate from heat released by the nuclear decay of radioactive elements deep within the Earth. The general tendency of the tectonic forces is to produce surface uplift and increase topographic relief. As a result of tectonic activity, new land masses are produced and mountain ranges are formed. The second group of landform-producing forces is the gradational forces. Powered largely by solar energy and given a downward directional component by gravity, these forces lower surface elevations and generally reduce topographic relief. Gradation operates chiefly through the erosive action of water, ice, and wind.

2. *Into what classification systems are landforms commonly organized?*

Two distinct approaches to landform classification are widely employed by geographers. The "descriptive approach" involves the division of the Earth's surface into regions of similar topographic appearance. One such system uses the four landform categories of plains, hills, mountains, and plains with areas of high relief. Descriptive systems stress surface form rather than process. The "genetic approach" to the classification of landforms involves the organization of either individual landforms or topographic regions into categories based on their processes of formation. These processes involve the tectonic and gradational forces. Consequently, the broad grouping of landforms into tectonic and gradational categories generally serves as the basis of classification in genetic systems.

Review Questions

1. Explain the differences between the geomorphic principles of catastrophism and uniformitarianism. Which principle is viewed as being more accurate today? Why?

2. Explain how the population distribution pattern within your own state or local area is related to its topographic features.

3. What are the two basic approaches to the systematic classification of landforms? What are the advantages and shortcomings of each?

4. Briefly describe or define each of the four major descriptive classes of landforms. List one or two examples of each in North America.

5. Describe the basic geographical pattern of the great world cordilleran belts. Do these belts generally lie near the centers, or the margins, of the continents? Why?

6. What are the two groups of geomorphic forces, and what is the ultimate energy source for each? How are the two groups of forces opposed to each other in terms of their influence on the topography?

7. Why are the Earth's surface features different from those of the other planets and larger satellites in our solar system?

Key Terms

topography
geomorphology
catastrophism
uniformitarianism
plain
plateau
tectonic forces
diastrophism
volcanism
gradational forces
weathering

CHAPTER 14

EARTH'S DYNAMIC LITHOSPHERE

OUTLINE

FOCUS QUESTIONS

1. What are the characteristics and locations of the internal layers of the Earth?
2. What are the major categories of rocks and minerals comprising the Earth's surface?
3. How do plate tectonic movements occur, and how do they create and destroy crustal material?

The tectonic and gradational forces act in concert to produce the surface features of the Earth. Each year, the gradational forces alone (ignoring human influences), erode some 1.0×10^{13} kilograms (11 billion tons) of material from the land and transport it to the oceans. If sediment removal proceeded unopposed at this rate, the continents would be reduced to sea level in a span of only 40 million years. A comparable mass of material, however, is lifted above the oceans annually by tectonic activity, so that the total mass of land tends to remain relatively constant.

Although the net gains and losses of mass from the Earth's land areas resulting from tectonic and gradational forces may be roughly equal, the geographical distributions of these opposing forces and the types of surface features they produce differ greatly. In a basic sense, the tectonic forces can be considered as primarily responsible for the large-scale patterns of surface features, such as the arrangements of the continents and ocean basins and the locations of the cordilleran belts. Gradational forces, on the other hand, can only operate where, and when, the tectonic forces have first created topographic features to erode.

The sizes, shapes, and locations of the continents, ocean basins, and cordilleran belts are among the most basic and important aspects of our planet's physical geography. They form the largest units of the Earth's surface features. In addition, as a result of the complex interactions that exist among the various Earth surface phenomena, they exert a vital influence on global patterns of climate, plant and animal distributions, and human activities. This chapter examines the composition of the Earth's large-scale geomorphic features as well as the processes responsible for their formation and distribution.

EARTH'S INTERIOR

The geographer is concerned primarily with the Earth's surface. The internal composition and structure of our planet is relevant to physical geography largely because it sheds light on surface processes and features. Events deep within the Earth produce the tectonic activity responsible for the formation of the continents, the oceans, and the atmosphere.

All knowledge of the Earth's interior must be obtained indirectly because humans have barely scratched the surface in their mining and drilling activities. Most information on the composition and structure of the interior has been gained through the analysis of shock waves produced by earthquakes and explosives as recorded by instruments called **seismographs.**

The Earth's interior has a layered structure produced by density differences resulting from variations in composition and temperature. It is dominated by a small group of chemical elements consisting mostly of iron, silicon, and oxygen, along with lesser quantities of magnesium, aluminum, and calcium. These are the heavy elements produced in greatest quantities by the fusion reactions in massive stars, which later spew them throughout the universe in supernova explosions (see the Focus Box on pages 12–13 in Chapter 2. The interior is differentiated into an iron-dominated central region called the **core,** and a larger silicon- and oxygen-dominated periphery containing the **mantle** and **crust** (see Figures 14.1 and 14.2).

The development of a layered interior apparently occurred during the Earth's formation. The process of planetary accretion was associated with very high temperatures, which, it is believed, were sufficient to partially melt the developing planet. The melting of the Earth's constituents allowed them to settle at different levels according to their respective masses to form the present layers of the interior. Fortunately, the separation process was not complete, or none of the heavy elements that are so essential to us today would exist at the surface.

The Core

The core is the innermost layer of the Earth and occupies one-fifth of our planet's total volume. It forms a dense metallic sphere with a radius of about 3,460 kilometers (2,160 miles). Its uppermost margin is located about 2,900 kilometers (1,800 miles) beneath the surface. Temperatures and pressures within the core are higher than anywhere else inside the Earth. All values are estimates, but temperatures are believed to range from 4,000° to 5,000°C (7,200° to 9,000°F), and pressures from 1.4×10^6 to 3.5×10^6 kilograms per square centimeter (20 to 50 million lbs/in^2). The composition of the core has not yet been ascertained, but, because of its mass, it is thought to consist largely of iron, perhaps alloyed with a small quantity of nickel.

Seismic data indicate that the core consists of two distinct layers. The larger **outer core** apparently is

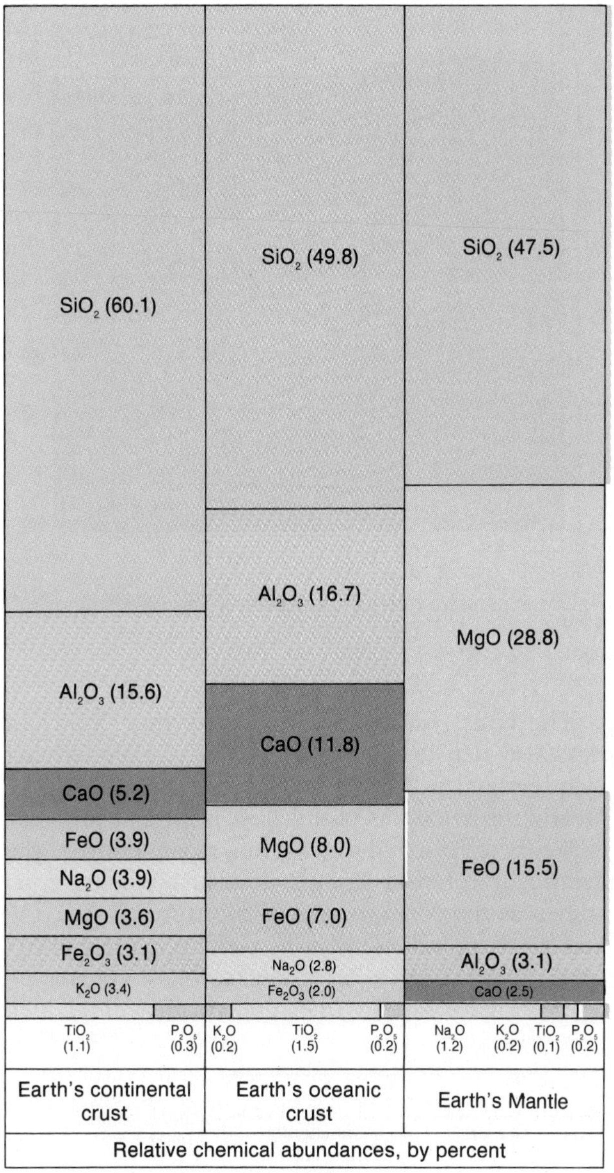

FIGURE 14.1 Proportional chemical constituencies of the mantle, the oceanic crust, and the continental crust.

liquid, since it cannot transmit the seismic s-waves that pass only through solids. Temperatures here must be high enough in relation to pressure to have caused melting to occur. The outer core is the only major portion of the Earth's interior to exist in the liquid state. Inside the outer core is a solid **inner core** with a radius of approximately 1,200 kilometers (750 miles). Temperatures here undoubtedly are even hotter than those of the outer core, but pressures are so high that liquefaction cannot occur.

The core, despite its remoteness, appears to play an important indirect role in tectonic processes affecting the surface. Apparently it contains a large proportion of the radioactive elements that power tectonic activity. The two most important of these elements, uranium and thorium, have high masses and would have tended to sink into the core during the period when the Earth was molten. Their gradual atomic decay through time has released great quantities of heat that eventually has made its way to the surface to help power tectonic processes. Fluid movements within the core also are believed to generate the Earth's magnetic field.

The Mantle

The mantle forms an extensive layer of rocky material extending from the top of the outer core almost to the surface of the Earth. It is nearly 2,900 kilometers (1,800 miles) thick and comprises about four-fifths of the Earth's total volume. In contrast to the metallic core, it is composed primarily of rock. The rock type apparently is peridotite, which is dense, dark in color, and consists largely of oxides of silicon, iron, and magnesium (see Figure 14.1). Recent research indicates that the boundary between the core and mantle is highly complex and irregular in shape. The mantle seems to penetrate several kilometers into the core in some places, and pockets of molten rock the size of small seas exist along the boundary zone in others.

The mantle, like the core, can be subdivided on the basis of its physical state. The lower mantle, below a depth of 700 kilometers (450 miles) beneath the surface, consists of immobile crystalline rock. Most of the upper mantle, in contrast, seems to be in a plastic state and is capable of undergoing a slow creeping flow in response to forces acting upon it. This plastic layer, termed the **asthenosphere,** plays a critical role in tectonic processes at the Earth's surface. The extreme upper mantle, above the asthenosphere, is a rigid solid. The uppermost portion of the mantle and the overlying crust to which it is attached are sometimes collectively termed the **lithosphere.**[1]

Heat from the decay of radioactive elements in both the core and mantle is conducted very slowly

[1] Note that this use of the term "lithosphere" is much more restrictive than its earlier application to denote the entire interior of the Earth.

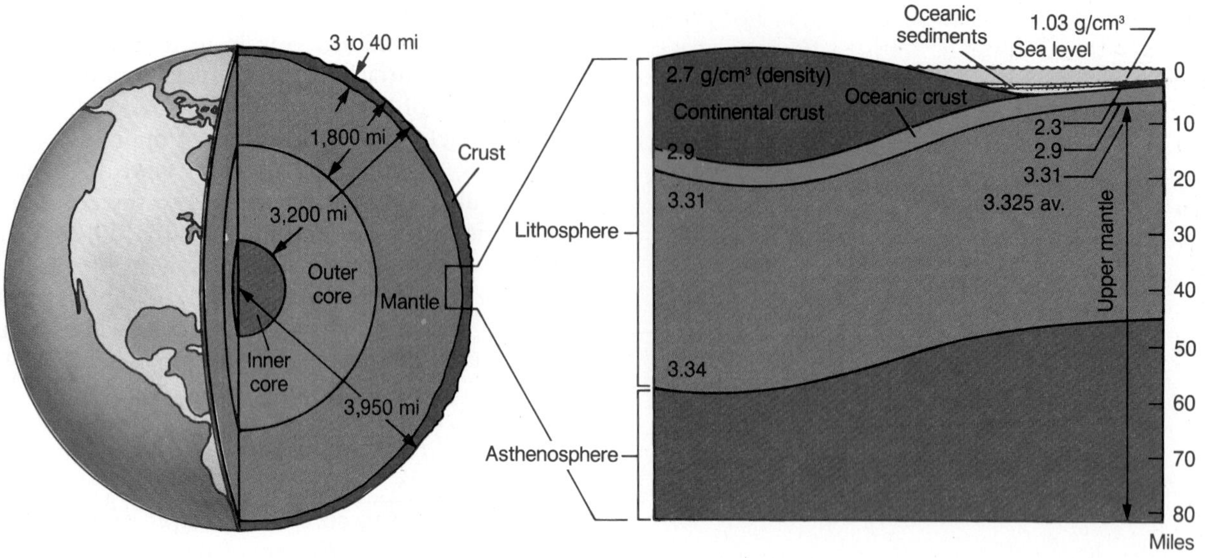

FIGURE 14.2 The internal layers of the Earth. On the right is an expanded view of a portion of the upper mantle and crust, with representative rock densities indicated.

surfaceward through the mantle. When it reaches the mobile asthenosphere, this heat generates slow convection currents within the rock material. The radioactive elements and the heat they produce are distributed unevenly within the Earth, just as they are at the surface. In places where they are relatively abundant, more heat is generated, and the mobile areas within or above them in the asthenosphere expand, increase in buoyancy, and become the sites of rising convection currents. In places where they are not as abundant, less heat is produced, and the affected mobile areas are likely to be the sites of descending currents of somewhat cooler and denser material. A continuous circulation system is thus established (see Figure 14.16). It has been found that the regions of mantle upwelling and sinking are not fixed, but instead slowly shift in various directions at mean rates of a several centimeters per year.

The Crust

The crust is the outermost layer of the Earth. Although by far the least massive of the three layers, its characteristics are overwhelmingly important to humans and to life in general and are closely interrelated with all the topics covered in this text. Landforms are merely the surface irregularities of the crust.

The crust consists of a relatively thin "skin" of rigid low-density rock that covers the underlying mantle to a depth of 5 to 70 kilometers (3-40 miles). Nearly three-fourths of the crust is formed from the elements oxygen and silicon. It is separated from the mantle by a rather sharply defined boundary zone known as the **Mohorovičić discontinuity** after the Yugoslavian physicist who first identified it in 1909 from seismic data (see Figure 14.3). Generally termed the **Moho,** this one kilometer (0.6 mile) thick

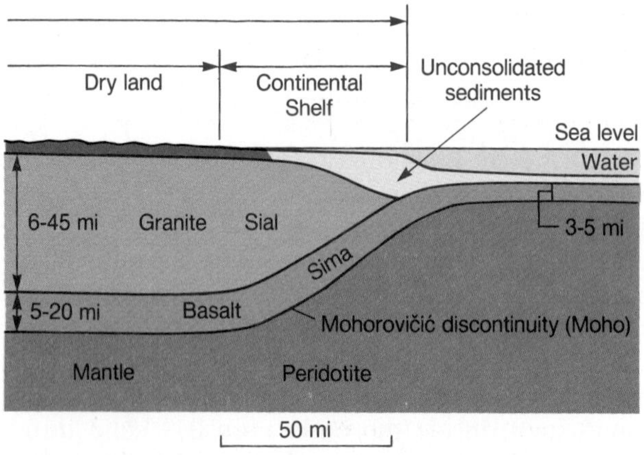

FIGURE 14.3 Cross-sectional view of a continental plate margin. The oceanic crust is shown in green and the continental crust in violet.

boundary zone tends to mirror the surface topography by lying deepest beneath high-elevation continental areas and most closely approaching the surface beneath deep ocean basins.

Like both the core and mantle, the crust is divided into two segments. The lowermost is the **oceanic crust**, a thin, continuous layer of relatively dense basaltic rock that reaches the surface beneath the ocean basins. Above this, and forming the continental land masses, is a discontinuous layer of very low density granitic rock referred to as the **continental crust.**

Approximately 65 percent of the Earth's surface, virtually all located beneath the ocean, is covered only by oceanic crust. This crustal layer generally is from 5 to 8 kilometers (3-5 mi) thick. It has a mean specific gravity of about 3.0 grams per cubic centimeter, a value some 10 percent greater than that of the continental crust.[2] In addition to oxygen and silicon, the oceanic crust contains a higher proportion of the heavier elements magnesium, iron, and calcium than does the continental crust (see Figure 14.1). The oceanic crust is sometimes referred to as the **sima,** a term derived from the words *si*licon and *ma*gnesium. Its thinness and high density cause it to rest at a relatively low level on the underlying mantle; for this reason it forms the ocean basin floors.

The continental crust is the uppermost of the solid layers of the Earth and is the only one that is discontinuous. It covers approximately 35 percent of the Earth's surface. In most places where it occurs, the continental crust extends from the surface to a depth of 10 to 70 kilometers (6-45 mi), making it much thicker than the oceanic crust. The continental crust contains a relatively high proportion of the elements aluminum, potassium, and sodium, which often combine with oxygen and silicon to form the rock called granite. The continental crust is sometimes referred to as the **sial,** a term derived from the words *si*licon and *al*uminum.

Because its thickness and low density of about 2.7 grams per cubic centimeter give the continental crust

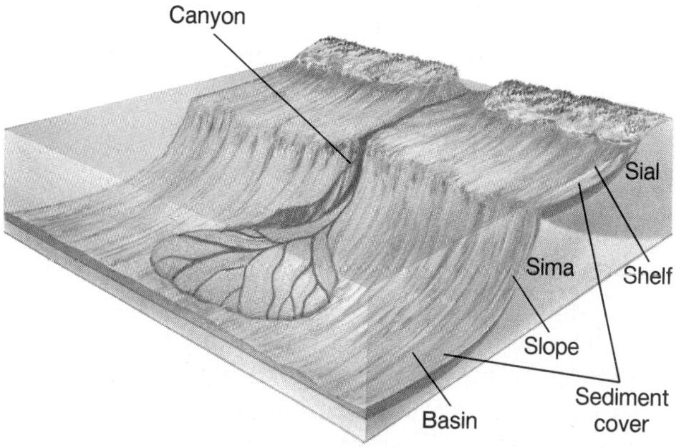

FIGURE 14.4 The tapered margins of the continental crust are overlain by the shallow ocean water covering the continental shelves. As the continental crust ends completely, ocean depths increase dramatically.

enough buoyancy to form the highest portions of the Earth's surface, it is the material of which the continents are composed. The underlying oceanic crust or sima stretches elastically to sag downward beneath the weight of the sial and the continents. The thickest areas of continental crust therefore underlie the highest mountains while the thinnest areas are beneath shallow ocean water, usually along the continental margins. Ocean depths increase dramatically as the blocks of continental crust taper off and end (see Figure 14.4).

While all of the oceanic crust is believed to have formed within the past 200 million years, much of the continental crust is very old. The most ancient portion of the crust to be dated so far is located in Canada's Northwest Territories and has an age of approximately 4.0 billion years. This is 87 percent of the 4.6 billion year age of the Earth itself. Because of its great age, many areas of continental crust have been subject to multiple episodes of tectonic activity and currently exhibit complexly deformed rock structures.

ROCKS AND MINERALS OF THE EARTH'S CRUST

The solid surface of the Earth is composed almost exclusively of rock materials and their weathering products. Rocks in different regions vary greatly in origin, composition, appearance, and in both physical and chemical characteristics. The distribution of

[2] **Specific gravity,** commonly stated in grams per cubic centimeter, is a frequently used measure of density. Fresh water has a specific gravity of exactly 1.0 g/cm^3; therefore, a given volume of oceanic crust weighs three times as much as the same volume of water. By comparison, the mean specific gravity of the Earth as a whole is about 5.5 g/cm^3, and that of the inner core averages 12.5 g/cm^3.

different rock types over the surface is in itself an important aspect of our environment. Two consequences of this distribution that are of great practical importance to humanity are the global patterns of soil types and that of mineral deposits. Of greatest relevance to this section of the text, however, is the fact that rock types and structures are crucial in determining the characteristics of landforms. In general, high-standing portions of the surface either have been uplifted recently by tectonic forces, or are composed of physically strong and chemically stable rock types that resist gradational forces. In contrast, lowlands frequently are underlain by rocks that are highly susceptible to weathering and erosion. A basic understanding of the major categories of surface rock types is therefore an important prerequisite to the study of landforms.

The Earth's crust contains only a small number of chemical elements in abundance. Only eight, in fact, individually account for more than 1 percent of the weight of the crust (see Table 12.2). Oxygen and silicon are about three times as abundant as all the rest combined. These two elements usually occur in chemical combination with metals to form the **silicate minerals,** the dominant mineral group of the crust and mantle. A **mineral** is a naturally occurring, solid, inorganic compound. It has a specific chemical composition and usually exhibits a crystalline structure that results from a distinctive atomic arrangement of its constituent elements. The more than 2,000 known minerals are divided into several major mineral groups dominated by the silicates, which comprise 92 percent of the Earth's crust.

Minerals can be considered the building blocks of rocks. A **rock** is simply a solid aggregate of minerals. Most rocks contain several different minerals, the individual crystals of which are sometimes readily visible (see Figure 14.5). In most areas the solid surface rock, or **bedrock,** is overlain by comparatively thin layers of weathered rock and soil. Most of the continental crust is composed of rocks with a high content of silicon and aluminum. These rocks are normally low in density and light in color. The oceanic crust, in contrast, consists almost entirely of denser, dark-colored rock containing a greater abundance of iron, magnesium, and calcium. This rock is sometimes brought to the surface by volcanic eruptions or is exposed by the long-term erosion of a land mass.

FIGURE 14.5 This closeup photo of a granite specimen clearly indicates its crystalline structure. (John S. Shelton)

Major Categories of Rocks

From the standpoint of the Earth scientist, the most meaningful way of categorizing rocks is by their method of formation. Three basic rock categories are recognized: igneous, sedimentary, and metamorphic. Within each category, a large variety of individual rock types exists. Differences among rock types are determined by mineralogical makeup and mode of origin as well as by the presence, size, and orientation of their constituent mineral crystals or rock particles.

Any rock type in any one of the three major rock categories discussed below can be converted by Earth processes to one or more types in either of the other two categories. These conversions occur because of changes in the environmental conditions that influence the rocks. A given rock type can form only in an environment where physical and chemical conditions favor its formation. If the rock-forming environment changes, or if the rock is transported to a different environmental setting, it is no longer in equilibrium with existing conditions. It therefore begins to change toward a condition of equilibrium with its new environment. For this reason, rocks that have formed deep within the Earth's interior, under conditions of high temperature and pressure, are chemically less stable, and consequently weather much faster, than most rocks that have formed on or near the surface.

The continuous cycle of rock formation and transformation is known as the **rock cycle.** Like the hydrologic cycle, it involves the same materials over

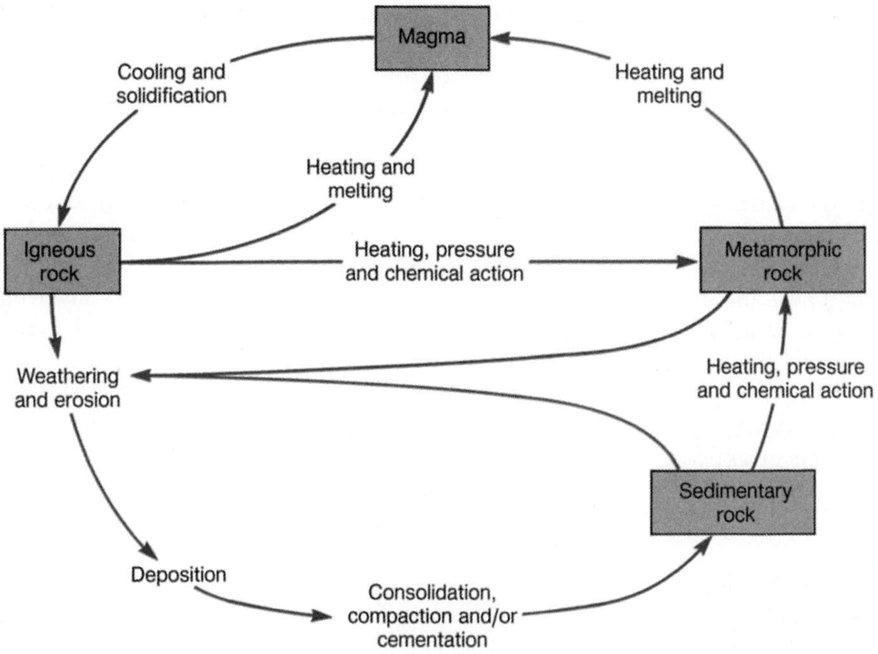

FIGURE 14.6 The rock cycle. Under proper conditions, a rock in any one category can be transformed to a rock in any other category.

and over again and contains several alternative routes that are capable of eventually returning participating materials to their point of origin. A diagram of the cycle appears in Figure 14.6. The energy that powers the rock cycle is derived from both the Sun and the interior of the Earth. As long as these two energy sources continue to operate, the materials of the Earth's crust will continue to be involved in a constant cycle of change.

Igneous Rocks

Igneous rocks (from the Latin *igneus*, "from fire") are formed directly from the solidification of molten rock and are produced by tectonic activity. If the Earth were at one time in a molten state, as is believed to have been true during the period of planetary formation, it must have consisted entirely of igneous rock when it solidified. This means that the constituents of all rocks must originally have been derived from igneous rock. The great majority of the Earth, including all of the mantle and oceanic crust and even most of the continental crust, still consists of igneous rock. Only within a few kilometers of the surface are sedimentary and metamorphic rocks found.

Most igneous rocks, including many now at the surface, formed in the lower crust or upper mantle. Although both of these segments of the Earth's interior consist primarily of solid rock, localized **magma** (molten rock) bodies form in the vicinity of zones of weakness where pressures are not sufficiently high to keep the rock from expanding and melting. Such masses of magma may be injected into overlying rock layers before solidification occurs, forming bodies of **intrusive igneous rocks.**

The most visible characteristic of intrusive igneous rocks is the large size of their constituent mineral crystals. These crystals, which have time to form because of the very slow rate of cooling and solidification of intrusive igneous rock bodies, normally are easily visible, and their sharp edges give the rock a rough, granular texture (see Figure 14.5). The best-known intrusive igneous rock type is **granite,** which is composed mostly of the minerals quartz, mica, and feldspar. Other common intrusive rocks include **diorite** and **gabbro.** Both of these rock types lack quartz and contain quantities of dark-colored minerals rich in iron and magnesium.

On occasion, magma makes its way to the Earth's surface, where it flows out as **lava** or is thrown out as **volcanic ejecta** (see Chapter 15). The solidification of this material forms what is termed **extrusive igneous rock** because it is extruded onto the surface while still molten. Because of the rapidity with which it cools, extrusive igneous rock generally has microscopic mineral crystals and, unless it contains air pockets, is relatively smooth. **Basalt,** a dense, black rock containing substantial quantities of iron and magnesium, is the best-known extrusive igneous rock type. The Hawaiian Islands and the Columbia Plateau of the northwestern United States consist largely of basalt. A less dense, lighter-colored extrusive igneous rock containing larger quantities of silicon and aluminum is termed **rhyolite,** while a third common rock type, **andesite,** is intermediate in composition. Some igneous rock types, including **obsidian** and **pumice,** have cooled so rapidly that mineral crystallization did not occur at all, and they are glassy in texture (see Figure 14.7 c and d).

Regional variations in igneous rock-forming processes have important topographic and economic implications. The types and relative proportions of minerals that crystallize during the formation of igneous rock masses strongly influence the nature of the landforms and soils that result from subsequent surface weathering processes. This mineralogical makeup is determined by the chemical constituency of the magma, which varies geographically. In addition, a chemical separation process takes place within subsurface magma bodies as the lighter, more silicon-rich materials tend to rise toward the top and the denser materials containing metallic elements sink toward the bottom. When mineral crystallization and rock formation occur later, this density layering process results in a vertical sequence of rock types within the igneous rock mass.

In most places where intrusive igneous rocks are exposed at the surface, they consist of rock types formed near the top of a magma body. This is because surface erosion rarely has proceeded far enough to expose the denser rock types often situ-

(a) (b) (c) (d)

FIGURE 14.7 Important igneous rocks. (a) Granite specimen with individual mineral crystals that are readily apparent. (b) Hand specimen of basalt. (c) Obsidian, or volcanic glass, is an extrusive igneous rock that solidified so quickly that mineral crystals did not have time to form. (d) Pumice specimen. This rock type contains so many gas pockets (or vesicles) that it can float on water. (a, c: Reprinted from Wicander/Monroe, *Historical Geology*, © 1989 West Publishing Co.; b:©JLM Visuals; d: John S. Shelton) ated several kilometers deeper. As a result, intrusive igneous rocks typically are composed of chemically stable, silicon-rich minerals. They therefore are less susceptible to surface weathering and erosion and are associated with less fertile soils than are the extrusive igneous rocks, which have solidified from a more thoroughly mixed magma body.

Intrusive igneous rock bodies (known as **plutons**) also serve as sites for the formation of veins of economically important metallic minerals. These veins form along fractures or other zones of weakness in the slowly cooling plutons. Mineral-bearing fluids enter and move along the fractures, depositing dissolved minerals because the reduction of pressure in these zones of weakness reduces the ability of the fluids to carry minerals in solution. Richly mineralized igneous rock bodies are the sources of much of the world's supplies of metals such as gold, silver, copper, and tin.

Figure 14.8 depicts the global distribution of surface rock types. It can be seen that intrusive and extrusive igneous rocks together cover about one-third of the Earth's land surface, and that the intrusive rocks are the more common of the two groups. The distributional pattern of intrusive igneous rocks is complex, but the largest areas consist of the cores of ancient plutons, such as those in eastern Canada, Scandinavia, and Africa. Only the erosion of great thicknesses of overlying rock masses has allowed these ancient rocks, which originally formed at great depth, to exist at the surface.

Extrusive igneous rocks are more commonly associated with oceanic crust than with continental crust. On land, they occur mostly within active volcanic zones located along the margins of the continents. They are quite common, for example, all along the borderlands of the Pacific. Because volcanic eruptions usually are rather localized phenomena, the resulting extrusive igneous rocks are scattered in numerous patches, many of which are too small to appear in Figure 14.8. In a few instances, though, large-scale fissure outpourings of lava have covered extensive areas. Notable in this regard are the Columbia Plateau of the western United States, the Paraná Plateau of southern Brazil, and the Deccan Plateau of India. These outpourings are discussed further in the next chapter.

Sedimentary Rocks

Weathering processes result in the production of loose surface materials sufficiently small and light to be picked up readily and transported to other sites by the gradational agents. Any such solid materials, of either organic or inorganic origin, that have been transported by water, ice, wind, or gravity, are collectively termed **sediments.** Familiar examples include mud or gravel deposits left by streams, beach sand and shells washed ashore by the ocean, unsorted rock debris deposited by retreating glaciers, and windblown leaves and dust.

Sediments are deposited where the energy of the transporting agents is insufficient to carry them farther. Because water is by far the most effective gradational tool, most depositional sites are associated with bodies of slow-flowing or standing water. Examples include valley floors, lake bottoms, and especially the ocean floor. The ocean can be thought of as the ultimate sediment trap. Once sediments reach it, as they do in tremendous quantities, there is normally no way for them to be removed because most sediments are denser than water and sink to the sea floor. The great majority of land-derived sediments reach the ocean by way of rivers, so that the greatest rates of sediment accumulation are along the continental margins, especially near the mouths of large rivers.

Where conditions remain favorable for the deposition of sediments over extended periods of time, they gradually accumulate to form horizontal layers termed **strata.** If accumulation is rapid, great thicknesses of material eventually may collect, resulting in the deep burial of the lowermost sediments. This generates a large amount of pressure, compacting these sediments. Frequently water circulating through the compacted sediments deposits chemical precipitates that cement the deposits together. The processes of compaction and cementation, along with chemical changes in the sediments themselves, result in their gradual conversion to **sedimentary rock.**

The sediments from which sedimentary rocks form may consist of rock particles of various sizes, of the remains or products of plants or animals, or of the products of chemical action or evaporation. As a group, sedimentary rocks are the softest and most pliable (or least brittle) of the three major rock categories. They also are the most widely distributed of the three at the Earth's surface. Although they comprise only about 5 percent of the volume of the continental crust, they form a relatively thin veneer over approximately two-thirds of the land surface and virtually all of the sea floor beneath the cover of marine sediments.

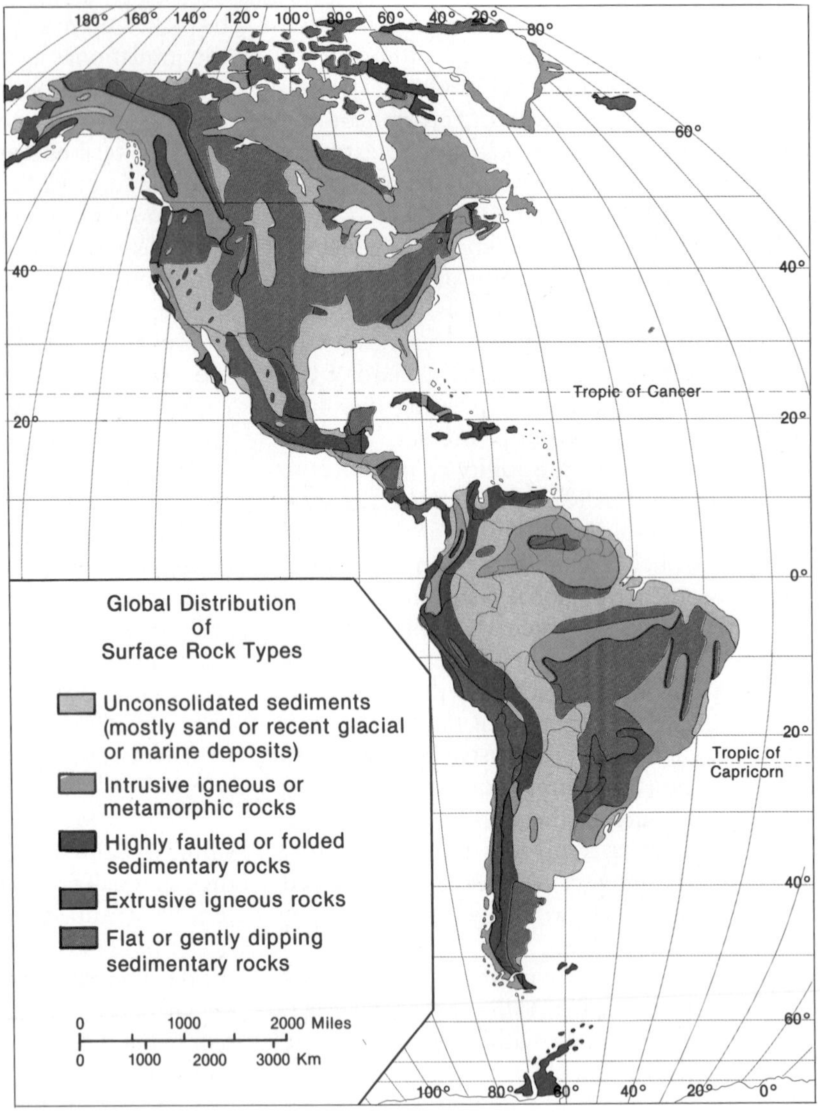

FIGURE 14.8 World distribution of rock types. Sedimentary rocks are most abundant, covering approximately two-thirds of the Earth's land surface and nearly all of the sea floor.

The boundaries or **bedding planes** between adjacent strata mark periods of gradual or sudden change from one rock-forming environment to another. Thick sequences of sedimentary rocks containing many hundreds of individual strata occur in various parts of the world (see Figure 14.9). Where deposition was continuous, these strata may provide information about past environmental conditions over a lengthy geological timespan. They bear eloquent testimony to the great age of the Earth and to the inconstancy of surface environmental conditions.

SEDIMENTARY ROCK TYPES Sedimentary rocks are categorized by their method of formation. Three basic categories are recognized: clastic, organic, and chemical. Each contains a number of individual rock types, the most important of which are listed in Table 14.1.

Clastic sedimentary rocks are composed of particles of preexisting rocks termed **clasts.** These particles were derived from the weathering of the parent rock and were subsequently transported, deposited, and **lithified,** or converted to rock. Clastic sedimentary rock types are differentiated on the basis of the sizes of their constituent clasts. **Conglomerate** contains the largest particles, which must be at least gravel size (see Figure 14.11a). **Sandstone,** one of the most abundant sedimentary rock types, consists

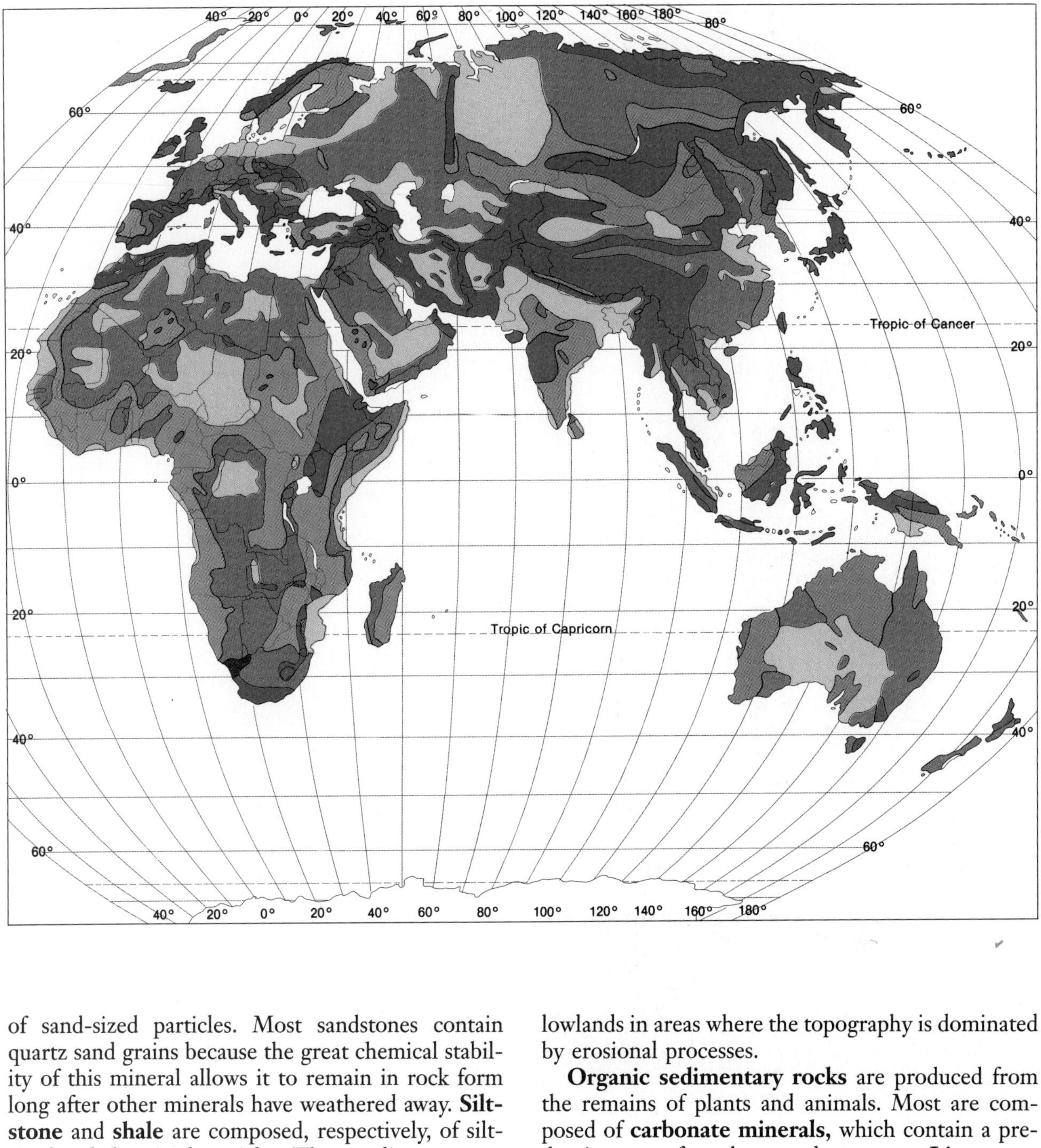

of sand-sized particles. Most sandstones contain quartz sand grains because the great chemical stability of this mineral allows it to remain in rock form long after other minerals have weathered away. **Siltstone** and **shale** are composed, respectively, of silt-sized and clay-sized particles. These sediments were nearly always deposited on the bottom of a standing body of water such as a lake or ocean. Shale is the chief sedimentary rock produced on the deep sea floor and comprises approximately half of all sedimentary rock. Both siltstone and shale, although chemically stable, are physically weak and susceptible to erosion so that they are frequently associated with lowlands in areas where the topography is dominated by erosional processes.

Organic sedimentary rocks are produced from the remains of plants and animals. Most are composed of **carbonate minerals,** which contain a predominance of carbon and oxygen. **Limestone** ($CaCO_3$) is by far the most abundant rock type within this category (see Figure 14.11b). It forms in clear, warm water as a result of the action of marine animals that synthesize calcium carbonate for their shells or exoskeletons. When these animals die, their remains accumulate on the sea floor and are buried, compacted, and lithified into limestone. Most atmo-

GEOTHERMAL HEAT AS AN ENERGY SOURCE

An important geological characteristic of many tectonically active regions is the presence of large volumes of hot ground water. In some areas, this water makes its way to the surface to form such natural phenomena as geysers, fumaroles, and hot springs. In addition to their scientific interest and economic value as tourist attractions, hot springs have been used from at least the days of the Roman Empire for heating, recreation, and health. During the twentieth century, geothermally heated water has become increasingly important for the production of electricity.

The chief advantage of geothermal power is that, when used within reason, it is a renewable resource with almost unlimited potential for development. The ultimate source of energy is the decay of fissionable materials in the Earth's interior. So far this energy source has lasted for several billion years, and apparently has the potential to last for several billion more. Unfortunately, technological limitations presently restrict the areas where geothermal energy can be economically utilized.

There are several site requirements for the development of a geothermal field to produce electric power. First, a large subsurface reservoir of superheated water must be present. Most geothermal electric power plants tap water with temperatures between 230° and 300°C (450° to 570°F). The water must be present at relatively shallow depths of 3,000 meters (10,000 feet) or less, and must have sufficient recharge ability to provide a continuous flow of hot water or steam for many years. The water also must be relatively free of excessive and corrosive chemicals. Finally, since the transportability of pressurized hot water is limited, the electrical generating plant must be constructed near the geothermal field.

Most currently developed geothermal fields are located in volcanically active areas, where hot igneous rock exists near the surface. The first geothermal electric power plant was constructed in Larderello, Italy, in 1904. Since then, other plants have been built in Japan, Iceland, the Philippines, New Zealand (see Figure 14.10), the Soviet Union, El Salvador, Nicaragua, Mexico, the United States, and several other countries. The field with the largest electrical output is The Geysers in California, but a plant currently under construction in the Soviet Union's Kamchatka Peninsula eventually may surpass it in production. Recent major finds of geothermal reservoirs in Central America make that region one of the most promising for future development.

FIGURE 14.9 Horizontally bedded sedimentary strata exposed on a cliff face in Arizona. (© JLM Visuals)

spheric carbon dioxide is eventually dissolved into the ocean waters and ends up in limestone. Most of the Earth's near-surface supply of carbon is locked up in this rock.

Although limestone is the chief organic sedimentary rock derived from animal remains, **coal** is the most common rock derived from the remains of plants (see Figure 14.11c). This rock type, composed largely of carbon, is formed from the burial, compaction, and gradual lithification of plant remains in shallow standing water or on water-saturated ground. Most large coal deposits formed about 250 million years ago, when large portions of the Earth were covered by swampy, forested lowlands. Coal is our leading fossil fuel resource and is widely distributed around the Earth.

FIGURE 14.10 Steam collection pipe system at a geothermal power plant in Warrakei, New Zealand. (J.N.A. Lott, McMaster University/BPS)

Despite its potential, a number of economic and environmental problems are associated with geothermal electric power production. Geothermal plants currently cost more than conventional plants to build. There also is a good deal of concern that overzealous development could damage or destroy natural geothermal phenomena in places such as Yellowstone Park, Hawaii, and New Zealand, as well as deplete the geothermal potential of individual fields. Obnoxious gases, such as hydrogen sulfide, are released from geothermal fields to the surrounding air, and the spent geothermal water may be a source of thermal and chemical pollution due to its dissolved salt content if released to surface lakes and streams. All in all, however, the potential benefits of geothermal power seem to far outweigh the problems associated with the continued development of this natural resource, and it appears to have a bright future.

Chemical sedimentary rocks are formed by the precipitation of chemicals in water. In order for precipitation to occur, the water must become locally supersaturated in these substances. It already has been noted that chemical precipitates play an important role in the formation of the coarser clastic rocks, notably conglomerate and sandstone, by serving as the cementing agents holding the clasts together. Under certain conditions, rocks are formed largely or entirely of these precipitates. The most common chemical precipitate is calcium carbonate ($CaCO_3$), which can accumulate to form chemical limestone deposits. Most currently forming chemical limestones are on the floors of shallow seas in the tropics.

Evaporites, another group of chemical sedimentary rocks, are produced by the evaporation of saline water bodies. One of the most common evaporites is **halite,** or rock salt (NaCl). In the geologic past, the slow evaporation of extensive cutoff arms of the sea in what are currently the western Gulf of Mexico region and in various parts of Eurasia and North Africa caused the deposition of thick layers of halites. The salt deposits were subsequently buried deeply by denser sediments. In some places, these sediments were sufficiently weak that the underlying buoyant salt layer was able to coalesce into giant "bubbles" that forced their way surfaceward to produce **salt domes.** Salt domes not only serve as sources of salt

TABLE 14.1 Major Sedimentary Rocks and Their Sediment Sources

ROCK CATEGORY	ROCK TYPE	SEDIMENT SOURCE
Clastic	Conglomerate	Rounded pebbles, cobbles, boulders
	Sandstone	Sand (usually quartz or feldspar)
	Siltsone	Silt
	Shale	Clay, mud
Organic	Limestone ($CaCO_3$)	Marine organisms
	Coal	Plants
Chemical	Limestone ($CaCO_3$)	Direct precipitation
	Dolomite [$CaMg(CO_3)_2$]	Direct precipitation; chemical alteration of limestone
	Chert (SiO_2)	Direct precipitation
Evaporite	Halite (NaCl)	Direct precipitation
	Gypsum ($CaSO_4 \cdot 2H_2O$)	Direct precipitation

FIGURE 14.11 Important sedimentary rocks. (a) Hand specimen of conglomerate. (b) Hand specimen of limestone. (c) Bituminous coal. (d) The cross-bedding displayed by this sandstone in Apache County, Arizona, indicates that it was originally deposited as sand dunes. (a,b,c: Reprinted from Wicander/Monroe, *Historical Geology* © 1989 West Publishing; d: E.D. McKee/USGS)

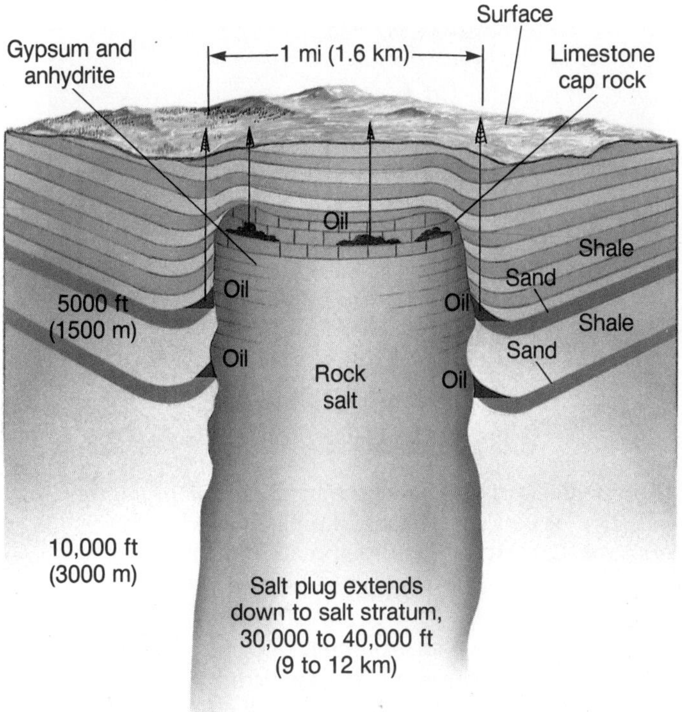

FIGURE 14.12 Profile view of a salt dome, showing the locations of oil-bearing strata.

but, more importantly, can trap commercially valuable petroleum and natural gas deposits, as shown in Figure 14.12.

DISTRIBUTION OF SEDIMENTARY ROCKS Sediments underlain by sedimentary rocks cover nearly all of the ocean floor and most of the continents. The chief land areas not covered by these rocks are ancient plutonic rock areas and areas that have experienced recent volcanism (see Figure 14.8). In addition, some sedimentary rock areas are overlain by thick unconsolidated glacial deposits.

The widespread presence of sedimentary rocks of marine origin in continental settings indicates that the land in these areas must have risen above sea level since the rocks were formed. Marine sedimentary rocks, for example, cover much of the midwestern United States. In most low-lying coastal regions where sedimentary rocks of recent origin are found, oceanic retreat apparently was accomplished by slight reductions in sea level or by surface uplift with little accompanying disturbance to the rock layers. The nearly horizontal sedimentary rocks of the Atlantic and Gulf coastal plain of the eastern United States serve as good examples. In other places, though, major tectonic upheavals have lifted sedimentary rock sequences great distances above sea level, often complexly deforming them while doing so. Most of the great cordilleran belts, containing some of the highest mountains on Earth, are composed of sedimentary rocks that formed on the ocean floor.

Most coarser grained clastic rocks, such as sandstone and conglomerate, were eroded from currently or previously existing cordilleran systems and were deposited near continental margins. These rocks have formed the surfaces of most of the world's coastal plains. Geologically older, better consolidated sedimentary rocks cover most interior lowland regions such as those of the central United States and the western Soviet Union. Formed largely when these regions were covered by shallow inland seas, they consist mostly of fine-grained clastic rocks and limestones.

Metamorphic Rocks

Metamorphic rocks (from the Greek *metamorphosis*, or "transformation") are rocks altered from their original form by heat, pressure, and/or chemical activity, while remaining in the solid state. Metamorphic changes in rocks produce textures, structures, mineral compositions, and general appearances that differ from those of the original rocks. The constituent chemical elements remain the same, but they are rearranged into different mineral combinations. Metamorphism results from the high temperatures and pressures generated well beneath the surface by tectonic processes. Most metamorphic rocks are produced by large-scale or **regional metamorphism** that occurs during mountain-building activity. These rocks, formed in the root zones of mountain systems, are exposed at the surface only after long periods of erosion.

All types of rock are subject to metamorphism with sufficient applications of heat and pressure; this includes igneous, sedimentary, and even metamorphic varieties. Each rock type has one or more metamorphic equivalents. Varying degrees of metamorphism exist, from slight alterations to complete recrystallization (see Figure 14.13). The type of metamorphic rock produced at any given site depends on the original type of rock and on the nature, intensity, and duration of the metamorphic processes. Most metamorphic rocks are classified as one of seven important types, listed in Table 14.2.

FIGURE 14.13 Important metamorphic rocks. (left) Marble—a nonfoliated rock produced by the metamorphism of limestone. (right) Hand specimens of banded gneiss. The application of heat and pressure during tectonic activity caused the alignment of their mineral crystals, giving the rocks a foliated texture. (left: Reprinted from Wicander/Monroe, *Historical Geology*, © 1989 West Publishing; right: Richard Jacobs/© JLM Visuals)

Metamorphic rocks are by far the least abundant of the three major surface rock categories. They are much more common deep within the crust where they have been produced by tectonic movements. Because great thicknesses of overlying materials must be eroded before rocks subjected to regional metamorphism appear at the surface, they are exposed mostly in ancient plutons. Most places where they occur were at one time the sites of cordilleran systems.

PLATE TECTONICS

In the preceding sections of the chapter, it was noted that tectonic forces have produced a large proportion of the Earth's surface materials and topographic features. It is now time to examine the mechanisms by which these forces operate.

Until a few decades ago, the processes by which tectonic forces operated were largely unknown. It had long been realized, of course, that these forces were powered by energy from our planet's interior, but the exact sources of this energy and the processes by which it produced tectonic features at the surface were matters of speculation. It is now known that the lithosphere, for all its seeming solidity, is divided into a number of segments, or **plates,** that are in constant motion. This motion is powered by the atomic decay of radioactive elements in the core and mantle. As the lithospheric plates move, collisions occur and existing plates are distorted or even broken apart. In addition, new crustal material constantly is being formed from magma rising from the mantle, and old crust is destroyed as it plunges back into the mantle. Each of these motions is associated with its own group of tectonic landforms.

The entire process of plate formation, movement, and destruction is known as **plate tectonics.** First proposed during the 1940s, plate tectonic theory gained widespread acceptance during the late 1950s and 1960s. Continuing research has provided a great deal of supporting evidence for the theory and has enabled us to refine our understanding of the various processes involved.

The development of plate tectonic theory has been the landmark occurrence in modern Earth science. It has undoubtedly resulted in the greatest increase in our understanding of landform-producing processes since the concept of uniformitarianism was accepted during the nineteenth century. It has shown that the Earth's surface does not consist merely of a

TABLE 14.2 Major Metamorphic Rocks and Parent Rock Types

METAMORPHIC ROCK	PARENT ROCK(S)
Quartzite	Quartz conglomerate, sandstone, or siltstone
Slate	Shale
Phyllite	Shale, slate
Schist	Shale, slate, phyllite
Gneiss	Granite, diorite
Marble	Limestone, dolomite
Anthracite coal	Bituminous coal

random assortment of land and water bodies, but instead forms an integrated system that is intimately related to materials and processes deep within the interior.

Evidence for Continental Drift

Many of us have at some time been struck by the thought that a map of the Earth resembles a jigsaw puzzle, with the continents representing the puzzle pieces separated by areas of ocean. On closer inspection, a number of the continental "pieces" do seem to fit together remarkably well. Especially striking is the match between the east coast of South America and the west coast of southern Africa. In a few other easily seen match-ups, the island of Madagascar fits neatly onto the east coast of Africa, the Arabian Peninsula into northeastern Africa, and New Guinea into northern Australia. Scholars as early as the sixteenth century noticed these coastal similarities and wondered whether they could be coincidental or if, indeed, the land masses had somehow separated and moved for long distances over the surface of the planet.

The matter remained largely in the realm of speculation until 1912, when German meteorologist and geophysicist Alfred Wegener presented a detailed theory for **continental drift** that not only traced past positions and subsequent motions of the continents, but also for the first time suggested a means by which this movement could have been accomplished. According to Wegener, all the larger land masses were at one time assembled into a single supercontinent that he called **Pangaea**. This supercontinent later separated into a number of smaller blocks that subsequently drifted to their present positions (see Figure 14.14). However, Wegener's theory that the

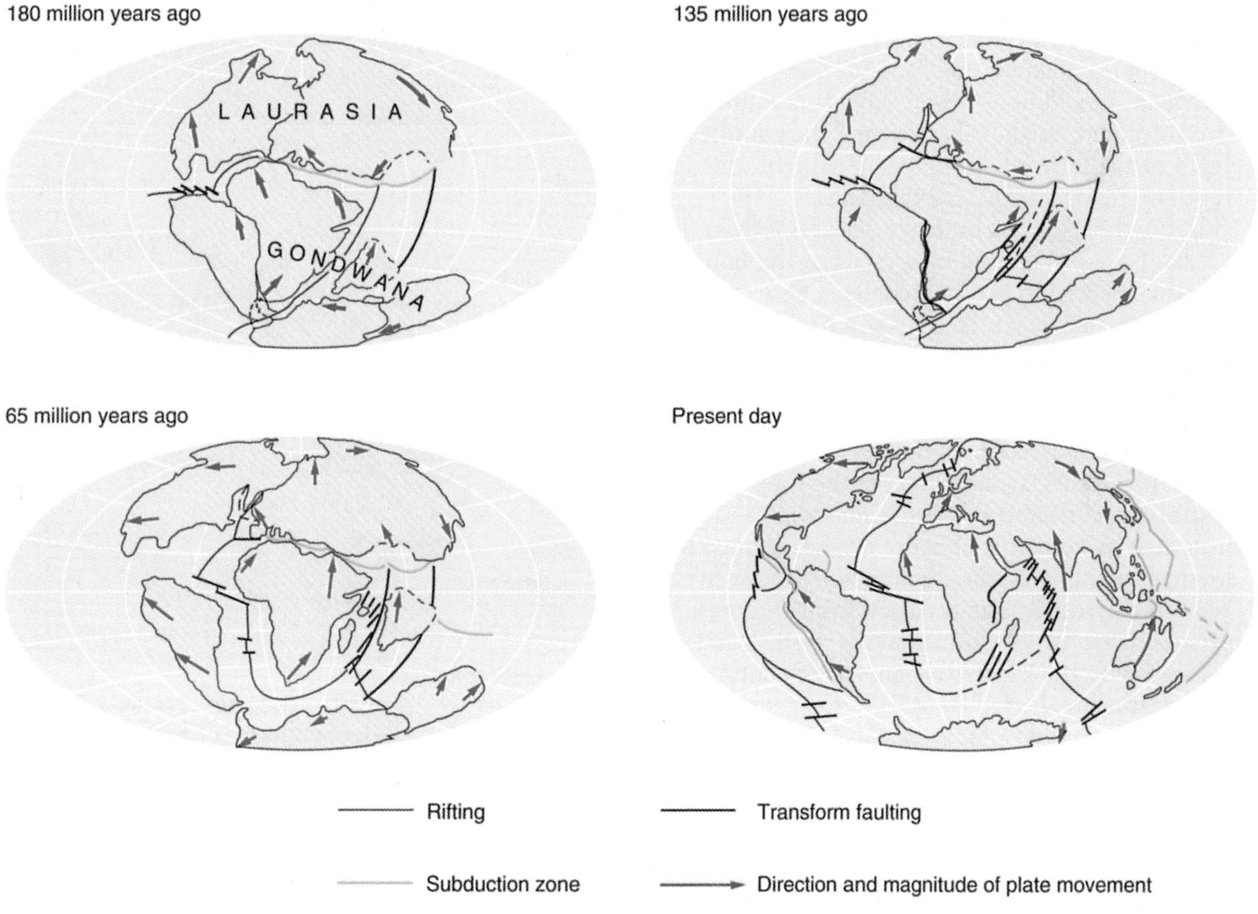

FIGURE 14.14 Map sequence showing the breakup of Pangaea and the movement of the continental blocks to their present positions.

continental crust moved about on the underlying oceanic crust was proven impossible. This lack of a workable explanation caused the idea of continental drift to fall into disrepute for a time. Wegener had nonetheless succeeded in focusing attention on the possibility that major shifts in the positions of segments of the crust somehow could have occurred.

A revival of research into continental drift took place following World War II. By this time, more sophisticated geologic instrumentation was being rapidly developed, and Earth scientists soon began amassing additional evidence in support of large-scale crustal movements. It was discovered, for example, that certain rock sequences extending to the coast of Brazil reappeared on the African coast precisely where they would be expected if the two continents were once joined. Fossil remains indicated that similar or identical plants and animals appeared during the same geological periods on land masses presently located great distances from one another. Evidences of a past ice age were unearthed in currently subtropical portions of South Africa and South America, and fossil beds of tropical plants were discovered in Antarctica and Britain. In addition, the present-day locations of active volcanoes, earthquake foci, and the cordilleran belts in general are all strongly aligned with the apparent boundaries of the crustal plates.

Some of the strongest evidence of all came from the analysis of the orientation of crystals of the mineral magnetite (Fe_3O_4) in oceanic basalts. These crystals, when free to move, act like tiny compass needles by aligning themselves along the axis of the Earth's magnetic field. When lava containing magnetite crystals solidifies, they are frozen into place so that their orientation provides a record of the direction toward the North and South Magnetic Poles at the time that solidification occurred. It has been found that while recent lava flows contain magnetite crystals aligned toward the present poles, the alignment in progressively older flows gradually rotates away from the present polar orientations. When magnetite crystal data are compared with the theorized patterns of continental drift since the breakup of Pangaea, a good correlation is obtained, providing almost conclusive evidence that these movements did in fact take place.

It also was discovered in the 1960s that the magnetic North and South Poles on occasion actually "flip," or undergo a reversal in polarity. This causes the magnetic North Pole to temporarily become the magnetic South Pole, and vice-versa. Any magnetite crystals located in basalt that solidified during a period of reversed polarity are aligned opposite from those of the present time. It has been found that the Earth's magnetic field has reversed polarity dozens of times over the past few tens of millions of years, and that the latest reversal occurred approximately 700,000 years ago.

In 1964, aircraft equipped with magnetic sensing devices flew over the Mid-Atlantic Ridge and discovered that the basalt on both sides exhibited a mirror-image pattern of stripes of normal and reversed magnetic polarity (see Figure 14.15). The Mid-Atlantic

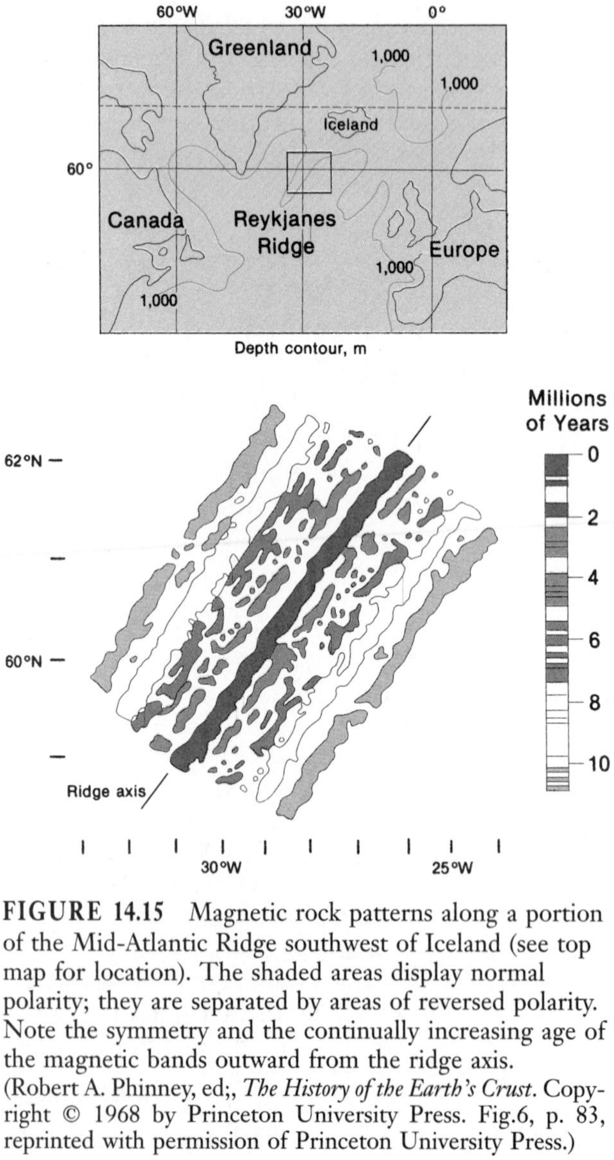

FIGURE 14.15 Magnetic rock patterns along a portion of the Mid-Atlantic Ridge southwest of Iceland (see top map for location). The shaded areas display normal polarity; they are separated by areas of reversed polarity. Note the symmetry and the continually increasing age of the magnetic bands outward from the ridge axis. (Robert A. Phinney, ed;, *The History of the Earth's Crust*. Copyright © 1968 by Princeton University Press. Fig.6, p. 83, reprinted with permission of Princeton University Press.)

Ridge is part of a system of oceanic ridges extending through the Atlantic, Pacific, and Indian Oceans that reaches heights of several thousand meters above the surrounding deep ocean floors. These ridges are the sites of active and nearly continuous submarine volcanism, which extrudes great quantities of lava that solidifies into basalt containing magnetite crystals. The inescapable conclusion was that these ridges are the boundaries between newly forming portions of two crustal plates that are diverging at rates of several centimeters a year. The movement is much like that of two belts located on adjacent rollers rotating in opposite directions.

Additional evidence for the divergent movement of the crust from the oceanic ridges is provided by the progressive increase in the age and thickness of the overlying cover of sediments and by the gradually lowering crustal temperatures outward from the ridges. Because most new crustal material seems to form and diverge along these oceanic ridges, continental drift appears to be attributable largely to the process of **seafloor spreading.**

Plate Tectonics Processes

It is now known that the moving plates are not composed exclusively of crustal material, but also include the rigid uppermost portion of the mantle. Taken together, the crust and uppermost mantle comprise the lithosphere, which averages about 100 kilometers (60 mi) in thickness. It therefore is more accurate to refer to these plates as **lithospheric plates** than to use the older term "crustal plates."

The asthenosphere is the layer upon which the lithospheric plates float. As noted earlier, the asthenosphere not only is plastic in constituency, it also contains a complex pattern of slow-moving convection currents produced by uneven heating from the decay of radioactive elements in the mantle and core (see Figure 14.16). Heat from the decay of these elements is conducted very slowly surfaceward through the mantle. When it reaches the asthenosphere, this heat generates slow convection currents within the rock material. The radioactive elements and the heat they produce are distributed unevenly within the Earth, much as they are at the surface. In places where they are relatively abundant, more heat is generated, and the mobile areas within or above them in the asthenosphere expand, increase in buoyancy, and become the sites of rising convection currents. In places where radioactive elements are not as abundant, less heat is produced, and the affected mobile areas are likely to be the sites of descending currents of somewhat cooler and denser material. A continuous circulation system thus is established.The lithospheric plates, which float on the upper portions of the convection currents, are carried slowly along like pieces of driftwood on a stream. Atomic energy, manifested in the form of convection currents in the asthenosphere, therefore provides the long-sought mechanism for continental drift. Continental drift, in turn, results in collisions between and separations of adjacent lithospheric plates, and these occurrences produce tectonic landforms.

The lithosphere is divided into six major plates as well as a number of smaller plates, as shown in Figure 14.17. The plates are moving in various directions at speeds of up to 20 centimeters (8 inches) per year. Movement in each case is away from the oceanic

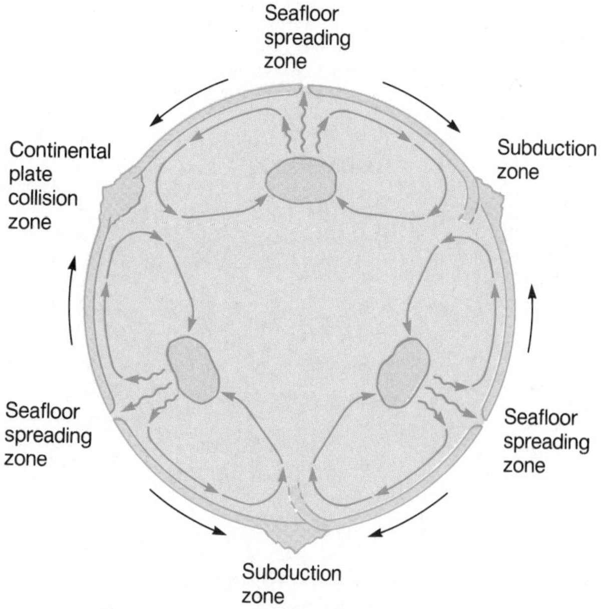

FIGURE 14.16 Diagrammatic view of the Earth's interior, showing the cause of plate tectonic movements. Convection currents in the asthenosphere, heated by the decay of radioactive elements in "hot spots" in the mantle, rise as plumes to the base of the lithosphere and then spread laterally, dragging the overlying lithospheric plates with them. The resulting plate collisions produce mountain ranges, while seafloor spreading zones in areas of plate divergence are the sites of new oceanic crust formation.

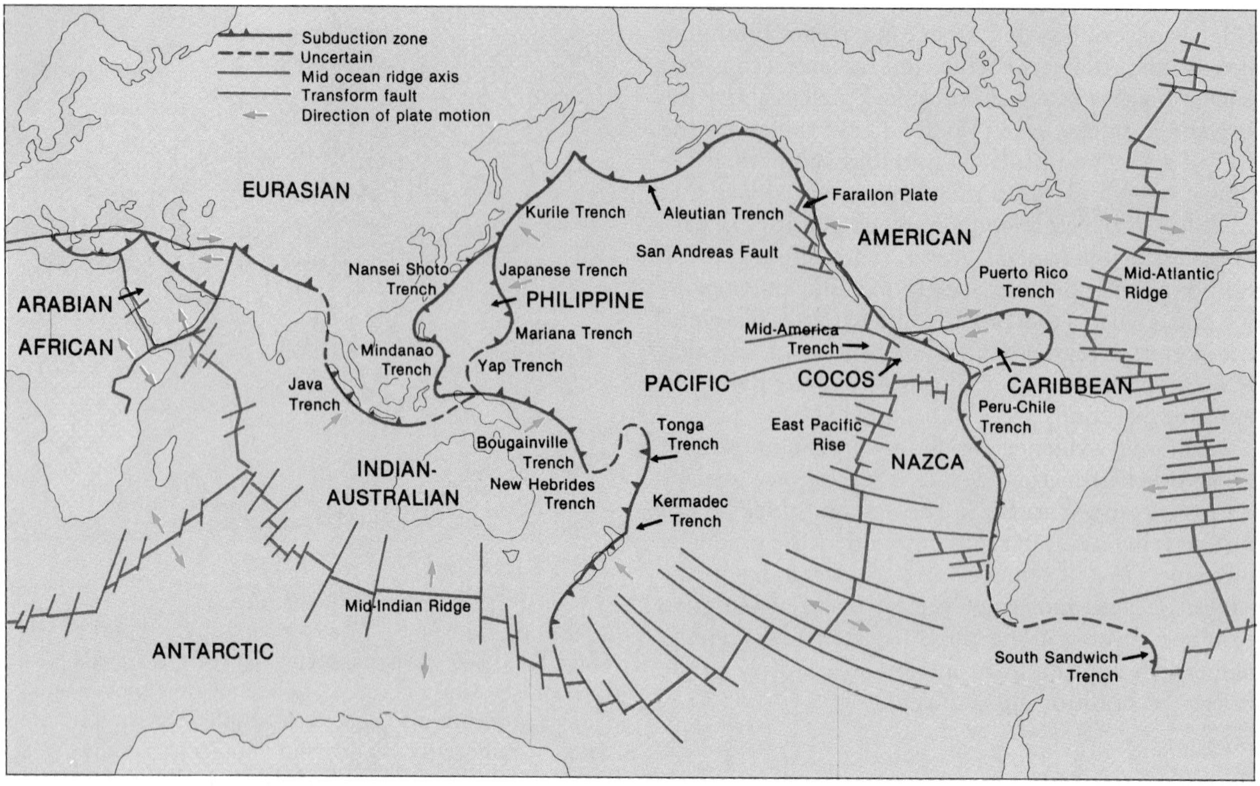

FIGURE 14.17 Global pattern of lithospheric plates and types of plate boundaries.

ridge along which the plate is being manufactured. For example, the American Plate, which contains the continents of North and South America as well as the western half of the Atlantic Ocean, is being manufactured along the Mid-Atlantic Ridge. As new crustal material is extruded along the ridge, the plate is carried westward by convection currents in the underlying asthenosphere. This westward motion causes the American Plate to collide with the Pacific and the Nazca Plates, which are located to its west. This collision has helped to crumple the forward edge of the American Plate to produce the Rocky Mountains cordilleran belt in North America and the Andes in South America. The movement also is gradually increasing the width of the Atlantic Ocean, which began forming about 150 million years ago when North and South America first began to separate from Europe and Africa as a part of the breakup of Pangaea. In similar fashion, the northeastward motions of the African and Indo-Australian Plates are thrusting up a great cordilleran belt extending from the East Indies westward to North Africa and Spain as they collide with the Eurasian Plate.

Plate Formation and Movement

The convection currents of the asthenosphere are responsible not only for the motions of the lithospheric plates, but for their creation and destruction as well. Internal heat seems to be carried surfaceward largely in the form of upwelling "plumes" that form in various locations within the asthenosphere. As these plumes approach the upper margin of the asthenosphere, they gradually change direction from vertical to horizontal. More than 120 active plumes have been tentatively identified. They are most numerous along and near the oceanic ridges, but a large number are scattered in seemingly random locations beneath other parts of the oceans and the continents. They usually are associated with surface volcanic activity, and most mid-oceanic volcanic islands or island groups such as the Hawaiian Islands, Azores,

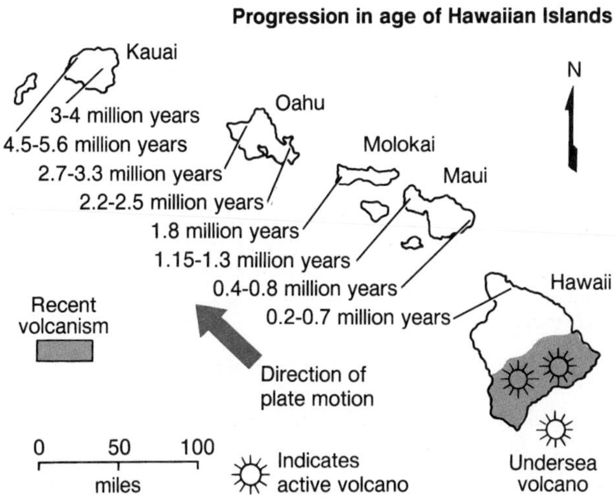

FIGURE 14.18 Ages of the Hawaiian Islands, as determined by radiometric dating techniques. The islands have been formed by the northwestward movement of the Pacific Plate over a stationary "hot spot" or mantle plume. The oldest islands lie at the western end of the chain, while active volcanism continues to form islands at the eastern end.

Iceland, and Réunion have formed on the sites of active plumes.

Isolated mantle plumes such as the one forming the Hawaiian Islands are able to protrude through a lithospheric plate without seriously disrupting it (see Figure 14.18). In places where several nearby plumes break through the crust, however, they may weaken it so much that it splits apart to produce two smaller plates.[3] This is one mechanism by which the lithosphere is believed to have been originally broken into a system of separate plates.

The splitting of a plate causes a substantial reduction in pressure on the underlying asthenosphere. The reduced pressure lowers the melting point of the rock comprising the upper mantle, allowing the selective melting of its lighter constituents, which have the lowest melting points. The zone of selective melting extends from a depth of some 50 to 70 kilometers (30 to 40 mi) below the surface upward to the base of the crust. Only 10 to 20 percent of the rock is able to melt, and it eventually collects in a magma chamber beneath the new crustal break. At this point the molten rock is likely to experience a further separation of its constituents on the basis of mass, with the lighter elements rising toward the top of the chamber and the heavier elements sinking toward the bottom. Eventually, part or all of this rock material makes its way to the surface through a system of vertical passageways lying along the fracture and is expelled onto the ocean floor during submarine volcanic eruptions. As it solidifies, this material forms new oceanic crust of basaltic composition. The combination of upward expansion due to heat and the expulsion of magma produces an oceanic ridge and apparently forces the divergence of the newly forming lithospheric plates on either side. The divergence of the plates also may be aided by gravity as they move downslope from the elevated oceanic ridge systems toward the deep oceanic trenches that usually border the far sides of the plates.

In some instances, a plume-induced break occurs in a portion of the lithosphere overlain by continental crust. In this case, subsequent divergence of the sides of the break will first stretch and weaken the crust, producing a deep, linear, steep-sided trough called a **rift valley.** An excellent example of such a trough, produced by the early separation stages of two segments of continental crust, is the East African Rift Valley (see Figure 15.12). If divergence continues, the two continental plate segments will become completely separated as ocean water fills in the gap that is being covered by newly manufactured oceanic crust. This stage of the divergence process is well illustrated by the Red Sea, located between Africa and the Arabian Peninsula. Continued divergence eventually may produce a full-sized ocean, as in the case of the Atlantic.

Three major zones of plate formation and divergence marked by oceanic ridges presently exist. They are the Mid-Atlantic Ridge, the East Pacific Rise and Pacific-Antarctic Ridge system, and the Y-shaped Indian Ocean Ridge system (see Figure 20.19). They total some 64,000 kilometers (40,000 miles) in length.

The rate of seafloor spreading varies considerably in different areas. Along the Mid-Atlantic Ridge it averages about 4 centimeters (1.5 inches) per year. The fastest rate of spreading is along the East Pacific Rise, where it averages about 15 centimeters (6 inches) per year. Even though spreading rates of a few centimeters a year may seem very slow, these movements can total thousands of kilometers in a few tens of millions of years. They therefore are capable

[3] By analogy, a single nail usually can be driven into a board with no adverse effects on its integrity. The placement of too many nails too closely together, however, is likely to cause the board to split.

YELLOWSTONE – A VOLCANIC "HOTSPOT" WITH AN EXPLOSIVE PAST AND AN UNCERTAIN FUTURE

The Yellowstone region, located mostly in the northwestern corner of Wyoming, is an area whose scenic beauty and unique geothermal features caused it to be designated in 1872 as the first national park in the United States. Yet, from a geological standpoint, the Yellowstone is one of the most unstable and dynamic segments of the North American Plate. Like the island of Hawaii, it is situated above a "hot spot" in the mantle and has been the site of repeated violent volcanic eruptions. Unlike the Hawaiian hot spot, however, the Yellowstone hot spot is located beneath continental crust, and the eruptive material is viscous and chemically acidic (that is, it has a high aluminum and silicon content). This type of magma is associated with explosive volcanic eruptions.

Large volumes of magma apparently are being continuously generated in the upper mantle and lower crust and are accumulating in one or more huge chambers deep be-neath the Yellowstone Plateau. Geologic evidence indicates that similar magma accumulations were released in three huge, but short-lived volcanic eruptions during the past two million years. Each eruption emptied the subterranean magma chambers; this caused the overlying surface to collapse, forming large calderas, or volcanic depressions. Recent studies have indicated that the erupted material was not primarily lava, but instead consisted largely of ash and hot gas. The resulting ash flows were larger in scale than those known anywhere else in the world, and upon cooling they formed extensive deposits of rhyolite and tuff.*

Yellowstone's most recent phase of volcanic activity began about 1.2 million years ago, and the climactic eruption that last emptied the magma chambers occurred about 600,000 years ago. This eruption expelled about 1,000 cubic kilometers (250 mi^3) of ash. (By comparison, the 1980 eruption of Mount St. Helens released about 4 cubic kilometers, or 1 mi^3, of ash.) The expulsion of this material created a caldera that was 45 kilometers (28 miles) wide and 75 kilometers (47 miles) long. It has been partially filled by sediments and by material from subsequent smaller eruptions.

The floor of the Yellowstone Plateau currently is being uplifted again at the relatively rapid rate of 1.4 centimeters (0.5 inches) per year. This uplift apparently is being caused by the injection of new magma into the magma chambers underlying the plateau. At some future time, the volcanic cycle will undoutedly be repeated as the chambers are emptied by another violent eruption.

*Robert B. Smith and Robert L. Christiansen, "Yellowstone Park as a Window on the Earth's Interior," *Scientific American*, 242, no. 2 (1980): 104-17.

FIGURE 14.19 Boiling geothermal pools, like Fireside Lake shown here, are common in Yellowstone National Park. Their heat is derived from a subterranean magma chamber. (M. Schneider/H. Armstrong Roberts)

of completely rearranging the surface of our planet over relatively short geologic timespans.

Plate Collisions and Destruction

Based on their relative motions, lithospheric plate boundaries are classified as divergent, transform, or convergent. Along **divergent boundaries,** the adjacent plates move away from one another. Along **transform boundaries,** the motions are lateral, so that the two plates slide past each other while moving in opposite directions. Neither of these two types of boundary movements is directly responsible for the production of a large proportion of Earth's continental landforms. Diverging boundaries do produce volcanic islands and rift valleys, and, as separation continues, seas and even oceans may eventually form. The chief material manufactured along these boundaries, though, is oceanic crust, which normally is located on the sea floor or buried beneath continental crust. Discussion of the various types of seafloor features that form on this layer can be found in Chapter 20.

Convergent boundaries have by far the greatest influence on the production and distribution of continental tectonic features. Not only are they responsible for producing most tectonic landforms, these movements actually form continental crust in the process of destroying oceanic crust. Without the convergence of lithospheric plates, no major mechanism would exist for the production of light, buoyant, and therefore high-standing masses of continental granitic rock, and the Earth would be almost completely covered by ocean.

Because the Earth is not increasing in size, the rate of lithospheric plate destruction must equal its rate of formation. This destruction occurs at the convergent plate boundaries, and the nature of the surface features produced depends primarily on whether the plates are covered by oceanic or continental crust. Three types of convergent boundaries exist; the first is formed by the collision of two oceanic plates; the second, by a collision between an oceanic and a continental plate; and the third, by a collision of two continental plates.

When two oceanic plates collide, one of the plates is bent downward by the other so that it is pressed gradually into the asthenosphere at an angle of 30° or more in a process called **subduction.** As the subducted plate reaches the base of the asthenosphere at depths exceeding 600 kilometers (400 miles), it is heated to the point that it loses its rigidity. The entire process from the time of initial plate contact to the loss of rigidity of the subducted plate may take some 10 million years.

The portion of the oceanic plate that is being subducted is likely to be covered by a great thickness of sediments. These sediments have accumulated on the seafloor since the time the plate was manufactured along an oceanic ridge. The sediments contain a large amount of water, and as they are subducted along with the plate, this water reduces the melting point of the lighter rock-forming materials both within the subducted plate and in the portions of the mantle with which the plate comes into contact. This results in the selective melting of these materials, which rise as magma toward the surface. The magma may either solidify beneath the surface to form large granitic plutons known as **batholiths** (see Chapter 15), or reach the surface in volcanic eruptions. In either case, the resulting igneous rock material becomes continental crust.

Because this type of plate convergence occurs between two oceanic plates, the initial tectonic features are produced on the seafloor. Among the most striking features are **submarine trenches** that form along the line where the denser oceanic plate is being subducted. These narrow chasms contain the deepest points on the ocean floor. Challenger Deep in the Marianas Trench reaches a depth of 11,033 meters (36,198 feet).

Several hundred kilometers from a trench, and typically paralleling it above the zone where the subducted plate extends deep into the asthenosphere, an **island arc** is produced near the margin of the overriding plate (see Figure 14.20). This occurs when the volcanic eruptions associated with the rising continental crust-producing magmas have accumulated in sufficient quantity to protrude above sea level. Examples of such island arcs include the Marianas of the western Pacific, the Aleutian Islands of Alaska, and the Lesser Antilles that separate the Atlantic from the eastern Caribbean Sea.

The gradual accumulation of lighter rock from both surface volcanism and subsurface granitic intrusions eventually forms continents. This process is accomplished largely by the collision and joining of numerous island arcs or other small segments of continental crust, each of which is termed a **terrane,** by the sweeping action of the lithospheric plate move-

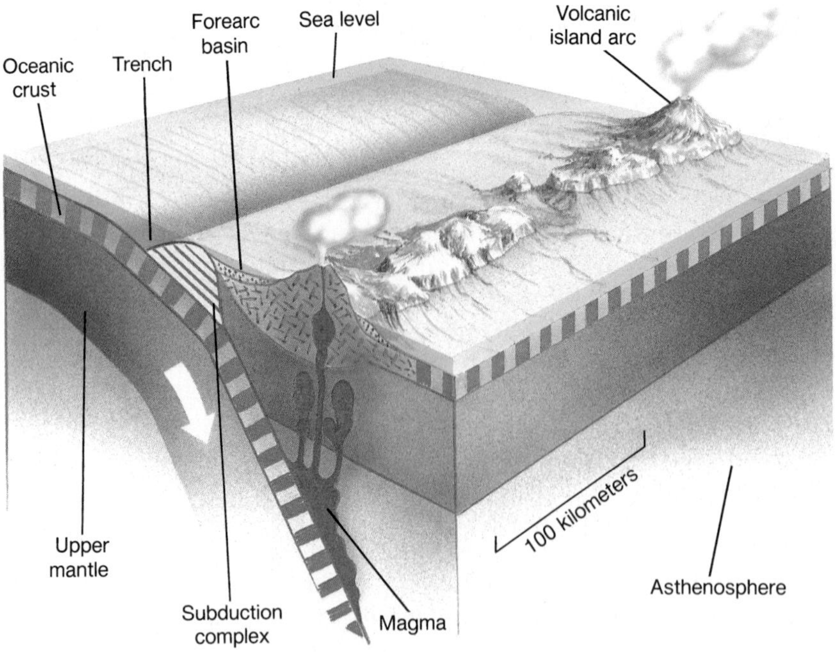

FIGURE 14.20 Diagrammatic view of the collision between two oceanic plates. Magma rising from the subducted plate forms an island arc composed of new continental crustal material. (Wicander/Monroe)

ments. The west coast of North America from California to Alaska, for example, has been formed by the accretion of numerous terranes onto the western margin of the American plate. In addition, some geologists believe that the Piedmont region of the eastern United States consists of several crushed-in layers of ancient island arcs that were collected at a time when a plate boundary existed along the axis of the Appalachians.

When an oceanic plate collides with a continental plate, the denser oceanic plate is always subducted (see Figure 14.21). A marginal volcanic zone forms as

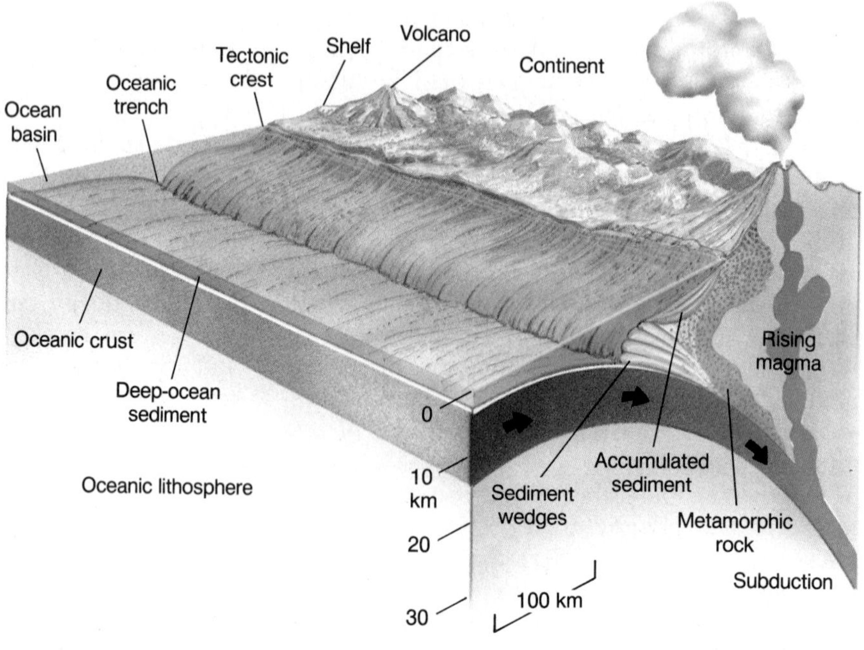

FIGURE 14.21 Diagrammatic view of the collision between a continental plate and an oceanic plate.

magma from the subducted oceanic plate rises into and sometimes through the adjacent margin of the overriding continental plate to form new continental crust. This category of plate collision is the most common of the three and is primarily responsible for the fact that the majority of the world's cordilleran belts are located along the margins of the continents. The Pacific Plate is currently being subducted by continental plates on all sides, producing its surrounding ring of volcanoes.

Behind the coastal volcanic zone, the overriding continental plate may experience extension, compression, or both. If it is being extended, the plate is stretched and down-warped, sometimes producing an inland sea. The Sea of Japan, between volcanic Japan and mainland Asia, is a good example. In the Great Basin region of the western United States, crustal extension has broken the crust into numerous north/south-oriented fault blocks, some of which have tilted to form mountain ranges. If compression occurs, a broad mountain area dominated by folding and faulting is formed. The Rocky Mountains to the east of the Great Basin region serve as a prime example.

Seaward of the volcanic zone, continuing subductions results in a scraping of the oceanic plate by the overriding edge of the continental plate. This tends to peel off the top of the oceanic plate, including its surface cover of sediments, and to pile this material onto the margin of the continental plate outside the volcanic zone (see Figure 14.21). This process can form a nonvolcanic island chain paralleling the coast, which may be pushed onshore later, as has occurred along the west coast of South America. The thickness and light weight of these materials often result in the formation of high-elevation surfaces.

When two continental plates collide, neither one is subducted, and little volcanism normally occurs. Instead, a broad zone of compression and diastrophism, often many hundreds of kilometers wide, is formed (see Figure 14.22). Frequently, one of the plates is thrust for long distances over the other. This produces a double thickness of continental crust, which, in combination with its light weight and buoyancy, results in the formation of a high, rugged cordilleran system.

The crumpling of the forward edges of the two colliding continental plates results in the production of extremely complex geological structures. The plate boundaries often lose their identities, and instead a broad and poorly defined zone of compression is formed. Mountains may form along a zone of weakness rather than along or near the actual line of plate convergence. Continued convergence eventually results in the fusion, or **suturing,** of the two plates into a single larger plate. This mechanism balances the process of plate separation and keeps the crust from being divided into progressively smaller units.

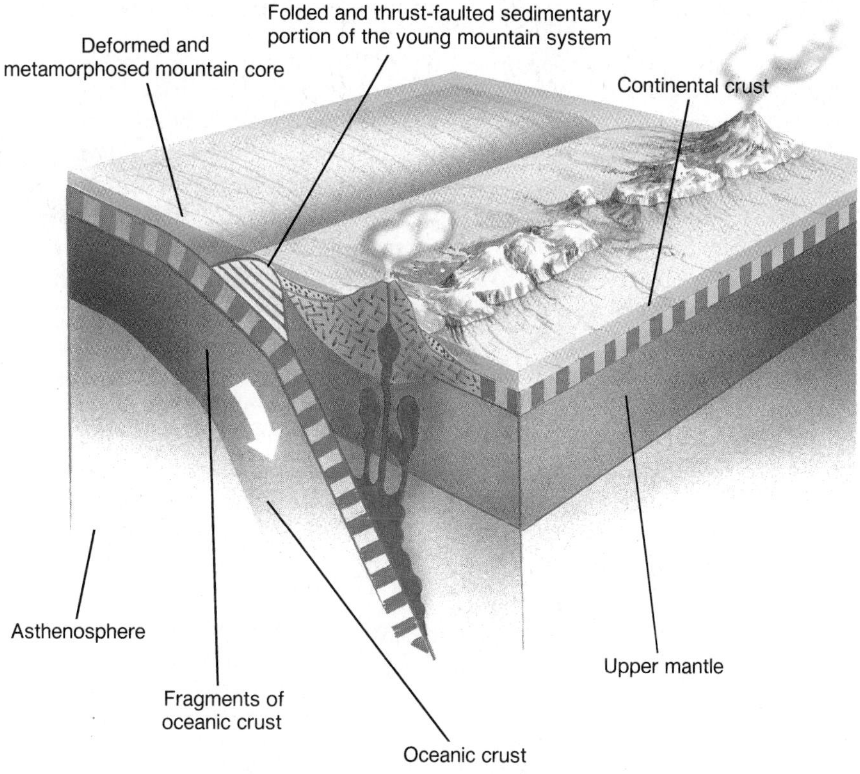

FIGURE 14.22 Diagrammatic view of a collision between two continental plates. The Himalayas have been formed by this type of collision (see Figure 14.23). (Wicander/Monre)

FIGURE 14.23 View of the Himalayas. These mountains are being formed by the northward movement of the Indian subcontinent into the Eurasian Plate. Mt. Everest, the world's highest peak, is on the right. (G.J. James/BPS)

The premier example of a continental plate collision is that occurring between the Indian subcontinent and southern Asia. This collision, which began 50 million years ago because of the northward movement of the Indo-Australian Plate, has produced the Himalayas—the world's loftiest mountain range (see Figure 14.23). Asia has proven to be the weaker of the two participants and has deformed the most. Faulting in Asia extends nearly 4,800 kilometers (3,000 mi) north of the collision zone, and the continent has been shortened latitudinally and stretched longitudinally by the collision. A similar type of collision at an earlier stage is occurring to the west, where the northward-moving African Plate is beginning to run into the European portion of the Eurasian Plate. The intervening Mediterranean Sea is the remnant of a much more extensive ocean that once separated these two plates, and continuing plate movement should eventually compress it into oblivion. These plate collisions are major generators of earthquakes—a topic explored in Chapter 15.

The process of continental plate collision results in the formation of great thicknesses of continental crust in the zone of convergence. For this reason, ancient plate boundaries, which now generally are located within the interiors of sutured lithospheric plates, are exceptionally strong and resistant to further deformation. These areas are termed **shields** because of their strength and their broadly convex surface profiles.

Shields form the stable core areas of continents. They have remained near or above sea level for long periods of geologic time, causing them to be subjected to slow erosion that has exposed the granitic and often metamorphosed "roots" of the ancient mountain systems. In many areas, the surfaces of the shields, which contain the oldest rocks on Earth, are exposed. In other localities, though, they lie beneath a veneer of much younger sedimentary rocks deposited when these shield areas were temporarily covered by shallow seas. At present, shields occupy more than a third of the continents. The Canadian Shield, for example, extends over most of eastern Canada, while the African Shield comprises the majority of that large continent. The locations of the major shield areas are depicted in Figure 14.24.

Isostatic Uplift

The lithospheric plates, which float on the asthenosphere to just the depth needed to support their weight, are said to be in **isostatic equilibrium** (from the Greek *isos*, "equal" and *stasia*, "standing") with their surroundings. When erosion removes material from the surface, the weight of the plate in that location is reduced. The reduction in weight results in the **isostatic uplift** of the region. This is a broad, gentle upwarping that, like the gradual rise of the underwater portion of an iceberg when its top melts, allows the plate to remain in a state of isostatic equilibrium.

Isostatic uplift occurs most rapidly in mountainous regions, where the reduction of mass by the erosional processes generally is more rapid than it is on flatter surfaces. As erosion proceeds, then, the surface lowering of a region resulting from the continuing removal of weathered materials is largely offset by its gradual uplift because of its loss of weight. This allows old mountain systems in tectonically stable regions like the Appalachians to maintain themselves for a much longer time than they otherwise could. Isostatic uplift also allows shield areas, with their great underlying thicknesses of buoyant continental crustal material, to remain above sea level for lengthy geologic timespans despite constant slow erosion.

Future of Plate Tectonics

Two hypothesized long-term trends in the Earth's plate tectonic activity are the gradual increase in the

50 pts T/F Matching
2 of 3 short answer

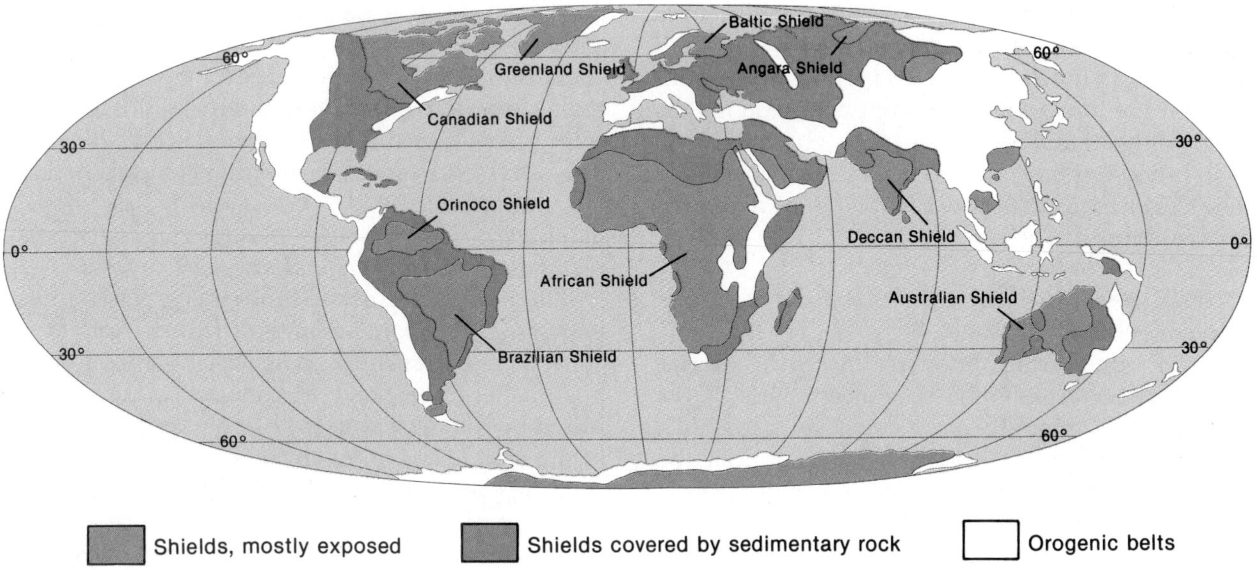

FIGURE 14.24 Major shield regions of the Earth.

volume and area of continental crust at the expense of the oceanic crust, and the slow worldwide decline in overall tectonic activity. Because only oceanic plates are subducted, they are destroyed within a relatively brief geologic timespan after their formation. They therefore consist of relatively recent rock. The oldest known oceanic rock, located in the northeastern Pacific, was formed approximately 180 million years ago. The continental plates are too buoyant to be subducted and "ride out" the plate movements on the surface indefinitely. For this reason, many consist of ancient rock more than 600 million years old. Continental plates do, however, lose surface material by the erosional action of water, ice, and wind, and a current area of investigation concerns the relative rates of formation and destruction of continental crust. Many researchers believe that the total mass of continental crust is increasing gradually with time.

The radioactive isotopes of uranium and thorium that power the movements of the lithospheric plates are declining gradually in abundance as they are converted to stable elements through the process of atomic decay. The output of heat from this source in the Earth's early history was perhaps three times the current amount. It is estimated that the Earth currently is losing heat at about twice the rate that it is being produced. This means that the interior must be slowly cooling.

In the future, the interior of the Earth will continue to cool slowly. The lithospheric plates will thicken and eventually the asthenosphere will become rigid, so that the surfaceward transport of heat by convection currents will cease. This will allow no further tectonic activity to occur. In that distant future period, probably several billion years from now, the contest between the tectonic and gradational forces finally will be decided, as gradation achieves the ultimate victory. The Earth in its old age will have its land masses eroded to low plains. These eventually will be covered by shallow seas as the waves accomplish the final step in the destruction of the continents.

SUMMARY

This chapter has examined the composition and internal structure of the Earth. Our planet contains only a small number of chemical elements in abundance, but they have been acted upon by a variety of processes in differing environmental settings. This has produced a large number of materials with diverse physical and chemical properties. Tectonic forces originating deep within the interior are responsible for the continual alteration of the Earth's surface. New crustal material is manufactured, while old material is destroyed. At the same time, the positions of the continents and oceans are constantly rearranged by the slow creep of huge volumes of plastic rock in the mantle. These motions are responsible for the formation of the continents and for the uplift of the cordilleran belts.

In order to summarize the major topics covered in the chapter more systematically, we return to the Focus Questions:

1. *What are the characteristics and locations of the internal layers of the Earth?*

The Earth is divided into three concentric layers on the basis of composition. The innermost layer, the core, is a dense metallic sphere composed largely of iron. The inner core is believed to be solid, while the outer core is liquid. The surrounding mantle is an extensive layer of dense rock containing large amounts of iron and magnesium. The lower and extreme upper portions of the mantle exist in a rigid solid state, but an intermediate portion, the asthenosphere, is plastic and contains slow-moving currents. The Earth's surface layer, the crust, consists of low-density rock composed largely of oxygen and silicon. A continuous lower portion, the oceanic crust, reaches the surface in the ocean basins. The uppermost portion, the continental crust, is discontinuous and forms the continents.

2. *What are the major categories of rocks and minerals comprising the Earth's surface?*

Most rocks comprising the Earth's crust consist of silicate minerals, which are composed largely of oxygen and silicon. This is especially true of the continental crust, which has a predominantly granitic composition. The oceanic crust consists largely of basaltic rock. It is denser than the continental crust and contains a greater proportion of ferromagnesian minerals.

Rocks commonly are divided into three categories according to their mode of formation. Igneous rocks have solidified directly from a molten state. Intrusive igneous rocks form below the surface and are exposed by the erosion of the overlying materials. Extrusive igneous rocks are formed at the surface by volcanic eruptions or fissure outpourings of lava. Sedimentary rocks are formed from the accumulation and lithification of sediments. Clastic sedimentary rocks, including conglomerate, sandstone, and shale, consist largely of particles of preexisting rocks. Organic sedimentary rocks, including limestone and coal, are formed from the remains of plants or animals. Chemical sedimentary rocks, including some limestones and salt, result from chemical precipitation. The least abundant surface rock category, the metamorphic rocks, are altered from their original states by mountain-forming tectonic stresses. Familiar examples include slate, marble, and gneiss.

3. *How do plate tectonic movements occur, and how do they create and destroy crustal material?*

The movements of the lithospheric plates apparently are caused largely by the frictional drag of convection currents in the underlying asthenosphere. These currents are produced by heat released by the nuclear decay of radioactive elements in the mantle and core. Upwelling mantle plumes result in the formation and divergence of new oceanic crust along the oceanic ridge systems. A collision between two plates having differing densities results in the subduction and destruction of the denser plate as it is forced downward into the mantle. The melting and surfaceward movement of some of the lighter and more volatile components of the subducted plate forms continental crustal material, which is volcanically erupted onto the surface. The collision of two continental plates often produces a high-standing cordilleran belt composed of greatly thickened and complexly deformed continental crust.

Review Questions

1. How has most information about the deep interior of the Earth been obtained?

2. Name the internal layers of the Earth and briefly describe their physical and chemical characteristics.

3. Explain how convection currents are formed within the mantle. What energy source powers these currents, and in what portion of the mantle do they occur?

4. Explain how the continental and oceanic crusts differ in terms of composition, thickness, density, age, and surface distribution. How does this distribution relate to the current global distribution of continents and ocean basins?

5. List and define the three major categories of rocks. Describe the environmental settings in which each is formed. What types of rocks dominate in the area in which your college or university is located?

6. How can intrusive igneous rocks be visually distinguished from extrusive igneous rocks? Why does this difference occur?

7. What are the chief sources of rock-forming sediments? How are these sediments lithified?

8. What is the rock cycle? Give an example of the operation of the cycle, beginning with igneous rock, and passing through the other two rock categories before returning to igneous rock. Explain how and why each change in your example occurs.

9. Discuss several types of evidence that lend support to the concept of continental drift.

10. What are lithospheric plates? Describe the current motion of the one on which you live and relate its movement to the major topographic features of the plate.

11. Explain how and why continental drift occurs.

12. Explain the process of subduction. Why is oceanic crust destroyed by the subduction process, while continental crust is not? How does the subduction of oceanic crust create new continental crust?

13. What are the three types of plate motions, based on the relative movements of the plates involved? Which one is the major producer of continental crust, and why?

14. Explain how the East African Himalaya Mountains, and the Ale been produced.

15. What are shield areas and how formed? Describe the geographic lo major shields. Why does it take so lo areas to be destroyed by erosion?

16. What will be the ultimate fate of the process of continental drift, and what impact will this have on the Earth's surface?

Key Terms

core
mantle
crust
asthenosphere
lithosphere
oceanic crust
continental crust
silicate mineral
mineral
rock
intrusive igneous rock
extrusive igneous rock
sediment
strata
sedimentary rock
metamorphic rock
plate tectonics
seafloor spreading
lithospheric plates
subduction
suturing
isostatic uplift

C H A P T E R 15

DIASTROPHIC AND VOLCANIC LANDFORMS

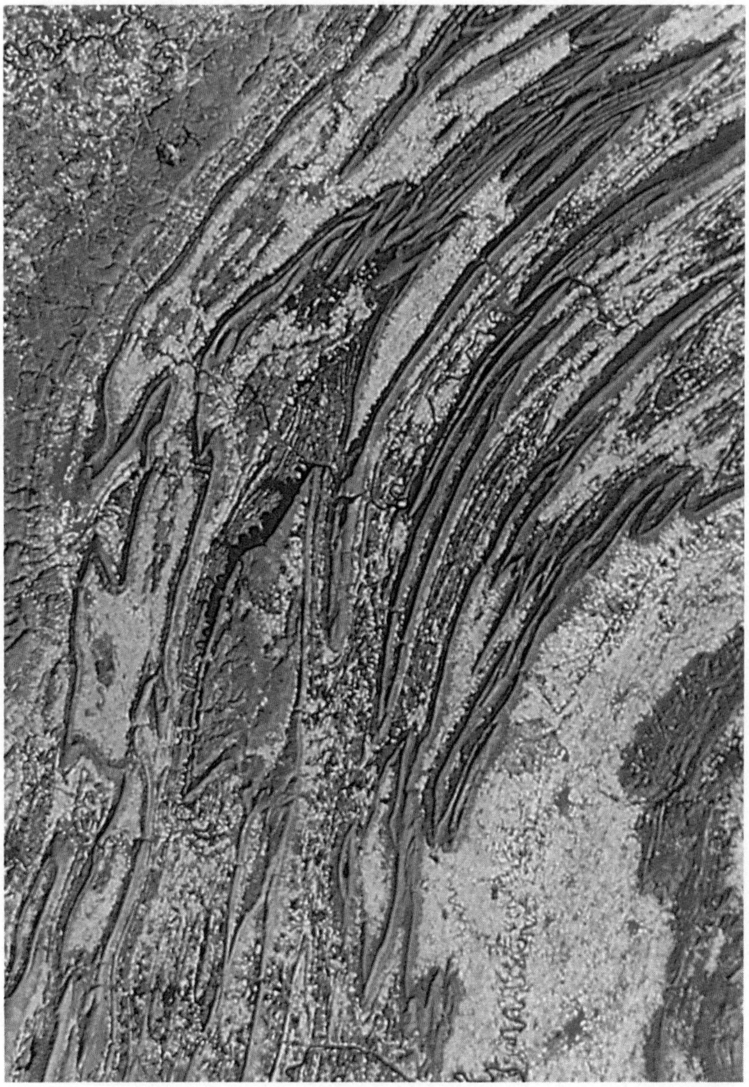

OUTLINE

FOCUS QUESTIONS

1. What causes rocks to fold, and what topographic features result from folding?
2. What are the major types of fault motions and their resulting topographic features?
3. What causes earthquakes? What controls the geographic distribution of earthquake hazards?
4. What causes volcanic eruptions, and what types of surface features do they produce?

Tectonic processes, as we saw in Chapter 14, control the formation and distribution of the Earth's large-scale surface features, including the continents, ocean basins, cordilleran belts, and shields. This chapter examines the surface features produced by tectonic activity at much closer range. The emphasis is on localized crustal motions and the types and distributions of the resulting tectonic landforms. Two basic categories of tectonic activity are recognized: diastrophism, or crustal displacements resulting from solid-state motions, and volcanism, or crustal displacements involving the transfer of molten rock-forming material.

FIGURE 15.1 *Diagrams*—Linear fold terminology and features. *Photo*—Tight folding in limestone and shale strata exposed in a railroad cut in the central Appalachians near Cumberland, Maryland. (John M. Morgan, III)

DIASTROPHISM

Diastrophism (from the Greek *diastrophe* "twisted or distorted") is the solid-state deformation of the Earth's crust resulting from tectonic forces. Deformation can occur either through the folding of rock material or through faulting, which involves the breaking and subsequent displacement of the rock. The crustal response to tectonic forces depends on both the characteristics of the rock material and on the nature and extent of the stresses involved.

Linear Folding—Processes and Features

We tend to think of the rocks at the Earth's surface as possessing great strength. Strength, though, is a relative term, and the tectonic forces are easily capable of deforming great thicknesses of crustal rock. In addition, the application of these forces deep in the Earth's interior occurs under conditions of high temperature and pressure that give many rock types the flexibility to fold without breaking.

Most folding takes place in response to lateral compression resulting directly or indirectly from lithospheric plate collisions. The rocks buckle and fold, and the entire area generally is uplifted as the relatively buoyant near-surface rock layers are compressed horizontally and increased in thickness. The resulting crustal shortening also increases the weight of the crust on the underlying mantle, causing it to press deeper beneath the surface than before.

Folding usually involves sedimentary rocks, which, as a group, are softer and more flexible than either igneous or metamorphic rocks. The ability of these rocks to deform plastically without breaking is further increased by the pressure and heat resulting from their burial at depths of often several kilometers at the time of folding and by the fact that the compression normally is applied at maximum rates of only a few centimeters a year.

The compression of rock strata produces a linear pattern of folding. Folded rock structures are highly similar in both cause and appearance to the patterns of folds produced when a carpet is pushed against a wall. Direct compression of this type produces a fold pattern with its long axis or **strike** perpendicular to the axis of compression (see Figure 15.1). This explains, for example, why the westward movement of the American Plate has resulted in the north-south orientation of the Rockies and Andes, and why the northward movement of India into Asia has produced the east-west strike of the Himalayas.

The areal extent and intensity of folding are controlled largely by the degree of compression and by the strength and thickness of the rock strata involved. With moderate compression, a wavelike series of open folds is produced. Each upfold is termed an **anticline,** and each downfold a **syncline.** Both the vertical dimensions and the angle of inclination or **dip** of the flanks of the folds increase as the amount of compression becomes greater. If compression is intense, the folds may be squeezed in on one another so that the fold structure closes and collapses, causing the axial plane to become nearly horizontal. This results in the overturning of rock strata on the fold flanks, and the folds themselves are termed **recum-**

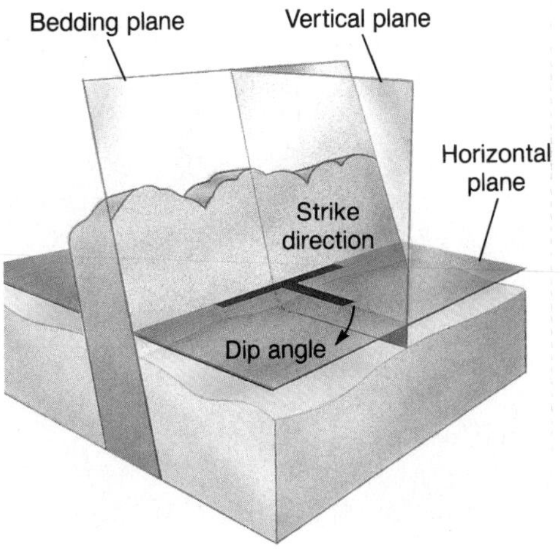

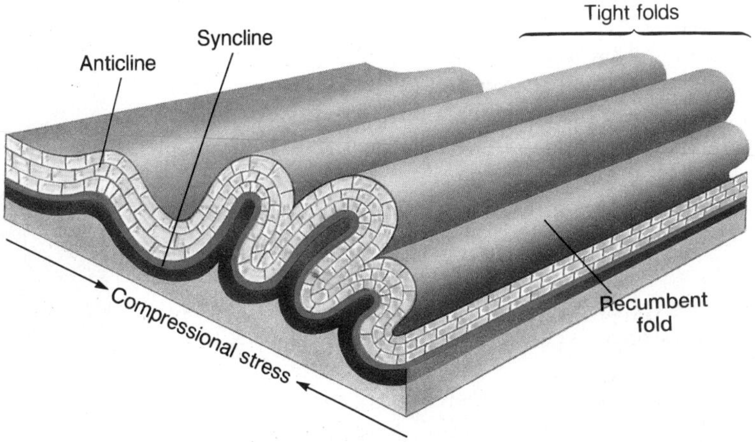

bent folds. Within an area of folded rock strata, the intensity of folding usually is greatest along one axis and gradually dies out to produce progressively more open folds away from this axis. If compression near the center is sufficiently intense, the rocks there may be metamorphosed to varying degrees.

When compressional folding occurs on horizontal rock strata, the resulting fold features are level and typically extend for long distances, sometimes many hundreds of kilometers. Frequently, though, rock strata that have previously been tilted are subjected to folding. In this case, a series of **plunging** anticlines and synclines is produced (see Figure 15.2).

The relationship between fold patterns and topographic features might be expected to be a straightforward one, with upfolded areas producing mountain ridges and downfolded areas producing valleys. This situation, however, rarely exists in reality because of the effects of **differential erosion,** or erosion occurring at differing rates. Even though folding occurs very gradually in human terms, the geologic timespan during which a given area is actively undergoing folding normally is quite short. On the other hand, once fold structures have been produced, they remain until the folded rocks have been eroded completely away. As a result, the great majority of areas that contain folded rock strata are not currently undergoing folding, but instead were folded at some time in the geologic past. Consequently, the forces of erosion have had enough time to modify the topography considerably since the folding occurred.

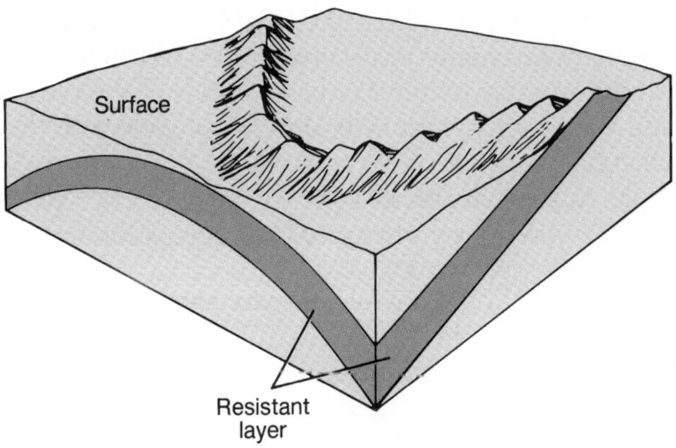

FIGURE 15.2 A plunging fold results in the formation of a U-shaped mountain ridge along the surface outcrop of a resistant sedimentary rock layer.

In a few regions folding has occurred recently enough, and on a large enough scale, to control the present pattern of surface features directly. Most such regions have dry climates, so that the lack of surface water reduces the effectiveness of the gradational processes. Included are parts of North Africa, the Middle East, New Zealand, and limited portions of western North and South America. Within these regions, the surface is dominated by anticlinal ridges and synclinal valleys produced directly by folding. These two types of features are said to display an **accordance** (conformity) between structure and topography. An **anticlinal ridge** is a ridge composed of upfolded rock strata, while a **synclinal valley** is a valley developed upon downfolded strata. It should be stressed, though, that in most areas ridges are not produced directly by upfolding, nor are most valleys produced by downfolding. How, then, are these topographic features produced?

Over extensive areas, folding and gradation exert a dual control on the topography. Folding initially deforms the rock strata so that they vary greatly in their angle of inclination, or dip. As erosion lowers the surface, rock strata are exposed that vary considerably in their susceptibility to weathering. Erosion occurs at different rates on these strata, with layers of resistant rock eroding slowly to form ridges or other uplands, while weak rock layers erode quickly, becoming valleys or lowlands. The resulting topography therefore is controlled by differences in erosional rates between stronger and weaker rock strata whose surface patterns have resulted from folding.

The stretching and compression of rock strata during the folding process may also influence their susceptibility to erosion. Upfolding stretches the rock, weakening it and often producing stress fractures that facilitate subsequent weathering and erosion. Conversely, downfolding compresses the rock, generally increasing its strength. Other factors being equal, then, anticlines contain somewhat weaker rocks than do synclines.

Let us imagine an area of horizontally bedded sedimentary rock being subjected to compressional folding, and trace the likely course of geomorphic events (see Figure 15.3). If folding occurs rapidly enough, a series of anticlinal ridges and synclinal valleys will be formed initially. These features, however, will be short-lived. The anticlinal ridges are eroded rapidly, both because they are the highest and most exposed features on the landscape, and because the expansionary forces that attended the folding weakened the rock. Sediment eroded from the ridges is partly deposited in the valleys, helping to protect them from erosion, while the remainder is transported out of the area by streams. The anticlinal ridges eventually are removed by erosion and, as time goes by, the greater weakness of the upfolded rock strata results in the formation of **anticlinal valleys** in their places. The more resistant downfolded rock strata, meanwhile, are being eroded more slowly, so that they eventually extend above their surroundings in the form of **synclinal ridges**. At this time, an inverse relationship exists between surface features and rock structure, so that a **reversal of topography** is said to have occurred since the time of initial folding. Because of the influence of folding on rock strength, a reversal of topography is more common than an accordance between structure and topography.

FIGURE 15.3 *Diagram*—Formation and evolution of anticlinal, synclinal, and homoclinal ridges and valleys. AM–anticlinal mountain; AV–anticlinal valley; SM–synclinal mountain; SV–synclinal valley; HM–homoclinal mountain; HV–homoclinal valley; WG-water gap. *Photo*—Sideling Hill, in western Maryland, is a synclinal ridge whose constituent strata have been exposed by a roadcut for U.S. Highway 40. (Ralph Scott)

In many folded areas, the erosional resistance of the rock strata is influenced more by rock type than by fold structures. Where differing rock types exist, some are likely to be more resistant to erosion than others because of their physical hardness and/or chemical stability. In areas containing folded or tilted rock strata of greatly differing resistances to erosion, the positions of the ridges and valleys will be controlled by the surface distribution of strong and weak rocks. Resistant strata will undergo reduced erosion rates and become ridges, and nonresistant strata will erode more rapidly to become valleys. As time goes by and erosion continues, the positions of the ridges and valleys will continually shift to coincide with the changing patterns of resistant and nonresistant rock types exposed at the surface. The ridges and valleys of the Appalachians are a good example of the results of this process (see Figure 15.4).

In areas of tilted or folded sedimentary rocks that differ greatly in their resistance to erosion, ridges and valleys tend to form along the surface outcrops of strong and weak strata. The upthrust edges of the resistant layers extend above the surrounding surface as **homoclinal ridges** (see Figure 15.3c). Likewise, the more rapid erosion of the dipping flanks of weak strata produces **homoclinal valleys**. If the strata

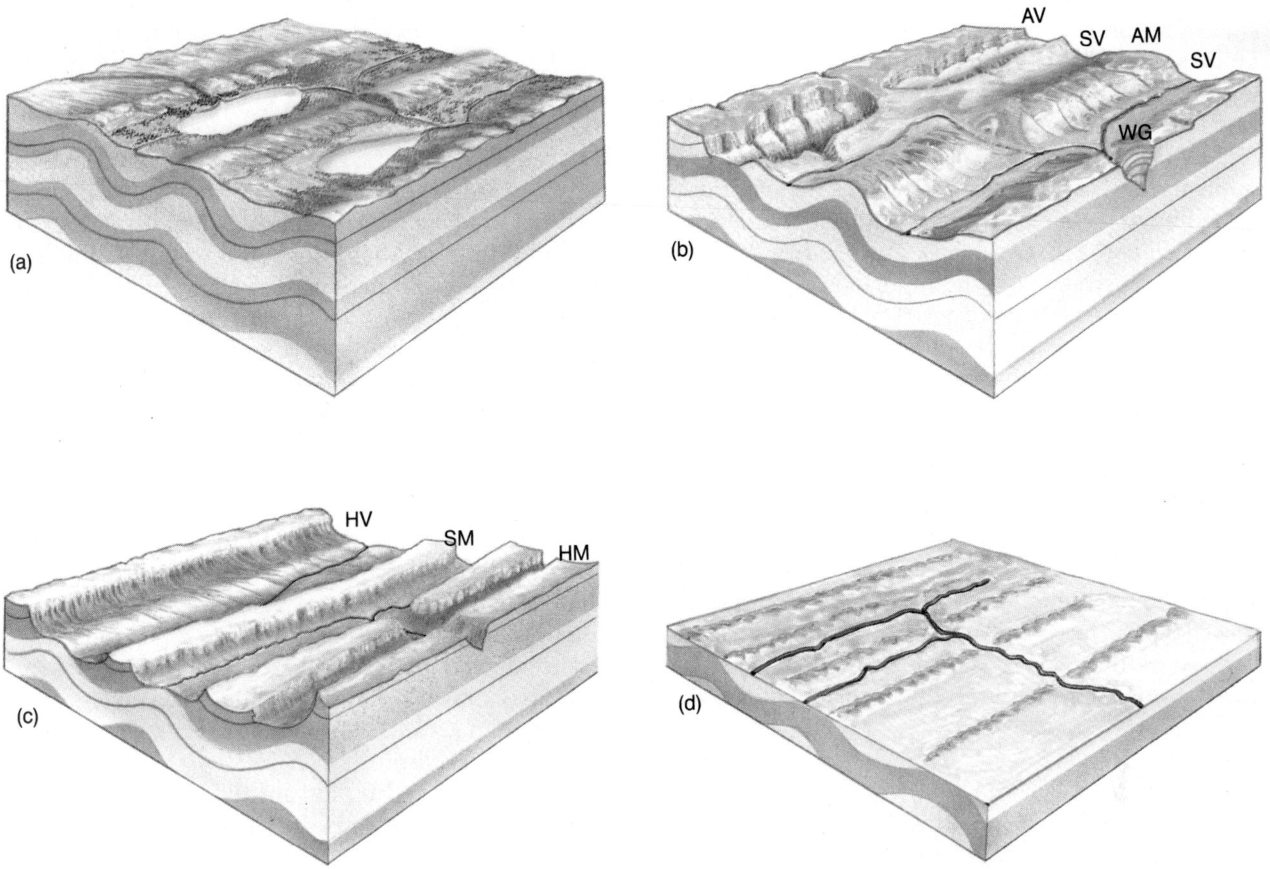

FIGURE 15.3 *continued*

were horizontal at the time of folding, the ridges and valleys developed on them will be linear and may extend for considerable distances. The erosion of tilted anticlines and synclines, however, will produce relatively short, zigzagged ridges and valleys. Such features are well developed, for example, in portions of the Pennsylvania Appalachians and in the Zagros Mountains of western Iran (see the chapter opener photo on page 373).

Regardless of the types of fold structures, if the folding occurred geologically recently and resulted from substantial crustal compression, the affected area normally will be uplifted to a high mean elevation. Erosion therefore should be vigorous, and if the rock layers within the area differ considerably in their erosional resistance, a rugged high-relief surface will result. Conversely, ancient folded rock areas that have been geologically stable for long periods usually display low relief because even the resistant strata have been worn down.

FIGURE 15.4 The Susquehanna River has cut water gaps through these linear Pennsylvania ridges composed of resistant sandstone. (John S. Shelton)

The major cordilleran belts are the primary global sites of recent folding; within many portions of these belts, active folding is still taking place. Indications are that average maximum rates of uplift within portions of these belts are approximately 6 to 7 millimeters (about 0.25 in) per year, and maximum rates of uplift in the California Coast Ranges may approach 12 millimeters (0.5 in) per year. Older mountain systems such as the Appalachians, Urals, Great Dividing Range of Australia, and the highlands of the British Isles contain highly folded strata but are not currently undergoing active folding. They are located on now-sutured lithospheric plate boundaries and owe their survival as low mountain systems to sporadic isostatic uplift of the crust. The ancient shield regions also contain complexly folded, faulted, and metamorphosed rock strata produced by plate collisions that occurred hundreds of millions or even billions of years ago. In many areas they are covered by more recent undeformed strata. Taken together, most land surfaces are underlain, at least at depth, with rocks that have been folded by compressional forces associated with lithospheric plate collisions.

Domes and Basins

Linear folding, as we have seen, is largely a product of horizontal compression resulting from lithospheric plate collisions. Folding, however, also can be produced by vertical forces. These forces consist of upward pressures from the Earth's interior and the downward pull of gravity. If the upward pressures exceed the gravitational pull, the affected area may be arched upward to produce a rounded anticlinal structure known as a **dome** (see Figure 15.5). If, on the other hand, internal pressures become insufficient to counteract gravity, the lack of support may cause the surface rock layers to sag, producing a synclinal **basin**.

Structural domes and basins underlie extensive portions of the Earth's surface. They display great variability in size, shape, age, and extent of folding. Some very large domes and basins cover hundreds of thousands of square kilometers. These structures are all very old, contain very gently arched or downwarped rock strata, and have rather obscure origins. Although they may be visible from very high altitudes, they usually are unrecognizable at the surface; as a result, their identification often requires the careful measurement of rock stratum dips over large areas.

Perhaps the most important practical topographic effect of large dome and basin structures occurs when they are near sea level. Doming may in this case produce sufficient uplift to raise a portion of the seafloor and produce a land area, as in the case of the Florida peninsula. Downwarping, conversely, may submerge an area of continental crust, as in the case of Canada's Hudson Bay.

Most domes and basins are old enough that the associated landforms are the direct result of differential erosion rather than of vertical bedrock movements. A dome or a basin normally will stand topographically higher than its surroundings if the rocks exposed in its center are more resistant than those outside the structure. One of the best-known exam-

FIGURE 15.5 A geologically recent dome in Wyoming probably caused by an igneous (laccolithic) intrusion. (John S. Shelton)

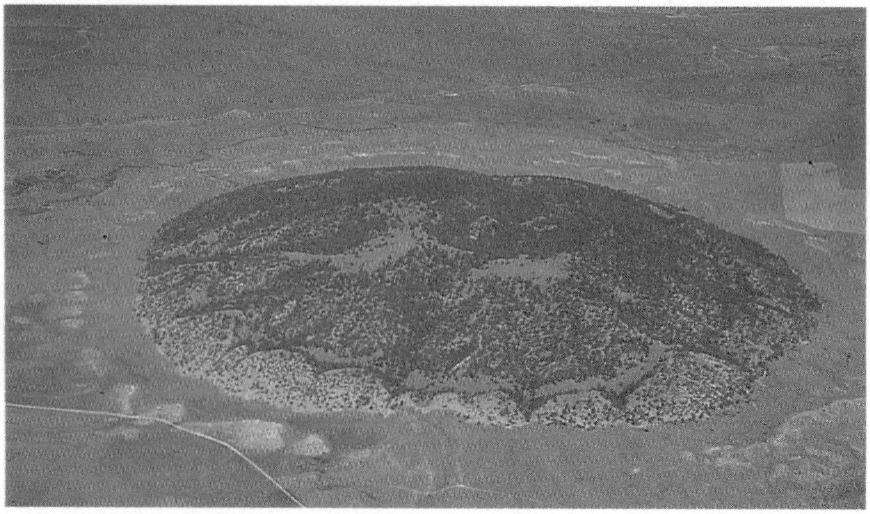

ples of a large dome with an elevated, mountainous center caused by the exposure of resistant rocks is the Black Hills of South Dakota (see Figure 15.24). If, on the other hand, weaker rocks are exposed in the center of either a dome or a basin, a lower surface will result. A good example is provided by the Nashville Basin in central Tennessee. Although this feature is a structural dome, it is a topographic basin because erosion has stripped away the overlying more resistant sandstone layers, exposing less resistant limestone in its center. Often surrounding the raised or lowered centers of eroded domes and basins are concentric rims of tilted sedimentary rock strata that stand at various levels as a result of differential erosion.

Faulting

Faulting involves the breaking or fracturing of rock under pressure, followed by the movement or offsetting of the two sides of the fracture relative to one another. The forces causing major folding and faulting are in most cases identical—they are largely tectonic stresses associated with lithospheric plate movements. The nature of the rock itself is the primary factor determining whether the rock will respond to these stresses by folding or faulting. If the rock is sufficiently flexible, it will fold; if it is relatively brittle, it will fracture and undergo fault displacements. While folding is most frequently associated with areas of sedimentary rock, faulting can occur in rocks of any type but is especially associated with the more brittle igneous and metamorphic rocks. Folding and faulting generally occur together in areas subjected to tectonic stress.

Although the stresses producing both fold and fault movements are applied very gradually, a critical difference exists between the two movements. Whereas folding slowly and continuously relieves the stresses, rocks subject to fault movements frequently store and then release these stresses suddenly, producing earthquakes. Individual fault movements are relatively small, ranging from less than a centimeter up to about 7.5 meters (25 ft), but they are repeated periodically when stresses build to the point that they exceed the resistance of the rock on opposing sides of the fault. For a fault that has long been active, the total displacement of the opposing sides over millions of years can attain values of a thousand or more kilometers horizontally and tens of kilometers vertically.

Fault Terminology and Types of Faults

A fault takes the form of a two-dimensional plane that typically extends from the Earth's surface down to a variable but often considerable depth (see Figure 15.6). The trace of the fault on the surface is termed the **fault line**. Fault lines may extend for distances of hundreds of kilometers, but lengths of a few tens of kilometers are more common. Most faults are nearly straight. This linearity, which results from the tendency of rock to fracture along straight lines, contrasts markedly with the irregularity of the features produced by most other geomorphic processes.

The two-dimensional surface of a fault as it extends into the ground is termed the **fault plane**. The distance below the surface to which faults penetrate varies greatly. Earthquake data indicate that the deepest active faults extend to the upper boundary of the asthenosphere at a depth of about 640 kilometers (400 mi), although most are far shallower.

Geologists recognize four general categories of faults based on the nature of the displacements that occur: normal, reverse, transcurrent (or strike-slip), and thrust faults. Actually, as is true of most other natural phenomena, a continuum of fault types exists. A given fault may have characteristics intermediate between two categories or may grade laterally from one type into another.

Normal faults are the most common of the four fault types. Relative motion is more vertical than horizontal, and an expansionary component is present so that the opposing sides also move apart, resulting in crustal extension (see Figure 15.6a). The **hanging wall** side of the fault, under which the fault plane extends, is down-dropped relative to the **foot wall** side (see Figure 15.6 top). Normal faults usually are produced by broad regional arching in areas of tectonic stress.

Reverse faults are so-named because the movement of the opposing sides is reversed from that of normal faults. Like normal faults, they have deeply dipping fault planes and undergo predominantly vertical motion (see Figure 15.6b). Unlike normal faults, though, reverse faults are produced by regional compression and their hanging wall sides are raised relative to their foot wall sides. Crustal shortening results and a net uplift of the surface normally occurs.

Transcurrent faults undergo a predominantly horizontal offsetting of their opposing sides. Their fault planes are vertical, or nearly so, and slippage is parallel to the strike, or surface orientation, of the

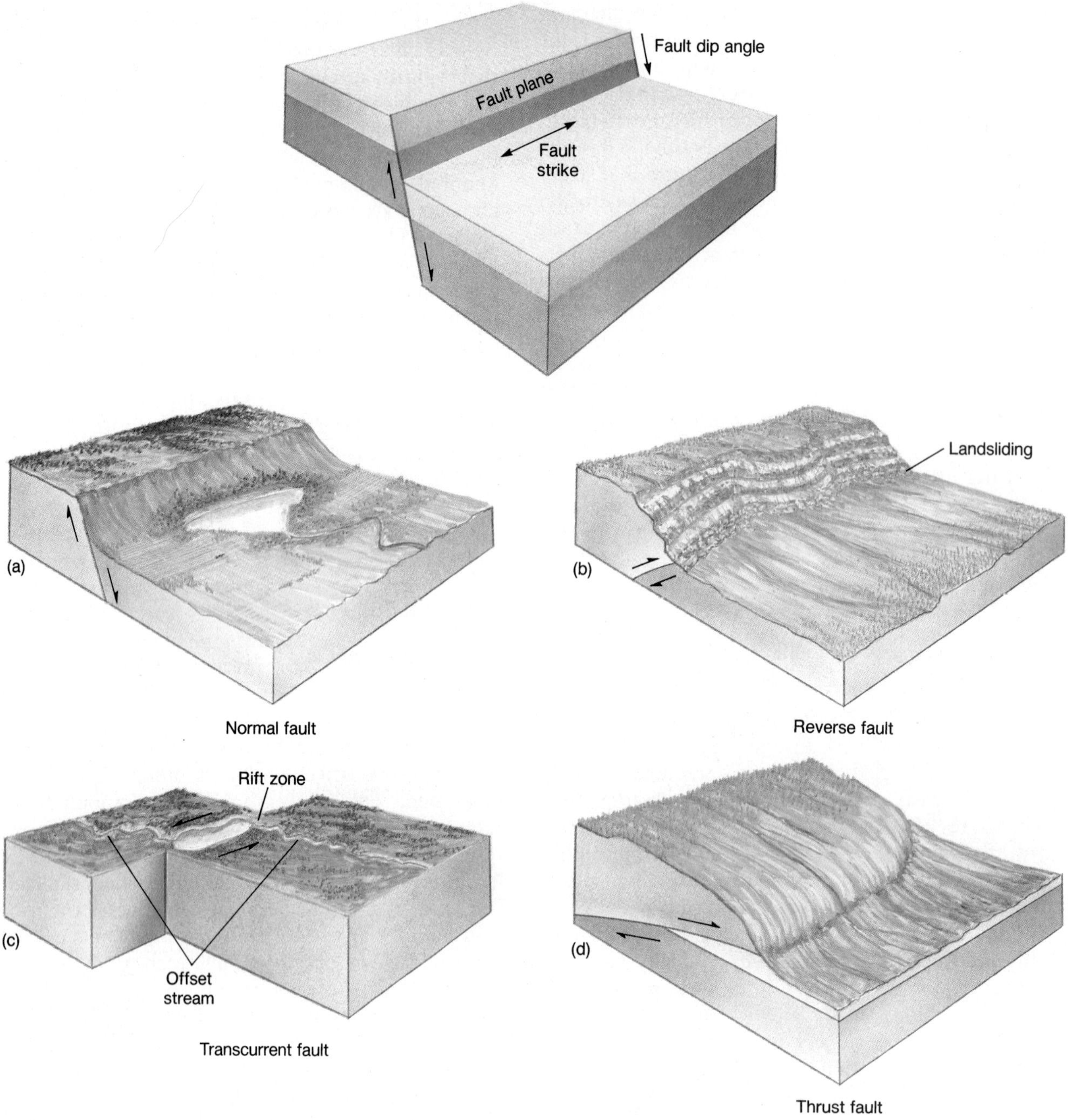

FIGURE 15.6 Fault terminology and profile views of the four basic fault types.

fault lines (see Figure 15.6c). For this reason, they are commonly also referred to as **strike-slip faults**. Transcurrent faults are most frequently located along transform plate boundaries where the relative motions of the opposing lithospheric plates are essentially parallel to one another. Most transcurrent faults are located on the floors of oceanic plates and are produced by seafloor spreading movements, but some, like California's San Andreas fault (see Figure 15.10), occur on land.

Thrust faults result from the extreme compression of rock strata produced by lithospheric plate collisions. The relative movement of the opposing sides is similar to that of a reverse fault. The fault plane, though, dips into the ground at a low angle, and motion is predominantly horizontal as one side is thrust over the other, sometimes for considerable distances (see Figure 15.6d).

Landforms Associated with Faulting

Fault movements are responsible for the production of a great variety of topographic features. These range in scale from large mountain systems and fault troughs with thousands of meters of relief down to microrelief features that may go completely unnoticed on the landscape. It should be stressed that large fault features are the net result of many individual fault displacements occurring over extended periods of time. Fault features may be produced directly through the movement of the blocks of land on opposing sides of the fault, or indirectly as a result of differential erosion. Indirectly produced fault features usually result either from the fault-induced juxtaposition of rocks of differing resistances to erosion or from the formation of easily eroded zones of crushed rock along the fault plane. In either case, the gradational agents of water, ice, or wind subsequently will erode the less resistant rocks at a more rapid rate.

High-angle faults chiefly involve vertical displacements of rock strata; consequently, if these faults are active for a long enough time, they can produce features with large vertical dimensions. The basic landform feature associated with high-angle faulting is the **fault escarpment** (or **fault scarp**). It ideally is a steep, linear, smooth-sided slope consisting of the exposed surface of the fault plane along the edge of the higher fault block (see Figure 15.7). Frequently, the rocks of the fault scarp do not have sufficient strength to maintain a steep slope. Consequently, landslides and subsequent erosion reduce the angle of inclination until a condition of stability is reached. Repeated movements along a major fault over a period of millions of years can produce fault scarps thousands of meters high extending for hundreds of kilometers. Most of the Earth's loftiest mountain ranges, including the Himalayas, are produced largely by high-angle fault displacements.

Topographic features produced by low-angle faulting normally have lower relief and are much less conspicuous on the landscape than are those produced by high-angle faults. Transcurrent faults especially may provide little surface indication of their existence in localities where the offsetting of the two sides is essentially parallel. Frequently, though, a zone of crushed rock is produced along the fault line as the opposing sides grind against one another. This forms a zone of weakness that can be exploited by the agents of weathering and erosion to produce a linear depression, perhaps occupied by a stream or lake. As a stream crosses a transcurrent fault line, its course may be offset by subsequent fault movements so that the stream channel is pulled apart, giving it a zigzagged course (see Figure 15.8).

FIGURE 15.7 A portion of the fault scarp that formed in Gallatin County, Montana, as the result of a major earthquake in August 1959. Note the man standing at the base of the scarp. (J.R. Stacy/USGS)

California's famed San Andreas Fault is the best-known example of a transcurrent fault. Extending through the state for some 880 kilometers (550 mi) from the Mexican border to Tomales Bay just north of San Francisco, it marks the boundary between the American and Pacific Plates (see Figure 15.10). Geologically, the portion of California west of the fault is not a part of North America, but is simply moving north along the margin of the continent at the present time. The San Andreas Fault apparently has been active for several tens of millions of years, and at least 550 kilometers (350 mi) of offsetting of the two sides has occurred (see Figure 15.8). It is believed that the continuing northward movement of the Pacific Plate eventually will carry western California to a position off the coast of Alaska. Because its speed of movement in this direction averages only several centimeters per year, the chief cause for concern to California's residents is the potential destruc-

THE LOMA PRIETA EARTHQUAKE–PRELUDE TO DISASTER?

The fault best known and most feared for its earthquake potential of any within North America is California's San Andreas Fault (see Figure 15.10). Minor slippages occur frequently along this great fault and its numerous offshoots, but the last large-scale movement took place in 1906. This involved a 3-to 6-meter (10-20 ft) displacement of the opposing sides of the northern portion of the fault. It produced the great San Francisco earthquake and fire that leveled much of the city and claimed 700 human lives.

Much more recently, on October 17, 1989, another slippage on the San Andreas Fault caused the Loma Prieta earthquake. This earthquake, which lasted less than 15 seconds, had its epicenter in the Santa Cruz Mountains about 100 kilometers (60 mi) south of San Francisco. It was responsible for 67 deaths and 3,500 injuries, left 12,000 homeless, and caused property damage estimated at nearly $6 billion (see Figures 15.9 and 15.16). The Loma Prieta earthquake had a Richter Scale magnitude of 7.1, making it only one-tenth as powerful as the great earthquake of 1906. Future earthquakes along the fault are inevitable, and Californians are increasingly concerned that one of catastrophic proportions may take place in the near future.

Two lines of evidence point to the possibility of a powerful earthquake along the San Andreas Fault within the next few decades. In the first place, more than 85 years have now elapsed since the San Francisco earthquake of 1906. Prior to that time, major slippages had been taking place at approximately 50-year intervals, so the next movement is now overdue. A second line of evidence comes from measuring instruments and geological observations that indicate that a large amount of strain has developed between the opposing sides of the fault in several areas. One ominous indication of this strain is the so-called "Palmdale Bulge," centered on the fault about 55 kilometers (35 mi) northeast of Los Angeles, where the Earth's surface has in recent years risen about 0.3 meters (12 in). The southern section of the San Andreas Fault apparently became tightly locked after its last large-scale movement, which produced the Fort Tejon earthquake of 1857; to date it has resisted any further

FIGURE 15.8 Stream channels have been offset 22 meters (70 ft) along this portion of the San Andreas Fault on California's Carrizo Plain. (John S. Shelton)

tion from earthquakes resulting from future fault displacements. This topic is examined in the Focus Box above.

Thrust faulting involves the forced movement of thick sheets of rock, called **nappes**, over other rock strata. This motion is the result of extreme compression, which produces crustal shortening to the extent of double-layering the crust in the affected area. In most cases of thrust faulting, older strata are thrust over younger rocks, reversing the normal stratigraphic positioning of younger rock layers atop progressively older layers.

Nappes that have been thrust over younger rocks for distances of tens of kilometers have been identified in a number of mountain systems. These include the western edge of the Appalachian Blue Ridge, parts of the northern Rocky Mountains, and several portions of the European Alps. In the southern Appalachians, thrust faulting produced by the collision

FIGURE 15.9 Most of the casualties resulting from the Loma Prieta earthquake occurred when the top section of Interstate Highway 880 collapsed onto the bottom section. (Reuters/Bettmann Newsphotos)

movements of consequence. Many experts believe that the southern portion of the fault is more likely to be the site of the next major earthquake than is the northern portion near San Francisco.

The tremendous population growth in western California during the twentieth century has greatly increased the potential for severe loss of life and property. Construction standards for new buildings have been improved substantially to make them more earthquake resistant, and civil defense authorities have long been preparing to cope with the consequences of a major earthquake. All that most Californians can do at present, though, is to wait and hope that the fault will provide some prior warning of its next movement.

between the American and African Plates some 300 million years ago drove a nappe more than 130 kilometers (80 mi) westward. In some parts of the Alps, compression has been so great that multiple thrust sheets have been pancaked atop one another, producing of some of the most complex, and most studied, geological structures on Earth.

Multiple Faulting

When an area is subjected to tectonic stresses, it is rare for only a single fault to develop; instead, faults, like folds, tend to form in multiples. An area under stress usually will develop a series of faults that parallel one another. The orientation of fault lines, as with fold axes, generally is perpendicular to the direction of stress application. The blocks of land between such multiple fault systems are relatively lacking in support and are likely to experience vertical displacement. This may be accompanied by tilting, so that a considerable variety of movements, and resulting topographic features, can be produced.

The basic features produced by the movements of blocks of land between parallel faults are horsts and grabens. A **horst** is a raised block of land bounded by two normal faults (see left half of Figure 15.11). In this case, the fault movements have resulted from crustal extension, while pressure from below the central block was sufficient to raise or maintain it above the level of the blocks on either side. A recently formed horst ideally will appear as a linear, steep-sided, flat-topped ridge. The forces of erosion soon modify this form, so that it eventually is reduced to a range of hills or low mountains. Horsts exist in a

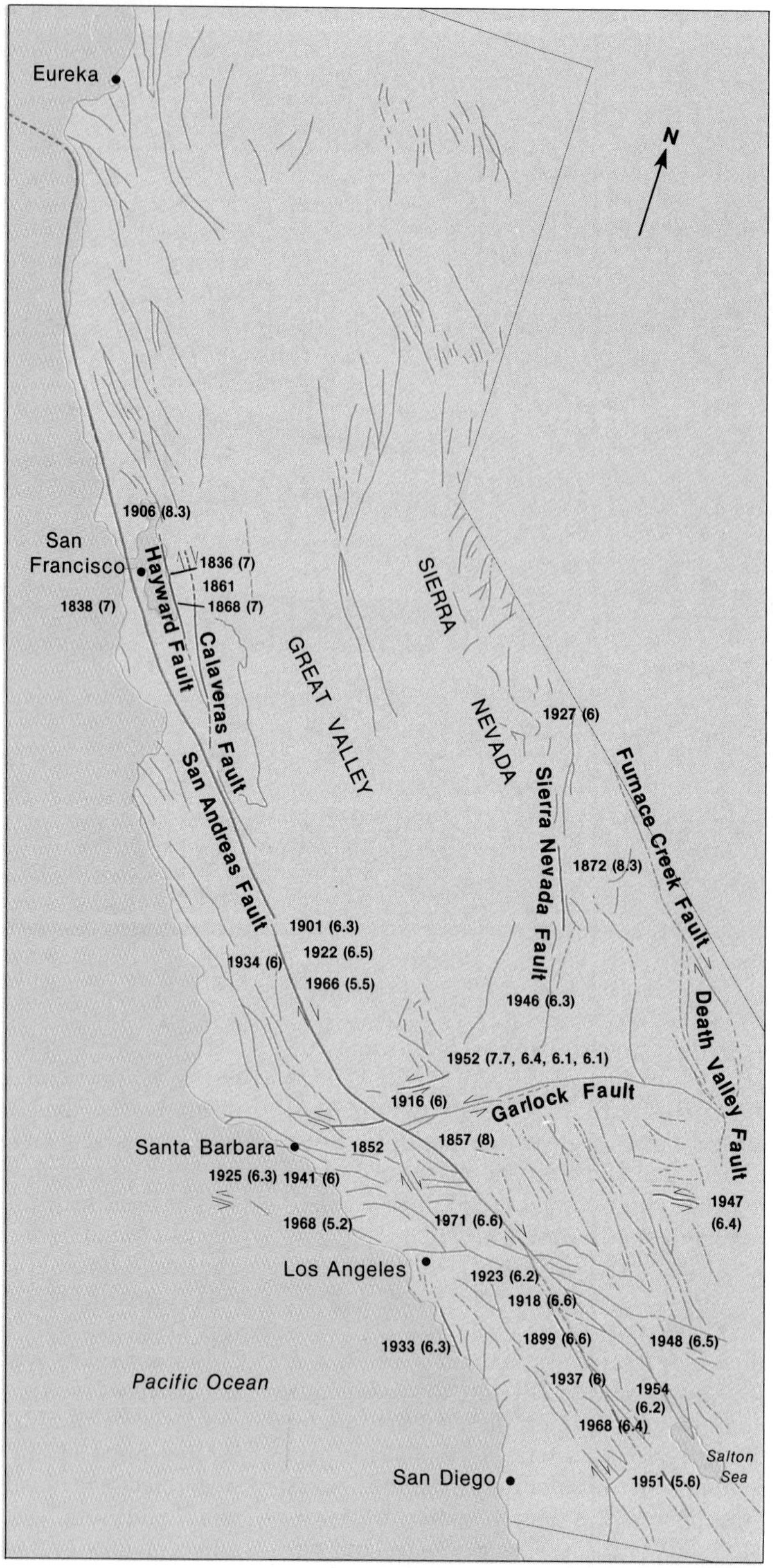

FIGURE 15.10 Map of California showing the location of the San Andreas Fault and its subsidiary faults. The dates and Richter scale magnitudes of major historical earthquakes are given.

variety of sizes; some well-known examples of large horsts include the Palestine Plateau, the Black Forest of Germany, the Vosges Mountains of eastern France, and the Sinai Peninsula.

A **graben** is structurally similar to a horst, except that the central block of land is downdropped between two normal faults (see right half of Figure 15.11). Ideally, a graben initially forms a linear, smooth-sided, flat-bottomed lowland. Once again, the forces of erosion soon modify it so that its original angularity and linearity are lost. Well-known examples of grabens include California's Death Valley, the Rhine Valley separating Germany and France, the Dead Sea Trough, and the trench occupied by Lake Baikal in the Soviet Union.

If the expansionary stresses that produce grabens continue, the separation between the opposing sides may increase until they assume the dimensions of **rift valleys**. Rift valleys are especially large, elongated grabens whose formation frequently marks an early stage in the separation of two lithospheric plate segments. The outstanding present-day example of this feature is the East African Rift Valley (see Figure 15.12). This immense and complex system, which extends more than 3,200 kilometers (2,000 mi) through East Africa from Mozambique northward to Djibouti, is bounded by fault scarps up to 900 meters (3,000 ft) high. On its floor lie some of the world's largest lakes, including Lakes Tanganyika, Malawi, Albert, and Rudolf.

In some parts of the world, large areas of continental crust have been faulted and broadly arched upward by internal pressures that are later relaxed. The faults divide the surface into large blocks that, upon relaxation of pressure from below, drop and tilt to varying degrees. The higher blocks tend to form steep-sided linear mountain ranges. They are separated by broad, flat lowlands developed on the down-

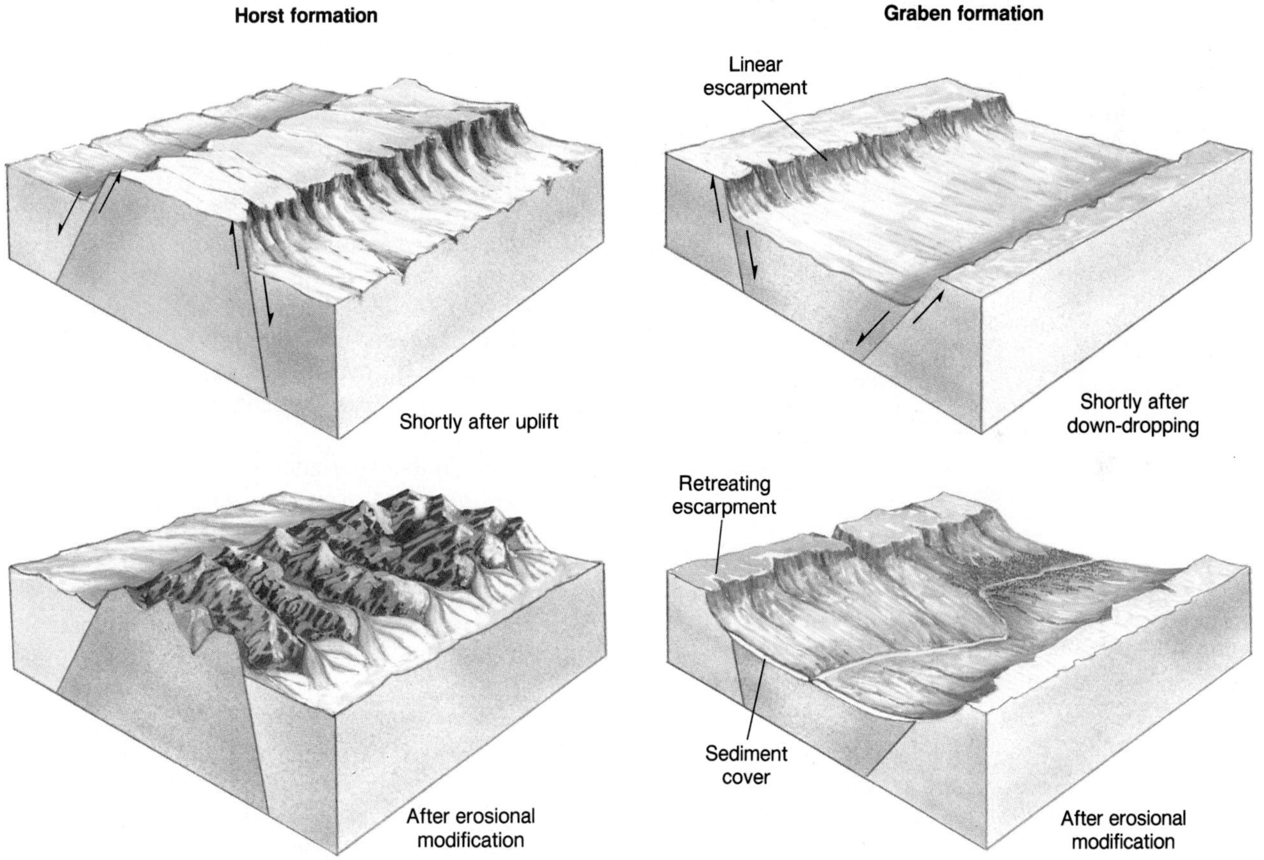

FIGURE 15.11 Block diagram depictions of a horst and graben shortly after formation (top views), and after erosional modification (bottom views).

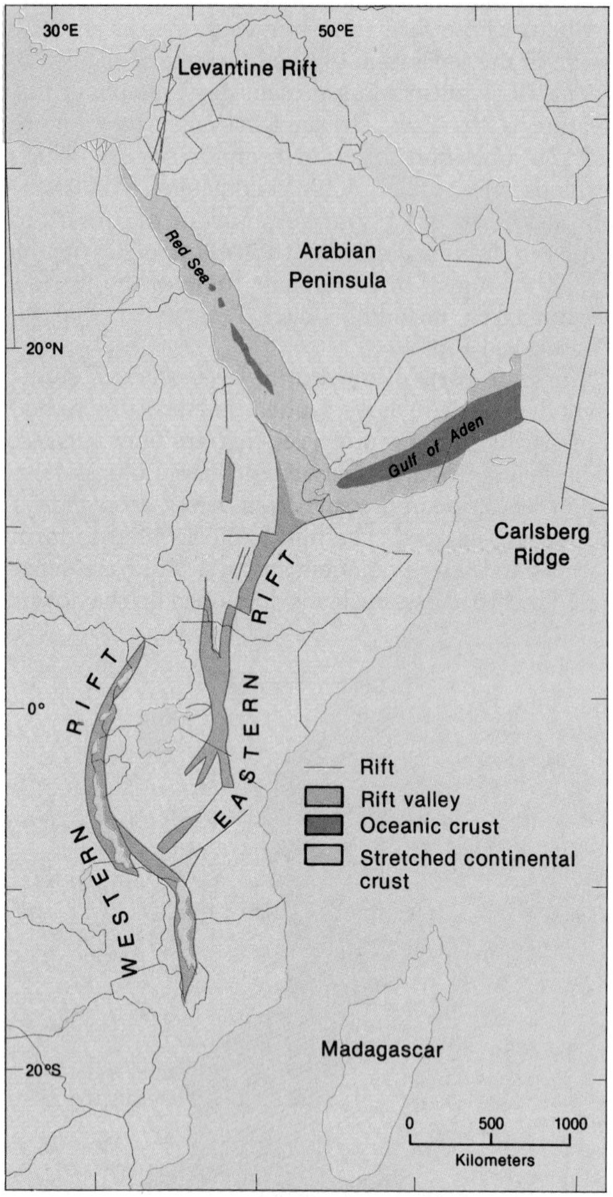

FIGURE 15.12 The East African Rift Valley, shown here, is apparently being formed by the gradual separation of eastern Africa from the rest of the continent as a result of divergent plate movements. The earlier separation of other blocks formed Madagascar and the Arabian Peninsula.

dropped blocks. The result is what has become known in the United States as **basin and range topography**.

Most multiple fault-block mountains and basins exist between split sections of the cordilleran belts. These include the Great Basin of the interior western United States and adjacent northern Mexico, the Iranian Plateau, the Tibetan Plateau, and the Bolivian Altiplano. The American Great Basin provides an excellent example of the basin and range topography developed in such areas. This region contains a large number of normal faults separating block-faulted mountain ranges and intervening broad basins (see Figure 15.13). The western edge of the Great Basin is bounded by the Sierra Nevada of eastern California. This mountain system consists of a giant tilted block of ancient igneous and metamorphic rock that originally formed as an igneous intrusion (or batholith). The fault motion producing the Sierra Nevada has lifted the surface in much the same way that a hinged box lid is raised. This has produced an asymmetrical mountain range, with a gentle west-facing slope and a precipitous east-facing escarpment. Active uplift continues at a rate of approximately a centimeter (0.4 in) per year. Glaciation within the Sierras has largely destroyed its linearity as an upthrust fault block and has carved a series of individual peaks.

Earthquakes

The most important influence of fault movements on the inhabitants of an affected region is the earthquakes they frequently generate. **Earthquakes** are sudden movements of the surface produced when the slowly accumulated strain along the opposing sides of an active fault is released as the two sides are torn apart. Individual fault motions are small and involve relative displacements ranging from a fraction of a centimeter up to 5 to 6 meters (15 to 20 ft). As a general rule, larger displacements produce more violent earthquakes.

The same faults experience slippage and produce earthquakes over and over again as the accumulated stress periodically exceeds the shear strength of the rock. The rocks on the opposing sides of a fault deform elastically in much the same fashion as a rubber band, which stretches in response to a buildup in strain, then suddenly snaps when the strain becomes too great. Although stress tends to accumulate at a constant rate along any given fault, elapsed timespans between successive earthquakes often vary because on some occasions the two sides of the fault may be more tightly locked together than on other occasions. In some cases, the opposing sides of a fault are not locked together at all, and strain is released gradually and continually by a slow **fault creep** that may be detectable only by the gradual offsetting of fea-

FIGURE 15.13 The Great Basin of Nevada and western Utah is in an area of crustal extension and consists of tilted and block-faulted mountain ranges, with broad intervening basins. (Visuals Unlimited/© A.J. Copley)

tures on opposite sides of the fault. Most earthquakes are undetectable except to delicate recording instruments called **seismographs**. Thousands of perceptible earthquakes, however, occur each year somewhere on Earth.

The shaking of an earthquake is actually produced by the passage of a shock wave resulting from the sudden fault movement. A variety of different types of shock waves are produced, but the two most important are the **P waves**, or primary waves, which result from compression, and the **S waves**, or shear waves, which exhibit a side-to-side motion.

Earthquake magnitudes are commonly measured by use of the **Richter scale**, devised by seismologist Charles F. Richter in 1935. It is a logarithmic scale, with each number having 10 times the earthquake wave amplitude (amount of side-to-side motion) and about 31.5 times the total energy release of the preceding number. A Richter value greater than about 2.5 is detectable without special instruments, one above 5 is considered damaging, and one of 8 or above is catastrophic. The most intense earthquake measured since accurate readings have been available, which occurred in central Chile in 1960, had a Richter scale magnitude of 8.9. By way of comparison, the great San Francisco earthquake of 1906 had an estimated magnitude of 8.3, and the Alaskan earthquake of 1964 was measured at 8.5.

The distribution of earthquakes is essentially identical to that of active faults—they are concentrated along lithospheric plate boundaries, especially those where convergence and active subduction are taking place. A map of earthquake surface locations, or **epicenters**, clearly outlines the major plate boundaries (see Figure 15.14). Because these boundaries frequently are the sites of active tectonic uplift, earthquakes on land generally are centered in mountainous regions within the cordilleran belts. Nearly 95 percent of recorded earthquakes occur around the rim of the Pacific and from southern Asia westward through southern Europe. Most other earthquakes take place along generally less active faults located within the lithospheric plates. Some of the movements along these faults also can produce devastating earthquakes on occasion. For instance, much of the city of Charleston, South Carolina, was destroyed by an 1886 earthquake. A series of quakes centered near New Madrid, Missouri, in 1811 and early 1812 sent huge waves rolling down the Mississippi River, changing its course in several locations. Although some regions are notably less prone to earthquakes than others, few if any parts of the world can be considered immune to a major quake (see Figure 15.15).

Individual earthquakes generally have only a minor impact upon landforms. They may offset surface features by a few meters, produce small fault scarplets (see Figure 15.7), and trigger landslides that may on occasion block streams to produce lakes. By far the most important effect, though, is on people, especially those occupying buildings constructed of massive materials and those living along coasts that may experience earthquake-produced tsunami (see

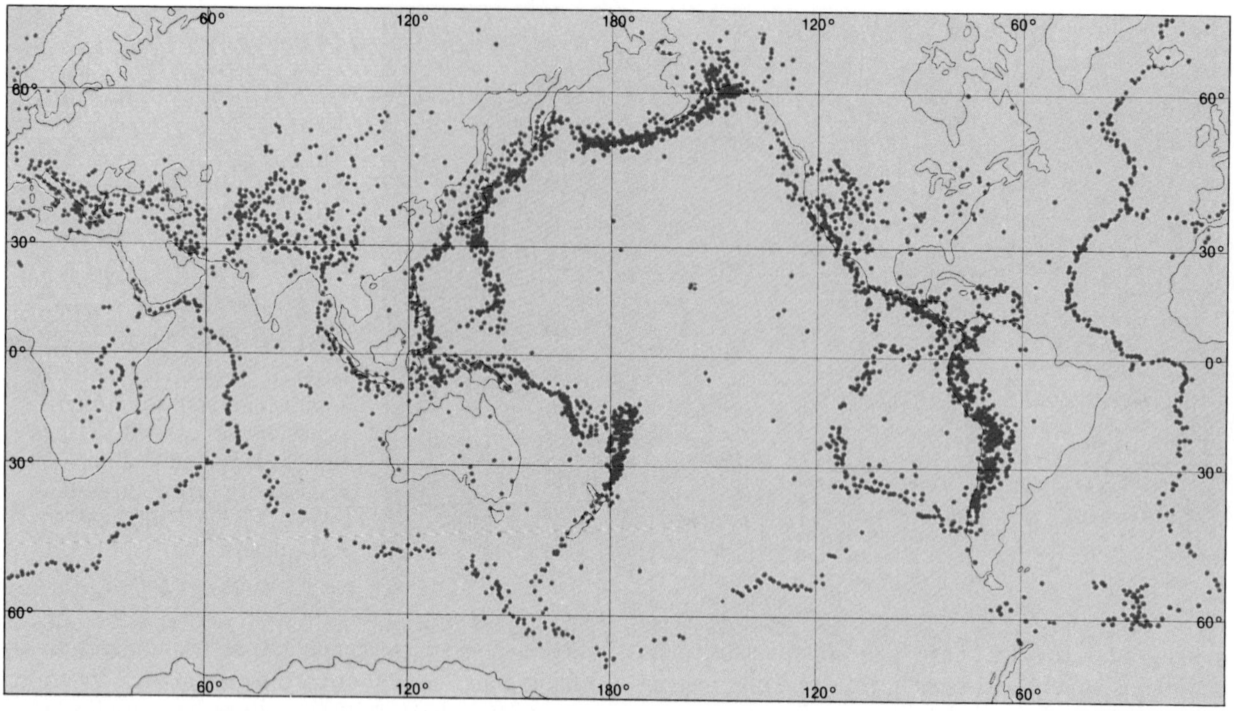

FIGURE 15.14 Global distribution of the epicenters of earthquakes between 1963 and 1977. (The **epicenter** is the surface location directly above the earthquake location, or **focus**.) Note that the dots collectively outline the lithospheric plate boundaries depicted in Figure 14.17.

the Focus Box on page 389). Major earthquakes rank with hurricanes and volcanic eruptions as producers of the most catastrophic natural disasters in recorded history (see Figure 15.16). Cities are especially vulnerable, since buildings may be shaken to the ground, and ruptured gas pipes and downed electric wires cause fires that cannot be reached to be extinguished. The greatest loss of life resulting from an earthquake in the twentieth century occurred in Tangshan, China, in 1976, where an estimated 655,000 persons lost their lives. A 1923 earthquake in Japan destroyed most of Tokyo and claimed 100,000 victims. As already noted, the growing potential for a major earthquake along the San Andreas Fault is currently of great concern to the residents of California.

VOLCANISM AND VOLCANIC LANDFORMS

Volcanism is the second major category of tectonic activity. In contrast to diastrophism, which involves solid-state movements of the Earth's crust, **volcanism** involves the surfaceward movement of fluid magma. The magma, which may be derived from either the lower crust or upper mantle, frequently solidifies into igneous rock before reaching the surface. Volcanism occurs where the lithosphere contains deep fractures or faults, a common situation along and near plate boundaries. These breaks in the lithosphere serve two essential purposes in promoting volcanic activity. First, they may relieve pressures sufficiently to permit the localized melting of the rock that forms the magma; second, they provide pathways for its surfaceward movement.

In the discussion of rock types in Chapter 14, intrusive igneous rocks, which solidify inside the Earth, were differentiated from extrusive igneous rocks, which reach the surface in a molten state before rapid solidification occurs. This section covers the processes by which these movements take place, the geomorphic features they produce, and the geographical distributions of volcanic processes and features.

TSUNAMI

Most ocean waves are produced by the frictional effects of the wind. The largest wind-generated storm waves, as noted in Chapter 11, are more than 15 meters (50 ft) high. The largest and most destructive waves of all, however, are seismic waves caused by sudden tectonically induced movements of the sea floor. These movements result most frequently from earthquakes, but they also may be caused by volcanic eruptions or large underwater avalanches. In the past these waves were commonly referred to as "tidal waves," but they have nothing to do with the tides and are more properly termed **seismic waves** or given their Japanese name **tsunami**. They are most numerous in the Pacific, which is ringed by active fault lines and volcanoes. Like wind-generated waves, they are capable of traveling great distances from their points of origin.

Tsunami are essentially shock waves produced by underwater displacements. They travel much more quickly than wind-generated waves, their speed increasing with water depth. For example, in water 300 meters (1,000 ft) deep, they travel at 185 kilometers per hour (115 mph), but in water 4,600 meters (15,000 ft) deep, they travel at 770 kilometers per hour (480 mph). This rapid movement provides relatively little time for threatened coastlines to make preparations.

A tsunami quickly loses height as it travels outward from its point of origin, and it passes unnoticed beneath ships in deep water. Upon approaching a coastline, however, the frictional drag of the increasingly shallow ocean bottom on the rapidly moving wave causes it to increase in height to as much as 30 meters (100 ft) or more. Its approach is first indicated by a marked fall of water levels along the coast over a period of several minutes. This is followed by a rapid rise of water that soon inundates the entire shore and may advance well inland.

Tsunami effects are most extreme on low-lying coasts and especially in funnel-shaped bays, where the force of the rising water is magnified. A dramatic example of the magnification of a tsunami in a narrow bay occurred in Lituya Bay, Alaska. On July 10, 1958, a tsunami generated by a nearby undersea earthquake destroyed trees to an elevation of 525 meters (1,720 ft) at the head of the bay.

One of the most destructive tsunami on record was produced by the explosion of the East Indian volcano Krakatau on August 27, 1883. This tsunami devastated the densely populated islands of Java and Sumatra, taking in excess of 36,000 lives (see the Case Study at the end of this chapter). The Hawaiian Islands also have been visited on occasion by destructive tsunami.

Extrusive Volcanism

Extrusive volcanism consists of processes by which molten rock reaches, and is extruded upon, the Earth's surface. Although it occurs less frequently than intrusive volcanism and generally involves smaller quantities of molten rock, it is much better known, more spectacular, and has a more direct influence on the topograpy.

Two basic groups of extrusive volcanic features exist—volcanoes and lava flows. The difference between the two is determined largely by the fluidity of the magma and the force with which it is expelled onto the surface. A comparison of volcanoes and lava flows provides an excellent illustration of the fact that features with similar modes of origin may differ greatly in their topographic expressions.

Volcanoes

A **volcano** is a hill or mountain constructed of materials from the interior of the Earth that have been ejected under pressure from a vent (see Figure 15.17). Large volcanic peaks are among the most beautiful and distinctive types of landforms. They typically have a conical and sometimes nearly perfectly symmetrical shape and may extend to great heights above the surrounding landscape. They often are surmounted by a cap of snow and ice and, if

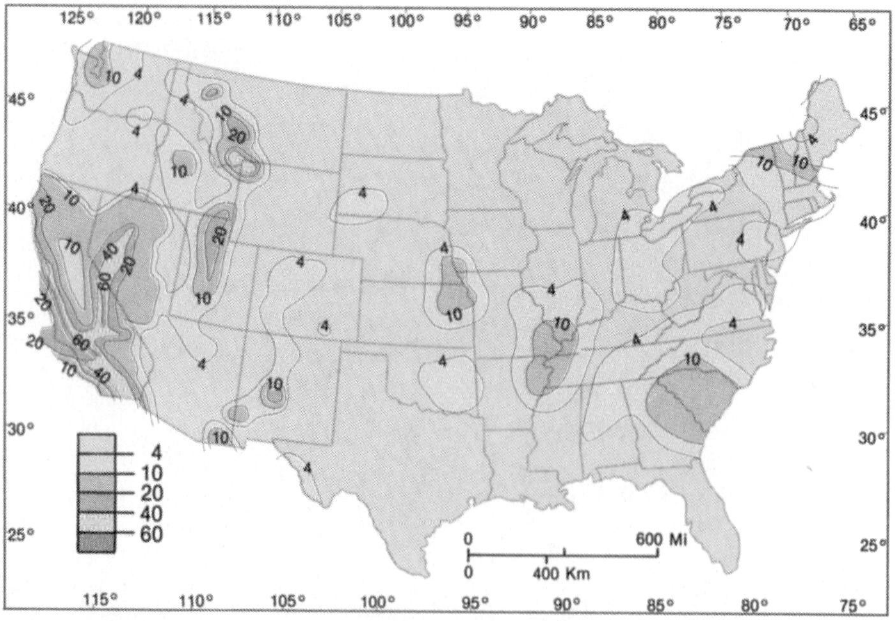

FIGURE 15.15 Map of estimated U.S. earthquake potential. (Modified from S.T. Algermissen and D.M. Perkins, "A Probabilistic Estimate of Maximum Acceleration in the Contiguous United States," USGS Open-File Rep. 76–416, July 1976.

active, may extrude steam, smoke, and sometimes ash or lava. A volcanic eruption, especially when viewed at night, may be the most spectacular and awe-inspiring of all geomorphic events (see the chapter opener photo). Humanity's fascination with volcanoes also is increased by their unpredictable nature and demonstrated potential for dealing death and destruction on a large scale.

On the other hand, because of their spectacular nature, it is tempting to overrate the geomorphic influence of volcanoes. These features are relatively restricted in geographic extent, and most of the Earth's surface does not contain any extrusive volcanic features at all. At present, approximately 540 active volcanoes are scattered throughout the world. Even in areas where they do occur, volcanic peaks generally are relatively localized, and other geomorphic processes, such as stream erosion, are likely to exert a much greater geomorphic influence on a regional scale.

Volcanic eruptions have taken approximately 300,000 human lives during the past four centuries.

FIGURE 15.16 A front-end loader tears down the remains of an apartment complex in San Francisco's Marina district the day after the Loma Prieta earthquake of October 17, 1989. (Reuters/Bettmann Newsphotos)

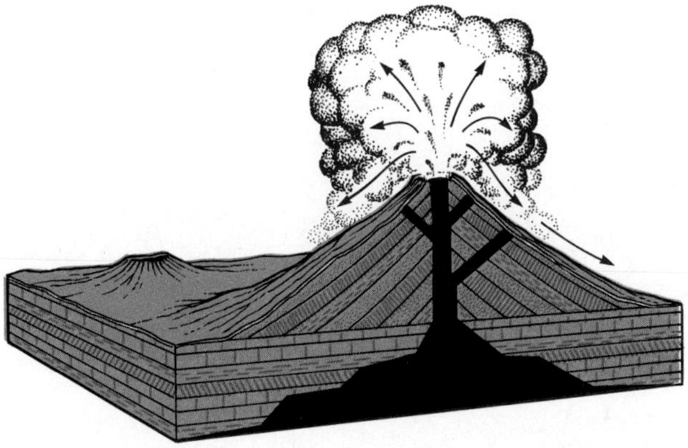

FIGURE 15.17 Cutaway view of a composite volcano, showing the magma chamber, system of vents, and layers of lava and ash comprising the cone.

The greatest known loss of life during that period occurred in the 1815 eruption of Tamboro volcano in the East Indies, which killed 92,000 people. Most eruptions occur from already existing volcanoes, but about 15 or 20 new volcanoes have formed during historic times. Two well-documented cases of volcanic births are of Parícutin in a Mexican cornfield in 1943 and Surtsey off the coast of Iceland in 1963. Volcanoes vary greatly in the timespan over which they are active; some erupt only once over a period of a few days or weeks, while others are intermittently active for hundreds of thousands of years.

The global zones of present-day volcanic activity are closely associated with those areas where folding, faulting, and earthquakes are currently taking place. As a result, most volcanoes are located within cordilleran belts on or near lithospheric plate boundaries. Nearly four-fifths of the active volcanoes on land are located in the "Ring of Fire" around the circumference of the Pacific. Portions of the Pacific margin where volcanoes are especially numerous include the southern and northern Andes, southern Mexico, the Aleutians, and the entire western Pacific arc from the Soviet Union's Kamchatka Peninsula southward to northern New Zealand (see Figure 15.18). Other notable centers of activity include the Mid-Atlantic Ridge—especially in the vicinity of Iceland—the East African Rift Valley, the Lesser Antilles, the Mediterranean Sea, the islands of Java and Sumatra in the East Indies, and scattered mid-oceanic "hot spots" situated above mantle plumes, such as Hawaii, Tahiti, the Galápagos, and Réunion. It also should be noted that great quantities of lava are extruded relatively quietly from fissures along the seafloor-spreading centers of the oceanic ridges.

The nature of the volcanic activity within any region is controlled largely by the type of magma involved. This, in turn, depends upon the type of plate tectonic movements and on the source of the magma. Highly fluid magmas containing considerable magnesium and iron are believed to be produced by the partial melting of rock within the upper mantle in areas of divergent or transform plate movements or above mantle plumes. They are extruded at temperatures averaging above 1,200°C (2,200°F). All magmas contain great quantities of dissolved gases that come out of solution as the magma approaches the surface. These gases remain dissolved longer in fluid magmas, and when they come out of solution, they do so quietly and easily. The result is that eruptions involving fluid magmas are relatively nonviolent and are characterized by the outpouring of large quantities of lava. At present, most such eruptions are occurring on the ocean floor or on mid-oceanic islands situated above mantle plumes, such as Hawaii. The basaltic rock formed from the solidification of these eruptions has produced the lower crust, or sima.

Volcanoes situated above areas of plate convergence and subduction apparently derive their magma from the melting of the lighter components of the subducted lithospheric plates. The result is the production of a somewhat cooler magma (averaging about 900°C or 1,600°F) that is considerably more chemically **acidic**, or rich in silicon dioxide. The result is a much more viscous or "pasty" type of magma. The high viscosity of this magma inhibits the release of its compressed gases. This usually results in violent volcanic eruptions powered by the explosive escape of the pent-up gases. It was this factor, for example, that caused the great violence of the initial eruption of Mount St. Helens in May 1980 (see Figure 15.19).

Acidic eruptions normally involve the expulsion of relatively modest quantities of lava but produce large amounts of **pyroclastic** (from the Greek *pyros*, "fire," and *klastos*, "broken") or fragmental material. Particle sizes range from fine ash, which may remain suspended in the atmosphere for years if it reaches the stratosphere, to blocks the size of houses. Most violent eruptions of acidic materials occur along continental margins or island arcs where convergence is taking place either between two oceanic plates or between an oceanic and a continental plate.

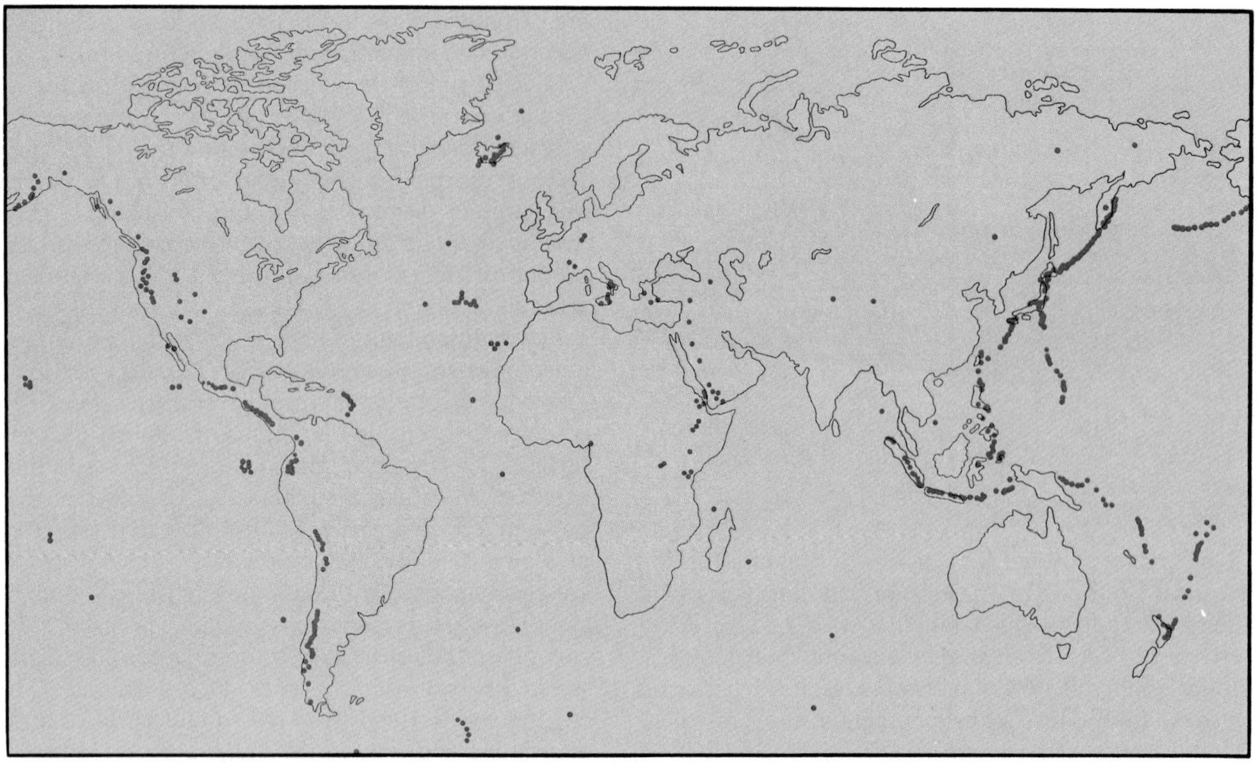

FIGURE 15.18 Global distribution of volcanoes that have erupted in geologically recent times.

The type and quantity of available magma, the pressure under which it is extruded, and the persistence of subsequent eruptions at the same site all affect the nature of the surface features produced by volcanic eruptions. The most abundant, most wide spread, and generally the smallest of the major types of volcanoes are **cinder cones**. They are conical hills or low mountains constructed of loose fragmental material thrown into the air from a central vent (see Figure 15.20a), and occur in areas underlain by viscous, acidic magmas. Most cinder cones are relatively steep, with slope angles determined by the maximum angle of repose of the pyroclastic fragments of which they are composed. This angle is usually near 40° and increases from the base toward the summit, giving the cinder cone a concave profile. The central vent forms a depression, or crater, which is filled by loose pyroclastic material. Cinder cones exist in a variety of sizes, but most are only a few hundred meters high. They generally form very rapidly, and most become extinct after only one period of eruption. Their composition of loose, easily weathered material causes cinder cones to be eroded rapidly by gradational forces, so that they are short-lived features on the landscape. Geologically recent cinder cones are found in several portions of the western United States, notably in central Arizona and in Idaho's Snake River Plain.

FIGURE 15.19 Aerial view of the Mount St. Helens eruption of May 18, 1980. (USGS/JLM Visuals)

FIGURE 15.20 (a) Sunset Crater, near Flagstaff, Arizona, is a geologically recent volcanic cinder cone with a basal lava flow. Its only eruption occurred in 1065 A.D. (b) The gently sloping profile of Mauna Loa volcano as seen from the north coast of Hawaii. (c) Mt. Mayon in the Philippines is a composite volcano with an almost perfectly symmetrical conical profile. (d) Oregon's Mt. Hood is a dormant composite volcano whose outlines have been strongly modified by glacial erosion. (a: John S. Shelton; b: B.F. Molnia; c: Camerique/H. Armstrong Roberts; d: D.R. Crandell/USGS)

Shield volcanoes, in contrast to cinder cones, are formed from highly fluid basaltic (or **basic**) magmas. They have broad, gently sloping surfaces and areconstructed primarily of solid basaltic rock rather than pyroclastic materials (see Figure 15.20b). They are located mostly in the ocean basins at "hot spots" above mantle plumes and form volcanic islands or island groups. Examples of shield volcanoes are Mauna Loa and Mauna Kea in Hawaii, Mt. Hekla in Iceland, and Mt. Etna in Sicily, as well as the volcanoes that have formed the islands of Réunion, Mauritius, Tahiti, Samoa, and the Azores.

The eruptions of shield volcanoes are relatively quiet and consist largely of outpourings of great quantities of fluid lava. Individual shield volcanoes may erupt sporadically over long geologic timespans before they become extinct, so that mountains of great size are eventually produced. The Hawaiian volcanoes Mauna Loa and Mauna Kea, for example, not only are very broad, but they also have elevations exceeding 4,000 meters (13,000 ft). If measured from their bases on the floor of the Pacific at a depth of about 5,500 meters (18,500 ft), they form the tallest mountains on Earth.

Most of the largest volcanic peaks are constructed of alternating layers of pyroclastic materials and lava. Because their surficial and eruptive characteristics are in some respects transitional between cinder cones and shield volcanoes, they are known as **composite volcanoes.** Nearly all are located within the cordilleran belts or on island arcs situated above convergent lithospheric plates. Most eruptions of composite volcanoes consist chiefly of pyroclastic materials derived from acidic magmas, but the composition of the magma can change in different eruptions, or even during different portions of the same eruption, allowing the release of lava flows of more basaltic composition. Some of the better known examples of composite volcanoes are Mt. Fuji in Japan, Mounts Shasta, Hood, Rainier, and St. Helens in the western United States (see Figures 5.19 and 5.20d), Mt. Vesuvius in Italy, Mt. Kilimanjaro in Tanzania, Mt. Mayon in the Philippines (see Figure 15.20c), and Mt. Egmont in New Zealand. Composite volcanoes combine the steepness and distinctiveness of cinder cones with the large size and geological persistence of shield volcanoes. Unlike cinder cones, they frequently have multiple craters and are asymmetrical in shape.

The sudden eruptions of composite volcanoes have been responsible for most major volcanic disasters. Eruptions are sometimes exceedingly violent because extreme gas pressures can build up when hardened lava from earlier eruptions has blocked the volcano's vent. When these pent-up gases are finally released, explosive eruptions occur that are capable of releasing prodigious quantities of ash. (The Case Study at the end of the chapter provides a good illustration.) Heavy ash falls can bury the surrounding countryside to depths of tens or even hundreds of meters, producing ash-covered plains or ash-filled basins. Fine ash and sulfuric acid droplets can be injected high into the stratosphere, blocking out enough solar energy to detectably lower worldwide temperatures for several years. Associated lava flows also can bury the towns and fertile agricultural lands that often surround volcanoes.

Two major volcanic eruptions that occurred within a week of each other were the June, 1991, eruptions of Japan's Mt. Unzen and of Mt. Pinatubo, in the Philippines. Prodigious ashfalls from the Mt. Pinatubo eruption forced the U.S. military to evacuate Subic Bay Naval Base and led to the decision to permanently abandon Clark Air Force Base. The volcano erupted great quantities of sulfur dioxide (SO_2) gas, which combined with water vapor to form an extensive cloud of microscopic sulfuric acid (H_2SO_4) droplets at an altitude of about 24 kilometers (15 miles). Atmospheric scientists believe that this cloud will reflect enough insolation over the next several years to lower the Earth's mean temperature by an estimated 0.6C° (1.0F°).

During some especially violent eruptions, avalanches of superheated gases and ash called **nuées ardentes** (French for "glowing clouds") are ejected from volcanoes. Despite their incorporated gases and temperatures of several hundred degrees Celsius, they are dense enough to rush down the flanks of the volcano under the influence of gravity, charring or setting fire to anything combustible and burning or suffocating all living organisms in their paths. On several occasions, nuées ardentes have almost instantaneously taken the lives of the inhabitants of entire cities located near the bases of erupting volcanoes. The Roman residents of Pompeii are believed to have been overwhelmed by such a phenomenon during the famed eruption of Mt. Vesuvius in A.D. 79, and during the 1902 eruption of Mt. Pelée on the island of Martinique in the West Indies, a nuée ardente killed all but two people in St. Pierre, a city of 30,000 inhabitants.[1]

In various parts of the world, large, roughly circular depressions of volcanic origin known as **calderas** are found. Most are believed to be formed by the collapse of a volcano into its own subsurface magma chamber after the chamber has been emptied of its contents by an eruption. Diameters of calderas range from about 3 kilometers (2 mi) up to 16 kilometers (10 mi) or more. One of the largest known is Buldir caldera, which occupies a submerged site between the islands of Buldir and Kiska in the Aleutians. It is 43 kilometers (27 mi) long and 21 kilometers (13 mi) wide. The best-known caldera in the United States is that formed by the prehistoric eruption and collapse (about 7,000 years ago) of Mt.

[1] One of these individuals owed his life to the fact that he was a prisoner in an underground jail cell at the time of the eruption. For many years thereafter, he made a living with a traveling circus as a survivor of the eruption.

FIGURE 15.21 Crater Lake, in Oregon, is a caldera formed by the explosion and collapse of a volcano. Wizard Island, a cinder cone, is visible on the far side of the lake. (Reprinted from Wicander/Monroe, *Historical Geology*, © 1989 West Publishing)

Mazama in Oregon. This 8 kilometer (5 mi) wide caldera is now occupied by Crater Lake, and its natural beauty and unique origin have caused the United States government to preserve it within a national park (see Figure 15.21). The violent eruption of Krakatau in the East Indies in 1883, described in the Case Study, formed an underwater caldera, and the Yellowstone Plateau, discussed in the Focus Box on page 364, consists of a huge multiple caldera system.

Lava Flows

Lava flows occur when subsurface gas pressures are insufficient to throw molten rock material into the air during an eruption; instead, it wells out of the ground in a fluid state. This may occur during the latter stages of a volcanic eruption due to the reduction in gas pressures at this time. Lava flows, sometimes on a massive scale, also may occur where magmas are so fluid that they offer little resistance to gases, allowing them to escape readily to the surface. This normally occurs when magmas are of basaltic composition.

At some sites beneath divergent plate boundaries and above mantle plumes, temperature and pressure conditions allow the partial melting of large quantities of rock within the upper mantle, producing reservoirs of basaltic magmas. If the magmas reach the surface, extensive outpourings of highly fluid **flood basalts** are produced. Most flood basalt flows occur on the ocean floor along the axes of the oceanic ridges. The movement of the lithospheric plates away from these ridges has spread their eruption products over the Earth to produce the lower crust, or sima.

FIGURE 15.22 The margin of a geologically recent basaltic lava flow on Idaho's Snake River Plain within the Columbia Plateau (John S. Shelton)

Of greater direct significance to the topography of the continents are the large outpourings of flood basalts on land above "hot spots" apparently associated with mantle plumes. These outpourings occur as **fissure eruptions** along a major fracture or fault rather than from a single vent, as in a volcano. Huge outflows of flood basalts have occurred in a number of widely separated locations, including the Columbia River Plateau and Snake River Plain of the northwestern United States (see Figure 15.22), the Deccan

Plateau of northern India, the Paraná Plateau of southern Brazil, and the Drakensberg Plateau of South Africa, as well as in portions of Antarctica, Bolivia, Ethiopia, Iceland, the Soviet Union, and northwestern Arabia. Much older sites of flood basalts have been identified in other locations such as the Piedmont of the eastern United States and in the British Isles. Iceland is the only location where such eruptions have taken place during historic times, and these eruptions occurred on a relatively limited scale. The last major Icelandic fissure eruption took place in 1783, when approximately 12.5 cubic kilometers (3 mi^3) of lava were released along the 32-kilometer (20 mi) long Laki fissure.

Fissure eruptions may largely or completely bury the original surface, producing a gently rolling landscape that, depending on its elevation, may be termed a **lava plain** or a **lava plateau**. Successive flows may in some cases accumulate to depths far greater than 1,000 meters (3,000 ft). The Columbia Plateau, for instance, consists of approximately 260,000 square kilometers (100,000 sq mi) of lava-covered surface, and in some locations the lava reaches depths of 3,000 meters (10,000 ft). The lavas of the Columbia Plateau have been extruded geologically recently in a large number of separate eruptions. The eruptive cycle probably is not yet completed, and future outpourings of flood basalts may one day again cover large portions of this region.

Intrusive Volcanism

Intrusive volcanism occurs when rising magma solidifies into igneous rock before reaching the surface. If solidification takes place far below ground, any effect on the surface is indirect and occurs at a much later geologic period when the deep-seated pluton is exposed by the erosion of the overlying rock. Near-surface intrusions as well generally have their greatest influence on the topography when exposed by erosion. Unlike plutonic intrusions, however, near-surface intrusions also may have a direct and immediate effect on the surface if they deform the overlying rock layers during their upward movement. The surrounding rock may be domed up and also may be metamorphosed by the heat and pressure of the molten rock. When intrusive igneous rock masses are exposed at the surface, their greatest effect on landforms results from differential erosion rates. In most but not all cases, the igneous rock is more resistant to erosion than the surrounding rock and produces high-standing features.

By far the largest of the plutonic rock features are **batholiths**. They are huge plutons with diameters of usually 50 kilometers (30 mi) or more and indeterminate depths that form by the slow cooling and solidification of magma deep within the crust (see Figure 15.23). Rock crystallization is sufficiently slow to allow mineral differentiation to occur. If only the upper portion of the batholith is exposed at the surface, as is normally the case, this differentiation causes it to consist largely of granite. The slow cooling and solidification also allow time for mineral-bearing fluids to deposit veins of metallic minerals in fractures, so that today valuable metal deposits are being mined within exposed batholiths in a number of regions. Batholiths currently are exposed in the cores of many mountain ranges and it is believed that they may underlie, at depth, all major mountain regions. The origin of batholiths has been the source of much controversy. Most researchers believe that the relaxation of pressure at some stage during the formation of a cordilleran system lowers the melting point of the rocks comprising its roots sufficiently to cause large-scale melting, thus producing the magma pool from which the batholith forms. Within the United States, batholiths are exposed in the Sierra Nevada, the Salmon River Mountains of central Idaho, the Front Range of the Colorado Rockies, the Adirondacks of upstate New York, and the Black Hills of South Dakota (see Figure 15.24).

Laccoliths are smaller bodies of relatively acidic magma that work their way surfaceward through vents and solidify as dome-shaped intrusions between flexible sedimentary strata (see Figure 15.23). If they approach the surface closely enough, they are capable of bulging up the overlying strata to create **laccolithic domes** (see Figure 15.5). In time, the igneous core of the laccolith may be exposed by erosion, forming a central mass of hills or mountains. Several mountain masses in southern Utah, including the Henry and LaSal Mountains, represent the cores of exposed laccoliths.

Sills consist of sheets of magma injected between sedimentary rock strata relatively near the surface. They are rather similar to laccoliths, but are flatter because of the greater fluidity of the magma involved. As a result, they do not noticeably dome the overlying rock layers. Sills vary greatly in breadth and in thickness, depending on the quantity of magma involved.

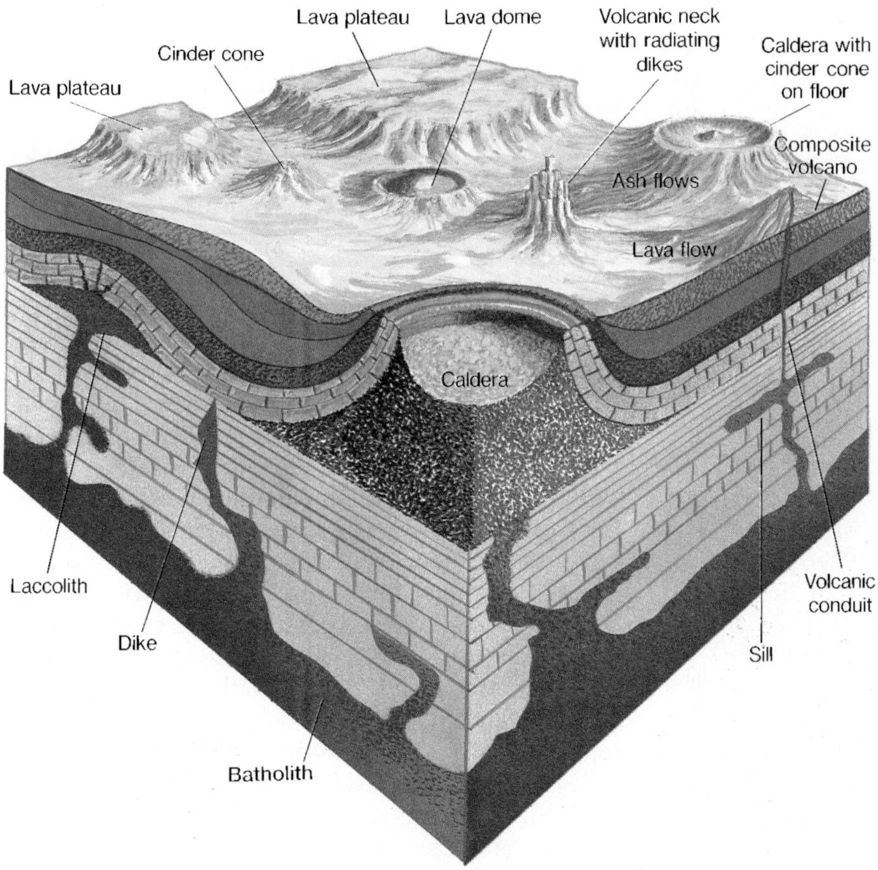

FIGURE 15.23 Block diagram showing various intrusive and extrusive volcanic features.

Larger sills may have horizontal dimensions of tens of square kilometers and thicknesses of more than 30 meters (100 ft). Exposures of the edges of large sills often display the columnar jointing typical of basalt, as in the well-known examples of the Giant's Causeway on the Irish coast and the Palisades along the lower Hudson River in New York.

Dikes are igneous rock masses formed from sheets of magma that have been injected discordantly, and often nearly vertically, through preexisting rock fractures. They are very much like sills, except that they cut across the structures into which they are intruded while sills parallel the structures. Dikes are extremely narrow compared to their length. They may extend for many kilometers, but usually are only a few centimeters to several meters thick. They are distributed widely, being found in "swarms" beneath some volcanoes or volcanic remnants, where they often radiate outward from the central vent that supplied their magma.

Volcanic Erosion and Remnants

Individual volcanic peaks are relatively short-lived features on the landscape. Once eruptions cease, the forces of gradation quickly lower and eventually remove them completely. Most symmetrical volcanoes currently on the Earth's surface have been active within the past few thousand years. Cinder cones and composite volcanoes, because of their steep slopes and loose pyroclastic composition, erode especially rapidly. The uneven distribution of more resistant lava flows on composite volcanoes typically results in an irregular erosional profile because the softer pyroclastic portions are removed more rapidly than the harder portions composed of solidified lava (see Figure 15.20d). Shield volcanoes, with their great masses, gentle slopes, and predominant composition of solid basalt, are more durable.

A volcano composed primarily of loose pyroclastic materials may leave as its last erosional remnant a **volcanic neck**. This is a solid spire of hardened lava

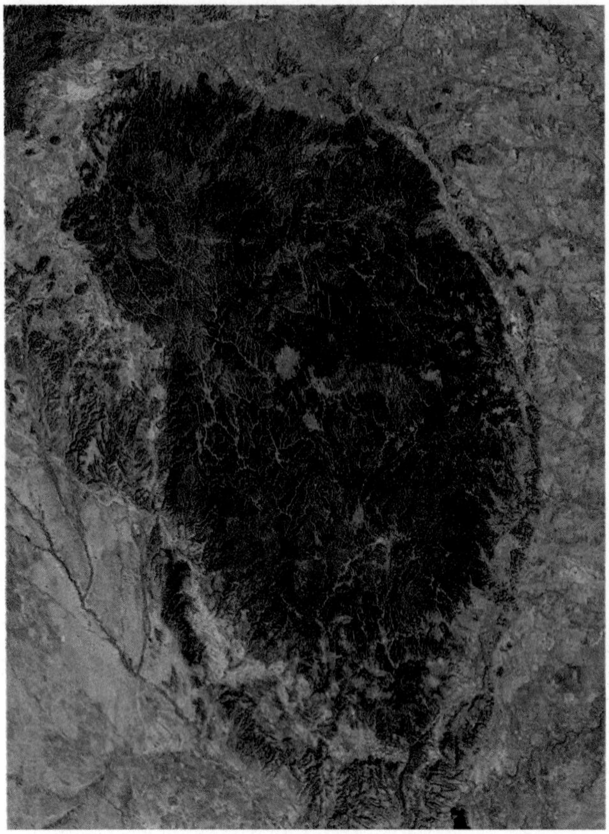

FIGURE 15.24 EROS satellite photograph of the Black Hills of South Dakota and Wyoming. The Black Hills formed from a batholithic dome whose igneous core has been partially exposed by the erosional removal of the overlying sedimentary strata. (USGS)

that was in the vent of the volcano at the end of its final eruption. The rapid removal of the unconsolidated material that once surrounded it and formed the volcanic cone may leave the volcanic neck standing conspicuously alone above the neighboring landscape. A well-known example of a volcanic neck in the western United States is Ship Rock, in New Mexico (see Figure 15.25). On occasion, volcanic necks may have exposed **dike ridges** radiating from them and appearing like stone walls. The degree of topographic expression of such volcanic remnants depends largely upon their resistance to erosion compared to that of the surrounding rocks. They stand out most conspicuously in areas where they have been intruded into weak sedimentary strata.

Flood basalts contain much more massive quantities of volcanic materials than do individual volcanic peaks and remain as landscape features for correspondingly longer timespans. Their basaltic composition, however, causes them to weather rapidly at the surface, so that in most areas they soon become covered by fertile soils. If the accumulation of basalt has formed a high-standing plateau, over time it is likely to become increasingly dissected by steep-sided river valleys or canyons. In arid regions, especially, the margins of lava plateaus are frequently abrupt and highly irregular and are marked by retreating escarpments.

SUMMARY

The lithospheric plate motions described in Chapter 14 are responsible for a variety of surface tectonic

FIGURE 15.25 Ship Rock is a volcanic neck with radiating dike ridges that stands conspicuously above the surrounding New Mexico desert. (John S. Shelton)

phenomena. These include earthquakes and volcanic eruptions—two of the most impressive and deadly manifestations of the forces locked in our planet's interior. Tectonic processes are divided into two general categories based upon the physical state of the rock material undergoing movement and deformation. Diastrophism involves solid-state motions and encompasses the processes of folding and faulting. Volcanism involves the movement of magma within the Earth and its expulsion onto the surface. Most tectonic landforms are produced along and near the boundaries between lithospheric plates and are associated with the major cordilleran belts.

We now return to the Focus Questions posed at the beginning of the chapter:

1. *What causes rocks to fold, and what topographic features result from folding?*

Most folding is caused by the linear compression of sedimentary rock strata due to lithospheric plate collisions. This results in the formation of elongated anticlinal and synclinal rock structures whose long axes, or strikes, lie perpendicular to the axis of compressive stress. Subsequent weathering and erosion of rocks of differing erosional resistance typically will produce shifting patterns of ridges and valleys. Rounded fold structures in the form of domes and basins are caused by the subsurface injection or removal of plastic or fluid substances such as magma or salt. Domal structures vary greatly in their size, age, and extent of rock deformation and are widely distributed over the Earth.

2. *What are the major types of fault motions and their resulting topographic features?*

Faulting involves the fracturing of rock, followed by the subsequent offsetting of the two sides. Individual fault movements are small, but they may be repeated at frequent intervals, eventually producing major topographic features. Two of the four basic fault types experience primarily vertical offsetting and may eventually produce mountain ranges or large escarpments. These include normal faults, which have an expansionary component, and reverse faults, which result from compression. Faults primarily involving horizontal offsetting typically produce lower relief topographic features, but are feared for their earthquake potential. They include transcurrent (strike-slip) faults, which involve lateral motion of the two sides of the fault, and thrust faults, which involve extreme compression resulting in the overlap of their two sides. Multiple faults often occur in areas subjected to tectonic stresses and the intervening blocks of land may be raised, lowered, or tilted.

3. *What causes earthquakes? What controls the geographic distribution of earthquake hazards?*

Earthquakes are sudden movements of the Earth's surface produced by fault displacements. They occur as the rocks on opposing sides of a fault plane are torn apart and rebound elastically after being subjected to gradually increasing strain. Earthquakes vary greatly in magnitude as a result of factors such as depth of the focal point, type of rock, and amount and nature of the stresses involved. They can be devastating events in densely populated regions, especially those containing large stone structures such as buildings and dams. The distribution of earthquakes coincides closely with that of the active lithospheric plate boundaries. The most severe earthquakes occur in regions where the opposing plates are "locked" together, so that a great deal of strain must be developed before displacement occurs.

4. *What causes volcanic eruptions, and what types of surface features do they produce?*

Volcanic eruptions involve the surface expulsion of molten rock in the form of either lava or pyroclastic fragments. The formation of magma bodies in the upper mantle or lower crust apparently results from the reduction in confining pressure along and near lithospheric plate boundaries. Surfaceward movement occurs because of the buoyancy of the magma, and the violence of the eruption itself results largely from the explosive release of pressurized gases originally dissolved in the rock. Ideally, a volcano takes the form of a symmetrical conical peak with a central vent. Three basic types of volcanoes are recognized. Cinder cones typically are small and steep and are composed primarily of loose pyroclastic material. Shield volcanoes, conversely, are very large, gently sloping, and composed of accumulated flows of basaltic lava. Composite volcanoes consist of alternating layers of ash or other pyroclastic material and of lava. Lava flows associated with volcanic eruptions usually are of limited extent, but extensive fissure outpourings of fluid basaltic lava in the geologic past have covered extensive regions.

Rising magma often solidifies before reaching the surface, resulting in the formation of intrusive volcanic features such as batholiths, laccoliths, sills, or dikes. These later may be exposed by erosion, and they often form high-standing surface features because of their resistance to erosion.

Review Questions

1. Under what tectonic and geologic conditions does folding take place? What is the relationship between the direction in which stress is applied and the orientation of the fold axis? Give three examples of folding and relate them to particular lithospheric plate movements.

2. Why is it rare to find a good accordance (correlation) between fold structures and topography within a folded region? What is the chief factor responsible for the existing patterns of ridges and valleys within a folded mountain region such as the Appalachians?

3. In what ways do the characteristics of domes and basins differ from those of linear folds? What are some of the causes of dome and basin formation? What determines whether the center of a dome or basin will stand higher or lower than its surroundings?

4. Contrast the geographic distributions of regions of linear folding with those of domes and basins.

5. What is faulting? What factors determine whether folding or faulting will take place in response to tectonic stresses?

6. List and describe each of the four basic fault types. Describe the types of lithospheric plate boundaries with which each normally is associated.

7. In what respects are horsts and grabens similar, and in what respects do they differ? Cite some well-known examples of each. How are grabens and rift valleys related?

8. What causes earthquakes? What factors determine their frequency and intensity? How are earthquake magnitudes measured?

9. Why do many experts expect a major earthquake in western California in the near future? Where will this earthquake most likely occur? Why there?

10. What is volcanism? What is the source of the magma involved in volcanic activity, and how does it reach the surface?

11. Describe, in general terms, the global distribution of active volcanism and relate this distribution to the global patterns of folding and faulting. Explain the distributional relationship between these patterns.

12. Describe the general characteristics of each of the three major types of volcanoes, and give two examples of each type. Which type has produced the most catastrophic eruptions from a human standpoint, and why?

13. What are flood basalts? In what major respect do they differ from lava flows produced by volcanoes? What effects do flood basalts have on the topography? What portion of the United States is covered by flood basalts?

14. Explain how a plutonic igneous intrusion such as a batholith can influence the surface topography. Cite one or more actual examples of such an influence.

15. What factor chiefly controls the rate and pattern of erosion of an extinct volcano? Name and explain the origins of two volcanic erosional features.

Key Terms

diastrophism
anticline
syncline
differential erosion
reversal of topography
homoclinal ridges and valleys
dome
basin
fault
normal fault
reverse fault
transcurrent (strike-slip) fault
thrust fault
fault escarpment (scarp)
horst
graben
rift valley
earthquake
extrusive volcanism
volcano
pyroclastic material
cinder cone
shield volcano
composite volcano
caldera
lava flow
fissure eruption
lava plain
intrusive volcanism
batholith
laccolith
sill
dike
volcanic neck

CASE STUDY

THE ERUPTION OF KRAKATAU VOLCANO

The spectacular eruption of Krakatau Volcano in the East Indies provides a graphic illustration of the awesome powers locked within the Earth. Although not the largest volcanic eruption on record, it was the most violent and certainly one of the most celebrated eruptions to have occurred during recent historical times. The eruption took place a little more than a century ago—from May through August 1883—but only in recent years has a full understanding of this event begun to emerge.

Among the characteristics of Krakatau's eruption that have resulted in its notoriety are the following:

1. Its climactic explosion on August 27 produced the loudest sound ever known. It was heard in places as far away as the Philippines, Sri Lanka, central Australia, and Rodriguez Island in the Indian Ocean. Rodriguez Island is located more than 5,000 kilometers (3,100 mi) from the site of the eruption. If Mount St. Helens had erupted as loudly in 1980, it would have been heard in New York City and in Miami, Florida.

2. The eruption destroyed most of Krakatau Island and ejected many cubic kilometers of ash and pumice into the air. Fine ash that reached the stratosphere produced numerous optical phenomena, including spectacular sunrises and sunsets around the world for months following the eruption.

3. The eruption produced a series of tsunamis that drowned an estimated 36,000 people. This is one of the greatest losses of life recorded for any single historical eruption.

Krakatau in 1883 was a large uninhabited island located in the Sunda Strait between Sumatra and Java (see Figure 15.26). It consisted of three overlapping volcanic cones, the highest of which reached an elevation of 813 meters (2,667 ft). The island was at the juncture of two intersecting submarine grabens in an area of crustal extension along the border of the Indo-Australian and Eurasian Plates. The most recent previous eruption of one or more of the island's volcanoes is believed to have occurred in 1680.

The 1883 eruption began rather suddenly on May 20, after a period of frequent minor earthquakes caused by the surfaceward movement of magma. The May eruptions were accompanied by loud explosions heard more than 150 kilometers (90 mi) away in Jakarta, on Java. Intermittent eruptions continued through June, July, and most of August. These eruptions were not unusually intense, and their main effect was to deposit some 50 centimeters (30 in) of ash on the island.

At approximately 1:00 P.M. on August 26, the character of the eruption suddenly changed. Intense explosions occurring at about 10-minute intervals produced a cloud of smoke and ash that rose well into the stratosphere and reached a peak altitude of about 25 kilometers (16 mi). Heavy ashfalls of up to 20 meters (60 ft) were experienced on the island and by nearby ships. At about 5:30 the following morning,

continued on next page

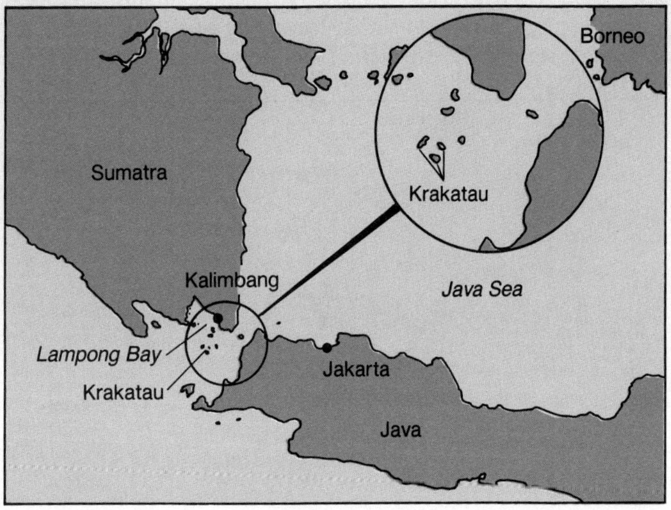

FIGURE 15.26 Krakatau's location at the "hinge" between the islands of Java and Sumatra.

the eruption increased greatly in violence, culminating in an enormous explosion at 9:58 A.M. that marked the destruction of the island. This was the explosion heard over a radius of thousands of kilometers. Great volumes of ash and pumice were hurled high into the air, and the eruption cloud reached an altitude of 40 kilometers (25 mi). The denser portion of this cloud of incandescent gases and volcanic fragments quickly fell back to the surface. Much of the material fell into the sea, but a considerable portion rushed off laterally as a massive *nuée ardente*. The finer ash stayed in the atmosphere, plunging the neighboring islands into darkness.

The momentum gained by its fall of many kilometers during its return to the surface imparted tremendous speed to the *nuée ardente* as it rushed away from the shattered remnants of Krakatau and out over the adjacent sea. As it traveled, a winnowing action took place, and a thick deposit of ash and pumice was left on nearby islands. Deposition also produced floating islands of pumice on the sea surface of the Sunda Strait, temporarily blocking shipping. The lighter ash and gases traveled at least 40 kilometers (25 mi) to the northeast and eventually reached the coast of Sumatra. They were still hot enough to cause burns to the residents of the town of Kalimbang as they rushed up from beneath the floorboards of their raised houses. The final eruption also produced a devastating tsunami that reached an estimated height of 130 feet (40 m) and obliterated coastal settlements on the nearby islands, taking some 36,000 lives.

The explosion produced a caldera up to 290 meters (950 ft) deep on the site where the island had been (see Figure 15.27). It resulted primarily from the collapse of the island into the magma chamber that fed the eruption. Field investigations show that approximately 95 percent of the erupted material came from the magma chamber and the remaining 5 percent was torn from the island by the explosion.

Two theories exist for the origin of the tremendous quantities of gas that were necessary to produce the final explosive destruction of Krakatau. One long-held idea is that seawater penetrated into the magma chamber, forming great volumes of water vapor that "blew the lid" off the chamber and hurled its contents into the air. Recent analyses of the composition of the ash produced by the eruption, however, have led to the formulation of a different hypothesis. The ash, rather than being of uniform composition, consists of a mixture of acidic (rhyolitic) and basic (basaltic) fragments. It is hypothesized that most of the eruption involved acidic materials, but that late in the eruption a body of gas-rich basaltic magma was injected into the base of the magma chamber. Pressures within the chamber were low enough to allow huge quantities of gases to come rapidly out of solution from this magma, causing the explosion and subsequent collapse of the chamber's roof.[1]

Differing theories also exist as to the origin of the tsunami produced

[1] Peter Francis and Stephen Self, "The Eruption of Krakatau," *Scientific American*, 249, no. 5 (November 1983): 172–87.

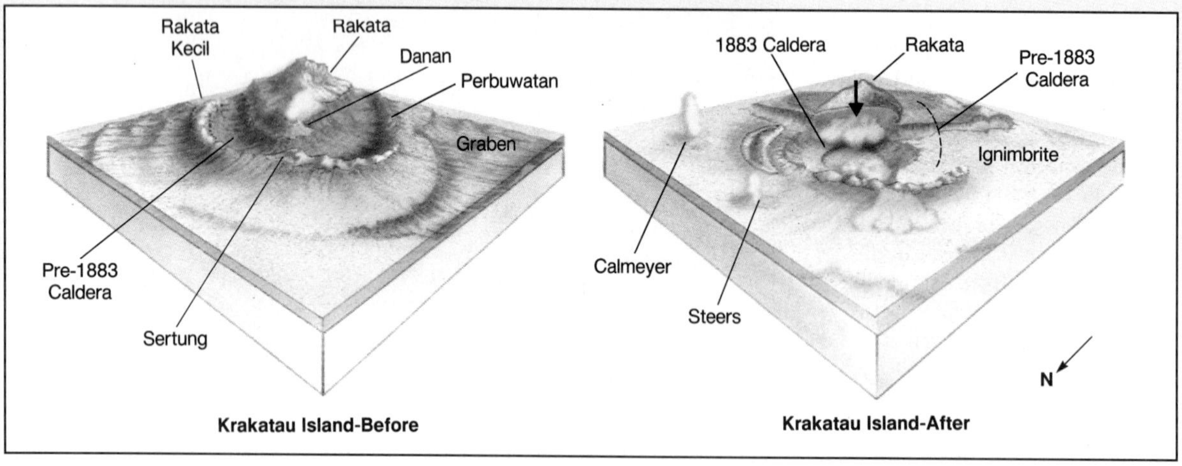

FIGURE 15.27 Block diagram depictions of Krakatau Island before and after the eruption of August 27, 1883.

during the final eruption. It may have resulted from the collapse of the magma chamber, but it also may have been caused by the splashdown of the vast quantities of pyroclastic material hurled into the air by the final explosion.

Currently, the remnant of Krakatau consists of the bisected cone of one of the three volcanic peaks that comprised the original island. The fertile soils and humid tropical climate have allowed a rapid return of life to the island. The area has been set aside by the Indonesian government as a nature preserve, and it is the last home of the nearly extinct Javan rhinoceros. Continuing sporadic eruptions have been occurring within the adjacent caldera formed by the 1883 eruption, and in 1928 a new island emerged that was named Anak Krakatau ("Child of Krakatau"). This island is slowly increasing in size as continuing eruptions add to its mass, and it seems to be gradually taking on the shape of the original island. Krakatau may thus eventually be resurrected, perhaps at some future time to be destroyed by another volcanic cataclysm.

CHAPTER 16

WEATHERING, MASS WASTING, AND KARST TOPOGRAPHY

OUTLINE

FOCUS QUESTIONS

1. How do the weathering processes operate, and in what environmental settings are they most effective?
2. What causes mass wasting, and what types of mass wasting are most widespread and important?
3. What are the causes and characteristics of karst topography, and how is it distributed around the world?

Tectonic uplift is accompanied, and followed, by the gradational processes that operate under the influence of gravity to lower the land surface. Gradation, unlike tectonism, is continuously active everywhere on land. As a consequence, a much larger total surface area is dominated by gradational than by tectonic features. Gradation primarily involves the transport of materials from higher to lower elevations by the three "tools of gradation": water, ice, and wind. Chapters 17 through 20 are devoted to the actions of these gradational agents and to the landform features they produce. In order for gradation to operate at a geomorphically effective rate, however, surface materials must first be reduced in size and erosional resistance through weathering processes. Earth materials, especially if weakened by prior weathering, do not always require transportation by water, ice, or wind, but can be eroded directly by the force of gravity. In addition, differential weathering processes can selectively dissolve some rock types to produce a distinctive landscape assemblage called **karst**. The three topics of weathering, gravity movements, and karst are covered in this chapter.

WEATHERING

Weathering is the combined action of physical and chemical processes that disintegrate and decompose bedrock. The term is used because agents associated with the weather—specifically air and water derived from precipitation—perform weathering operations. Weathering processes are of vital importance because they produce essential weathering products, and because they greatly increase the effectiveness of subsequent gradational processes. Weathered rock material is the major component of most soils, a commodity without which advanced plant and animal life on land would never have evolved. Weathering also produces loose sediments that are later converted to the sedimentary rocks that comprise the majority of the Earth's surface rock cover.

The role of weathering in soil and bedrock formation already has been covered; our main concern with weathering here is with its influence on gradational landforms. Weathering is an essential prerequisite for gradation because it reduces the size or alters the chemical composition of rock materials in such a way that erosional agents can remove and transport them much more readily. The tools of erosion, even ice, are largely ineffective when operating on solid bedrock. The gradational process therefore would be slowed immeasurably if bedrock were not first prepared for removal by weathering activities. Weathering sets the pace of gradation; because its progress is slow and steady, the gradational processes also operate in a slow and steady manner.

Weathering is not always considered part of the gradational process itself because no significant material transport is involved. The products of weathering remain at or near their original sites awaiting transport by water, ice, wind, or gravity. Weathering is normally, but not necessarily, followed by erosion by one of these agents.

Because weathering involves contact between rock materials and constituents of the atmosphere, it is most effective on near-surface materials. In general, the nearer rock materials lie to the surface, the more they have been altered by weathering. Weathering at any given site should not be thought of as a single process, but rather as a series of processes, each of which alters the rock further and generally weakens or disassembles it. Waves of sequential weathering processes work their way slowly down through the rock material; the position of the lowermost wave sometimes is called the **weathering front**.

The speed and types of weathering at any given site are controlled by five factors: rock type, rock structure, topography, vegetation, and, most important, climate. The significance of rock type is related primarily to the rock's physical strength and chemical stability. Physically weak and chemically unstable rocks are most readily attacked by weathering agents. Rock structures such as joints or bedding planes may provide avenues for attack by increasing the amount of surface area within reach of the weathering agents. Conversely, massive, solid rock weathers slowly because only its outer surface is presented to attack. The influence of topography on weathering primarily involves the slope angle and the resulting depth of soil cover. A deep soil layer, such as often covers areas with gentle slopes, may insulate the underlying bedrock from some types of weathering, but may also supply chemicals that promote other weathering reactions. Vegetation generally promotes weathering both through physical stresses produced by the growth of roots and by the production of organic acids. Climate exerts a strong control on weathering by determining temperature and moisture as well as by influencing the topography and vegetation.

The mix of these five factors can vary significantly over small distances, resulting in differing weathering rates. This, in turn, is likely to lead to differing erosion rates and to the production of a variety of topographic features.

Types of Weathering

For purposes of classification and discussion, weathering activities are commonly divided into the two categories of physical weathering and chemical weathering. Although in this section the various processes in each category are discussed individually, it should be realized that several processes normally act in conjunction or in sequence at a given site.

Physical Weathering

Physical weathering involves processes that reduce the size of rock masses without altering their chemical composition. Often described as producing rock disintegration, it converts masses of rock into more numerous smaller pieces of the same kind of rock that may be more readily removed by erosion. The most important physical weathering processes are joint and fracture formation, frost wedging, salt crystal growth, thermal expansion and contraction, and biological forces.

JOINT AND FRACTURE FORMATION The first weathering process to affect many rocks, and the deepest acting, is the formation of joints or fractures. Such breaks often are produced by tectonic stresses or result from cooling contractions. The formation of joints or fractures by any cause facilitates subsequent weathering by greatly increasing the amount of surface exposed to physical and chemical attack (see Figure 16.1).

Horizontal fracture systems also may be produced by the process of **unloading**. This refers to rock breakup caused by the gradual decrease in gravitational pressure as erosion removes the overlying material. Chemical weathering that occurs along the fracture planes facilitates the process. The resulting vertical expansion causes the rock to **exfoliate,** or to separate into exfoliation sheets (from the Latin *folium,* "leaf") paralleling the surface. Exfoliation sheets may vary in thickness from less than a centimeter to several meters; when the rock is exposed at the surface, they tend to peel or break off somewhat like the layers of an onion. The exfoliation of rock layers, when operating on a small scale, acts to produce highly rounded boulders. On a larger scale, mountain-sized **exfoliation domes** can be produced by this process, especially when it operates on large masses of exposed granite (see Figure 16.2). Half Dome in Yosemite National Park, California, and Stone Mountain near Atlanta, Georgia, are well-known examples of exfoliation domes.

FROST WEDGING **Frost wedging** results from the growth of ice crystals, derived either from precipitation or sublimation, within rock fractures or hollows. This process can generate pressures of more than 100 kilograms per square centimeter (1,400 lbs/in^2) of rock surface. The force is concentrated at the apex of the fracture and often is sufficient to cause further splitting and widening.

Frost wedging can be highly effective in regions where daily freeze and thaw cycles occur during a large part of the year and where little soil cover exists to insulate the solid bedrock from the effects of atmospheric temperature fluctuations. In particular, locations within the subpolar regions and at high elevations above the tree line but below the zone of perennial ice and snow are susceptible to this process. Within the middle latitudes, frost wedging may be effective during the winter in areas where bedrock outcrops occur at the surface. In the low latitudes it is ineffective, except at high elevations, because of a lack of subfreezing temperatures.

Active frost wedging in mountainous regions can litter the surface with angular rock fragments of all sizes, producing **boulder fields** (see Figure 16.3). Most of the rocks typically break off nearby cliffs. The lower portions of large, steep-sided cliffs may be buried beneath cone-shaped accumulations of such rocks. These accumulations, known as **talus cones,** are common in mountainous portions of western North America (see Figure 16.4).

SALT CRYSTAL GROWTH In arid regions, saline ground water is drawn surfaceward by capillary tension and the salts are eventually deposited at or near the surface as the water evaporates. Runoff from precipitation can then dissolve the salt and carry it into bedrock fractures, intergranular voids, or other openings. The subsequent evaporation of the water allows salt crystals to form and grow, generating expansionary pressures similar to those associated with

FIGURE 16.1 This granite outcrop in Wisconsin displays two clearly defined joint systems. They will serve as pathways for attack by subsequent physical and chemical weathering activities. (© JLM Visuals)

frost wedging. Fractures thus can be enlarged and rocks eventually split apart. Crystalline rocks, such as granite, may have individual mineral crystals pried loose by growing salt crystals, and granular rocks such as sandstone can have their constituent grains detached one by one. In portions of the southwestern United States, this process has occurred on a large enough scale to produce hollows in sandstone cliffs that became the habitation sites of cliff-dwelling Amerindian tribes.

THERMAL EXPANSION AND CONTRACTION The daily heating and cooling of bedrock exposed to sunlight at the surface stresses the rock material because thermally produced changes in volume occur. Rock is a poor conductor of heat, so that the expansion resulting from heating by direct sunlight, for example, is concentrated in the near-surface layers, causing them to expand more than the underlying layers. Conversely, nightly cooling causes the surface layers to contract more than the deeper layers.

Thermal expansion and contraction once was thought to be the primary cause of the granular disintegration of exposed bedrock often observed in arid regions. Recently, its effectiveness has been questioned because laboratory experiments subjecting rocks to repeated heating and cooling have produced no perceptible effects. The more extreme heating

FIGURE 16.2 Enchanted Rock, Texas, is a classic exfoliation dome. (David Butler)

FIGURE 16.3 A block field produced by active frost wedging in Greenland. (R. B. Colton/USGS Photo Library)

FIGURE 16.4 Talus cones lie beneath towering peaks in the Canadian Rockies near Moraine Lake in Banff National Park, Alberta. (Ralph Scott)

FIGURE 16.5 Bingham open pit copper mine, near Salt Lake City, Utah, is one of the world's largest human excavations. (U.S. Dept. of Agriculture, ASCS Western Aerial Photo Lab, Salt Lake City, Utah)

produced by fires, however, has been observed to produce rapid rock disintegration. Repeated heating and cooling, when combined with the presence of air and water over a prolonged period in an outdoor environment, may well play a significant role in the weathering of exposed rock surfaces.

BIOLOGICAL FORCES Biological forces are of limited significance with respect to physical weathering. Probably the most important is the growth of plant roots, especially those of trees, in rock crevices. The buckling of concrete sidewalks by tree roots is a familiar illustration of the power that growing plants can bring to bear on rock material.

If the effects of human beings are included within the category of physical weathering, it attains substantially more importance. Humans, especially during the present century, have developed the technological capacity to alter landforms on a major scale (see Figure 16.5). Examples of human activities resulting in the mechanical breakup of bedrock include drilling and blasting associated with mining and quarrying, building and road construction, and warfare.

Chemical Weathering

Chemical weathering involves chemical changes in the rock material that result in the formation of new substances. In general, the new substances are softer, more finely divided (i.e., they consist of smaller pieces), more soluble, and larger in volume than the original rock. These changes make the rock material more susceptible to further weathering and erosion. The breakup of rock by chemical weathering often is referred to as rock decomposition.

Chemical weathering often, but not always, follows physical weathering processes such as joint formation and frost wedging that open the way for chemical attack. As a result, weathering advances most rapidly along fractures or other lines of weakness such as bedding planes. In contrast to the angularity of rocks that have been influenced primarily by physical weathering, those dominated by chemical weathering typically are more smoothly rounded. Rounding occurs because rock corners and edges have large surface areas in proportion to their masses, allowing chemical attack to alter them relatively rapidly.

As a rule, rocks formed under the highest temperatures and pressures are the least stable chemically. In particular, the chemically basic igneous rocks derived from constituents of the upper mantle and lower crust tend to be chemically unstable. Sedimentary rocks, which form much nearer the surface, are chemically more stable but are physically much weaker, and have a tendency to separate into their original constituents when acted upon by weathering agents. From a broad perspective, then, weathering processes occur because rock materials tend to undergo alterations that increase their physical and chemical stability with environmental conditions at or near the surface. Most areas therefore display a weathering "profile" ranging downward from highly altered materials, approaching an equilibrium with surface conditions, to completely unweathered deep-lying bedrock in equilibrium with the interior environment in which it was formed (see Figure 16.6).

Several important categories of chemical weathering activities exist. Most involve chemical reactions with water or with gases dissolved in water; consequently, the availability of water is critical to chemical as well as physical weathering processes. The five most important categories of chemical weathering processes are hydration, hydrolysis, oxidation, carbonation, and solution.

FIGURE 16.6 A vertical progression of chemical weathering intensity is evident in this view of a freshly exposed bank in North Dakota. (Visuals Unlimited/ ©Bonnie L. Heidel)

HYDRATION **Hydration** involves the absorption, or adhesion, of water molecules to the molecules of a mineral. It is not a permanent chemical combination and can be reversed with the application of heat. An important example is the conversion of anhydrite ($CaSO_4$) in the presence of water to gypsum ($CaSO_4 \cdot 2H_2O$). Hydration increases the mass and volume of a mineral and softens it. The alternating hydration and dehydration of the minerals in some rocks in response to changing temperature conditions produces shrink-swell pressures that can dissociate the rock constituents. In some rocks it produces the exfoliation of thin sheets, and in others it aids in granular disintegration (see Figure 16.7).

HYDROLYSIS **Hydrolysis** is a permanent chemical combination with water. Like hydration, it results in the expansion and weakening of the rock material. It is especially effective on igneous rocks, because hydrolysis particularly affects feldspars and micas, which these rocks contain in quantity. The hydrolysis of igneous rocks by ground water, especially in the humid tropics, can occur at considerable depths.

OXIDATION **Oxidation** involves the chemical combination of minerals with oxygen that normally has first been dissolved in water. It is most effective on iron compounds, especially in warm climates, and primarily affects igneous rocks containing iron. The rusting of automobiles, iron railings, or other objects when exposed to moisture is a familiar example of oxidation. The presence of large quantities of iron oxides in the soils of the humid tropics is responsible for their distinctive red color.

FIGURE 16.7 Granite exposed in this roadcut in California is undergoing granular disintegration. (Ralph Scott)

CARBONATION **Carbonation** consists of chemical reactions of rock material with carbonic acid (H_2CO_3), which is produced when carbon dioxide is dissolved in water. The carbon dioxide involved generally is produced by plant decay, so that carbonation is most active in humid, well-vegetated environments. Carbonation results in the conversion of oxides to more soluble carbonates. It is highly effective on the feldspar minerals that comprise a large proportion of most igneous rocks. Carbonation also is the process by which limestone and dolomite are converted to calcium bicarbonate, a physically weak and highly soluble mineral that can be removed readily by ground water flow.

SOLUTION **Solution** refers to the dissolving of solid rock material in water. The water often contains substances such as carbonic acid that enhance the solution process. In effect, the rock is converted from a solid to a dispersed molecular state and therefore loses all resistance to erosion. Solution is commonly considered a chemical weathering activity, but it is not necessarily accompanied by permanent chemical change because the dissolved substances eventually may be precipitated back into their original chemical forms.

All substances are soluble in water to some degree. Solution is most effective by far, however, on highly soluble rock types, notably the evaporites halite and gypsum. As noted above, solution also plays a role in the weathering of limestone and dolomite following their conversion to calcium bicarbonate. Although limestone and dolomite are less soluble than the evaporites, they are much more widely distributed. In areas where they are the dominant bedrock type, they may be removed wholesale by erosion following the carbonation/solution process to produce a karst landscape. This subject is covered later in the chapter.

Global Distribution of Weathering Activities

Chemical and physical weathering differ considerably with regard to the climatic and lithologic conditions under which they operate most effectively. As a result, major differences exist in the relative importance of these two groups of processes in different parts of the world.

For the world as a whole, chemical weathering is more active than physical weathering. This is because chemical weathering operates at a much faster rate than physical weathering under favorable conditions, and because it produces by far the most complete transformation of the rock material. The end product of physical weathering is merely smaller rock particles of the same composition as the original. The end products of most chemical weathering processes, on the other hand, are completely new substances. In fact, the complete chemical weathering of rock-forming silicate minerals forms the clay minerals that are essential components of most soils.

Because most chemical weathering processes generally operate most efficiently under warm, wet conditions, the most rapid chemical weathering occurs in the humid tropics (see Figure 16.8). Here, the deepest and most highly weathered soils exist and the fastest topographic modification takes place. Conversely, chemical weathering processes are least active in polar and arid regions, and landscapes in these environments evolve relatively slowly. Chemical weathering is also aided by the presence of fragmented, soluble, or chemically unstable rocks, so that geology as well as climate is important in controlling the rate of chemical weathering within a region.

Physical weathering occurs most rapidly in regions that favor the operation of its primary mechanism—frost wedging. It therefore is most vigorous in the subpolar regions and in high mountain areas in the low and middle latitudes (see Figure 16.9). Physical weathering also attains relatively great significance in deserts, largely because the lack of water inhibits chemical weathering activities. Desert regions are frequently littered with fragmented rocks resulting from

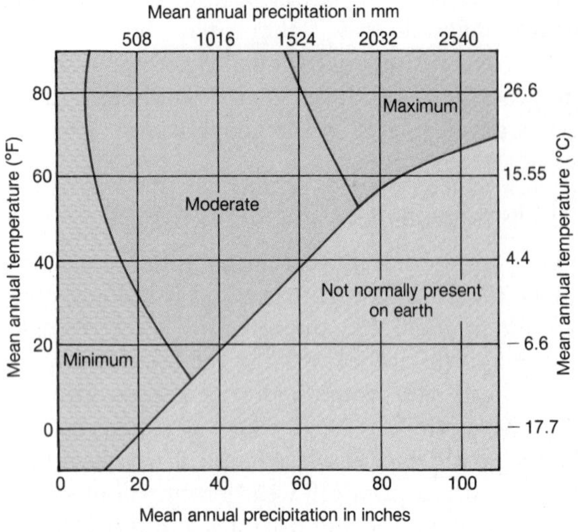

FIGURE 16.8 Relationship between climate and chemical weathering rates.

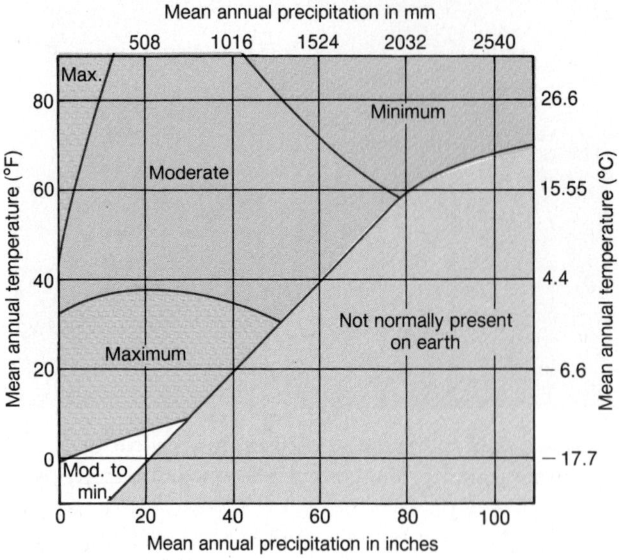

FIGURE 16.9 Relationship between climate and physical weathering rates.

the effects of salt crystal growth, frost wedging, and thermal expansion and contraction. Physical weathering processes also are aided by certain lithologic characteristics. These include large surface exposures of brittle, fractured, thinly bedded, or physically soft rock. In most regions, both physical and chemical weathering processes work together so intimately that it is difficult and rather arbitrary to separate them, and the existing landscape represents their collective contributions.

The zone of weathering varies in depth from the surface in some polar and arid environments to well over 30 meters (100 ft) in portions of the tropics. In most middle latitude areas, the weathering front is some 3 to 9 meters (10 to 30 ft) below the surface. The great majority of the rock materials that have been weathered and eroded at some time during the Earth's long history have been deposited and relithified and now exist in the form of sedimentary and metamorphic rock.

Regional variations in the rates, types, and products of weathering are largely responsible for differences in the geomorphic characteristics of regions with differing climates. For example, humid tropical environments dominated by chemical weathering processes typically display a deep soil cover and rounded topographic features. This stands in sharp contrast to the stark, angular features of many arid regions that result from the lack of a soil and vegetation cover and from the dominance of physical weathering activities (see Chapter 19). The correlation between climate and topography would be still greater were it not for the topographic variability imposed by differing rock types and especially by variations in the nature and degree of activity of the tectonic processes. These two topographic controls operate independent of climate, so that a newly formed volcano, for example, looks much the same in Alaska as in Mexico.

In concluding this section, it should again be stressed that weathering is not a landscape producer in itself; it is merely the necessary preparatory step in the gradation process. Because weathering activities normally occur near the surface and involve little movement of materials, features produced by differential weathering alone are not usually apparent unless some erosional mechanism exposes them. Such exposure is generally accomplished by the selective relocation or removal of the more highly weathered materials by water, ice, wind, or gravity, leaving the less weathered materials behind. Differential weathering therefore facilitates differential erosion. In general, high-standing gradational landforms are composed of materials that are relatively resistant to weathering, and low-standing features are composed of easily weathered materials. An example is provided by the folded Appalachians. Here, the ridges are composed of physically hard and chemically resistant sandstone, while the valleys are composed of soluble limestone or physically weak shale.

MASS WASTING

Once weathering processes have weakened surface materials, these materials become much more susceptible to the influence of outside forces, including gravity. Gravity is the direct motivating force in the process of gradation. Most of the work of gradation is accomplished by water, ice, and wind, which, impelled by gravity, transport surface materials downslope. A vital role in the gradational process, though, is performed by gravity acting alone. The downslope movement of surficial material under the direct influence of gravity is termed **mass wasting**. Although no medium such as water or ice serves as a transporting agent, one or both of these substances are present and act as destabilizing factors in most individual mass movements.

Mass wasting has a much greater impact on the Earth's surface than is often realized. It performs the vital role of transferring the products of weathering from their original sites, primarily on hillsides or other upland surfaces, to the valley or lowland sites where they are in position to be picked up for long-distance transport by the erosional tools of water and ice.

Mass movements exhibit a tremendous range of physical attributes. They vary in size from the motion of a single grain of sand to the massive displacement of many cubic kilometers of rock. The materials involved may consist of unweathered or weathered rock, soil, vegetation, water, ice, or, more commonly, a mixture of several of these ingredients. The speed at which mass movements travel varies from the imperceptible to velocities exceeding 45 meters per second (100 mph). Distances range from fractions of a centimeter to many kilometers.

The larger and faster types of mass movements not only are awesome phenomena that are capable of sweeping away everything in their path, they also are likely to occur suddenly and unexpectedly. As a result, they can produce great property destruction and loss of life. In fact, mass movements comprise one of the major categories of natural hazards that have afflicted humanity for millennia. Even minor mass movements are quite capable of producing damage such as cracked road surfaces and building foundations, broken water mains, bent trees and telephone poles, and highway debris following rainstorms in mountainous terrain.

Factors Influencing Mass Wasting

The susceptibility of a slope to mass wasting is determined by the relative effectiveness of two opposing groups of forces. These two groups are the forces that act to maintain slope stability, and the downslope forces that act to produce slope failure, or mass wasting. Three factors collectively aid in the maintenance of slope stability (see Figure 16.10). The first is the **normal stress,** or the inward compression of the slope materials caused by their weight. The second is the **cohesion** of the slope materials, which is a measure of the extent to which the individual particles comprising the slope adhere to one another. The third is the **internal friction** along any potential failure surface, resulting largely from the roughness of that surface. Collectively, these attributes determine the **shear strength** of the hillslope materials, which is a measure of their ability to resist downslope movements.

Opposed to the shear strength is a single force—the downslope force caused by the component of

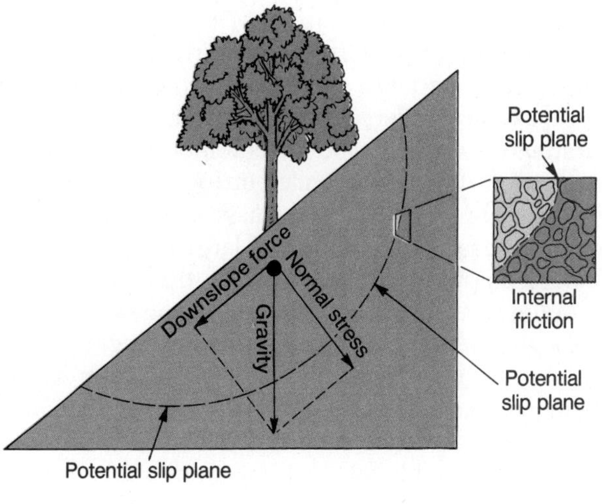

FIGURE 16.10 Factors influencing slope stability. Stabilizing factors include normal stress, the cohesion of the slope material, and internal friction along the potential slip plane. The destabilizing factor is the downslope force caused by the pull of gravity. The overall stability of the slope is determined by the relative magnitudes of these opposing sets of forces.

gravity acting parallel to the slope. This force becomes greater as slope angles increase. The steepest angle that can be maintained on any given slope without failure of the slope materials is called the **angle of repose**. Its exact value depends upon the type, amount, and condition of the slope materials. The angle of repose represents a threshold at which the shear strength of the materials is exactly counterbalanced by the downhill pull of gravity. As long as the angle of repose is not exceeded, mass movements will not occur. Factors that foster mass movements cause slope angles to exceed the angle of repose either by increasing the steepness of the slope or by producing conditions that at least temporarily reduce the shear strength of the slope materials.

FIGURE 16.11 Large rotational slump resulting from wave undercutting at Pacific Palisades, California. (John S. Shelton)

Oversteepening of Slopes

As a general rule, the steeper a slope, the less stable it is. Steep slopes therefore are likely to experience more frequent and more rapid mass movements than are gentle slopes. The angle of repose for slopes covered by unconsolidated materials normally ranges between 25° and 40°. At progressively steeper slope angles, there is a general decline in the thickness of the weathered mantle as mass wasting and other erosional processes become more efficient in transporting it downslope. Slopes steeper than 40° usually have solid rock surfaces.

A number of processes may increase slope steepnesses past the angle of repose. Probably the most important is undercutting by water, so that mass wasting of various types is common along stream banks. Likewise, wave undercutting, especially during storms, often produces mass movements along oceanic coasts or the shores of large lakes. The California coastline and portions of the Great Lakes shoreline are noted for such occurrences (see Figures 10.9 and 16.11). Oversteepening leading to slope failure also may result from valley glaciation within mountainous regions and from roadcuts that have been made at too steep an angle. The latter factor is largely responsible for the frequent mass movements that occur along roads constructed through mountainous terrain.

Weaknesses in Slope Material

Weak or loose slope materials are much more prone to mass wasting than is solid bedrock. Steep or even vertical cliffs can be maintained in a relatively stable condition in massive, solid rocks such as granite, limestone, or basalt. Fractures, steeply inclined bedding planes, and other zones of weakness, though, can greatly reduce the angle of repose. Poorly consolidated rocks, such as shale and many sandstones, also are unable to maintain steep slopes. In particular, the presence of a strong **caprock** (such as basalt) overlying a much weaker rock (such as shale) may lead to unstable slope conditions, since the more rapid erosion of the weak rock will likely lead to the undermining of the caprock. Many steep slopes in western North America are destabilized by this set of conditions.

The presence of a deep weathering mantle also favors mass movements. Many mass movements, in fact, involve only the displacement of weathered materials; the underlying bedrock is largely or entirely unaffected. This factor brings into consideration the indirect influence of climate, which is crucial in determining the types and rates of weathering processes (see Figure 16.12). The humid tropics are particularly susceptible to mass wasting of chemically weathered materials. Mass movements involving physical weathering products, conversely, achieve great geomorphic significance in alpine and high latitude environments. Mass wasting is so important in the periglacial environment that it is sometimes regarded as the leading geomorphic process operating in that region (see Chapter 18).

Triggering Mechanisms

The slope instability factors just described are considered "passive" factors, because over time they

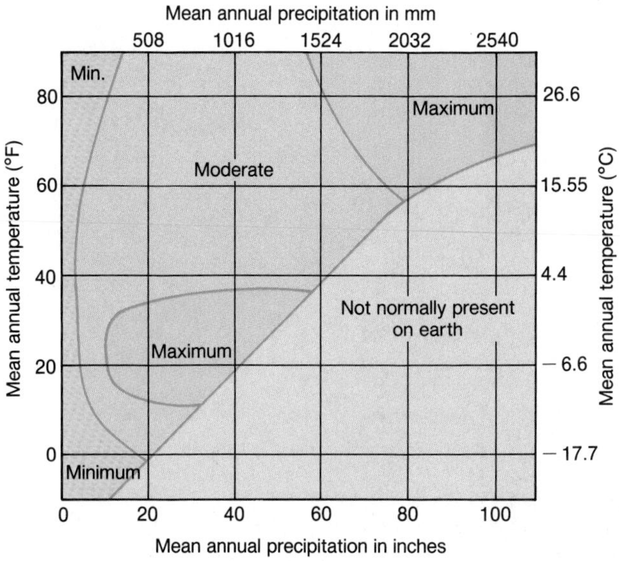

FIGURE 16.12 Relationship between climate and mass wasting potential.

tend to make a slope gradually more susceptible to mass movement. Most mass movements are actually triggered by an "active" factor that temporarily reduces the shear strength of the slope materials below the critical angle of repose. Although no clear-cut triggering mechanism exists for some mass movements, most are initiated either by strong vibrations or by the temporary presence of large amounts of water. Vibrations that trigger mass movements often result from earthquakes, which are relatively common in tectonically unstable mountainous regions prone to mass movements. If a slope is unstable enough, a lesser disturbance such as a loud clap of thunder may suffice to initiate slope failure. **Avalanches,** which are rapid movements of masses of snow and ice down steep mountain slopes, may be triggered by the sound of a gunshot or even a shout.

Large quantities of water that may produce slope failure usually are provided by heavy storms. The water adds weight to the slope and lubricates and buoys the individual slope particles so that their cohesion is reduced. Water is especially effective on slopes with a high clay content, since clay is not only capable of holding large quantities of water but also consists of platy minerals that readily slide over one another when wet. (This accounts for the slickness of mud, as opposed to the firmness of dry soil.) Several types of mass movements occur only when the water content of the slope materials is unusually high.

Types of Mass Wasting

Mass movements, like most other natural phenomena, exist in a continuum of types, and one type often grades into another with distance. Although this means that no mass movement classification system can describe adequately the existing spectrum of types, a number of systems have been devised. Mass movements usually are classified according to whether their motion consists primarily of sliding, falling, or flowing; whether the material involved is predominantly rock, earth, or **debris** (a mixture of soil, rock, and vegetation); or whether their motion is relatively slow or rapid. We shall use the last of these criteria to divide mass movements into two groups and will describe briefly the major types of mass movements included within each (see Figure 16.13).

Rapid Mass Movements

Rapid mass movements are those in which visually perceptible motion occurs. As a group, they are much briefer, more destructive, potentially more dangerous, but less geomorphologically significant than the slow mass movements. They tend to occur on relatively steep slopes and are likely to contain unweathered rock as a component of their material. Rapid mass movements in which motion occurs primarily by sliding are sometimes collectively referred to as **landslides**. This term is not employed here because of its lack of specificity.

ROCKFALLS **Rockfalls** are among the smallest, most common, and most rapid mass movements. They simply involve the falling of individual rocks from nearly vertical cliffs. Rockfalls usually result from frost wedging in alpine environments and are common in mountainous portions of western North America. Accumulations of angular rock fragments from innumerable individual rockfalls form the talus cones described earlier in the chapter.

ROCKSLIDES **Rockslides** are produced by the sliding of masses of detached rock down moderate to steep slopes. They exhibit tremendous variability in size, ranging from individual rocks to entire mountainsides (see Figure 16.14). Large rockslides are usually triggered by earthquakes. They may begin as coherent (solid) masses but quickly become fragmented by the violence of their motion. On long, steep slopes in high mountain areas such as the Andes

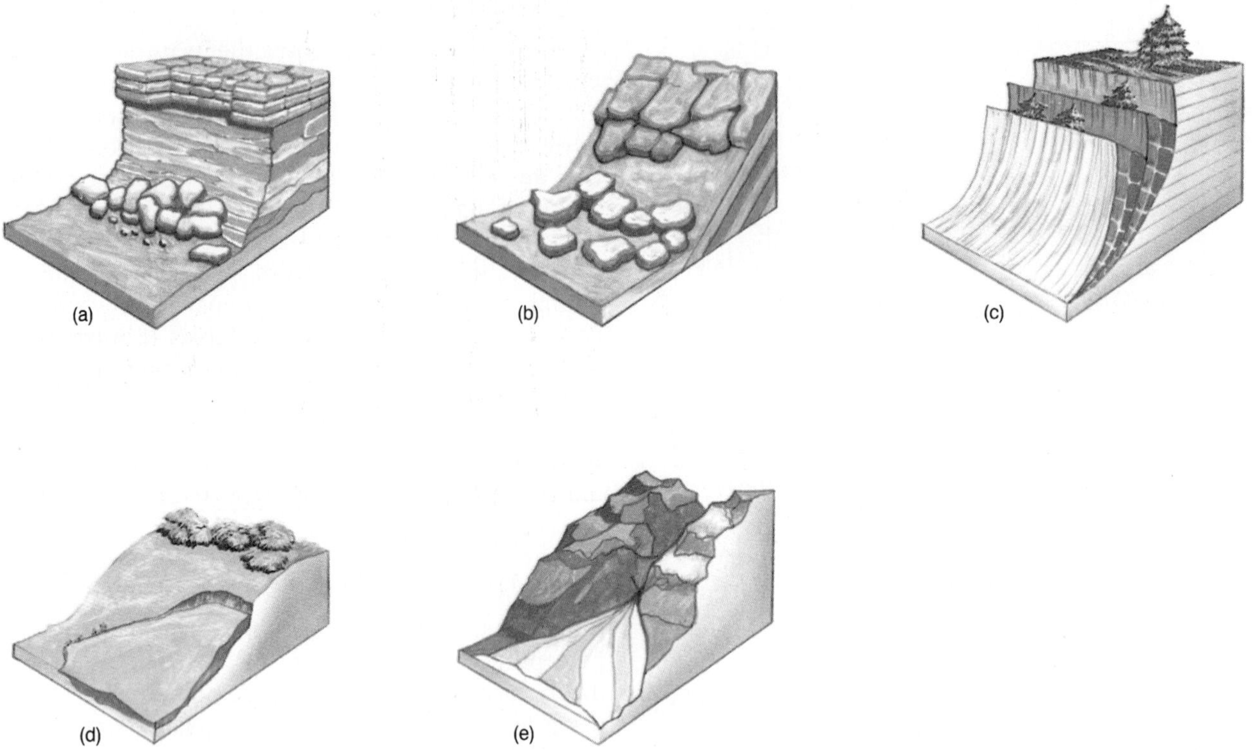

FIGURE 16.13 Some major types of mass movements. (a) *Rockfall* The relatively free falling of a detached mass of bedrock of any size from a cliff or steep slope. (b) *Rockslide* The usually rapid sliding of a detached mass of rock down a steep bedrock slope. (c) *Slump* The usually slow downward slippage of a coherent body of rock or regolith along a curved slip-plane. (d) *Earth flow* The slow flowage of a mass of nearly saturated soil or regolith down a moderate to steep hillside. (e) *Mudflow* The perceptible and often rapid downhill flow of mud, usually mixed with rocks and other debris, along a linear track.

FIGURE 16.14 A rockslide in central Utah. (©JLM Visuals)

or Himalayas, they can attain speeds well in excess of 160 kilometers per hour (100 mph).

The world's largest known single mass movement of any kind was the prehistoric Saidmarreh rockslide, which occurred in southwestern Iran. It consisted of approximately 21 cubic kilometers (5 mi^3) of limestone and marl that became detached from an anticlinal ridge and slid into the adjacent valley and well up the ridge on the valley's opposite side. It left a scar some 14 kilometers (9 mi) long, and the resulting debris covered an area of 166 square kilometers (64 mi^2).

DEBRIS SLIDES, DEBRIS AVALANCHES, AND DEBRIS FLOWS These mass movements all involve the rapid transport down moderate to steep slopes of a mixture of soil, rock material in various stages of weathering, and vegetation. They differ from one another primarily in their water content and their consequent fluidity of motion. The debris slide is the driest of the three, while the debris flow is the most fluid. They are all similar in shape, tending to be elongated, narrow, and to follow hillside depressions or hollows (see Figure 16.30). When they reach the base of the slope, they typically spread out in fan-shaped deposits. All three normally are triggered by exceptionally heavy rainstorms and can inflict considerable destruction.

One of the major natural disasters of recent times resulted from a huge debris flow that began when an earthquake triggered an avalanche on Mt. Huascarán, Peru's highest peak, in 1970. Accelerating down the steep mountainside and picking up debris as it went, it may have attained speeds as great as 90 meters per second (200 mph). It overwhelmed and destroyed several towns near the base of the mountain, taking an estimated 70,000 lives. The Case Study at the end of this chapter discusses the causes and consequences of the occurrence of a large group of debris avalanches in central Virginia as a result of torrential rainfalls associated with Hurricane Camille in 1969.

MUDFLOWS **Mudflows** are the most fluid of the major types of mass movements. They consist of a perceptible and often rapid downhill flow of mud, usually mixed with rocks and other debris. Mudflows generally follow preexisting channels and are long and relatively narrow. Most occur in arid and semiarid regions, where they are triggered by heavy thunderstorms, but they also may form in alpine and ash-covered volcanic regions. In all three of these environments, the surface is covered by loose, fine-textured materials that are not stabilized by vegetation.

FIGURE 16.15 A prehistoric viscous mudflow northeast of Doublespring Pass, Idaho. (John S. Shelton)

In interior portions of the western United States, mudflows caused by mountain thunderstorms may suddenly and unexpectedly debauch from canyons onto flat farmland in adjacent basins, burying large areas in mud. In southern California, slopes denuded of vegetation by summer brush fires often become the sources of destructive mudflows following the onset of winter rains.

Slow Mass Movements

Slow mass movements advance at a visually imperceptible rate and may take days, weeks, or even years to become evident. Although they are not nearly as spectacular or dangerous as the rapid mass movements, they are more geographically widespread and are responsible for the downhill transport of a greater total volume of weathered materials.

SLUMP A **slump** is the intermittent movement of a mass of earth or rock along a curved slip plane. It is characterized by the backward rotation of the slump block so that its surface may eventually tilt in the opposite direction from the slope from which it became detached (see Figure 16.13c). Slumps are most likely to occur on steep slopes with deep, clay-rich soils after saturation by heavy rains. The movement generally extends over many days or weeks and is nearly impossible to control or halt once it has begun.

LAND SUBSIDENCE—ITS CAUSES AND ENVIRONMENTAL IMPACT

Land subsidence refers to the vertical settling or sinking of the ground due to compaction or the removal of supporting materials. Although subsidence usually is slow and relatively unspectacular, it exists in many forms and has numerous causes. Some forms of subsidence are of natural origin, but many are caused or abetted by human activities. Human-induced subsidence occurs in at least 38 states within the United States, and the total area affected exceeds 40,000 square kilometers (15,500 mi^2). Conservative estimates of its annual costs within the United States exceed $100 million.

Four categories of human activities result in significant subsidence in the United States. Probably the oldest of these is underground mining, especially of coal. Approximately 32,000 square kilometers (12,000 mi^2) of land have been undermined by mining operations, and a third of this land already has experienced subsidence. Annual costs, resulting largely from building damage due to settling and cracking, may run as high as $30 million.

The withdrawal of ground water and petroleum from unconsolidated sediments has caused well-publicized problems in a number of major cities around the world, including Venice, Mexico City, and Bangkok. Within the United States, petroleum removal from the Wilmington oil field in California produced as much as 9 meters (27 ft) of subsidence in the city of Long Beach between the early 1940s and the mid-1960s.

The drainage of excess water from organic soils prior to agricultural and suburban development is probably the most costly cause of subsidence. Drainage not only leads to the compaction of the soil, but also often allows its organic materials to be consumed rapidly by aerobic bacteria. Resulting subsidence rates can exceed 8 centimeters (13 in) per year. In the Sacramento Delta of California, some 775 square kilometers (300 mi^2) of land have subsided below sea level, necessitating the construction of an expensive levee system.

The most spectacular form of subsidence is associated with the sudden appearance of collapse sinkholes in limestone areas. Sinkhole development, which is discussed later in this chapter, is a natural occurrence in limestone regions, but water extraction for irrigation or the disruption of ground water flow patterns sometimes lowers the water table sufficiently to trigger surface collapse. The development and rapid growth of a large sinkhole during a period of dry weather on May 8–9, 1981 (see Figure 16.16), in Winter Park, Florida, made headlines around the county. This sinkhole eventually attained a diameter of more than 100 meters (330 ft) and a depth of 35 meters (115 ft). It swallowed up a home, six vehicles, part of a city swimming pool, and portions of two streets.

FIGURE 16.16 This collapse sinkhole formed in May 1981 in Winter Park, Florida, as the result of a drought that lowered water table levels. It grew to be 100 meters (330 ft) wide and 35 meters (115 ft) deep, and swallowed up several buildings and vehicles. (USGS)

Slumps are common along the California coast, where slopes frequently have been oversteepened by wave undercutting (see Figure 16.11). They also occur along the side slopes of river gorges in various parts of the western United States. Small slumps capable of blocking traffic are frequent occurrences on steep road cuts following heavy rains.

EARTH FLOW **Earth flows** involve the slow flow of nearly saturated soil down moderate to steep slopes. They are more fluid, shallower, and usually smaller in size than slumps. An earth flow typically has a depressed upper portion and a bulging "toe" at its base where the material has come to rest, perhaps beneath a mat of vegetation (see Figure 16.17). Earth flows commonly occur in humid climates on grassy, soil-covered slopes following heavy rains. Large numbers may form on a single hillside under favorable conditions.

CREEP **Creep** is the slowest and least noticeable, but the most widespread and geomorphologically important, of all the categories of mass movements. It consists of the very slow downhill movement of soil or rock material over a period of years. Creep involves the entire hillside and probably occurs, to some extent, on any sloping, soil-covered surface. Because the near-surface layers creep fastest, it is commonly responsible for the downhill tilting of buried or partially buried objects such as trees, telephone poles, stone walls, and inclined rock strata (see Figure 16.18).

FIGURE 16.17 Earth flows that occurred on a grassy hillslope near San Francisco following heavy rains. (Martin Miller/©JLM Visuals)

Creep is caused by any disturbance to the surface mantle, because such disturbances almost invariably result in a net downhill displacement of surface particles. Some types of disturbances contributing to creep include **frost heaving** (the loosening of soil by growing frost crystals), wetting and drying, heating and cooling, plant root growth, raindrop impact, and the trampling of livestock. Creep is especially active on slopes covered by deep soils with a high water content, and is therefore much more significant in humid than in arid or semiarid regions. It is thought to be largely responsible for the generally rounded topographic features of humid landscapes. The subdued profiles of the Appalachian Mountains, for example, stand in stark contrast to the angular features of mountain ranges in arid portions of the West. This contrast is probably more attributable to variations in rates of chemical weathering and creep than it is to geologic age. Creep may be the chief mechanism by which weathered materials are transported down hillsides in soil-covered areas to the valleys of the streams that ultimately carry these materials to the sea.

KARST TOPOGRAPHY

In some areas underlain by thick sequences of limestone and dolomite, the weathering processes of carbonation and solution can themselves be considered responsible for the major characteristics of the topography. Even here, a transporting agent (in the form of ground water) is needed to remove the weathered materials from their original sites. Because these materials have first been converted to an individual molecular state, however, carbonation and solution often are considered the primary geomorphic mechanisms in the development of the resulting topography.

Cause and Global Distribution of Karst Topography

Ground water in humid regions tends to become somewhat acidic from the solution of atmospheric carbon dioxide as well as from the acquisition of organic acids from the soil. It becomes, in effect, a weak carbonic acid solution. Carbonic acid reacts with carbonate rocks to produce calcium bicarbonate, which is highly soluble in water and is readily transported away by ground water flow. An area with landform features produced largely by the differential solution and subsequent erosion of limestone and dolomite is called a **karst** area. The term is derived from the

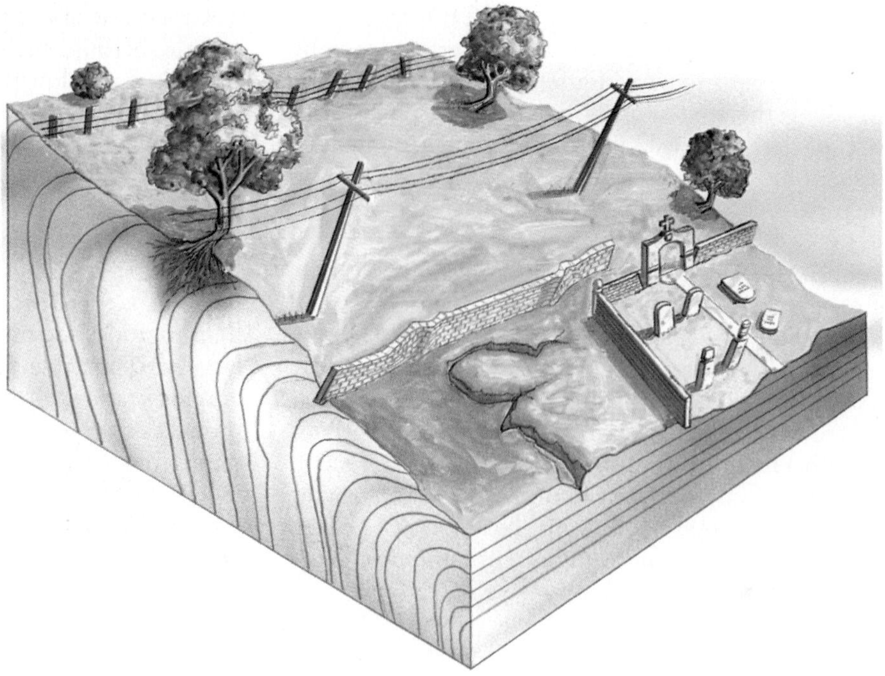

FIGURE 16.18 The effects of creep include the bending of near-surface inclined rock strata and tree trunks, as well as the tilting of such features as telephone poles, fences, walls, and tombstones.

Kras region of Yugoslavia near the Adriatic Coast, where some of the earliest research on this topographic assemblage was conducted.

A combination of lithologic, topographic, and climatic factors determines the favorability of a given site for the formation of karst topography. There are four prerequisites for optimum karst development. First, thick sequences of relatively pure limestone or dolomite must be present at or just beneath the surface. Second, the rock should be highly jointed in order to provide ingress for surface water, but should otherwise have low porosity and permeability. This is because karst features develop as a result of differential solution rates. A third and very crucial factor is a moderate to abundant supply of precipitation. This provides the water needed for the solution process. Plentiful precipitation also favors the growth of vegetation and the formation of a soil humus layer that lowers the pH of the downward-moving water, thereby greatly increasing its solution potential. The final prerequisite for maximum karst development is for the zone of ground water saturation to be situated well below the surface. This zone marks the lower limit of downward percolation of the acidic water that accomplishes the most rapid solution. A near-surface location therefore inhibits the vertical development of solution features.

Karst topography is geographically widespread, although individual sites are sometimes highly localized. Conditions generally are favorable for karst development in humid portions of the middle latitudes, and several major karst regions are located there, especially in the United States and Europe. It reaches optimum development, however, in the humid tropics, where heavy precipitation and organic acids produced by the decomposition of the dense vegetation combine to favor the rapid solution of limestone and dolomite. Major areas are depicted in Figure 16.19. The United States probably contains more karst than does any other country. Important areas include Kentucky and Indiana, the central Appalachians, Florida, the Ozark Plateau, western Texas and eastern New Mexico, and the Black Hills of South Dakota. Karst also is well developed in southern Europe, especially in the southern Alps and in a number of regions bordering the Mediterranean Sea. Australia contains an extensive but little-publicized karst region in the Nullarbor Plain, located in the south-central part of the continent. The most important areas of tropical karst include South China and adjacent portions of Vietnam, Thailand, and Burma; parts of the East Indies; a large area in eastern Brazil; Mexico's Yucatán Peninsula; and the West Indian islands of Jamaica, Cuba, and Puerto Rico.

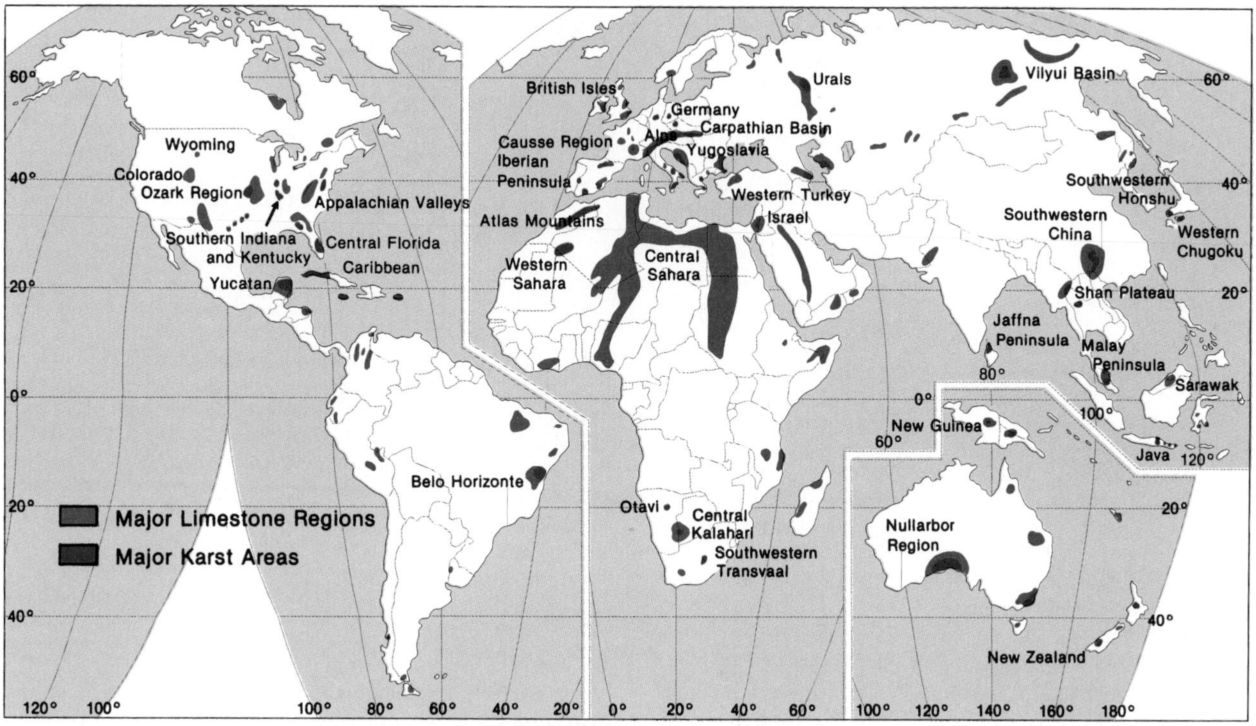

FIGURE 16.19 Global distribution of major limestone and karst regions.

Karst Features

Karst regions vary greatly in their topographic characteristics (see Figure 16.20). Some may be characterized as plains, some are mountainous, but most frequently they are hilly. The larger relief features tend to be very abrupt and steep-sided, giving the landscape a picturesque and sometimes spectacular appearance, but typically reducing accessibility and land-use potential.

The solution rates of limestone and dolomite can vary greatly over very small distances because of the influence of fracture and bedding patterns, slight variations in chemical composition, and access to acidic ground water. As a result, even in areas of low relief, differential solution produces an extremely rough and intricately sculpted surface topography on exposed bedrock outcrops. Rock outcrops are common and frequently extensive in karst regions, and these outcrops typically exhibit gullies, grooves, solution pits, and various other small-scale differential solution features in abundance. In addition, loose stones and boulders often litter the surface. The reddish, clay-rich soils of karst areas are fertile but discontinuous and often quite thin, sometimes existing only in patches within low-lying sites.

Sinkholes, also termed **sinks** or **dolines,** are the most widespread and undoubtedly the best known of the major karst features. They are rounded and often nearly circular depressions formed by subsurface solution. Most develop on bedrock joint intersections, are rather small, and are produced by the gradual subsidence of the weathered surface mantle into an enlarging solution hollow (see Figure 16.21). Sinkholes also can form rather suddenly and spectacularly by the collapse of the surface into an underground void. Such **collapse sinkholes** tend to be larger, deeper, and steeper than the more common **subsidence sinkholes** (see Figure 16.16). Sinkholes commonly range up to 30 meters (100 ft) in diameter and to 10 meters (30 ft) in depth, but within a given area most are likely to be considerably smaller and shallower than this.

Although most individual sinkholes are highly rounded, they may overlap where they are especially numerous to form irregularly shaped compound sinkholes. On occasion, an unevenly floored linear

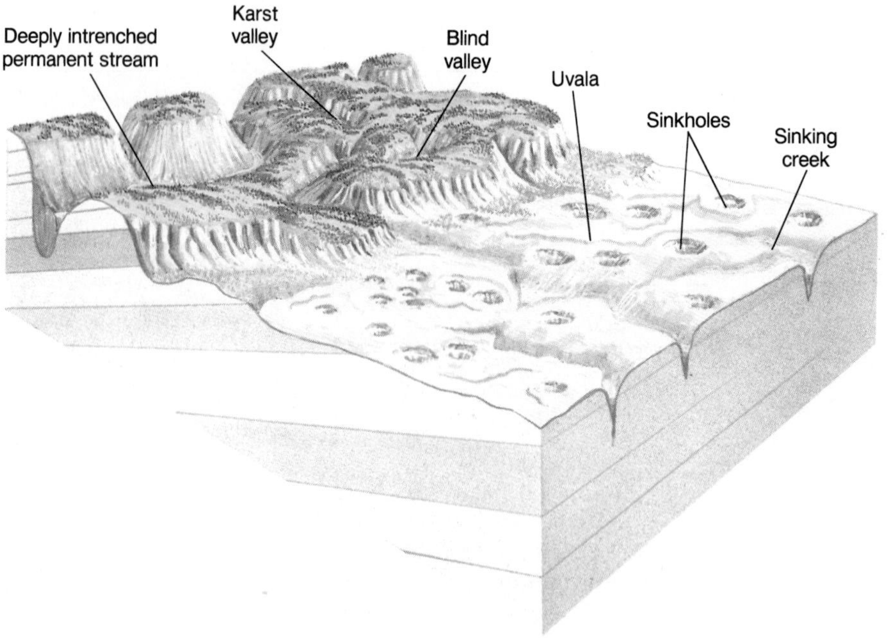

FIGURE 16.20 Important surface karst features.

depression termed a **uvala** will be formed by the coalescence of a line of sinks above the course of an underground stream. Sinkholes typically are dry if their bottoms are situated above the water table, but are partially filled with water if they extend below it. The rounded lakes of the Central Florida Lake District, for example, occupy large sinkholes in an area with a high water table. Sinkholes are most abundant and conspicuous in low-relief areas. **Karst plains,** such as those in portions of Indiana and Kentucky, are plains covered with sinks that may number a hundred or more per square kilometer (see Figure 16.22).

Well-developed karst areas display an almost complete lack of organized surface drainage. The underlying carbonate rocks are honeycombed with solution channels that receive surface water through **swallow holes** typically located at the bottoms of sinkholes. The surface is almost like a sieve, and runoff cannot travel far before it enters a sinkhole and disappears underground. One consequence of the thin soil cover and rapid underground diversion of surface water is the development of a drought-prone surface incapable of retaining moisture for long after a rainfall.

In areas where sinks and swallow holes are less numerous, surface streams may form, only to disappear suddenly below ground to become **sinking creeks** as the water enters large swallow holes in the stream bed. A sinking creek may flow beneath the

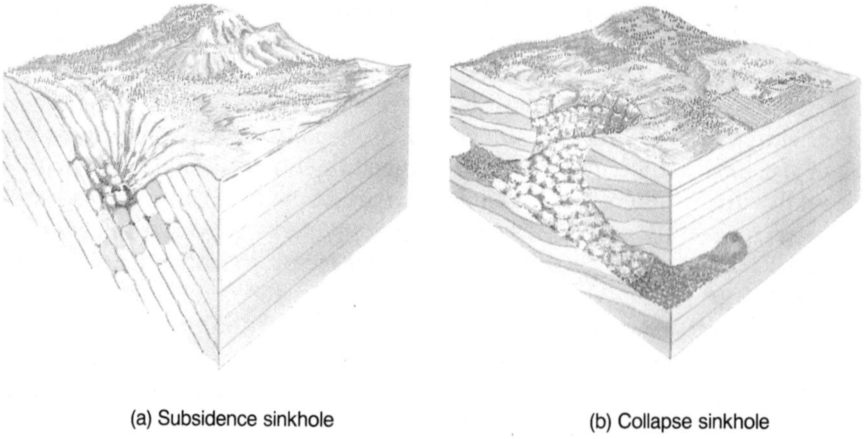

FIGURE 16.21 Surface appearance and subsurface structure of collapse and subsidence sinkholes.

FIGURE 16.22 This karst plain in western Illinois contains numerous water-filled sinkholes. (B. F. Molnia)

surface for a considerable distance before suddenly reappearing as a spring. Some springs in karst areas have very large volumes of flow. A well-known example is Silver Springs, located near Ocala, Florida. It has a flow volume of approximately 15 cubic meters (500 ft^3) of water per second and has been developed as a tourist attraction.

Caves are produced by a variety of geomorphic processes, but they are most numerous and complex by far in karst areas, where they are formed by subsurface solution and erosion. A **cave** is any natural cavity within or beneath the Earth's surface. It may range from a small bedrock hollow set into a cliff, to an elaborate cave system consisting of an interconnected maze of rooms and passageways. Although normally predominantly horizontal, caves may develop on more than one vertical level.

Within karst areas, caves are formed largely by ground water solution along joints, fractures, or bedding planes located at, or just below, the water table. Because these zones of weakness usually exhibit a geometric pattern, cave systems likewise tend to be organized into distinctive, and often rectilinear, patterns (see Figure 16.23). The world's largest known cave system, Mammoth Cave, contains approximately 530 kilometers (330 mi) of interconnecting passageways. As solution progresses and the water table falls, the original water-filled caves are gradually drained, while new levels may be formed below them. In this way, a multiple-level cave system is believed to develop from the upper levels downward.

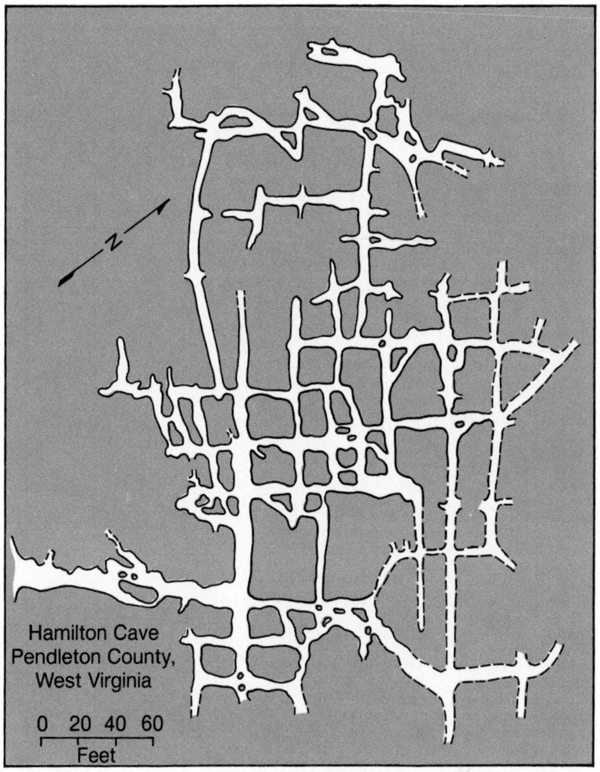

FIGURE 16.23 Joint-controlled system of passageways in Hamilton Cave, West Virginia.

Variations in the quantity and acidity of the available ground water as well as in the solubility and structure of the rock determine the rate at which solution occurs. Where solution is rapid and the overlying rocks are strong, large chambers can be formed. The largest known chamber—Sarawak Chamber in Good Luck Cave, Sarawak, is 700 meters (2,300 ft) long, 400 meters (1,300 ft) wide, and 280 meters (900 ft) high. In a large cave system a number of chambers will be connected by a series of passages formed along joints and bedding planes. These passages vary greatly in width, but usually are rather narrow. Even single passageways vary greatly in dimension from place to place.

Adding greatly to the mystery and beauty of limestone caves is the great variety of **travertine** features that often develop within them (see Figure 16.26). These features consist of calcium carbonate deposits left by the slow dripping of water from the roof and walls of the cave. The deposition of cave travertine features results from the gradual diffusion to the atmosphere of the carbon dioxide dissolved in the suspended droplets. The loss of the carbon dioxide

KARST SEQUENCE OF DEVELOPMENT

It has been theorized by some geomorphologists that karst landscapes tend to display an orderly sequence of topographic characteristics over time. This sequence would result from the slow progress of solution and from resulting changes in the depth of the water table. Support for this concept and for the existence of other cycles of landscape evolution has considerably diminished in recent decades. This is due to a growing awareness that such sequences would take so long to be completed that they would invariably be interrupted by events such as tectonic deformation and climate changes. Nonetheless, evolutionary landscape sequences continue to serve as useful models for stressing the inherent logic in the operation of geomorphic processes.

We can imagine, as a starting condition in an idealized karst sequence of development (see Figure 16.24), that a relatively smooth limestone surface has been lifted recently above sea level by tectonic activity. As solution and erosion begin, the surface becomes irregularly etched by small-scale solution features, sinkholes develop, and the water table begins to fall (View a). The lowered water table allows for the extensive removal of subsurface limestone, causing drainage gradually to be diverted to underground routes. Sinking and rising creeks as well as cave systems develop, and karst features in general reach a

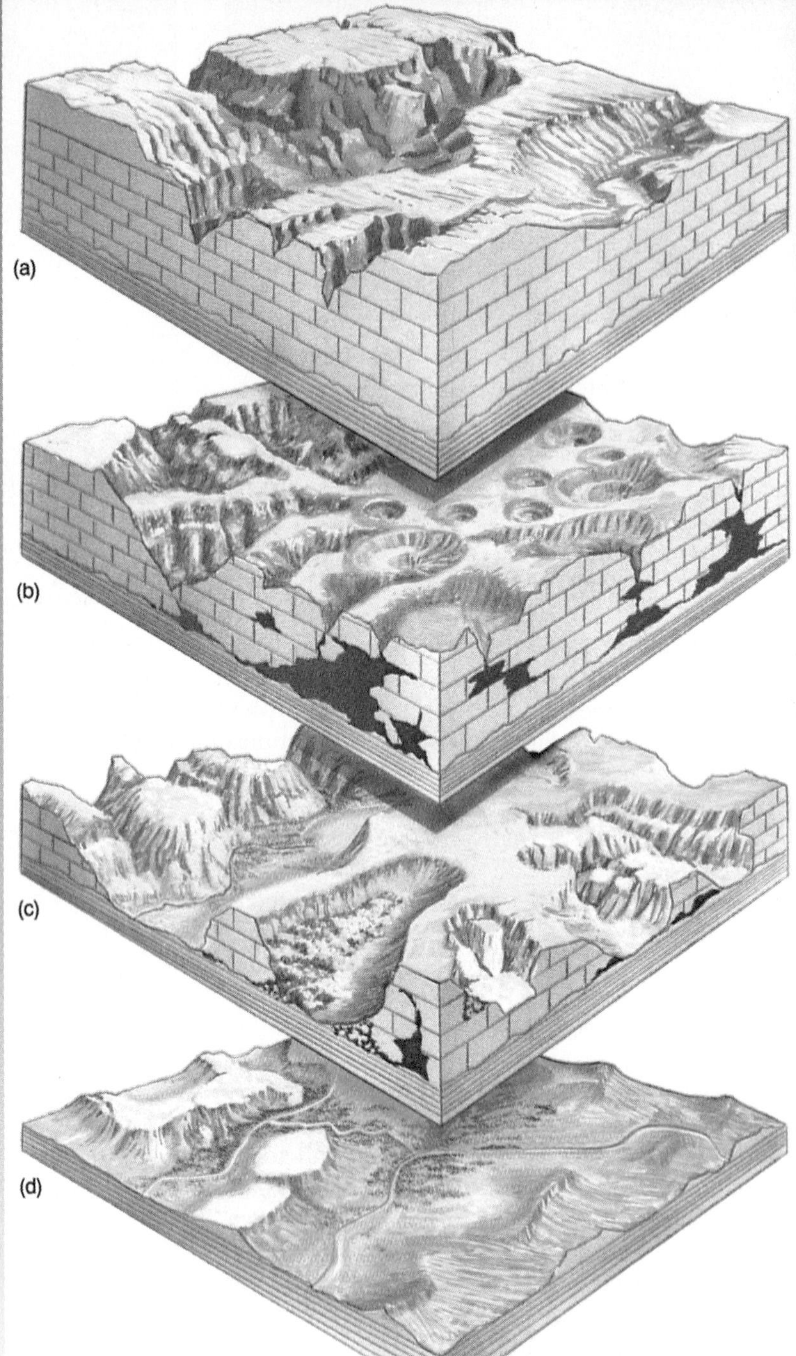

FIGURE 16.24 The karst topography sequence of development.

maximum (View b). A continuation of underground solution results in the progressive undermining of the near-surface rock, causing the gradual collapse of extensive portions of the original surface (View c). Uncollapsed remnants form natural tunnels and bridges, as well as haystack hills. The solution process is all the while slowly and unevenly dissolving away the limestone near and above the level of the lowered water table, so that eventually a new karst plain with a near-surface water table is formed (View d). If tectonic uplift or a fall in sea level once more lowers water table levels, the cycle can be repeated. In the meantime, the dissolved calcium carbonate is transported by rivers to the sea. There it is eventually reconstituted, by chemical and biological processes, into new limestone. This limestone may be raised once more above sea level by tectonic activity to produce a new karst landscape.

greatly reduces the ability of water to hold calcium carbonate in solution, thereby forcing its deposition.

The most important travertine features include the iciclelike **stalactites** that hang from the cave roofs, upward-growing **stalagmites** on the cave floors, **pillars** or **columns** formed by the coalescence of stalactites and stalagmites, and travertine **curtains** or **ribbons** formed by water dripping along an incline. All these features come in an infinite variety of sizes and shapes, and they may be stained various hues by chemical impurities in the water.

Although cave systems are located in most areas underlain by massive limestone and dolomite, they are readily accessible only if they are currently above the level of the water table. As a consequence, most well-known cave systems are associated with hilly or mountainous regions situated well above sea level and with deep water tables. Important regions include the Kentucky-Indiana area, the Ozarks, the central Appalachians, and the Black Hills of the United States (see Figure 16.27); the Dinaric Alps of Yugoslavia; Mexico's Yucatán Peninsula; and the mountains of south-central China.

Cave systems are inherently unstable and geologically short-lived natural phenomena because their continued growth eventually leads to loss of roof support and collapse. Cave systems usually collapse in segments, rather than all at once. Uncollapsed portions may form **natural tunnels,** if broad, or **natural bridges,** if relatively narrow. Probably the most famous limestone bridge in the world is Natural Bridge, in Rockbridge County, Virginia.

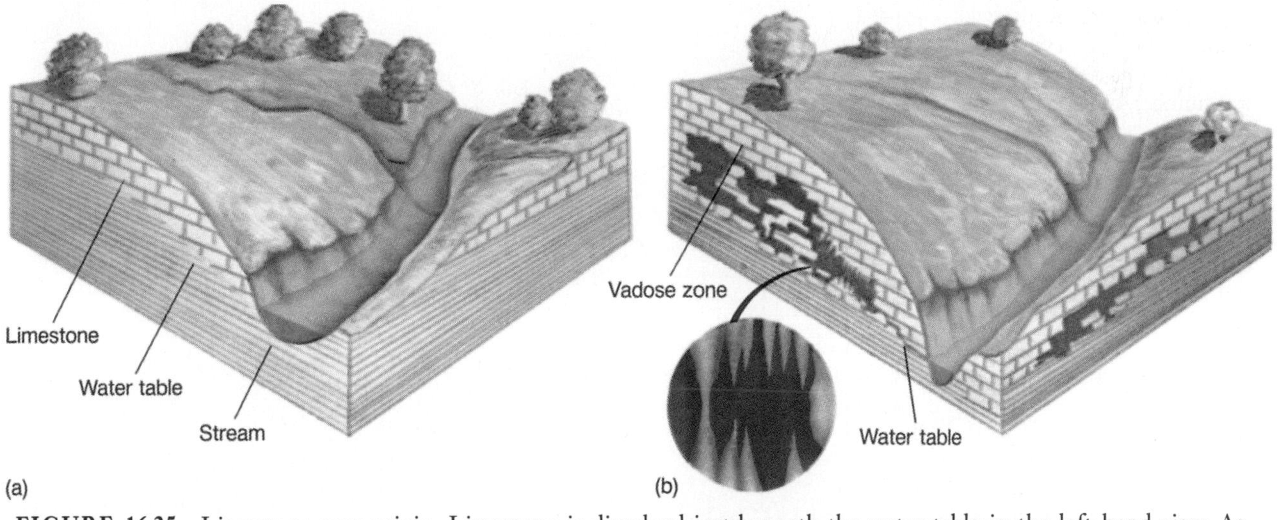

FIGURE 16.25 Limestone cave origin. Limestone is dissolved just beneath the water table in the left-hand view. As the stream erodes more deeply, the water table falls, and the area of dissolved limestone becomes a dry cave.

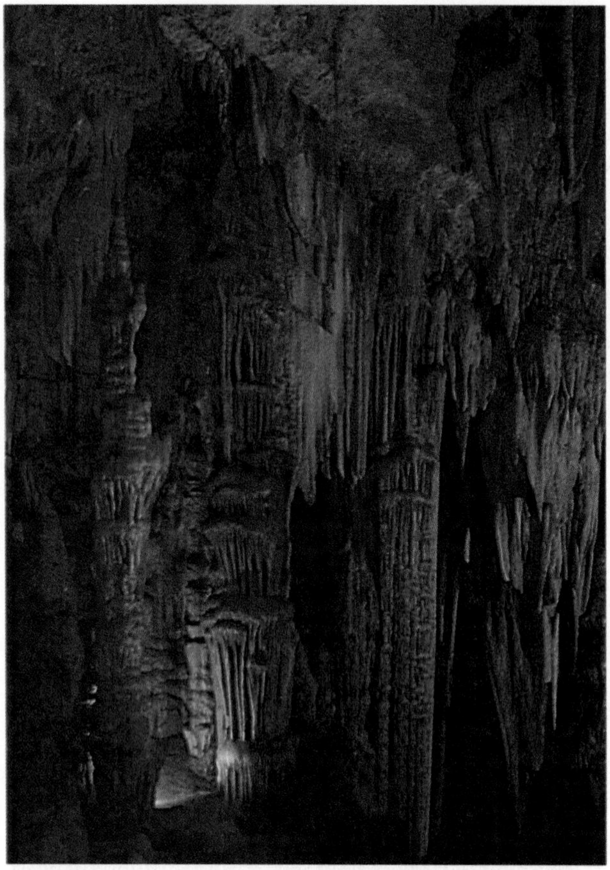

FIGURE 16.26 A view of various travertine features within Luray Cavern, Virginia. (Ralph Scott)

Another type of erosion remnant consists of hills that remain on a karst surface as indicators of a former higher topographic level. These hills vary in steepness and height in different areas, but reach optimal development in the humid tropics. They are known regionally by a variety of terms, including **haystack hills**—an allusion to their steep-sided and highly rounded forms. A landscape dominated by numerous steep-sided, conical hills and intervening star-shaped depressions, which is especially well developed in portions of the West Indies, is known as **cone karst**. In parts of Southeast Asia, such remnant limestone peaks attain mountainous proportions, producing a spectacular and highly distinctive landscape referred to as **tower karst** (see Figure 16.28). This landscape has served as a source of inspiration to Chinese artists for many centuries.

SUMMARY

This chapter begins our examination of the gradational processes and their effect on the Earth's surface. Most gradational landforms are produced by erosion and deposition resulting from the movement of water, ice, and wind. Before these agents can operate effectively, however, the surface rock must first be acted upon by weathering processes which greatly reduce its resistance to erosion. Solar energy and gravity are the forces that power the gradational

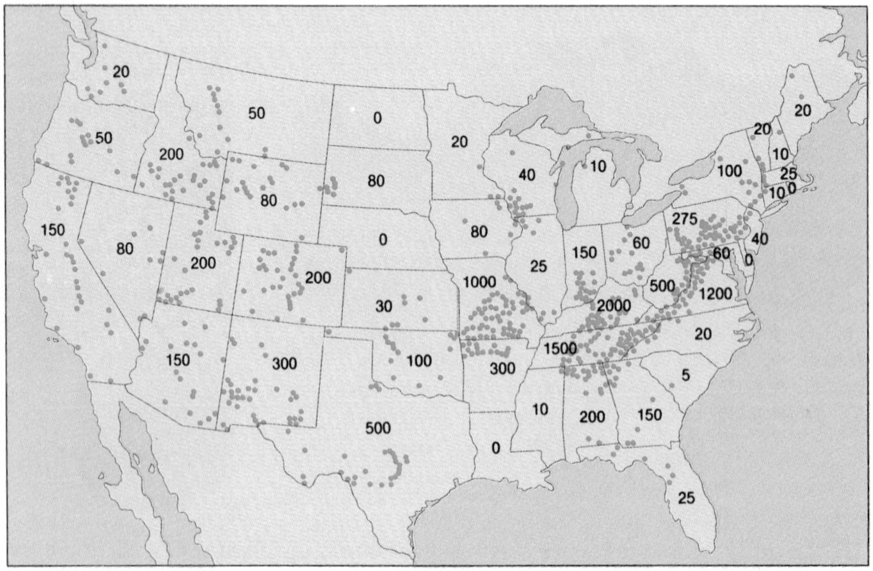

FIGURE 16.27 Distribution of karst cave areas in the coterminous United States. Numbers indicate the estimated total number of caves in each state.

FIGURE 16.28 Tower karst in the Li River Valley of southern China. (R. Krubner/H. Armstrong Roberts)

tools, and gravity also is capable of producing a variety of mass movements of the surface on its own. Another set of gradational features not normally considered a product of the "traditional" operation of the tools of gradation are the karst features of limestone regions.

In order to summarize more systematically the three chapter topics of weathering, mass wasting, and karst, we return to the Focus Questions:

1. *How do the weathering processes operate, and in what environmental settings are they most effective?*

Weathering involves the actions of physical and chemical processes that reduce the size of rock materials and often produce softer, more easily eroded substances. Physical weathering involves only a physical reduction in rock size and is most important in cold and dry environments. The chief physical weathering processes include the formation of joints and fractures caused by tectonic stresses, cooling contractions, and unloading; frost wedging caused by ice crystal growth; the growth of salt crystals in rock fractures; thermally induced changes in volume; and the mining and construction activities of humans.

Chemical weathering has greater overall geomorphic importance than physical weathering; it is most significant in warm and moist environments. Chemical weathering processes alter the chemical composition of the rock and produce new substances. The processes of hydration and hydrolysis involve chemical combinations with water, while oxidation and carbonation involve reactions with oxygen and carbon dioxide that generally have been first dissolved in water. Solution, another important chemical weathering process, causes the molecules of rock material to become detached from one another and to be suspended in water.

2. *What causes mass wasting, and what types of mass wasting are most widespread and important?*

Mass wasting, or mass movement, is the downward movement of surface materials under the direct influence of gravity. Because it is gravitationally induced, it is most likely to occur in areas that have steep slopes or physically weak surface materials, or in places where subsurface support has been removed. Factors resisting mass movements are the compression caused by the weight of the slope materials, as well as their cohesion and internal friction. Most mass movements are triggered by heavy rainfalls or earthquake vibrations.

Rapid mass movements such as rockfalls, rockslides, and debris avalanches are brief, spectacular, and often highly destructive. They are most common in geologically youthful mountain regions along or near lithospheric plate boundaries. Slow mass movements, especially creep, apparently have a greater total geomorphic impact because they are more widespread and collectively involve the downward transport of a greater volume of surface material.

3. *What are the causes and characteristics of karst topography, and how it is distributed around the world?*

Karst topography is created by the differential solution and erosion of soluble rocks, notably the carbonate rocks limestone and dolomite. Carbonate rocks are distributed widely around the world, but are especially susceptible to karst processes in areas of heavy precipitation and abundant vegetation. These conditions are associated with the production of large quantities of acidic surface water. Most important karst regions are located in humid portions of the tropics and middle latitudes.

The karst sequence involves three steps. The rock is first attacked by the chemical weathering process of carbonation, which changes the carbonate minerals to highly soluble bicarbonates. The bicarbonates are then dissolved by acidic water and, finally, are removed by surface or ground water flow. A karst landscape therefore is characterized by a variety of differential solution features. Small-scale surface features include grooves, solution pits, and loose rocks in more rugged areas. The most important larger surface features are circular depressions known as sinkholes. These features may coalesce to form linear valleys termed uvalas. Subsurface solution features also are very prominent. Most drainage takes place through underground routes; this frequently results in the formation of cave systems containing a variety of travertine features.

Review Questions

1. What is weathering? Explain the role that weathering plays in the gradational process.

2. What is the difference between physical weathering and chemical weathering in terms of their effects on the rock? Give two examples of each weathering category.

3. How do the global distributions of physical and chemical weathering rates vary? Why do they do so? For the world as a whole, which of the two categories of weathering processes has the greatest geomorphic significance? Why?

4. What are exfoliation domes? What weathering process or processes are primarily responsible for their production? What are talus cones? What weathering process or processes are primarily responsible for their production?

5. What is mass wasting? Discuss the importance of mass wasting in the gradational process.

6. Explain the significance of the angle of repose in mass wasting. Describe several ways by which human activities can cause slope angles to exceed the angle of repose.

7. Why can creep be considered the most important of the individual types of mass movements?

8. Describe the physical characteristics of a location that would favor each of the following types of mass movements:

A. rockfall
B. rockslide
C. mudflow
D. slump
E. creep

9. Describe the environmental conditions that favor the development of karst topography. Why is karst more widespread in the humid tropics and middle latitudes than in the subtropics and high latitudes?

10. Describe the surface appearance of sinkholes. In what two ways do they form?

11. Discuss the ways in which drainage in a karst area differs from the drainage elsewhere. What karst features develop as a result of these drainage characteristics?

12. How are karst caves believed to have formed? What distinctive features do they contain that make them more "scenic" than caves formed by other processes? How are these features produced?

Key Terms

physical weathering
unloading
exfoliation
frost wedging
chemical weathering
mass wasting
shear strength
angle of repose
rockslide
mudflow
slump
creep
karst
sinkhole
sinking creek
cave
tower karst

C A S E   S T U D Y

DEBRIS AVALANCHES PRODUCED BY HURRICANE CAMILLE

When conditions become highly favorable for their formation, mass movements sometimes occur in large numbers within a restricted area, resulting in considerable destruction and even loss of life. A major example of such an occurrence was the formation of several hundred debris avalanches in western Virginia as a result of torrential rains associated with Hurricane Camille in 1969.

THE STORM

The storm that was to become Hurricane Camille first appeared about August 5, 1969, as a weak tropical disturbance in the eastern Atlantic off the West African coast. After drifting westward for several days, the disturbance began to intensify rapidly on August 14 in the western Caribbean Sea. Camille, as the storm was named, attained hurricane strength early on August 15, and later that day crossed the western tip of Cuba with 50 meter per second (115 mph) winds. Emerging into the warm waters of the Gulf of Mexico, the storm continued its rapid intensification, with wind speeds approaching a phenomenal 90 meters per second (200 mph) by the evening of August 16. Camille moved ashore in the vicinity of Gulfport, Mississippi, on the evening of August 17, accompanied by winds of at least 83 meters per second (185 mph) and a 7 meter (23 ft) storm surge. The combination of winds and the storm surge resulted in some 171 deaths in Mississippi and Louisiana, but timely warnings and a massive evacuation are credited with having saved as many as 100,000 lives.

By the evening of August 19, as the weakened remnants of the storm were crossing the central Appalachians, its rainfalls briefly intensified to unprecedented levels, producing widespread flooding and debris avalanches in several Appalachian counties of West Virginia and Virginia.

The heaviest rainfall was centered in Nelson County, Virginia (see Figure 16.29). It is a rural county of about 10,000 inhabitants situated in the west-central part of the state on the eastern flanks of the Blue Ridge Mountains. Excessively heavy rain began falling in Nelson County about 9:30 P.M. and continued until about 3:30 A.M.—a period of six hours. The National Weather Service later conducted bucket surveys (field surveys of water levels in containers known to be empty preced-

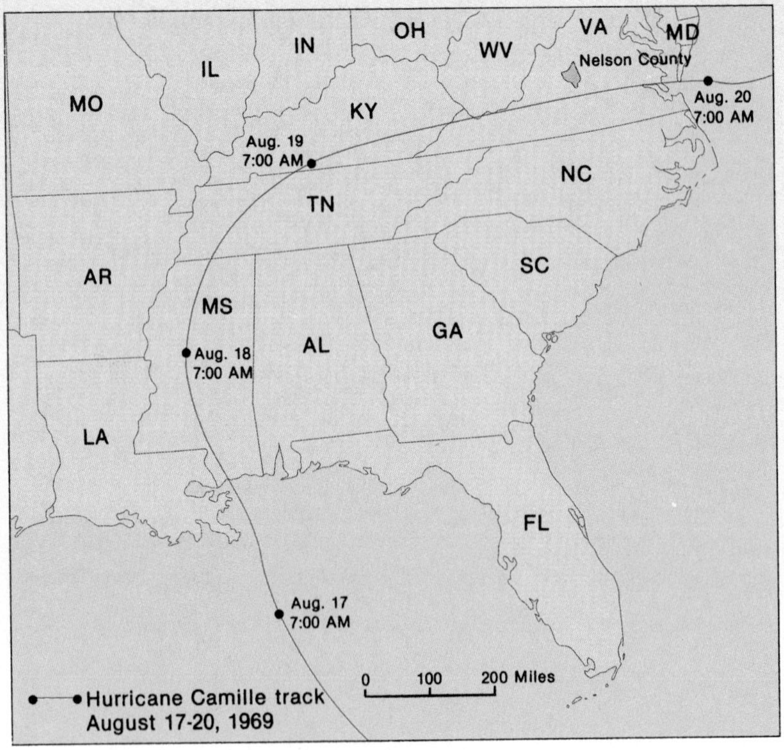

FIGURE 16.29 Track of Hurricane Camille and location of Nelson County, Virginia.

ing the storm) and at one point confirmed a storm rainfall total of 80 centimeters (31.5 in). Unofficial indications existed that in a few sites the storm total may have exceeded the annual mean of 100 centimeters (40 in). These figures are far larger for a six-hour timespan than any officially recorded rainfall totals anywhere in the United States.

The cause of the sudden and rather localized increase in storm rainfall intensities is apparently related to a combination of three circumstances. First, the entire region was covered by an unusually moist maritime tropical air mass. Second, the center of the storm passed just south of Nelson County, producing an easterly wind flow. Because the county is on the easternmost flanks of the Appalachians, a significant orographic lifting factor was produced. Finally, an area of prefrontal thunderstorms from an approaching cold front reached the area at the same time as the remnants of Hurricane Camille.

THE DEBRIS AVALANCHES

Even before the onset of the storm, soils were very moist from one of the wettest summers on record. As the torrential rainfalls caused the saturated soils to lose coherency, small roadbank slumps began by 10 P.M. and the first debris avalanches by about 11 P.M. By 1 A.M., debris avalanches were occurring in large numbers, and new ones continued to form until about 4 A.M.

The sites on which the debris avalanches occurred were the mountainside hollows that serve as conduits for subsurface drainage. The highly porous soils on these slopes normally absorb water as rapidly as it is received and direct it as ground water into the hollow systems. The water eventually surfaces at the foot of the slopes to form small streams.

The debris avalanches apparently were initiated by the coalescence of extraordinarily large amounts of subsurface water into the hollow systems. This water may have literally lifted the soil in each hollow from the underlying bedrock, allowing the entire detached mass to slide and flow down the hollow and into the stream valley below.

Witnesses reported that the debris avalanches occurred in groups. The noise and vibration from one avalanche was apparently sufficient to trigger several others nearby in a chain reaction sequence. In all, more than 275 occurred, including many with multiple branches totaling more than 1,000 meters (3,300 ft) in length. The debris avalanches left narrow, elongated scars on the hillsides, while the debris—a chaotic jumble of soil, rocks, giant boulders, and trees—was deposited in the adjacent flat stream valleys (see Figure 16.30). Many houses in the affected area were swept away by either the debris avalanches or the flooded streams, and approximately 100 people died in Nelson County alone.

In the years following the storm, the debris avalanches have been healing rapidly. The debris itself was quickly cleared, with government assistance, from the agriculturally productive valleys. The hill-

FIGURE 16.30 Debris from the three debris avalanche scars on the hillslope has been deposited in the foreground. This photo was taken in Nelson County, Virginia, a few months after the area was devastated by torrential rains from the remnants of Hurricane Camille. (Donald Poole)

side scars remain, but are much less conspicuous than they once were because they are being recovered slowly by soil that has been washing and creeping onto them from upslope. Vegetation in this humid temperature setting returns quickly once a soil cover is preset, and most scars already contain small trees.

Debris avalanches are undoubtedly recurrent phenomena within these mountainside hollow systems. Creep and debris avalanching apparently play complementary roles in transporting weathered materials down the steep slopes to the stream valleys below. Creep performs the initial task of filling the hollow systems with weathered materials from upslope. These materials are then periodically flushed down the hollow systems by debris avalanches during exceptionally heavy storms.

What may at first seem to be a single, random occurrence thus becomes a step within an organized gradational sequence. It is another indication that the various Earth phenomena we have been studying are organized into complex interrelated systems. One of the physical geographer's basic tasks is to discover the existence of such systems, how and where they operate, and the role they play in the development and evolution of our planet's surface features.

CHAPTER 17

FLUVIAL PROCESSES AND LANDFORMS

OUTLINE

FOCUS QUESTIONS

1. How do fluvial drainage systems develop, and what geographical patterns do they exhibit?
2. How do streams erode, and what are the most important landforms produced by stream erosion?
3. Why do streams deposit sediments, and what are the most important landforms produced by stream deposition?

The surface features of the land are more the work of streams than of any other geomorphic agent. In Chapter 10, discussion centered on the hydrologic aspects of streams, especially on their widespread distribution and the role they play in the hydrologic cycle. In the present chapter, we examine the physical characteristics of streams and stream systems, with particular emphasis on their role as shapers of landforms. The term **fluvial** (from the Latin *fluvius*, "river"), which commonly is used to identify stream-related processes and features, will appear frequently during the discussion.

FIGURE 17.1 Even in desert regions, fluvial features often dominate the landscape. Here, dendritic stream channels cover the surface of a portion of the Mojave Desert in southeastern California. Vegetation in the moister channels serves to accentuate the drainage pattern. (John S. Shelton)

GEOMORPHIC IMPORTANCE OF STREAMS

Streams are a dynamic component of the physical environment and can undergo major changes over relatively short periods of time. They perform two vital natural functions. First, they remove excess water from the land. The water transported by streams to the ocean comprises about one-fifth of all the precipitation that falls on the Earth's land surface. By flowing in response to gravity, streams gain enough kinetic energy to perform their second function. This is the erosion, downhill transport, and eventual deposition of weathering products supplied to the streams from adjacent slopes largely through mass wasting processes. The redistribution of these materials by streams produces most fluvial landforms.

Streams have had a profound impact on human activities. They have served vital functions as transportation routes and irrigation sources since the dawn of recorded history. They have formed the fertile agricultural floodplains on which we raise many of our food crops. From a broader perspective, fluvial processes acting over lengthy geologic timespans are responsible for most of the plains on which the majority of our planet's human inhabitants reside. Two examples of human interactions with stream systems are provided in the Case Study at the end of the chapter and the Focus Box on Rio Grande boundary changes.

Satellite imagery indicates that fluvial landform processes dominate perhaps three-fourths of the Earth's land surface (see Figure 17.1). This geomorphic dominance results from the nearly global distribution as well as the great gradational power of stream systems. Only the 10 percent of the land surface covered by continental glaciers does not contain at least intermittently active streams. Certain other nonglaciated regions contain some fluvial features, but are dominated by landforms produced by other geomorphic processes. Chief among these are the coastal margins (discussed in Chapter 20), some tectonically active areas (Chapter 15), and areas covered by shifting sands (Chapter 19).

Most of the areas just cited have limited practical importance to humans as sites for settlement—ice-covered areas are too cold and have no soil, tectonic areas are rugged and often prone to natural disasters, and sand-covered areas are dry and largely sterile. Only the coastal margins favor human settlement and they too are shaped by water.

The geomorphic effect of flowing water, though important nearly everywhere, displays significant spatial variations in its nature and intensity (see Table 17.1). These variations are related chiefly to climatic and vegetative patterns and are influenced strongly by local soil and bedrock characteristics. One might expect fluvial effects on the landscape to be least important in arid regions and to increase directly with precipitation averages, making the wettest regions the areas with the best-developed and fastest forming fluvial features. Figure 17.2 indicates that this is not the case. Flowing water actually is most effective as a landscape modifier in semiarid and subhumid regions. Very dry regions are minimally influenced be-

TABLE 17.1 Rates of Regional Erosion in the United States*

DRAINAGE REGION	DRAINAGE AREA (1,000 km²)†	RUNOFF (1,000 m³/s)	LOAD (t/km²/yr) Dissolved	Solid	Total	EROSION, (cm/1,000 yr)	AREA SAMPLED (PERCENT)
Colorado	629	0.6	23	417	440	17	56
Pacific Slopes, California	303	2.3	36	209	245	9	44
Western Gulf	829	1.6	41	101	142	5	9
Mississippi	3,238	17.5	39	94	133	5	99
South Atlantic and Eastern Gulf	736	9.2	61	48	109	4	19
North Atlantic	383	5.9	57	69	126	5	10
Columbia	679	9.8	57	44	101	4	39
Total/Average	6,797	46.9	43	119	162	6	

*After Sheldon Judson and D. F. Ritter, "Rates of Regional Denudation in the United States," *Journal of Geophysical Research*, 69, (1964): 3399.
†Great Basin, Saint Lawrence, and Hudson Bay drainage not considered.

cause of a lack of water, but humid regions with abundant precipitation have only a moderate rate of development of fluvial features. This results largely from the protective influence of vegetation, which generally increases in density with increased precipitation. The climatic zone of maximum fluvial effectiveness receives enough rainfall to produce considerable storm runoff, but is dry enough for the year as a whole to support only a limited cover of protective vegetation.

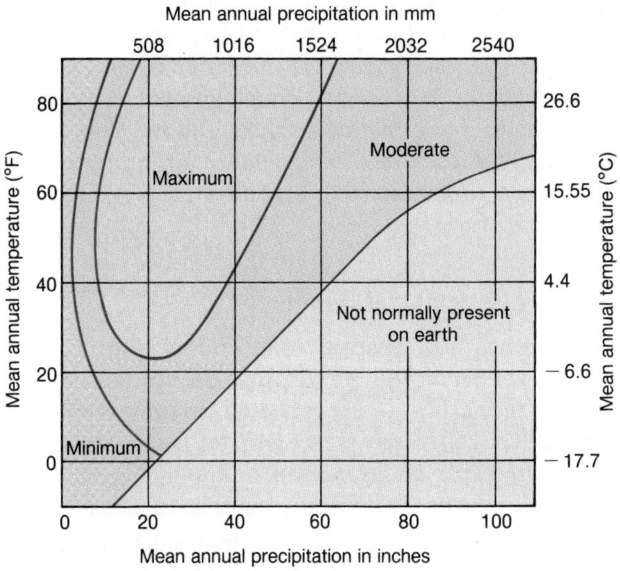

FIGURE 17.2 Relationship between climate and the effectiveness of fluvial geomorphic processes.

Although the topography of most of the Earth's land surface results primarily from fluvial processes, streams at any given time occupy and actively modify only a very small proportion of the total surface area. All stream action and energy are concentrated in a small area, producing features that stand in marked contrast to the remainder of the surface. The differential operation of fluvial processes therefore produces well-developed features, while their uniform operation would not. The majority of the surface in most areas consists of **interfluves,** or areas between streams, and their topographic characteristics are largely determined by what nearby streams have *not* modified.

The subsequent material is divided into two sections. We first analyze the causes and characteristics of stream flow—how and why streams develop, the mechanics of fluvial erosion and deposition, and the forms assumed by individual streams and the drainage systems of which they are a part. Second, we examine the major landforms produced by fluvial erosional and depositional processes.

CAUSES AND CHARACTERISTICS OF STREAM FLOW

Streams are formed by water that reaches the Earth's land surface and cannot be absorbed immediately. The surface in each locality has a maximum rate at which it can absorb water. This is known as the **infiltration capacity** and is related primarily to the texture, degree of compaction, and existing water

FIGURE 17.3 Newly developed rills leading into small gullies on agricultural land in Tennessee. (Tim McCabe/Soil Conservation Service)

FIGURE 17.4 Major gully development in the upper basin of the Rio Reventado in Costa Rica. Gullies result from the removal of the protective cover of dense natural vegetation in a region subject to heavy downpours. (H.H. Waldron/USGS)

content of the surface materials. If precipitation exceeds the infiltration capacity, water begins to collect on the surface. Depressions are filled first, and then, assuming that some slope exists, the excess water begins to flow downhill. In the absence of existing channels, the initial flow of water is disorganized and spreads over the surface in a nonchannelized form known as **overland flow.** It may flow as a sheet of water on a smooth surface or, more commonly, as uneven threads around obstacles like plants and stones. Because this flow is not concentrated, it is very shallow and friction with the underlying surface is great. As a result, even on relatively steep slopes, its flow is slow, and its capacity for eroding and transporting sediments or debris of any kind is highly limited. Overland flow dominates near the crests of interfluves, producing zones of no erosion there.

Proceeding downhill, the flow increases in volume as the total upslope area being drained becomes progressively larger. **Channelization,** the beginning of a stream, occurs where the water depth and speed enable the flow to overcome the cohesion of the soil particles. On unprotected slopes, such as those cleared for agriculture, a nearly parallel series of small shoestring channels called **rills** forms first (see Figure 17.3). Farther downhill, the rills grow through coalescence into steep-sided **gullies** (see Figure 17.4). Neither rills nor gullies are normally considered to be stream channels, because they carry water only infrequently. The continued deepening of a gully, though, eventually may enable it to reach the water table and to begin tapping ground water supplies. At this point the gully has been converted to a stream channel and constitutes an extension of the stream system it joins.

Rills and small gullies normally do not form on undisturbed, well-vegetated slopes because the soil is highly absorbent and is held firmly in place by the vegetation mat. Under these conditions, water from even heavy storms typically is absorbed as fast as it is received. This water travels downslope as throughflow (see Chapter 10), until it surfaces as a spring or seeps out along the side of a stream bank at a point where the surface intersects the water table. Deeper ground water flow travels much more slowly than either surface water flow or throughflow. Thus, in a humid climate, a stream supplied by ground water is likely to maintain a stable base flow even during dry periods.

Drainage Systems and Patterns

Streams are not separate, independent entities, but instead are organized into **drainage systems** consisting of numerous interconnected stream segments that collectively remove excess water from a drainage basin. In form and function a drainage system is much like a tree. It normally consists of a master or trunk stream at its base that branches uphill into a progressively larger number of increasingly small segments, or **tributaries.** The tributaries extend into all portions of the drainage basin. Like the branches of a tree, the branches of a drainage system conduct water and other materials (in this case, stream sediments) from one part of the system to another.

Around the peripheries of drainage basins of any size are highland rims that serve as **drainage divides.** These divides separate the flow and direct it toward adjacent drainage systems. In some areas, the divides are low and indistinct, while in others they form lofty mountain systems. The largest and most important drainage divide in the United States is the Continental Divide, which separates runoff flowing east into the Mississippi River system and ultimately the Gulf of Mexico from that flowing west through the Colorado and Columbia River systems to the Pacific (see Figure 17.5).

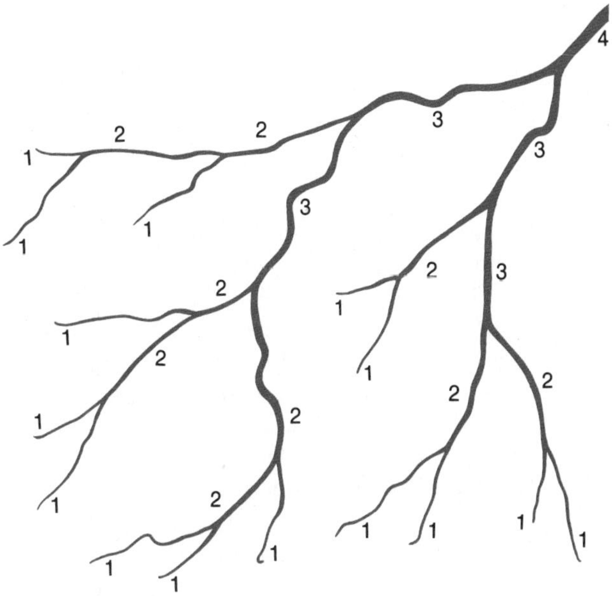

FIGURE 17.6 The Horton-Strahler stream ordering system. A stream of a given numerical order is formed by the juncture of two streams of the next lower order.

Stream Orders

Stream size, a basic and important concept, can be described in several ways. Figures such as length, width, discharge, and area drained all can be used as indices of stream size. A general idea of the regional significance of a stream, though, can be conveyed in comparatively simple fashion by the concept of **stream order.** The stream ordering system was first proposed by the American geomorphologist Robert Horton in the 1940s. In applying the system, streams are ranked from 1st order to approximately 12th order based on the number and nature of their tributaries—factors that correlate closely with the size and regional significance of the stream (see Figure 17.6). By definition, a 1st order stream is a small stream with no tributaries. It is analogous to the outermost twigs on a tree. In humid climates, 1st order streams generally are fed directly by subsurface water

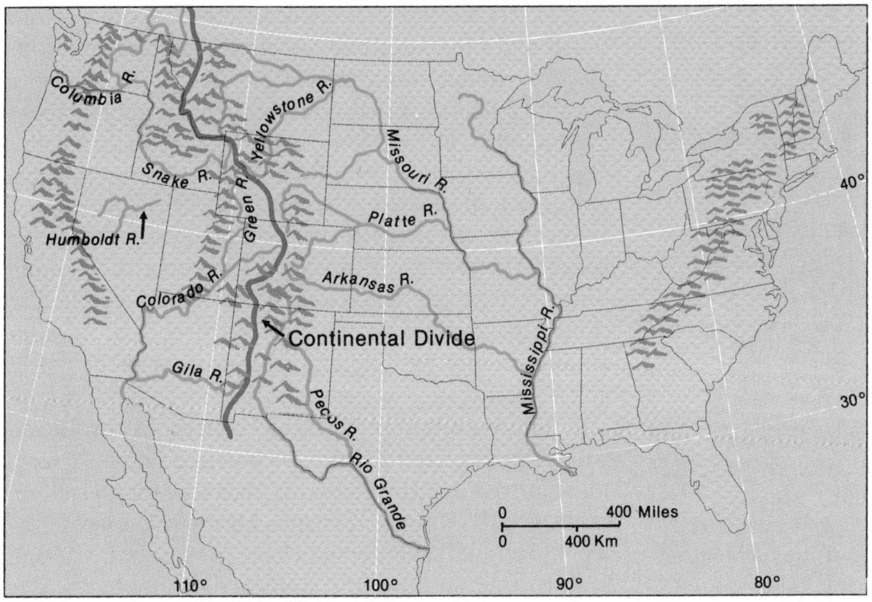

FIGURE 17.5 The Continental Divide (in red) separates streams flowing into the Pacific from those flowing into the Atlantic via the Gulf of Mexico.

outflow from adjacent interfluves. A 2nd order stream is formed when two 1st order streams join. Subsequent orders follow the same pattern, increasing by whole number units only when two streams of the next lower order join. Within a given drainage system, as the order becomes higher, the total number of stream segments of that order becomes smaller, but the physical dimensions of the streams increase. The numerical value of the highest order depends, of course, on the size of the stream system, and on factors such as slope steepness, vegetation density, and rock type, that influence surface erosional resistance and infiltration capacity. The Mississippi River, the largest in the United States, is generally considered a 10th order stream.

Horton and other researchers have discovered statistical relationships between stream order and a host of other stream and drainage basin characteristics. These include mean stream length, width, discharge, depth, area of drainage basin, number of tributaries, and slope gradient (see Figure 17.7). Of these, all except the slope gradient increase in value with an increase in stream order; within a given drainage system, all these factors usually can be related numerically to one another with a reasonable degree of accuracy. The relationships among these characteristics, and the differences they display between regions, provide another indication that drainage systems develop in response to natural controls.

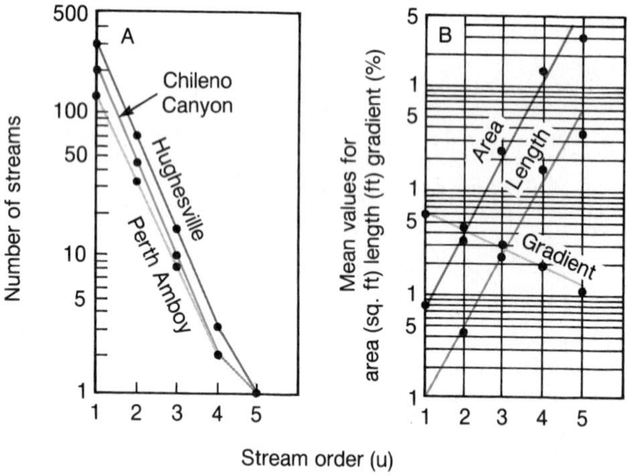

FIGURE 17.7 Quantitative relationships between stream order and stream frequency, length, gradient, and drainage basin area. (Stanley A. Schumn, "The Evolution of Drainage Systems and Slopes in Badlands at Perth Amboy, New Jersey," Bulletin of the Geological Society of America, vol. 67, 1956, pp. 603-604.)

Drainage Patterns

The drainage pattern is the surface pattern collectively formed by the streams in a drainage system. All the stream systems in a given region typically exhibit essentially the same type of pattern. The existing pattern is largely a response to differences in rock types and structures as well as to preexisting surface features. It therefore provides an indication of the region's geologic and geomorphic history as well as its present characteristics. A large number of drainage pattern types exist, and, as might be expected, some regions exhibit patterns that are poorly defined, are intermediate between two types, or are complex because of multiple influences. The types described here and illustrated in Figure 17.8 are among the most widespread.

The **dendritic pattern** is the most widely distributed of the drainage pattern types. It develops on gently sloping surfaces that are homogeneous in their resistance to erosion and therefore exert no significant structural control on stream location. It is especially likely to occur in areas of horizontally bedded sedimentary rocks and glacial deposits. The dendritic pattern is treelike in form, with a large degree of variability in the orientation of tributaries, and with stream junctures normally forming angles well under 90°.

The **trellis pattern** typically develops in regions of parallel ridges and valleys where resistant and nonresistant sedimentary strata alternate. It consists of a parallel pattern of major streams that occupy valleys eroded into the weaker rocks, while much smaller and shorter tributaries flow down the steep ridge flanks to join the major streams at nearly right angles. In places, a stream may cut through an adjacent ridge to join the stream in the next valley. In the United States, trellis drainage patterns are well developed in the Ridge and Valley portion of the central Appalachians and in the Ouachita Mountains of Arkansas and Oklahoma.

The **rectangular pattern** usually is produced when the drainage pattern is controlled by intersecting fault or fracture systems in areas of crystalline rocks such as granite. Streams alternate between unusually straight segments, where they flow along one fault or fracture line, and sudden right-angled bends,

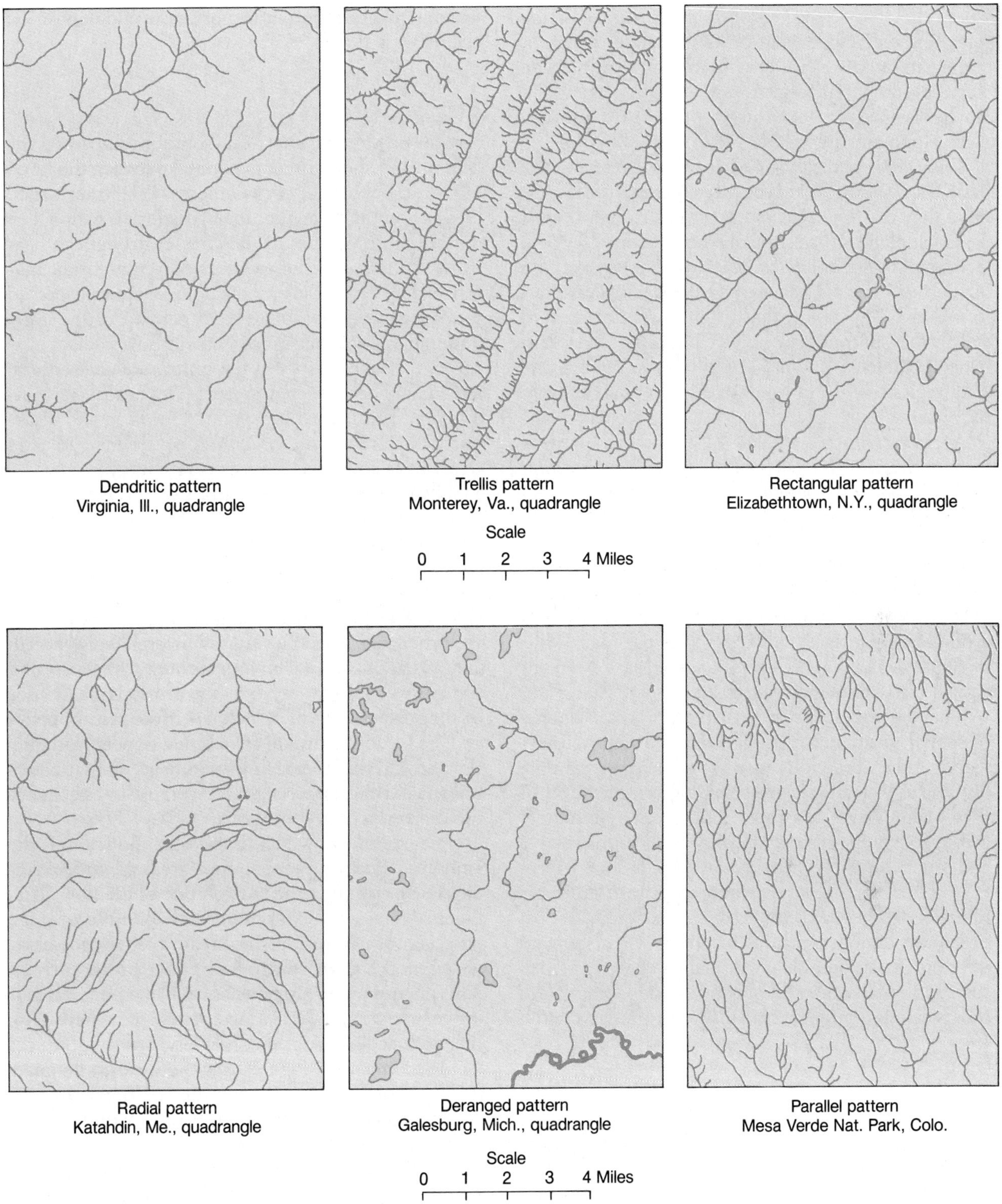

FIGURE 17.8 Map depictions of major drainage pattern types.

where they are redirected onto another intersecting line. The rectangular pattern is similar to the trellis pattern in its angularity, but different in that it has no single predominant direction of flow. A rectangular pattern is well developed in portions of the Adirondack Mountains of upstate New York.

The **radial pattern** is perhaps the simplest of the basic drainage pattern types. It consists of the outward flow of streams in all directions from a central peak or upland. It is especially well developed in the vicinity of large, isolated volcanic peaks such as those in the northern Cascades of Oregon and Washington. Conversely, a converging flow of streams toward a central basin produces a **centripetal pattern.** Centripetal patterns are typical of both karst regions and arid regions with interior drainage; consequently, they are widespread.

A more complex variety of the radial pattern, termed the **annular pattern,** develops when an uplifted region is surrounded by inclined bands of sedimentary rock strata that vary in their resistance to erosion. This structural pattern produces an alternation of divergent stream segments and of radial segments that follow the outcrops of the less resistant strata. An annular drainage pattern often develops around large igneous masses, such as the Black Hills of South Dakota and Wyoming, that have been intruded into sedimentary strata.

The **deranged pattern** displays a great degree of spatial disorganization. Streams are highly irregular in their directions of flow, sometimes exhibiting sharp, V-shaped bends. In addition, they do not flow smoothly downhill, but pass alternately through depressions occupied by lakes and swamps, and steep segments marked by rapids and waterfalls. The existence of a deranged pattern indicates that it is of geologically recent origin and that an organized drainage pattern has not yet had time to develop. It is common in glacially scoured regions such as the Canadian Shield and Scandinavia, where the previous drainage systems were erased by the ice.

Stream Erosion, Transportation, and Deposition

It is through the ability of streams to erode, transport, and deposit sediments that fluvial landforms develop. Because the characteristics of fluvial features are determined by the nature of the activities that produce them, we now examine the processes by which stream erosion, transportation, and deposition occur.

Erosional Processes

Stream erosion results from solution and from the friction and shear forces generated between the flowing water (including any sediments being transported by this flow) and the nonmoving materials with which it comes into contact. If the stresses brought to bear by this action exceed the cohesive strength of the materials, these materials will be eroded and transported downstream. Approximately 95 percent of the gravitational energy that causes stream flow is dissipated through friction between the water molecules of the stream and is converted to heat. The critical remaining 5 percent is used for sediment erosion and transport. The great erosional power of streams indicates that this is still a large amount of energy.

The erosive energy of a given stream is related to its velocity and volume of flow and to the amount of friction generated with its **banks** (sides) and **bed** (bottom). An increase in any of these factors will increase the erosive power of the stream. At the same time, a change in any of these three factors will affect the other two, since they are all interrelated. A fourth factor that must be taken into account when assessing a stream's ability to erode is the erosional resistance of the surface on which it flows. If the stream banks and bed consist of solid rock, erosion is normally very slow. If the stream flows within accumulations of unconsolidated materials, however, it can effect very rapid erosional modifications.

The speed at which the water flows and the amount of friction it encounters are both critical factors in determining the **turbulence** of the flow. Turbulence refers to the tumbling eddy motions of small parcels of water within the stream that are superimposed on the main downhill direction of the current. This phenomenon already has been encountered as it influences wind-flow characteristics (see Chapter 6). Just as friction with the Earth's surface produces wind turbulence, which causes the wind to be gusty and to have the capacity to erode and transport sediments, so too the friction of a stream with its banks and bed causes the stream to flow in a turbulent manner and gives it the ability to erode and transport sediments with a higher specific gravity than water. Water flows at a much slower average speed than the wind, but it is a much denser medium and has correspondingly greater erosional and transportational capabilities. An increase in either streambed roughness or speed of flow will increase turbulence and give the stream a greater erosional ability. In fact, it has been determined that the erosive power of a stream increases as approximately the cube of its increase in speed of flow. This means that if the speed of flow of a stream doubles, its capacity to erode sediments increases by 2^3, or eight times.

In order for stream erosion to occur, the water must flow rapidly enough that the friction of the moving water, coupled with the lift produced by eddies within the flow, can dislodge the materials over which the water passes. The minimum stream velocity needed to erode a given streambed particle is called its **critical erosion velocity;** this is illustrated graphically in Figure 17.9. This velocity has been found to be lowest for sand-sized particles with diameters of about 0.5 millimeters (0.2 in). Smaller and lighter silt and clay particles have higher critical erosion velocities because of their cohesiveness and because they collectively produce such a smooth streambed that turbulence is reduced. Particles larger than 0.5 millimeters, on the other hand, require greater flow speeds for erosion because of their weight. As the settling velocity line in Figure 17.9 indicates, the coarser the texture of the stream-borne particles, the higher the flow speeds needed to continue to transport them. Note the large gap between the critical erosion velocity line and the settling velocity line on the left side of the graph. This indicates that small streambed particles are difficult to erode but that, once eroded, they can continue to be transported at very low flow speeds. Conversely, high velocity flow is required for both erosion and transport of coarse materials. Most streams do not maintain a high flow velocity for very long; as a result, most transported stream sediments consist of sand-sized or finer particles.

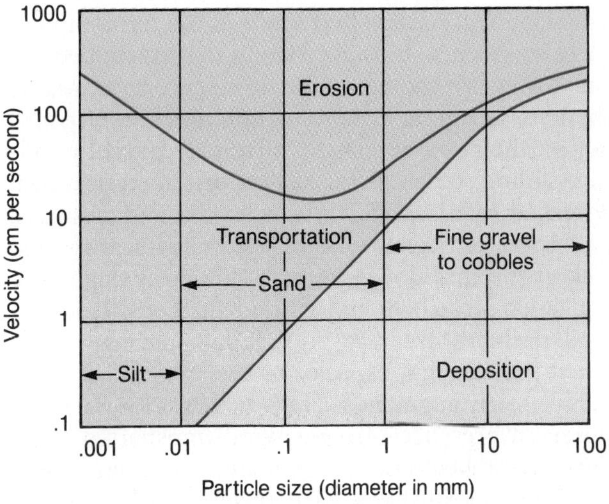

FIGURE 17.9 Stream sediment erosion, transportation, and deposition regimes as related to particle size and water flow velocity.

During times of high flow volumes, streams have the capacity to erode and transport a much greater sediment load than at times of normal flow. This occurs for several reasons: the quantity of water available to erode and transport sediments is greater; the "wetted perimeter," or total surface area covered by the stream, increases, giving the stream access to more sediments; and, most important, the speed of flow increases. As a result, more stream erosion and fluvial landscape modification generally occur during floods than during the much longer periods of normal or low flow conditions. Floods of progessively greater magnitude are increasingly infrequent events within a given stream system, but major floods are capable of modifying the landscape to a much greater degree than is minor overbank flooding. This raises the question as to whether the greatest total geomorphic work is accomplished by streams during very rare extreme floods, or collectively during the relatively frequent minor floods. Flood magnitude/frequency studies have indicated that it is the minor to moderate floods, with mean recurrence intervals of about three to four years, that accomplish the most sediment erosion and transportation.

The actual process of sediment erosion by streams is accomplished in three different ways. **Hydraulic action** consists of the direct sweeping away of loose materials by the friction and turbulence generated by the moving water itself. Although very effective on unconsolidated materials, it has a negligible effect on solid rock. **Abrasion** involves the scraping of particles carried in the water against the stationary materials of the banks and bed. This process can slowly erode even solid rock if the stream-borne particles are harder than the rock they are scraping. Abrasion also occurs within the flow itself as the transported sediments are repeatedly scraped or knocked together by the turbulence of the flow. This process results in the gradual rounding of the sediments. **Corrasion** is the dissolving of rock materials so that they enter the flow in a molecular state. On the average, about 30 percent of the total sediment load of streams is transported in solution.

Most streams simultaneously erode both vertically and laterally. The relative proportion of erosion that takes place in each of these directions is critical to the nature of the resulting topography, since a predominance of vertical erosion will produce a high relief surface containing deep, steep-sided valleys while a predominance of lateral erosion will result in the formation of broad, gently sloping valleys. In general, vertical erosion is dominant where the stream gradient is steep and lateral erosion dominates where it is gentle, although other factors such as volume of flow, channel form, and surface erosional resistance also are important. A gentle gradient reduces the stream's velocity, thereby lessening the rate of downcutting sufficiently that valley-widening processes such as sheet flow, tributary inflow, and mass wasting can keep up with deepening so as to maintain a gentle valley side-slope.

Variations in stream velocity, turbulence, and resulting erosional capacity can exist over short distances. Erosion therefore does not occur simultaneously everywhere along the course of a stream, but is at any given time concentrated at certain points. These points are located where active currents are in contact with bed and bank materials, especially if these materials are weak or unconsolidated.

Transportation

The same water turbulence that lifts sediments from the streambed keeps them from settling back to the bed, allowing the current to transport them downstream. As illustrated in Figure 17.9, the critical erosion velocity is always greater than the settling velocity for a given particle. As a consequence, as long as the stream velocity and turbulence that eroded the sediments are maintained, stream sediments will continue to be transported downstream without being deposited.

Streams transport sediments in a number of ways, depending upon the sizes and physical state of the sediments as well as the speed and turbulence of the flow (see Figure 17.10). A substantial proportion of the total sediment load is carried in solution. The **dissolved load** tends to be especially large in relation to the total load in areas of readily soluble rock such as limestone, in areas where the velocity of flow and resulting turbulence are low, and in humid areas where vegetation is abundant and the water is mildly acidic. Most dissolved materials are transported entirely through the stream system, and eventually may be precipitated to form chemical sedimentary rock on the sea floor.

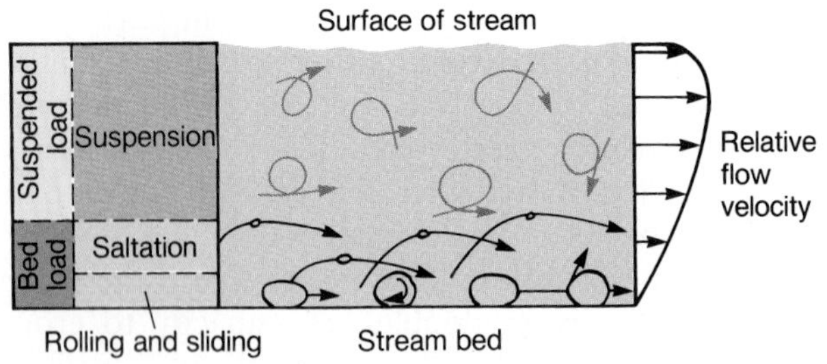

FIGURE 17.10 Methods of stream sediment transport. The relative flow velocity profile at right indicates that water flows fastest just below the surface of a stream and most slowly along its bed.

More than half the total sediment load collectively transported by streams travels as **suspended load.** Suspended materials consist of generally fine-textured particles that are carried within the flow itself. They are kept from settling by the upward component of the turbulent eddies generated by friction. Streams flowing through areas in which unconsolidated materials are plentiful carry the greatest volumes of suspended sediments. The greatest quantities of such materials are picked up and transported during floods (see Figure 17.11).

The coarsest particles that a stream is capable of moving are carried as **bed load.** As the term implies, these sediments are moved along the stream bed because they are too heavy for turbulence to transport them for any great distance within the flow. Depending on their size and shape, they may travel by rolling, sliding, or bouncing (**saltation**) along the bed. The bed load, unlike the suspended and dissolved loads, moves intermittently. It typically is transported a short distance downstream, temporarily deposited, and later picked up and carried further. Many bed load particles are transported only during floods, when the carrying capacity of the stream is temporarily greatly augmented. This accounts for the often observed presence of giant flood-transported boulders on the bed of a small stream that seems utterly unable to move them. As it travels, the bed load is gradually rounded and reduced in size by abrasion.

The form of a stream channel is controlled largely by the quantity and coarseness of the sediment load

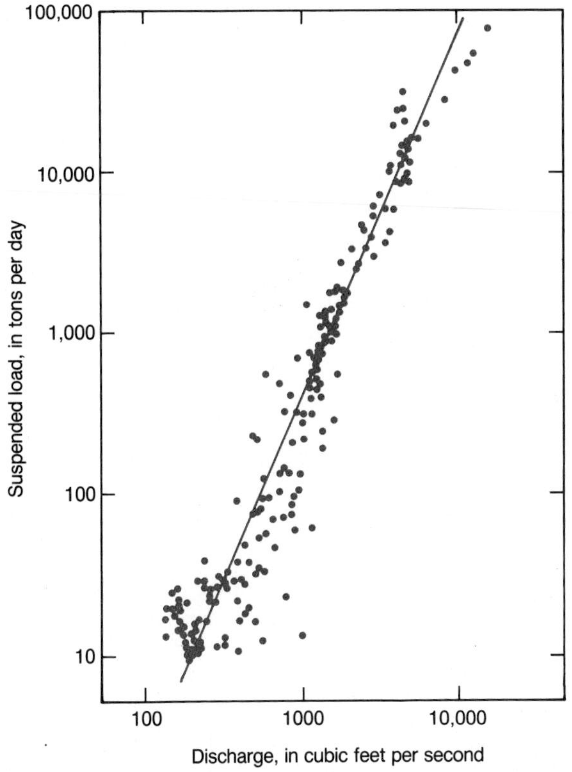

FIGURE 17.11 Relationship between stream discharge (volume of flow) and suspended sediment load in Brandywine Creek near Wilmington, Delaware. Each point represents a separate measurement. A modest increase in discharge can be seen to produce a large increase in suspended load.

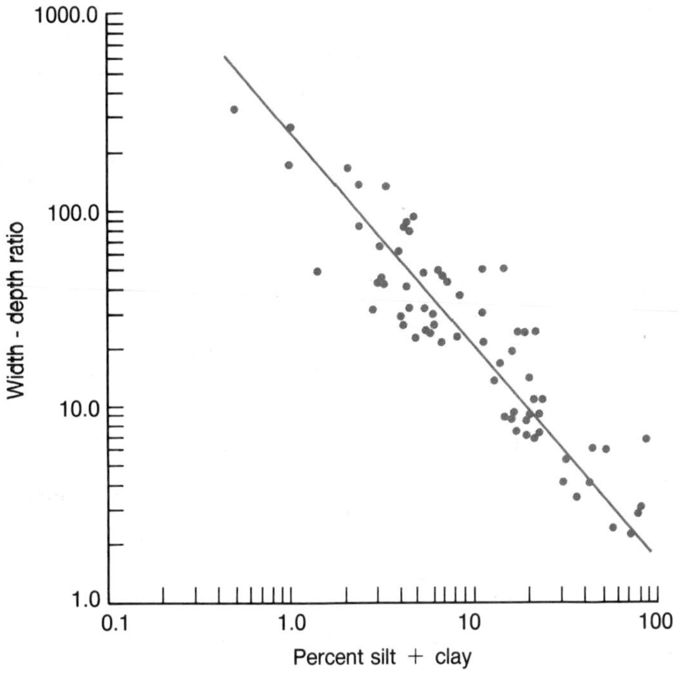

FIGURE 17.12 Relationship between stream width/depth ratio and total percentage of silt and clay in the stream sediments. The graph indicates that streams with broad, shallow channels transport small proportions of silt and clay, while streams with deep, narrow channels transport large quantities of these fine-textured sediments.

it transports (see Figure 17.12). If the stream carries a limited quantity of predominantly fine-textured sediments, it tends to be deep and narrow, to have a gentle downhill gradient, and to display a high degree of sinuosity (i.e., to have numerous curves or **meanders**). If the sediment load is large in quantity and coarse in texture, the stream will be broad and shallow, steeper in slope, less sinuous in form, and likely will develop a braided flow pattern (see Figure 17.19).

These characteristics of channel form bring about an adjustment between the quantity and texture of the sediment load and the stream's ability to transport that load. The basic mechanics of the adjustment process are relatively simple. If the sediment-carrying capacity of a stream exceeds its load, the excess energy is used to erode the stream channel vertically. This, in turn, lowers the elevation of the stream and reduces its downhill gradient, reducing its speed of flow, turbulence, and sediment-carrying capacity. In the absence of any disrupting factors, downcutting will continue until the carrying capacity of the stream is reduced to the point that it becomes just sufficient to transport its sediment load. No further significant downcutting will then occur, because there is no longer a supply of excess stream energy.

By the same token, if the stream's sediment supply exceeds its carrying capacity, the excess sediment will be deposited. This will **aggrade** (build up) the channel, increasing the stream's downhill gradient, which in turn will increase its flow speed, turbulence, and sediment-carrying capacity. Deposition and aggradation will continue until the ability of the stream to transport sediment has been increased sufficiently that it can carry all of its load.

The extent to which a given stream has adjusted its downslope gradient to achieve a balance between its sediment-carrying capacity and sediment load can be estimated roughly by examining its **longitudinal profile.** This is a graphic depiction of the downhill gradient of the stream from its head to its mouth (see

Figure 17.13). A stream that has not achieved balance between flow and load typically has an irregular longitudinal profile. Such a profile is likely to contain **knickpoints,** or points at which an abrupt increase in slope steepness occurs. Most knickpoints occur where the stream passes over a resistant bedrock outcrop. They are typically the sites of rapids or of waterfalls.

On the other hand, a stream that has become **graded**—that is, has attained a balance between flow and load—ideally displays a smoothly concave longitudinal profile on a bed of uniform resistance. The development of a graded profile means that the stream flows along a gradually declining gradient from head to mouth. Streams are able to continue to transport their sediment loads at progressively reduced slope angles because their increased volumes of flow result in a gradual reduction in the amount of friction per unit volume of water. In addition, the size (or caliber) of the streambed particles decreases gradually downstream because of the effects of abrasion. As an illustration of the influence of discharge on speed of flow, the lower Amazon River, with an extremely gentle gradient of only 4.75 centimeters per kilometer (3 in/mi), flows at nearly 2.4 meters per second (8 mph) during times of high water. This is faster than many mountain streams with much steeper gradients.

Deposition

Streams deposit sediments when a reduction in their speed, turbulence, and/or volume of flow make them incapable of transporting all of their sediment load. As their carrying capacity declines, they first deposit the coarsest materials, which have the greatest settling velocities, and then deposit progressively finer sediments. The finest sediments of all, including the stream's dissolved load, generally are transported completely through the stream system to the water body into which the stream empties. This association between stream-transporting ability and the size of the particles being deposited causes fluvial deposits to be **sorted,** or separated by size. A high degree of sorting, in fact, is one of the most important and easily recognizable characteristics of fluvial deposits (see Figure 17.14).

The accumulation of thick deposits of sediments is common in the lower reaches of stream systems, and especially at the mouths of large rivers that empty into the sea. Deposition also occurs locally and temporarily at numerous points along the courses of nearly all streams. Sites favoring deposition include shallow stretches along the stream course such as the insides of bends, places where the channel widens, and around the bases of obstructions to flow such as boulders lying on the streambed. Human actions can greatly influence the pattern of stream deposition, with dams and reservoirs acting as sediment traps. The effect of channel modifications on the depositional pattern of the Mississippi River is discussed in the Case Study at the end of the chapter.

Accumulations of fine-textured stream-deposited sediments are called **alluvium.** The chief source of the alluvial deposits of most streams is soil eroded from areas upstream. Thick accumulations of allu-

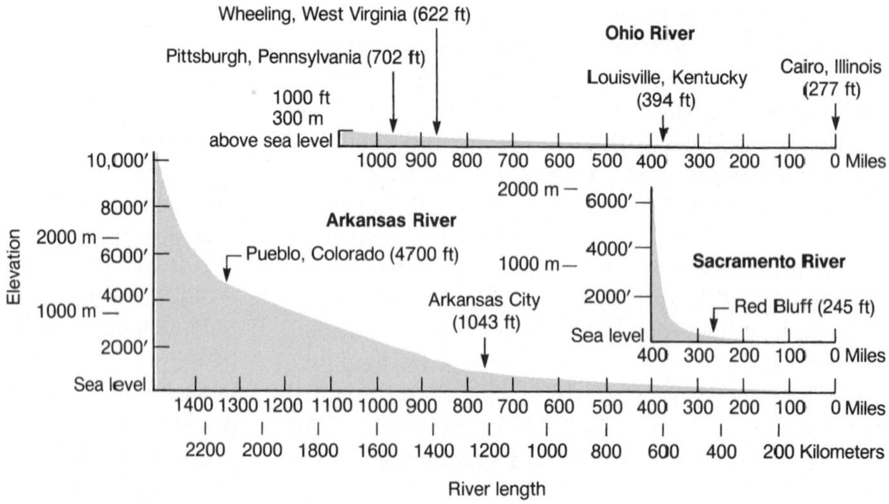

FIGURE 17.13 Measured stream longitudinal profiles. All three streams have concave profiles, although the degree of concavity differs for each as a result of differing slope steepness. Vertical scale exaggeration is approximately 275:1.

FIGURE 17.14 Closeup view of rounded and sorted fluvial deposits on a stream bank near Lafayette, Indiana. The dime provides scale. (B.F. Molnia)

vium cover the lower valleys of most large river systems, and their generally high content of soluble plant nutrients often makes them agriculturally productive.

FLUVIAL LANDFORMS

Streams are not the most powerful of the natural forces shaping the surface features of the Earth. That distinction, among the major groups of forces, goes to the tectonic and glacial processes. The restricted size of the areas affected at any given time by tectonic activity and glaciation, however, reduces their global significance, so that streams are the leading geomorphic agents in terms of their overall influence on the Earth's surface topography. In many parts of the world, the three processes of weathering, mass wasting, and fluvial erosion act in sequence to erode the surface and produce most gradational landforms. Weathering acts first, producing unconsolidated surface materials; mass movements next transport these materials to stream valleys; and streams then transport them, eventually, to the sea. During the last stage of the sequence, a variety of fluvial erosional and depositional features is likely to be formed.

In a direct sense, the dominant fluvial feature is the **valley.** A valley is a linear lowland produced by stream erosion. Most streams, of course, produce valleys, making them common and familiar surface features. The dimensions of a valley are determined largely by the size (or, more precisely, the volume of flow) of the stream that produced it, the rate of stream erosion, and the length of time over which the valley has developed. Valleys contain a variety of both erosional and depositional features, as we will see shortly. From an indirect standpoint, a second major category of fluvial landforms is erosional hills and mountains. These features are the upland erosional remnants located between stream valleys in a fluvially dominated landscape.

In a very general sense, fluvial landforms can be divided into two groups: those formed primarily by erosion and those formed primarily by deposition. It is actually rather arbitrary, however, to classify many fluvial features as either erosional or depositional because they are composite in nature. In other words, they are produced by both erosional and depositional processes operating in different places or at different times. Nonetheless, these two groups of processes are used here as a convenient way of organizing the discussion of fluvial landforms.

Fluvial Erosional Features

Streams erode when their volume, speed, and turbulence of flow give them the ability to remove and transport a greater quantity of sediment than they already contain. Excess erosional capacity is most common in the upper reaches of drainage systems. In these areas the downhill gradient is normally considerably steeper than it is farther downstream, the sediment load is usually much smaller in volume, and significant tectonic uplift is more likely to have occurred in the recent geologic past. The two major categories of features produced by fluvial erosion, as already noted, are valleys and erosional hills and mountains.

Erosional Valleys

Nearly all river valleys are produced largely or entirely by fluvial erosion, but the term **erosional valley** as used here refers to a valley still being actively deepened by a river that is downcutting toward base level. **Base level** is the level or elevation at any point along a stream's course that provides the stream with a gradient just sufficient for it to transport its sediment load past that point with no erosion or deposition. It is, in other words, the level of flow/load equilibrium for the stream. Any erosion or deposition can be viewed as a response by the stream to a changing or as yet unattained base level. If base level is achieved all along its course, the entire stream system is considered to be **graded.**

A great deal of controversy has occurred among fluvial geomorphologists regarding the validity of the concepts of stream grade and base level. One of the chief difficulties with these related ideas lies with the great variations in time scales over which they may be viewed (see the discussion of changes in fluvial landscapes over time in the Focus Box on page 455). As a result, both temporary and long-term base levels can be postulated.

A temporary base level may be achieved quickly through the reworking of stream sediments. During floods, for example, a stream gains excess erosional capacity and rapidly erodes sediments lying within its channel. The base level in this case has been temporarily lowered by the stream's increased sediment-carrying capacity, and the stream therefore reduces its gradient in order to restore a balance between flow and load. When the flood subsides, the stream is located at too gentle a gradient to transport its sediment load with the now reduced volume of water it contains. It therefore deposits enough sediment to aggrade the channel back to its original level. Because of nearly constant fluctuations in their volumes of flow, most streams are continuously eroding or depositing at various points along their courses in order to adjust to shifting temporary base levels.

On a long-term basis, an ungraded stream may have to erode through great thicknesses of rock over lengthy geologic timespans in order to reach a stable base level. A classic example is the Colorado River, which, over many millions of years, has eroded 1.6 kilometers (1 mi) into the surface of the Colorado Plateau of northern Arizona to produce the Grand Canyon (see Figure 17.15). The river in this area apparently has not yet achieved its base level elevation, in part because tectonic uplift is still taking place, and active downcutting continues. Superimposed on this long-term erosional trend, though, the Colorado River locally erodes and deposits sediments at various points along its course in response to short-term variations in its volume of flow and sediment load.

If a stream is downcutting rapidly and the near-surface rock material is strong enough to maintain steep slopes, the erosional valley will be deep, steep walled, and relatively narrow (see Figure 17.16). The width of such a valley is restricted because valley-widening processes depend largely upon the slow progress of weathering to weaken valley-side materials so that they may be removed by mass wasting. Valley deepening, on the other hand, can be effectively accomplished by stream particle abrasion on unweathered materials. Even the deepest and most steep-walled stream valleys rarely have depths that exceed their widths. If it is exceptionally deep and narrow, a valley is referred to as a **gorge.** A steep-walled but broader and often flat-bottomed valley is called a **canyon.**

FIGURE 17.15 The Grand Canyon, 1.6 kilometers (1 mi) deep and about 30 kilometers (20 mi) wide, is one of the world's most impressive fluvial erosional features. (Bob Thomason/Tony Stone Worldwide)

FIGURE 17.16 The deep, steep-sided valley of a California stream that is still actively downcutting toward base level. (Visuals Unlimited/©Albert Copley)

The side slope characteristics of stream valleys and especially of rapidly expanding erosional valleys depend largely on the nature of the materials of which they are composed. If the materials are homogeneous, a smooth side slope is produced. If the materials consist of alternating layers of varying resistance to erosion, a highly irregular slope may be produced (see Figure 17.15). This slope normally will consist of protruding ledges and cliffs formed on the resistant strata, and inset slopes of intermediate steepness on the weaker strata.

Erosional valleys exist mostly in uplands such as plateaus or in hilly or mountainous regions. Most large river systems have their headwaters in such regions and eventually enter plains farther downstream. Erosional valleys are especially steep, narrow, and spectacular in arid or semiarid regions. Here they are less masked by vegetation, and much less moisture is available for chemical weathering to facilitate the valley-widening processes.

Erosional Hills and Mountains

Erosional hills and mountains are the highland areas that remain on the landscape following the fluvial dissection of a plateau, fault block, or other raised surface. We have in previous chapters given the credit for the widespread distribution of such features to the tectonic forces that produced the initial surface uplift, but perhaps equal credit should be given to the gradational forces that have carved these raised blocks into their present intricate patterns of hills or mountains and intervening valleys.

The pattern of hills or mountains within a region is largely determined by the type of drainage pattern that develops. The drainage pattern, in turn, is controlled by the nature of the uplift and by the rock types and structures within the area. If, for example, a plateau is formed by the broad uplift of a mass of homogeneous or horizontally bedded rock, a dendritic drainage pattern will develop. The randomness in location of the stream segments in this pattern will cause the intervening hills or mountains to display a random pattern also. The Appalachian and Ozark Plateaus are both good examples of rugged but disorganized upland surfaces formed by the development of dendritic drainage patterns in regions of uplifted horizontal sedimentary rocks (see Figure 17.17).

In places where folding or faulting has exposed strata that differ in their resistance to erosion, streams develop linear drainage patterns such as the trellis pattern as they preferentially carve valleys in the weaker rock layers. This action will produce a parallel system of resistant linear ridges and alternating valleys such as that found in the Appalachian Ridge and Valley region and in portions of the Coast Ranges of California.

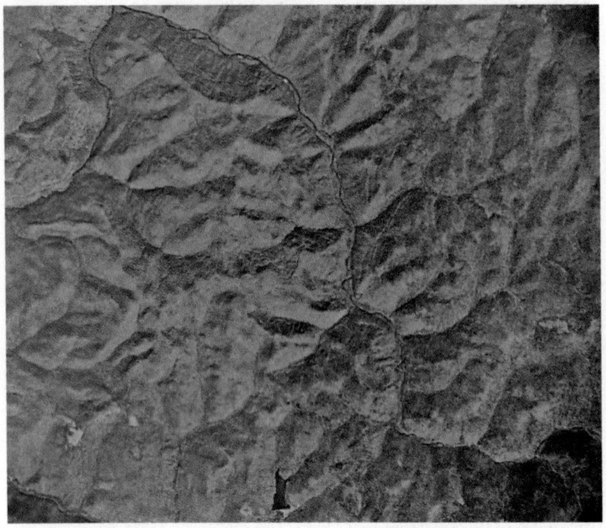

FIGURE 17.17 Aerial view of maturely dissected topography displaying dendritic drainage on the Appalachian Plateau of central West Virginia. (USGS)

Fluvial Depositional Features

Deposition occurs on the largest scale in the lower portions of drainage systems, where slopes are normally gentle, the streams and their sediment supplies are large in volume, and a stable long-term base level is more likely to have been achieved than farther upstream.

The dominant landform resulting from fluvial deposition is the floodplain. Floodplains, in turn, contain a number of important secondary features such as meander deposits, natural levees, deltas, and alluvial terraces.

Floodplains are broad, flat-floored valleys covered by alluvium and subject to flooding at times of high water. Floodplains comprise the floors of valleys originally formed by vertical erosion and are normally in the process of being slowly widened by lateral erosional processes. In one sense, then, the floodplain, like any stream valley, can be considered more a product of erosion than of deposition. On the other hand, the stream occupying a floodplain typically has reached its long-term base level and has essentially ceased downcutting. Furthermore, nearly all floodplains are covered with stream-deposited alluvium, and the small-scale features within the floodplain are alluvial and depositional in origin. A floodplain therefore is a depositional feature located within a larger erosional feature. Because the erosional processes that result in valley formation already have been discussed, the emphasis in this section is on the depositional aspects of floodplains.

Floodplain formation begins when a stream reaches its long-term base level of erosion and active downcutting ceases. This allows the valley floor to be gradually widened through the lateral gradational processes of weathering, mass wasting, and bank undercutting by the stream. During times of major floods, the stream temporarily erodes loose deposits and may actively scour the bedrock floor of its valley. When water levels subsequently fall and the stream loses its erosional capacity, it deposits a covering of alluvium over the width of the valley, bringing its surface back to base level.

One of the most important characteristics of a floodplain, from both a geomorphic and a human standpoint, is the fact that it is subject to occasional flooding. Streams flowing within floodplains are not normally incised into the surface, but instead flow close to the level of the surrounding land. During times when they carry large amounts of runoff, they can readily overtop their banks. As a result, overbank floods occur in a typical floodplain section about two years out of three. The relative flatness of the surface usually causes these floods to be extensive but rather shallow. Large quantities of sediments are transported by a stream during floods and are deposited in a layer over the surface, especially as the water recedes. Such deposits add valuable topsoil eroded from upstream portions of the drainage basin. The short-term economic hardships of floods for floodplain residents must therefore be weighed against the long-term benefits of naturally maintained soil fertility.

Extreme floods, which fortunately are quite rare on any given stream, may cover the floodplain so deeply and with such swiftly flowing water that erosion rather than deposition predominates. These floods help to maintain a long-term surface level equilibrium because they offset the slow buildup of sediments from "normal" floods by periodically flushing out the excess.

Many floodplains are covered by only a relatively thin veneer of alluvium, and streams retain the capability of temporarily exposing part or all of the underlying bedrock floor during extreme floods. Other floodplains, however, currently are covered so deeply by alluvium that bedrock exposure is impossible. In some cases, this has occurred because the floodplains are located in slowly subsiding tectonic basins. This is true, for instance, of the Mississippi, Amazon, and Ganges River Valleys. The slow subsidence of their valleys has caused these rivers to deposit large quantities of sediment in order to maintain a graded state. For example, the Mississippi River Valley near Vicksburg, Mississippi, contains an average of more than 45 meters (150 ft) of sediment, and in places sediment is twice as thick. Pleistocene sea level changes and human activities also have increased deposition in the floodplains of many rivers.

Because a stream occupying a floodplain flows near the surface in unconsolidated alluvium, it can shift its course much more readily than can a stream occupying a narrow erosional valley. Over long periods of time, then, the stream is likely to flow over all portions of its floodplain in turn (see Figure 17.18). On occasion, it will impinge on portions of the sidewalls of the floodplain, undercutting the bank and widening the floodplain. Fluvial features therefore are not restricted to the immediate vicinity of the present channel, but are distributed widely over the floodplain surface. Most of the modest surface relief found within floodplains is associated with these features.

When a stream flows through a region of unconsolidated and easily eroded materials like sand, the weakness of the banks causes the stream to become wide and shallow. Under such conditions, an excess of sediment will likely enter the stream, choking the channel and dividing it into a shifting series of deeper and shallower channel segments. The deepest and most actively flowing segments will consist of a number of interweaving channels separated by elongated **bars** of sand or gravel that are deposited at places where the speed of flow diminishes. During times of low water, these bars may protrude well above the water. Because they are composed of loose sediments and are located within the stream channel, they are highly vulnerable to erosion and are subject to constant and rapid fluctuations in size, shape, and position. A stream that exhibits a pattern of multiple interwoven channels separated by intervening bars is said to have a **braided flow pattern** (see Figure 17.19).

Meanders and Associated Features

The development and migration of broad looping bends, or **meanders,** in the courses of streams occupying floodplains is responsible for many floodplain features (see Figures 17.18 and 17.20). Stream meanders display a distinctive cycle of formation, growth, and disappearance. A meander begins as a slight bend in the channel, perhaps initiated by differing degrees of cohesiveness of the materials on the opposing stream banks or by the inflow of water from a tributary. The diversion of flow results in the erosion of the unconsolidated material on the outside of the bend, allowing it to become more pronounced (see Figure 17.21). As the bend increases in amplitude, centrifugal force becomes the primary factor furthering its growth by deflecting the main current toward the outer bank, where erosion occurs (see Figure 17.22). As it grows outward, the meander slowly migrates down-valley.

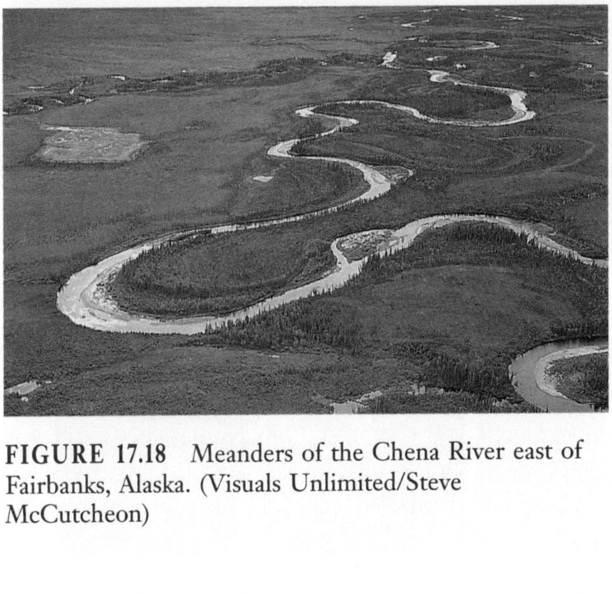

FIGURE 17.18 Meanders of the Chena River east of Fairbanks, Alaska. (Visuals Unlimited/Steve McCutcheon)

FIGURE 17.19 A semiarid climate and plentiful sand supply have caused Nebraska's Platte River to develop a braided flow pattern. (John S. Shelton)

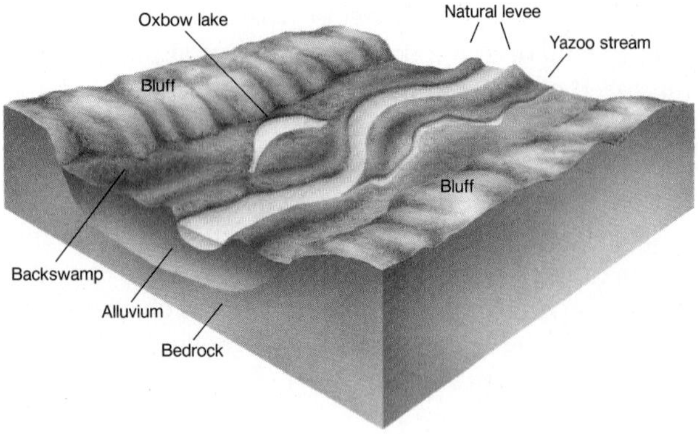

FIGURE 17.20 Some major floodplain features.

The reduction in the speed and depth of flow on the inner side of the developing meander causes the deposition of sediments, which form a series of **point bars.** Centrifugal force, acting on the current as it passes through the meander, causes the sides of the meander to converge gradually to form a narrowing **meander neck.** Eventually, the sides intersect to produce a **meander cutoff.** When this happens, the meander is abandoned by the active current as the river takes the shorter and steeper route through the cutoff.

The abandoned meander eventually is separated from the stream by bank deposits along its new course so that it becomes an arcuate lake called an **oxbow lake.**[1] Lakes are relatively ephemeral features on the landscape, and an oxbow lake gradually fills with fine sediments and organic matter to form first a swamp and eventually dry land. The meander scar may long be visible on the landscape, especially from the air, because its vegetation usually differs from that of its surroundings (see Figure 17.23).

Over long periods of time, the development and migration of meanders allows the stream to flow over all parts of its floodplain in turn (see Figure 17.24). Many floodplains have well-defined margins in the

[1] An oxbow, which gives this lake type its name, is the U-shaped part of an ox yoke that passes under and around the animal's neck and helps attach it to a cart or plow.

FIGURE 17.21 Stages in the "life cycle" of a stream meander. As centrifugal force causes the meander to grow outward, its neck is gradually constricted, leading to the eventual cutoff of the meander (View 3) and the formation of an oxbow lake.

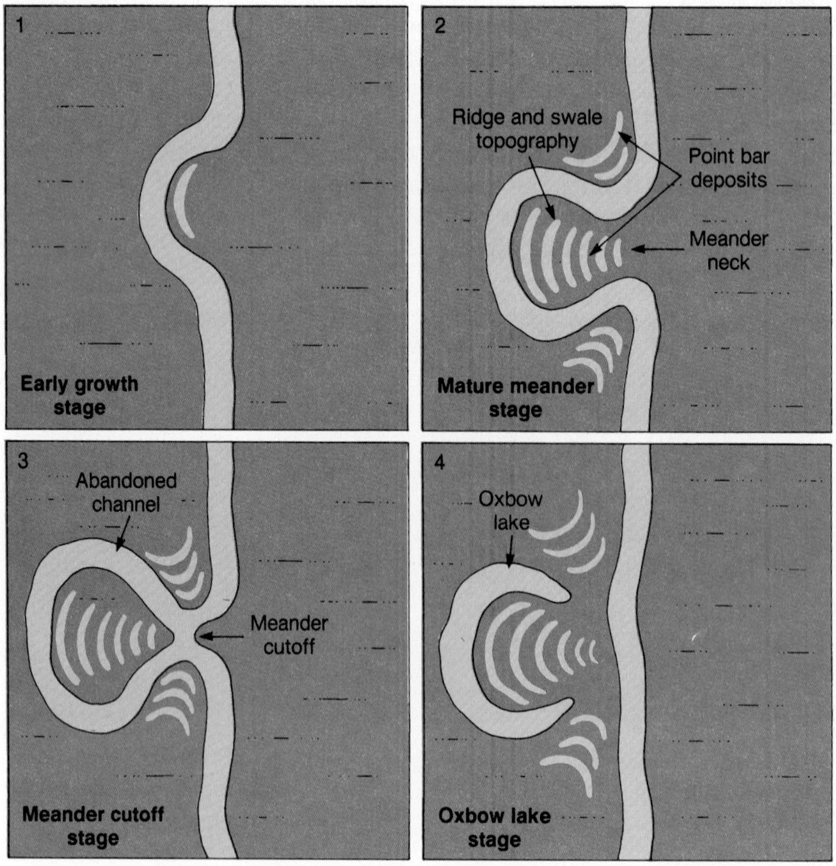

FIGURE 17.22 The growth of stream meanders tends to undercut, steepen and often to destabilize their outer banks, while deposition on the insides of the meanders leaves low-lying sand or gravel bars. (©JLM Visuals)

form of steep **bluffs** that rise abruptly above the flat valley floor.

Natural Levees

When a river within a floodplain spills out of its channel during a flood, the water outside the main channel is subjected to a sudden increase in friction due to the shallowness of the water and the irregular surface provided by submerged vegetation; its speed of flow consequently is diminished greatly. This reduction in flow reduces the water's sediment-transporting capacity and results in the deposition of the excess sediment on the floodplain surface. The coarsest materials, which have the largest settling velocities, sometimes are deposited atop the stream banks where they accumulate to form natural levees.

FIGURE 17.23 A variety of meander scars, as well as ridge and swale topography produced by the Tallahatchie River in Mississippi. Dark green areas are densely vegetated swamps occupying abandoned meanders. (USGS)

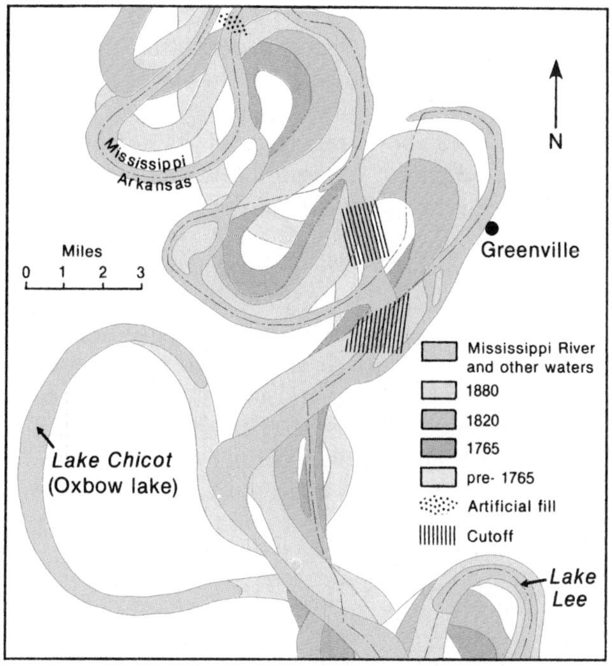

FIGURE 17.24 Map showing historical changes in the course of a portion of the Mississippi River on the Arkansas-Mississippi border. The boundary between these two states no longer follows the river's course because the channel has changed several times since the boundary was established in the early twentieth century.

Natural levees take the form of low, often broad ridges of fine sand and coarse silt that parallel each side of the stream channel. They are sometimes the highest natural features within the entire floodplain and, as such, increase channel stability by helping to keep the stream within its established channel. Natural levees along the lower Mississippi River, for example, reach heights of up to 5 meters (15 ft) above the main floodplain surface. If the river is aggrading its channel, the natural levees may enable the water surface to stand above the level of the surrounding floodplain. This is an unstable situation, because a breach in any portion of the levee will allow the river to abandon its channel and form a new channel at a lower level in another portion of the floodplain. Humans, who usually have a vested interest in keeping rivers within their existing channels, may reinforce natural levees in an attempt to avoid such breaches, which usually occur during floods (see the Case Study at the end of the chapter). In places the water level of China's Huang Ho River flows more than 10 meters (33 ft) above the level of its surrounding floodplain because it has long been kept from shifting its course by a combination of natural and artificial levees.

The tendency of rivers in floodplains to meander and to form natural levees, then, leads to the development of a variety of relatively short-lived features on the floodplain surface. In general, these features are most recently formed and abundant near the current river courses. Their presence gives the relatively flat floodplain floor a complex microrelief that shows up well on topographic maps employing small contour intervals. They are also often clearly visible on aerial photographs because of the vegetative patterns they produce (see Figure 17.23).

Deltas

Deltas are deposits of alluvium formed when streams enter standing water bodies. They exist on a wide variety of scales, from that of a small brook entering a pond to that of a giant river such as the Amazon or Mississippi entering the sea. Deposition takes place because the sudden cessation of flow that occurs as the stream enters the standing water body causes it to lose its sediment-carrying capacity. This causes all but dissolved sediments and clay-sized particles to settle quickly to the bottom.

The outward spread of sediments that produces the fanlike shape of a typical delta is the result of active deposition and a lack of confining channel banks. Deposition causes the stream to aggrade itself repeatedly out of its channel in order to shift to a lower route. This process eventually causes a fan-shaped deposit to be built up. Frequently, the stream splits into several **distributary channels** that divide the river into a number of smaller channel segments.

In general, deltas are proportional in size to the sizes and sediment loads of the streams that produce them. All of the world's major deltas are produced by large rivers that transport heavy sediment loads to the sea. Some large rivers, though, carry little sediment and have not formed deltas. For example, Canada's St. Lawrence River cannot obtain the necessary sediment load for delta formation because the Great Lakes, which it drains, act as settling basins. In contrast, many of the large rivers of Asia carry heavy sediment loads and have developed extensive deltas. The Ganges delta, with a surface area of 78,000 square kilometers (30,000 sq mi), is the world's largest.

Delta formation is also greatly facilitated if the standing water body that the river enters is shallow and has weak wave and current action. If the water is very deep, great quantities of sediment must accumulate before the deposits reach the surface to form the delta. This is one reason why few major deltas occur along the west coasts of North and South America. Similarly, if wave and current action are strong, stream deposits will be carried away or spread out along the coastline rather than accumulating in one place to produce a delta. For this reason, most of the best-developed deltas are formed in bays, partially enclosed seas, or other protected waters.

The term delta was taken from the Greek letter delta (Δ), which originally was used to describe the shape of the "classic" delta of the Nile River (see Figure 17.25). Other deltas, like those of the Ganges or Niger Rivers, are more fanlike in shape. A few deltas, most notably that of the Mississippi River, contain long fingerlike projections of land. These projections result from a combination of very weak currents that are unable to smooth off the outer perimeter of the delta and a slow subsidence of the surface that leaves only the crests of natural levees along the major delta channels above the water level. This form has been termed a **bird's-foot delta.** Finally, a number of deltas are located in semienclosed water bodies (**estuaries**) either in drowned river valleys or in submerged tectonic troughs. Examples of **estuarine deltas** include the deltas of the Yangtze, Susquehanna, and Colorado Rivers.

Deltas generally form seaward extensions of existing floodplains. Like floodplains, delta surfaces are low, flat, and often swampy and are composed of fine-textured sediments that normally contain a large proportion of organic matter. When adequately drained, they may form highly productive agricultural land and support dense populations.

Terraces

As noted earlier, the formation of a floodplain indicates that a stream has reached a stable base level and that downcutting is no longer significant. If the base level is lowered for any reason, a stream occupying a floodplain will renew downcutting and may eventually carve a new and narrower floodplain within the old floodplain surface (see Figure 17.26). The now

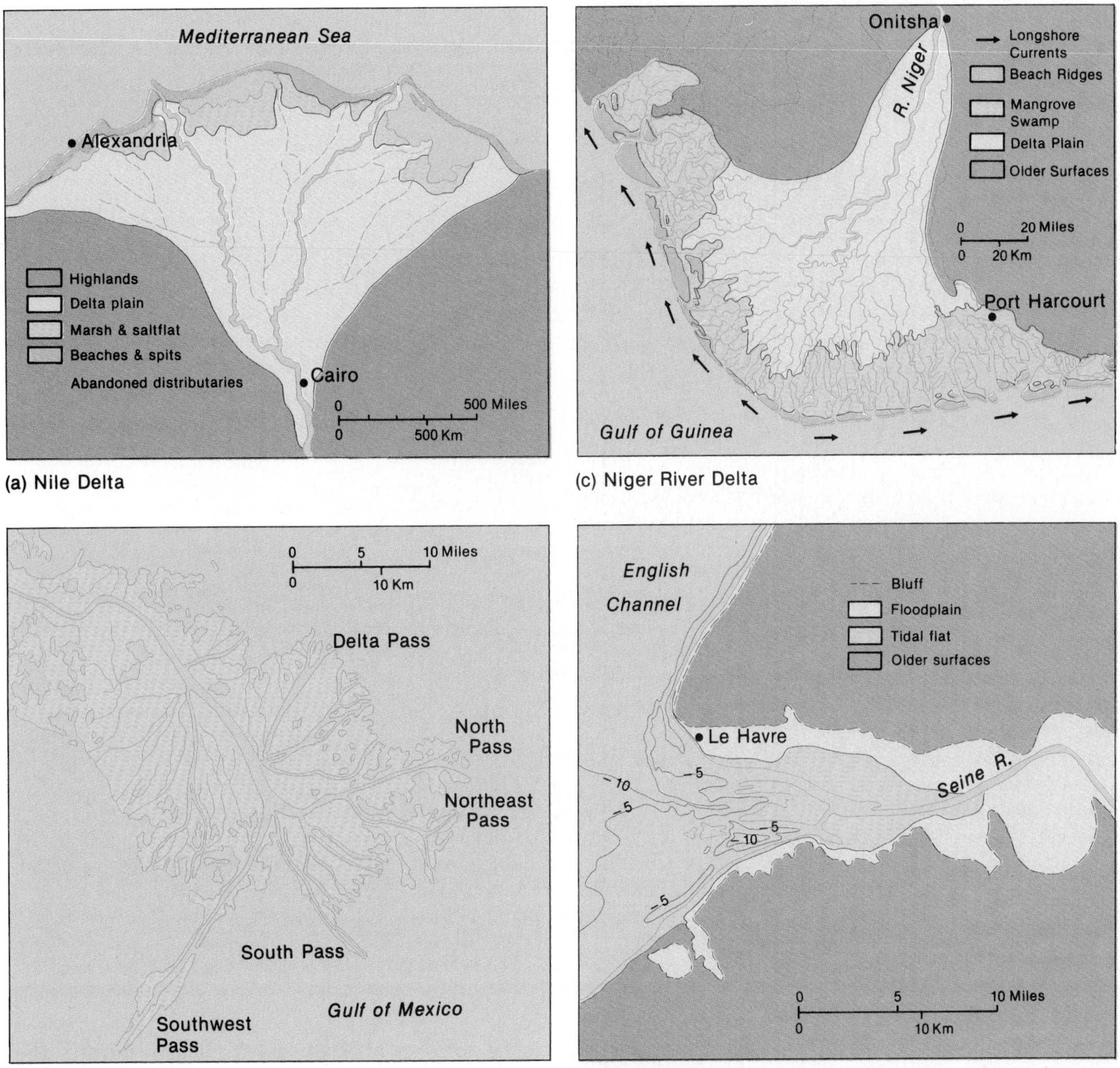

(a) Nile Delta

(c) Niger River Delta

(b) Mississippi Delta

(d) Seine Delta

FIGURE 17.25 Maps of four deltas with differing forms. Delta forms are controlled by such factors as the rate of sediment deposition, the bottom topography of the water body in which the delta is being formed, and the patterns of coastal currents. The Nile and Niger Rivers have produced "classic" deltas, while the Mississippi has produced a bird's-foot delta and the Seine an estuarine delta.

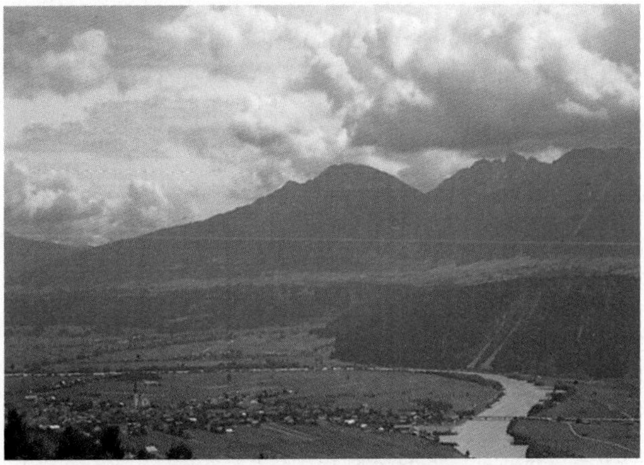

FIGURE 17.26 Fluvial terraces have formed along the margins of the Inn River Valley, in Austria, as tectonic uplift periodically caused the river to downcut into the floor of its floodplain. (Jerry Scott)

INTERNATIONAL BOUNDARY CHANGES ALONG THE RIO GRANDE

Rivers have long served an important political function as boundaries between nations. The tendency of some rivers to shift their courses, however, can cause major diplomatic problems for the countries involved. A good example is provided by the Rio Grande, which forms the international boundary between the United States and Mexico over a distance of roughly 1,600 kilometers (1,000 mi).

Soon after the 1848 Treaty of Guadalupe Hidalgo established the middle course of the Rio Grande as the international boundary, it became evident that the river occasionally experienced major changes in its course during floods. This raised a sticky political problem: Should these changes cause the international boundary to shift, or should the original boundary be maintained, leaving increasingly numerous enclaves of each country on the "wrong" side of the river?

Most changes in the Rio Grande's course during the nineteenth and early twentieth centuries were more of a nuisance than a major problem, because the lands involved were barren and unpopulated. In 1905, Mexico and the United States signed the Banco Treaty in which they agreed to exchange approximately equal parcels of land from time to time in an attempt to offset, as much as possible, shifts in the river's course.

A more serious situation had developed, however, in the densely populated area where the Rio Grande separated El Paso, Texas, from the neighboring Mexican city of Ciudad Júarez. Southward shifts of the river occurred in 1852 and again in 1864, placing tracts of Mexican land on the American side of the river. Beginning in the late 1800s, Mexico repeatedly requested that the lands involved be returned. The U. S. government's position at the time was that, since the shifts in the river's course resulted from natural processes of erosion and deposition, the lands involved now belonged to the United States.

In 1910, the United States finally agreed to settle the problem by arbitration, and a joint Mexican-American commission was appointed to decide the matter. The commission eventually decided that Mexican land lost by the 1852 shift should remain as part of the United States, but that the larger "El Camizal" tract lost in 1864 should be returned to Mexico. The United States refused for a long time to accept this finding, although it had originally agreed to abide by the commission's decision. The transfer of the land finally took place in 1963 as the result of an act signed by U. S. president John Kennedy and Mexican president Lopez Mateos.

elevated remnants of the original floodplain form **terraces** that usually end abruptly in steep slopes well back from the current river channel. In most cases a pair of terraces will be present, one on each side of the valley. The height of the terraces depends upon the extent to which base level has been lowered. The width of the new floodplain, and therefore the distance that separates the opposing terraces, is related to the amount of time that has elapsed since the river eroded to its new base level. If the new level is maintained long enough, lateral erosion eventually will widen the new level sufficiently to destroy the terraces. In some cases, multiple terrace levels exist within a valley. This indicates that the base level is being sporadically lowered, with intervening periods of stability.

The formation of alluvial terraces illustrates the complexity of some fluvial landform features and the problems involved with categorizing individual fluvial landforms as either erosional or depositional. Alluvial terraces form only after fluvial erosion has first produced a valley, then deposition has covered its floor with alluvium, and finally one or more episodes of erosion have partially removed the alluvial fill. Purely erosional **bedrock terraces** also can be produced by sporadic river downcutting in a valley not floored with alluvium.

The primary cause of terrace formation is an increase in the gradient of the stream course. This can be accomplished by tectonic or isostatic uplift of the surface or by a decline in sea level. River terraces are

CHANGES IN FLUVIAL LANDSCAPES THROUGH TIME

Streams and their surrounding landscapes can undergo profound changes through time. For instance, the geological structure of the Earth's shield areas and of the Piedmont of the eastern United States indicates that these regions once contained lofty mountain ranges. Today, however, they exhibit a rather subdued topography and in places contain broad, gently sloping river valleys.

The idea that the topographic features of a region could be used to determine its geomorphic age was first suggested by the famous American geographer William Morris Davis in the late nineteenth century. Davis saw an analogy between the progressive changes in the fluvial landscape with time and the changing physical features of human beings with increasing age. He developed an elaborate fluvial "cycle of erosion" that classified streams as being in youth, maturity, or old age on the basis of their flow character-

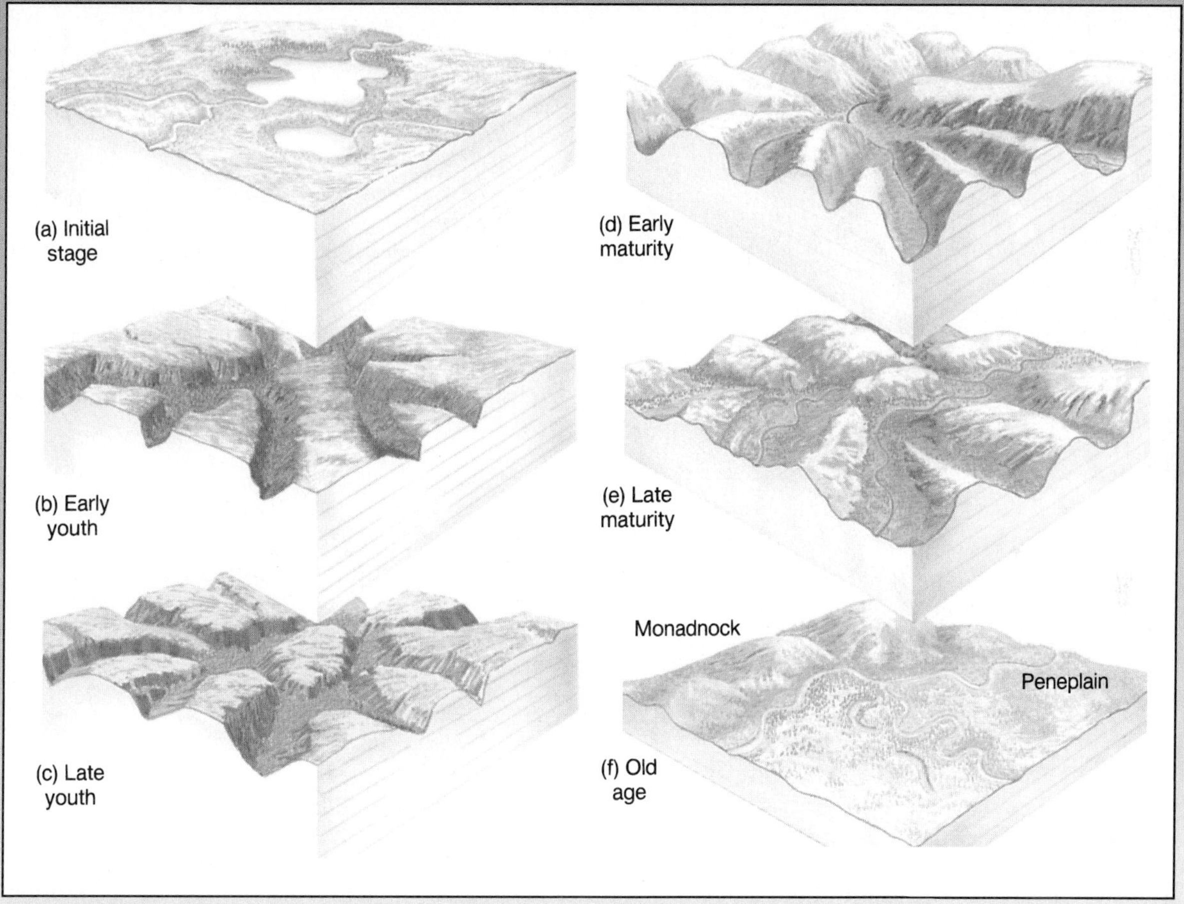

FIGURE 17.27 Davis's fluvial "cycle of erosion."

continued on next page

(a)

(b)

(c)

FIGURE 17.28 These three photos provide examples of landscapes in different stages of Davis's fluvial cycle of erosion.
(a) Youth. The narrow gorge of the Colorado River is incised into the Colorado Plateau upstream from the Grand Canyon.
(b) Maturity. The maturely dissected topography of the southern Appalachians in western North Carolina.
(c) Old Age. A resistant granite monadnock on a peneplain; Stone Mountain, Georgia. (a: John S. Shelton; b: Visuals Unlimited/©Joel Arrington; c: © JLM Visuals)

istics and the character of the surrounding landscape (see Figure 17.27). The validity of this cycle has been hotly debated ever since its initial publication.

DAVIS'S INITIAL ASSUMPTION

In order to trace the fluvial cycle of erosion from the earliest possible point, Davis had to begin with a surface that was totally unaffected by stream action. He therefore envisioned an irregular seafloor being uplifted rapidly by tectonic forces to a considerable elevation. Erosion of this surface by flowing water, of course, begins with the first rainfall. Overland flow occurs initially, but very quickly a drainage network starts to form.

YOUTH

During youth, the surface still retains many characteristics of its previous seafloor setting. Early in the youthful period, streams are highly ungraded. Surface irregularities cause the stream courses to contain lakes or swamps, rapids, and even waterfalls. Stream gradients, although very irregular, are on the average quite steep. This gives the streams the ability to downcut at a relatively rapid rate.

By middle youth, the major streams in the drainage network have formed deep, narrow, and steep-sided gorges (see Figure 17.28a). Lakes and waterfalls gradually disappear, and rapids become less numerous as the stream courses are gradually smoothed by erosion and deposition. The entire uplifted surface increases in ruggedness and relief as gorges are incised into it by the vigorously flowing streams.

Late in the youthful period, the major streams approach their long-term base level of erosion, and the rate of downcutting tapers off considerably. This allows the walls of the gorges to be worn back to less steep angles by lateral erosional forces.

MATURITY

Davis considered a stream mature when it reaches long-term base level and becomes graded. This achievement marks a major change in the geomorphic activity of the stream. Until this time it has had excess energy and has been involved primarily in downward incision. From maturity onward, it remains essentially stable in elevation, deposits as well as erodes sediments, and is involved chiefly in valley widening rather than deepening.

With the stabilization of valley bottom levels, floodplain formation begins. At this point, the relief of the surrounding landscape is at a maximum. Nearly the entire surface is in slope (see Figure 17.28b), and the last remnants of the original upland surface are being consumed by erosion. As the mature stage progresses, the river floodplain gradually widens, and various floodplain features such as meanders and natural levees begin to develop. The widening floodplain reduces the areal extent of the remaining upland surface as its elevation and ruggedness gradually diminish.

OLD AGE

By the time it has reached old age, the river flows sluggishly through a broad floodplain that may be covered by a considerable depth of alluvium. A variety of floodplain features develop and are erased in constantly changing patterns. The combination of long-continued weathering, fluvial erosion, and other gradational processes has converted the surrounding region to a **peneplain,** or a plain formed by the long-continued erosion of an upland surface. A few isolated highland remnants, called **monadnocks,** may survive as the last vestiges of the former uplands (see Figure 17.28c), but in time even they are destroyed by the inexorable progress of erosion. This leaves an essentially timeless landscape characterized by a low surface relief and slow-flowing, freely meandering streams.

REJUVENATION

Unlike a human being, a fluvial landscape can remain in old age indefinitely. On the other hand, old age or any of the earlier developmental stages can at any time be interrupted by surface uplift or a fall in sea levels. Such an event will result in a reduction in base level and a **rejuvenation** of the stream system. The increased elevations will again initiate a period of rapid downcutting, resulting in the return of the stream system to a youthful condition. If the change in elevation is of limited extent, the chief topographic result will be the formation of terraces. If a major uplift occurs, the streams, assuming they maintain their original courses, will incise themselves well below their former surface level. If this occurs to a meandering stream, the increase in river flow speed will result in the development of **entrenched meanders** (see the chapter opener photo on page 433). Rejuvenation will cause the stream system and the surrounding landscape to pass through the life cycle a second time, and any later uplifts will produce additional rejuvenations of the system.

VALIDITY OF THE CYCLE OF EROSION

As already noted, the accuracy and even the basic validity of Davis's cycle of erosion have long been debated among geomorphologists. Probably the chief point in favor of the concept stems from the simple fact that streams transport sediments in only one direction—downhill. In the absence of uplift, therefore, the landscape *must* eventually be worn down. From what we know of fluvial erosion processes, Davis's basic erosional sequence of initial vertical incision to a base level elevation followed by the lateral expansion of the surface at this elevation seems logical. Landscapes apparently representing all the stages of Davis's cycle have been identified.

Although there have been many criticisms of the cycle of erosion concept, the most fundamental is the lengthy geological timespan that would be necessary for the cycle to be completed. Most geomorphologists believe that the Earth is far too active tectonically to allow the cycle to run its entire course. They therefore maintain that the concept is of only theoretical applicability. Many geomorphologists also contend that streams can quickly adjust their courses by means of changes in their gradient

continued on next page

or flow pattern. This factor allows streams to remain in **dynamic equilibrium** with respect to changes in their environment.

It is likely that both the cyclical and the dynamic equilibrium concepts of fluvial landscape development possess validity, but over differing magnitudes of geological time. More research is needed to improve our understanding of the entire process of long-term landscape evolution. Regardless of the final verdict on the merits of Davis's fluvial cycle, it has played a key role in our understanding of geomorphic processes by stressing that streams and their surrounding landscapes do not display a random assortment of features. Rather, landforms are organized into a logical system that progresses toward an ultimate state of equilibrium with the existing set of environmental conditions.

common in areas undergoing tectonic uplift; this includes portions of California and the Alps. Another important cause of terrace formation is a reduction in the sediment load supplied to a stream. During the waning stages of the Pleistocene epoch, for example, streams draining the melting ice sheets of North America and Eurasia received tremendous influxes of glacial sediments. Because they were unable to transport all of this load, they deposited much of it in their valleys to form thick alluvial fills. When the glaciers disappeared, the sediment supply was greatly reduced, and the rivers actively eroded into their alluvial deposits, leaving the remnants standing as terraces well above the present valley floors.

River terraces have become favored settlement sites because of their relatively flat surfaces, fertile alluvial soils, and proximity to water. In addition, their high-standing position makes them immune to the floods that periodically inundate the lower floodplain.

SUMMARY

Streams perform two vital natural functions. The first is the removal of excess water from the land. The second is the removal of surface weathering products supplied to streams largely from adjacent slopes by mass wasting processes. Because of the gradational power and the nearly global distribution of streams, most of the Earth's land surface is dominated by fluvial landform features.

We return now to the Focus Questions:

1. *How do fluvial drainage systems develop, and what geographical patterns do they exhibit?*

Streams develop in places where there is a sufficient downhill flow of water to erode a channel. Each stream channel conducts water from a drainage basin, and the coalescence of stream channels forms a drainage system. Drainage systems display spatial patterns of organization dictated largely by existing patterns of rock types and structures as well as surface features. The dendritic pattern, the most widespread type, develops on homogeneous surfaces that exert no strong topographic or structural control on the locations of the stream segments. The trellis pattern develops where rock strata of differing resistance to erosion alternate. It consists of a parallel pattern of larger streams joined at nearly right angles by short tributaries. The rectangular pattern usually develops in areas where crystalline rocks contain intersecting fracture systems. The streams follow these lines of weakness and therefore contain linear segments joined by sudden sharp-angled bends. The radial pattern is a diverging flow of streams from a central upland, while the reverse centripetal pattern is a converging flow into an enclosed basin. The deranged pattern exhibits a high degree of disorganization, with the stream segments changing direction frequently, and often containing lakes, swamps, and rapids.

2. *How do streams erode, and what are the most important landforms produced by stream erosion?*

Streams erode by the processes of solution, hydraulic action, and abrasion. All rock materials are soluble to some degree, and on average about 30 percent of the sediment load of streams is carried in solution. Hydraulic action resulting from stresses generated by the force of the flowing water itself is highly effective in eroding the unconsolidated materials that form the banks and beds of most streams. Abrasion, the scraping action of stream-borne particles, can slowly erode solid rock. Once sediments have been entrained into the flow of a stream, they continue to be transported as long as the turbulence generated by the flow keeps them from settling to the streambed.

The primary fluvial erosional feature is the valley. A stream with excess erosive energy will downcut rapidly, producing a deep, narrow valley. Downcutting, however, reduces the slope angle and therefore the gravitational energy available to the stream. This eventually causes valley-widening processes to dominate in tectonically inactive areas. The fluvial erosion of uplifted blocks of land also carves them into hills and mountains. The topographic pattern displayed by these upland features largely depends upon the drainage pattern that develops, while their relief is determined by the rate of downcutting.

3. *Why do streams deposit sediments, and what are the most important landforms produced by stream deposition?*

A stream deposits sediments when a reduction in its speed or volume of flow, or a change in the geometry of its bed, reduces the turbulence of the flowing water to the point that it can no longer transport its entire sediment load. When this occurs, the coarsest sediments are deposited first, followed by progressively finer sediments as the stream's carrying capacity diminishes. As a result, fluvial sediments are sorted, or separated by size. The basic fluvial depositional feature is the floodplain, which is a broad, flat-floored valley covered by alluvium and subject to occasional flooding. Floodplains exhibit a variety of depositional features formed largely by lateral stream migrations. These include bars, natural levees, and various meander features such as oxbow lakes and meander scars. When a stream deposits sediments in a standing water body, a delta often forms. Deltas vary greatly in size and shape, but most display a fan-like pattern caused by shifts in the stream's course through time as well the formation of a branching pattern of distributary channels.

When a stream occupying a floodplain renews downcutting, it incises itself into the floor of the floodplain. The remnants of the original floodplain often are preserved along the sides of the valley as terraces. Some streams have formed multiple terraces at varying levels above the present valley floor.

Review Questions

1. What parts of the world are, and are not, dominated by fluvial landforms?

2. Describe the process by which gullies form on a slope that has been cleared of vegetation. How is this process likely to be different if the slope has a dense cover of vegetation?

3. Briefly describe the likely general appearance of the surrounding landscapes of areas that have each of the following drainage pattern types: radial, trellis, deranged, dendritic, annular.

4. How do streams erode loose sediments? How do streams erode solid bedrock?

5. Describe each of the three methods by which streams transport sediments. Which one moves the most sediments worldwide? Which moves the least?

6. What causes streams to deposit sediments? Describe the physical appearance of stream deposits.

7. What is a graded stream? How can the stream's longitudinal profile indicate whether it is graded?

8. Define the concept of base level. What is the difference between short-term and long-term base level? Explain how the concepts of base level and grade are related.

9. What is the major landform feature produced by fluvial erosion? Describe the characteristics of such a feature in which vertical erosion is occurring rapidly.

10. What is a floodplain? How does it differ from an erosional valley? Describe several important floodplain features.

11. How do stream meanders form? Describe the meander development cycle that eventually results in the production of an oxbow lake.

12. Explain why deltas form. Why are bird's-foot deltas like that of the Mississippi River relatively rare?

13. What does the development of a series of alluvial terraces within a valley indicate about the recent geomorphic history of the valley? Explain.

14. Discuss the differences between the "cycle of erosion" and the "dynamic equilibrium" theories of stream development.

Key Terms

fluvial
interfluve
infiltration capacity
drainage system
stream order
dendritic pattern
trellis pattern
rectangular pattern
radial pattern
deranged pattern
critical erosion velocity
hydraulic action
abrasion
suspended load
bed load
meanders
graded stream
base level
floodplain
braided flow pattern
point bars
oxbow lake
natural levee
delta
terrace

CASE STUDY

WILL THE MISSISSIPPI CHANGE ITS COURSE?

Humans have been altering the flow of rivers for thousands of years in an effort to increase their usefulness and to reduce the danger of flooding. Most early alterations involved small-scale diversions of flow for irrigation, but modern technology has expanded greatly the magnitude and variety of alterations that can be undertaken. A few examples of modern alterations of rivers include channel straightening and the construction of dams, reservoirs, locks, and levees.

Because all the components of the environment are complexly interrelated, purposeful human alterations occasionally produce unfavorable and unforeseen consequences. It would seem that our technological abilities sometimes surpass our understanding of the more subtle interactions among the Earth's physical systems. The history of control efforts of the Mississippi River provides a good example of how desirable human adjustments to the flow of a major river have produced a highly unstable situation that offers the prospect of eventual disaster for much of southern Louisiana.

The Mississippi River system, one of the world's largest and most important, drains 41 percent of the United States as well as a portion of Canada. Its drainage basin is exceeded in size only by those of the Amazon, Nile, and Congo Rivers. The heavily farmed Mississippi Basin is the agricultural heartland of the nation, and the river system itself provides the means for transporting a tremendous volume of agricultural, industrial, and petrochemical products.

The task of developing and controlling the Mississippi River has been assigned largely to the U. S. Army Corps of Engineers. The two major responsibilities of the Corps have been to reduce the danger of destructive floods and to develop the system's navigation potential. On the whole, they have been very effective in accomplishing these goals. The Corps maintains the river throughout its length by means of a complex system of locks, dams, diversion channels, and artificial levees. Channels have been straightened and shortened in various places, and some channels even have been paved in order to increase their stability. The environmental consequences of all these activities have been modest in the upper portion of the drainage basin, but are far more profound in the Lower Valley, where deposition predominates.

The Lower Mississippi is an alluvial river. It transports 58 million kilograms (64,000 tons) of sediment per day. Some of this material is deposited within the river channel, but most is carried into the Gulf of Mexico. Accelerated erosion from upstream agricultural activities has augmented the sediment load so that it currently is twice as large as it was 50 years ago. In the past, the Mississippi repeatedly changed its course due to channel aggradation, and the deposits thus were spread over a large area to form an extensive delta. Because of the artificial stabilization of its channel, however, widespread sediment distribution can no longer occur.

The major problem resulting from channel stabilization is that the present channel of the Mississippi has aggraded itself well above the level of the surrounding floodplain. This has caused the river to display an increasing tendency to

continued on next page

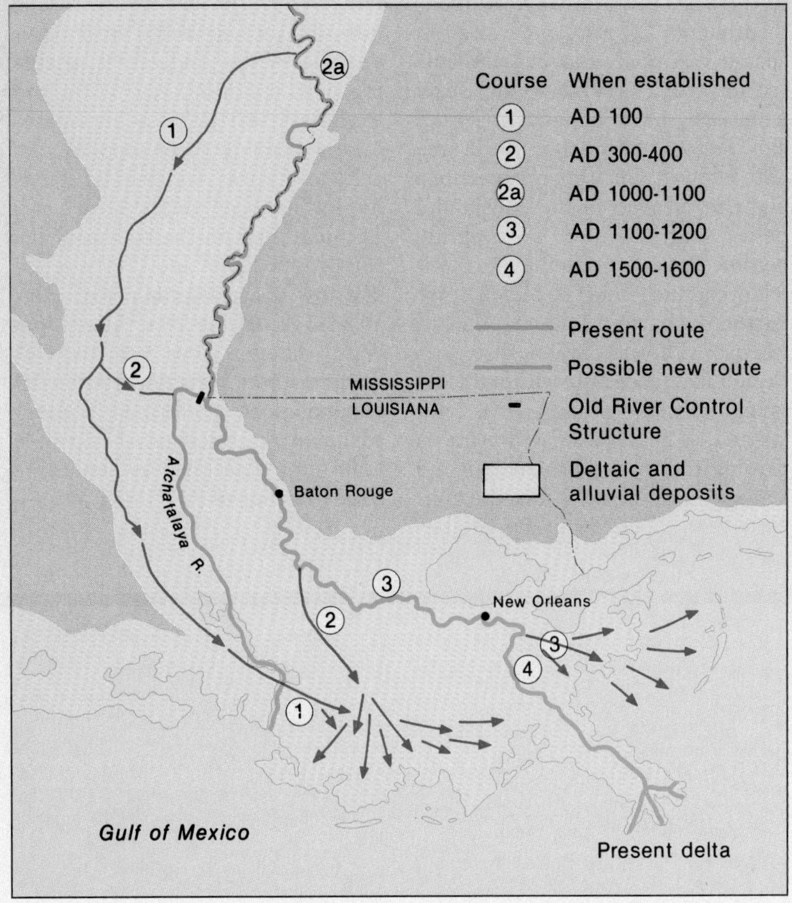

FIGURE 17.29 Map showing previous channels (in red) and the present channel of the Mississippi River. The location of the Old River Control Structure (ORCS) is indicated, as well as the possible new route of the river down the Atchafalaya.

change its course to a more westerly route that is lower in elevation and shorter in its path to the Gulf. Continuing alluviation within the present channel, which is occurring at a substantial rate because of the large influx of sediment from upstream agricultural sources, is building the river ever higher. This continues to make the situation progressively more unstable and necessitates the constant heightening and strengthening of the levees along the river's current channel.

The site at which the Mississippi River would abandon its present course if it had its own way is known. It is located at the juncture of the Mississippi with the Red and Atchafalaya Rivers near the southwestern corner of the state of Mississippi about 80 kilometers (50 mi) northwest of Baton Rouge (see Figure 17.29). The Atchafalaya is a distributary river that carries excess water from the Mississippi southward to the Gulf along a route located well to the west of the course of the Mississippi River itself. In 1839, a logjam on the upper Atchafalaya was cleared in order to open this river to navigation. The resulting increase in flow caused the Atchafalaya to become gradually deeper and wider and slowly to divert more and more water from the Mississippi. During this century the process accelerated, and in 1953 the Corps of Engineers notified Congress that, unless stopped, the Mississippi's flow would be completely diverted into the Atchafalaya by 1990. The consequences of such a diversion would be enormous. Baton Rouge, New Orleans, and many other cities and towns along a more than 480-kilometer (300 mi) stretch of the Mississippi would be left high and dry, while towns along the Atchafalaya—notably Morgan City—would be flooded out of existence.

In order to prevent such an occurrence, Congress hurriedly authorized the Old River Control Project at an initial cost of $70 million. This project had as its goal the regulation of the flow of water from the Mississippi into the Atchafalaya through a system of 84 floodgates. The project was completed in 1963 and seemed at first to be highly successful. In 1973, however, a major flood caused the collapse of part of the structure and nearly destroyed it. At present, such a state of imbalance has developed between the two channels that the undermining of the Old River Control Structure

continued on next page

(ORCS) by scour has become a very real danger. Three times each week, several hundred water depth soundings are taken in order to ascertain that no new scour channels are forming, and a backup control structure is being built. Even if a future major flood does not destroy the ORCS and produce a permanent shift in the Mississippi's course, floodwaters would be diverted down the Atchafalaya in order to save Baton Rouge and New Orleans. This places the future of rapidly growing Morgan City in grave doubt.

Another major problem also has been created by the tight control of the course of the Mississippi River. In the past, an extensive network of distributary rivers and swamps spread vast quantities of sediments over a broad area of southeastern Louisiana. These deposits offset the slow tectonic subsidence of the area and allowed for the maintenance and slow growth of the Mississippi delta. With the flow now tightly confined to a 1.6-kilometer (1 mi) wide channel, most sediments are currently shot out into the Gulf of Mexico. The deposition that does occur takes place only within the existing channel where it is not wanted. The lack of sediment is causing the slow subsidence of the remainder of the delta into the gulf. Each year, about 130 square kilometers (50 mi^2) of land are lost. The intrusion of salt water into other portions of the gradually submerging delta is killing vegetation and freshwater fish and is polluting domestic drinking water supplies.

Nature's price for channel stability thus has been the threat of a catastrophic shift in the course of the Mississippi River and the progressive loss of large amounts of valuable land from the lower Mississippi delta. These events provide eloquent testimony that environmental manipulations performed without adequate knowledge of possible long-term consequences may come back to haunt us.

CHAPTER 18

GLACIAL AND PERIGLACIAL LANDFORMS

OUTLINE

FOCUS QUESTIONS

1. What are glaciers, and why do they form?
2. What major categories of glaciers exist, and where are they found?
3. In what ways do glaciers modify the landscape?
4. What was the Pleistocene, and why did it occur?

Ice is the most powerful of the tools of erosion. Its gradational effect on the Earth as a whole, though, is greatly limited at present by its restricted areal extent. Ice currently covers only a tenth of the world's land area, and it is only there that it can directly influence the surface. During the very recent geologic past, however, in the epoch called the **Pleistocene** (sometimes referred to as the "Great Ice Age"), the global coverage of ice was as much as three times that of today. The effects of the ice on the landscape were profound, and the Pleistocene glaciation occurred so recently that, as yet, many landforms have been modified relatively little. The importance of ice as a geomorphic agent therefore is much greater than its current global distribution would indicate, and its significance is increased by the possibility that some day the glaciers may form again.

In this chapter, we examine the geomorphic influences of ice on the land and discuss the surface characteristics of landscapes dominated or strongly modified by glacial and ground ice. The chapter is divided into four major sections. The first provides an overview of glacial geomorphology and stresses the distribution, mechanics, and surface features of glaciers. The second section covers the subject of mountain or alpine glaciers. The third section examines continental glaciers as well as the causes and global impact of the Pleistocene. Finally, the periglacial environment, where topographic features are associated largely with the presence of ice in the underlying soil and rock, is discussed briefly.

GLACIER FORMATION, DISTRIBUTION, AND FEATURES

A **glacier** is a mass of freshwater ice, formed on land, that is or has been in motion. Glaciers, like rivers, move largely in response to gravity and perform the basic hydrologic function of removing excess precipitation from the land. The chief source of glacial ice is compacted snow, and in order for a glacier to develop, the accumulation of snow must exceed losses by melting and sublimation over an extended period. These losses are collectively referred to as **ablation**. Therefore, not only cold temperatures, but also significant snowfalls are required for glacier formation. Because of a lack of snowfall, some very cold but dry areas, such as parts of Alaska and much of northeastern Asia, apparently have never been glaciated.

When snowfall accumulation exceeds ablation over a prolonged period, the accumulated excess is converted gradually to glacial ice. A fresh snowfall contains mostly air and has a specific gravity of approximately 0.10, a tenth that of water. After a short period, it changes form to a much more compact granular state called **firn** that has a specific gravity of 0.55 or more. Further compaction, melting, and refreezing result in the expulsion of the remaining air and a slow conversion to glacial ice, with a specific gravity of about 0.90.

Present Extent of Glaciation

Glaciers are divided into two general categories on the basis of location, size, and shape. **Continental glaciers**, as the name implies, cover extensive land areas and form in the high latitudes. **Alpine glaciers**, also called "mountain glaciers," form at high elevations in mountainous regions.

The great majority of the Earth's currently glaciated surface is covered by the two surviving continental glaciers, or ice sheets—the Antarctic and Greenland ice sheets (see Figures 18.1 and 18.2). The Antarctic ice sheet is by far the larger; it covers 12.5 million square kilometers (4.8 million mi^2), an area larger than the United States and Mexico combined. It has a volume of approximately 24 million cubic kilometers, (5,750,000 mi^3), an average thickness of about 2,160 meters (7,090 ft), and a maximum thickness of 4,000 meters (13,000 feet or 2.5 miles).

The Greenland ice sheet is considerably smaller, with an area of 1.8 million square kilometers (696,000 mi^2), a volume of 2.3 million cubic kilometers (550,000 mi^3), and a maximum thickness of 3,350 meters (11,000 ft). These two ice sheets, if melted, would release enough water to supply all the world's rivers at their present discharge rates for nearly 900 years and, in doing so, would raise sea levels at least 45 meters (150 ft).

In contrast to continental glaciers, alpine glaciers are of limited areal extent, with a total area of approximately 508,000 square kilometers (196,000 mi^2) and a combined volume of 80,000 cubic kilometers (19,200 mi^3). Several thousand are scattered throughout the higher mountain ranges of the world. The more significant of these glaciers exist in Alaska, the Canadian Rockies, the Canadian Arctic Islands, the Andes, the European Alps, and several mountain ranges of Asia, including the Himalayas, Karakoram,

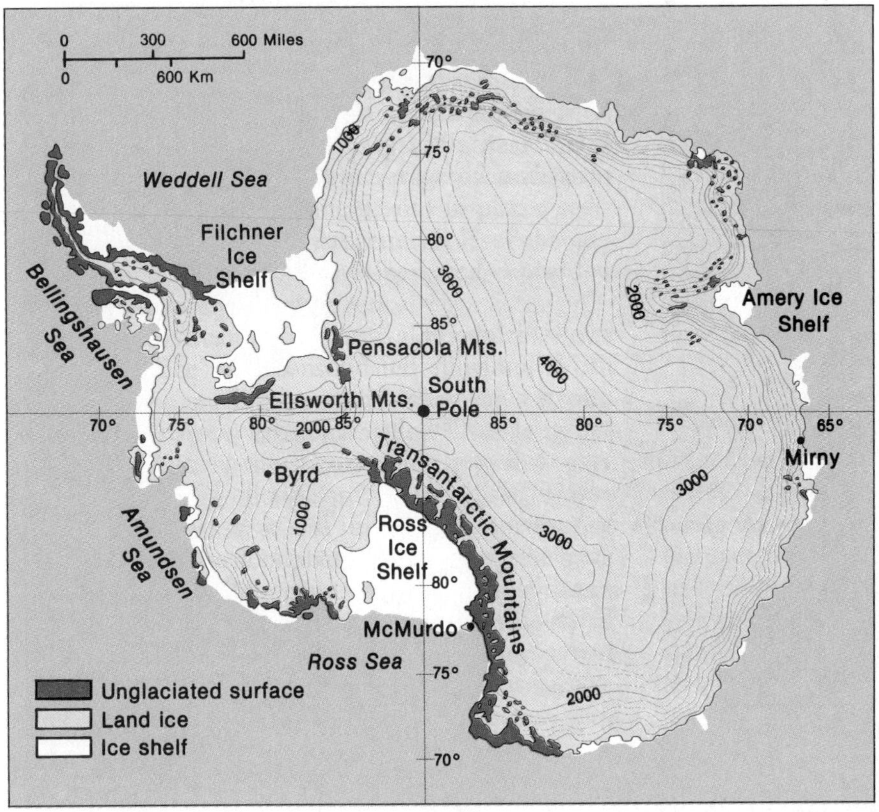

FIGURE 18.1 The Antarctic ice sheet. Antarctica is almost completely buried by an ice sheet that averages over 2,100 meters (7,000 ft) thick. It reaches a maximum thickness of 4,000 meters (13,000 ft) at a location about 480 kilometers (300 mi) inland from the Soviet base of Mirny. Contour lines give surface elevations in meters.

Tien Shan, and Caucasus (see Table 18.1). A few small glaciers are found in the southern U. S. Rockies, the Cascades, and in East Africa. Australia is the only continent with no glaciers at all.

Glacier Motion

Glacier motion is caused by the force of gravity, which compacts, deforms, and may locally melt the ice as it is drawn constantly downward. Unlike streams, the motion of a glacier is not always downhill; rather it is down-ice—that is, in the direction that the ice surface slopes.

In discussing glacial motion, we must distinguish between absolute and relative motion. The **absolute motion** is the actual forward movement of the particles of ice in the glacier. This motion nearly always is imperceptible and usually totals less than 300 meters (1,000 ft) per year. Stagnant or "dead" glaciers, located in areas where rapid losses by ablation are occurring, do not move at all, but decay in place. The absolute motion of a glacier may be far from steady. Most glaciers experience surges, or sudden

TABLE 18.1 Present-Day Ice-Covered Areas (km^2)

Antarctic	12,588,000
Greenland	1,802,600
Northeastern Canada	153,200
Central Asian ranges	115,000
Spitsbergen group	58,000
Soviet Arctic islands	55,700
Alaska	51,500
South American ranges	26,500
West Canadian ranges	24,900
Iceland	12,200
Scandinavia	3,800
Alps	3,600
Caucasus	1,800
New Zealand	1,000
USA (excluding Alaska)	500
Others	about 100

Total volume of present ice: 28 to 35 million km^3

SOURCE: C. Embleton and C. A. King, *Glacial Geomorphology*, (New York: Halsted Press, 1975), p. 25.

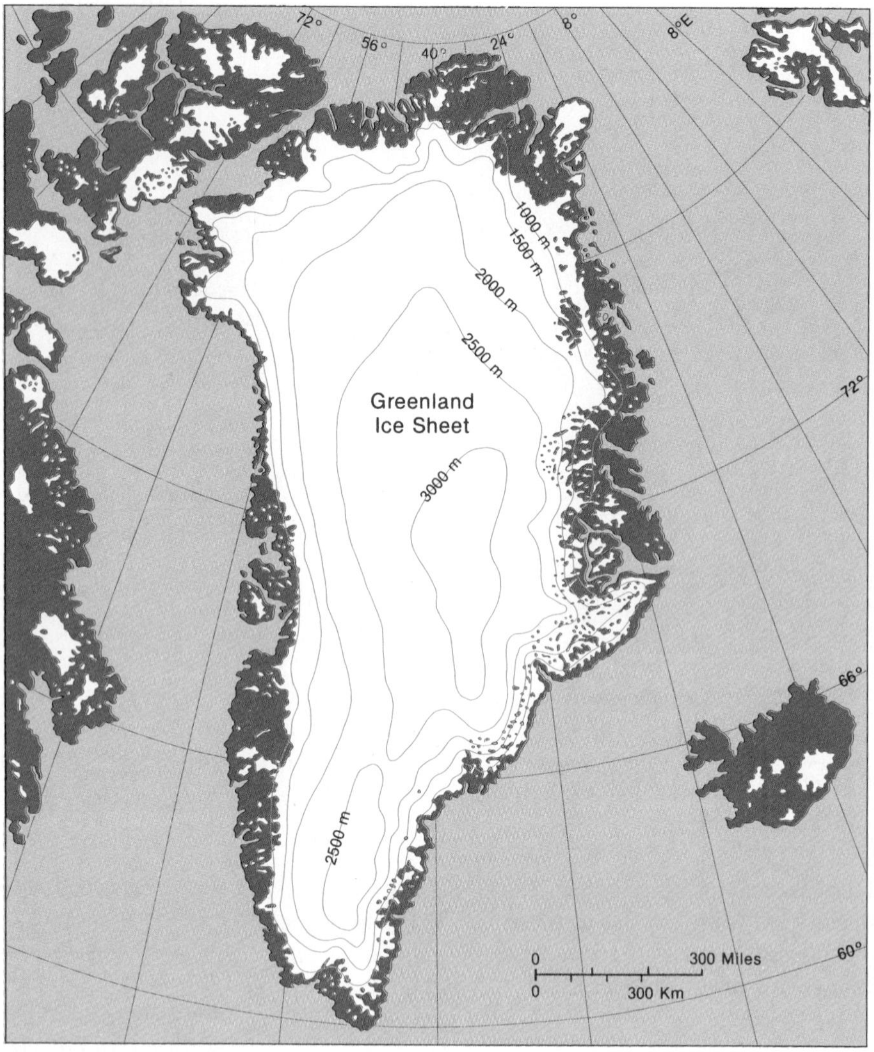

FIGURE 18.2 The Greenland ice sheet covers the interior of the world's largest island and has a maximum thickness of approximately 3,350 meters (11,000 ft).

increases in their rates of advance. The causes of these surges are not always apparent, but are sometimes associated with one or more seasons of heavy snowfall in the ice accumulation zone.

The **relative motion** of a glacier refers to the advance or retreat of the ice front. It is determined by the relationship between the absolute motion of the glacial ice and its rate of ablation. A glacier will advance if its absolute motion exceeds the ablation rate, will retreat if its ablation rate exceeds the absolute motion, and will remain stationary if these two factors are equal. Most glaciers have been retreating over the past few centuries, apparently as a result of a gradual global warming trend, even though the ice itself is still advancing in an absolute sense. Glacier motions are watched carefully, not only because of their geomorphic significance, but also because they are considered sensitive climatic indicators.

Every glacier has a **zone of accumulation** and a **zone of ablation**. If the glacier is to maintain a constant mass, the formation of new ice in the former zone must balance the losses in the latter. Separating the two zones is a so-called **equilibrium line**, where gains and losses are exactly equal (see Figure 18.3). In general, a size correlation exists between the zones of accumulation and ablation, so that larger glaciers, with more extensive zones of accumulation, typically extend to lower elevations than do smaller neighboring glaciers.

The precise mechanisms of glacial movement are complex, but may be divided into two major components: basal sliding and internal flow (see Figure

18.4). At any point on its surface, the absolute motion of a glacier is the sum of the velocities of basal sliding and internal flow.

Basal sliding involves a true sliding on the bedrock surface underlying the glacier. Where the bedrock floor is uneven, a process termed **regelation** makes ice motion possible. In order to envision this process, imagine a securely anchored rock outcrop, a few meters high, protruding from the bedrock floor into the ice. As the ice is pushed toward the outcrop, the pressure generated lowers the melting point of the ice, so that it melts and the water flows around the obstruction. As it reaches the lee side, the decreased pressure raises the melting point, so that refreezing, or regelation, occurs. Basal sliding, which is normally considered to include the regelation process, occurs only with **warm ice glaciers**, which have an underlying thin film of meltwater acting as a lubricant to aid slippage. Most alpine glaciers are of this type.

The second component of glacier motion, internal flow, is believed to occur primarily as a smooth flowage of horizontally oriented ice crystals that shear over those below. It takes place at depths greater than about 50 meters (160 ft) because the weight of the overlying ice makes the ice below this level lose its rigidity and deform in a plastic fashion. Internal flow, unlike basal sliding, has little or no effect on the underlying surface. It is the only mechanism of movement employed by **cold ice glaciers**, which are frozen fast to the underlying bedrock. Cold ice glaciers exist largely in the colder high latitude environments where continental glaciers form.

Because the nonmoving rock floor and walls of a glacier (if it is laterally confined) resist ice flow, the surface ice moves fastest. In addition, if the glacier takes the form of a stream of ice, as do most alpine glaciers, flow rates are greater near the center than near the margins. In these respects, the flow pattern of a glacier is very similar to that of a river. Major differences are the much slower flow velocities of glaciers and the predominance of laminar (smooth) rather than turbulent flow.

Glacier Erosional Processes and Features

The great power of glaciers as eroders and transporters of regolith stems from several factors. One is the fact that, unlike water and wind, glaciers exist in the solid state. This produces the physical hardness and firmness needed to dislodge vast quantities of weathered rock material effectively. A second factor is the tremendous pressure that hundreds or even thousands of meters of ice exerts on the surface. Although water depths and pressures equal or exceed those of

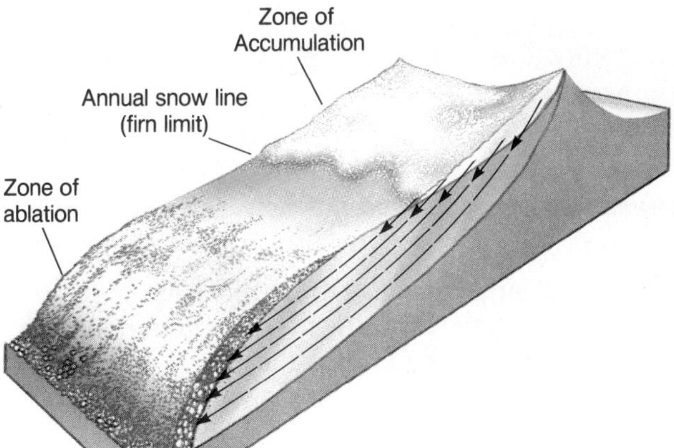

FIGURE 18.3 Zones of ice accumulation and ablation on a glacier. Rock debris is buried beneath newly forming ice in the zone of accumulation, producing a clean surface, but eventually resurfaces in the zone of ablation due to the melting of the overlying ice.

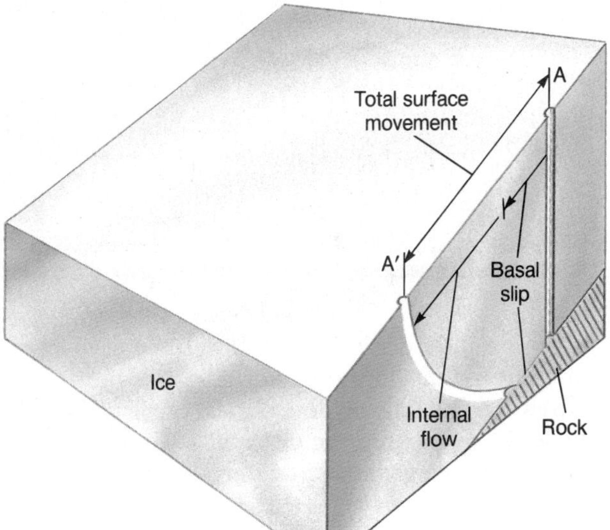

FIGURE 18.4 Mechanisms of glacier movement. The motion of warm ice glaciers is accomplished by both basal slip and internal flow. Cold ice glacier motion takes places only by internal flow.

ALASKAN GLACIERS – SURGING AND RETREATING

The Coast Mountains of southeastern Alaska and adjacent northwestern Canada, nurtured by copious year-round snowfalls, contain one of the world's most extensive systems of alpine glaciers. These glaciers, which are more accessible than those in many other areas, are of considerable interest to scientists both as climatic indicators and as outdoor laboratories for the study of glacial processes. Although concentrated within a restricted geographical area, the glaciers of southeastern Alaska tend to behave in highly individualistic fashion. In recent years some have suddenly surged forward, while others have undergone rapid retreat. The recent actions of the nearby Hubbard and Columbia glaciers illustrate both extremes of motion.

For most of the 90 years over which its movements have been recorded, the Hubbard Glacier has flowed into Yakutat Bay, near the northwest tip of the Alaskan Panhandle, at a relatively steady rate of about 150 meters (500 ft) per year. This motion has not been entirely counterbalanced by a loss of ice at the glacier's snout, so the glacier has been slowly advancing into the bay. During the winter of 1986, however, the Hubbard Glacier suddenly began to surge—moving as fast as 14 meters (46 ft) per day. A tributary glacier, the Valerie Glacier, also began to surge forward at speeds of up to 34 meters (112 ft) per day. This movement soon blocked off Russell Fjord at the head of the bay and turned it into a lake, temporarily trapping numerous seals and other sea animals (see Figure 18.5).

Geomorphologists believe that the surge may have been caused by entrapped subglacial meltwater, which provided a sliding plane for the ice and perhaps helped to buoy it up from the surface in places. In support of this theory, it has been noted that some neighboring glaciers, such as the Variegated Glacier, surge at regular intervals. Their surges are preceded by a substantial reduction in meltwater outflow, followed by a flood of meltwater at the end of the period of surge.

The recent behavior of the Hubbard Glacier stands in marked contrast to that of most of Alaska's glaciers, which are in retreat. The rapid melting of Alaskan glaciers, in fact, generally is believed to be a significant contributor to the gradual global rise in sea levels.

The Columbia Glacier, one of the last of Alaska's great tidewater glaciers and an important tourist attraction, recently has started to retreat at a rate exceeding 300 meters (1,000 ft) per year. The retreat seems to be taking place because the glacier's snout has now shrunk past its shallow water sill into much deeper water, where large chunks of ice can **calve** (break off). In the late 1970s, the water at the glacier's edge was only 17 meters (50 ft) deep. Currently, it is nearly 300 meters (1,000 ft) deep. It is believed that, over the next few decades, the snout of the Columbia Glacier will leave the sea and retreat several kilometers inland.

glaciers on most ocean floors and in some deep lakes, the bottom water in these cases usually is nearly stationary and has little or no erosional capacity. Indeed, the deep ocean basins are major depositional sites. A third factor aiding glacial erosion is the great extent of ice coverage within a glaciated area. While only a tiny proportion of the surface in a fluvially dominated landscape is covered by stream water at any one time, continental glaciers bury the surface, modifying the entire landscape. Even alpine glaciers, which are confined largely to mountainside depressions or valleys, may cover a significant proportion of the surface.

On the other hand, glaciers do have some erosional limitations. Probably the most important is the reduced rate of chemical weathering of the underlying bedrock, which is inhibited by cold temperatures and sometimes a lack of liquid water. The prior weathering of the bedrock is crucial in preparing material for glacial transport. Hard as ice is, most unweathered rock is harder yet and will not be affected significantly by direct glacial erosion. Frequently, however, pieces of bedrock become frozen into the base or sides of a moving glacier. They then scrape over the surface, abrading it in much the same man-

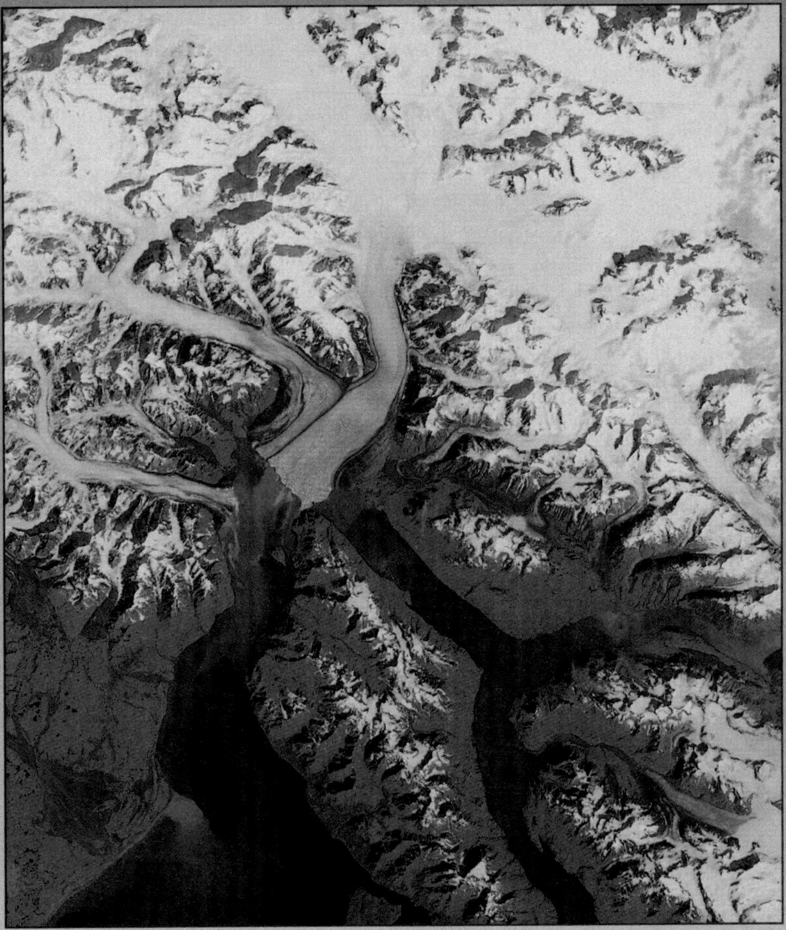

FIGURE 18.5 A surge in the flow of Alaska's Hubbard Glacier, which occurred in early 1986, temporarily blocked Russell Fjord (the inland water body to the right of the lower portion of the glacier), trapping a large number of marine animals. (USGS)

ner as the hard grains of sand on a sheet of sandpaper scrape a piece of wood.

A second factor limiting the effectiveness of glacial erosion is the slow rate of ice movement over the underlying surface. Ice is by far the slowest moving of the tools of erosion. In addition, much of the ice motion occurs internally; it therefore does not affect the underlying surface. Indeed, polar glaciers, frozen rigidly to their bases, perform little or no erosion.

Glacial erosion occurs by the processes of scraping and plucking. **Scraping** is accomplished primarily by rocks frozen into the base of the ice and dragged over the bedrock surface. Larger rocks may produce scratches called **striations**, which are sometimes enlarged to form grooves if rock fragments are dragged repeatedly over the same point (see Figure 18.6). Striations are valuable indicators of the direction of past ice flow. The abrasion performed by silt-sized particles, on the other hand, smooths and even polishes some rock surfaces. Because these smaller particles are much more widespread than the larger rock fragments along the glacier base, the overall effect of glacial abrasion is to smooth and round rock surfaces.

FIGURE 18.6 Closeup of glacial striations on bedrock in Jasper National Park, Alberta. (Richard Jacobs/©JLM Visuals)

Plucking involves the excavating of angular, often relatively large, rock fragments from the bedrock. The regelation process described earlier seems to be the key mechanism in making plucking effective. Ice melting on the up-glacier side of an obstacle such as a hill or boulder flows around the obstruction, freezes in rock fractures on the down-glacier side, and plucks out rock fragments as the ice moves on. The process is aided greatly by the presence of weathered or highly jointed bedrock. Some previously glaciated areas contain numerous bedrock features with their up-glacier sides smoothly rounded and their down–glacier sides irregularly plucked. These features are sometimes referred to as **roche moutonées**, a French term meaning "rock sheep."

Glaciers are most effective, by far, in eroding unconsolidated materials. This most commonly occurs when a glacier advances over terrain that has not been glaciated in the recent past. The hundreds of meters of glacial ice involved readily bulldoze the soil and weathered rock or incorporate it into the ice itself, leaving the underlying surface stripped to unweathered bedrock.

Because the sediment-carrying ability of the ice is virtually unlimited, glacially transported material is carried in an unsorted state. Temperate glaciers not bounded by rock walls carry most of their debris along or near their bases. Alpine glaciers contain much basal material but also transport large amounts of material, commonly dislodged by frost wedging, on their surfaces.

Glacier Depositional Processes and Features

Glaciers transport debris in a conveyor-belt fashion. Even if the forward edge of the ice is retreating, the absolute glacier motion is still down-ice, and this motion does the transporting. The general term for all deposits of glacial origin is glacial **drift**. This term is a holdover from the time when it was believed, for lack of a better explanation, that glacial deposits had been rafted to their present sites by icebergs during the biblical Flood. Drift consists of two general types, defined by the mode of deposition. Material deposited directly by glacial ice is known as **till**. It is unsorted and angular, with constituent particles ranging in size from clay to occasional large boulders. In contrast, **glacio-fluvial sediments** are deposited by glacial meltwaters and are partially stratified, or sorted.

Till is deposited primarily by the processes of dumping at the glacier terminus, or **snout**, and by lodging or plastering on the surface beneath the glacier. The deposits produced by either process are termed **moraines**. If the snout remains stationary for an extended period, so much till is conveyed to that point that a ridgelike **terminal moraine** is produced. This is a loose accumulation of till with its long axis oriented perpendicular to the direction of ice flow (see Figure 18.7). It is often arcuate, because most glacier snouts are convex in shape. If the glacier front is retreating, a series of such moraines may be formed, with each moraine marking the site of a temporary halt in the glacier's retreat. In this case, only the outermost moraine is referred to as the terminal moraine, and the others are termed **recessional moraines**. Terminal and recessional moraines, where undisturbed, provide precise information on past glacial positions and are invaluable to researchers attempting to reconstruct the chronology of events in a previously glaciated area. Between these moraines, a thinner covering of **ground moraine** is deposited by the more rapidly retreating ice. It has been found that, although continental glaciers may transport some rock debris for hundreds of kilometers, most glacial deposits are derived from relatively nearby sources.

Glacio-fluvial materials normally are deposited at or beyond the margins of the ice, although in some cases subglacial deposits may occur. In general, the farther these deposits, called **outwash**, have been transported by water, the better sorted they are. Glacial meltwater streams usually are overloaded with

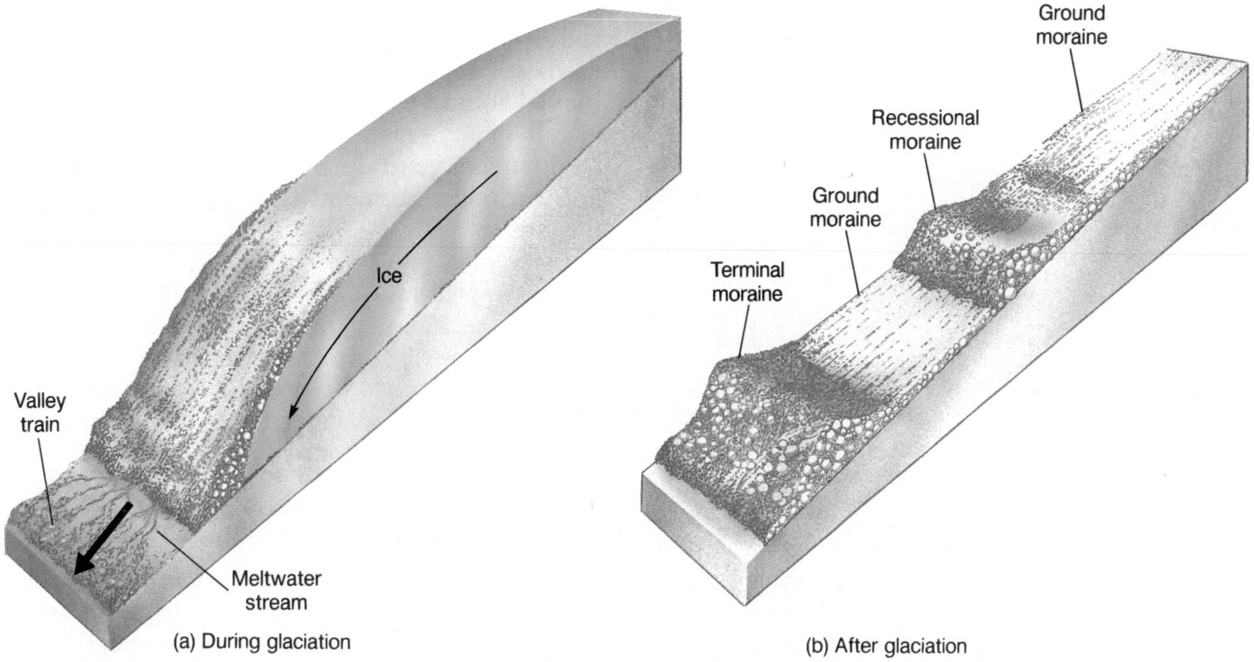

FIGURE 18.7 Block diagrams depicting the formation of a terminal moraine, as well as the appearance and relative positions of terminal, recessional, and ground moraines after the melting of the glacier that formed them.

sediment and frequently aggrade their courses, typically developing braided channels as they do so. If multiple streams issue from the ice front, a broad **outwash plain** may be formed. If only one fairly channelized stream emerges, a single long **valley train** of deposits is produced, perhaps extending for tens of kilometers from the ice front.

ALPINE GLACIERS AND ASSOCIATED LANDFORMS

Alpine glaciers form in mountainous regions at high elevations and flow downhill under the influence of gravity. Because global temperatures tend to increase equatorward, alpine glaciers become progressively less numerous with decreasing latitude and their zones of formation occur at progressively higher elevations. The lower limit of ice accumulation lies essentially at sea level within 10 degrees of the poles and increases to elevations of 4,600 to 5,000 meters (15,000 to 16,000 ft) in the Central Andes and the Ruwenzori Range of East Africa near the equator. The zones of ablation of well-fed glaciers can extend significantly below these limits, though, so that the lower limit of glaciation depends not only on temperature, but also on snowfall amounts and on the elevation and size of the glacial accumulation area.

Characteristics of Alpine Glaciers

Most alpine glaciers originate in relatively small mountainside hollows called **cirques**. If conditions are marginal for ice accumulation, the glacier may remain as a small, rounded **cirque glacier**. If conditions favor ice accumulation, however, the glacier will outgrow its cirque and flow down the mountainside as a long, narrow tongue-like mass of ice. Upon reaching the base of the mountain, continued glacier expansion usually will result in the ice advancing down a preexisting valley as a **valley glacier** (see Figure 18.8). Large valley glaciers, like those of the Alps, typically are fed by several glacier tongues, so that a dendritic pattern of intersecting glaciers is developed.

The surface features and appearance of an alpine glacier can vary considerably from point to point (see Figure 18.9). In the zone of accumulation, the ice usually is covered by snow and cannot be seen, while

FIGURE 18.8 Alaska's Kaskawulsh Glacier is a major alpine glacier with numerous tributary glacial tongues. The glacial zones of accumulation (snow-covered portion) and ablation are readily discernible. (John S. Shelton)

in summer, at least, the surface of the glacier in the zone of ablation is visible. Where the glacier passes over convexities on its bed, the stretching of the near-surface ice causes it to fracture, forming parallel open fissures called **crevasses**. The deepest known crevasses extend downward approximately 50 meters (160 ft). Below this level, the weight of the overlying ice gives it enough plasticity to close any gaps that may form.

Lateral moraines frequently take the form of thick stripes of rock debris along the sides of alpine glaciers. This material accumulates both from glacial erosion and from weathering processes, especially frost shattering, which dislodge rocks from the slopes above. Where two glacial tongues merge, their innermost lateral moraines coalesce to form a single **medial moraine** within the glacier. A large alpine glacier often contains several medial moraines, the total number generally indicating the number of individual valley glaciers that have combined to form the glacier.

Proceeding into the zone of ablation, the glacier takes on a progressively dirtier, more debris-laden appearance, because the gradual melting of the ice produces a continually increasing proportion of morainic material. Near the glacial snout, debris may completely cover the surface, making it impossible to accurately determine the most forward position of the ice.

Alpine Glacier Erosional Landforms

Erosion by alpine glaciers tends to increase the relief and ruggedness of the topography. It has been largely responsible for producing the spectacular alpine scenery that annually draws millions of tourists to highly glacially modified mountain systems such as the Alps and northern Rockies (see Figure 18.10). A number of landform features are produced by alpine glacier erosion, but two of the most important are cirques and glacial troughs.

Cirques are bowl-shaped depressions eroded into the rock faces of mountainsides in glaciated areas (see Figures 18.11 and 18.12). They vary greatly in de-

FIGURE 18.9 Major features of an alpine glacier. The equilibrium zone separates the zone of accumulation from the zone of ablation.

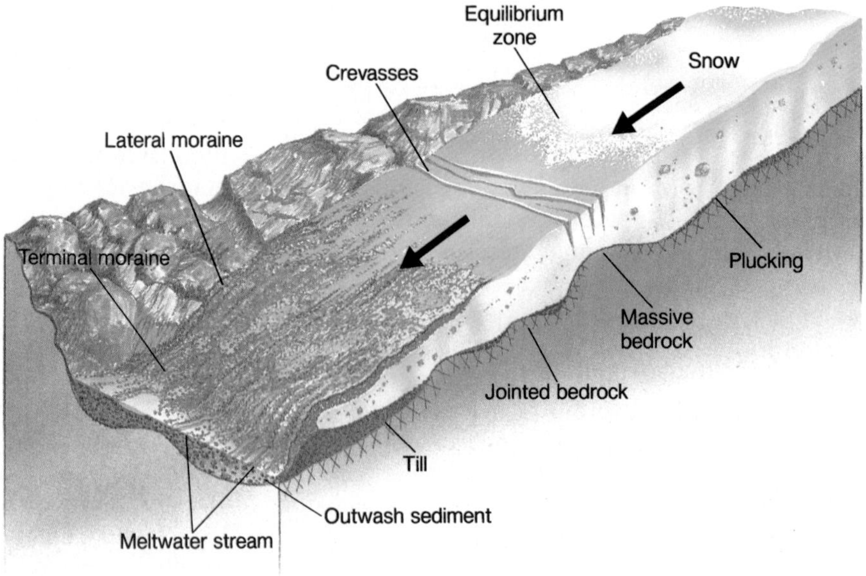

FIGURE 18.10 St. Mary Lake occupies a glacial trough in Montana's Glacier National Park. A horn and a knife-edged arête further enhance this area's spectacular alpine scenery. (Ralph Scott)

gree of development and perfection of shape, but well developed cirques consist of three basic sections: a steep headwall, a smooth central basin, and a slightly raised lip or threshold (see Figure 18.13). Some uncertainty exists as to the precise mechanism by which cirques develop, but they generally are believed to form where mountainside snow patches provide water for accelerated physical and chemical weathering. As the hollow enlarges, sufficient snow collects to prevent complete summer melting, and a cirque glacier forms. The cirque is subsequently enlarged and molded by frost shattering on the headwall and by erosion of the cirque floor by scraping and plucking. Many cirques that formed during the Pleistocene are now devoid of ice and instead contain small lakes called **tarns**.

As growing cirques erode headward from opposite sides of a linear mountain or upland area, they eventually may reduce the intervening mass to a narrow, serrated ridge called an **arête**. Headward cirque growth from three or more sides can produce a sharp-crested pyramidal peak termed a **horn**. The Matterhorn of Switzerland is the most famous example of this process, but most isolated peaks in glaciated mountain regions have formed in essentially the same manner. The cirque growth process therefore is a highly effective erosional mechanism and is largely responsible for the upland features in extensively glaciated mountains.

The valley features in glacially modified mountainous regions result largely from the development of glacial troughs. A **glacial trough** is a deep valley, usually of fluvial origin, that has been subsequently occupied and modified by a valley glacier (see Figure 18.14). Glacial troughs are noted for their broad, rounded U-shapes, as opposed to typical fluvially eroded mountain valleys, which are more V-shaped. In contrast to river valleys, they also possess markedly ungraded longitudinal profiles and greater overall linearity.

The U-shape of a glacial trough results from the large size of a valley glacier as compared to a river

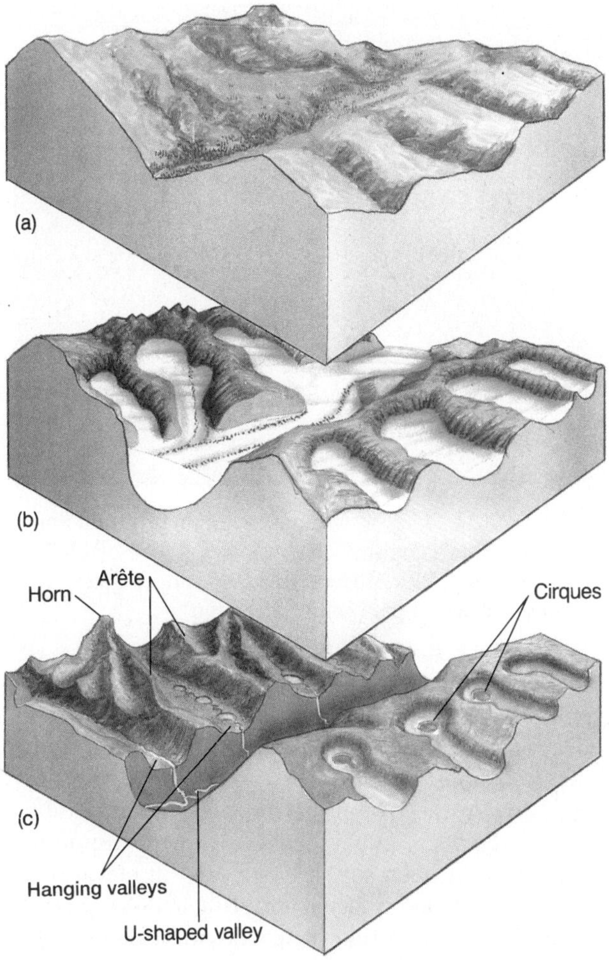

FIGURE 18.11 The glacial modification of a mountainous landscape is represented in this sequence of diagrams. Part (a) shows a fluvially dissected landscape prior to the onset of glaciation. Note the smoothly rounded appearance of the slopes. Part (b) depicts the same landscape at the height of glaciation. In Part (c), the glaciers have retreated, leaving in their wake a rugged landscape containing numerous glacial erosional features.

FIGURE 18.12 Classic cirque development producing "biscuitboard topography" in the Uinta Range of Utah. These cirques are Pleistocene features, and no glaciers currently occupy them. (John S. Shelton)

draining the same area. This size is necessary because glacial ice flows much more slowly than river water. Indeed, a valley glacier usually fills its entire trough, often to a considerable depth. The deepening and steepening of the trough by its glacier considerably increase the local relief because nearby mountain peaks have not been similarly reduced in elevation.

Smaller tributary glaciers entering a valley glacier also carve U-shaped valleys, but do not usually erode nearly as deeply. When the ice melts, their valleys intersect the main glacial trough high on its sides, forming **hanging valleys**. As streams issuing from these hanging valleys cascade into the major trough, spectacular waterfalls may result. Yosemite Falls, in California's Sierra Nevada, is a famous example.

The ungraded longitudinal profile of a glacial trough results from variations in the rate of vertical erosion by the ice in different parts of the valley. Erosion is aided by factors such as a narrowing of the valley (causing a thickening of the glacier), the influx of a tributary glacier, and the presence of relatively weak rock. Such areas are eroded to form localized basins within the trough; they usually are occupied by lakes after the ice melts.

At higher latitudes, glacial troughs may be eroded below sea level and occupied by ocean water when the ice melts. Such drowned troughs are called **fjords**. They are typically long, winding, steep-walled, and relatively narrow (see Figure 18.15). Fjords formed because the great thickness of the ice enabled it to erode deeply even though it was partially buoyed by the ocean water into which it advanced. Sea level rises subsequent to the melting of the glaciers resulted in further deepening. As a consequence, some fjords in Norway and southern Chile attain depths of more than 1,300 meters (4,000 ft). They therefore provide excellent protected routes and harbors for coastal shipping and fishing interests. Fjords are best developed along the coasts of Norway, western Canada, southern Alaska, Greenland, southern New Zealand, and southern Chile.

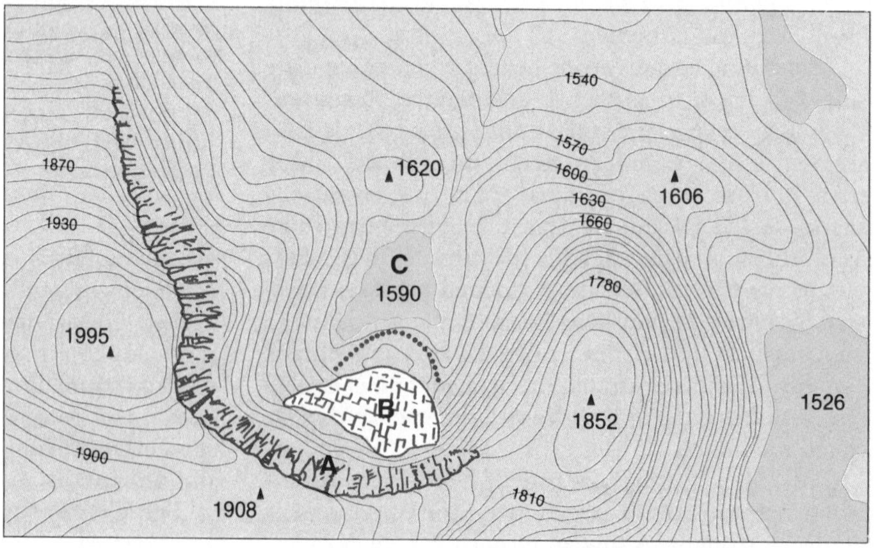

FIGURE 18.13 A topographic map depiction of a well developed cirque in Norway. Note the precipitous headwall (A), the remnant cirque glacier (B), and the central basin containing a tarn, or small lake (C). Elevations are given in meters.

FIGURE 18.14 The valley of McDonald Creek in Glacier National Park, Montana, is a U-shaped glacial trough. (©JLM Visuals)

FIGURE 18.15 Geirangerfjorden, in Norway, is a fjord with numerous waterfalls issuing from hanging valleys on its sides. (Tony Stone Worldwide)

Alpine Glacier Depositional Landforms

The depositional features of alpine glaciers typically are more poorly defined and less obvious than are the erosional features for several reasons. First, they have much less relief than the erosional features. In addition, they consist of unconsolidated materials and therefore do not have the permanency of the erosional features, which generally are composed of solid bedrock. Finally, because most glacial deposits are located in valleys, the rivers that reoccupy the valleys after the departure of the ice rework and often largely remove these materials.

The major till features are moraines of various types. Fresh moraines consist of ridges of loose, unsorted rock debris. Lateral moraines form terracelike deposits along valley margins. Those produced by large valley glaciers are sometimes more than 100 meters (330 ft) in height and often become progressively larger and better defined down-valley. Medial moraines, which are more likely to have been removed by fluvial processes, may form ridges well within the glacial trough. Terminal and recessional moraines form arcuate ridges, bulging down-valley, that are oriented perpendicular to the lateral and medial moraines. Many formerly extensive alpine glaciers have disappeared completely, so that morainic deposits occur from their lower valleys all the way up to their cirque floors.

Glacio-fluvial deposits already have been discussed. Of those produced by alpine glaciers, the most significant are the valley trains deposited by debris-laden glacial meltwater streams.

CONTINENTAL GLACIERS AND THE PLEISTOCENE

Continental glaciers differ from alpine glaciers in several major respects. The most important difference is in their size and thickness. While alpine glaciers extend, at most, a few tens of kilometers and frequently have widths of only a kilometer or two, continental glaciers have horizontal dimensions of hundreds or even thousands of kilometers. Alpine glaciers lie cupped in mountainside hollows or in deep valleys, while continental glaciers usually cover the entire surface deeply.

The underlying surface exerts much less control on the movement of continental glaciers than it does for alpine glaciers. The movement of the entire ice sheet is not unidirectional, but rather is radial, with the ice spreading outward in all directions from one or more centers of accumulation. The cause of this radial spread is not so much the downhill force of gravity as it is the down-ice pressure exerted by the great weight of the ice itself. Radial expansion gives the glacier a rounded, domelike overall appearance. Examined in more detail, however, the ice front advances as a series of lobes whose positions are dictated by the locations of the centers of ice accumulation and by the topography and geology of the underlying surface.

The geographic setting of the zone of ice accumulation of alpine and continental glaciers also differs in one major respect. While high elevations are responsible for the cold climates that support alpine glacier growth, high latitudes are responsible for the low temperatures that support the growth of continental glaciers. Since larger areas of the Earth are at high latitudes than at high elevations, continental glaciers are much more extensive than are alpine glaciers.

The Pleistocene Epoch

The geologic record indicates that for most of its existence the Earth has had no continental glaciers. There may always have been alpine glaciers in the highest mountains, but continental glaciers seem to be a rare phenomenon. Scientists have evidence of only five relatively brief periods of continental glaciation in the past billion years. The first four occurred hundreds of millions of years ago. The fifth—the **Pleistocene**—began only about two million years ago and is still continuing.[1]

Over the past one million years, there have been approximately 10 major glacial periods, or stages, and as many as 40 more minor periods of glacial advance. The **Wisconsinan Stage** (or **Würm Stage**, as it is known in Europe), is the most significant because of its relative recency. The Wisconsinan Stage began some 50,000 to 100,000 years ago and generally is considered to have ended about 10,000 years ago, when the retreating ice left the north-central United States. The last ice may not have melted in central Canada until about 4,000 years ago. The recency of this period is such that only a superficial modification of the landscape has occurred in many of the glaciated areas; as a result, these areas tend to be dominated by relict glacial features.

Between the various glacial stages were lengthy interglacial stages, when the ice sheets generally melted back past their present positions. Evidence suggests that during at least one interglacial stage the ice may have disappeared largely or entirely from Greenland and Antarctica. It is likely that we are currently in an interglacial stage.

Geographic Extent of Pleistocene Glaciation

During the Pleistocene glacial stages, ice covered as much as 30 percent of the Earth's present land area. Glacial ice now covers some 14 million square kilometers (5.5 million mi^2), but during the Pleistocene it covered up to 44 million square kilometers (17 million mi^2). Then as now, the Antarctic ice sheet was the largest of all (see Figure 18.16). During the Pleistocene, it enlarged and spread farther out over the adjacent sea than at present. The Greenland ice sheet was considerably larger than it is today, extending to the coast and probably forming ice shelves in protected coastal waters. The major addition of ice,

[1] Many scientists consider the Pleistocene to have ended about 10,000 years ago. There is increasing evidence, however, that the cyclical pattern of continental glacier formation and retreat will continue in millennia to come. Even if this does not occur, the continuing existence of massive glaciers in Antarctica and Greenland indicates that the present glacial stage is not completed.

FIGURE 18.16 Nunataks, or isolated mountain peaks, protrude above a sea of ice on the edge of the Antarctic Plateau. (Bruce F. Molnia)

however, consisted of two giant ice sheets that formed over the northern half of North America and over northern Europe and adjacent northwestern Asia.

The North American or **Laurentide ice sheet** spread outward from two or more centers in interior Canada to cover virtually all of Canada, including Hudson Bay and most of the Canadian Arctic Islands (see Figure 18.17). The United States was covered as far south as a line extending westward from Long Island through northern Pennsylvania, then along the courses of the Ohio and Missouri Rivers, and finally westward near the international border to northern Washington. Surprisingly, much of Alaska remained unglaciated, apparently because of a lack of snowfall and the presence of orographic barriers. It should be stressed that the line just described is a composite maximum line of glacial advance. During each major glacial stage, the ice advanced to a different boundary. The total area covered by the Laurentide ice sheet was approximately 10.2 million square kilometers (4 million mi^2), about 80 percent as large as the Antarctic ice sheet is today. Its thickness is unknown but was probably comparable to that of the Antarctic ice sheet. It was at least thick enough to cover completely the 1,918 meter (6,288 ft) crest of Mt. Washington, New Hampshire, which was located near the presumably thinner margin of the ice.

The Eurasian or **Fennoscandian ice sheet** radiated from a center in eastern Sweden to cover a max-

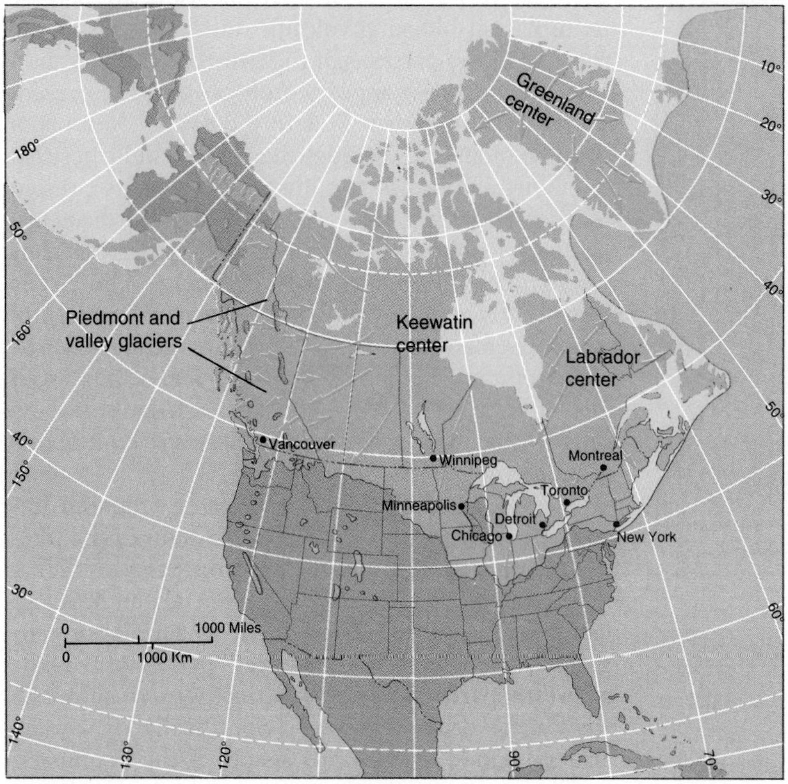

FIGURE 18.17 Centers of ice formation and maximum extent of Pleistocene glaciation in North America. Areas of both glacial ice and sea ice are shaded in blue.

imum of 5.1 million square kilometers (2 million mi^2). Its southernmost boundary extended through southern England, eastward through the Netherlands, then east-southeastward across central Europe to the southern Ukraine, and finally east-northeastward to a point in north-central Siberia (see Figure 18.18). The northern terminus of the ice extended some distance past the present continental margin into the Arctic Ocean. The North Sea and the Baltic Sea also were under glacial ice, which reached far enough westward to cover Ireland.

The Pleistocene glacial stages also were periods of formation or expansion for smaller glaciers in many other parts of the world. A subcontinental glacier may have formed over eastern Siberia, which was excessively cold during the Pleistocene but was too dry to support the growth of a more massive glacier. Mountain glaciers in many areas also expanded greatly, and many presently unglaciated mountains supported extensive Pleistocene glaciers. For example, the Canadian Rockies developed a system of coalescent alpine glaciers, sometimes termed the **Cordilleran ice sheet**, that merged with the Laurentide ice sheet east of the continental divide. The Alps developed an ice cap above which only the higher peaks protruded. Glaciers formed in the mountains of southeastern Australia where none exist today. The southern Rockies, which contain only a few cirque glaciers at present, were extensively glaciated. Glaciers may have descended to 2,000 meters (6,500 ft) in the Andes at the equator, where today the glacial limit is at least 4,300 meters (14,000 ft).

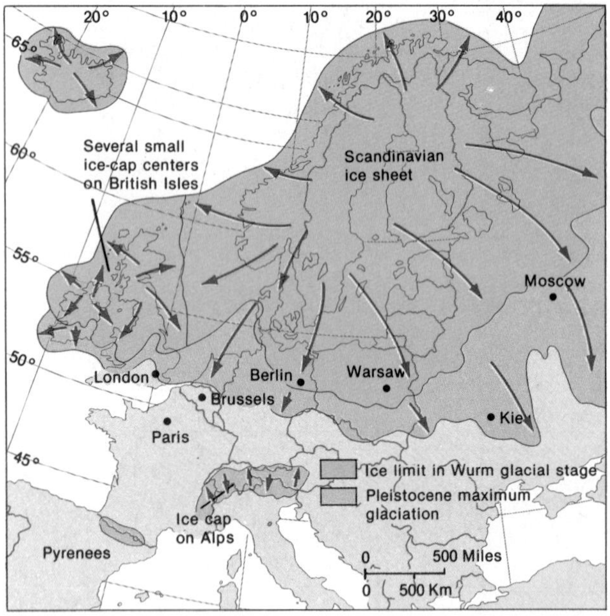

FIGURE 18.18 Areas of continental glacier coverage and directions of ice flow during the Pleistocene in Europe. Both the maximum extent of the ice sheets and the limits of ice advance during the last (Würm) glacial stage are shown.

Causes of the Pleistocene

The cause of the Pleistocene has produced one of the greatest controversies in geomorphology ever since nineteenth-century Earth scientists confirmed the past existence of an ice age. There are actually two facets to the problem, and their respective solutions may or may not be related.

The first question is why the Pleistocene occurred at all. More specifically, why, after many tens of millions of years of nearly global warmth, did the Earth become cold enough within the past two million years to permit the growth of massive continental glaciers? The answer to this question still is not known, although a number of hypotheses have been developed. These include the possibility of a recent decrease in the output of solar radiation, an increase in insolation-blocking volcanic dust, and a change in the circulation patterns of the ocean currents. Whatever the answer, it appears probable that the present position of the North Pole in the Arctic Ocean is of great significance. Geological evidence indicates that the North Pole has not always been in its present location; in fact, it is believed to have long been situated in the North Pacific in waters so open that they would never have been able to freeze. The Arctic Ocean, however, not only has a restricted pattern of water circulation with the adjacent warmer Atlantic and Pacific that allows its surface to freeze, but also is virtually surrounded by massive high-latitude land masses that provide potential sites for continental glacier growth.

The second question is of greater potential import, because its answer relates strongly to the likelihood of future periods of glaciation. It is this: What is the cause of the alternation of glacial and interglacial stages within the Pleistocene? Scientific investigation over the past two decades has shown that the answer apparently is related directly to variations in the Earth's orbital geometry. Specifically, these variations include (1) the precession of the equinoxes, which determines the time of year that the Earth is at aphelion and perihelion, and (2) long-term variations in the tilt of the Earth's axis. The influence of these factors is discussed in more detail in the Focus Box on page 480. An important implication of these findings is that we are probably now in an interglacial period and that the ice sheets of North America and Eurasia may well re-form in the near geologic future.

Landforms Produced by Continental Glaciers

Continental glaciers produce many of the same types of landforms as do alpine glaciers. Important differences exist, however, in the sizes, locations, and relative abundance of most of these features. As in the case of alpine glacial features, those produced by continental glaciation can be categorized as erosional or depositional in origin (although some are produced by both processes). The areas that undergo net erosion by a continental glacier generally are located beneath the central portion of the ice sheet, in the zone of ice accumulation. Surrounding this, in the zone of ablation, are the major areas of deposition.

Continental Glacier Erosional Landforms

A basic difference exists between the erosive actions of alpine and continental glaciers. Alpine glaciers, as we have seen, sharpen the topography and usually increase local relief. Continental glaciers have just the opposite effect. Because they cover everything, they subject the entire surface, not just lowland areas, to erosion. Surface features such as hills or mountains are especially susceptible to erosion because they protrude well into the faster moving zones of the ice. Lowland areas, on the other hand, are somewhat protected and may even be sites of local deposition within an area dominated by erosion. The surface as a whole therefore is smoothed and rounded by the ice.

The extent of actual surface lowering by continental glacier erosion has long been a subject of controversy. The central question has involved how effectively the ice erodes unweathered bedrock. Most researchers now believe that the ice is relatively ineffective on most solid bedrock and that glacial erosion largely involves the removal of the soil and regolith down to the weathering front.

The zone of erosion of the Laurentide ice sheet included the Canadian Shield area, the Canadian Maritime Provinces, and northern New England. The Fennoscandian ice sheet zone of erosion included Scandinavia and adjacent areas currently beneath the shallow waters of the North Sea and Gulf of Bothnia.

For the most part, the surfaces in the zones of erosion of both the Laurentide and Fennoscandian ice sheets were shield areas of ancient crystalline rocks, with predominantly low surface relief. These areas were further smoothed by the ice, so that today **ice-scoured plains** are the most extensive features of continental glacier erosion. These areas now consist of vast expanses of irregularly rolling surface, largely stripped of soil and covered by myriad lakes and swamps (see Figure 18.19). The glaciers originally removed virtually all soil and weathered bedrock from these areas. Most areas today, though, have a thin covering of soil produced either by deposition as the ice melted, or by postglacial weathering.

Among the most important features of ice-scoured plains are their numerous lakes and swamps. These

FIGURE 18.19 View of the rocky, lake-strewn surface of an ice-scoured plain in southern Alaska. (©JLM Visuals)

EARTH'S ORBITAL GEOMETRY AND CLIMATIC CHANGE

Ever since nineteenth-century scientists discovered that the Earth has been experiencing a recent series of glacial and interglacial periods, a search has been underway for the mechanisms causing these changes. During the 1920s and 1930s, the Serbian mathematician Milutin Milankovitch argued that the glacial and interglacial stages resulted from long-term variations in the Earth's orbit. As late as the 1970s, most experts believed that changes in insolation resulting from these orbital variations would be in-

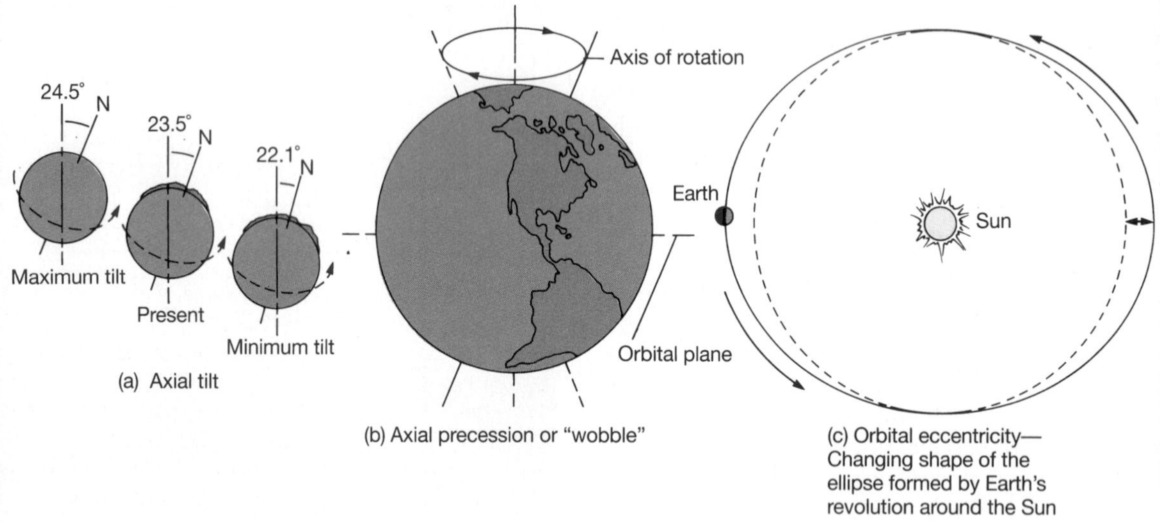

FIGURE 18.20 The three Earth orbital variations believed to be collectively responsible for the sequence of Pleistocene glacial and interglacial stages.

features have formed because erosion and deposition by the glaciers obliterated the preglacial drainage network. In addition, areas of deeply weathered rock were scooped out by the ice, forming countless rock basins that subsequently filled with water to form lakes. Wandering through these plains in a haphazard fashion from lake to lake are stream systems displaying markedly ungraded long profiles and deranged drainage patterns. The presence of glacial lakes has earned Minnesota the nickname "The Land of 10,000 Lakes." The glacially derived lakes of Canada are so large and numerous that they reputedly contain more than half the world's fresh surface water. The Great Lakes were either formed or greatly enlarged when the ice excavated areas of weathered rock.

The extensive glacial erosion of areas such as the Canadian Shield has greatly reduced their economic potential. The soil that once covered the surface was removed, and most weathered rock material was transported southward, where it now enhances the agricultural productivity of the midwestern United sufficient to produce the necessary climatic changes. In the past two decades, however, knowledge of the chronology of the Pleistocene glacial and interglacial stages has been refined to the point that it has been possible to compare the timing of these stages with those of terrestrial orbital variations. The correlation has been so close that it is now almost certain that long-term variations in the Earth's orbit are the chief cause of the glacial and interglacial stages.

Three orbital variations are involved (see Figure 18.20). First are changes in the tilt of the Earth's axis with respect to the plane of the ecliptic. The tilt currently is approximately 23.5°, but it varies between 22.1° and 24.5° over a 41,000-year cycle. An increase in tilt concentrates insolation toward the poles and decreases insolation received at the equator. A decrease in tilt, which is currently taking place, has the opposite effect.

A second orbital variation is the precession of the equinoxes. The Earth's axis has been found to gyrate or "wobble" like that of a slowly spinning top. In addition, the elliptical orbit of the Earth is slowly rotating. The net result of these two independent movements is that the time of year that the Earth is at different points in its orbit—at aphelion and at perihelion, for example—gradually changes, completing a full cycle each 22,000 years.

A final orbital variation is the change in the eccentricity of the Earth's orbit over cycles of 100,000 and 400,000 years. When the eccentricity of the orbit increases, the Earth takes a less circular orbit around the Sun, and consequently there is a greater annual variation in distance between the Earth and Sun.

It has been found that, of these motions, the precession of the equinoxes and the changes in axial tilt have the greatest influence on temperature changes and on continental glacier formation and retreat. The influence of the Earth's orbital eccentricity is of secondary importance and mainly influences the amplitude of the precession cycle. Although human activites may well enter into the picture from this point onward, the net effect of these natural orbital variations over the next 10,000 or so years would seem to be the gradual ending of the present relatively warm interglacial stage and the beginning of a new period of continental glaciation.

States. In addition, much land is agriculturally unusable because of glacially induced poor drainage. As a consequence, most of interior eastern Canada today is an almost uninhabited rock, lake-strewn wilderness.

Continental Glacier Depositional Landforms

Continental glaciers produce a variety of depositional features (see Figure 18.21). Some are similar in origin to those produced by alpine glaciers, but occur on a grander scale; others are unique to continental glaciation. Most recognizable depositional features were produced within the past 20,000 years and are the product of the Wisconsinan Stage.

The most extensive depositional feature is the **till plain**, a thick accumulation of ground moraine covering the preglacial surface over an extensive area (see Figure 18.22). Depths of till often exceed 30 meters (100 ft), with thicker accumulations in originally lower areas such as valleys. In central Ohio, for example, till thicknesses over preglacial uplands average about 15 meters (50 ft), increasing to about 60 meters (200 ft) over lowland areas. Surface relief therefore is reduced by glacial deposition, although the preglacial topography is still generally reflected in a subdued fashion on the surface of the till plain.

Most of the till forming a till plain is fine in texture and consists of sand, silt, and clay. Occasional rocks or boulders are mixed in, however, and in some areas they are quite numerous. The entire mixture is unsorted, since it was deposited directly by the melting ice. In some localities, the separation of huge blocks of ice from nonmoving or "dead" glaciers during the melting process has produced pitted till plains, with the pits, called **kettles**, often currently occupied by small lakes.

Till plains cover much of the north-central portions of both the United States and Europe. The general flatness of the terrain, the high mineral content of the still lightly leached till, and the favorable present-day climate have combined to produce fertile soils. The American Midwest and the North European Plain are renowned for their agricultural productivity and currently support substantial human populations.

Systems of terminal and recessional moraines form the major relief features on till plains in many

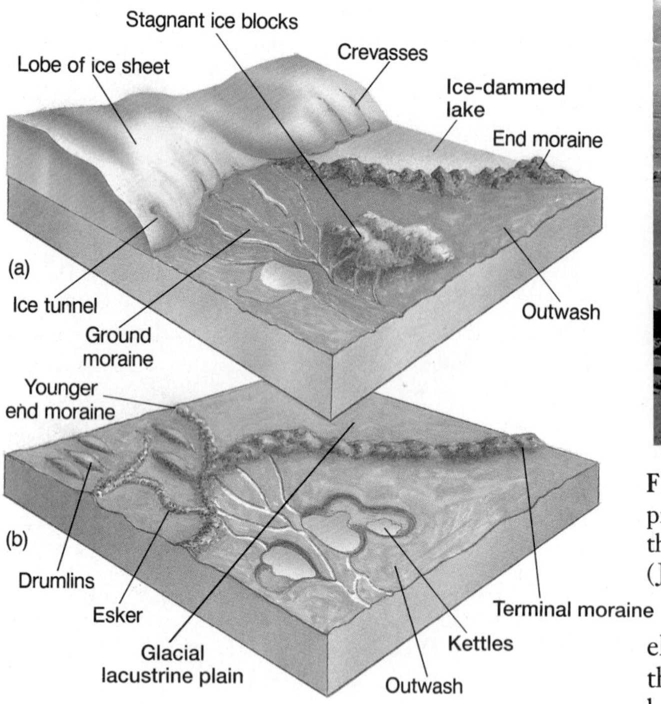

FIGURE 18.21 Block diagrams of landscapes during and after glaciation, showing the origin of various continental glacier depositional features.

FIGURE 18.22 The checkerboard pattern of productive farmland attests to the fertility of the soil on this flat till plain southeast of Aberdeen, South Dakota. (John S. Shelton)

areas of both the United States and Europe. These moraines take the form of long belts of low, gently sloping ridges or hills. They are arcuate in shape, bulging southward; this reflects the fact that the ice sheets, along whose melting edges the moraines were deposited, were lobate in form. Well-developed recessional moraine systems are especially numerous in the vicinity of the Great Lakes and in the area to their south and west in Ohio, Indiana, Illinois, Minnesota, and the Dakotas. In Eurasia, a series of moraines extends some 4,800 kilometers (3,000 mi) from northern Germany through Poland and the western Soviet Union to north-central Siberia. Both Long Island, New York, and the Danish Peninsula consist largely of morainic deposits. **Interlobate moraines**, originally deposited at the juncture of two adjacent lobes of ice, also are relatively numerous in the Great Lakes area. In contrast to the east-west oriented terminal and recessional moraine systems, interlobate moraines typically have a north-south orientation.

Another, smaller feature frequently encountered on till plains is the **drumlin**. Drumlins are smooth, elongated hills composed of till. When well formed, they are oval in shape, resembling an egg that has been cut in half lengthwise and placed with its flat surface down. They generally occur in groups and are aligned with their tapered ends pointing down-ice (see Figure 18.23). Some question exists as to their exact method of formation, but many apparently were produced by the plastering of till around subglacial obstructions lying in the path of the ice in places where it was overloaded with debris. A typical drumlin is perhaps 30 meters (100 ft) high and 600 meters (2,000 ft) long. Especially large groups or "swarms" are located in the Lake Ontario Plain of upstate New York, southeastern Wisconsin, northern Lower Michigan, the Niagara Peninsula of Ontario, and in Ireland and Scotland.

Till plains, moraines, and drumlins are all features produced by direct glacial deposition. Glaciofluvial features also are of great importance in areas of Pleistocene continental glaciation. Some of these features formed above, within, or beneath the glacier, but most formed beyond the melting edge of the ice.

Foremost among the glacio-fluvial features in importance and areal extent are outwash plains and lacustrine plains. **Outwash plains** may be regarded as the glacio-fluvial counterparts of till plains. They consist of extensive areas covered by stratified drift deposited by meltwaters flowing outward from glaciers. Not only are the deposits vertically stratified, they also tend to be horizontally graded, with coarser

FIGURE 18.23 These elongated hills east of Rochester, New York, are drumlins. Agricultural fields, which have followed the contour of the land, typically run parallel to the orientation of the drumlins. (John S. Shelton)

sediments deposited near the ice front and progressively finer sediments away from the ice. Most outwash plains have rather smooth surfaces, but pitted outwash plains, containing numerous water-filled kettles, have formed where large blocks of ice became detached from the glacier and settled onto the outwash before melting.

Outwash plains are scattered throughout the areas of Pleistocene glacial deposition, primarily in the Great Lakes region and Upper Midwest of the United States and in the Northern European Plain. Areas of outwash plains are interspersed with moraine systems, till plains, and lacustrine plains in complex patterns because of the numerous changes in position of the ice front. Where the ice front remained stationary for an extended period, however, an orderly sequence of deposits resulted, with till plain deposits beneath the ice, a terminal moraine along the forward edge of the ice, and outwash plain deposits beyond the ice margins.

Outwash plains formed where the surface sloped away from the glacier front; this allowed sediments to be transported and deposited by braided meltwater streams. In places where the surface sloped toward the ice front, however, drainage was impeded and ice-dammed lakes frequently formed. Very fine sediments accumulated on the floors of these lakes, and their eventual drainage produced **glacial lacustrine plains.** These areas now have exceptionally smooth, level surfaces and, like the outwash plains, have been converted largely to productive agricultural land. Many of the Pleistocene lakes were very large, and the lacustrine plains they created currently cover extensive areas. In North America, the largest lake of all was Lake Agassiz. Late in the Wisconsinan glacial stage it covered more than 260,000 square kilometers (100,000 mi^2) in eastern North Dakota, Manitoba, Saskatchewan, and Ontario. Most of this area has now drained, although it is still dotted with thousands of smaller lakes. The largest surviving remnant is Lake Winnipeg.

The Great Lakes, extensive at present, were much larger still during the late Pleistocene, when ice blocked their current outlet through the St. Lawrence Valley (see Figure 18.24). The Great Lakes went through a highly complex sequence of changes during the Wisconsinan Stage, but at their maximum extent they coalesced into a single giant inland sea, covering nearly twice their present area. Northern Eurasia also contained several large ice-dammed lakes. The largest was in western Siberia on the present-day site of the huge Vasyugansk Swamp that occupies the watershed between the Ob and Irtysh Rivers.

River systems also underwent profound changes as a result of glacial blockage and rerouting. The upper Missouri River, for example, originally flowed north to Hudson Bay. Glacial blockage diverted it southward and added it to the Mississippi River drainage basin. The present courses of both the Ohio and Missouri Rivers developed roughly parallel to the line of maximum glacial advance and served as lateral drainage routes for glacial meltwaters. During the late Wisconsinan Stage, when the St. Lawrence River was blocked by ice, the Great Lakes drained southward through several routes into the Mississippi River system (see Figure 18.24).

The rapid melting of great volumes of ice during the Pleistocene summers resulted in the production of enormous quantities of debris-laden glacial meltwater. Where valleys were present to channelize the flow, large meltwater rivers were formed. These rivers, in most cases overloaded with glacial debris, formed large glacial spillways floored by extensive valley train deposits, often extending hundreds of kilometers from the ice front. The Mississippi River Valley contains deep accumulations of glacially derived alluvium and can be regarded as a largely buried valley train of giant proportions.

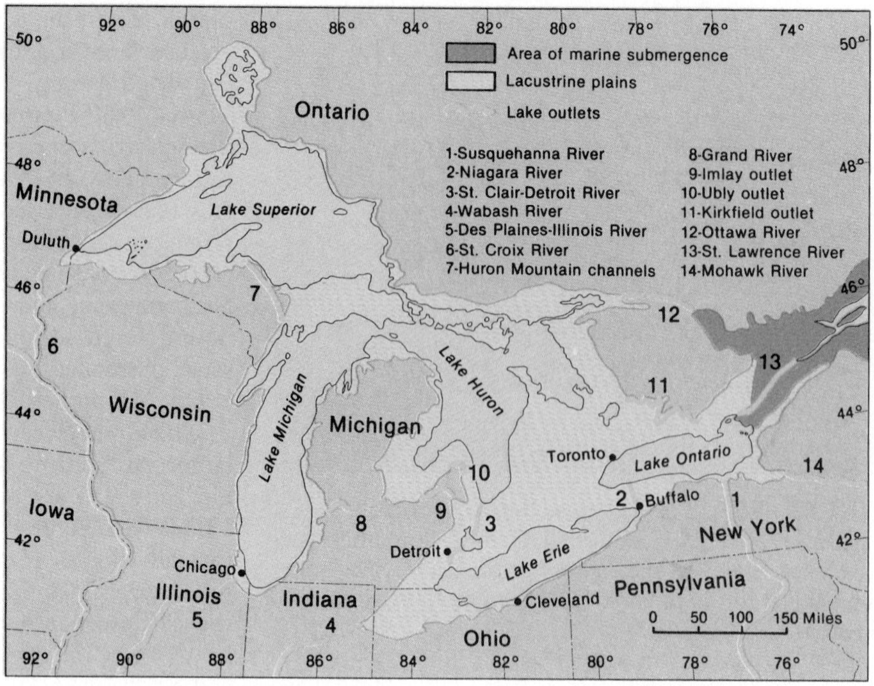

FIGURE 18.24 This map shows the expansion in areal extent of the Great Lakes during the late Pleistocene, when their present drainage route through the St. Lawrence River was blocked by ice. Major Pleistocene drainage outlets are numbered.

Eskers are among the more unusual features associated with continental glaciation. They are long, meandering ridges of stratified drift, usually located on till plains. Most apparently were deposited by subglacial streams flowing through tunnels in the ice near the margin of the glacier. Because the surrounding ice locked the stream in place, it was unable to shift its channel as it deposited sediment and aggraded its bed. This produced an inverted stream channel deposit—higher, rather than lower, than its surroundings. Eskers are especially numerous in New England, near Lake Michigan, and throughout the Canadian Shield and Scandinavia. Most formed during the final melting of the Wisconsinan ice sheets, some 6,000 to 10,000 years ago.

Another feature of glacio-fluvial origin is the **kame**, a mound of slightly sorted water-deposited debris. The exact origin of kames can vary, but they sometimes form when glacial meltwater, flowing on the ice surface, enters a crevasse and descends to the base of the glacier. When the water reaches the bedrock floor of the glacier its velocity is reduced, and the debris it carries is deposited to form the kame.

Indirect Pleistocene Effects

The effects of the Pleistocene were not limited solely to those areas covered by glaciers or glacially derived meltwaters; probably every part of the Earth experienced changes in climate during this period. Pleistocene temperatures over most of the Earth are estimated to have averaged some 3 to 6 C° (5-10 F°) lower than at present. Pleistocene temperature fluctuations were responsible for major redistributions of plants and animals, including the extinction of numerous species. The cooler world climate during the Pleistocene glacial stages also influenced the locations of the global air pressure and wind belts. These belts, with their associated climatic conditions, generally shifted equatorward during cold periods, giving many areas markedly different precipitation patterns than they have at present.

The cooler, wetter Pleistocene climate resulted in the formation of large lakes in a number of currently dry areas with interior drainage. The currently arid Great Basin of the western United States, for instance, contained a large number of extensive Pleistocene lakes that have now either dried completely or have shrunk to small, often highly saline remnants (see Figure 18.25). The largest lake of all was Lake Bonneville, which at one point occupied more than 51,000 square kilometers (20,000 mi^2) in Utah, Nevada, and Idaho. Its chief remnant is the Great Salt Lake, with an area of 2,400 square kilometers (940 mi^2). Lake Bonneville fluctuated considerably in extent, and a series of elevated terracelike abandoned

FIGURE 18.25 Present-day lakes of the western United States (shown in light blue) occupy only a small proportion of the area covered by lakes during the cooler and wetter Pleistocene epoch (dark blue).

shorelines can be found at sites as much as 300 meters (1,000 ft) above the present surface of the Great Salt Lake. Numerous other large lakes existed in western Nevada, southern Oregon, and southeastern California, as well as in other parts of the world.

The drying of the climate in some areas apparently has continued into recent times. Large portions of the northern Sahara, for example, are covered by giant drainage networks that rarely, if ever, carry water at present. Extensive areas in extreme North Africa were forested only a few thousand years ago. The disappearance of these forests has been attributed primarily to human activities, but the present climate apparently is too dry to permit their return.

Another Pleistocene effect with worldwide implications was changing sea levels. The alternate formation and melting of the ice sheets involved the withdrawal and addition of vast quantities of ocean water. During the height of the glacial stages, enough water was locked in the glacial ice to lower global sea levels some 90 to 150 meters (300 to 500 ft). As a result, the Earth's appearance was altered; the surface was covered by less water and more land, and the shapes and locations of coastlines differed greatly from those of the present. For example, the British Isles were connected to mainland Europe, and North America and Eurasia were joined across what is now the Bering Strait. It is believed that the Bering Strait land bridge allowed Asiatic nomads who were the forebears of native Americans to migrate from eastern Asia to the Americas.

Conversely, during the interglacial stages, sea levels were substantially higher than at present. A series of seven Pleistocene marine terraces, each representing a past position of the coastline, has been recognized along the Atlantic and Gulf coastal plain of the Untied States. The continued slow melting of the world's remaining glaciers is partially responsible for the gradual rise in ocean levels observed on most present coastal margins.

A related Pleistocene influence is the **isostatic rebound** of previously glaciated areas. The great weight of the continental glaciers depressed the underling surface much as the pressure of a person's thumb might depress the surface of a rubber ball. When the pressure is removed, the ball springs back to its original shape. In similar fashion, the recent melting of the Laurentide and Fennoscandian ice sheets has allowed the areas formerly covered by the ice to rebound slowly to their original elevations. This process has been continuing for thousands of years and is now perhaps two-thirds completed. The areas that are rebounding are increasing in elevation with respect to sea level, causing their coastlines to advance seaward. As a result, countries such as Canada and Finland gain land each year. The displaced ocean water, which is relocated southward, has contributed to the current slow rise in sea level over much of the world. The resulting widespread problems of coastal flooding and beach erosion are discussed in the Case Study at the end of Chapter 20.

THE PERIGLACIAL ENVIRONMENT

The term **periglacial** literally means "near glacial." It is the climate of periglacial areas, however, rather than their location, that is near to glacial in its characteristics. Although conditions are warm enough, or dry enough, to prevent a permanent ground cover of ice and snow from forming, the climate is still severely cold. Landscape development in the periglacial environment is associated largely with the alternate freezing and thawing of the near-surface layers.

Location and General Appearance

At present, periglacial processes are most active in tundra climate areas. The tundra zone, as described earlier, includes areas with a warmest monthly mean temperature of 0°C to 10°C (32°F to 50°F). The absence of periglacial conditions in areas averaging below 0°C (32°F) in all months of the year results from the permanently frozen and generally snow- and ice-covered nature of the surface. Conversely, areas with temperatures averaging above 10°C (50°F) in summer generally are forested. Trees reduce the effectiveness of the freeze-thaw cycles and hold the soil firmly in place. The periglacial realm therefore is restricted at present largely to the far northern portions of North America and Eurasia, to some unglaciated fringe areas of Antarctica, and to scattered high mountain and plateau areas including, notably, much of the Tibetan Plateau.

Periglacial conditions dominated much larger areas during the Pleistocene. As the Pleistocene glaciers advanced and retreated, periglacial environments occupied shifting zones generally adjacent to the ice margins. These included, at different times, most of Canada and Alaska, the northern United States, much of northern and western Europe, much of northern Asia, and scattered areas in southern South America, New Zealand, and southeastern Australia, as well as extensive upland areas in other locations. In some of these areas, relict periglacial features still are discernible.

The periglacial landscape, especially in colder locations, often consists of a lightly vegetated surface covered with a thin, stony soil or regolith. Physical weathering processes are important, and in more rugged areas the surface is often littered with a jumble of angular rock fragments of various sizes. These rocks generally have been freed by frost shattering from adjacent hillsides. In flatter areas, which are extensive, the surface material is mostly fine in texture and a thin soil cover has formed. Drainage tends to be very poor, and irregularly shaped lakes and swamps, generally of glacial origin, are common in many lower and flatter sites. The entire area is frozen and largely snow-covered for 8 to 10 months of the year. Vegetation, which is more abundant toward the warmer margins, is of the tundra type and consists of a mixture of grasses, small flowering plants, heaths, mosses, and lichens.

Major Periglacial Features

Periglacial areas display a variety of distinctive surface features. The most prominent of these are patterned ground features and mass movement features. Both owe their existence largely to the alternate freezing and thawing of the surface.

Permafrost, or perennially frozen ground, underlies extensive land areas of the higher latitudes, including not only glaciated Antarctica and Greenland, but also much of northern North America and Eurasia. It extends well beyond the periglacial margins and into much of the taiga forest zones of these continents (see Figure 18.26). Permafrost penetrates to an average depth of about 300 meters (1,000 ft), but

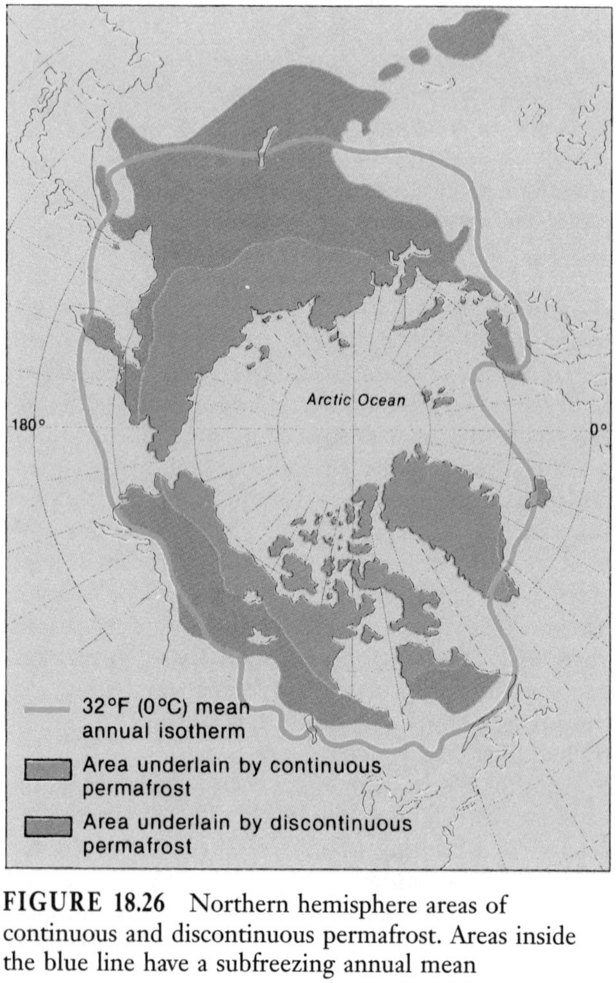

FIGURE 18.26 Northern hemisphere areas of continuous and discontinuous permafrost. Areas inside the blue line have a subfreezing annual mean temperature.

is as much as 600 meters (2,000 ft) deep in parts of northeastern Siberia.

Frozen ground serves as a barrier to the downward percolation of water. Although the upper meter or so of soil generally thaws in summer in unglaciated areas, the underlying permafrost traps water in the surface layer, resulting in nearly saturated conditions in the summer melt season. Poor drainage in many areas is reinforced by the Pleistocene disruption of the surface drainage network, as well as by cool summer temperatures which reduce evapotranspiration rates.

Repeated freezing and thawing of the surface, with its attendant expansion and contraction, produces a churning action in the upper soil and rock material, often with rather striking results. The surface material tends to develop polygonal patterns of a symmetrical nature, often covering vast areas (see Figure 18.27). These so-called **patterned ground** features may be sorted or unsorted. Sorted patterns display a gradation from the finest material in the center of each polygon to the coarsest material, often in the form of large rocks, filling the cracks between the polygons. The sorting mechanism apparently is volume changes associated with the repeated freezing of near-surface water. This pushes rocks first to the surface and then laterally into the cracks. Nonsorted patterns show no significant size gradation of particles, but are evidenced by the slightly raised or lowered edges of the polygons or by variations in vegetation.

FIGURE 18.27 Patterned ground is caused by the contraction and cracking of the near-surface soil layers as they freeze. The cracks are occupied by vertical ice wedges that push the cracks farther apart as they expand. (Visuals Unlimited/©Steve McCutcheon)

Mass movements also play a crucial role in periglacial landscape evolution. The chief factors favoring mass movements are the existence of saturated soil conditions in summer, a scarcity of stabilizing vegetation, and most important, the action of ground ice. The volumetric expansion associated with the freezing of water loosens the regolith. On flatter surfaces this produces patterned ground features, which are actually small-scale mass movement phenomena. On sloping ground it results in a net downhill motion, generally consisting of a form of creep.

On many slopes, the upper surface of the permafrost, during the summer melt season, serves as a sliding plane for the overlying saturated masses of soil. The result is a type of earth flow termed **solifluction**, which may cause an entire hillslope to be covered by hundreds of solifluction lobes bulging downhill. The combined actions of creep and solifluction typically produce a steady transport of regolith down slopes that in most other environments would be too gently inclined to experience mass movements of any significance. Other, more rapid mass movements also occur on steeper slopes under periglacial conditions, but are less widely distributed. Taken as a whole, mass movements exceed fluvial processes in importance in the periglacial environment, so that landscape evolution is accomplished more by slope retreat through mass wasting than by fluvial erosion.

SUMMARY

This chapter has examined the influence of accumulations of ice on the Earth's land surface. Ice in the form of glaciers is the most powerful of the three tools of gradation. It also is an effective landscape modifier in periglacial environments, where it exists in a less concentrated form as ground ice. Despite its power as a geomorphic agent, ice is geographically restricted to low-temperature environments at high latitudes and elevations. In the geologically recent past, however, colder worldwide temperatures prevailed, and ice was of much greater geomorphic importance than at present. Many relict glacial and periglacial features still exist in areas that formerly served as sites of ice accumulation.

We now return to the Focus Questions posed at the beginning of the chapter:

1. *What are glaciers, and why do they form?*

Glaciers are large accumulations of freshwater ice derived primarily from the compaction of snow. They form on land areas where snowfall receipts exceed melting rates over an extended timespan.

2. *What major categories of glaciers exist, and where are they found?*

Glaciers can be divided into two basic types—alpine glaciers and continental glaciers. Alpine glaciers, which are much smaller than continental glaciers but are more widely distributed, generally assume the form of long, relatively narrow tongues of ice occupying depressions in high mountain regions. Continental glaciers, which form over extensive high latitude land areas, generally are domelike and move outward in all directions from their centers of accumulation. At present, continental glaciers occupy the continent of Antarctica and the interior of Greenland.

3. *In what ways do glaciers modify the landscape?*

Glaciers modify the landscape by eroding and depositing surface materials. Erosion is accomplished by the entrainment or "bulldozing" of weathered surface materials, and, to a lesser extent, by the scraping of rocks frozen into the base of the ice against the surface. The chief erosional landforms of alpine glaciers are cirques and glacial troughs; the ice-scoured plain is the chief erosional landform of continental glaciers.

Most glacial deposition occurs after the ice melts sufficiently to become overloaded with sediment. Sediment is transported in a conveyor-belt fashion from the glacial erosional area to the glacial depositional area. The chief depositional features of alpine glaciers are lateral and terminal moraines. Major depositional landforms produced by continental glaciers are terminal and recessional moraines and till plains. In addition, glacial meltwaters can carry large amounts of glacially derived materials away from the ice front to produce glacio-fluvial features such as lacustrine plains, outwash plains, and valley trains.

4. *What was the Pleistocene, and why did it occur?*

The Pleistocene was a geologically recent period of colder temperatures and widespread glaciation. There have been about 10 periods or "stages" of extensive glaciation over the past two million years, separated by relatively warm interglacial stages. During the glacial stages continental glaciers formed over northern North America and northwestern Eurasia, leaving a major imprint on those areas today. Alpine glaciers also increased greatly in number and extent. The cause of the Pleistocene is not known, but the present location of the North Pole in the restricted waters of the Arctic Ocean is probably an important factor. The cyclic alternation of glacial and interglacial stages within the Pleistocene is apparently the result of long-term variations in the Earth's orbit around the Sun. There is growing evidence that the Pleistocene has not yet ended, and that the Earth is currently in an interglacial period.

Review Questions

1. What is a glacier? What conditions are necessary for a glacier to form?

2. What areas currently are covered by continental glaciers? By alpine glaciers?

3. What is the difference between the absolute motion and the relative motion of a glacier? What factors are responsible for the glacier's rate of motion in each case?

4. How does the erosional effectiveness of glacial ice compare with that of flowing water?

5. How do glaciers transport and deposit regolith? How do the characteristics of direct glacial deposits differ from those of fluvial deposits? What are glacio-fluvial deposits?

6. At what specific sites within a mountainous region do alpine glaciers generally form? What determines the paths they take as they flow?

7. What are some of the common surface features of alpine glaciers? How does the surface of an alpine glacier change in appearance between the zones of accumulation and ablation?

8. Describe the overall impact of extensive alpine glaciation on a mountainous region such as the Alps. What are some of the more important erosional and depositional landforms?

9. How do the physical characteristics of alpine and continental glaciers differ?

10. What was the Pleistocene? Briefly describe three hypotheses regarding its origin.

11. Describe the maximum geographical extent of Pleistocene glaciation in North America and Eurasia.

12. Describe the present surface characteristics of the areas that experienced a predominance of continental glacial erosion during the Pleistocene. Where are these areas located?

13. What are till plains? What are the present surface characteristics, geographic locations, and economic significance of the major till plains produced by the Pleistocene glaciers of North America and Eurasia?

14. What are the two most extensive types of glacio-fluvial features produced by continental glaciers? What are their surface characteristics? Where are they found in relation to till plains?

15. What changes did the Pleistocene glaciers produce in the locations of major rivers and lakes in North America?

16. What changes occurred during the Pleistocene in areas of the low and middle latitudes far removed from the ice sheets?

17. Under what circumstances do periglacial conditions exist? What parts of the world currently have periglacial conditions?

18. What is permafrost? What environmental problems are caused by the presence of permafrost?

19. Describe the general mechanism involved in the formation of most small-scale periglacial landforms. What specific types of landforms are produced by this mechanism?

Key Terms

glacier
continental glacier
alpine glacier
till
moraine
crevasse
cirque
horn
glacial trough
fjord
Pleistocene
Laurentide ice sheet
ice-scoured plain
till plain
drumlin
glacio-fluvial features
outwash plain
esker
periglacial
patterned ground

CASE STUDY

A GLACIAL CATASTROPHE: THE SPOKANE FLOOD

Geomorphologists and other Earth scientists are on the whole committed to the principle of **uniformitarianism**, which is often described by the expression, "the present is the key to the past." The idea is that natural Earth phenomena of the geologic past were produced by the same physical processes that operate at present. On occasion, however, landforms are found that differ so greatly in appearance or magnitude from those with which we are familiar that at first they seem to defy the uniformitarian concept. An area containing such landforms exists in the Columbia Basin of southeastern Washington. After more than 50 years of study, Earth scientists have concluded that many of its features were produced by the largest verified floods in the history of the Earth.

The Columbia Basin is a low-elevation portion of the Columbia Plateau. The climate is arid, and the topography varies from flat to hilly. Despite its present aridity, however, a large portion of this area shows strong evidence that short-lived floods of gigantic proportions occurred during the Pleistocene. An area of approximately 28,500 square kilometers (11,000 mi^2) is covered by a variety of fluvial erosional and depositional features, some on a scale unknown anywhere else in the world. Among these features are numerous dry canyons (called **coulees**) containing streamlined mesas and buttes, rockrimmed basins, low basaltic ridges apparently eroded into segments by rushing floodwaters that overtopped them, dry waterfalls, giant gravel bars, and, perhaps most amazingly, ripple marks 15 meters (50 ft) in height and spaced 150 meters (500 ft) apart (see Figure 18.28).

Debate on the origins of this area, called the Channelled Scablands because of the abandoned river canyons scarring its surface, continued for many years following its first description in a 1923 study by J. H. Bretz.[1] Many geomorphologists originally believed that it resulted from a long-continued flow of glacial meltwater. It gradually was realized, though, that features such as the overtopped drainage divides and giant ripple marks could be explained only as the result of one or more gigantic, but very brief, floods. The problem was then to determine how and where such floods could have originated.

In 1942 an answer finally was found. It had been known for some time that, during the Wisconsinan glacial stage, much of what is now western Montana was covered by a glacially dammed lake that was given the name Lake Missoula. J. T. Pardee, a geomorphologist employed by the U. S. Geological Survey, hypothesized that the lake was blocked by a lobe of ice that extended down the Purcell Trench in northern Idaho.[2] The melting or

continued on next page

[1]J. H. Bretz, "The Channelled Scablands of the Columbia River Plateau," *Journal of Geology*, 31 (1923):617-49.

[2]J. T. Pardee, "Unusual Currents in Glacial Lake Missoula," *Bulletin of the Geological Society of America 53* (1942):1569-1600.

FIGURE 18.28 Giant ripple marks produced by the Spokane Flood. (John S. Shelton)

breaching of this ice lobe allowed the estimated 1,250 cubic kilometers (300 mi^3) of water in Lake Missoula to be released within a period of a few hours or days, creating the catastrophic flood that carved the Channelled Scablands (see Figure 18.29). This origin now is generally accepted.

Recent fieldwork in the area indicates that Lake Missoula drained suddenly and catastrophically on numerous occasions during the Late Pleistocene. Geologist R. B. Waite has presented evidence of 40 catastrophic floods during the Late Wisconsinan period (between about 17,000 and 12,000 years ago) alone.[3] He believes that, for a time, the lake filled and then suddenly drained at intervals of only a few centuries or even decades. The drainage apparently did not occur due to a warming of the climate, but took place when the water level in Lake Missoula rose to approximately 90 percent of the height of the ice dam (the Pend Oreille lobe of the Cordilleran ice sheet). At this point, the ice was buoyed up, water began flowing under the dam, and subglacial tunnels were rapidly carved that allowed the lake to drain completely. Each breached ice dam was soon replaced by fresh ice as the glacier continued to move south, while Lake Missoula, fed by vast quantities of meltwater from the Cordilleran ice sheet to its north, refilled in a relatively short period.

Estimates of the volumes of flow and duration of the floods are imprecise at best, but it is believed that Lake Missoula may have drained at an approximate rate of 40 cubic kilometers per hour (9.5 mi^3/hr) over a period of 40 hours. This is about 10 times the present combined volume of flow of *all* the world's rivers.

[3]R. B. Waite, Jr., "About Forty Last-Glacial Lake Missoula Jokulhlaups through Southern Washington," *Journal of Geology* 88 (1980):653-79.

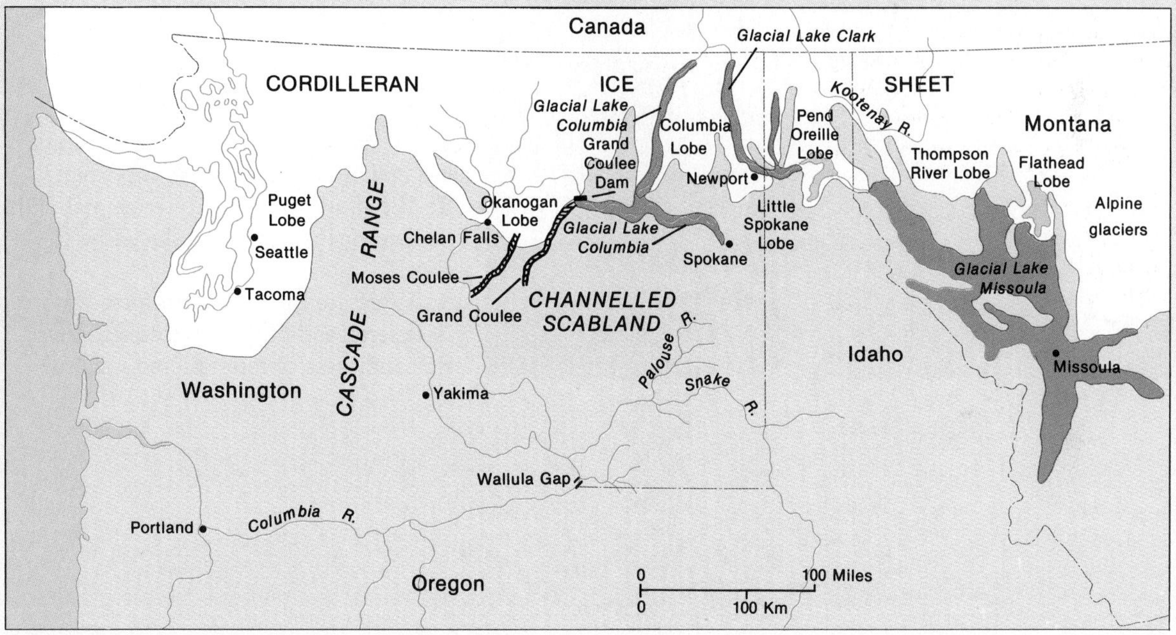

FIGURE 18.29 Location of glacial Lake Missoula and the Channelled Scablands at the margin of the Pleistocene Cordilleran ice sheet.

CHAPTER 19

EOLIAN PROCESSES AND DESERT LANDSCAPES

OUTLINE

FOCUS QUESTIONS

1. In what geomorphic environments are eolian processes most important?
2. What are the primary erosional and depositional features produced by the wind?
3. What are the major landform features of deserts, and how do they differ from the landforms of more humid regions?

This chapter examines the geomorphic aspects of two related topics. We begin with a discussion of the surface effects of the wind. The wind is the weakest of the tools of gradation; it attains importance as a landscape modifier only in areas devoid of a protective vegetation cover. Most such areas have arid or semiarid climates, although sandy beaches and other localized environments also can be strongly influenced by the effects of the wind. Geomorphic processes performed by the wind are called eolian processes after Aeolus, the Greek god of the winds.

The second section departs from the single geomorphic approach of the last several chapters and examines the landscape characteristics of desert regions. Desert landscapes are not dominated by a single geomorphic process; instead, they are polygenetic in origin, with tectonic, fluvial, and eolian processes being most important. The specific actions and the extent of the contributions of each of these processes differ from place to place, causing deserts to vary greatly in appearance. As a group, though, the surface characteristics of deserts differ in several basic respects from those of humid landscapes.

THE WIND AS A GEOMORPHIC AGENT

The ability of desert windstorms to erode and transport enormous quantities of loose sand and dust has long been demonstrated to travelers in arid lands, and until fairly recently the wind was believed to be the dominant geomorphic agent within desert regions. This assessment of the wind's surface influence has been downgraded considerably, however, because the erosive power of the wind on consolidated materials has been found to be minor. Its weakness results largely from the gaseous condition of the atmosphere and the air's very low density ratio of about 1:2,000 as compared to rock. (Water and ice, by contrast, have density ratios of about 1:3 when compared to rock.) As a result, the wind is restricted primarily to the redistribution of loose, fine-textured surface materials. Where such materials are plentiful and are not immobilized by moisture or vegetation, the wind can perform substantial erosional and depositional work and produce a variety of distinctive surface features.

Wind Sediment Transport Mechanisms

The wind transports sediments in much the same manner as does flowing water, although some important differences exist. Both wind and water are fluids whose flow over a surface generates turbulence. The upward component of the turbulent eddies allows both fluids to pick up loose sediments and to carry them in suspension as an important mechanism of transport. Although the wind typically flows much faster than water, its ability to carry suspended particles is reduced by its lower density. As a result, normally only silt and clay particles are carried in suspension by the wind. These fine particles, however, can be lifted high into the air and quickly transported for great distances before being deposited.

Where winds are exceptionally strong and large quantities of loose soil are available, **dust storms** are sometimes produced. These storms can reduce surface visibilities to only a few meters and can last for hours. Tremendous volumes of valuable topsoil can be lost from agricultural regions. The best-known dust storms in the United States occurred during the 1930s in the "Dust Bowl" region of the southern Great Plains (see Figure 19.1). Although improved agricultural practices and moister climatic conditions have greatly reduced the frequency of such storms within this region, severe dust storms are still common in portions of Africa, Asia, and Australia.

If sediments are too coarse textured to be carried in suspension, wind—like water—can transport them by surface creep or by saltation over the surface. **Surface creep** refers to the slow shifting of surface materials by rolling or sliding; it accounts for approxi-

FIGURE 19.1 The awesome front of a dust storm approaching Springfield, Colorado, in the spring of 1937. This storm produced nearly total darkness for a half-hour period. (AP/Wide World Photos)

mately one-fifth of all eolian surface sediment transport. **Saltation** is an asymmetrical bouncing motion of surface grains. It is caused by both the aerodynamic lift resulting from the increase in wind speed with height above the ground and the impact of other saltating grains. Saltation moves materials rapidly and accounts for about four-fifths of all eolian surface sediment transport. It is through saltation that sand grains are aggregated into shifting sand dunes. Because saltation rarely lifts sand grains more than a short distance above the surface, true **sandstorms** consist of a stinging mass of wind-whipped sand grains extending only a meter or two above ground level.

It has been found that, in order to initiate the eolian transport of particles of any given diameter, a critical wind speed threshold is needed to produce the necessary frictional pull or **drag velocity** (see Figure 19.2). Once particles begin saltating over the surface, though, their impacts with other particles can sustain the motion at somewhat lower wind speeds.

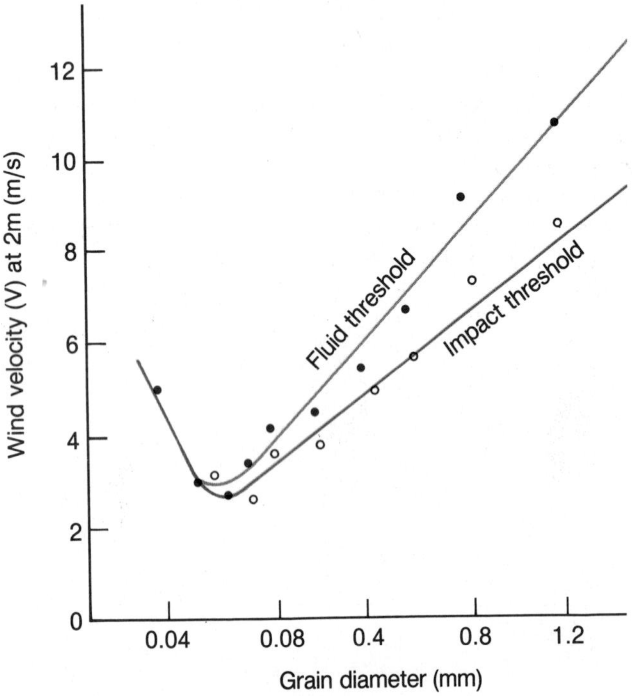

FIGURE 19.2 The relationship between wind drag velocity and particle diameter. The fluid threshold is the minimum wind speed needed to initiate grain movement by the force of the wind alone. The impact threshold is the minimum wind speed needed to sustain the movement of a grain that has been first set in motion by the impact of another particle.

Despite the general similarities between the transport of sediments by water and by wind, several important differences exist between the two. Wind transport is much more rapid but is restricted to finer textured sediments than is water transport. In addition, the wind does not carry a dissolved sediment load. Finally, the wind is much less restricted than water in the directions it can travel and in the sites in which it can erode or deposit sediments. Water always transports sediments downhill, while the wind can transport sediments either uphill or downhill (although downhill movement predominates). Fluvial processes are confined largely to the lowland sites where water flow is concentrated. In contrast, the wind flows over the entire surface and is capable of eroding or depositing materials almost anywhere within a suitable region.

Eolian Environments

In order for eolian processes to be effective, surface materials must be fine textured, dry, and loose. In addition, the affected area must experience at least occasional strong winds. A vegetation cover greatly diminishes the effectiveness of the wind by producing a large amount of near-surface friction, by covering the soil, and by holding the soil particles in place. For these reasons, areas subject to significant wind erosion must be at least partially devoid of vegetation.

The most extensive geomorphic environments meeting these criteria are deserts. Every continent contains large deserts where the wind plays an important role in landscape development. In most deserts, wind erosion has predominated over deposition. The surface in such areas is covered largely by stones of various sizes because the finer sand, silt, and clay particles have been winnowed out by wind action. Approximately one-fourth of the world's deserts, though, have been the sites of net deposition and are covered partially or completely by sand.

Another well-known but much less areally extensive category of eolian environment consists of the coastlines of large water bodies, including the ocean, seas, and large lakes. Along coastlines, waves and currents supply weathering products susceptible to wind action, and vegetation is discouraged by storms and tides. The coastal environment is described in detail in Chapter 20.

Three types of more localized environments also may contain large quantities of loose, fine-textured sediments capable of removal or redistribution by the wind. One includes areas underlain by weakly cemented sandstone. In such areas, weathering activities detach the individual grains to form loose sand.

The valleys of streams draining arid and semiarid regions also may be important sites for eolian processes. These streams support relatively little streambank vegetation. They also fluctuate greatly in volume of flow, often exposing extensive strips of sandy alluvium during periods of low water. Strong winds are able to remove the silt and clay particles and to collect the sand grains into dunes. The floodplains of several large rivers of the American Great Plains, including the Missouri, Platte, and Arkansas, are important sites for these eolian activities.

The third local environment is a legacy of the Pleistocene, when tremendous quantities of fine-textured glacial sediments were transported southward from the margins of the North American and Eurasian ice sheets by glacial meltwaters. These meltwaters were not always organized into discrete streams, but often spread over large areas in braided or overland flow patterns. They frequently were overloaded with sediments, which they deposited over extensive areas. The wind removed great volumes of predominantly silt-sized materials from these deposits, and their downwind deposition produced extensive windblown sediment or **loess** (from the German *löss*, or "loose") accumulations. Most glacial outwash deposits currently are stabilized by vegetation.

Over the last few centuries, eolian processes unfortunately have been augmented considerably by human activities. Especially affected are semiarid and seasonally dry regions in Africa and Asia that are used for agricultural or pastoral pursuits. The activities of the rapidly growing human populations within these regions have led to the extensive removal of the already marginal vegetation cover. The loss of protective vegetation, in turn, has caused the removal by the wind of tremendous quantities of irreplaceable soil. The human element in eolian sediment transport is explored further in the Case Study at the end of the chapter.

Wind Erosion

The wind erodes materials from the surface in two different ways. **Abrasion** involves the impact of saltating sand grains and often is described as a "sandblasting" effect. **Deflation,** on the other hand, involves the removal of loose surface materials directly by the wind.

The effect of wind abrasion on solid rock is relatively minor and generally superficial. The wind does not normally produce major bedrock erosional features; it merely modifies their surfaces by pitting, etching, smoothing, or polishing. These activities are most effective on weak sedimentary rocks such as soft shales and poorly consolidated sandstones. On such rocks, the wind tends to etch out micropatterns that reflect differences in rock hardness. **Ventifacts** are desert rocks with prominent wind-polished, grooved, or faceted surfaces. Sand grains large enough to perform abrasion are transported only a small distance above the surface. As a consequence, features standing more than a few meters above the ground are not significantly affected by this process (see Figure 19.3).

Deflation is a much more important mechanism of wind erosion than abrasion, but it requires the presence of unconsolidated surface materials such as sand or dry soil. In many instances the entire surface is rather evenly lowered by deflation; as a result no striking surface features are produced, and it is difficult to estimate the quantity of materials removed. On occasion, though, remnants of the original surface, perhaps anchored by the roots of a tree or shrub, are left to provide an indication of the depth of surface removal.

FIGURE 19.3 The effect of wind abrasion on solid rock is generally considered to be relatively minor. Repeated sandstorms, though, are sometimes capable of producing "mushroom rock" formations in soft rock outcrops because erosion is most vigorous just above the ground. The example here is in southeastern Iran. (Rodman Snead/©JLM Visuals)

One widespread category of features resulting from the differential erosion of surface materials is **deflation hollows** or "blowouts". These are shallow, often circular depressions that can vary greatly in size. The largest are more than 1.6 kilometers (1 mi) in diameter and up to 15 meters (50 ft) deep, although most are much smaller. Many thousands of such depressions dot portions of the southern Great Plains of the United States. These features, which may have been formed in part by solution and past bison wallowing, become the sites of shallow ponds after heavy rainfalls.

Deflation is the only natural mechanism that can remove sediments from enclosed desert basins. Large desert basins, such as the Qattara Depression in northwestern Egypt and several basins in interior Australia, may owe their origin primarily to deflation; in effect they may be gigantic blowouts.

In many arid regions the eolian removal of sand-sized and finer sediments has left behind deposits of pebbles, stones, and boulders that are too large and heavy to be blown away. In time, the effects of gravity, salt crystal growth, and ephemeral floodwaters can arrange these deposits into a **desert pavement** of close-fitting pebbles or stones that protect the underlying surface from erosion (see Figure 19.4). A stony desert from which the fine particles have been removed by the wind is called a **reg** in Arabic, and this term has been applied to similar deserts in other parts of the world. Regs cover large areas in the southwestern United States, the Sahara, and interior Australia.

FIGURE 19.4 A close-up view of a desert pavement surface in Nevada. The finer surface materials have been removed by the wind. (John S. Shelton)

Wind Deposition

Wind, like water, deposits sediments when its speed of flow and resulting turbulence become insufficient for it to continue to transport these materials. The transport of sand requires higher wind speeds and greater turbulence than does the transport of finer sediments, and this factor leads to the sorting of eolian deposits on the basis of mass. Where both sand and silt are derived from the same source region, the coarser sand particles usually are deposited relatively near the source of supply, while the windblown silt is deposited as loess over a larger region farther downwind. Not only are these two eolian deposits geographically separated from one another, but each has distinctive physical characteristics, typical landform assemblages, and human implications. They therefore are treated separately in the following discussion.

Topography of Sand-Covered Surfaces

Sand consists of loose grains of rock ranging from 0.0625 to 2.0 millimeters (0.0025 to 0.08 in) in diameter. Most eolian sand deposits are composed largely of the stable mineral quartz and have average grain diameters of about 1 millimeter (0.04 in). Such grains, when dry and loose, begin to move over the surface when wind speeds exceed 4.5 meters per second (10 mph).

Because sand does not normally travel far from its region of origin and can be moved by the wind only when it is dry and loose, it accumulates primarily in deserts. The impression that many people have of most deserts being blanketed by sand, however, is incorrect. Sand does not cover the majority of any large desert. Rather, it has accumulated within certain areas by the action of the wind, and, to a lesser extent, of ephemeral streams. Sandy deserts, sometimes referred to as **ergs** (another Arabic term), comprise at most one-fourth of the world's deserts. They contain not only local sand, but also sand derived from surrounding nonsandy desert areas. The quantities of sand within such an accumulation area vary greatly. In many instances, sand exists only in scattered patches; in other cases, it may deeply cover the entire surface (see Figure 19.5).

The major sandy deserts of the world are depicted in Figure 19.6. The largest ergs are located in the central Sahara, eastern Saudi Arabia, west-central Australia, southeastern Iran, northwestern India and adjacent Pakistan, and western China. In addition,

FIGURE 19.5 Giant sand dunes in the Sahara (Hammada du Guir, Morocco). (R. Krubner/H. Armstrong Roberts)

smaller sandy deserts are scattered throughout other arid lands, including the southwestern United States. A well-known sand accumulation in the United States is located within White Sands National Monument near Alamogordo in southern New Mexico. It is unusual in that it contains snow-white gypsum sand derived from deposits on a nearby dry lake bed. Great Sand Dunes National Monument in south-central Colorado is another sand accumulation site that has been preserved as a tourist attraction.

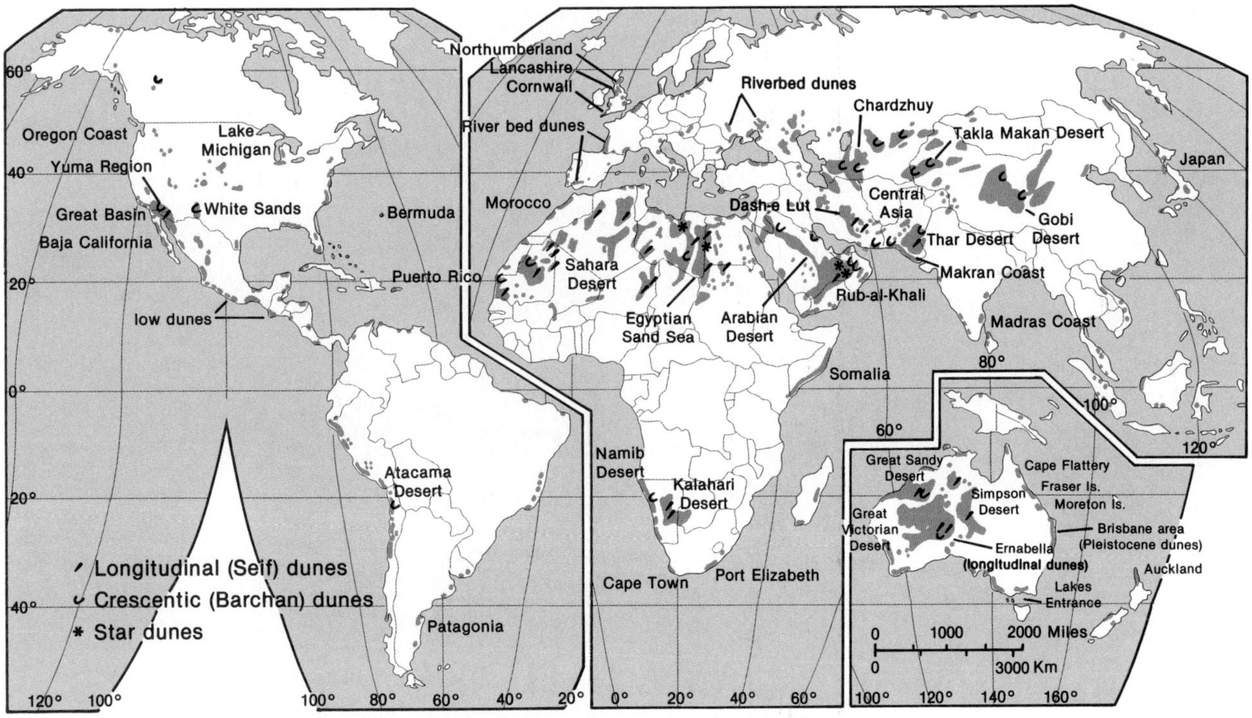

FIGURE 19.6 Global distribution of sand-covered regions and their characteristic dune forms.

The most characteristic and important surface features associated with sand-covered regions are dunes. A **dune** is a mound or ridge of wind-deposited sand. Most sand-covered regions exhibit dunal topography. Sand dunes vary greatly in size and shape; small dunes, which are most common, usually are 3 to 15 meters (10 to 50 ft) in height. Where sand is plentiful and winds are strong, though, dunes can attain heights of 180 meters (600 ft) and extend for many kilometers. The dimensions and forms of sand dunes are determined chiefly by the quantity of available sand, the speed and direction of the wind, and the presence and extent of stabilizing vegetation. Areas with winds that are relatively constant in direction generally have the most organized dune patterns.

Dunes form and grow largely because of the frictional resistance they offer to the wind. They are self-generating features because, as they grow, they become more effective barriers to the wind, reducing its speed and causing it to deposit any sand it may be transporting. Because other surface features also offer frictional resistance to the wind, dunes are best developed on flat, unvegetated surfaces.

The sides of most dunes are asymmetrical, with a gentler windward-facing **back slope** up which the sand grains are blown, and a steeper leeward-facing **slip face** down which the grains tumble until they land within a niche in which they can lodge (see Figure 19.7). The slip face normally rests at the angle of repose of dry sand, which is approximately 34° from the horizontal. When a slope receives so many additional sand grains through the process of surface creep that its steepness exceeds the angle of repose, a sand slip occurs, and a mass of sand slides down the slip face, generally coming to rest at its base. Through the mechanism of individual sand grains saltating up the back slope and masses of sand sliding down the slip face, the dune gradually changes its location and shape.

Dunes are described as "active" if they are not covered by vegetation and are currently moving and changing shape. Conversely, they are "inactive" if they are stabilized by a partial or complete cover of vegetation. Stabilized dunes often are relict features of a past drier climate. For example, the Sand Hills of central Nebraska is an area of sandy Pleistocene outwash deposits, reworked by the wind, that is currently covered by sod and grasses. The Palouse region of Washington is another important area of stabilized sand dunes (see Figure 19.8). In addition, large areas of stabilized dunes exist along the equatorward margins of the Sahara and Kalahari Deserts. They provide evidence that the subtropical deserts, at least in Africa, have shifted poleward in post-Pleistocene times.

Local wind systems frequently organize sand dunes into complexes that give a distinctive pattern to the topography. Dunes tend to be similar within a given area because the controlling factors usually are fairly constant. A basic distinction is normally made between longitudinal dunes, which are elongated parallel to the prevailing winds, and transverse dunes, which are elongated perpendicularly to the prevailing winds. The following paragraphs describe, and

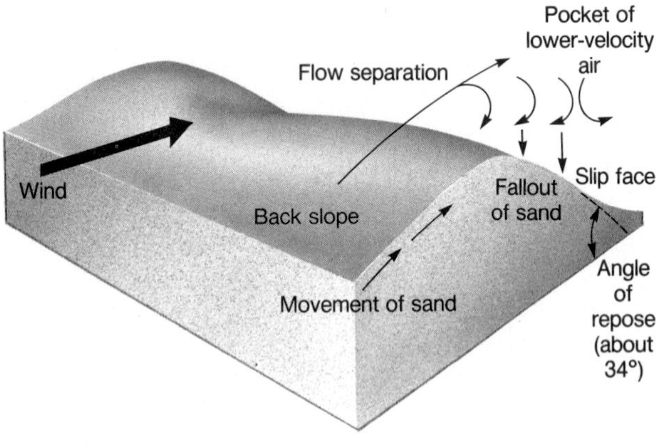

FIGURE 19.7 Profile view of a sand dune. The steeper slip face forms at the angle of repose of dry sand.

FIGURE 19.8 A sea of stabilized Pleistocene sand dunes in the Palouse region of Washington State. (H. Armstrong Roberts)

Figure 19.9 illustrates, some of the major categories of sand dunes on the basis of form.

Probably the most common and familiar of the dune forms are the **barchans.** They are relatively small, crescent shaped, and are oriented with their long axes perpendicular to the prevailing winds and their tapered ends or "horns" pointing downwind. When well formed, they are remarkably similar in shape to French crescent rolls. Barchans are found in portions of all large deserts. They develop in areas of relatively constant wind direction and restricted sand supply and are the most mobile of the major dune types.

Longitudinal dunes are long ridges of sand generally aligned parallel to the direction of the prevailing winds. They may become very large, attaining heights of more than 100 meters (300 ft) and lengths of many kilometers. Longitudinal dunes are especially well developed in central Australia, where they cover nearly one-fourth of the continent and are remarkable for their even spacing. Most of the Australian dunes are presently stabilized by vegetation and apparently formed during a past drier period. Currently active longitudinal dunes cover extensive desert areas in Saudi Arabia, Egypt, and Iran.

Transverse dunes are elongated perpendicular to the prevailing winds. Common in regions of abundant sand, they typically assume a wavelike pattern with undulating crest lines, and have been likened in appearance to a frozen picture of a storm-tossed sea. They are much less symmetrical and continuous than longitudinal dunes. Some transverse dunes develop a clearly distinguishable linked barchan form and may separate into individual barchan dunes if they extend into an area containing less sand. Transverse dunes frequently form on sandy beaches that have a prevailing onshore wind flow.

Sand dunes do not always develop the symmetrical forms just described. Complex dunes of almost any shape may form, especially in areas of variable winds. These include the so-called **star dunes** of Egypt and the Arabian peninsula, which are among the highest in the world. These consist of pyramidal mountains of sand with sharp crest lines on the various radiating ridges that form the "points" of the star. At the other extreme, some deserts are covered by flat **sand sheets** that display little or no tendency toward dunal formation. One of the best examples of these sand sheets covers a large portion of the Libyan desert.

Most sand dunes, unless stabilized by vegetation, are free-moving forms that gradually change their sizes, shapes, and locations. In contrast, fixed dunes, sometimes called **sand shadows,** frequently form in the lee of an obstruction to the wind such as a building or a rock outcrop.

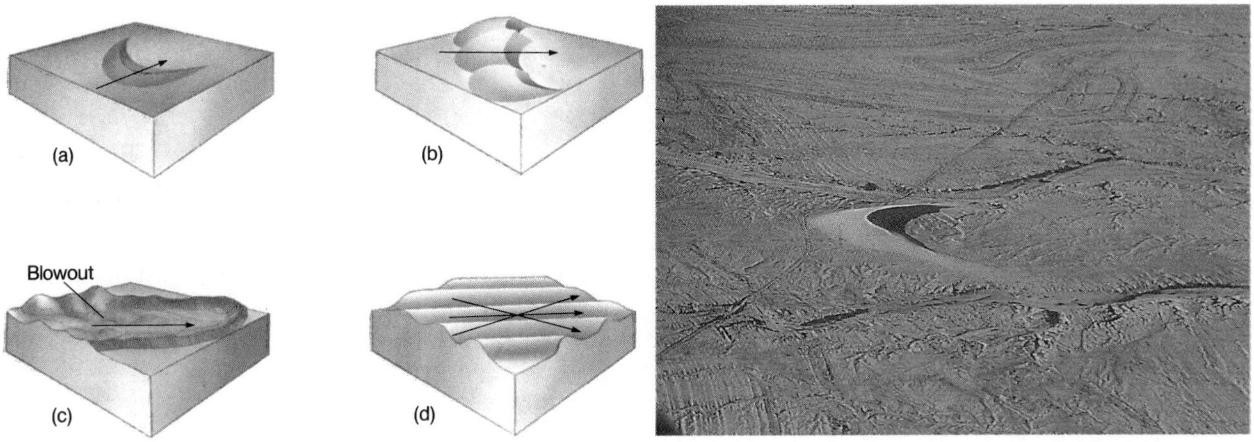

FIGURE 19.9 *Art:* (a) Barchan dune: A crescent-shaped dune with its horns pointing downwind. Barchans exist on hard, flat floors of deserts where there is constant wind and a limited sand supply. (b) Transverse dune: A dune that forms a wavelike ridge transverse to the wind direction. They exist in areas with abundant sand and little vegetation. (c) Parabolic dune: A U-shaped dune with its open end facing upwind. Some form by the piling of sand along the margins of a growing blowout. (d) Longitudinal dune: A linear dune oriented parallel to the wind. They exist in deserts with a limited sand supply and strong winds from one prevailing direction. *Photo:* A solitary barchan dune on the California desert west of the Salton Sea. (John S.Shelton)

Superimposed on all sand deposits are small **sand ripples** that contain only 2 to 5 centimeters (1–2 in) of surface relief. They are highly similar in cause and appearance to the small ripples on water that are superimposed on the larger wave forms on a windy day. Sand ripples are erased and re-formed during each windy period and develop an undulating pattern elongated perpendicular to the direction of the wind.

Topography of Loess-Covered Surfaces

Loess consists predominantly of silt-sized eolian deposits. It is typically buff-colored and is composed of a well-sorted mixture of quartz and calcite ($CaCO_3$) particles. Loess is derived from the same general areas as sand. The three most important loess sources are deserts, Pleistocene outwash deposits, and the floodplains of rivers in semiarid regions.

Despite its general similarity in origin to sand, loess differs markedly in most other geographic and geomorphic characteristics. Because it must be stabilized by moisture and a cover of vegetation in order to accumulate, it is not associated with desert environments. At present, most loess deposits are overlain by both vegetation and a surface soil layer; consequently they are not as easily recognizable as sand deposits, which often are devoid of both soil and vegetation. Soils derived from loess, in contrast to those developed from sand, often are highly fertile and agriculturally productive.

Uneroded loess deposits usually do not form distinctive topographic features because the wind spreads the loess as a smooth blanket over the surface. As a result, some of the most level terrain on Earth occurs in undissected loess regions. These regions become much more topographically distinctive, however, when subjected to stream erosion. Loess particles are angular and tend to interlock and develop a columnar structure. At the same time, the unconsolidated nature of loess readily permits streams to erode into the deposits and form nearly vertical banks or cliffs. As a result, eroded loess areas typically are dominated by deep, steep-walled valleys subject to rapid lateral and headward expansion (see Figure 19.10).

Loess, predominantly of Pleistocene origin, at present covers approximately 10 percent of the Earth's land surface and 30 percent of the Untied States. Depths range from as much as 90 meters (300 ft) immediately downwind from a major source to only a few centimeters. In general, the deposits are wedge-shaped, tapering off in a downwind (usually easterly) direction. The major loess-covered regions of the world are shown in Figure 19.11. Among the most important are portions of the North European Plain from Belgium eastward to the Ukraine, parts of the western Asiatic Soviet Union, northern China, the Pampas of Argentina, and the Great Plains, Midwest, and Mississippi River Valley in the United States. It is notable that most large deposits exist within the semiarid and subhumid areas of the middle latitudes that collectively comprise most of the world's major grain-producing regions.

FIGURE 19.10 This view of a roadcut in Chinese loess illustrates the ability of loess to maintain vertical banks. (Rodman Snead/©JLM Visuals)

Most loess deposits within the United States were derived from braided river courses such as that of the Missouri River, dry lake beds, and glacial outwash plains at times when these surfaces were largely devoid of vegetation. Maximum thicknesses of 15 to 30 meters (50 to 100 ft) occur on the eastern bluffs of the Lower Mississippi River and just east of the major Pleistocene outwash plains in Nebraska, Iowa, and northern Kansas. A smaller area of loess occurs in the Palouse region of southeastern Washington. The loess deposits of Europe and the southern Soviet Union also had a glacial eolian origin, but those of northern China and Argentina apparently were derived largely from deserts to their west.

DESERT LANDSCAPES

A **desert** is a large area with little or no vegetation. This broad definition can be taken to include the

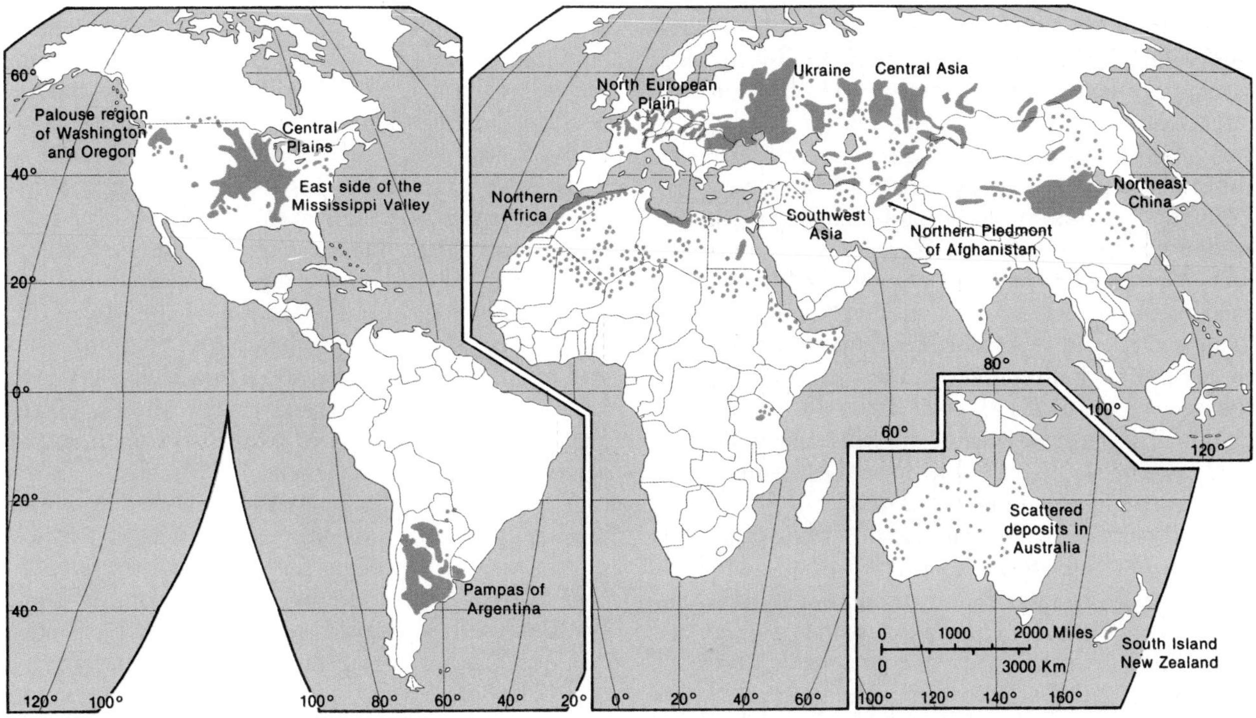

FIGURE 19.11 Global distribution of major loessal regions.

so-called polar deserts, where plant life is absent because of a combination of cold temperatures and a surface cover of ice and snow. In this section, however, the concept is restricted to deserts resulting from aridity.

The causes, climatic characteristics, and global distributions of arid regions were discussed in Chapter 9. It should be recalled that arid regions are those in which potential evapotranspiration rates exceed precipitation receipts on a long-term basis. Most are located in the subtropics and owe their dryness to the subsidence of air within the subtropical high pressure belts. Additional deserts exist in the middle latitudes. Middle latitude deserts sometimes are described as topographic or rainshadow deserts because their aridity results largely from the moisture-blocking influences of mountain barriers. In all, some 19 percent of the Earth's land surface consists of arid deserts.

Surface Characteristics of Deserts

The first-time visitor to a desert is struck immediately by the great contrast between the desert landscape and those of more humid environments. Perhaps the most fundamental impression is of the starkness of the land (see Figure 19.12). The surface is not covered by a mantle of soil and vegetation, but instead is likely to consist largely of a loose jumble of rocks of various sizes. The dominant colors are the buffs, grays, and browns of rock in various states of

FIGURE 19.12 Tilted fault-block mountains and salt-encrusted playas are common geomorphic features in the Great Basin of the western United States. (John S. Shelton)

weathering rather than the soft greens of live vegetation. The topography often appears highly angular, with a prevalence of horizontal and vertical surfaces and a lack of the smoothly rounded slopes typical of many humid landscapes. Even the clear weather and unlimited visibilities that prevail in most desert regions add to the impression that the universe consists predominantly of rock and sky.

The underlying factor controlling all the differences between desert and nondesert regions is the reduced availability of surface moisture. The low precipitation totals received in deserts not only discourage the growth of vegetation, but also profoundly affect the processes of landscape modification. Chemical weathering processes, which are highly dependent upon the presence of water, are especially affected. Although still important, they progress much more slowly than in humid climates and generally do not reach the clay mineral production stage. Physical weathering activities are less affected, because more of them require either no water or else just the short-term presence of water. The major physical weathering processes in deserts are salt crystal growth, frost wedging in colder areas, and perhaps thermal expansion and contraction related to daily heating and cooling.

The reduced rates of weathering cause bedrock to be exposed much more extensively in deserts than it is in most humid environments. Based on the nature of their surface materials, about one-fourth of the world's deserts may be classified as ergs, or sandy deserts. Over half are **regs,** or stony deserts. A third category of desert contains a solid bedrock surface, sometimes covered by a scattering of boulders. Known as **hamada,** they are less extensive than the other two categories. True desert regions generally contain only relict soils (left from a previous more humid period) or, locally, soils developed on alluvial deposits.

The angular topography that characterizes many desert regions has been attributed largely to the virtual absence of creep as a downhill transporter of weathered materials. Instead, most downhill movement of weathering products occurs by surface water transport during heavy rainstorms or by rapid mass movements such as rockslides and mudflows.

Despite its restricted availability within desert regions, water is still the most effective erosional agent for deserts as a whole, with wind relegated to a distant second place. Because less water is available for weathering and erosional activities, the overall rate of landscape evolution is much slower in arid than in humid environments. On the other hand, weathered surface materials within deserts are highly mobile because they are not stabilized by vegetation. This allows flowing water and strong winds to be effective transporting agents and results in the continual rearrangement of surface materials.

A factor increasing the long-term importance of fluvial processes in current desert regions is the relatively frequent shifts in position of the global climate belts. Most deserts apparently have not been arid for long periods of geologic time; many therefore contain relict fluvial landscape features that remain well preserved because of the slow operation of current gradational processes.

A basic fluvial characteristic of desert regions is the small quantity of water normally carried by stream systems. Streams serve the hydrologic purpose of removing excess precipitation from the land. Within deserts such an excess exists only rarely and locally. Water tables in desert regions generally are well below the surface. As a result, sites of ground water outflow are rare, and most streams are fed only by surface runoff. This causes nearly all desert streams to be ephemeral, carrying water for only short periods at infrequent intervals.

Major exceptions to the ephemeral nature of most desert streams are the **exotic rivers** that cross some deserts. These rivers originate in humid regions and pass through deserts on their way to the ocean. Although they have no tributaries and lose large volumes of water to seepage and evaporation as they traverse the desert, the larger rivers have sufficient flow to survive the passage. A large proportion of the human inhabitants of desert regions reside along the banks of exotic rivers because of the irrigation water, fertile alluvial soils, and means of transport they provide (see Figure 19.13). Some examples of important exotic rivers are the Colorado and Rio Grande in North America, the Nile and Niger in Africa, and the Indus in Asia.

Except in regions adjacent to exotic rivers or coasts, surface runoff in deserts does not reach the ocean, but instead flows into enclosed basins, where it eventually evaporates. Such desert regions are characterized by **interior drainage.** Runoff usually travels only a short distance before either sinking into the ground or entering an enclosed basin. Because weathering products are not transported out of the area, depositional features of various kinds—dunes, mass movement debris, and stream deposits—are especially abundant.

FIGURE 19.13 Irrigation water from the Colorado River makes intensive agriculture possible in this semidesert area near Bakersfield, California. (P. Degginger/H. Armstrong Roberts)

Desert regions are notoriously susceptible to destructive flash floods, which play a vital role in landscape development. Despite the overall paucity of precipitation, desert thunderstorms can produce localized torrential downpours. Flooding also is promoted by the impermeable nature of most desert surfaces. These surfaces are likely to contain a **duricrust layer** formed by the near-surface precipitation of salts, or to consist of a solid bedrock surface beneath a thin cover of sand or stones. No deep soil layer exists to absorb and hold moisture. The lack of vegetation and of soil organisms allows for little working and loosening of the weathered surface materials, so they are typically highly compacted. Finally, major drainage channels capable of carrying off large volumes of water are few and far between. This lack of surface drainage capacity causes runoff to flow in sheets or shallow rivulets over much of the surface during a heavy storm.

Because runoff can flow with little friction or absorption over the surface, it can travel very rapidly, especially on moderately to steeply sloping terrain. Dry stream channels therefore can be quickly transformed into raging torrents. Desert storms typically are highly localized and short in duration, though, so floods can subside almost as quickly as they arrive.

Although drainage routes in deserts usually are dry and large streams are rare, small stream channels often are closely spaced and highly visible. On a very localized scale, the surface in many areas is covered by an interweaving network of miniature drainage lines that may have widths of only 0.3 to 0.6 meters (1 to 2 ft) and vertical dimensions of 2.5 centimeters (1 in) or less. Somewhat larger and slightly deeper channels are termed **washes** in the western United States. On a still larger scale, ephemeral stream channels with typical widths of several meters and depths of half a meter (2 ft) or more are called **wadis** in parts of the Mideast and **arroyos** in the American West. They generally are steep-sided and flat-bottomed features (see Figure 19.14). At the largest scale of magnitude are the gorges and canyons of larger, often exotic rivers like the Colorado.

Desert Landform Assemblages

There is no predominant desert landscape. Deserts vary greatly in their landform assemblages because in different regions they have been subject to differing combinations of geomorphic controls. Factors such as temperature, precipitation amounts and intensities, and wind speeds and directions are all significant. Probably the most critical controls, however, are lithologic factors and recent climatic and tectonic events. Lithologic features such as differing rock strata, bedding planes, and fractures often are strikingly exposed because of the general absence of soil and vegetation. Recent tectonic processes are especially important in deserts because the gradational processes that modify and eventually eliminate tectonic features have been slowed greatly by a lack of water. Because tectonic features are so important, the nature and recency of their actions strongly influence the three categories of desert landform assemblages discussed in this section. These are desert plains, deserts containing plateaus, and mountainous deserts.

FIGURE 19.14 A dry arroyo in Arizona. (©JLM Visuals)

Desert Plains

Many deserts have developed in stable continental interiors where little tectonic activity has occurred for long periods of geologic time. These areas have low to moderate elevations and limited surface relief, so that they are classified topographically as plains. Important examples include the Australian, Arabian, Turkestan, and Kalahari Deserts, as well as most of the Sahara.

Desert plains contain some of the most uniformly flat surfaces on Earth, chiefly because vertical erosion is restricted by interior drainage. Because no fluvially transported materials can leave the area, local base levels prevail. Sediments are deposited in low-lying sites, eventually filling them so that an almost perfectly level surface is produced.

If at least some slope is available, the surface is likely to be covered by a braided pattern of rills or shallow washes. The finer textured, damper soils within the channels encourage the growth of vegetation, frequently causing them to become covered or paralleled by growths of small shrubs. These plants accentuate the drainage pattern so that it is readily visible from the air.

Surface materials vary considerably in different desert plains. A reg surface of closely fitting stones is most common but extensive desert plains are covered by shifting sands. Sandy regions, at least, display highly irregular although usually rather small-scale surface features in the form of dunes. Also in contrast to other desert plains, sand-covered surfaces usually are free of visible drainage channels because of the high porosity and shifting nature of the sand. Still other desert plains are bedrock-surfaced hamada or contain stony residual soils (see Figure 19.15).

Deserts Containing Plateaus

Tectonic forces have raised some arid and semiarid regions well above sea level with little deformation so that they now form plateaus. The Colorado Plateau of northern Arizona and southern Utah is an excellent example of such a region. Other desert plateaus are located in northern Mexico, southern Argentina, the Middle East, western Arabia, Southwest Africa, and western China.

Typically the majority of a desert plateau surface is flat, making it topographically indistinguishable from a desert plain. Higher elevations and some orographic influence are likely to make plateaus somewhat cooler and less arid than their low elevation counterparts, so that they usually are somewhat better vegetated. Most flat upland surfaces are developed on sedimentary rocks such as sandstone or limestone, which are resistant to weathering and erosion in dry climates. They may even be resistant enough to form a **cap rock** layer that protects the weaker underlying strata from erosion. If a cap rock

FIGURE 19.15 A nearly level desert surface covered by stony residual soils and scattered bedrock outcrops in Joshua Tree National Monument, California. (Tony Craddock/Tony Stone Worldwide)

layer has been uncovered by the erosion of overlying weaker rocks, the surface is described as a **stripped plain**. Such plains may contain flat-topped remnants of former higher surfaces.

Among the most notable features of desert plateaus are their steep, clifflike escarpments. These escarpments usually exist either where a river has incised a deep gorge or canyon into the surface or at the eroded margins of the plateau. In either case, the likely presence of a resistant cap rock stratum and the reduced importance of chemical weathering help to maintain slope steepness. Along escarpment faces, differing rates of erosion of resistant and nonresistant strata often are clearly evidenced by the development of a steplike series of alternating cliffs and slopes.

Escarpments in all climates typically retreat in an irregular fashion, so that they have highly uneven margins. Uneroded segments of a retreating scarp-frequently become isolated by the erosion of the surrounding rock. These outliers are termed **mesas** if they still retain a flat top and **buttes** if erosion has reduced them to narrow spires with no flat upper surface. Isolated mesas or buttes also may exist as the last remnants of plateaus that have otherwise been completely consumed by erosion (see Figure 19.16).

Mountainous Deserts

Mountainous deserts contain the greatest variety and complexity of surface features. The major mountainous deserts of the world include the Great Basin of the western United States and adjacent northern Mexico (see Figure 19.17); South America's Atacama Desert; large portions of southwestern Asia, including much of Iran, Afghanistan, and Pakistan; limited areas of the central Sahara and central Australia that contain isolated mountain masses; and portions of western China and Mongolia.

Structural features in mountainous deserts typically are exposed clearly. Rock strata, folds, fault lines, large-scale and small-scale differential weathering features, and many other phenomena are easily seen and often sharply delineated, making such regions ideal geomorphological laboratories.

Two important gradational activities that are confined largely to arid regions and are of great importance in shaping the landscapes of mountainous deserts are back-wasting and interior drainage. Due apparently to the reduced chemical weathering and the virtual absence of creep, the erosional reduction of highland masses in arid regions takes place by **back-wasting** (see Figure 19.18). The significance of back-wasting is that mountain masses, while being reduced in size and areal extent, maintain their steepness and ruggedness.

Interior drainage is critically important in mountainous deserts because it allows no weathering products to leave the region (except in limited quantities by wind transport); consequently there is no significant reduction in mean surface elevations. Instead, as

FIGURE 19.16 Monument Valley, in northeastern Arizona, contains a large number of residual sandstone buttes and mesas. Seen here is one of the Mittens. (E.D.McKee/USGS)

highland areas are gradually worn back, the adjacent lowlands that receive the products of erosion are aggraded. This results in a gradual reduction in surface relief unless tectonic activity produces further uplift.

The Great Basin of the western United States, which lies between the Rocky Mountains and the Sierra Nevadas has been studied more intensively than any other mountainous desert region. It consists largely of linear, north/south-oriented, block-faulted mountain ranges separated by broad flat basins. The

FIGURE 19.17 California's Death Valley is a desert graben surrounded by rugged mountains. (© JLM Visuals)

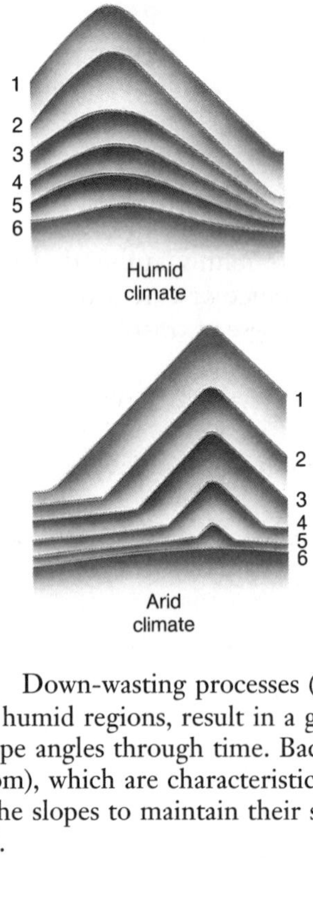

FIGURE 19.18 Down-wasting processes (top), which predominate in humid regions, result in a gradual reduction of slope angles through time. Back-wasting processes (bottom), which are characteristic of arid regions, cause the slopes to maintain their steepness as they are eroded.

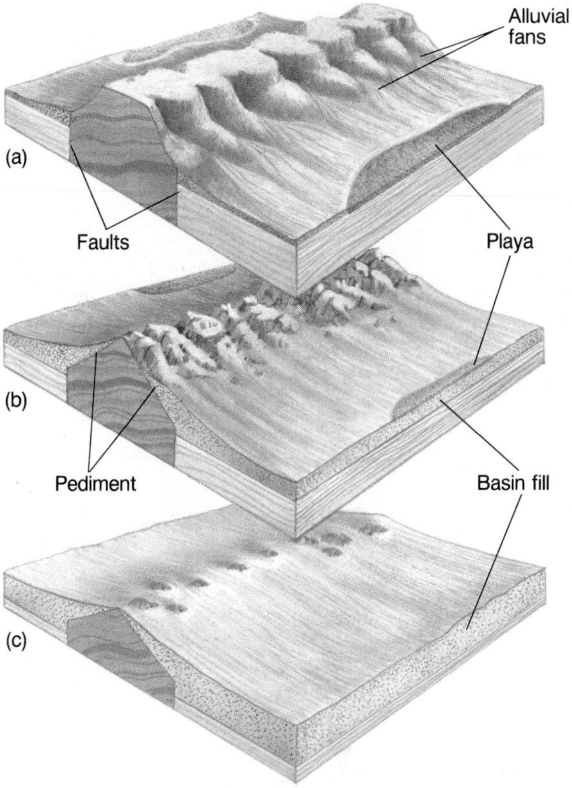

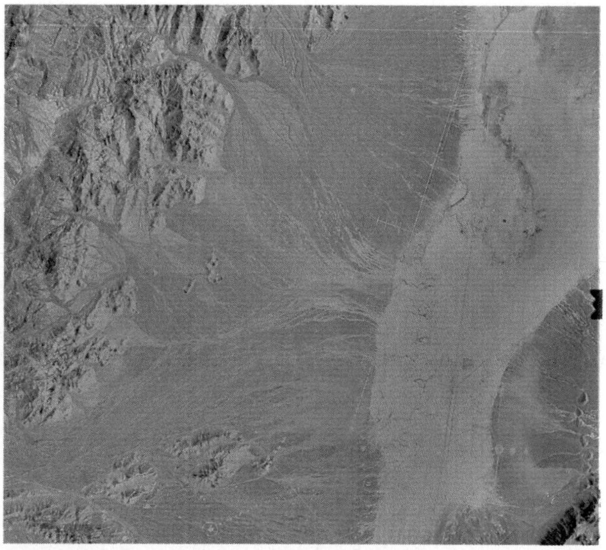

FIGURE 19.19 Artwork: Three stages in the erosion of an upfaulted block (a horst) in an arid climate. As the block is eroded back, the remaining upland segments retain their steep slopes. Interior drainage exists, so sediments are transported only into the adjacent basins. Photo: Aerial view of residual mountains, a pediment surface covered with coalescing alluvial fans (forming a bajada), and a playa near Las Vegas, Nevada. (USGS)

discussion of mountainous desert features that follows stresses the landforms of this region. Other regions in different lithologic, tectonic, and climatic settings may contain features that vary to some degree in appearance or extent of development, but all mountainous deserts display a basic similarity of landform types.

Mountains in most desert regions rise abruptly from a level or gently sloping surface. This, as we have seen, appears to result largely from backwasting processes and the absence of soil creep. In some localities, however, the sudden breaks in slope along mountain fronts have been buried by such features as talus cones or alluvial fans, which are formed from weathered materials transported down the mountain flanks by mass wasting or storm runoff. The mountain masses themselves typically are steep, rugged, and rocky. If composed of sedimentary strata, the rock may be largely intact, with the surface supported by a physically strong limestone or sandstone cap rock layer. If it is composed of an intrusive igneous rock such as granite, though, weathering processes operating along fractures frequently have reduced the mountain, at least near the surface, to a mass of blocks or boulders.

The surface at the foot of the mountain often is covered by deposits eroded from the adjacent uplands. If these materials have been transported away from the mountain front, the mountains are typically surrounded by a planed rock surface called a **pediment.** The origin of pediments has been the subject of much controversy, but they apparently are produced by the combined erosional activities of backwasting, sheet floods following heavy rains, and the lateral shifting of ephemeral streams from the adjacent mountains. The lack of vegetation to trap and hold soil also aids in the formation of a bare rock surface. Pediments increase in size with time as the mountain fronts are worn back at a constant angle. The pediments serve as surfaces of transportation because their slope angles of 4° or less are just sufficient for floods to carry weathered materials from the adjacent mountains across them. Some pediments have their solid rock surfaces exposed to view, but more frequently they are hidden beneath a veneer of loose rock debris that has been temporarily deposited

THE DESERT AND OFF-ROAD VEHICLES

The desert environment is highly fragile and is easily disturbed by intensive human use. In many areas, this sensitivity presents no major problems, because the desert offers little inducement to human occupation. Such a situation, though, is not true of the deserts of southern California, which are within easy driving distance of a large and highly mobile population. Every weekend, these once empty lands are invaded by thousands of sightseers, picnickers, fossil and Indian artifact hunters, plant and animal collectors, and recreational vehicle owners. Although all of these groups leave their mark on the land, the recreational vehicle owners are of greatest concern to environmentalists.

In recent years there has been a dramatic increase in the popularity of off-road vehicles (ORVs) of all types, including motorcycles, dune buggies, and especially the relatively inexpensive all-terrain vehicles. There currently are more than 13 million ORVs in use in the United States, with well over a million of these in southern California alone. The major environmental problems associated with their use in the California desert have been soil erosion, changes in hydrology, and damage to plants and animals.

Vehicle tracks made on desert surfaces can last for years and sometimes even for centuries (see Figure 19.20). Passing vehicles destroy the structure binding the thin surface soil, leaving it as loose dust. At the same time, they compact the underlying materials, greatly reducing their ability to absorb water. Runoff from ORV sites is as much as eight times higher than that from adjacent undisturbed sites. Because of their lack of permeability, vehicle tracks, especially those extending downhill, tend to be transformed into gullies by desert rainstorms. What water remains in the soil is more tightly held and less available to plants than before. Diurnal soil temperature ranges also increase, placing additional thermal stresses on plants and animals that must already cope with extremely high daily temperature variations.

The most obvious change in the appearance of many areas frequented by ORVs is the reduction of vegetation. Plants are killed both directly, by being run over, and indirectly, as a consequence of soil erosion, compaction, and desiccation. Animals also are killed or displaced. One 1975 survey by the U.S. Bureau of Land Management following a cross-country motorcycle race between Barstow and Las Vegas found a 90 percent reduction in the local population of small mammals.

The ongoing controversy between environmentalists and recre-

while being transported across the pediment. The thickness of the rock debris covering the pediment surface tends to increase gradually away from the mountain front until finally, in block-faulted regions like the Great Basin, the pediment slopes steeply downward to form the basement rock of a sediment-filled basin (see Figure 19.19).

The enclosed desert basins into which mountain-derived sediments are transported are termed **bolsons.** Their flat surfaces often are developed upon hundreds of meters of debris, which is coarsest in texture at their margins and finest in their centers. Because they contain no drainage outlets, the centers of bolsons become the sites of temporary lake basins called **playas** following rainstorms. Playa lakes usually are extensive, very shallow, and have rapidly shifting boundaries as water is added by inflow or removed by evaporation and seepage. Playas often are completely dry or consist of mudflats that may develop cracked surfaces as they dry out.

The water of playa lakes is highly saline because salts brought in by the runoff are left behind when the water evaporates. When the playas are dry, the centers of their bolsons may be covered by glistening white surfaces of precipitated salts (see Figure 19.12). Salts in some localities have accumulated in sufficient quantity to be mined commercially. The mining of nitrates, for example, is a major industry in the Atacama Desert of northern Chile, and borates have long been mined in California's Death Valley.

Another common feature of mountainous deserts is the **alluvial fan.** Alluvial fans consist of gently sloping fan-shaped deposits of alluvium. They form where ephemeral mountain streams carrying large ational vehicle owners over the use of ORVs in California's deserts boils down to a difference in environmental perception between the two groups. The environmentalists contend that the desert contains a valuable and fragile ecosystem, and that vehicular use in these areas should be restricted or prohibited. The recreational vehicle owners tend to view the desert as an empty wasteland that is of little or no practical use other than for recreation. They point out that they are using these unoccupied lands so that they will not disturb the residents of populated areas. As citizens and taxpayers, they feel that they have a right to use these public lands just as campers and hikers do.

It is obvious that some desert land must be set aside for ORV use. The chief problems are how much land is needed, where it should be located, and how environmental damage can be minimized. One promising trend has been the opening of privately owned ORV sites with established trails that are regularly tended and watered to keep down dust.

In 1980, after four years of hearings, the California Desert Conservation Area Plan was enacted into law. It should provide the framework for an effective desert land-use policy that will have something to offer all groups concerned with the use, and protection, of the desert.

FIGURE 19.20 Denudation and erosion of private land by off-road vehicles. This area is near Corona, on the east side of the Los Angeles Basin. (Howard Wilshire)

sediment loads enter the relatively flat adjoining pediment surface. Here the reduction in slope steepness and especially the removal of confining channel walls greatly reduces the speed of flow of the stream and allows it to spread out, causing it to deposit most of its sediment load. In origin and general surface configuration alluvial fans are similar to deltas, but, unlike them, they are formed entirely on land, usually along the base of a mountain front. Alluvial fans also form in humid regions on sites such as the foot of a river bluff or terrace. They are best developed in mountainous deserts, however, because of the frequency of sudden changes in slope, and are highly visible there because of the lack of vegetation (see Figure 19.21).

Alluvial fans vary greatly in size, but their dimensions are generally proportional to the areas drained by the mountain streams forming them. Along a linear mountain front, individual fans may coalesce to form a ramplike alluvial apron called a **piedmont alluvial slope** (or **bajada**). Large fans are favored settlement sites in arid regions because they generally contain near-surface supplies of ground water. In addition, the lower portions of fans consist of fine-textured sediments suitable for cultivation, and their gently sloping surfaces allow for good drainage of water and air, thus reducing frost hazards. A portion of Salt Lake City, Utah, and a number of cities and towns in California's Central Valley have been built on alluvial fans.

The lack of external drainage that characterizes most mountainous deserts means that the fluvial transport of weathering products is limited to the

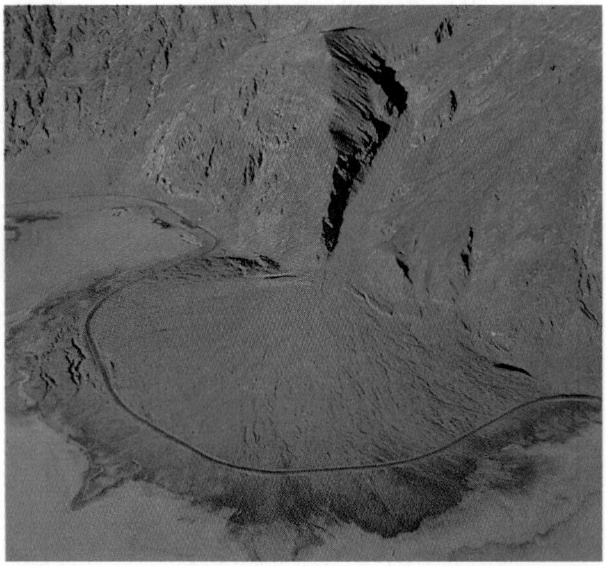

FIGURE 19.21 Aerial view of a "classic" alluvial fan in California's Death Valley. A road winds around the outer portion of the fan. (Martin Miller/©JLM Visuals)

FIGURE 19.22 View of "The Olgas," a conglomerate inselberg in Australia. (E.D. McKee/USGS)

route between the mountain masses and the nearby bolsons. As time passes, the mountain masses are reduced in stature and areal coverage, the pediments at their bases become increasingly extensive, and the bolsons become progressively more deeply buried in debris. In the arid equivalent of "old age" in the erosional cycle of a once mountainous desert, nearby pediments coalesce as the intervening mountain masses are consumed to produce a **pediplain.** This erosion surface differs from its humid landscape counterpart, the peneplain, in that it has been produced largely by back-wasting and sheet floods rather than by the lateral shifting of rivers. It also is likely to lie at a considerable elevation because it has developed in response to local base levels. These base levels actually have been raised as the adjacent bolsons have been progressively filled with debris because of the lack of exterior drainage.

Standing conspicuously above some pediplain surfaces in arid, semiarid, and savanna landscapes are isolated steep-sided mountain remnants called **inselbergs** (German for "island mountains"). They may be considered the dry-land equivalent of monadnocks. Like monadnocks, they have survived for a longer period than the surrounding uplands largely because of their greater resistance to weathering. Inselbergs are especially common in the Kalahari, in parts of Algeria, northwestern Nigeria, and in western Australia (see Figure 19.22).

SUMMARY

This chapter deals with a geomorphic process—the wind—and an important surface environment—the desert—that are associated with areas largely devoid of vegetation. In vegetation-free settings, weathering products frequently lie loose on the surface and are eroded easily by even low-energy gradational forces.

In order to review these subjects more systematically, we return to the Focus Questions:

1. *In what geomorphic environments are eolian processes most important?*

Eolian processes are important in regions that contain loose, dry surface materials of sand size or smaller. The frictional drag and turbulence of the wind enable it to move these materials through the processes of surface creep, saltation, and aerial transport. The most geographically extensive of the eolian environments are the deserts. Areas that experienced desert conditions in the recent geological past or those located downwind from deserts also may contain wind-produced landforms. Coastlines with sandy beaches also are important eolian settings. Other areas in which wind-produced features are important include those underlain by weakly cemented sandstone, the floodplains of streams draining arid and semiarid regions, and regions containing fine-textured glacial deposits. Human activities, especially those associated with agricultural pursuits, also have allowed the wind to erode and redistribute vast quantities of soil.

2. *What are the primary erosional and depositional features produced by the wind?*

Wind erosion may act rather evenly to produce a general surface lowering, leaving little in the way of distinctive features. More commonly, though, it selectively removes finer or weaker materials while leaving those of greater resistance. Deflation hollows of various sizes exist in many areas as a result of wind erosion, and many desert regions contain reg or desert pavement surfaces from which the finer materials have been selectively winnowed.

Wind deposits of sand-sized particles differ considerably from those of silt-sized particles. Sand typically is deposited in dunes within collection areas known as ergs. Several major types of dune forms exist, including barchans, longitudinal dunes and transverse dunes. Dunes not stabilized by vegetation migrate slowly over the surface, mostly within desert regions, in response to prevailing local wind patterns. Silt-sized eolian deposits, commonly termed loess, typically are situated downwind of sand deposits because their smaller size allows them to travel farther. Most loess deposits were derived originally from Pleistocene glacial outwash. At present most are stabilized by vegetation, and their surfaces often have weathered to form fertile and agriculturally productive soils. Loessal surfaces are predominantly flat and nearly featureless, but the angular nature of the loess particles has allowed streams in some areas to erode steep-walled gullies.

3. *What are the major landform features of deserts, and how do they differ from the landforms of more humid regions?*

The basic difference between deserts and most humid climate regions is the absence of a vegetation cover. The general lack of surface water also results in a greatly reduced rate of chemical weathering and of soil development. The effects of these factors are that bedrock features generally are exposed at the surface and that the topography tends to be angular. Streams in desert regions normally carry water only infrequently, but fluvial geomorphic processes dominate in most arid regions because of the lack of surface vegetative protection. Desert rainstorms, when they do come, can supply copious amounts of precipitation that can produce spectacular and geomorphically effective floods. Many surfaces are covered by shallow washes, or drainage channels, that may be visually enhanced by vegetation patterns. Approximately one-fourth of the world's deserts are ergs, covered by shifting sand, but most are stony regs or bouldery hamada.

Despite the slow rate of landform evolution in arid regions, many deserts are situated in tectonically inactive regions and therefore consist of rather monotonous plains. Flat upland surfaces likewise may have arid or semiarid climates, and often are bordered by steep-faced escarpments. Mountainous deserts like those in parts of the southwestern United States contain the greatest diversity of features. They are characterized by rocky and rugged mountain masses that tend to retain their steepness as they are worn back. The mountain masses typically are surrounded by gently sloping, planed-off pediment surfaces, sometimes partially overlain by alluvial fan deposits. These features merge downhill with sediment-filled basins of interior drainage called playas.

Review Questions

1. How does the wind erode and transport materials that are much denser than air? How does the eolian transport of dust differ from that of sand?

2. In what environmental settings are eolian processes most important? What basic similarities of all these settings enable the wind to be an effective geomorphic agent?

3. What is the difference between abrasion and deflation? Which is more important as a gradational mechanism? Why? Name and briefly describe the major surface features associated with each process.

4. Are most desert regions covered with sand? Why or why not? Where are some of the most extensive ergs located?

5. What is a sand dune? Why and how do dunes form? What determines whether they are "live"?

6. What is loess? Why are loess deposits not normally found in the same places as sand deposits? Why does loess weather into better agricultural soils than sand?

7. Describe the basic differences between the surface appearance of a desert landscape and a humid landscape.

8. Explain why water, which is scarce in desert regions, still is normally the most effective desert erosional agent.
9. What is interior drainage? Why is it prevalent only in arid regions?
10. List some major desert plains regions. Why are desert plains more geographically extensive than mountainous deserts? What three types of surface materials (based on size) cover large portions of desert plains?
11. What is the difference between back-wasting and down-wasting? Which is more common in deserts? Why?
12. What is a pediment, and what function does it serve in the mountainous desert erosional cycle? How does a bolson differ from a pediment?
13. Explain the similarities and differences in the forms and functions of alluvial fans and deltas.
14. What human activities are most responsible for the process of desertification?

Key Terms

eolian
surface creep
saltation
sandstorm
abrasion
deflation
deflation hollow
desert pavement
loess
erg
dune
barchan
longitudinal dune
transverse dune
desert
exotic river
interior drainage
back-wasting
pediment
bolson
playa
alluvial fan
pediplain
inselberg
desertification

CASE STUDY

THE EXPANDING DESERTS

The deserts, which already occupy approximately one-fourth of the Earth's land surface, have in recent decades been growing at an alarming pace. The expansion of desertlike conditions into areas where they did not exist previously is called **desertification**.

The spread of deserts into nondesert lands has caused a great deal of human suffering, including economic hardship, malnutrition, starvation, and mass migration. Desertification has been occurring for a long time, but recently was brought forcibly to public attention by the Sahelian drought of 1968–73. (The Sahel is the name given to the vast semiarid region of Africa lying just south of the Sahara.) This drought resulted in the deaths of up to 250,000 people, the loss of millions of livestock, and the southward displacement of hundreds of thousands of people. More recent major droughts in Ethiopia and neighboring countries in East Africa have taken hundreds of thousands of additional human lives.

Most regions undergoing desertification lie on the fringes of existing deserts. Included are the areas to the north and south of the Sahara, the "horn" of East Africa, regions to the northwest and southeast of the Kalahari, and large portions of the Middle East and southwestern Asia, as well as smaller areas in Australia, eastern Brazil, southern Argentina, the central Andes, northern Mexico, and the western United States (see Figure 19.23).

The advance of the desert does not take place in a smooth, continuous manner, but instead occurs in a highly irregular, piecemeal fashion (see Figure 19.24). It may be halted during a period of wet years, only to progress rapidly during a succeeding dry period. Often the less dry areas desertify the fastest, because they are used more intensively than are drier, less productive areas. Each year some 60,000 square kilometers (23,000 mi^2) of land turn into desert. This is an area the size of West Virginia! Additional vast areas annually suffer from accelerated erosion and loss of agricultural supporting capacity. It has been estimated that the total area threatened by this process covers 37,600,000 square kilometers (14,500,000 mi^2). This is one-fourth of the land surface of the Earth, or an area four times the size of the United States.

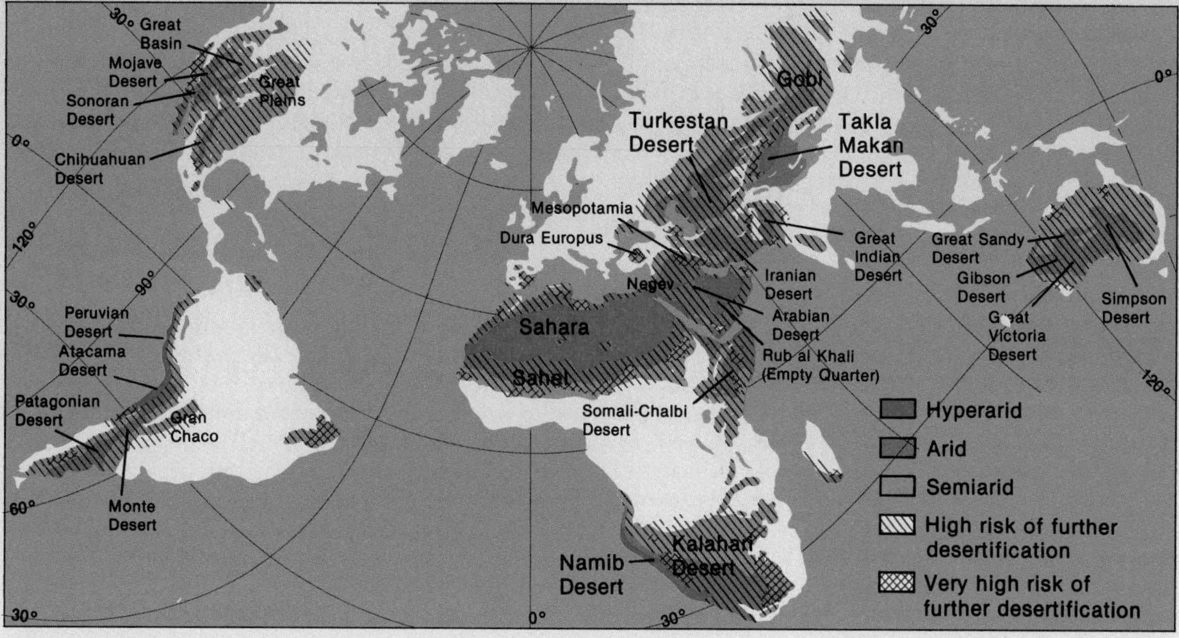

FIGURE 19.23 Global distribution of existing desert areas and of areas where desertification is taking place.

CAUSES OF DESERTIFICATION

Desertification is accomplished primarily through the loss of stabilizing natural vegetation and the subsequent accelerated erosion of the soil by wind and water. In some cases, the loose soil is blown completely away, leaving a reg or hamada surface. In other cases, the finer particles are removed, while the sand-sized particles accumulate to form mobile dune systems.

Even in areas that retain a soil cover, the reduction of vegetation generally results in the loss of the soil's ability to absorb substantial quantities of water. The impact of raindrops on the loose soil transfers fine clay particles into the soil pore spaces, sealing them and producing a largely impermeable surface. Water absorption is greatly reduced and, as a consequence, runoff is increased, resulting in accelerated erosion rates. The gradual desiccation of the soil caused by its diminished ability to absorb water results in the further loss of vegetation, so that a cycle of progressive surface deterioration is established.

In some regions, desertification is occurring largely as the result of a trend toward a drier climate. The continued slow global warming following the end of the last glacial stage of the Pleistocene may have much to do with this process. As the Earth warms, the subtropical highs slowly expand in size and shift poleward. This has produced a well-documented increase in aridity for areas such as North Africa and the southwestern United States over the past few thousand years. (The process may well accelerate in subsequent decades if an intensified greenhouse effect resulting from human-produced air pollution and forest destruction occurs.)

There is little doubt, however, that desertification in most areas results primarily from current human activities. The semiarid lands bordering the deserts exist in a delicate ecological balance and have limited potential to adjust to increased environmental pressures. Most of these areas are located in less-developed countries that currently are experiencing rapid population increases of 2.5 to 3.5 percent each year. The expanding populations are subjecting the land to increasing pressures to provide them with food and fuel. The inequitable distribution of property in many of these countries compounds the problem, since the large landholdings of the wealthy often are effectively off-limits to agricultural development, thereby increasing pressures on the remaining land.

In wet periods, the land may be able to respond to these stresses.

continued on next page

During the short-term droughts that are common and expectable phenomena along the desert margins, though, the pressure on the land often is far in excess of its diminished supporting capacity, and desertification results.

Four specific activities have been identified as the major contributors to desertification. They are overcultivation, overgrazing, firewood gathering, and overirrigation. The cultivation of crops has expanded into progressively drier regions as population densities have increased. These regions are especially drought-prone, so that crop failures are common. Because crop production generally necessitates the prior removal of the natural vegetation, crop failures leave extensive tracts of land devoid of a plant cover and susceptible to wind and water erosion.

The raising of livestock is a major economic activity in semiarid lands, where grasses generally are the dominant natural vegetation. This is especially true in parts of Africa, where the number of cattle a family owns are a measure of its wealth and status. Livestock numbers have been increasing at about the same rate as the human population, and in many regions they now far exceed the carrying capacity of the land (see Figure 19.24). The consequences of an excessive number of livestock are the reduction of the vegetation cover and the trampling and pulverization of the soil. This is usually followed by the drying of the soil and accelerated erosion by wind and water.

Firewood is the chief fuel used for cooking and heating in many developing countries. The increased pressure of expanding populations has led to the removal of woody plants so that many cities and towns, especially in the Sahel, are surrounded by large areas completely denuded of trees and shrubs. Journeys to obtain firewood in these areas may take two or three days. The reduction in vegetation also decreases local evapotranspiration rates, resulting in the addition of less water vapor to the atmosphere. As plants are removed, the underlying lighter-colored soil is increasingly exposed, thereby increasing surface albedos. This may result in lowered temperatures, which in turn increase the stability of the air and decrease precipitation.

The final major human cause of desertification, soil salinization resulting from overirrigation, was the subject of a Focus Box study in Chapter 12. Soil salinization has been an especially significant contributor to desertification in the Middle East, southwestern Asia, and North Africa, and is a growing problem in the western United States.

IS DESERTIFICATION IRREVERSIBLE?

The extreme seriousness of the desertification problem results from the vast areas of land and the tremendous numbers of people affected as well as the great difficulty of reversing or even slowing the process. Once the soil has been removed by erosion, only the passage of centuries or millennia will enable new soil to form. In areas where considerable soil still remains, though, a rigorously enforced program of land protection and cover-crop planting may make it possible to reverse the present deterioration of the surface. If the process of desertification is to be halted, the basic rule that must be followed is that the intensity of use of marginal lands must not exceed their carrying capacity in even the *poorest* years.

FIGURE 19.24 Desertification in progress in eastern Africa. The barren land on the right has been overgrazed by cattle. The well-vegetated land in the left distance has been left ungrazed. (Breck Kent/©JLM Visuals)

CHAPTER 20

COASTAL LANDFORMS AND SEAFLOOR TOPOGRAPHY

OUTLINE

FOCUS QUESTIONS

1. What geomorphic processes are chiefly responsible for the production of coastal features?
2. What are the most important types of erosional and depositional coastal landforms?
3. What are the major causes of coastal retreat and advance, and what categories of coastlines have resulted from these coastal changes?
4. What are the major topographic divisions of the ocean floor?

Water is the most geomorphically effective of the gradational tools. This chapter examines two important groups of geomorphic features in which water plays a dominating role: coastal landforms and the seafloor. Coastal landforms actually are polygenetic in origin, since terrestrial, marine, and atmospheric processes all contribute to their development. Because some knowledge of each of these processes is necessary for an understanding of coasts, this subject has been placed in the final chapter of the book.

Coastal processes often are powerful and can perform relatively rapid geomorphic work, but they are highly restricted in areal extent at any given time. Unlike fluvial, glacial, and eolian processes, they affect the land only along its margins. The coastlines of the world, where coastal processes perform all their work, are several hundred thousand kilometers long, but are generally quite narrow. Their extent is somewhat enlarged if the shorelines of large lakes, such as the Great Lakes, are included in the inventory. Lakeshores are subject to many of the same geomorphic processes as oceanic coasts, but the processes operate on a more limited scale.

The coastal zone, despite its restricted extent, is especially critical for human beings. It contains some of the world's highest population densities and provides sites for some of the most intensive human activities. Many of the world's largest cities, including Tokyo, New York, Shanghai, Rio de Janeiro, Los Angeles, and St. Petersburg, have been built on the coast. The coastal zone also is one of the most dynamic and environmentally sensitive of all natural habitats, so the potential exists for major coastal alterations as a result of human actions. Although a thorough discussion of the multiple influences of humans on coastal processes and features is beyond the scope of this text, some aspects of this topic are examined briefly in the Case Study at the end of the chapter.

The second portion of the chapter briefly investigates the major features of the seafloor. The ocean covers 71 percent of the Earth's surface, making the seafloor by far the most extensive of all geomorphic environments. Yet, it is still poorly known, and detailed mapping of its features has only just begun. Studies already have revealed, however, that seafloor features are both diverse and often strikingly different from those on land.

COASTAL PROCESSES

Coastal processes, like the gradational processes already discussed, can locally erode, transport, or deposit materials, depending upon the balance between available energy and the sediment supply. Where a large amount of energy is available and the sediment supply is restricted, erosion is dominant; where sediment is abundant and the available energy is restricted, deposition predominates. The physical characteristics of a given coast, then, control the relationships between energy and materials.

Most coastal sediments undergo multiple episodes of erosion, transportation, and deposition in response to the shifting balance of forces that influence the coastal environment. In the absence of tectonic uplift or a decline in sea level, though, the ultimate tendency of coastal processes is toward erosion because of the inexorable downward force of gravity. A continual net seaward transport of sediments takes place on a global scale, and the deep seafloor is the ultimate repository for sediments derived from the land. Terrestrial sediments must, of course, be transported out to sea through the coastal zone, and the geomorphic effect of coastal processes is to produce a continuous, gentle graded transportation slope seaward that facilitates this process. As a consequence, steep coasts tend to be eroded back, while low-lying coasts are likely to experience net deposition over the long term.

Waves, currents, and tides, the origins and characteristics of which were discussed in Chapter 10, are the most active agents shaping coastlines. Of these three oceanic movements, waves generate the fastest water motions and are by far the most powerful and effective geomorphic agent. Their strength results largely from the fact that all the energy they contain is released suddenly as they break, often in contact with the coastline (see Figure 20.1). Waves and currents are set in motion by the wind. Through its influence on the ocean water, the wind, whose relative weakness as a direct geomorphic agent was discussed in the last chapter, attains considerable additional importance as a shaper of landforms.

The combined motions of waves, currents, and tides are responsible for a variety of coastal gradational processes. Because they are produced by water movement, these processes are very similar to those performed by streams. Most important is **abrasion,** caused by the impact of materials carried in the water against those on the shore. Breaking waves some-

FIGURE 20.1 Large wave breaking on the California coast near San Francisco. (B.J.O'Donnell/BPS)

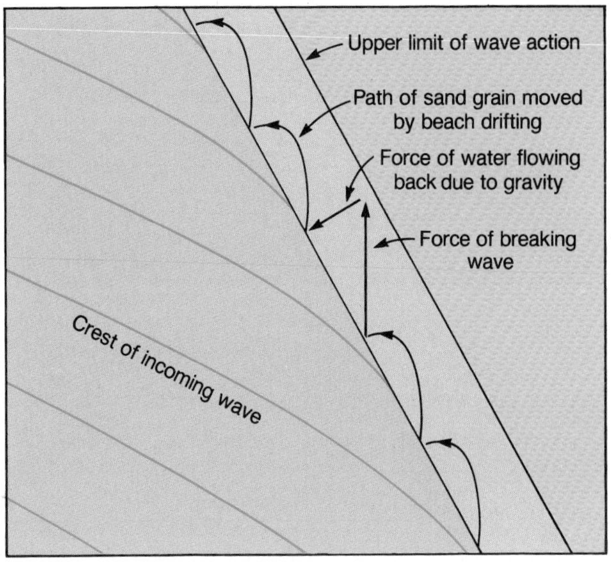

FIGURE 20.2 Beach drifting occurs as wave swash carries sand grains obliquely up the beach. The return backwash then carries them back toward the ocean directly down the slope of the beach, resulting in a net lateral transport.

times are capable of lifting even heavy rocks from the seafloor, and large storm waves have been known to hurl rocks for considerable distances. A related erosional process is **attrition**—the wearing down of loose rock materials by impact or rubbing action. Attrition is especially effective shoreward of the zone of wave-break because of the ceaseless alternation between the forward motion of wave **swash** and the return flow of wave **backwash.**

Solution is locally important on coasts containing soluble rock types such as limestone, although it affects all rock types to some extent. Finally, **hydraulic action,** or the direct impact of waves on the coast, can at times be a highly effective erosional mechanism. Pressures generated by the impact of large waves can be enormous. Water and/or compressed air can be forced repeatedly into rock fractures, eventually splitting the rocks apart. The greatest effect of hydraulic action, though, is upon unconsolidated shoreline materials. These materials may be removed rapidly during periods when large waves strike the coast.

One of the most obvious characteristics of a sandy beach is the smoothness of the surface in the zone of wave action. As waves alternately push fine-grained materials up the beach and pull them back, they produce a smooth, gently inclined surface. If waves approach the shoreline at an angle, sand grains move down the beach. This movement, termed **beach drifting,** occurs because the wave swash carries the sand grains up the beach at the oblique angle at which the waves intersect the shore, while the backwash carries them directly downhill in response to gravity (see Figure 20.2). The resulting zigzag movement of the particles transports them laterally along the beach, and in time may move large quantities of materials for substantial distances.

Localized **longshore currents** also may be produced by the lateral movement of water associated with breaking oblique waves. These currents consist of a steady flow of water paralleling the shore in the down-beach direction of oblique wave approach. They tend to augment the sediment-transporting effects of beach drifting by carrying fine particles somewhat farther offshore.

Although waves normally approach the coastline from a specific direction at any given time, coastal irregularities bend or refract the wave fronts as they approach the shore. **Wave refraction** occurs because shallow water with a depth of less than half the wavelength exerts a frictional drag on the waves that reduces their rate of landward propagation. If a portion of an incoming wave is thus slowed, the axis of the wave rotates so that it is directed more toward the shallow water (see Figure 20.3). Along an irregular coastline, this process concentrates wave energy on

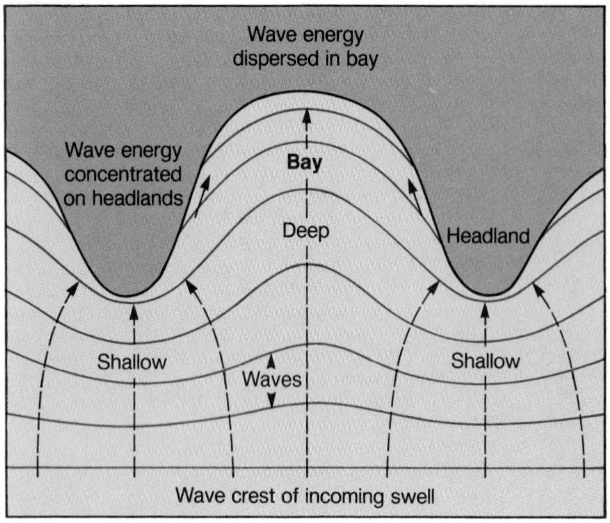

FIGURE 20.3 Wave refraction is caused by the reduced speed of landward propagation of the waves as their bases rub against the seafloor in shallow areas. This process acts to straighten shorelines by concentrating wave erosional energy on headlands. The dashed lines indicate the direction of movement of the wave fronts.

seaward-extending promontories or **headlands,** while reducing wave energy in coastal indentations such as bays or inlets. The net geomorphic effect of this process is to straighten the coastline by encouraging the erosion of headlands as well as the deposition of sediment in indentations.

The motion of waves is always toward the coast, and this generates a net shoreward flow of water near the ocean surface. This flow is counterbalanced by a seaward movement of subsurface water, called **undertow.** Undertow serves as the major transporter of fine sediments into deep water. When large waves produce a vigorous landward flow of water, the return seaward flow may be concentrated in **rip currents.** These currents consist of narrow zones of rapidly seaward-moving water that extend all the way to the surface. Rip currents sometimes carry unwary swimmers well away from the shore.

The effects of eolian and fluvial processes on the coast also are important. The wind forms dunes on many sandy beaches and can erode or redistribute fine-textured, dry coastal sediments that are not stabilized by vegetation. Streams erode valleys that reach the sea to form coastal indentations; if sea levels rise to flood the lower portions of these valleys, extensive **estuaries** can be produced. Streams that enter the ocean are vital to coastal processes because they supply most of the sediments that are distributed along the coast by waves and currents to form beach deposits. Low-lying coasts frequently are major sites of fluvial deposition and delta formation.

The rates at which coastlines are modified by erosion and deposition vary greatly. Modification rates depend primarily on wave size, the strength of the coastal materials, the angle of seaward slope of the coastal surface (determining how far offshore the waves will break), and the shape of the coastline (which may locally focus or diminish wave energy by refraction). Just as rivers in flood produce the most rapid fluvial landform alterations, large storm waves produce the most rapid coastal modification. Depending upon local conditions, inland floods can cause either rapid erosion or deposition; similarly, coastal storm waves can either transport sediment onshore, resulting in deposition, or carry it out to sea, resulting in erosion. At any given site, the erosional or depositional energy of the sea is concentrated in the zone of wave impact. Over a 12- or 24-hour period, this zone shifts between high and low tide levels. Areas with large tidal ranges therefore have coastal processes spread over larger vertical and generally larger horizontal distances than do areas with small tidal ranges.

If erosion occurs, it will do so most rapidly in unconsolidated coastal materials such as glacial or alluvial deposits or beach sand. Individual severe coastal storms, particularly hurricanes and intense winter frontal cyclones, can produce major erosion and even change the configuration of a coastline by eroding and redistributing sediments. Hurricanes crossing North Carolina's Outer Banks, for example, have altered the pattern of tidal channels approximately 30 times in the last 400 years by cutting new channels and filling others. Erosion also can be relatively rapid on a long-term basis along steep coastlines composed of weak sedimentary rock such as chalk or some sandstones. The famed Cliffs of Dover on the southeastern coast of England are being eroded at rates of up to 1.6 kilometers (1 mi) per thousand years. Conversely, coasts composed of massive resistant rock such as granite can withstand even intense wave action for long periods with little erosion.

COASTAL LANDFORMS

In this section, features produced directly by coastal marine processes are stressed. It should be noted, though, that many coastal features, even though they have been modified by coastal processes, originally were produced by tectonic activity or by the actions of streams, ice, wind, and gravity. A common method of categorizing coastal features, and the one employed in this section, is by determining whether they are predominantly erosional or depositional in origin.

Erosional Features

Coastal erosion occurs when the near-shore waters have the capacity to transport a greater quantity of sediment than is locally available. If coastal erosion has long been active in an area of consolidated materials, the coast generally is steep and rugged (see Figure 20.4). The existence of such a coast indicates that terrestrial gradational processes such as streams, which erode the surface from the top, have been unable to keep pace with the undercutting action of marine erosion. A steep coast allows large waves to break at or near the shore, concentrating their energy on the land rather than dissipating it farther out to sea.

The most characteristic landform developed along an erosional coast is a **sea cliff** (see Figure 20.5). Sea cliffs are produced by wave undercutting in areas of strong rock. The undermining of sea cliffs by wave hydraulic action, abrasion, and solution causes periodic slope failure as a portion of the cliff falls or slides into the ocean, where it is consumed by the waves. Along the California coast, where sea cliffs are well developed, most failures occur as rotational slumps (see Figure 16.11). In other places, they may take the form of rockslides or rockfalls. These mass movements temporarily stabilize the sea cliff, but the process of undermining immediately begins anew.

Cliffed coasts frequently are highly irregular in shape. This sometimes reflects differences in the erosional resistance of the rocks comprising the cliff. In other cases it results from differences in the elevation of the land along whose margin the sea cliff has developed. For instance, a ridge cut into by a sea cliff would contain much more rock for the sea to erode than would a valley. For this reason, the ridge would tend to be eroded back more slowly and would form a coastal promontory or **headland,** while the valley would be the site of a coastal indentation. Such indentations, depending upon their sizes and shapes, may be referred to as **bays, coves,** or **inlets.**

The process of wave refraction concentrates wave attack on headlands; consequently, the cliffs at their margins are likely to be especially bold and steep. A narrow headland attacked from opposite sides by the sea eventually may be cut through to form a **sea arch.** The collapse of the arch will produce a steep-sided island remnant called a **stack.** These features are common along portions of the west coast of the United States (see Figure 20.6).

Conversely, coastal indentations between headlands are subject to much less vigorous wave attack. The quieter environment encourages the local dep-

FIGURE 20.4 A cliffed coast in Oregon with a pocket beach. (David Butler)

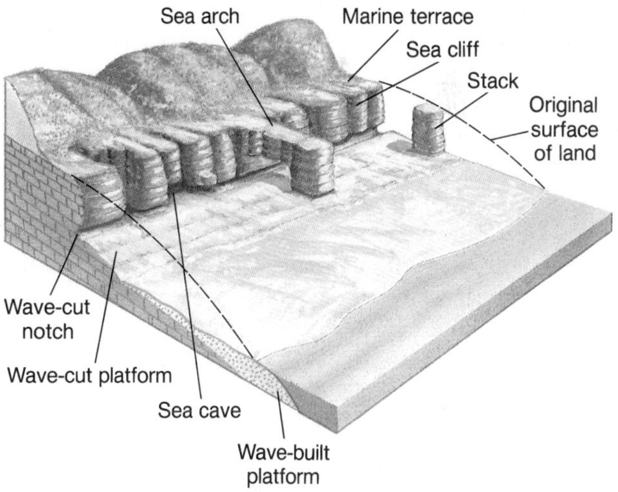

FIGURE 20.5 Features produced by coastal erosion.

FIGURE 20.6 A sea arch frames an offshore stack on the California coast near Santa Cruz. (Camerique/H. Armstrong Roberts)

FIGURE 20.7 A wave-cut platform exposed at low tide along the California coast. (John S. Shelton)

osition of sediments, often derived from the headlands, so that crescent-shaped **pocket beaches** may develop even along a coastline that is, as a whole, being worn back by erosion (see Figure 20.4).

Coastal erosion of solid rock surfaces is effective only down to the lowest level of wave attack at low tide. As a sea cliff retreats, then, it leaves behind a planed-off rock platform at a level representing the base of erosionally effective wave action. This surface is termed a **wave-cut platform.** A wave-cut platform serves as a surface of transportation between an erosional site—the sea cliff—and a depositional site—the deeper water farther offshore. The platform may be bare, or it may be covered by a veneer of rock fragments derived from the sea cliff and in the process of being transported seaward. In form and function, then, a wave-cut platform is analogous to the pediment surface that develops at the foot of a desert mountain range (see Figure 20.7).

Sediment frequently is deposited at the normally steeper outer edge of the wave-cut platform. In time, depending upon the steepness of the coast and the amount of available wave energy, these sediments may accumulate to form a **wave-built platform** consisting of mud, sand, gravel, or cobbles. The wave-built platform therefore extends the gradual slope of the continental margin farther seaward.

Depositional Features

Deposition occurs when the coastal waters are supplied with a greater quantity of sediment than they can transport and remove from the coastal zone. These materials are derived from various sources. Some are picked up well offshore and carried to the coastline by large storm waves. Others are eroded from sea cliffs or other coastal features. In most areas, though, the majority of sediments are transported to the coast by streams from inland locations. Waves and longshore currents then sort and spread these sediments laterally along the coastline. Conditions favor deposition where sediment supplies are plentiful and both onshore and offshore slopes are gentle. Gentle slopes cause large, erosionally effective waves to break and expend their energy well away from the shoreline.

The most characteristic feature of a depositional coast, and undoubtedly the best-known coastal landform of all, is the beach. Beaches annually attract tens of millions of tourists to coastal resorts within the United States alone, and add billions of dollars to coastal economies (see Figure 20.8). A **beach,** by definition, is the gently sloping shore of a body of water that is washed by waves or tides and is normally covered by loose deposits of sand or gravel. Beach sed-

FIGURE 20.8 Sandy beaches annually draw millions of tourists to North American beach resorts. (Stefan Schluter/The Image Bank)

iments often accumulate on a wave-cut platform during periods when conditions favor deposition. The depth of accumulation is highly variable, but is usually quite shallow.

The most familiar beaches are composed of whitish or buff-colored sand-sized quartz grains, but beach deposits can vary considerably in both composition and size. Some portions of the Hawaiian Islands, for example, have beaches of black lava sand. In some areas, shell fragments are an important component of beach sand, and on many tropical islands, especially in the Pacific, beaches are formed from crushed coral. Where coastal slopes are fairly steep and wave action is too vigorous for sand-sized deposits to be retained, beaches typically are covered by gravel or cobbles. In some locations, the type of beach materials may change from one time of year to another because of seasonal variations in climate. Some California beaches, for example, are covered with sand during the tranquil summer months, but winter storms transport the sand seaward, exposing an underlying cobble beach. With the return of summer, the sand once again is carried onshore where it re-forms the higher summer beach.

The California example illustrates the point that even where beach deposits cover the coast for long periods, they are highly vulnerable to erosion and may be removed quickly under extreme conditions. Beach materials are transient because they are normally fine textured, are located in a high-energy environment subject to changeable conditions, and are usually not stabilized by vegetation. Material continuously arrives and departs and the beach remains relatively unchanged only if the opposing tendencies offset one another.

Because beaches are highly sensitive to coastal processes, they normally develop in discontinuous fashion along an irregular coastline. Smooth, regular coastlines, conversely, are more likely to be continuously fronted by beaches. The Pacific coast of the United States from California to Washington is predominantly high, rugged, and irregular. As a result, beaches have developed only in protected sites, especially within bays or coves. Conversely, the Atlantic and Gulf Coast from Long Island to Texas is predominantly low-lying and gently sloping, and beaches have developed along virtually its entire length.

The gentle seaward slope and large sediment supply of a coast along which deposition is occurring allow waves, and especially large storm waves, to redistribute the bottom sediments to a considerable distance offshore. Just seaward of the **shoreline,** where land and water meet, there typically exists a trough of relatively deeper water excavated by moderate-sized waves breaking near the shore (see Figure 20.9). Farther seaward, however, the water often becomes shallower again as the bottom rises to form a **sand bar** paralleling the coastline. A sand bar is a largely submerged ridge, usually composed of sand, formed in the near-shore waters by currents and larger waves. The top of the bar may be exposed at the surface at low tide, but it is submerged at high tide. It is located in the zone between which large storm waves and moderate-sized "normal" waves break. While the beach contains an accumulation of sediments thrown forward by normal waves, the sand bar consists largely of an accumulation of sediments thrown forward by storm waves. Bars also serve as

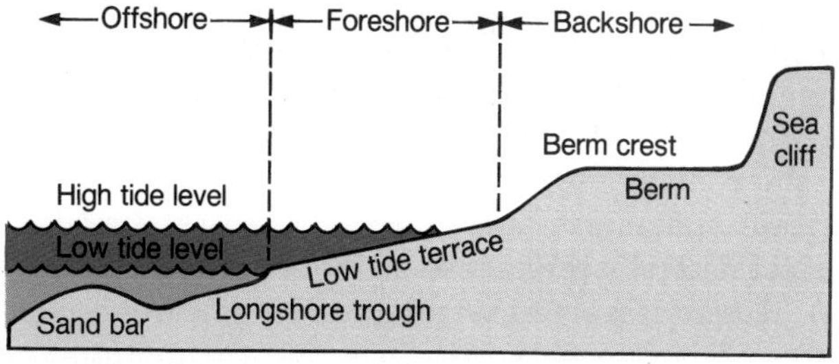

FIGURE 20.9 Cross-sectional profile of a typical beach.

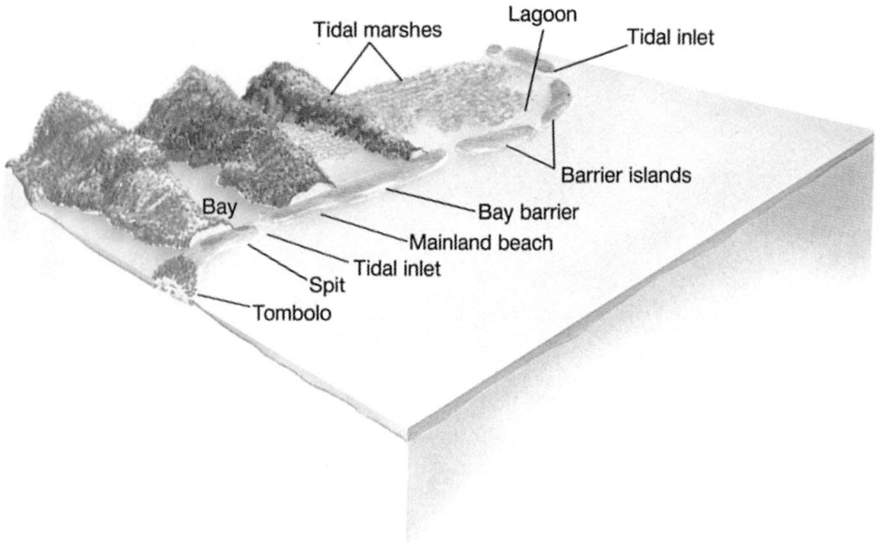

FIGURE 20.10 Depositional coast features.

temporary repositories of sediments eroded from the beach during periods of larger waves, such as those occurring in winter in portions of the middle latitudes.

Continued deposition by storm waves, coupled with a reduction in sea levels, may convert bars to **barrier islands** that lie continuously above the ocean surface even at times of high tide (see Figures 20.10 and 20.11). Like the bars from which they are formed, barrier islands are long and narrow and parallel the coastline. They often form in chains along low-lying sandy coasts and are separated from one another by **tidal inlets** through which the sea flows (see Figure 20.12). These inlets enable water levels seaward and landward of the barrier island chain to equalize during storms and as the tides change. The water between the barrier island chain and the mainland, referred to as a **lagoon,** is generally rather tranquil because it is protected from storms and large waves. The protected waters of coastal marshes and lagoons generally receive a continuous flow of fine sediments and organic materials from rivers entering them from the landward side. As a result, they typically contain a great quantity and variety of aquatic plants and animals and comprise one of the world's most naturally productive and sensitive ecosystems.

Barrier islands vary greatly in length, width, and distance from the coast. Some are situated only a short distance offshore, but others are 30 kilometers (20 mi) or more from the mainland, as along the North Carolina coast. Small barrier islands may be only 100 meters (300 ft) or so wide, but larger ones may exceed 2 kilometers (1.2 mi) in width. These ribbons of land may extend for many kilometers. Padre Island off the coast of Texas is one of the world's longest, with a length of about 160 kilometers (100 mi). Many barrier islands, as discussed in the Case Study, have been developed extensively as coastal resorts and contain large human populations.

Bars and barrier islands, like beaches, are highly unstable features and are subject to rapid changes in

FIGURE 20.11 A barrier island fronted by sandy beaches and backed by mudflats and a lagoon near Cape Lookout, North Carolina. (P. Godfrey)

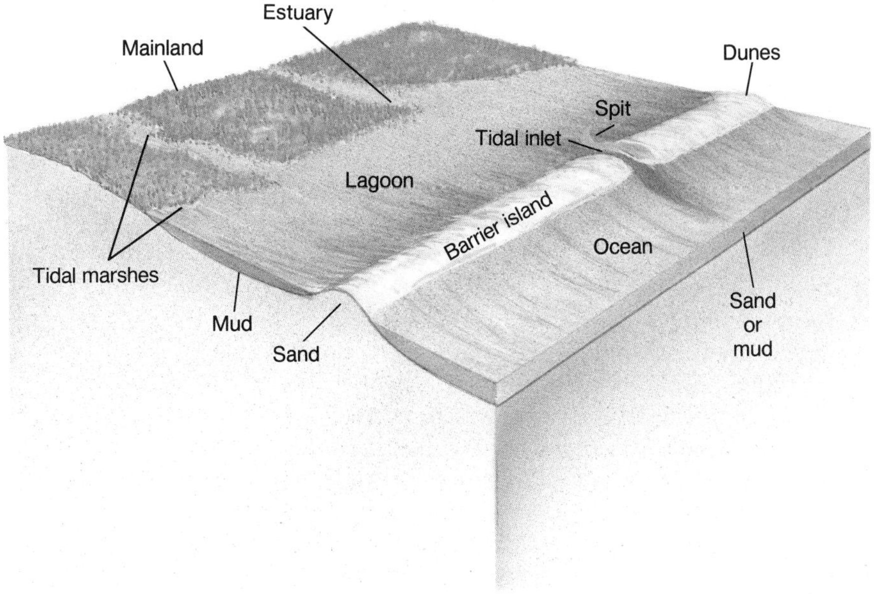

FIGURE 20.12 Block diagram of a barrier island system separated from the mainland by a coastal lagoon.

size, shape, and position. Bars and barrier islands tend to grow in the direction taken by the longshore drift of sediments. If the coastline is irregular, with headlands and bays, they generally are attached at one end to a headland. In this case, the strip of land is a peninsula, rather than an island, and is more appropriately termed a **barrier beach.** Along less rugged, straighter coasts, they usually are entirely separated from the land. A barrier beach attached to the land at one end may extend partially across a bay to form a hook-shaped **spit** ending in open water. Cape Cod, Massachusetts, is an example of a giant coastal spit (see Figure 20.13).

The protected lagoons on the landward sides of barrier beaches generally are tranquil enough that fine silt and clay particles can settle to the bottom. Plants such as marsh grasses that grow in the shallow water near the shore aid in trapping the sediment. Thus lagoons tend to fill gradually with organic matter and fine-textured river sediments. The continued filling of lagoons by sediments and organic material may eventually convert them to **mudflats,** which

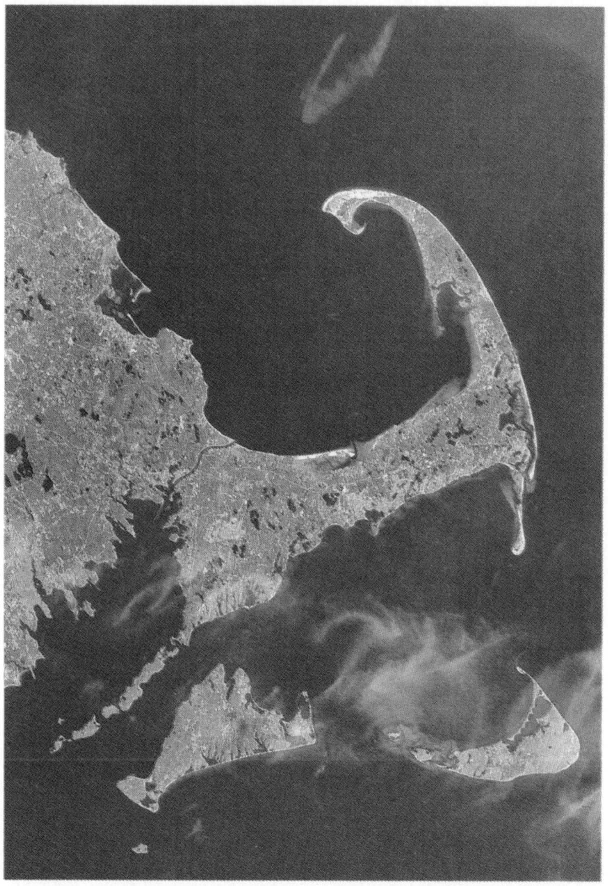

FIGURE 20.13 Cape Cod, Massachusetts, is a classic example of a large coastal spit. It contains a number of smaller spits. (GEOPIC™, copyright Earth Satellite Corporation)

typically are submerged only at times of high tide. Coastal mudflats may be diked, flushed of salts, and drained for settlement and agriculture; this has occurred on a large scale on the "polderlands" of the Netherlands. When mudflats are colonized by marsh grasses, they become **salt marshes.** A decline in sea level will convert such marshes to meadows and eventually to dry land.

TYPES OF COASTS

A number of coastal classification systems have been devised, but no one system has proven completely satisfactory. The basic problem is the great variety of processes that affect coasts and the resulting complexity of forms exhibited by the coastlines of the world. In order to classify coasts adequately, a system would have to be as complex as the interaction of the processes that determine coastal characteristics.

One very basic division of coasts, which is used as a starting point in some classification systems, is based on whether coasts are retreating (losing land) or advancing (gaining land). This system, on a very general level, is employed in this section.

Retreating Coasts

Coastal retreat caused by the erosion of the land and the deposition of the eroded materials on the seafloor is a slow process. In theory it could continue, in the absence of tectonic or other uplift, until all land areas were completely removed and a global ocean prevailed. There is another basic process of coastal retreat, though, that has much less potential for continental inundation, but that can operate much more rapidly in the short term. This is submergence caused by a rise in sea level. Submergence currently is causing a relatively rapid retreat of most coastlines because of the post-Pleistocene rise in sea level. The sea level rise results largely from the isostatic rebound of previously glaciated regions in the high latitudes, which displaces ocean water equatorward and increases its depth in the low and middle latitudes. Other related causes of the rise in sea level are the continuing melting of glaciers and the thermally-induced expansion of the ocean water.

Rates of coastal submergence have been accelerating in recent decades. Along the east coast of the United States, for instance, sea level currently is rising at a rate of up to 0.3 meters (1 ft) or more per century. Depending upon the steepness of the coast, this generally translates to a horizontal shoreline retreat of 30 to 300 meters (100 to 1,000 ft) per century. The effects, of course, are most evident along low-lying coasts. It has been speculated that the recent acceleration in the rate of sea level rise is an early indication that the widely predicted global warming trend resulting from the increase of atmospheric carbon dioxide levels has begun (see Chapter 5).

Erosion and submergence, although they both result in coastal retreat, produce coastlines that differ in basic characteristics. Each type of coast will now be examined briefly.

Erosional Coasts

Erosional coasts, as noted earlier, are generally steep, irregular, and rugged. They often contain bold cliffed headlands and intervening indentations, sometimes occupied by pocket beaches. The sea cliffs frequently exhibit a variety of secondary erosional features such as sea caves, arches, and stacks, and they are commonly fronted by wave-cut and wave-built platforms. Beyond the outer platform margins, water depths increase rapidly, so that the coastal zone, both seaward and landward of the shoreline, is narrow.

Submergent Coasts

Submergence brings the sea into contact with a land surface that has been shaped by terrestrial processes such as rivers or glaciers. Unless this inherited surface is exceptionally flat, the result is usually a highly irregular coastline.

The development of a coastline on a surface that has been dominated by fluvial processes produces what is frequently called a **ria coast** (from the Spanish *rio,* or "river"). Submergence results in the drowning of the lower portions of the river valleys to produce an extremely irregular coastline dominated by numerous bays or estuaries. The higher interfluves, conversely, stand out as broad headlands separating the coastal indentations. Where drowning is more extensive, the remnants of headlands may exist only as strings of aligned islands. With time, erosion of the headlands and deposition within bays straightens the coast. This process has not proceeded far since the Pleistocene, though, because submergence has been taking place for only about 15,000 years and is an ongoing process.

The east coast of the United States from New York south to the Carolinas is a prime example of a ria coast. Numerous large embayments, including Delaware Bay, Chesapeake Bay, the Potomac and James River estuaries, and Albemarle and Pamlico Sounds, were produced by the drowning of the lower portions of large river valleys (see Figure 20.14).

In high latitude regions where Pleistocene valley glaciers once reached the sea, coastal submergence has resulted in the development of **fjord coasts.** Their characteristics and global distributions are discussed in Chapter 18. Like ria coasts, fjord coasts contain deep indentations. The fjords, however, tend to be much deeper, steeper walled, and more consistent in width than the drowned river valleys of ria coasts. Both ria and fjord coasts contain protected natural harbors that are well suited for commercial maritime activities.

Advancing Coasts

Because of recent rises in worldwide ocean levels, advancing coasts are much less common than retreating coasts. They occur only where surface uplift or coastal addition operate quickly enough to more than offset the loss of land from erosion and sea level rises.

Advancing coasts can be divided into two categories. First are coasts emerging from the sea because of tectonic uplift or isostatic rebound. A decline in water levels would also produce an emergent coast, and although this is occurring along the shores of many lakes, it is not currently taking place along oceanic coasts. The second category of advancing coasts consists of outbuilding coasts. These coasts are being built outward by geological or biological processes at a sufficiently rapid rate to allow the position of the coastline to advance seaward.

Emergent Coasts

Most actively emerging coastlines are located in the high latitudes of the northern hemisphere and around the rim of the Pacific. The high latitude locations most notably include the Canadian Shield region and Scandinavia. These two areas are still rebounding from the released weight of the Pleistocene ice sheets. The coastal margins of the Pacific are rising as a result of tectonic uplift caused by the overriding of the Pacific Plate by lighter plates containing continental (sialic) crustal rock. These areas include the west coasts of North and South America.

Emergent coastlines associated with tectonic uplift are relatively straight and regular compared to

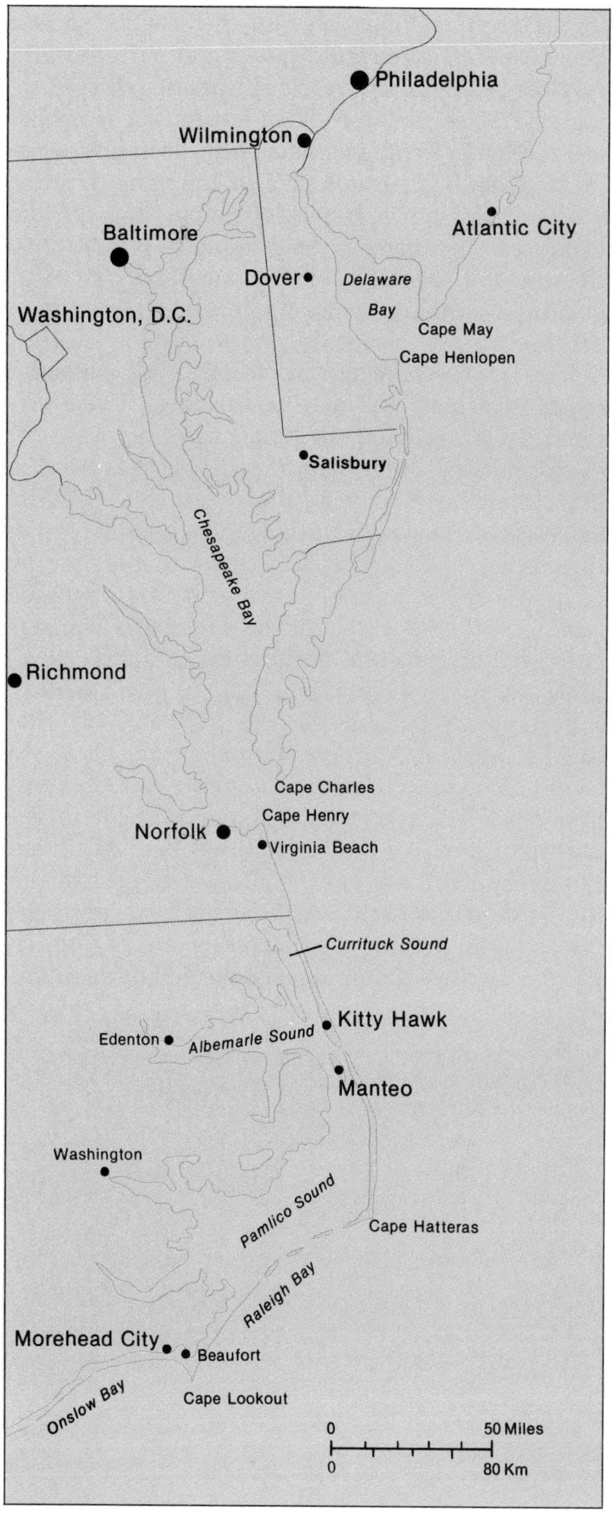

FIGURE 20.14 The U.S. Mid-Atlantic Coast is a ria coast. Its numerous bays and sounds were produced by the flooding of the lower portions of river valleys by the sea.

submergent coastlines because the seafloor surface being exposed is generally quite smooth. In contrast, surfaces emerging as a result of isostatic rebound are highly irregular because glacial features such as fjords (as in Norway and Labrador) and drumlins (as in Finland) are being exposed. If an emerging coastline is steeply sloping, as is most commonly the case, no offshore bars or barrier islands develop. If it is gently sloping, and specifically if it is supplied with large quantities of sediments from rivers or other sources, bars and barrier islands do tend to develop.

The most characteristic features of emergent coasts in general are **marine terraces.** These elevated flat surfaces are the remnants of old wave-cut platforms, wave-built platforms, or beaches that are now located inland well above sea level. Coastal emergence seems in most cases to be sporadic rather than steady, and each terrace represents the position of the shoreline during a previous period of coastal stability. Along portions of the southern California coast, multiple marine terraces extending to elevations as great as 425 meters (1,400 ft) give a steplike appearance to the landscape (see Figure 20.15). Elevated terraces also are common along the shores of lakes that formerly contained more water. Well-developed lake terraces are found adjacent to the northern shores of the Great Lakes because of isostatic rebound. Lake terraces also are associated with the western United States, particularly within the Great Basin, because the climate of this region has become warmer and drier since the end of the Pleistocene and numerous lakes have either disappeared or diminished greatly in size (see Figure 18.25).

FIGURE 20.15 Multiple marine terraces resulting from sporadic tectonic uplift are evident on the west side of San Clemente Island, California. (John S. Shelton)

Outbuilding Coasts

Outbuilding coasts are advancing because new material is being deposited on the coastline at a rate rapid enough to counteract processes of erosion and submergence. The types of materials and the modes of deposition vary, leading to the formation of several distinct types of outbuilding coasts.

DEPOSITIONAL COASTS Depositional coasts are experiencing the active deposition of loose clastic sediments. The sediments are transported to the coast largely by rivers and commonly are redistributed by waves and currents. The chief factors favoring the development of this type of coast are an abundant sediment supply and a gently sloping coastal margin. Where wave and current action is sufficiently vigorous to spread these materials along the coastline, depositional coasts are characterized by beaches and by offshore bars and barrier islands.

VOLCANIC COASTS Volcanic coasts exist where volcanic eruptions have raised new land above the sea. This may occur along the margins of already existing landmasses, such as those around the rim of the Pacific, or it may result in the formation of new volcanic islands. Volcanic coasts usually are very steep, but typically are fairly regular in form. A recently formed volcanic island is likely to be nearly circular, and a larger volcanic land mass along which several volcanoes have erupted typically has a coastal margin consisting of multiple arcs. After the volcanoes become dormant or extinct, their steep flanks are eroded rapidly and made more irregular by wave action.

CORAL COASTS Coral coasts are composed of limestone deposited by coral polyps and other reef-building marine organisms. The coastline itself is fronted by a **reef,** which is a ridgelike accumulation of the exoskeletons of large colonies of these small animals. Reefs grow upward through time, since the living organisms construct their limestone structures only on the upper and outer portions of the reef. Reef-building organisms require clear, shallow, and warm water in order to survive. As a result, coral coasts occur in the tropics between approximately

30°N and 25°S. In addition, they favor islands rather than mainland coasts containing rivers that discharge large quantities of sediments.

Depending upon their relationship to land, coral reefs can be divided into three types. **Fringing reefs** are attached to the coast and indicate that local land levels have been rising relative to the sea. **Barrier reefs** are separated from the coast by a lagoon that is too deep for coral growth (see Figure 20.16). In form, they are very similar to barrier beaches and, like barrier beaches, they protect the coast from storms. The largest and best-known barrier reef is the Great Barrier Reef off the east coast of Australia. Many volcanic islands, especially in the western Pacific, are surrounded by barrier reefs.

Atolls, the final type of reef, differ from the other two in that they are not associated with other land masses. They typically consist of circular reef growths surrounding a central lagoon. Small low-lying islands of sand derived from crushed coral may surmount high points on the reef to give the complex a "pearl necklace" appearance from the air. It is believed that most or all atolls originated as fringing or barrier reefs surrounding volcanic islands. They later evolved into atolls as the central islands eroded and subsided beneath the ocean while the continued upward growth of the reefs allowed them to remain at or near the ocean surface. Most atolls are located in the western Pacific.

Compound Coasts

Compound coasts add complexity to the variety of coastal types by displaying evidence of both submergence and emergence. They are common around the world because of the numerous Pleistocene and Holocene (post-Pleistocene) fluctuations of sea level. A good example of a compound coast is the Mid-Atlantic Coast of the United States, where recent drowning of a generally emergent coastline has been taking place. As a consequence, the coast contains both large estuaries, formed by the submergence of the lower portions of river valleys, and a well-developed system of barrier islands.

SEAFLOOR TOPOGRAPHY

The ocean covers some 71 percent of the Earth's surface, so that the seafloor is nearly three times as extensive as all other topographic surfaces combined. Our knowledge of the seafloor, although still far from complete, is increasing rapidly. Like our knowl-

FIGURE 20.16 A volcanic island surrounded by a barrier reef in the Society Islands of French Polynesia. (Marcel Isy-Schwart/The Image Bank)

edge of the Earth's interior, it has been obtained largely through indirect means because environmental conditions make direct observations extremely difficult.

As data continue to accumulate, it becomes increasingly evident that the topography of the ocean floor is highly diverse and is at least as rugged and varied as that of the land. The fact that major differences exist between the topography of the land and the seafloor should not be surprising, since the forces that produce them differ greatly. Of the three gradational tools—wind, ice, and water—that largely sculpt the land surface, two—wind and ice—are not present, and the action of water is much modified and reduced by its general lack of movement along the seafloor. The dominant geomorphic processes therefore are largely tectonic and gravitational in nature. The weakness of the gradational forces allows the larger tectonic features to be maintained until they are buried in sediment or subducted into the Earth's interior.

While erosion exceeds deposition in quantity on the continents, deposition greatly exceeds erosion on the ocean floors. Sediments transported from the land drift to the seafloors, especially near the margins of the continents, where eventually they can accumulate to great depths. The deep seafloors are the ultimate sediment traps; because of the density of the underlying simatic rock, they are unlikely ever to be uplifted sufficiently to become land. The oceanic rocks and their overlying sediments do not last forever, though, because they are recycled by plate tectonic movements (see Chapter 14).

The ocean basins occupy a geologically different portion of the Earth's crust than do the continents. Unlike the continents, they do not contain a thick upper zone of relatively low density granitic sial rock; instead, they consist of a much thinner layer of dense basaltic simatic rock that is covered by ocean water because it floats at a lower level on the underlying mantle.

If the ocean is considered to "belong" over the sima-surfaced portions of the crust, however, it is overfull of water. The ocean actually extends well beyond the margins of the basins and onto the geological rims of the continents. This allows the floor of the ocean to be divided into two segments—the continental margins and the ocean basins. Both segments have several major subdivisions and important accompanying topographic features.

The Continental Margins

The continental margins are composed geologically of the tapering edges of the sialic continental platforms. The crust in these areas is neither quite light enough nor thick enough to extend above sea level and therefore is overlain by shallow ocean water. In total, the continental margins comprise slightly over one-fifth of the ocean area (see Table 20.1). This

TABLE 20.1 Relative Extent of Various Features of the Main Oceans and Adjacent Seas

WATER BODY	CONTINENTAL MARGIN[1]		OCEAN BASIN			PERCENT OF TOTAL OCEAN
	Continental Shelves and Slopes	Continental Rise and Partially Filled Sedimentary Basins	Deep Seafloor	Oceanic Ridges	Other Areas	
Pacific and adjacent seas	13.1%	2.7%	43.0%	35.9%	6.3%	50.1%
Atlantic and adjacent seas	17.1	8.0	39.3	32.3	2.7	26.0
Indian and adjacent seas	9.1	5.7	49.2	30.2	5.8	20.5
Arctic and adjacent seas	68.2	20.8	0	4.2	6.8	3.4
Total ocean	15.3	5.3	41.8	32.7	4.9	100.0

[1]The continental margin has a total area of about 74.5 million square kilometers (28.2 million mi^2).

SOURCE: David A. Ross, *Introduction to Oceanography*, 3d ed., (Englewood Cliffs, N.J.: Prentice-Hall, 1982), p. 106. Ross obtained the data from Menard and Smith, 1966.

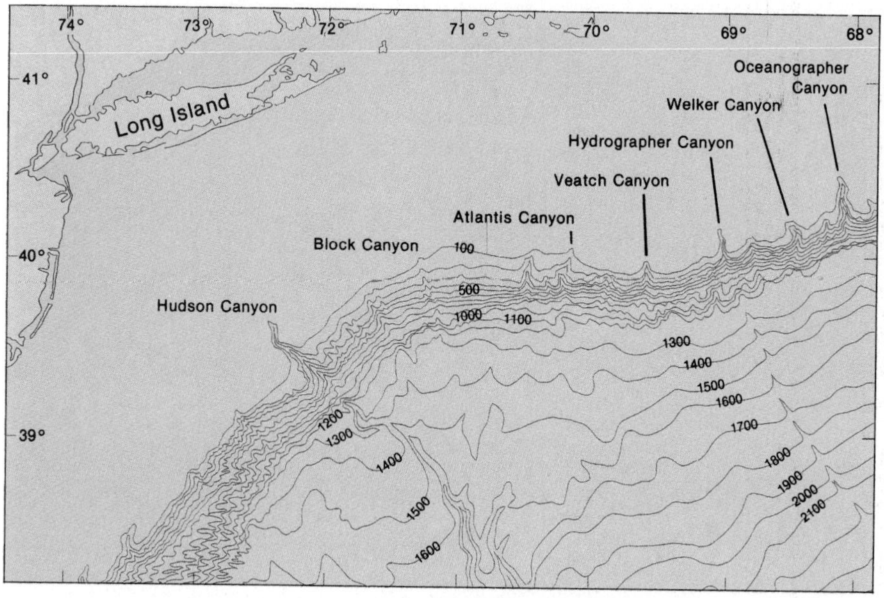

FIGURE 20.17 The continental shelf, continental slope, continental rise, and a submarine canyon system are all evident in this bathymetric map of the Atlantic off the east coast of the United States. Water depths are in fathoms. (One fathom equals six feet, or about two meters.)

zone may be subdivided into three sections: the continental shelf, the continental slope, and the continental rise (see Figure 20.17).

The **continental shelves** consist of the shallow portions of the seafloor that adjoin and surround most parts of the continents. They are quite shelflike in nature, as they are relatively smooth and level, and form the upper edges of the continental platforms.

Continental shelves are extensively developed along most continental margins and, in total, cover nearly one-sixth of the seafloor (see Figure 20.18). They range up to 1,300 kilometers (800 mi) in width and average about 75 kilometers (47 mi) wide. They tend to slope seaward, but at a nearly imperceptible mean inclination of only 0°07′. Their average depth below sea level is 60 meters (200 ft), and at their sharply defined outer edges they average about 180 meters (600 ft) in depth.

The shelves apparently originate largely from the combined effects of erosion and deposition. Lower sea levels during the Pleistocene caused most shelf areas to be exposed as land during that period, and they were subjected to subaerial gradational processes. As sea levels rose and the shelves were flooded, they began receiving large quantities of sediments from the adjacent continents. These sediments buried most surface irregularities, producing a relatively smooth offshore platform.

The most impressive and controversial features of the otherwise relatively smooth continental shelves are the deep **submarine canyons** that are entrenched into their surfaces in places. These V-shaped canyons generally are cut entirely through the thick accumulations of shelf sediments and extend well into the underlying bedrock. Some of the largest rival or exceed in size the deepest river canyons on land. Like river canyons, they are winding and contain tributary systems of smaller branching canyons. Most submarine canyons are located immediately seaward of large rivers on the continents. They seem to exist on all continental shelves, and about 20 are located off both the east and west coasts of the coterminous United States.

The headward portions of most submarine canyons apparently formed first as river valleys on land during periods of lowered sea levels. They subsequently were enlarged and extended seaward into deep water by **turbidity currents**, which are rapid flows of dense, sediment-laden water down the steep continental slopes at the outer edges of the shelves. Fluvially derived sediments apparently accumulate on the shelves until they attain unstable slopes. They are then carried by turbidity currents down the continental slope and onto the deep seafloor. If this idea is correct, turbidity currents are the chief mechanism responsible for completing the gradational cycle by transporting terrestrial sediments to the deep seafloor. The submarine canyons serve as chutes to facilitate these movements.

The **continental slope** is the long slope that separates the continental shelf from the deep seafloor. It is the surface manifestation of the two major divi-

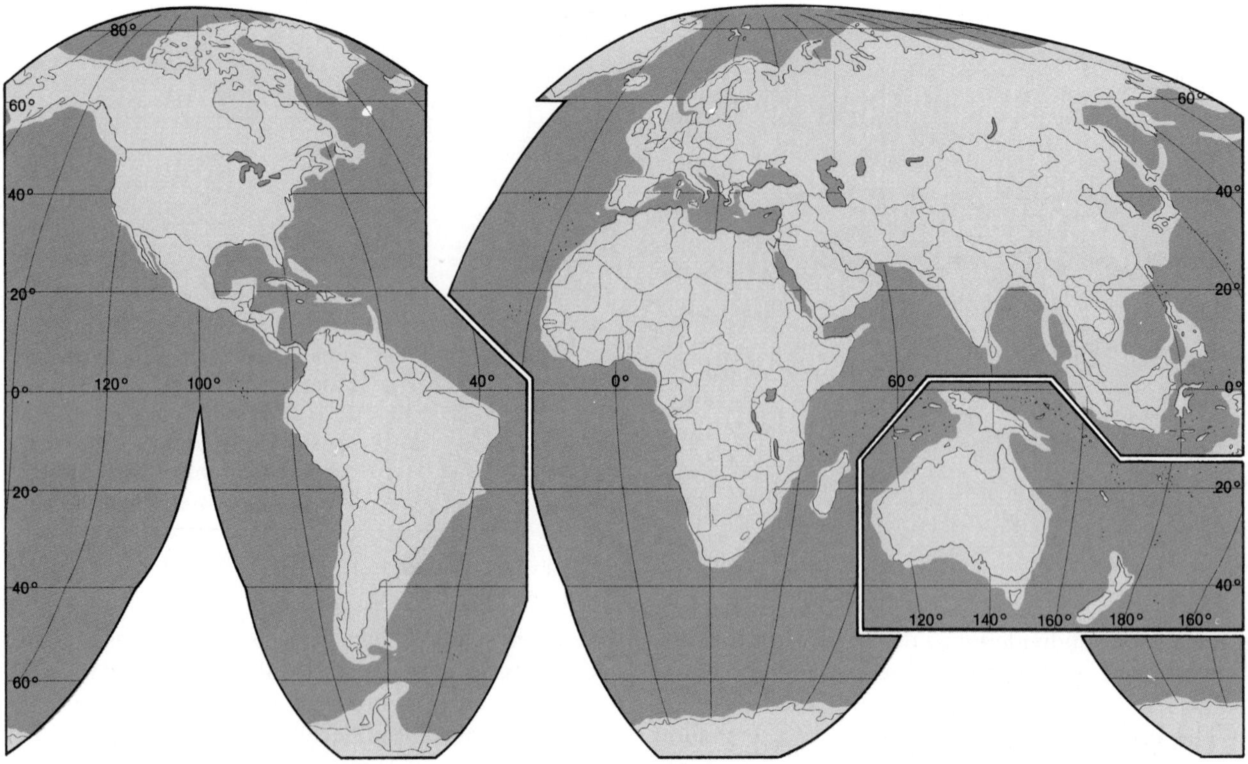

FIGURE 20.18 Continental shelf areas, shown here in light blue, comprise nearly a sixth of the total ocean floor.

sions of our planet's crust—the sial and the sima. The slope has an average width of 20 kilometers (12 mi) and an average inclination of 4° or more. In all, the seafloor drops an average of 3,000 to 3,650 vertical meters (10,000 to 12,000 ft) across this slope.

The surface of the continental slope is covered largely by fine muds of terrestrial origin. Much of this material apparently was transported directly to the upper portions of the slope by rivers and winds during the Pleistocene, when lowered sea levels caused it to lie just offshore. At regular intervals, the slope is trenched by the submarine canyons down which turbidity currents convey sediments from the adjacent continental shelf.

The **continental rise,** which occupies about 5 percent of the seafloor, is a broad, very gentle depositional ramp located at the base of the continental slope. It consists of a wedgelike apron of sediments that have spread out from the lower ends of the submarine canyons. The continental rise represents the final depositional site for most continental sediments that are transported to the sea. In some areas, sediments have accumulated on the continental rise to depths as great as 10 kilometers (6 mi). Sediment depths are especially great seaward of large rivers such as the Mississippi, Ganges, and Indus. Slope angles on the continental rise generally are less than half a degree (0.5°). Widths are highly variable—from 100 to 1,000 kilometers (60 to 600 mi).

The Ocean Basins

The ocean basins begin at the seaward margins of the continental rise and collectively comprise nearly four-fifths of the ocean floor (see Figure 20.21). In most places the ocean basin topography is rough and irregular, with various troughs, basins, ridges, and plateaus, as well as hills and mountains existing either individually or in groups or ranges (see Figure 20.19). One basic reason for the generally rugged surface of the deep seafloor is that insufficient sediment is available to cover surface irregularities produced by tectonic movements. Only limited quantities of the finest terrestrial sediments are deposited so

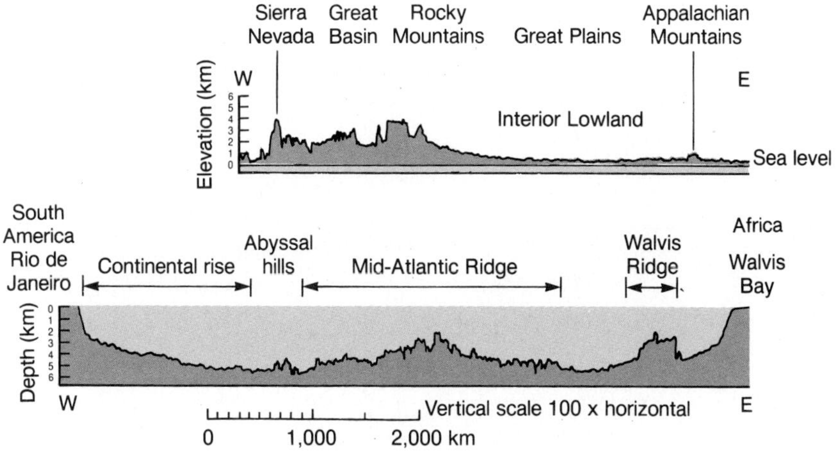

FIGURE 20.19 Comparative profile views of transects of the United States and of the Atlantic seafloor illustrate the fact that the ocean bottom topography is at least as rugged as is the land.

far from land, and marine sediments from sea plants and animal remains or direct chemical precipitation are in most places quite limited in volume. The major topographic features of the ocean basins are abyssal plains, trenches, the oceanic ridge complex, and seamounts.

Abyssal plains are deep, smooth-surfaced basins covered with a mixture of terrestrial and marine sediments and located at mean depths of 4.0 to 5.6 kilometers (2.5 to 3.5 mi) below sea level. Most occur near the western margins of the ocean basins, adjacent to continental east coasts that have large sediment-laden rivers entering the sea. The smoothness of these plains seems to result largely from the fact that enough terrestrial sediments reach them to have buried most surface irregularities. Deep sea cores taken from abyssal plains show that they contain interlayered deposits of unsorted sandy terrestrial sediments and well-sorted marine clays.

The systems of oceanic ridges and deep sea trenches were discussed in Chapter 14. They are opposite types of features both tectonically and topographically. The trenches are narrow slits in the ocean floor where old, deeply sediment-covered oceanic crust is subducted into the asthenosphere and destroyed. They are the sites of the world's greatest ocean depths. Conversely, the oceanic ridges are broad mountainous zones of newly formed crustal material with little or no sediment cover. The total oceanic ridge complex has a combined length of some 64,000 kilometers (40,000 mi), rises from 1,500 to 3,000 meters (1 to 2 mi) above the deep seafloor, and covers nearly a third of the total ocean floor. Together, the trench and ridge systems form an organized pattern that divides the remainder of the seafloor into a number of separate compartments.

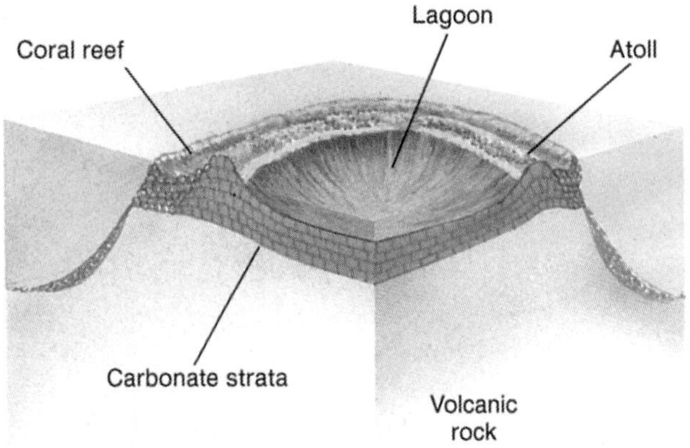

FIGURE 20.20 Cutaway profile view of a guyot surmounted by an atoll. These features are most common in the Pacific.

Seamounts, which were produced by undersea volcanic eruptions, are individual circular or elliptical submarine mountains rising more than 900 meters (3,000 ft) from the deep seafloor. Most apparently form near the oceanic ridges and are then shifted laterally by seafloor spreading motions. As they move away from the zone of volcanism, they become extinct, cool and contract, and gradually subside. Although most seamounts are located in the Pacific, they occur in all oceans. More than 1,400 are known to exist, and hundreds or even thousands of others await discovery. A few of the highest and most recent break the ocean surface to form volcanic islands, but the tops of most are situated well below sea level.

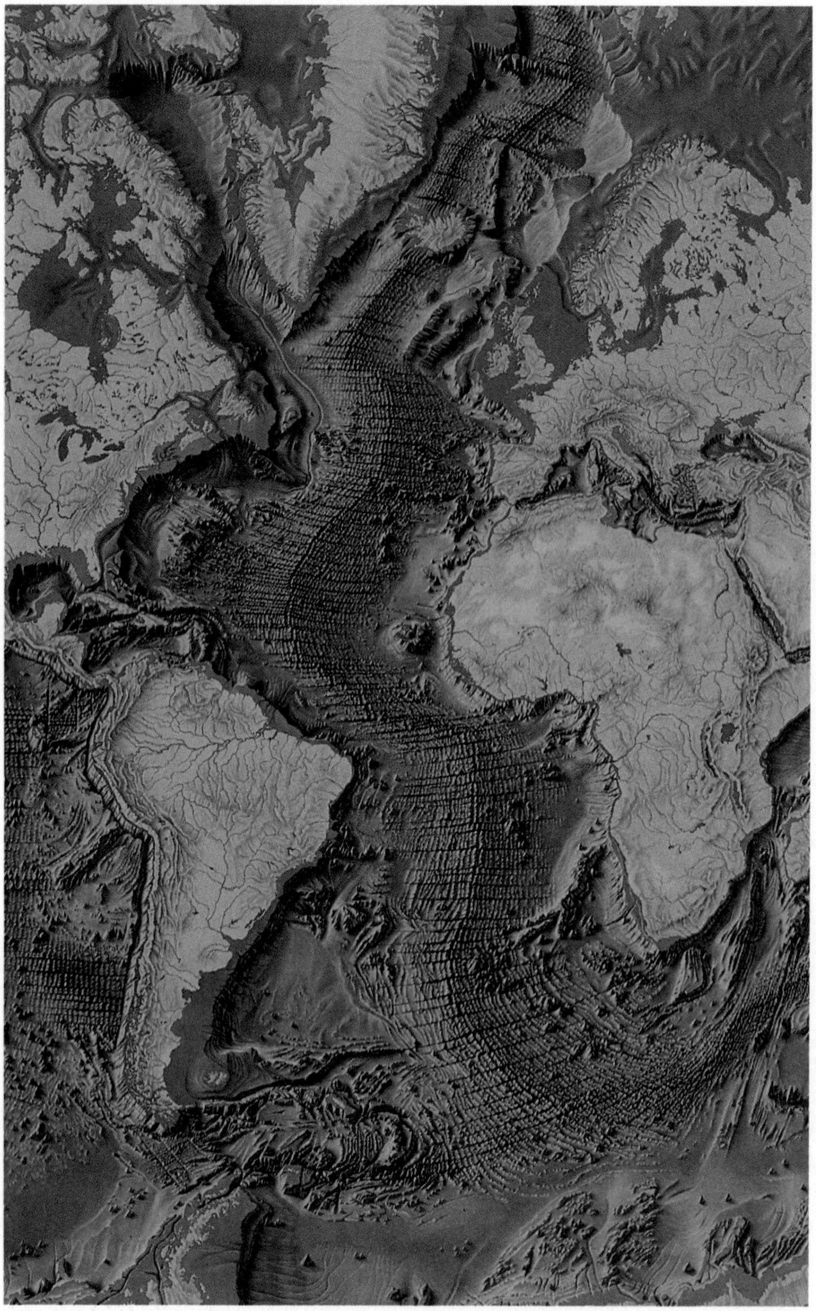

FIGURE 20.21 The physiography of the ocean floors. (Copyright © 1980 Marie Tharp)

Many seamounts are a flat-topped variety termed a **guyot.** Guyots apparently once extended above the water surface, where they were planed off by subaerial erosional processes. In some cases, coral colonies became established around these islands and subsequently were able to grow upward rapidly enough to offset the subsidence of the seamounts and remain near the surface. They are now the sites of coral atolls (see Figure 20.20).

SUMMARY

This chapter has dealt with two categories of geomorphic features that differ greatly in their physical

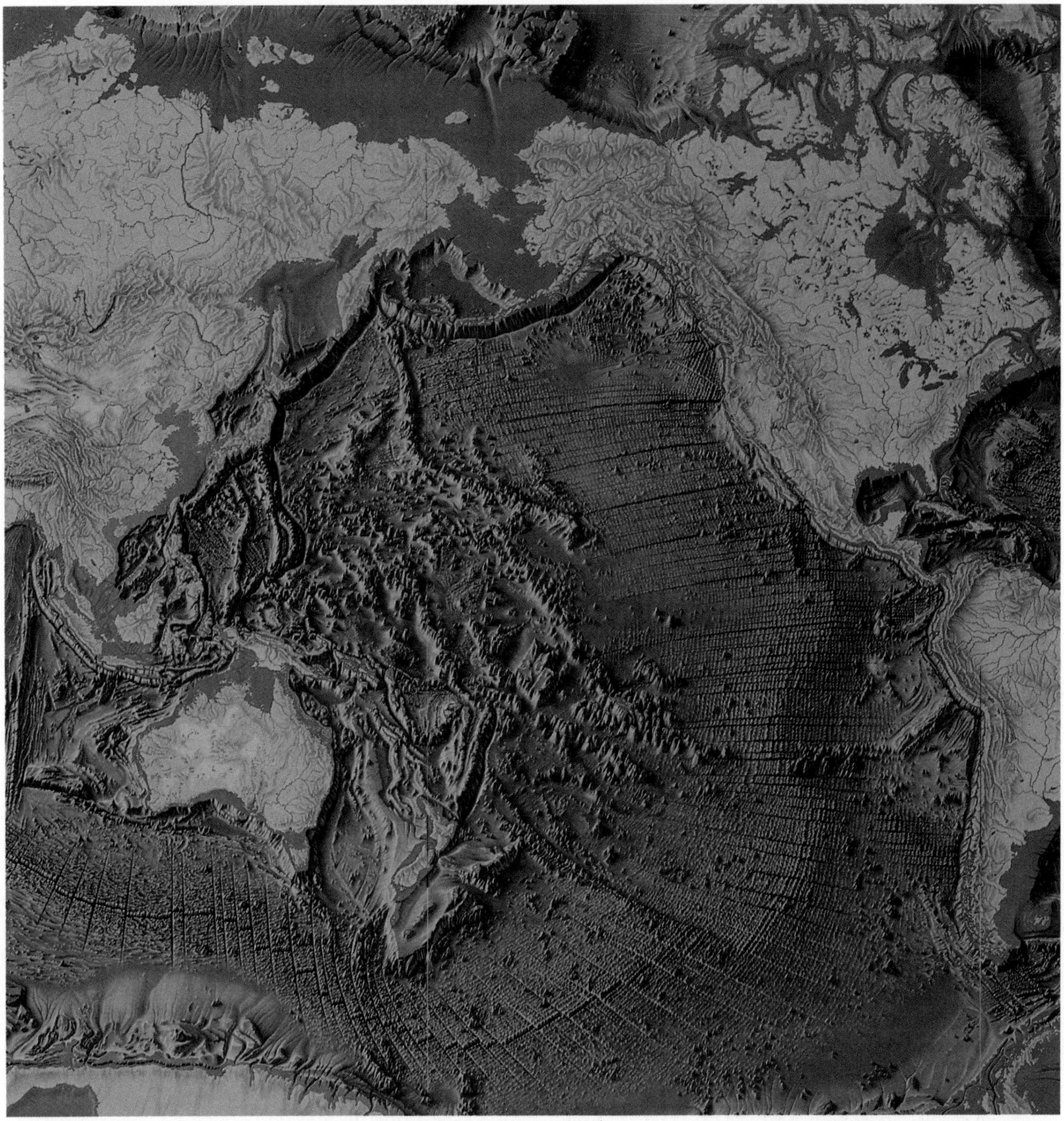

FIGURE 20.21 *continued*

characteristics and human impact, but share the common bond of being dominated by marine processes. The coastline, located at the interface of air, land, and sea, is influenced by a complex combination of atmospheric, terrestrial, and marine processes. Despite its limited extent, this highly dynamic physical environment is crucial to human beings. The ocean floor, on the other hand, occupies more than 70 percent of the Earth's surface, but has so far been little used, or even explored, by humans. Water here plays a much less active role in the production of topographic features. In fact, because of its lack of

SEAFLOOR METAL FACTORIES

Recent marine investigations have led to the discovery that rich deposits of metallic minerals currently are being formed on the ocean floors. The minerals are being deposited by "geysers" of superheated water rich in dissolved metallic solutions. The geysers apparently are widespread in volcanically active areas, especially in the vicinity of seafloor spreading centers. They are known to exist on the East Pacific Rise off the west coast of Mexico and in the Red Sea, and also were discovered recently on a portion of the Mid-Atlantic Ridge. They probably exist elsewhere as well, such as in the vicinity of volcanic "hot spots" like the Hawaiian Islands and the Azores, and in active subduction zones, such as those near Japan, Indonesia, and the Lesser Antilles.

The geysers are convectionally driven hydrothermal features powered by heat from the mantle (see Figure 20.22). They exist because the rocks of the ocean floor, especially in volcanically active areas, are "leaky," allowing sea water to circulate freely through them. Cold, dense sea water enters bedrock fissures on the seafloor and descends until it encounters hot igneous rock. There, it is heated, expands, and rises back through the fissured rocks, leaching large quantities of minerals as it does so. When it is discharged into the ocean water in geysers, the super-heated water is almost immediately cooled, greatly reducing its ability to hold dissolved minerals in solution. These minerals consequently are precipitated onto the seafloor in the vicinity of the geyser vents.

The most potentially valuable mineral deposits are associated with "black smokers," the hottest type of geyser, with water temperatures higher than 315°C (600°F). Their name is derived from their surrounding clouds of precipitating polymetallic sulfide crystals. These crystals build tubular vents around the geysers and eventually accumulate to form substantial ore bodies on the adjacent seafloor. Some of the geyser water minerals apparently precipitate at a later time, helping to form the manganese nodules that cover vast portions of the oceanic abyssal plains.

Seafloor geysers are important contributors to the total dissolved mineral content of the ocean waters. Prior to their discovery in the 1970s and 1980s, most oceanic minerals were believed to have been washed into the oceans by rivers. It is currently thought that the geysers constitute at least as important a source of these minerals.

The geysers also seem to have been a key mechanism in the production of mineral deposits on land. These deposits, which often are similar in appearance to those forming on the seafloor, are major com-

movement, water on the ocean floor serves to protect features produced by tectonic and gravitational processes while permitting the deposition of vast quantities of both terrestrial and marine sediments.

We now return to the Focus Questions posed at the beginning of the chapter:

1. *What geomorphic processes are chiefly responsible for the production of coastal features?*

Atmospheric, terrestrial, and marine processes all contribute to the production of coastal features. The most powerful and effective geomorphic agent for most coastlines, however, is wave action. Waves erode shoreline materials through the processes of abrasion, attrition, solution, and hydraulic action. Waves also deposit sediments derived both from the land and from farther offshore along the coast. The action of currents is important in transporting fine-textured sediments along the coastline.

2. *What are the most important types of erosional and depositional coastal landforms?*

Coasts dominated by long-term erosion typically are steep and rugged. The undercutting action of waves along an elevated coastline characteristically produces sea cliffs, which gradually retreat as they are undermined and collapse. The process of wave refraction causes erosion to be most active around headlands, while coastal indentations may be the sites of localized deposition. Headland erosion leaves wave-cut platforms that can serve as future sites for the accumulation of beach sediments.

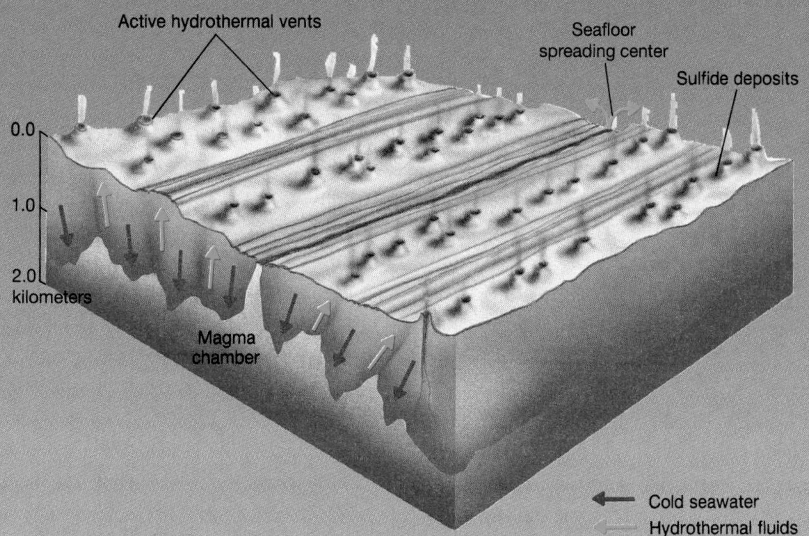

FIGURE 20.22 Diagrammatic view of the distribution of mineral-depositing hydrothermal vents along a seafloor spreading center.

mercial sources of gold, silver, copper, zinc, and other metals. The metallic ore bodies of eastern Quebec, for example, although found in rocks 2.6 billion years old, resemble polymetallic sulfide deposits currently forming on the seafloor around black smokers.

Valuable as these mineral deposits are, their commercial exploitation is not likely in the near future. Most of the deposits occur on the deep seafloor, far from land. No technology currently exists for breaking, lifting, and sorting these hard-rock deposits. Indeed, we are at present only in the early stages of discovering the existence and plotting the distribution of these minerals. Marine polymetallic sulfide (MPS) deposits, however, may constitute an economically important source of metals in the twenty-first century.

The dominant coastal depositional feature is the beach. Beaches may front mainland shorelines, but often are best developed on offshore barrier islands that are separated from the mainland by coastal lagoons.

3. *What are the major causes of coastal retreat and advance, and what categories of coastlines have resulted from these coastal changes?*

Coastal retreat can be caused by either erosion or submergence. Erosion typically produces a rugged coast characterized by slowly retreating headlands. Submergent coastlines are geographically widespread because of the nearly global post-Pleistocene rise in sea levels. The nature of the previously dominant geomorphic process largely determines the present surface characteristics of a given submergent coastline. Ria coasts, probably the most distinctive of the submergent coastal types, are formed from the partial drowning of coasts dominated by river valleys. They are characterized by numerous bays or estuaries.

Coastlines advance seaward either because of tectonic or isostatic uplift, or by outbuilding resulting from the addition of material by geological or biological processes. Large areas in the high latitudes of the northern hemisphere are advancing because of isostatic rebound. Emergent coastlines around the Pacific are rising because of crustal compression caused by the subduction of the Pacific Plate. Outbuilding coasts include those coasts subject to rapid fluvial or marine deposition, volcanic eruptions, and the growth of reef-building organisms.

4. *What are the major topographic divisions of the ocean floor?*

The ocean floor consists of two basic segments—the continental margins and the ocean basins. The continental margins have formed on the tapering edges of the continental plates and are major depositional sites for sediments derived from the land. Immediately bordering the continents are the ramplike continental shelves. They generally are broad, covered shallowly by water, and slope gently seaward. In places they are trenched deeply by submarine canyons. The continental shelves are bordered on their seaward sides by long and often steep continental slopes, which merge at their bases with the gently sloping and deeply sediment-covered continental rises.

The ocean basins are compartmentalized by their extensive systems of oceanic ridges and trenches. These systems separate deep and generally flat-floored abyssal plains. Scattered through the abyssal plains are individual rounded submarine mountains, termed seamounts, that are products of undersea volcanic eruptions.

Review Questions

1. What is the justification for making coastal landforms the last landform group to be covered in this text?

2. By what processes do waves and currents erode coasts? Briefly describe each process.

3. Explain how the process of wave refraction causes wave energy to be concentrated on headlands.

4. How are sea cliffs formed? How is their steepness maintained? What features may be produced as a sea cliff is eroded back?

5. What coastal environmental conditions favor coastal erosion? What conditions favor coastal deposition?

6. What is the most characteristic feature of a depositional shoreline? Explain how it forms. From what sediment sources are its materials derived?

7. What similarities and differences exist between bars and barrier islands? Where and why do they form? What effects do they have on the characteristics of the mainland coast?

8. What are the two chief mechanisms by which coastlines retreat? Why are more shorelines around the world currently retreating than advancing?

9. Explain how the Chesapeake Bay formed.

10. What are the two basic causes of present-day coastal advance? In what geographical areas of the world is each occurring?

11. What are marine terraces, and how are they produced? In what part of the United States are marine terraces well developed, and why?

12. Under what environmental conditions do coral reefs develop? Describe the three categories of coral reef coasts.

13. What are compound coasts? Why are they common at the present time?

14. What are the two basic divisions of the ocean floor? In what physical and geological respects do they differ?

15. How are submarine canyons produced and what is their geomorphic function? How do they differ in origin from deep sea trenches?

16. What are seamounts and how are they formed? What is their relationship to coral atolls?

17. Why do barrier beaches migrate under natural conditions? What human activities can accelerate the process of barrier beach migration?

Key Terms

beach drifting
wave refraction
undertow
sea cliff
wave-cut platform
wave-built platform
beach
sand bar
barrier island
tidal inlet
lagoon
spit
ria coast
marine terrace
reef
atoll
continental shelf
submarine canyon
turbidity current
continental slope
continental rise
abyssal plain
seamount

CASE STUDY

BARRIER BEACH MIGRATION AND COASTAL DEVELOPMENT

Increases in affluence and leisure time, coupled with the development of efficient modern transportation networks, have resulted in a great increase in tourism in the United States during the past few decades. The chief focus of the nation's multi-billion dollar tourist industry is the string of beach resorts that have been developed on the more than 300 barrier islands along the Atlantic and Gulf Coasts. Such cities as Atlantic City, New Jersey; Ocean City, Maryland; Virginia Beach, Virginia; Myrtle Beach, South Carolina; and Daytona Beach, West Palm Beach, Fort Lauderdale, and Miami Beach in Florida, are only some of the largest and best known of scores of tourist-oriented cities and towns along the East Coast.

Unfortunately, the barrier beaches upon which these cities and towns have been built are unstable and, geologically speaking, short-lived features. Furthermore, the human alteration of these strips of sand has in many places upset the delicate balance of nature and further destabilized conditions. A combination of ongoing natural processes and human influences has reduced the size of barrier beaches of the Atlantic and Gulf Coasts and has caused them to migrate slowly landward. This raises serious concerns for the future of the countless billions of dollars worth of nonmovable buildings and other property on these beaches and for the livelihoods of their hundreds of thousands of human inhabitants.

CAUSES OF BARRIER BEACH MIGRATION

The basic cause of barrier beach migration is the nearly global rise in ocean levels. Barrier beaches occupy a fixed position relative to the coast just shoreward of the zone of impact of storm waves. When sea levels rise—as they are doing now all along the U.S. Atlantic and Gulf Coasts—this zone shifts landward, resulting in the shoreward migration of the barrier beaches. Because slopes along barrier beach coasts tend to be quite gentle, a small increase in sea level produces a large horizontal migration of both the shoreline and the barrier beaches. For example, the approximate 0.3 meter (1 ft) per century rise in sea level experienced recently along much of the East Coast has produced a landward migration of about 60 meters (200 ft) in the shoreline and the offshore barrier beaches.

Barrier beach migration does not occur as a steady movement, but takes place sporadically during major storms. Large storm waves pound at the seaward sides of the barrier beaches, eroding the sand ridges behind the shoreline. If these ridges are eroded through, or if waves become high enough to breach them, waves can wash completely over the barrier beach from one side to the other. This **overwash** process causes sand to be removed rapidly from the seaward side of the barrier beach and deposited on the landward side. By this means, the material comprising the barrier beach slowly rolls over on itself in a manner similar to that of a moving tank tread (see Figure 20.23). Unfortunately, fixed structures are unable to move with the land, so a building originally located well away from the shoreline behind a protective beach ridge may have the shoreline migrate toward it until it is finally undercut by waves and destroyed.

A number of human activities further destabilize barrier beaches, hastening their erosion and landward migration. One is the destruction of dune grasses that help to hold the sand in place and assist in dune formation. These grasses may be removed purposefully for "aesthetic" reasons or be destroyed by trampling or by the use of off-road vehicles. The sand dunes themselves, which form the backbone of barrier beaches, may be flattened in order to improve the view of the ocean for residents of homes situated behind them. Of more widespread significance is the upstream damming of rivers that supply sand to the coast. The reservoirs behind the dams serve as sediment traps, greatly reducing the quantity of sand available for barrier beach construction. This factor is believed to play a major role in the rapid erosional losses currently being experienced on both barrier islands and mainland beaches in many areas. If the widely anticipated global warming of the greenhouse effect be-

continued on next page

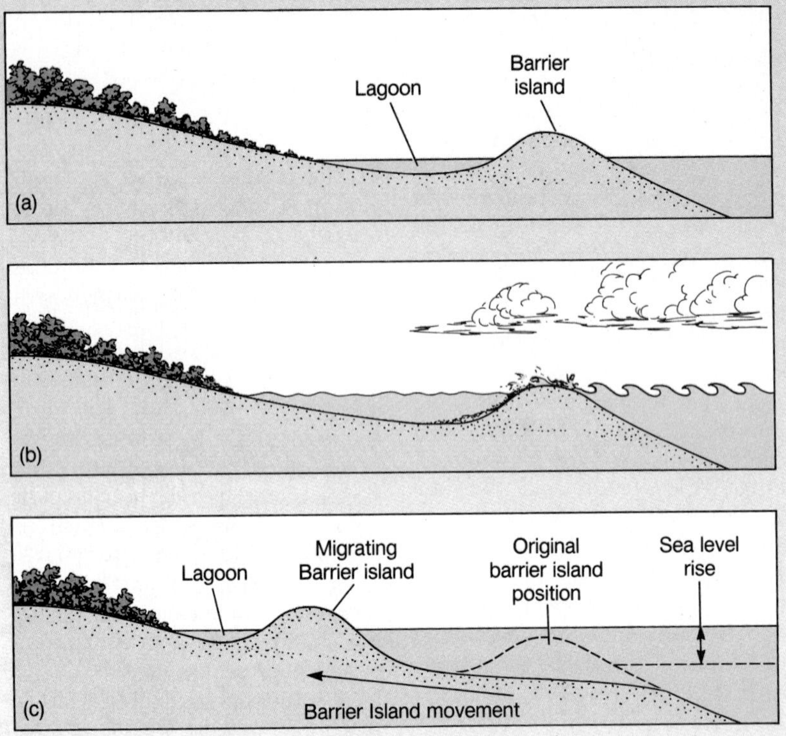

FIGURE 20.23 The landward migration of barrier island systems, shown here in a time sequence, results largely from the post-Pleistocene rise in sea level. The landward movement actually occurs when storm waves wash sand from the seaward side of the islands and deposit it in the lagoon (center view).

comes a reality, increased glacial melting and thermally induced sea water expansion will produce even more rapid rises in ocean levels than already experienced, further compounding the problem.

SOLUTIONS

The seriousness and complexity of the problem of barrier beach erosion and migration is such that many differing types of solutions have been tried. One group involves the construction of structures designed either to protect the shoreline from wave attack or to encourage the local deposition of sand. These include **seawalls,** which are rock or wooden barriers built parallel to the coast to absorb and dissipate wave energy and protect the land behind them. Unfortunately, seawalls do not protect the beach in front of them and often deflect wave energy downward so that they eventually are undermined and destroyed.

Another type of structure is the **groin,** a narrow seaward extending barrier, generally built of rock, that is designed to interfere with beach drifting and longshore currents and to encourage the deposition of sand. Series of groins have been constructed in numerous locations along the East Coast of the United States. Although they may locally solve the problem of beach erosion, at least temporarily, their interception of sand often causes the accelerated erosion of beaches immediately down the coast. In addition, both seawalls and groins are expensive to build and negatively affect the appearance of the beach and shoreline they are designed to protect. They are at best only temporary solutions to the problem of beach erosion because they do not address the underlying factor of rising sea levels.

Another very temporary but aesthetically more pleasing solution is to haul in replacement sand from elsewhere or to pump sand onto the beach from further offshore. A number of East Coast cities, including Ocean City (Maryland) and Miami Beach have recently employed this method of beach restoration. In only a decade or two, however, or perhaps much sooner if these sites are visited by major storms, this sand will be eroded away. This is especially true where pumping has deepened the water just offshore, allowing larger waves to reach the beach.

A different type of solution, which is favored by various government agencies and many conservation groups, is not to fight the sea, but rather to limit development on barrier beaches and to let nature take its course. In recent years, many barrier beaches along the Atlantic and Gulf Coasts have been set aside as National Seashores, National Wildlife Refuges, and state parks. Coastal changes within these areas are expected and are adjusted to by those responsible for their care. This policy is undoubtedly wisest in areas where no major development has yet occurred, but it is too late for many other areas where enormous investments of money, energy, and materials already have been made.

The future of the developed portion of our barrier beaches is very uncertain. In most cases, development has occurred recently enough that beach migration has progressed only to the point of being an expensive annoyance rather than a major disaster. Technological developments and the expenditure of large sums of money may postpone indefinitely the day of reckoning for some areas, but future coastal storms are certain to cause great destruction in others (see Figure 20.24). In the meantime, increased understanding of coastal processes and greater restraint in coastal development are necessary.

FIGURE 20.24 A beachfront home destroyed by Hurricane Hugo near Charleston, South Carolina in September, 1989.

APPENDIX A

METRIC AND IMPERIAL MEASUREMENT EQUIVALENCIES

Throughout this book, a dual system of numerical values has been used because both our traditional English (or Imperial) system and the metric system are widely employed. In most cases, metric values are given first, and Imperial equivalents are provided in parentheses. Although the metric system is undoubtedly a superior system and is used in most parts of the world, the conversion to metric in the United States is proving to be slower and more difficult than earlier supposed, and it appears that both systems will be in widespread use for some time. For convenience, conversion formulas for units of measurement especially relevant to the subject of physical geography are provided here.

Length

- 1 kilometer = 1000 meters = 0.6214 mile = 3281 feet
- 1 mile = 5280 feet = 1.609 kilometers = 1609.3 meters
- 1 centimeter = 0.3937 inch
- 1 inch = 2.540 centimeters
- 1 micrometer (μm) = 10^{-6} meter = 10^{-4} centimeter = 3.937×10^{-5} inch

Area

- 1 square kilometer = 0.3861 square mile = 247.1 acres
- 1 square mile = 2.590 square kilometers = 640 acres

Volume

- 1 cubic kilometer = 0.2399 cubic mile
- 1 cubic mile = 4.1684 cubic kilometers
- 1 cubic meter = 1.308 cubic yards = 61,024 cubic inches
- 1 cubic yard = 0.7646 cubic meters = 46,656 cubic inches
- 1 liter = 1.0567 quarts
- 1 quart = 0.946 liters

Mass

- 1 metric ton = 1000 kilograms = 2205 pounds
- 1 short ton = 2000 pounds = 907.2 kilograms
- 1 kilogram = 1000 grams = 2.2046 pounds
- 1 pound = 0.4536 kilograms

Speed

- 1 meter per second = 3.281 feet per second = 3.6 kilometers per hour = 2.237 miles per hour
- 1 mile per hour = 1.467 feet per second = 1.6093 kilometers per hour = 0.8684 knots
- 1 kilometer per hour = 0.6214 miles per hour
- 1 knot = 1.152 miles per hour = 1.853 kilometers per hour

Temperature Scale Conversion Formulas

- $°C = \frac{5}{9}(°F - 32°)$ or $°C = \frac{°F - 32°}{1.8}$
- $°F = \frac{9}{5}°C + 32°$ or $°F = 1.8\,°C + 32$
- $°K = °C + 273.15°$

Air Pressure

- Mean sea level air pressure = 1013.2 millibars = 29.92 inches of mercury = 101.3 kilopascals = 1 atmosphere

A P P E N D I X B

WEATHER MAP SYMBOLS

Surface weather maps are prepared daily by the U. S. National Weather Service in Washington, D. C., and it is possible to purchase a subscription to a weekly booklet entitled "Daily Weather Maps" from the superintendent of documents. Each map in these booklets depicts the surface weather conditions for the coterminous United States as well as adjacent portions of Canada and Mexico at 7:00 A.M., Eastern Standard Time. An example of the maps is shown in Figure 8.10. Also included are smaller maps depicting upper air windflow patterns, 24-hour high and low temperatures, and precipitation totals for a large number of reporting stations. The ready availability and high quality of these maps have led many individuals interested in the weather as well as numerous educational institutions to subscribe to them.

The weather information contained on the Daily Weather Maps is conveyed by the use of standardized symbols. The two major types of features shown are isobars, which are drawn at four-millibar intervals, and weather fronts, the symbols for which are shown in Table B.1. The positions of high and low pressure centers also are shown.

A large number of individual reporting stations are displayed on the maps, and their information provides the basis for the positions of the isobars and fronts. The sample report in Table B.2 illustrates the format used for displaying reporting station weather data. A translation of the information in the report appears in the table and further information about the symbols is contained in Figure B.1.

TABLE B.1 Weather Front Symbols

TABLE B.2 Standardized Weather Map Symbols

INFORMATION	MEANING (SAMPLE STATION CONDITIONS IN PARENTHESES)
	Circle represents station's location. Proportion blackened in represents amount of cloud cover (overcast).
	Shaft extending from station circle indicates direction from which wind is blowing. Lines on shaft indicate wind speed (northwest wind at 18–22 knots).
32	Current temperature in degrees Fahrenheit (32° F).
30	Current dew point in degrees Fahrenheit (30° F).
3/4	Visibility in miles, if under 10 miles (3/4 mile).
* *	Present weather of significance (continuous light snow).
247	Last three digits of air pressure reading to tenths of a millibar, with decimal point omitted (1024.7 mb).
+28/	Air pressure change in 3 hours preceding observation, including pressure tendency over the period (pressure has risen steadily a total of 2.8 millibars).
.4	Time precipitation last began or ended (precipitation began as rain between 3 and 4 hours ago).
.45	Amount of precipitation during past 6 hours in inches. If frozen, water equivalent is given (0.45″).
	Type of high cloud, if any (dense cirrus).
	Type of middle cloud, if any (altocumulus).
- - -	Type of low cloud, if any (fractostratus).
6	Fraction of sky covered by low- or middle-level clouds (seven-tenths to eight-tenths of sky covered by fractostratus clouds).
2	Height of cloud base or ceiling (ceiling 300–599 feet).

CLOUD ABBREVIATION		C_L	DESCRIPTION (Abridged From W M O Code)		C_M	DESCRIPTION (Abridged From W M O Code)		C_H	DESCRIPTION (Abridged From W M O Code) 3
St or Fs-Stratus or Fractostratus	1		Cu of fair weather, little vertical development and seemingly flattened	1		Thin As (most of cloud layer semi-transparent)	1		Filaments of Ci, or "mares tails," scattered and not increasing
Ci-Cirrus	2		Cu of considerable development, generally towering, with or without other Cu or Sc bases all at same level	2		Thick As, greater part sufficiently dense to hide sun (or moon), or Ns	2		Dense Ci in patches or twisted sheaves, usually not increasing, sometimes like remains of Cb; or towers or tufts
Cs-Cirrostratus	3		Cb with tops lacking clear-cut outlines, but distinctly not cirriform or anvil-shaped; with or without Cu, Sc, or St	3		Thin Ac, mostly semi-transparent; cloud elements not changing much and at a single level	3		Dense Ci, often anvil-shaped, derived from or associated with Cb
Cc-Cirrocumulus	4		Sc formed by spreading out of Cu; Cu often present also	4		Thin Ac in patches; cloud elements continually changing and/or occurring at more than one level	4		Ci, often hook-shaped, gradually spreading over the sky and usually thickening as a whole
Ac-Altocumulus	5		Sc not formed by spreading out of Cu	5		Thin Ac in bands or in a layer gradually spreading over sky and usually thickening as a whole	5		Ci and Cs, often in converging bands, or Cs alone; generally overspreading and growing denser, the continuous layer not reaching 45° altitude
As-Altostratus	6		St or Fs or both, but no Fs of bad weather	6		Ac formed by the spreading out of Cu	6		Ci and Cs, often in converging bands, or Cs alone; generally overspreading and growing denser; the continuous layer exceeding 45° altitude
Sc-Stratocumulus	7		Fs and/or Fc of bad weather (scud)	7		Double-layered Ac, or a thick layer of Ac, not increasing; or Ac with As and/or Ns	7		Veil of Cs covering the entire sky
Ns-Nimbostratus	8		Cu and Sc (not formed by spreading out of Cu) with bases at different levels	8		Ac in the form of Cu-shaped tufts or Ac with turrets	8		Cs not increasing and not covering entire sky
Cu or Fc-Cumulus or Fractocumulus	9		Cb having a clearly fibrous (cirriform) top, often anvil-shaped, with or without Cu, Sc, St, or scud	9		Ac of a chaotic sky, usually at different levels; patches of dense Ci are usually present also	9		Cc alone or Cc with some Ci or Cs, but the Cc being the main cirriform cloud
Cb-Cumulonimbus									

R_t	TIME OF PRECIPITATION 4	h	HEIGHT IN FEET (Rounded Off)	HEIGHT IN METERS (Approximate) 5	N	SKY COVERAGE (Total Amount) 6	N_h	SKY COVERAGE (Low And/Or Middle Clouds) 7
0	No Precipitation	0	0 - 149	0 - 49		No clouds	0	No clouds
1	Less than 1 hour ago	1	150 - 299	50 - 99		Less than one-tenth or one-tenth	1	Less than one-tenth or one-tenth
2	1 to 2 hours ago	2	300 - 599	100 - 199		Two-tenths or three-tenths	2	Two-tenths or three-tenths
3	2 to 3 hours ago	3	600 - 999	200 - 299		Four-tenths	3	Four-tenths
4	3 to 4 hours ago	4	1,000 - 1,999	300 - 599		Five-tenths	4	Five-tenths
5	4 to 5 hours ago	5	2,000 - 3,499	600 - 999		Six-tenths	5	Six-tenths
6	5 to 6 hours ago	6	3,500 - 4,999	1,000 - 1,499		Seven-tenths or eight-tenths	6	Seven-tenths or eight-tenths
7	6 to 12 hours ago	7	5,000 - 6,499	1,500 - 1,999		Nine-tenths or overcast with openings	7	Nine-tenths or overcast with openings
8	More than 12 hours ago	8	6,500 - 7,999	2,000 - 2,499		Completely overcast	8	Completely overcast
9	Unknown	9	At or above 8,000, or no clouds	At or above 2,500, or no clouds		Sky obscured	9	Sky obscured

FIGURE B.1 Daily Weather Map Information (Abridged from Daily Weather Maps, U. S. Dept. of Commerce, National Oceanic and Atmospheric Administration, and the Environmental Data Service.

WW PRESENT WEATHER (Descriptions Abridged from W. M. O. Code)

8

	0	1	2	3	4	5	6	7	8	9
00	Cloud development NOT observed or NOT observable during past hour	Clouds generally dissolving or becoming less developed during past hour	State of sky on the whole unchanged during past hour	Clouds generally forming or developing during past hour	Visibility reduced by smoke	Haze	Widespread dust in suspension in the air, NOT raised by wind, at time of observation	Dust or sand raised by wind, at time of observation	Well developed dust devil(s) within past hour	Dust storm or sand storm within sight of or at station during past hour
10	Light fog	Patches of shallow fog at station, NOT deeper than 6 feet on land	More or less continuous shallow fog at station, NOT deeper than 6 feet on land	Lightning visible, no thunder heard	Precipitation within sight, but NOT reaching the ground	Precipitation within sight, reaching the ground, but distant from station	Precipitation within sight, reaching the ground, near to but NOT at station	Thunder heard, but no precipitation at the station	Squall(s) within sight during past hour	Funnel cloud(s) within sight during past hour
20	Drizzle (NOT freezing and NOT falling as showers) during past hour, but NOT at time of observation	Rain (NOT freezing and NOT falling as showers) during past hour, but NOT at time of observation	Snow (NOT falling as showers) during past hour, but NOT at time of observation	Rain and snow (NOT falling as showers) during past hour, but NOT at time of observation	Freezing drizzle or freezing rain (NOT falling as showers) during past hour, but NOT at time of observation	Showers of rain during past hour, but NOT at time of observation	Showers of snow, or of rain and snow, during past hour, but NOT at time of observation	Showers of hail, or of hail and rain, during past hour, but NOT at time of observation	Fog during past hour, but NOT at time of observation	Thunderstorm (with or without precipitation) during past hour, but NOT at time of obs.
30	Slight or moderate dust storm or sand storm, has decreased during past hour	Slight or moderate dust storm or sand storm, no appreciable change during past hour	Slight or moderate dust storm or sand storm, has increased during past hour	Severe dust storm or sand storm, has decreased during past hour	Severe dust storm or sand storm, no appreciable change during past hour	Severe dust storm or sand storm, has increased during past hour	Slight or moderate drifting snow, generally low	Heavy drifting snow, generally low	Slight or moderate drifting snow, generally high	Heavy drifting snow, generally high
40	Fog at distance at time of observation, but NOT at station during past hour	Fog in patches	Fog, sky discernible, has become thinner during past hour	Fog, sky NOT discernible, has become thinner during past hour	Fog, sky discernible, no appreciable change during past hour	Fog, sky NOT discernible, no appreciable change during past hour	Fog, sky discernible, has begun or become thicker during past hour	Fog, sky NOT discernible, has begun or become thicker during past hour	Fog, depositing rime, sky discernible	Fog, depositing rime, sky NOT discernible
50	Intermittent drizzle (NOT freezing) slight at time of observation	Continuous drizzle (NOT freezing) slight at time of observation	Intermittent drizzle (NOT freezing) moderate at time of observation	Continuous drizzle (NOT freezing), moderate at time of observation	Intermittent drizzle (NOT freezing), thick at time of observation	Continuous drizzle (NOT freezing), thick at time of observation	Slight freezing drizzle	Moderate or thick freezing drizzle	Drizzle and rain, slight	Drizzle and rain, moderate or heavy
60	Intermittent rain (NOT freezing), slight at time of observation	Continuous rain (NOT freezing), slight at time of observation	Intermittent rain (NOT freezing) moderate at time of obs.	Continuous rain (NOT freezing), moderate at time of observation	Intermittent rain (NOT freezing), heavy at time of observation	Continuous rain (NOT freezing), heavy at time of observation	Slight freezing rain	Moderate or heavy freezing rain	Rain or drizzle and snow, slight	Rain or drizzle and snow, moderate or heavy
70	Intermittent fall of snow flakes, slight at time of observation	Continuous fall of snowflakes, slight at time of observation	Intermittent fall of snowflakes, moderate at time of observation	Continuous fall of snowflakes, moderate at time of observation	Intermittent fall of snowflakes, heavy at time of observation	Continuous fall of snowflakes, heavy at time of observation	Ice needles (with or without fog)	Granular snow (with or without fog)	Isolated starlike snow crystals (with or without fog)	Ice pellets (sleet, U. S. definition)
80	Slight rain shower(s)	Moderate or heavy rain shower(s)	Violent rain shower(s)	Slight shower(s) of rain and snow mixed	Moderate or heavy shower(s) of rain and snow mixed	Slight snow shower(s)	Moderate or heavy snow shower(s)	Slight shower(s) of soft or small hail with or without rain, or rain and snow mixed	Moderate or heavy shower(s) of soft or small hail with or without rain, or rain and snow mixed	Slight shower(s) of hail, with or without rain or rain and snow mixed, not associated with thunder
90	Moderate or heavy shower(s) of hail, with or without rain or rain and snow mixed, not associated with thunder	Slight rain at time of observation; thunderstorm during past hour, but NOT at time of observation	Moderate or heavy rain at time of observation; thunderstorm during past hour, but NOT at time of observation	Slight snow or rain and snow mixed or hail at time of observation; thunderstorm during past hour, but not at time of observation	Moderate or heavy snow, or rain and snow mixed or hail at time of observation; thunderstorm during past hour, but NOT at time of obs.	Slight or moderate thunderstorm without hail, but with rain and/or snow at time of obs.	Slight or moderate thunderstorm, with hail at time of observation	Heavy thunderstorm, without hail, but with rain and/or snow at time of observation	Thunderstorm combined with dust storm or sand storm at time of obs.	Heavy thunderstorm with hail at time of observation

9

ff	(MILES) (Statute) Per Hour	KNOTS
◎	Calm	Calm
	1 - 2	1 - 2
	3 - 8	3 - 7
	9 - 14	8 - 12
	15 - 20	13 - 17
	21 - 25	18 - 22
	26 - 31	23 - 27
	32 - 37	28 - 32
	38 - 43	33 - 37
	44 - 49	38 - 42
	50 - 54	43 - 47
	55 - 60	48 - 52
	61 - 66	53 - 57
	67 - 71	58 - 62
	72 - 77	63 - 67
	78 - 83	68 - 72
	84 - 89	73 - 77
	119 - 123	103 - 107

10

Code Number	a	BAROMETRIC TENDENCY	
0		Rising, then falling	
1		Rising, then steady; or rising, then rising more slowly	Barometer now higher than 3 hours ago
2		Rising steadily, or unsteadily	
3		Falling or steady, then rising; or rising, then rising more quickly	
4		Steady, same as 3 hours ago	
5		Falling, then rising, same or lower than 3 hours ago	
6		Falling, then steady; or falling, then falling more slowly	Barometer now lower than 3 hours ago
7		Falling steadily, or unsteadily	
8		Steady or rising, then falling; or falling, then falling more quickly	

11

Code Number	W	PAST WEATHER	
0		Clear or few clouds	Not Plotted
1		Partly cloudy (scattered) or variable sky	
2		Cloudy (broken) or overcast	
3		Sandstorm or dust-storm, or drifting or blowing snow	
4	≡	Fog, or smoke, or thick dust haze	
5	,	Drizzle	
6	•	Rain	
7	*	Snow, or rain and snow mixed, or ice pellets (sleet)	
8	▽	Shower(s)	
9		Thunderstorm, with or without precipitation	

12

13 Temperatures and dew points are coded in °C but plotted in °F on U.S. Surface Charts. Temperatures below 0°C are coded by dropping minus sign and adding 50. Thus, -5°C is coded as 55.

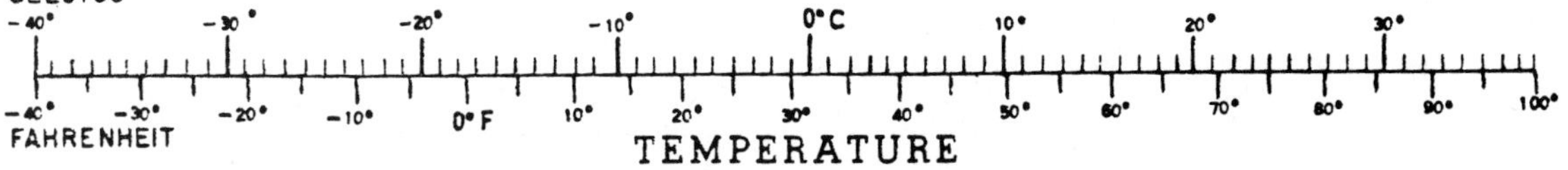

APPENDIX C

KÖPPEN CLIMATIC CLASSIFICATION SYSTEM

The Köppen climatic classification system was first published by the German biologist and climatologist Wladimir Köppen in 1901. It has been revised and refined on several occasions, most notably by Köppen's students R. Geiger and W. Pohl. The version currently in most widespread use, which is presented in simplified form here, was first published in 1953 and is frequently referred to as the Köppen-Geiger system. For convenience, we refer to it simply as the Köppen system.

ADVANTAGES AND DISADVANTAGES

The popularity of the Köppen system is based on several advantages that it offers the user. Foremost is its relative simplicity. Each climate type is defined on the basis of mean monthly or annual temperature and precipitation data, which are readily available from thousands of locations around the world. The definitions are precise and the formulas relatively simple, allowing for easy determinations of climate types. Because the climates are defined precisely, no decisions on the part of the user are necessary, nor is any previous knowledge of climate required. A major attraction for geographers is that the system relates well to global distributions of natural vegetation and soils. This factor is especially important when these two subjects are to be studied along with climate in a single academic course. Finally, the system has the advantage of widespread familiarity and use.

Among the several shortcomings often cited for the Köppen system is its relative lack of detail. The system divides the world into approximately 17 major climate types. For most of these climates, secondary and tertiary divisions are available, but compared to several alternative classification systems, the Köppen system is rather generalized, and its climatic regions contain distinctive unaddressed regional variations. Another criticism is that many important climatic factors are not taken into account. These include the radiation balance, evapotranspiration rates, precipitation type and intensity, winds, and temperature extremes. Associated with this factor is the Köppen system's lack of direct applicability to most practical uses in areas such as agriculture, forestry, and hydrology. Finally, the correlation with natural vegetation—the reason for which the system was initially devised—has always been imperfect.

Despite its shortcomings, the Köppen system remains an effective tool for organizing the study of general world climate patterns. Because such a study is precisely the goal of this textbook, the system is included here for those wishing to employ it.

CLASSIFICATION PROCEDURE

The Köppen system uses various uppercase and lowercase letters to represent mean monthly and annual temperature and precipitation data. Five major climatic categories, represented by the uppercase letters A–E, are recognized. They broadly differentiate world climatic zones in an ascending latitudinal sequence. Descriptions and definitions are as follows:

- A: Tropical climates with abundant precipitation during at least a portion of the year. Annual precipitation exceeds evapotranspiration, so that an overall moisture surplus exists. By definition, the mean temperature of each month must exceed 18° C (64.4° F).
- B: Arid and semiarid climates of the low and middle latitudes in which potential evapotranspiration exceeds precipitation. The boundaries between the B climates and adjacent humid climates take into account annual mean precipitation (P), annual mean temperature (T), and seasonal distribution of precipitation. It is calculated by the following formulas, with the first formula in each pair employing metric values (centimeters of precipitation and degrees Celsius).

The second formula (in parentheses) uses inches for precipitation and degrees Fahrenheit.

$$(P \leq C + 14) \quad (P \leq 0.44T - 3)$$

■ when at least ten times as much rain falls in the wettest summer month as in the driest winter month.[1]

$$(P \leq C) \quad (P \leq 0.44T - 14)$$

■ when at least three times as much rain falls in the wettest winter month as in the driest summer month.

$$(P \leq C + 7) \quad (P \leq 0.44T - 8.5)$$

■ when seasonal precipitation distribution is too even to meet either of the first two formula criteria.

Note: If the value of P exceeds the formula values, the climate is not a B climate. The B climate boundaries may also be determined by using Figure C.2. The B climate classification takes precedence over all the others. Therefore, when classifying a climate, it should first be determined if a B classification is warranted before proceeding further.

■ C: Mild and humid (mesothermal) climates primarily in the lower middle latitudes. The mean temperature of the coldest month must be between 18° C (64.4° F) and −3° C (26.6° F), and at least one month must average 10° C (50° F) or higher.

■ D: Humid continental climates with cold winters (microthermal), located within the upper middle latitudes and subpolar regions of the northern hemisphere. The coldest month must average below −3° C (26.6° F) and the warmest must average 10° C (50° F) or higher.

■ E: Cold climates of the high latitudes. All months must average below 10° C (50° F).

Note: Approximately 20 percent of the Earth's land surface has an A climate, 26 percent a B climate, 16 percent a C climate, 21 percent a D climate, and 17 percent an E climate.

Secondary and sometimes tertiary letters are employed in association with each of the five primary letters to allow for further differentiation into individual climate types. The most important of these secondary and tertiary letters are as follows:

[1]"Winter" refers to the colder half of the year. This includes the months October through March in the northern hemisphere and April through September in the southern hemisphere. The "summer" months are the opposite six months for each hemisphere.

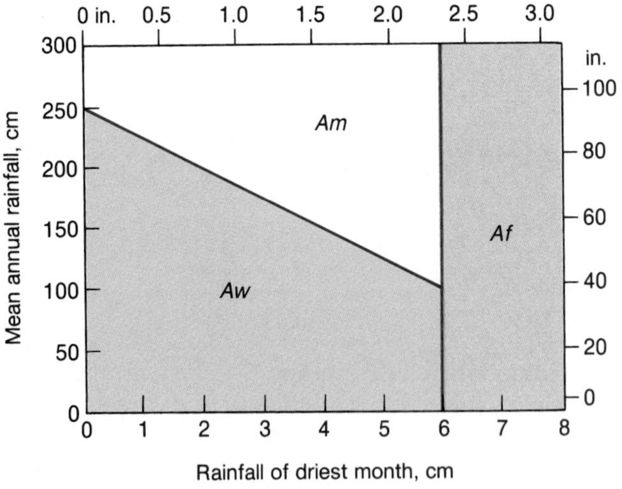

FIGURE C.1 Boundaries between the A climate subtypes.

A Climates

■ f: Mean precipitation of each month is at least 6 centimeters (2.4 in).

■ m: Short dry season compensated for by surplus precipitation in wetter months.

■ w: Well-defined dry season producing a period of water shortage.

The boundary between Am and Aw climates may be determined from Figure C.1.

B Climates

■ W: Arid. Annual precipitation totals less than half the amounts calculated from the B climate formulas.

■ S: Semiarid. Annual precipitation totals more than half the amounts calculated from the B climate formulas.

The boundaries between the BW and BS climates may be determined from Figure C.2.

■ h: Annual mean temperature is above 18° C (64.4° F).

■ k: Annual mean temperature is below 18° C (64.4° F).

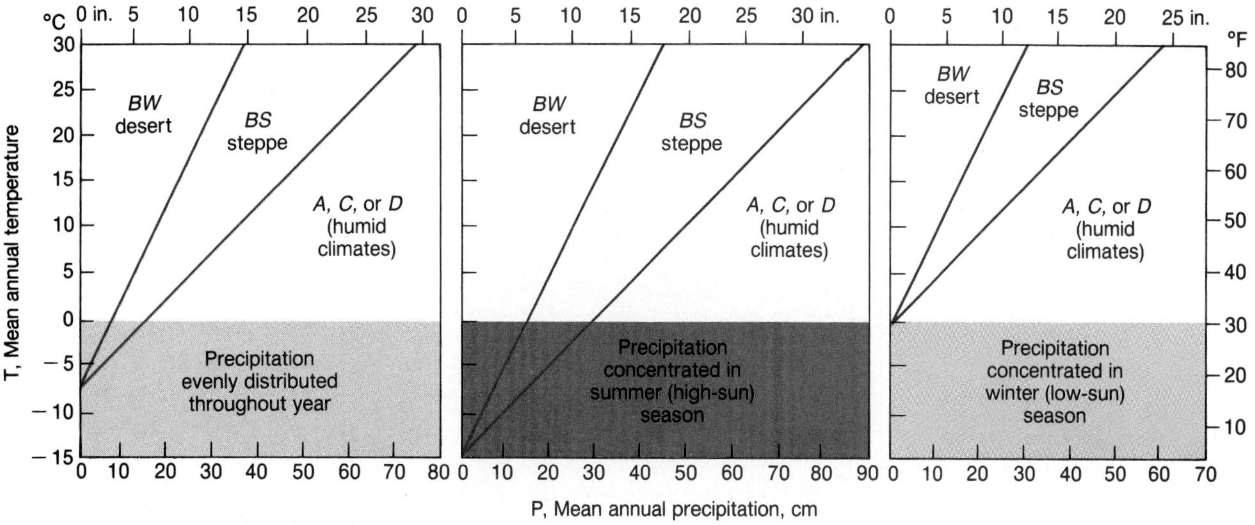

FIGURE C.2 Boundaries between the B climate subtypes for stations with differing annual temperature and precipitation means.

C Climates

- s: Summer dry season. Precipitation during wettest winter month is at least three times that of driest summer month; precipitation in driest summer month is less than 3 centimeters (1.2 in).
- w: Winter dry season. Wettest summer month receives at least ten times the precipitation of the driest winter month.
- f: No pronounced dry season. Criteria for neither s nor w are met.
- a: Hot summer season. Temperature of warmest month must average at least 22° C (71.6° F).
- b: Long mild summer. Warmest monthly mean is below 22° C (71.6° F) and four or more months are above 10° C (50° F).
- c: Short cool summer. Warmest monthly mean is below 22° C (71.6° F) and one to three months are above 10° C (50° F).

D Climates

- s, w, f: Defined as in the C climates.
- a, b, c: Defined as in the C climates.
- d: Extremely cold winter. Mean temperature of coldest winter month is −38° C (−36.4° F) or lower.

E Climates

- T: Tundra or Periglacial climate. Warmest monthly mean is between 0° C (32° F) and 10° C (50° F).
- F: Polar or Icecap climate. Mean temperature of every month is below 0° C (32° F).

MAJOR KÖPPEN CLIMATE TYPES

The names of the major Köppen system climate types are listed here. The world distribution of these climates is depicted in Figure C.3.

- Af: Tropical Rainforest climate
- Am: Tropical Monsoon climate
- Aw: Tropical Savanna climate
- BWh: Tropical Desert climate
- BSh: Tropical Steppe climate
- BWk: Mid Latitude Desert climate
- BSk: Mid Latitude Steppe climate
- Csa, Csb: Mediterranean climate
- Cwa, Cwb: Subtropical Monsoon climate
- Cfa: Humid Subtropical climate
- Cfb, Cfc: Marine climate
- Dfa, Dwa: Humid Continental–Warm Summer climate

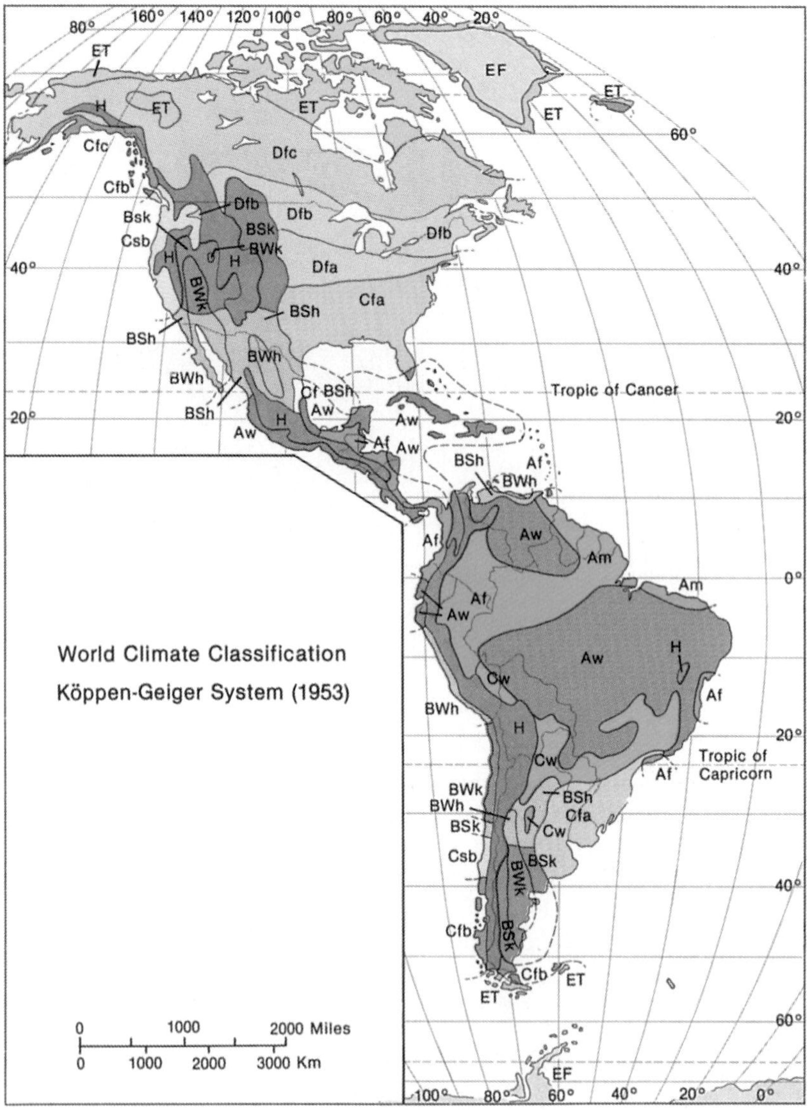

FIGURE C.3 Global distribution of major Köppen climate types.

- Dfb, Dwb: Humid Continental–Cool Summer climate
- Dfc, Dwc, Dfd, Dwd: Taiga (or Subarctic) climate
- ET: Tundra climate
- EF: Polar (or Icecap) climate

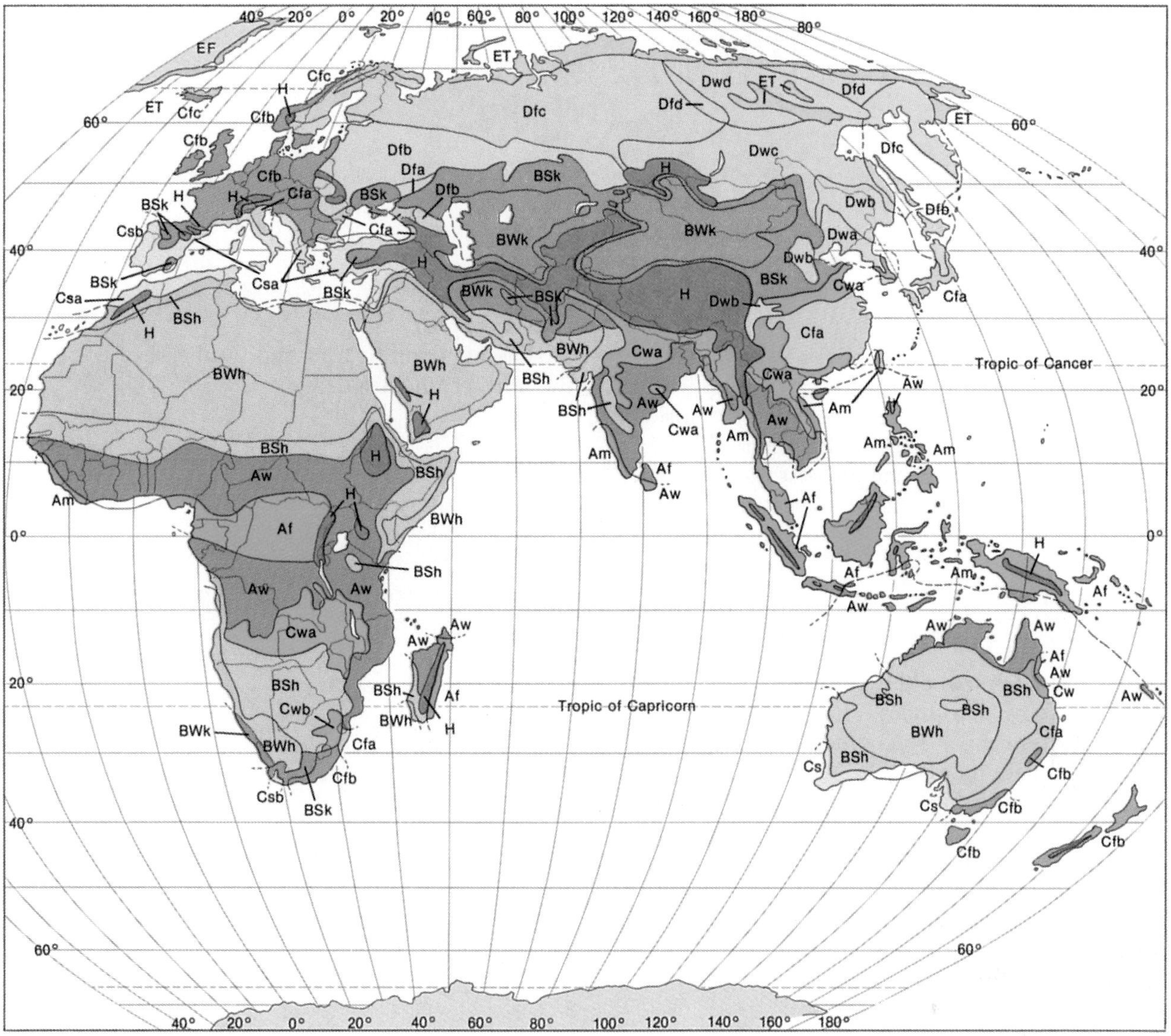
Tropic of Cancer
Tropic of Capricorn
EF
ET
Dfc
Dfb
Dfa
Dfd
Dwd
Dwc
Dwb
Dwa
BSk
BSh
BWk
BWh
Cfa
Cfb
Cfc
Csa
Csb
Cwa
Cwb
Cw
Cs
Af
Am
Aw
H

APPENDIX D

ANSWERS TO PROBLEMS

Chapter 2

1. 15° N
2. 170° E
3. 5 p.m., Monday
4. 4 a.m., Friday
5. a. The Sun's noon altitude is higher in Florida.
b. Maine has the longer daylight period.
c. Alaska experiences the greatest number of hours of twilight because it is located at the highest latitude.

Chapter 3

1. a. 1:506,880
b. 1 centimeter would represent 5.07 kilometers
2. a. 1 inch = 15.78 miles
b. 1 centimeter = 10 kilometers
3. a. 800 square miles
b. 2,070 square kilometers
4. a. 4 times as much paper
b. a sheet measuring 40 inches × 20 inches
5. a. 26 contour lines (27 if sea level is counted as a natural contour line)
b. 650 feet

Chapter 5

1. a. 10° C
b. 283° F
2. −459.4° F
3. 89.6° F
4. 39° F
5. 47° F
6. a. The average environmental lapse rate is 6.4C°/1000 m.
b. −7.2° C

Chapter 6

1. ENE, or 70°
2. SW, or 210°
3. The high will be centered southeast of you (in the northern hemisphere).

Chapter 7

1. 14,750 calories
2. 26,800 calories
3. 61% relative humidity
4. Unsaturated air is stable; saturated air is unstable.

GLOSSARY

ablation The loss of ice from a glacier as a result of melting and sublimation.

abrasion Erosion accomplished by the scraping of water-borne particles against other materials.

abyssal plain A deep, smooth-surfaced submarine basin that is covered with a mixture of terrestrial and marine sediments.

acid rain Rainfall of more than usual acidity. Acid rain is believed to result largely from the human release of sulfur dioxide (SO_2) gas, especially from coal-fired electric generating plants, and has proven harmful to plant and animal life as well as to a variety of material substances.

adiabatic temperature changes Temperature changes produced within a parcel of air because of the air's expansion or contraction. Adiabatic cooling normally occurs as a parcel of air rises and expands; adiabatic warming normally occurs as air descends and contracts. No energy is actually gained or lost by the air during the process.

advection The process of energy transfer through the horizontal movement of a fluid from one place to another. A type of convection.

advection fog A fog type that forms when moist air flows over a colder surface that cools it below the dew point.

air mass A large body of air with relatively uniform horizontal temperature and moisture characteristics.

air mass thunderstorm A thunderstorm that has developed within a single unstable air mass. Air mass thunderstorms typically show a lack of linear organization, and over land areas form most frequently during the afternoon hours.

air pressure The force exerted by the air on its surroundings because of the pull of gravity. Air pressure at sea level averages 1.0 kilograms per square centimeter (14.7 lbs/in^2).

albedo The short wave energy reflection rate of an object or substance, normally stated in percent.

alfisol A soil order containing fine textured soils with a moderately high cation exchange capacity that are low in humus and have a yellowish-brown A horizon. They are common in forest-grassland transitional environments in the low and middle latitudes.

alluvial fan A gently sloping fan-shaped deposit of alluvium that is formed when a stream flowing within a steep, narrow channel suddenly enters an open lowland and loses its capacity to transport its sediment load.

alluvial soil A soil formed from alluvium, which is stream deposited sediment.

alluvium An accumulation of stream-deposited sediments, typically found on the surface of a floodplain.

alpine glacier A relatively small glacier or glacier system that forms at high elevations in a mountainous region and moves downhill.

alpine meadow The name given to areas covered by tundra vegetation in high mountain (alpine) environments.

altitude Distance above sea level; generally used for a location above the Earth's surface.

altitude (solar) The angular distance in degrees of the Sun above the nearest horizon.

anemometer An instrument consisting of several rotating cups that measures wind speeds.

aneroid barometer A barometer that measures air pressure by means of the force exerted on a collapsible box, called a sylphon cell, from which some of the air has been removed.

angle of repose The steepest angle that can be maintained on a given slope without slope failure (mass movement) occurring.

annual A plant that has a lifespan of one year or less, usually because it is killed by cold or dry weather at the end of the growing season.

annual temperature range The number of degrees between the warmest and the coldest monthly mean temperatures for the year.

annual variability of precipitation The average annual percentage deviation of a site's annual precipitation total from the long-term mean total.

Antarctic Circle The most equatorward line of latitude in the southern hemisphere (66½° S) at which the Sun remains constantly above the horizon on the date of the southern hemisphere summer solstice, and constantly below the horizon on the date of the southern hemisphere winter solstice.

anticline An upward fold in one or more rock strata.

anticyclone A high pressure system, characterized by an outward spiraling windflow pattern.

aphelion The point of the Earth's orbit that is farthest from the Sun. It occurs about July 4th, when the two bodies are approximately 94.5 million miles apart.

aquiclude A relatively impermeable rock or sediment layer that acts as a barrier to the penetration of water from an adjacent layer.

aquifer A porous and permeable rock or sediment layer that contains large quantities of ground water and permits its ready movement.

Arctic Circle The most equatorward line of latitude in the northern hemisphere (66½° N) at which the Sun remains constantly above the horizon on the date of the summer solstice, and constantly below the horizon on the date of the winter solstice.

arête A narrow, serrated mountain ridge in a region of present or past glaciation. It is formed by the headward growth, and eventual meeting, of cirques from opposite sides of the ridge.

argon An inert gas comprising almost 1% of the atmosphere.

arid climate A climate with a potential evaporation rate that greatly exceeds its mean precipitation receipts, so that desert conditions prevail.

aridisol A soil order containing predominantly shallow, stony, and mineralogically immature soils that have developed in arid environments.

aspect (of a slope) The compass direction in which the slope faces.

association See *vegetative association.*

asthenosphere The plastic upper portion of the mantle. Convection currents in the asthenosphere are believed to produce lithospheric plate motions.

atmosphere The gaseous envelope surrounding the Earth.

atoll An often ring-shaped reef that forms an island or chain of islands. Atolls are believed to form around oceanic volcanic islands. They persist because of the continued growth of the reef-building organisms after the volcanic island has disappeared due to erosion or submergence.

aurora australis Luminous bands or streamers in the night sky of the southern hemisphere apparently caused by the disruption of the stream of ionized solar particles by the Earth's magnetic field.

aurora borealis Luminous bands or streamers in the night sky of the northern hemisphere apparently caused by the disruption of the stream of ionized solar particles by the Earth's magnetic field.

autumnal equinox The day that the Earth's axial inclination is exactly perpendicular with respect to the Sun, following a period in which one's hemisphere was inclined toward the Sun. Marks the official beginning of autumn.

avalanche The rapid movement of a detached mass of snow and ice down a steep mountain slope.

average environmental lapse rate The average rate at which the temperature declines with altitude in the troposphere. This value is approximately 6.4C°/1000 meters, or 3.5F°/1000 feet.

axis The line, connecting the North and South Poles, about which the Earth rotates.

azimuth system A directional system that states directions in degrees clockwise from due north. Values range from 0° to 360°.

azimuthal map A map that retains true directions outward from one central point.

backwash The gravity-induced return flow of wave swash to the ocean.

back-wasting Processes of weathering and erosion that result in a mountain mass maintaining its steepness and ruggedness while being reduced in size and areal extent. Back-wasting is characteristic of arid regions and contrasts to the down-wasting processes typical of humid regions.

bajada See *piedmont alluvial slope.*

bar A linear deposit of sand or gravel that is often found within a stream channel at a place where the speed of flow is diminished.

barchan A relatively small, crescentic sand dune that is oriented with its tapered ends pointing downwind.

barometer An instrument designed to measure air pressure.

barrier island An elongated sandy island paralleling the coastline that is formed by the continued deposition of sand by storm waves. This same feature, if connected to the mainland to form a peninsula, is termed a barrier beach.

basalt A dense, black colored extrusive igneous rock containing substantial iron and magnesium.

base level The elevation at any point along a stream's course that provides the stream with a gradient just sufficient for it to transport its sediment load past that point with no erosion or deposition. If a stream is at base level throughout its course it is considered a graded stream.

batholith A large mass of intrusive igneous rock, usually granitic in composition. Because of its resistance to erosion, it may form a mountain mass if exposed at the surface.

beach The gently sloping shore of a body of water that is subject to wave or tide action. It is normally covered by loose deposits of sand or gravel.

beach drifting The gradual movement of sediments along a beach in a zig-zag motion. This occurs as the sediments are repeatedly pushed up the beach by swash from oblique waves and then carried directly downslope by the wave backwash.

bed load Sediments that are moved along the bed of a stream by sliding, rolling, or bouncing (saltation).

bedding plane The boundary between adjacent sedimentary rock strata.

Bergeron process A process of precipitation formation, based on the fact that the dew point is slightly lower for water droplets than for ice crystals. This encourages the evaporation of the water droplets in a mixed cloud, and the subsequent sublimation of the moisture onto ice crystals.

big bang theory At present, the most widely accepted theory of origin of the universe. It holds that all matter in the universe was initially produced, and hurled outward, by the explosion of an inconceivably tiny and compact entity sometimes referred to as the "cosmic egg."

biogeography The branch of geography concerned with the spatial characteristics of plants and animals.

biological community An assemblage of plants and animals whose members occupy an environmental setting of restricted size.

biomass A measure of vegetation density, defined as the mass of vegetation per unit of area.

biome A global-scale community of plants and animals that exists in relative equilibrium with its environment.

biosphere The sphere of living organisms inhabiting the surface and near-surface portions of the Earth.

bird's-foot delta A delta that contains long finger-like deltaic projections. It develops when a stream deposits sediments in a quiet, partially enclosed water body that lacks the wave and current action to spread out the deltaic materials.

bolson An enclosed desert basin into which ephemeral streams transport water and sediments. Following rainfalls, they become the sites of temporary lakes termed playas.

braided flow A stream flow pattern that contains multiple interwoven channels separated by bars. This pattern tends to develop when the stream is supplied with an excessive quantity of coarse sediments.

broadleaf Usually used to describe trees with leaves of considerable width, such as oak, maple, or elm, as opposed to needleleaf trees.

butte A small, isolated, and steep-sided remnant of a former more elevated topographic surface. Buttes are often left as erosional outliers of retreating escarpments in arid regions.

caldera A roughly circular volcanic depression caused either by the explosion or collapse of a volcano.

calorie The amount of heat needed to raise the temperature of one gram of liquid water by one Celsius degree (1.8F°).

canyon A deep, steep-walled valley of considerable width, usually carved by a stream that is rapidly downcutting.

cap ice The most extensively distributed type of sea ice. It is relatively thick and continuous and typically has a rough surface.

cap rock A resistant surface rock layer that serves as a cap to protect the underlying less resistant strata from erosion. Most cap rock layers exist in arid regions.

capillary tension The electromagnetic attraction of a thin molecular film of water to solid surfaces, such as soil grains and plant root linings.

carbon dioxide An atmospheric gas essential to plant photosynthetic processes and largely responsible for the long wave radiation absorbing capacity of the atmosphere. It has the chemical formula CO_2.

carbonation A chemical weathering process involving the reaction of rock material with carbonic acid (H_2CO_3) produced when carbon dioxide is dissolved in water.

cartographer A professional map maker.

cartography The art and science of map making.

catastrophism The widely held view of a few centuries ago that the Earth's major surface features were produced in the recent past by cataclysmic events.

cation exchange capacity (CEC) The quantitative measure of the ability of a given soil to exchange cations with plants.

cave A natural cavity within or beneath the surface. Caves can be formed by a number of processes, but are especially numerous and important in karst areas, where they are produced by subsurface solution.

celsius The metric system temperature scale. It is based on the boiling point of water (under standard temperature and pressure conditions) at 100° C, and the freezing point of water at 0° C.

Channelled Scablands A region, located in Washington state, where the surface topography has been produced largely by repeated Pleistocene floods of massive proportions. These floods resulted from the breaching of an ice dam that blocked the outflow of glacial Lake Missoula.

chaparral The name used in California for a dense cover of woody shrubs and small trees ranging up to 6 meters (20 ft) in height. It is a variety of the Mediterranean Woodland and Shrub vegetation subassociation.

chemical sedimentary rock A sedimentary rock formed by the direct precipitation of chemicals in water.

chemical weathering Weathering processes involving chemical changes in the rock material that form new substances.

chinook A warm, dry wind that is adiabatically heated as it descends the eastern slopes of the Rocky Mountains in winter.

chlorofluorocarbon gases (CFC's) A group of synthetic gases that are apparently depleting the ozone layer, causing an increase in the quantity of short wave radiation reaching the Earth's surface.

cinder cone A conical volcanic hill or low mountain constructed of loose pyroclastic material thrown into the air from a central vent.

cirques Bowl-shaped mountainside hollows sculpted by glacial scour and frost action, and in which alpine glaciers typically have their heads.

cirrus From the Latin word meaning "curl of hair," cirrus clouds are the highest of the ten basic cloud types, are composed of ice crystals, and have a wispy appearance.

clastic sedimentary rock A sedimentary rock composed of particles of pre-existing rocks. Based on the sizes of the constituent rock particles, the four major categories are conglomerate, sandstone, siltstone, and shale.

clay-humus complex A complex combination of tightly bound clay and humus particles that is capable of attracting and exchanging cations needed by vegetation.

climagraph A graph of monthly mean temperature, precipitation, and sometimes other climatic data for a given reporting station.

climate The long-term or overall condition of the atmosphere with respect to the weather elements and weather systems.

climatology The branch of physical science that deals with the study of climate.

climax community The stable biological community that eventually occupies an ecosystem at the end of a successional sequence. It is theoretically the best suited to survive under existing environmental conditions of all possible communities available to form the ecosystem.

cloud A visible accumulation of minute water droplets or ice crystals, suspended in the air, that have condensed or sublimed around dust particles because the air has cooled below the dew point.

coal An organic sedimentary rock composed primarily of carbon, and formed from the burial, compaction, and lithification of plant remains in a poorly drained environment.

coalesence A process of precipitation formation in clouds, operating through the merging of droplets that come in contact with one another.

cold front A frontal boundary along which an advancing mass of cold air is underrunning relatively lighter warm air, forcing it to rise.

cold ice glacier A dry-base glacier that is frozen fast to the underlying bedrock surface. Glacial movement must therefore take place solely by internal deformation.

colloids Very fine clay particles that are capable of remaining suspended indefinitely in water and are soft and gelatinous in nature.

composite volcano A volcano constructed of alternating layers of pyroclastic materials and lava. Most large volcanoes are of this type.

condensation The phase change of a substance from a vapor to a liquid.

condensation nuclei Atmospheric particulates that serve as the collection centers for water molecules, enabling clouds and ultimately precipitation to form.

conduction The transfer of thermal energy by physical contact between two objects or substances having different temperatures.

cone karst A landscape dominated by numerous steep-sided, conical hills and intervening star-shaped depressions.

conformal map A map that has the property of showing the correct shapes of the areas it depicts.

conglomerate A clastic sedimentary rock comprised largely of rounded gravel or larger sized clasts.

conifer A tree that produces its seeds within cones.

continental arctic air mass A very cold, dry air mass that has formed over a snow and ice covered surface.

continental climate Climatic characteristics typical of a large land mass, especially including relatively large daily and annual temperature ranges and low amounts of precipitation, fog, cloudiness, and humidity.

continental crust The relatively light, discontinuous layer of granitic rock that comprises the uppermost portion of the crust and forms the continental masses.

continental drift The somewhat outdated term for the movements of the lithospheric plates as a result of convection currents in the asthenosphere.

continental glacier A glacier that forms at high latitudes and covers an extensive land area. Continental glaciers currently cover Antarctica and interior Greenland, and existed during the Pleistocene in northern North America and northwestern Eurasia.

continental polar air mass A relatively cold, dry air mass that has formed over a high latitude land mass.

continental rise The broad, gently inclined submarine depositional ramp located at the base of the continental slope.

continental shelf The shallow portion of the ocean floor that adjoins and surrounds most parts of the continents.

continental slope The long slope separating the continental shelf from the deep sea floor.

continental tropical air mass A relatively hot, dry air mass that has formed over a low latitude land mass.

contour lines Lines of constant elevation on a topographic map that are generally numbered from mean sea level.

convection The process of energy transfer through the physical movement of a fluid from one place to another. On the Earth, the chief fluids involved are air and water.

convectional lifting The rising of air because it is warmer, and therefore lighter, than the surrounding air.

cordilleran belts The belts of high, rugged mountains that are being actively formed by present-day tectonic processes.

core The innermost layer of the Earth, believed to be primarily metallic in composition. The core consists of two segments—a solid inner core and a liquid outer core.

Coriolis effect The apparent deflection of moving fluids, especially winds and ocean currents, on the Earth's surface, occurring as a result of the Earth's rotation. The direction of deflection is to the right in the northern hemisphere and to the left in the southern hemisphere.

cosmic radiation Charged atomic particles that enter the Earth's atmosphere at nearly the speed of light and are capable of destroying or mutating plant and animal cells.

creep The very slow downhill movement of soil or rock material over a period of years. Creep is most important in humid climate regions, and involves the entire mass of near-surface weathered material on most slopes.

crest (of a flood) The peak flow of a flood, measured in terms of either discharge or water level.

crevasse A large crack or fissure in the near-surface portion of a glacier, caused by the brittle fracturing of the ice as it passes over a convexity in its bed.

critical erosion velocity The minimum velocity of flow that will enable a stream to erode streambed particles of a given size.

crust The relatively thin "skin" of rigid low density rock that forms the outermost lithospheric layer of the Earth. The crust varies in thickness from 5 to 70 kilometers (3–40 mi) and consists of two segments, the sima and the sial.

cumulonimbus The cloud type responsible for thunderstorms, hail, and tornadoes; probably the world's leading precipitation producing cloud.

cumulus From the Latin word meaning "pile" or "heap," cumulus clouds are the most common of the ten basic cloud types. They have a puffy appearance, with flat bases and uneven sides and tops.

cyclone A low pressure system or storm center, characterized by an inward spiraling windflow pattern.

cyclonic lifting Atmospheric lifting associated with the convergence of air into a low pressure system.

daily mean temperature The average temperature of the day, calculated by adding the daily high and low temperatures and dividing by two.

daily temperature range The number of degrees between the high and the low temperatures for the calendar day.

Davis, William Morris An American geomorphologist best known for developing the concept of the fluvial landscape "cycle of erosion," patterned after the three human life cycle stages of youth, maturity, and old age.

daylight saving time A variation of standard time in which the time is set ahead by one hour in order to achieve greater correlation between the normal human activity period and the daylight period.

debris avalanche The rapid movement of a mixture of soil, rock, and vegetation down a pre-existing mountainside hollow. Movement is nearly always triggered by heavy rainfall; it typically begins as a slide and becomes a flow.

deciduous A leafy plant that loses all its foliage during a given season, usually as a mechanism for coping with cold or dry conditions.

deep ocean current An ocean current flowing at considerable depths that is produced by density differences resulting from variations in temperature and salinity from the surrounding water. It often displays significant vertical as well as horizontal components of motion.

deflation The erosion of loose surficial material directly by the force of the wind. It occurs when the wind speed exceeds the drag velocity of the materials.

deflation hollow A shallow, often roughly circular depression of variable dimensions that has been formed by deflation.

delta An often fan-shaped deposit of alluvium formed when a stream enters a standing body of water and loses its ability to transport its sediment load.

dendritic stream pattern A treelike drainage pattern that develops on homogeneous surfaces that exert no significant structural control on stream location. It is the most widespread of the drainage pattern types.

desert A region of substantial size that is largely or entirely devoid of vegetation, usually as a result of aridity.

desert pavement A desert surface of close-fitting pebbles or stones that protects the underlying surface from erosion. It results largely from the eolian removal of fine textured particles.

dew Water droplets that have condensed on surface objects because the overlying air has cooled below the dew point.

dew point The temperature at which a given parcel of air will become saturated with water vapor.

diastrophism Solid-state movements of the Earth's crust, including folding and faulting, that result from tectonic forces.

differential erosion Erosion occurring at differing rates within a given area because of the uneven operation of gradational forces.

dike A sheet of basalt that was injected as magma into a pre-existing rock fracture that is discordant with (cuts through) the host rock structure.

dike ridge A wall-like basaltic ridge formed by the erosional exposure of a dike injected into weaker sedimentary strata.

dip The angle at which a rock stratum or any planar feature is inclined from the horizontal.

discharge The volume of stream flow past a given point during a specified timespan, normally measured in cubic feet per second (cfs) or in cubic meters per second (cms).

doldrums Another name for the equatorial belt of variable winds and calms.

doline See *sinkhole.*

dome A rounded anticlinal structure. Smaller domes are usually formed by the surfaceward movement of material such as magma or salt.

Doppler radar A type of radar that holds great promise for tornado identification and prediction. It can remotely measure and display the speed of winds moving toward or away from the radar site, allowing for the detection of developing rotary motions in thunderstorms.

drag velocity The critical wind velocity needed to overcome frictional resistance and initiate the eolian transport of surface materials of a given diameter.

drainage basin The total surface area drained by a given stream system.

drift See *glacial drift.*

drizzle Liquid precipitation droplets with diameters less than 0.5 millimeters (0.02 in).

drumlin A smooth, elongated hill composed of glacial till. They generally occur in groups or "swarms," and are oriented so that their tapered ends point in the direction the ice was moving.

dry adiabatic rate The rate at which unsaturated air cools as it rises (or warms as it descends) through the troposphere. This rate is 5.5F°/1000 feet, or 10C°/1000 meters.

dune A mound or ridge of wind-deposited sand. Dunes assume a variety of forms and, unless stabilized by vegetation, tend to migrate slowly over the surface because of wind action.

duricrust A hard, nearly impermeable surface or near-surface layer, typically formed by the precipitation of salts in an enclosed basin in an arid climate.

dust The collective term for all liquid and solid particulates suspended in the atmosphere.

earth flow The slow flowage of a mass of nearly saturated soil down a moderate to steep slope.

Earth system The Earth and its atmosphere, treated as a unit.

easterly wave A north-south oriented low pressure trough in the trade wind belt that drifts westward and may have the potential to develop into a hurricane.

ecosystem A functioning biological entity consisting of all the organisms in a biological community, as well as the environment in which and with which they interact.

ecotone A transition area between two adjacent vegetative communities, such as forest and grassland, and therefore often exhibiting a complex ecology.

elevation The distance above sea level of a point on the Earth's surface.

El Niño The anomalous warm ocean current that sometimes flows eastward across the southern Pacific to reach the coast of South America. It recently has been discovered to exert a major impact on global climate patterns.

eluviation The removal of solid or dissolved materials from a soil horizon by downward percolating water.

entisol A soil order containing poorly developed soils, mostly of recent origin, that display little or no profile development.

entrenched meander A stream meander that has been incised deeply into the surface, perhaps as the result of the renewal of stream downcutting following a period of uplift.

eolian The term used for geomorphic processes performed or products produced by the wind.

ephemerals Small, short-lived desert plants that pass through their life cycles quickly following a substantial rainfall.

ephemeral stream A stream that flows only during wet periods when it is supplied by surface or soil water.

epicenter The point on the Earth's surface directly above the focus of an earthquake.

epiphyte An "air plant" that grows on other plants, but derives its moisture and nutrients from the air and is therefore not parasitic. Epiphytes are especially diverse and abundant in the tropical rainforests.

equal area map A map that has the property of showing the correct proportional sizes of the areas it depicts.

equator The line of fastest rotation on the Earth. It is a full circumference, located midway between the North and South Poles, and is the starting (0°) line of latitude.

equatorial belt of variable winds and calms The zone of light and variable winds associated with the Intertropical Convergence Zone.

equatorial lows Weak low pressure centers that develop along the ITCZ and drift westward, attended by extensive cloudiness and rainfall.

erg A sandy desert.

erosion The removal of Earth material from a given site, normally by the action of water, ice, wind, or gravity.

esker A long, meandering ridge of stratified drift, apparently produced by the deposition of sediments by a subglacial stream.

estuary A semi-enclosed water body located either in a drowned river valley or a submerged tectonic trough.

evaporation The phase change of a substance from a liquid to a vapor.

evaporite A sedimentary rock composed of minerals precipitated from a saline water body.

evapotranspiration The total water evaporated from both the Earth's surface and from vegetation.

evergreen A leafy plant that retains its foliage throughout the year.

exfoliation The separation of rock into sheets paralleling the surface as a result of unloading and chemical weathering.

exotic stream A large perennial stream, flowing through an arid region, that has its headwaters in a humid region from which it has derived most of its water.

extrusive igneous rock Igneous rock that was extruded onto the Earth's surface in a liquid state (as lava). Subsequent rapid cooling allowed little or no time for mineral crystals to form.

Fahrenheit The temperature scale most commonly used by the general public in the United States. It is based on the boiling point of water (at standard temperature and pressure) at 212° F, and the freezing point of water at 32° F.

fault escarpment (or fault scarp) Ideally, a linear, smooth-sided slope consisting of the exposed surface of the fault plane on the higher side of a fault.

faulting The fracturing of rock, followed by the movement of the two sides of the fracture relative to one another.

fault line The trace of a fault on the Earth's surface.

fault plane The two dimensional surface of the fault. It primarily exists below the surface, but a portion may be exposed along a fault escarpment.

Fennoscandian ice sheet The continental ice sheet that covered most of northwestern Eurasia during the glacial stages of the Pleistocene.

fetch The distance the wind blows over water. It is generally of concern when calculating expected wave heights.

field capacity The maximum quantity of water the soil is capable of retaining against the pull of gravity.

fjord A glacial trough that has been partially occupied by sea water to form a deep oceanic coastal inlet.

flood The inundation of the land along the banks of a stream that occurs when the stream's discharge exceeds its carrying capacity.

flood basalt A huge outpouring of fluid lava from a fissure that can cover an extensive region to form a lava plain or plateau.

floodplain A broad belt of low, flat land adjacent to a stream, formed by fluvial erosional and depositional processes, that is inundated periodically during floods.

fluvial An adjective used to identify stream-related processes and features.

fog A visible accumulation of minute water droplets or ice crystals suspended in air immediately overlying the surface. Fog normally forms as the air is cooled below the dew point, forcing the condensation or sublimation of water vapor on dust particles.

folding The bending of rock layers subjected to tectonic stresses.

foot wall The side of a fault plane that extends beneath the hanging wall side.

forest An area in which trees are spaced sufficiently close together that their foliage overlaps to form a more or less continuous canopy.

front See *weather front.*

frontal cyclone A low pressure system that has developed on a weather front. Most frontal cyclones have an associated warm front and cold front. Frontal cyclones are major precipitation producers for the middle and high latitudes.

frontal lifting The lifting of relatively light warm air by denser cold air along a weather front.

frontal thunderstorm A thunderstorm that has developed in association with a weather front. Frontal thunderstorms typically develop in a linear pattern, and are most frequently associated with cold fronts.

frontal wave A poleward bend in a weather front, normally associated with an area of weak low pressure. Frontal waves often develop into frontal cyclones.

frost Ice crystals that have sublimed on surface objects because the overlying air has cooled below the dew point.

frost wedging The process of rock breakup resulting from the growth of ice crystals within rock fractures or hollows.

galeria Strip growths of forest in the moist soils along riverbanks in savanna grassland regions.

geography The study of the distributions and interrelationships of Earth phenomena.

geomorphology The systematic study of landforms, including their origin, characteristics, and distribution.

geothermal heat Heat from the Earth's interior, believed to result largely from the radioactive decay of uranium and thorium. It serves as the energy source for plate tectonic processes.

glacial drift A general term for all deposits of glacial origin.

glacial lacustrine plain A plain covered by fine textured lake sediments. These sediments were deposited at a time when the surface was covered by a lake whose drainage outlet was blocked by a glacier.

glacial trough A deep and generally "U" shaped valley, usually of fluvial origin, that was subsequently occupied and modified by a valley glacier.

glacier A mass of fresh water ice, formed on land, that is or has been in motion as a result of gravity and/or internal pressure.

glacio-fluvial material Rock material that was partially sorted by meltwater after being deposited by a glacier. It therefore displays some stratification.

global radiation balance The long-term global balance between incoming short-wave insolation and outgoing long-wave terrestrial radiation.

gorge A narrow, deep, and steep-walled valley that is usually carved by a stream that is rapidly downcutting.

graben A block of land bounded by two normal faults that has been downdropped relative to the blocks on either side.

gradational forces Solar powered forces that, in conjunction with the force of gravity, act to lower the Earth's surface. The three major "tools" of gradation are water, ice, and wind.

graded stream A stream that has adjusted its flow parameters to the quantity and texture of its sediment load, so that it is neither eroding nor depositing materials (at least on a long-term basis).

granite A low density intrusive igneous rock consisting primarily of quartz, feldspar, and mica, that comprises much of the continental crust.

grass A large and biologically diverse family of plants with bladelike, opposing leaves connected to a jointed stem by a sheathlike attachment. The cereal grains that form the basic foodstuffs of humans and livestock are grasses.

great circle A circle, drawn on a sphere, that bisects the sphere and is a full circumference of the sphere.

greenhouse effect The heat-trapping ability of the atmosphere, especially of the gases carbon dioxide and water vapor. The process is analogous to the heat-trapping ability of the glass in a greenhouse.

Greenwich meridian The starting line of longitude; also known as the prime meridian.

ground water Gravitational water that completely fills the pore spaces or other voids in the soil or rock below the water table.

growing season The period of time during the year when weather conditions continuously favor the growth of vegetation. In the United States, it is generally considered as the period between the last killing frost of spring and the first killing frost of fall.

guyot A flat-topped seamount that apparently once extended above the ocean surface, where it was planed off by subaerial erosion.

gyre The geometrical pattern formed by surface ocean current systems, consisting of a giant elliptical loop.

habitat A specific site that has environmental conditions favoring its habitation by a given species or group of species.

hail A precipitation form consisting of relatively large, rounded particles of ice that typically exhibit internal layering. Hail forms as strong updrafts and downdrafts in a thunderstorm carry ice pellets alternately above and below the freezing level in a thunderstorm.

halophytes Salt-tolerant plants that normally occupy lowland basins of interior drainage in arid regions.

hamada A desert area with a solid bedrock surface, sometimes covered with a scattering of boulders.

hanging valley A small valley that enters a larger valley high up on its flank, and therefore "hangs" above the larger valley. This markedly ungraded situation generally results from either glaciation or fault displacement.

hanging wall The side of a fault that extends above the foot wall side.

haystack hill A rounded hill, somewhat resembling a haystack, that remains on a karst surface as an erosional remnant of a former higher topographic level.

heat of condensation The quantity of heat energy, totaling approximately 590 calories per gram of H_2O, that is released when water vapor condenses to liquid water. It is the direct energy source for many atmospheric processes.

heat of fusion The quantity of heat energy, totaling approximately 80 calories per gram of H_2O, that is released when liquid water freezes.

heterosphere The compositionally layered upper atmosphere, located above an altitude of 80 kilometers (50 mi).

high See *high pressure system.*

high pressure system An area of higher barometric pressure than the surrounding areas, characterized by an anticyclonic windflow pattern, and typically associated with fair weather.

histosol A soil order consisting of dark colored organic soils of poorly drained areas. Most histosols are highly acidic and low in fertility.

homosphere The compositionally homogeneous lowest 80 kilometers (50 mi) of the atmosphere.

horn A sharp-crested, pyramidal peak, representing the remnant rock mass between three or more coalescing cirques.

horst A raised block of land bounded by two normal faults.

human geography The branch of geography concerned with the spatial aspects of Earth phenomena that are produced or are primarily controlled by humans.

humidity A term referring to the water vapor content of the air. Several types of humidity measurements exist.

humus Finely divided, decomposed organic matter with a brownish-black color that is a major supplier of plant nutrients.

hurricane A tropical cyclonic storm that forms over water and has sustained winds of at least 33.5 meters per second (75 mph). Hurricanes are the world's most destructive storm type.

hydration A chemical weathering process involving the absorption, or adhesion, of water molecules to the molecules of a mineral.

hydraulic action The direct removal of loose materials by the friction and turbulence generated by flowing water in a stream or by oceanic wave action.

hydrograph A graph of stream discharge at a given gaging station over a period of time.

hydrologic cycle The continuous cycle of phase change and movement of water between the Earth's surface and the atmosphere.

hydrolysis A chemical weathering process consisting of a permanent chemical combination of minerals with water.

hydrosphere The aqueous envelope of the Earth, including water in the atmosphere, at the Earth's surface, and in the near-surface layers of the ground.

hygrophyte A plant especially adapted to grow in a very moist or waterlogged surface environment.

ice-scoured plain An extensive, lake-strewn, low relief surface that was subjected to glacial scour by a continental ice sheet. Ice-scoured plains presently occupy much of interior eastern Canada and Scandinavia.

iceberg An islandlike mass of fresh water ice produced by the breakup of a glacier as it enters the ocean or other large standing water body.

igneous rock A rock that has formed directly from the solidification of magma or lava resulting from tectonic activity.

illuviation The deposition, usually in a lower soil horizon, of solid or dissolved materials that were removed (eluviated) from another horizon.

inceptisol A soil order containing chemically immature soils with weak profile development.

infiltration capacity The maximum rate at which a given surface can absorb water. It is determined primarily by the texture, degree of compaction, and existing water content of the surface materials.

inselberg An isolated, steep-sided remnant of a previous mountain mass or other upland surface in an arid or seasonally dry region. Inselbergs are sometimes considered the arid climate counterparts of monadnocks.

insolation A shortening of the term "incoming solar radiation," referring to the total quantity of solar radiation reaching the Earth system.

interception The process by which vegetation intercepts precipitation before it reaches the ground. This process becomes ineffective once the vegetation has been fully wetted.

interfluve The generally well drained higher ground between streams.

interior drainage Drainage that does not reach the ocean, but instead enters an enclosed basin where it is absorbed into the ground or evaporates. Interior drainage is characteristic of most arid regions.

International Date Line This line is based on the 180th meridian and is located in the central Pacific. Time varies by a full 24 hours on either side, with the west side being a day later than the east side.

Intertropical Convergence Zone (ITCZ) The zone of low barometric pressure and wind convergence centered near the equator, caused primarily by the prevailing high temperatures.

intrusive igneous rock Igneous rock that has formed inside the Earth. The slowness of the solidification process usually allows visible mineral crystals to form.

invertebrate An animal without a backbone or spinal column.

island arc An arc-shaped group of islands paralleling, at a distance of several hundred kilometers, a submarine trench. It is formed when magma from the melting subducted plate that produced the trench has been erupted in sufficient mass to protrude above sea level.

isobars Lines connecting places having the same barometric pressure.

isostatic rebound The buoyant uplift of a portion of continental crust that had formerly been depressed by the weight of a continental glacier.

isostatic uplift The broad, gentle upwarping of a region that has become buoyant because of its reduction in mass as a result of erosion or deglaciation.

isotherms Lines connecting places having the same temperature.

jet streams Relatively narrow bands of strong westerly winds centered in the upper troposphere above the middle latitudes and the subtropics of both hemispheres. They act as steering currents to direct the movements of surface weather systems.

jungle A nearly impenetrable growth of low growing plants that is common in warm, moist areas where sunlight can reach the near-surface vegetation layers.

kame A mound of slightly sorted glacio-fluvial debris, often originating as a crevasse filling.

karst Topography developed largely by the differential solution and subsequent erosion of highly soluble rocks, especially limestone and dolomite.

karst plain A limestone plain covered with large numbers of sinkholes and sometimes other solution features.

Kelvin temperature scale A temperature scale with its starting temperature (0° K) at absolute zero—the temperature at which all molecular motion ceases. Like the Celsius scale, it employs 100 units between the freezing and boiling points of water (273° K and 373° K, respectively).

kettle A depression within a till plain that may be occupied by a small pond. Kettles are formed by the weight of detached blocks of ice from a stagnant glacier.

knickpoint A point where the downhill gradient of a stream abruptly increases.

Köppen climate classification A system of classifying global climates according to their basic temperature and precipitation characteristics. (Details of the system, developed by Wladimir Köppen about 1918, appear in the Appendix.)

laccolith A dome-shaped igneous intrusion of relatively acidic magma that is injected into a sequence of near-surface sedimentary strata.

lagoon A shallow, protected body of salt water located between a barrier island chain and the mainland. It may consist of an extensive mudflat at the time of low tide.

lake A landlocked body of water with a horizontal surface level.

land and sea breeze A diurnally reversing coastal windflow pattern established in some areas by land and water temperature differences. Sea breezes are generated during the day, and land breezes at night.

landforms Individual Earth surface features.

lapse rate See *average environmental lapse rate.*

laterite A hard, iron-rich subsurface layer that sometimes develops in humid tropical regions following the removal of vegetation.

latitude Angular distance in degrees north or south of the equator.

Laurentide ice sheet The continental ice sheet that covered most of Canada and the northern United States during the glacial stages of the Pleistocene.

lava Molten rock material formed by the expulsion of magma onto the Earth's surface.

lava plain (or lava plateau) An extensive, gently rolling plain or plateau surface formed by consecutive fissure outpourings of basaltic lava.

leaching The removal of dissolved materials and fine textured particles from the upper soil by downward percolating soil water. These materials may be deposited deeper within the soil, or they may be removed completely from the soil.

leap year A year containing 366 days (including February 29).

leeward coast A coast with a prevailing land to water windflow.

legumes A group of plants that host bacteria in their root nodules that extract nitrogen from the atmosphere and convert it to a form available to vegetation.

liana Any of a variety of climbing plants that root in the ground. Large woody lianas are characteristic of the tropical rainforest.

limestone An organic sedimentary rock composed of calcium carbonate ($CaCO_3$). Most limestone formed in quiet tropical sea floor environments either by direct chemical precipitation or through the accumulation of the remains of marine organisms.

lithosphere The solid portion of the Earth, composed largely of rock and metal. More recently, this term has also been used to refer to the rigid uppermost portion of the mantle and the overlying crust that forms the lithospheric plates.

lithospheric plate A discrete segment of the Earth's crust and uppermost mantle that moves as an entity when transported by convection currents in the mantle.

loam A soil of intermediate texture that generally contains substantial proportions of sand, silt, and clay.

local time The time of day based on the east-west position of the Sun at a given locality.

loess Silt-sized eolian deposits that are typically buff-colored and are composed of a well sorted mixture of angular quartz and calcite ($CaCO_3$) particles.

longitude Angular distance in degrees east or west of the prime meridian, stated to values of 180° E or W.

longitudinal dune A long ridge of sand that is typically aligned parallel to the direction of the prevailing winds in an arid region.

longshore current A current paralleling the shoreline, often generated by the momentum of waves approaching the shoreline obliquely.

low See *low pressure system.*

low pressure system An area of lower barometric pressure than the surrounding areas, characterized by a cyclonic windflow pattern, and typically associated with cloudiness and precipitation.

magma Molten rock material inside the Earth. If it reaches the surface in a molten state, it is termed lava.

magnetic North Pole The northern locus of the Earth's magnetic field, and the place to which a compass needle points. It is located in the Canadian Arctic.

magnetic South Pole The southern locus of the Earth's magnetic field, located near the coast of Antarctica.

mantle The most extensive of the internal layers of the Earth, occupying an intermediate position between the core and the crust. The mantle is approximately 2900 kilometers (1,800 mi) thick, and is apparently composed primarily of peroditite rock.

map A pictorial representation of the geographic location of selected surface features at a reduced scale.

map scale The ratio between the separation of places on a map and on the Earth's surface.

marine terrace An elevated surface of low relief that represents the remains of an old wave-cut or wave-built platform. These features occur in areas where the land has been uplifted or sea level has dropped.

Mariner's Compass A directional system that states directions by employing combinations of the words north, south, east, and west (or letters N, S, E, or W).

maritime climate Climatic characteristics typical of an area over or near a large water body, especially including relatively small daily and annual temperature ranges and high quantities of atmospheric moisture.

maritime equatorial air mass A very warm, moist air mass that has formed over a water body near the equator.

maritime polar air mass A relatively cold, moist air mass that has formed over a high latitude water body.

maritime tropical air mass A relatively warm, moist air mass that has formed over a water body in the low latitudes.

mass wasting The downslope movement of surficial material under the direct influence of gravity.

meander A broad looping bend in the course of a stream occupying a floodplain. Meanders typically display a distinctive cycle of formation, growth, and disappearance.

mercurial barometer A barometer that measures air pressure by the height of the mercury column in a glass tube from which the air has been removed.

meridian Another term for a line of longitude.

mesa An isolated, steep-sided and flat-topped remnant of a previous more elevated topographic surface. Mesas are often left as erosional outliers of retreating escarpments in arid regions.

mesopause The boundary between the mesosphere and the thermosphere. At this level, temperatures stop falling with increasing altitude.

mesophyte A plant adapted to an environment with a moderate water supply that is available throughout the year. It is intermediate in characteristics between a xerophyte and a hygrophyte.

mesosphere The thermal layer of the atmosphere between the stratosphere and the thermosphere, in which temperatures fall with increasing altitude.

mesquite A drought-resistant, spiny, deciduous tree or shrub with an extensive root system that has recently proliferated in the American Southwest.

metamorphic rock Rock subjected to solid-state alteration from its original form by heat, pressure, and/or chemical activity. Each igneous and sedimentary rock type has one or more metamorphic rock equivalents.

meteorology The branch of physical science that systematically studies the weather.

Milky Way galaxy The organized aggregation of stars in which our solar system is located.

mineral A naturally occurring, solid, inorganic compound that has a specific chemical composition and usually exhibits a crystalline structure that results from a distinctive atomic arrangement of its constituent elements.

minute An angular distance of 1/60th of a degree; used for accurately stating latitude and longitude.

Moho (or Mohorovičić discontinuity) The relatively sharp boundary between Earth's mantle and crust.

mollisol A soil order containing soils with a thick, dark colored, humus-rich A horizon and a high cation exchange capacity. Most mollisols are highly fertile and agriculturally productive and are associated with major grain producing regions.

monadnock An isolated hill or mountain that stands as an erosional remnant of a previous higher and more mountainous topographic surface.

monsoon winds A seasonally reversing regional-scale windflow pattern caused by land and water temperature differences. They are especially well developed in southern and eastern Asia and in parts of North Africa.

monthly mean temperature The average temperature for the month, calculated by averaging the daily mean temperature values for each day of the month.

moraine A glacial depositional feature composed of till. Several types of moraines exist, depending on their sites of deposition. These include lateral, medial, terminal, recessional and ground moraines.

mountain and valley breezes A diurnally reversing wind-flow pattern established in some mountain areas as a result of surface heating and cooling.

mudflow The perceptible and often rapid downhill flow of mud, usually mixed with rocks and other debris.

muskeg A swampy area within the Northern Coniferous Forest that is covered with mosses, swamp grasses, shrubs, and scattered hygrophytic trees such as larches.

mutation A sudden and inheritable alteration in the characteristics of a plant or animal caused by a molecular rearrangement of its genetic material.

nappe A thick sheet of rock that has been thrust by extreme compressionary forces over other strata.

natural levees Low and often broad ridges of stream deposits that parallel each side of a stream channel within a floodplain. They are formed by deposition resulting from the sudden increase in friction as a flooding stream overtops its banks.

neap tides Tides with exceptionally small vertical ranges, occurring approximately twice a month, caused by the perpendicular alignment of the Moon and Sun with respect to the Earth.

needleleaf Used to describe trees with needle-shaped leaves, such as pine, spruce, or fir, as opposed to broadleaf trees.

nimbus A Latin term meaning "rain cloud", used in conjunction with the two cloud types responsible for most of the world's precipitation—nimbostratus and cumulonimbus.

nitrogen A relatively inactive gas that comprises approximately 78% of the lower atmosphere.

nitrogen fixation The process by which atmospheric nitrogen is extracted by legumes and converted to a form available to vegetation.

normal fault A fault in which the vertical displacement exceeds the horizontal and an expansionary component is present.

North Pole The northern end of the Earth's axis, located at 90° N.

northern hemisphere The half of the Earth north of the equator.

nuée ardente A dense cloud of superheated gases and fine particles that rushes down the flanks of an erupting volcano, charring or setting fire to anything combustible and burning or suffocating all living organisms in its path.

oasis A desert locality where the water table intersects the surface, resulting in the growth of significant vegetation.

oblate ellipsoid A three-dimensional ellipse that is flattened at the poles. The Earth assumes this shape because the centrifugal force of rotation bulges out the equatorial regions.

occluded front A frontal boundary separating, at the surface, two relatively cold air masses, with a warmer air mass aloft. It is formed by the overtaking of a warm front by a cold front.

ocean The interconnected body of salt water that dominates the Earth's surface. It is commonly divided into segments named the Pacific, Atlantic, Indian, and Arctic Oceans.

ocean current The continuous movement of an ocean water stream of similar temperature and density.

oceanic crust The thin, continuous layer of dense basaltic rock that comprises the lowermost portion of the crust and forms the oceanic basins.

organic sedimentary rock Sedimentary rock produced from the remains of plants and animals. Most are composed of carbonate minerals, especially limestone ($CaCO_3$).

orographic lifting The lifting of air over a topographic barrier.

outwash Glacio-fluvial materials that have been transported and deposited beyond the edge of a glacier by meltwaters.

outwash plain A relatively extensive and flat glacial outwash surface deposited by multiple meltwater streams issuing from the margin of a glacier.

overland flow A non-channelized, sheetlike flow of water that is common on agricultural fields and near drainage divides.

oxbow lake An arcuate lake that occupies an abandoned stream meander.

oxidation A chemical weathering process involving the chemical combination of minerals with oxygen that has often first been dissolved in water.

oxisol The deepest and most highly chemically weathered of the soil types. They are quite limited in natural fertility and have developed on ancient surfaces within the humid tropics.

oxygen The second most abundant gas in the atmosphere, necessary to animal respiratory processes, and derived from carbon dioxide by means of plant photosynthesis.

ozone A tri-atomic form of oxygen (O_3), occurring in very small quantities in the atmosphere, that helps to protect the Earth's surface from cosmic ray bombardment.

ozone layer The layer of the upper atmosphere, extending from about 15 to 65 kilometers (10–40 mi) altitude, in which ozone is most abundant.

P waves Primary earthquake shock waves that result from compression.

pack ice A type of sea ice that is relatively thin and has numerous breaks or gaps through which sea water appears. It is generally peripheral in location to the thicker and more continuous cap ice.

parallelism The idea that the Earth's axis is constantly oriented in the same direction with respect to the plane of the ecliptic.

parallels Another term for lines of latitude.

parent material The solid inorganic and organic material from which the soil develops. The inorganic parent material may be either residual (originally derived from that location) or transported (derived from elsewhere and deposited in that location).

patterned ground Polygonal patterns that develop on the surface in many periglacial regions. They are caused by the alternate freezing and thawing of the ground, producing a churning action that cracks the surface and shifts surface materials in a selective fashion.

ped A soil structural aggregate formed by the tendency of individual soil particles to adhere to one another. Different soil types tend to develop peds of differing sizes and shapes.

pediment A gently sloping bedrock erosional surface that is commonly found at the foot of a mountain mass in an arid region.

pediplain A desert erosional plain formed by the coalescence of pediments as the intervening mountain masses are worn away. It is sometimes considered the arid climate counterpart of a peneplain.

pedologist A soil scientist.

perennial A plant that has an expectable lifespan of at least several years. It is therefore not normally killed by adverse weather conditions at the end of the growing season, although it may experience a period of dormancy.

perennial stream A stream that flows continously because it has a dependable source of ground water.

periglacial A geomorphic environment that develops in a severely cold climate. Although the surface is not glaciated, landscape development is associated with the alternate freezing and thawing of the near-surface layers.

perihelion The closest approach between the Earth and Sun. It occurs about January 3rd, when the two bodies are approximately 147.5 million kilometers (91.5 million miles) apart.

permafrost The condition of perennially frozen subsoil and bedrock that underlies large areas in arctic, subarctic, and alpine environments.

permeability The ability of the soil or rock to transmit the flow of water.

photosynthesis The formation of carbohydrates in the chlorophyll-containing tissues of plants exposed to light.

physical geography The branch of geography concerned with the spatial aspects of natural Earth phenomena, particularly with weather and climate, plants and animals, soils, and landforms.

physical weathering Weathering processes that reduce the size of rock masses without altering their chemical composition.

piedmont alluvial slope A ramplike alluvial apron, typically found along the base of a desert mountain range, that is formed by the coalescence of numerous alluvial fans. Also called a bajada.

place A specific location on the Earth's surface that contains a unique assemblage of spatial attributes.

plain An area of restricted local relief that has a predominantly flat to gently rolling surface.

plane of the ecliptic The angular plane on which the Earth's orbit around the Sun is located.

planimetric map A map that provides systematic information about only the horizontal characteristics of the areas it depicts.

plant association A global-scale community of plants that exists in relative equilibrium with its environment.

plant community A plant assemblage whose members occupy a specific environmental setting of restricted size.

plate See *lithospheric plate.*

plateau A relatively level upland surface that is frequently bounded on at least one side by an abrupt descent to lower elevations. Also termed a tableland.

plate tectonics Encompasses the processes involved in the formation, movement, and destruction of lithospheric plates.

playa A shallow desert lake, often of considerable areal extent, that temporarily occupies the center of a bolson following a rainfall.

Pleistocene The geologic epoch, extending from about 2,000,000 B.P. to 10,000 B.P., during which continental and alpine glaciers alternately advanced and receded over large portions of the world, especially in North America and Eurasia.

plunging fold A fold whose constituent rock strata were subjected to prior tilting so that the fold "plunges" into the ground in one direction.

pluton A body of intrusive igneous rock that has generally solidified well beneath the surface.

pocket beach A small, cresent-shaped beach located on a coastline that is, as a whole, being worn back by vigorous wave attack. These beaches, common along the U. S. West Coast, often develop in areas of relatively weak rock.

point bar A streambed bar of sand or gravel that develops on the inside of a meander. It is caused by the diminished flow at this point.

polar easterlies The belts of easterly winds blowing from the polar high pressure centers of each hemisphere toward the subpolar lows.

polar front The major weather front of the middle latitudes, separating air masses of subtropical and subpolar origins. It is a zone along which frontal cyclones frequently form.

polar front zone The zone of highly variable winds associated with the subpolar low pressure belts in each hemisphere.

polar highs High pressure caps centered near the North and South Poles, resulting largely from cold surface temperatures.

polar zones of variable winds and calms The zones of light and variable winds associated with the centers of the polar highs.

pore spaces The voids between the solid soil particles that contain soil liquids and gases.

porosity The volume of soil pore spaces divided by the total volume of the soil.

potential evapotranspiration The total quantity of water that would evaporate and transpire from a given site if the available moisture supply were unrestricted.

prairie Tall grassland generally associated with subhumid portions of the middle latitudes. Most prairie grasslands have been destroyed by human agricultural activities.

precipitation Water in liquid or solid form falling through the atmosphere toward the Earth's surface. The major forms are rain, snow, sleet, hail, and drizzle.

pressure gradient The rate of change in air pressure per unit of distance. When uncompensated for by other factors, it causes the wind to blow.

primary succession Vegetative succession that begins on a "new" surface completely devoid of vegetation, such as a lava flow or a previously glaciated surface.

prime meridian The starting line of longitude. A separation line between the eastern and western hemispheres.

pyroclastic Fragmental material erupted from a volcano.

radar An instrument used to determine the approximate size, shape, and intensity of areas of precipitation. It emits pulses of microwave energy that are reflected from the precipitation and displayed on a screen.

radiation The process of energy transmission by electromagnetic waves.

radiation fog A fog type that commonly forms over land areas on calm, clear nights because surface radiation chills the overlying air below the dew point.

radiosonde A weather instrument package that is sent aloft by means of a hydrogren or helium filled balloon and automatically radios back upper atmosphere data on temperature, air pressure, and humidity.

rain gage A funnel-shaped, tubular instrument used for the collection and measurement of precipitation.

rainshadow effect The reduction of precipitation by the blocking effect of a topographic barrier.

recumbent folds Folds that have been compressed so tightly that the fold structures have closed in on one another and collapsed, resulting in a nearly horizontal axial plane.

reef A ridge-like accumulation of the exoskeletons of large colonies of coral polyps or other reef-building marine organisms. The reef itself may be entirely beneath the water or may protrude above the water surface to form a coastal shelf or island. A fringing reef is attached to the mainland, while a barrier reef is separated from the mainland by a lagoon.

reg A stony desert from which the finer materials have been removed by the wind.

regelation A process of glacial motion resulting from the melting of ice as it is pressed into the up-ice side of a bedrock obstruction, followed by the refreezing of the water after it has flowed around the obstruction to the down-ice side.

region A portion of the Earth that displays relative similarity in one or more attributes.

regional metamorphism Large-scale metamorphism that occurs in conjunction with tectonic activity; especially found in the root zones of mountain systems.

regolith Weathered rock material, commonly located between unweathered bedrock (below) and soil (above).

relative humidity The percentage ratio of the amount of water vapor actually in the air to the maximum amount of water vapor the air could hold at that temperature and pressure.

relief (topographic) Differences in elevation within an area. The topographic relief of a mapped area is computed by taking the elevational difference between the highest and lowest point.

remote sensing The collection of information at a distance from a location by means of mechanical devices that receive emitted or reflected electromagnetic radiation.

reversal of topography Occurs when intense folding or faulting causes rock strata to overturn, so that older strata overlie younger strata.

reverse fault A fault in which the vertical displacement exceeds the horizontal and a compressionary component is present, with some overlapping of the two sides.

revolution The yearly orbital motion of the Earth around the Sun.

ria coast A highly irregular erosional coast dominated by numerous bays or estuaries that were produced by the gradual submergence of river valleys. The Middle Atlantic coast of the U. S. is a classic example.

Richter scale The most commonly used scale for measuring earthquake magnitudes. Each number on the scale has ten times the earthquake wave amplitude of the preceding number.

rift valley A deep, linear, steep-sided trough produced by the subsidence of a strip of land between two faults.

rift valley A large-scale graben, sometimes formed during an early stage in the separation of two continental plate segments.

rills A series of nearly parallel small shoestring channels that result from the early stages of fluvial erosion of an unprotected slope.

riparian Riverbank vegetation that often flourishes in the moist soils along streams in arid and semiarid regions.

rip current A narrow zone of water moving rapidly seaward along a shoreline.

roche moutonée A bedrock outcrop in glaciated terrain that exhibits a smoothly rounded up-ice side and an irregularly plucked down-ice side.

rock A solid aggregate of minerals. Different rocks are defined on the basis of their proportional mineral constituency. Rocks may also consist of organic materials.

rock cycle The continuous cycle of rock formation, erosion, deposition, and reconstitution.

rockfall The falling of an individual rock or small rock mass down a nearly vertical cliff.

rockslide The sliding of a mass of detached rock down a moderate to steep slope.

rotation The spinning of the Earth on its axis. This motion is the basis of our calendar day.

S waves Earthquake shear waves that are characterized by a side-to-side motion.

saline water intrusion The intrusion of salt water into a fresh water aquifer, often as the result of the excessive withdrawal of well water.

salinization The increase in soil salinity, usually in poorly drained arid areas, that occurs as a consequence of irrigation for a prolonged period. The salt is derived either from the deeper soil layers or from the irrigation water itself.

saltation An asymmetrical bouncing motion exhibited by relatively coarse textured sediments being transported either by a stream or by the wind.

salt dome A large dome-shaped intrusion of salt (halite) that has risen surfaceward through surrounding rock strata because of its buoyancy.

sand bar A largely or entirely submerged ridge composed of sand that is formed in the nearshore oceanic waters by currents and larger waves.

sandstone A clastic sedimentary rock comprised largely of sand-sized clasts that are usually composed of silica.

sandstorm A stinging mass of wind-blown sand grains extending to only a meter or two above the ground.

savanna A tropical grassland that contains scattered trees or shrubs. Savannas are dominant in forest/grassland transitional areas in regions with alternating wet and dry seasons.

sclerophyll A plant with small, hard, thick, leathery leaves that serve to minimize transpirational losses.

scrub A dense growth of shrubs and small trees, typical of many tropical and subtropical areas with alternating wet and dry seasons.

sea cliff A cliff formed along or just behind a shoreline by wave undercutting in areas of competent (strong) materials.

sea-floor spreading The process of formation of new oceanic crust by the extrusion of lava along divergent lithospheric plate boundaries.

sea ice Ice that forms on the ocean surface from the freezing of sea water. Most exists in the Arctic Ocean and near the coast of Antarctica.

seamount A circular or elliptical submarine mountain rising more than 900 meters (3,000 ft) from the deep sea floor that was produced by an undersea volcanic eruption.

second An angular distance of 1/60th of a minute, or 1/3600th of a degree.

sedimentary rock Rock that has formed from the lithification of sediments and typically accumulates in layers (strata).

sediments Solid materials of either organic or inorganic origin that were transported to their depositional sites by water, ice, wind, or gravity. The lithification of sediments forms sedimentary rocks.

seismograph An instrument used to measure earthquake shock waves.

semiarid climate A climate with a potential evaporation rate that slightly exceeds its mean precipitation total, so that it is moderately dry.

shale The most abundant sedimentary rock type, composed of clay-sized particles, and often containing fossils. Most shales formed from muds deposited on the deep ocean floor.

shear strength The measure of the ability of surficial materials to resist downslope movements.

shield An ancient block of continental crust that has been tectonically stable for a lengthy geologic timespan.

shield volcano A broad, gently sloping volcano constructed primarily of basalt derived from fluid lava.

sial Derived from the words "silicon" and "aluminum," this term refers to the granitic rocks that comprise most of the continental crust.

silicate minerals A mineral family consisting primarily of silicon and oxygen. It is the dominant mineral group of the crust and mantle.

sill A relatively thin sheet of magma injected more or less horizontally between near-surface sedimentary rock strata.

sima Derived from the words "silicon" and "magnesium," this term refers to the basaltic rocks that comprise the oceanic crust.

sinkhole A rounded depression formed by subsurface solution, especially in regions underlain by limestone. Sinkholes can develop either by gradual subsidence or by the sudden collapse of the surface. Also called a doline.

sinking creek A stream that enters a swallow hole in a karst region and continues its course beneath the surface.

sleet A precipitation form consisting of small ice pellets produced by the freezing of descending raindrops.

slump The intermittant and relatively slow movement of a mass of earth or rock along a curved slip-plane.

small circle A circle, drawn on a sphere, that does not bisect the sphere and does not form a full circumference of the sphere.

soil A loose mixture of weathered rock material, organic matter, water, and air, that can support plant growth. Most undisturbed soils tend to develop distinctive layers (horizons) that parallel the surface and differ in their physical, chemical, and biological characteristics.

soil horizons Soil layers, roughly paralleling the surface, that differ in their physical, chemical, and biological characteristics as a result of their formation under differing environmental conditions.

soil pH The acidity or alkalinity of the soil, determined by its supply of exchangeable $H+$ ions.

soil profile Vertical slice through the soil in which the various horizons are exposed for analysis.

Soil Taxonomy A nickname for the U. S. Comprehensive Soil Classification System—the system currently in official use in the United States.

soil texture The size of the individual soil particles. Different texture categories are determined by the soil's proportions of sand, silt, and clay-sized particles.

soil water Water that adheres in a molecular film to the soil particle surfaces, and partially occupies the pore spaces between the soil particles.

solar day The period of time needed for the Sun to make one apparent 360° circuit of the sky.

solar noon The exact time that the Sun reaches its zenith (high point in the sky) at any given locality.

solar system The Sun and the other astronomical bodies, notably the nine planets, that revolve around the Sun because of its gravitational attraction.

solifluction A form of earthflow that occurs on hillsides in periglacial environments as the surface layers thaw.

source region A large land or water body over which air masses form and from which they derive their temperature and moisture characteristics.

South Pole The southern end of the Earth's axis, located at 90° S.

southern hemisphere The half of the Earth south of the equator.

specific heat The quantity of heat energy that must be absorbed or released by a particular substance in order to undergo a given temperature change.

specific humidity The ratio of the mass of water vapor in a parcel of air to the total mass of the air, including the water vapor. Usually stated in grams per kilogram.

spit A typically hook-shaped barrier beach that is attached to the mainland at one end and often extends partially across a bay on the other end.

spodosol A soil order containing acidic soils with a light colored eluvial E horizon and an illuvial B horizon. They are typically associated with a subarctic climate and with taiga vegetation.

spring The site of an outflow of subsurface water.

spring equinox See *vernal equinox.*

spring tides Exceptionally high and low tides, occurring approximately twice a month, caused by the parallel alignment of the Moon and Sun with the Earth.

squall line A line of heavy showers or thunderstorms associated with strong, gusty winds. Squall lines frequently develop in advance of strong cold fronts, in which case they are known as pre-frontal squall lines.

stable air Air with a vertical temperature profile that causes it to resist any lifting influences.

stack A small, steep-sided island located near the shoreline that is an erosional remnant of a retreating sea cliff.

stalactite An icicle-like travertine feature that hangs downward from the roof of a cave.

stalagmite The upward growing counterpart of a stalactite, formed of travertine, and found on the floor of a cave in a limestone region.

standard time system The currently used time system for most of the world. Time is based on the east-west position of the Sun at meridians with longitudes divisible by 15, and changes by whole hour increments as time zones are crossed.

stationary front A relatively stationary frontal boundary between two differing air masses. Such a front eventually will either dissipate, or will begin to move as a warm front or cold front.

steppe Short grasslands generally associated with semiarid climates in the low and middle latitudes. Grass heights are usually less than 0.6 meters (2 ft).

storm surge The very high ocean water levels that accompany the landfall of a hurricane.

strata Horizontal layers, usually of sedimentary rocks.

stratopause The boundary between the stratosphere and the mesosphere. At this level, temperatures stop rising with increasing altitude.

stratosphere The thermal layer of the atmosphere above the troposphere, in which temperatures increase with altitude.

stratus From the Latin word meaning "layer," stratus clouds are low, dense, gray clouds that usually cover the entire sky.

stream A channelized body of water that flows downhill in response to gravity.

stream order A numerical system for classifying the approximate size of a stream, based on the number and nature of its tributaries. Two streams of a given order must join to form a stream of the next higher order.

stream stage The measured height of the water surface in a stream.

stream terrace A terrace consisting of the remnant of a former floodplain surface that was developed at a topographic level higher than the present stream valley level.

striations Sets of parallel bedrock scratches caused by the passage of glacial ice containing embedded rock fragments.

strike The compass orientation of a horizontal line in the plane of an inclined rock stratum or outcrop.

strike-slip fault See *transcurrent fault.*

stripped plain A plain formed when the erosional stripping away of weak rock strata leads to the uncovering of a horizontally bedded resistant rock layer.

structure (of soil) The characteristic shape, size, and organization of the soil aggregates or peds.

subduction The angular descent of the margin of an oceanic lithospheric plate into the asthenosphere, where it is eventually destroyed. Subduction results from a collision between two plates, with the denser plate being subducted.

sublimation The direct phase change of water vapor to ice, or vice-versa. Approximately 670 calories of energy are involved for each gram of H_2O that sublimes.

submarine canyon A deep, V-shaped canyon that is entrenched into the continental shelf. Such canyons are believed to have been formed initially by rivers when sea levels were lower, and to have been subsequently enlarged by the action of turbidity currents.

submarine trench A linear trench on the sea floor formed along the line where an oceanic lithospheric plate is being subducted. The world's greatest ocean depths occur in these trenches.

subpolar jet A jet stream that is produced by temperature contrasts between the low and high latitudes. Its position is closely related to the surface position of the polar front.

subpolar lows Zones of rising air and low pressure centered in the subpolar latitudes of both hemipsheres, caused primarily by the Earth's rotation.

subsidence The vertical settling of the surface as a result of the removal of subsurface support by mining, solution, or other processes.

subsoil The soil horizon that underlies the topsoil. It usually contains little organic matter and is frequently an illuvial horizon.

subtropical belts of variable winds and calms The narrow belts of light and variable winds associated with the centers of the northern and southern hemisphere subtropical highs; also referred to as the "horse latitudes."

subtropical highs Zones of subsiding air and high pressure centered in the subtropics of both hemispheres, and caused primarily by the Earth's rotation.

subtropical jet A jet stream that is centered above the surface positions of the subtropical highs.

summer solstice The day that one's hemisphere reaches its maximum inclination toward the Sun. The Sun is directly overhead at a latitude of 23½°. Marks the official beginning of summer.

surface creep The slow transport of wind-blown surface materials by rolling or sliding movements. This process accounts for about 20 percent of all eolian surface sediment transport.

suspended load The portion of the sediment load of a stream that consists of fine textured solid particles transported within the flow, and kept from settling by turbulence.

suturing The fusion of two lithospheric plates into a single plate following their collision.

swamp A poorly drained area, frequently covered with hygrophytic vegetation.

swash The forward (up-beach) motion of water from a breaking wave.

syncline A downward fold in one or more rock strata.

system A set of interrelated components through which energy flows to produce orderly changes. A closed system has a limited supply of energy from an internal source. An open system has an unlimited supply of energy from an external source.

taiga The Russian name for the northern coniferous forest, the most poleward of the forest subassociations.

talus cone A conical accumulation of angular rocks resting against a cliff or mountainside, derived largely from the process of frost wedging.

tarn A small lake occupying the enclosed basin of a cirque that no longer contains a glacier.

tectonic forces Landform producing forces originating inside the Earth, primarily as a result of the motions of tectonic plates.

temperature A measure of the thermal energy contained by an object or substance. This energy level is related to the rate of motion of its constituent molecules.

temperature inversion An increase in temperature with altitude in the troposphere. This situation represents a reversal of the normal decline in temperatures with altitude.

terrace A steplike surface that is relatively flat and linear. It is bounded on one side by a slope to a lower level and on the other by a slope to a higher elevation.

terranes Small pieces of continental crust, often of uncertain origin, that may be swept up by the movements of lithospheric plates and accreted onto the margin of a continent.

terrestrial radiation The long wave electromagnetic radiation emitted by the Earth system, which we sense as heat.

thermal high A high pressure system produced by the compaction of air as a result of cold surface temperatures.

thermal low A low pressure system produced by the expansion of air as a result of hot surface temperatures.

thermosphere The uppermost of the four thermal layers of the atmosphere, in which temperatures rise with increasing altitude.

throughflow The lateral flow of water near, but slightly below, the soil surface. It results from a temporary condition of saturation in the lower soil.

thrust fault A fault in which extreme compression causes one side to be thrust at a low angle over the other, often for a considerable distance.

thunderstorm A storm associated with cumulonimbus clouds, thunder and lightning. It is the world's most common storm type, and is often associated with heavy rains and strong local winds.

tidal bore A wave produced along the advancing front of the rising tide. It forms as the tide advances into a linear water body of restricted size, such as a river or bay.

tidal currents Currents in restricted bodies of water influenced by the tides that flow landward during the period of the rising tide and seaward during the period of the falling tide.

tidal inlet An oceanic inlet between barrier islands that connects the coastal lagoon to the open sea.

tides The rhythmic rise and fall of the ocean surface level produced by the gravitational influence of the Moon and Sun.

till Rock material deposited directly by glacial ice, and therefore displaying a lack of sorting.

till plain An extensive low relief surface covered by a thick accumulation of ground moraine. Much of the Midwest United States and the North European Plain are till plains.

topographic map A map that provides systematic information on surface elevations for the areas it depicts by means of colors, hachures, contour lines, or some other method.

topography The combination of landform characteristics and distributions within a region.

topsoil The uppermost mineral horizon of the soil; often referred to as the A horizon.

tornado A tubular vortex of whirling air surrounding a central core of extremely low pressure, which extends downward from the base of a cumulonimbus cloud.

tower karst Steep-sided remnant limestone peaks of mountainous proportions.

trade winds The major wind belt of the tropics, blowing from the subtropical high pressure centers of each hemisphere toward the Intertropical Convergence Zone. Prevailing wind directions are northeasterly in the northern hemisphere and southeasterly in the southern hemisphere.

transcurrent fault A fault in which the two sides are predominantly horizontally offset. Also called a strike-slip fault.

translocation The transfer of dissolved materials and fine textured particles from one place to another in the soil as a result of soil water transport.

transpiration The evaporation from plant surfaces of water previously absorbed through the plant root systems.

transverse dune A linear sand dune with an undulating crestline that is elongated perpendicular to the prevailing wind direction.

travertine Calcium carbonate ($CaCO_3$) deposits generally found in caves. They assume a wide variety of forms, including stalactites, stalagmites, columns, and ribbons.

tree A large woody plant normally containing a single primary stem or trunk that develops many branches.

treeline The cold climate limit of tree growth at high latitudes and elevations. It normally coincides roughly with the 10° C (50° F) isotherm for the mean temperature of the warmest month.

trellis stream pattern Consists of a parallel pattern of major streams, which are joined at nearly right angles by much smaller and shorter tributaries. This pattern typically develops in regions containing alternating bands of resistant and nonresistant sedimentary strata.

Tropic of Cancer Located at 23½° N, it is the farthest north line of latitude that the Sun's vertical rays reach during the course of the year.

Tropic of Capricorn Located at 23½° S, it is the farthest south line of latitude that the Sun's vertical rays reach during the course of the year.

tropical depression A weak closed low pressure center that has formed over a tropical water body and has maximum sustained winds of less than 18 meters per second (40 mph). It may have the potential to develop into a tropical storm and, eventually, into a hurricane.

tropical storm A storm that has developed over a tropical water body and has maximum sustained winds of between 18 and 33.5 meters per second (40–74 mph). Tropical storms are given names, are capable of causing destruction if they strike land, and have the potential to strengthen into hurricanes.

tropopause The boundary between the troposphere and the stratosphere. At this level, temperatures stop falling with increasing altitude.

tropophyte A plant adapted to alternating favorable and unfavorable climatic conditions for growth. Most tropophytes grow in areas with alternating warm and cold or wet and dry seasons.

troposphere The lowermost of the four thermal layers of the atmosphere, in which temperatures normally decline with increasing altitude. Most weather phenomena exist in this layer.

turbidity current A rapid submarine flow of dense, sediment-laden water down the continental slope at the outer edge of the continental shelf.

turbulence The chaotic eddying or swirling motions imparted to wind or flowing water by the effects of friction.

twilight The condition of semi-darkness caused by sunlight being scattered by air molecules and dust particles in the upper atmosphere when the Sun is less than 18° below the horizon.

typhoon The term used for a hurricane in the western North Pacific.

ultisol A soil order consisting of highly chemically weathered soils of limited natural fertility that have developed in warm, moist climates.

undertow The seaward movement of subsurface water near the shoreline. It occurs as a counterbalance to the landward flow of water carried by surface waves.

uniformitarianism The geologic principle stating that the Earth's surface features have evolved from the long continued application of currently operative geomorphic processes.

unloading A physical weathering process involving rock breakup resulting from the gradual decrease in gravitational pressure as erosion removes the overlying material.

unstable air Air with a vertical temperature profile that produces a tendency to rise, especially if the air is given an initial lift.

upwelling The flow of deep ocean water to the surface.

urban heat island The islandlike area of relatively warm temperatures associated with a city. The heat island results both from the nature of the city surface materials and from human activities that release heat.

valley train A linear train of glacio-fluvial materials deposited by a meltwater stream issuing from the margin of a glacier.

vegetative association See *plant association.*

vegetative succession The gradual change in the composition of a plant community in response to changing environmental conditions following a disturbance such as a fire.

ventifact A desert rock with prominently wind polished, grooved, or faceted surfaces.

vernal equinox The day that the Earth's axial inclination is exactly perpendicular with respect to the Sun, following a period in which one's hemisphere was inclined away from the Sun. Marks the official beginning of spring.

vertebrate An animal with a backbone or spinal column.

vertisol A soil order consisting of soils that contain large amounts of clay that swells when wetted and shrinks when dried. This results in a churning action that inhibits profile development.

volcanic neck A solid spire of hardened lava that was in the vent of a volcano at the end of its final eruption.

volcanism Processes involving the transfer of molten rock material either from one subsurface location to another, or its expulsion onto the surface.

volcano A hill or mountain constructed of materials from the interior of the Earth that have been ejected under pressure from a vent.

warm front A frontal boundary along which an advancing mass of warm air is overrunning and displacing relatively denser cold air.

warm ice glacier A wet-base glacier that has an underlying film of meltwater acting as a lubricant to aid slippage over its bedrock base.

water table The upper boundary of the zone of permanently saturated soil or rock; the boundary between soil water and ground water.

water vapor Water (H_2O) in the invisible gaseous state. It comprises an average of about 1.4% of the atmosphere and is the source of moisture for clouds and precipitation.

wave A rhythmic vertical motion of the near-surface portion of any water body, usually generated by the wind. The water molecules move in circular or elliptical paths, with little or no net forward motion.

wave-built platform A platform of unconsolidated marine sediments that is generally deposited in the nearshore waters immediately seaward of a wave-cut platform.

wave-cut platform A rock platform that has been planed off at the base of erosionally effective wave action along an eroding coast.

wave refraction The bending of the angle of approach of incoming waves. Refraction occurs as the waves obliquely approach the shoreline and encounter a shallow bottom that slows and rotates them more directly toward shallow water.

weather The short-term condition of the atmosphere with respect to temperature, air pressure, wind, humidity, clouds, and precipitation.

weather front A boundary zone along which two or sometimes three air masses of differing temperature and moisture characteristics meet, often producing an extensive area of cloud cover and precipitation.

weathering The combined action of physical and chemical processes that disintegrate and decompose bedrock. An essential preliminary for soil formation and gradation.

westerlies The major wind belts of the middle latitudes, blowing from the subtropical high pressure centers of each hemisphere toward the subpolar lows.

wet adiabatic rate The rate at which saturated air cools as it rises (or warms as it descends) through the troposphere. This rate averages approximately 3.2F°/1000 feet, or 6.0C°/1000 meters.

wilting point The soil moisture level at which plants begin to wilt. Wilting occurs because the remaining soil water is so tightly bound to the soil particles that the plants are unable to extract it.

wind shear Local variations in wind speed and direction that are sometimes dangerous to aircraft.

windward coast A coast with a prevailing water to land windflow.

winter solstice The day that one's hemisphere reaches its maximum inclination away from the Sun. The Sun is directly overhead at a latitude of 23½° in the opposite hemisphere. Marks the official beginning of winter.

xerophyte A plant especially adapted to grow in a region deficient in water.

zone of aeration The zone above the water table, where pore spaces or other openings are jointly occupied by soil water and gases.

zone of saturation The zone below the water table, where pore spaces or other openings are completely occupied by ground water.

Index